SCHAUM'S OUTLINE OF

THEORY AND PROBLEMS

of

BEGINNING PHYSICS II
Waves, Electromagnetism, Optics, and Modern Physics

ALVIN HALPERN, Ph.D.

Professor of Physics
Brooklyn College
City University of New York

ERICH ERLBACH

Professor Emeritus of Physics
City College
City University of New York

SCHAUM'S OUTLINE SERIES

McGRAW-HILL

New York San Francisco Washington, D.C. Auckland Bogotá
Caracas Lisbon London Madrid Mexico City Milan Montreal
New Delhi San Juan Singapore Sydney Tokyo Toronto

ALVIN HALPERN, Ph.D., Professor of Physics at Brooklyn College of the City University of New York. Dr. Halpern has had extensive teaching experience in physics at all college levels. elementary through doctoral. He was chairman of the physics department at Brooklyn College for ten years, and Vice President for Research Development at the Research Foundation of CUNY for four years. He presently is Acting President of the Research Foundation and University Dean for Research.

ERICH ERLBACH, Ph.D., is Professor Emeritus of Physics at The City College of the City University of New York. He has had over 35 years of experience in teaching physics courses at all levels. Dr. Erlbach served as chairman of the physics department at City College for six years and served as Head of the Honors and Scholars Program at the College for over ten years.

Schaum's Outline of Theory and Problems of
BEGINNING PHYSICS II: Waves, Electromagnetism, Optics, and Modern Physics

1 2 3 4 5 6 7 8 9 10 11 12 13 14 15 16 17 18 19 20 PRS PRS 9 0 2 1 0 9 8

ISBN 0-07-025707-8

Sponsoring Editor: Barbara Gilson
Production Supervisor: Sherri Souffrance
Editing Supervisor: Maureen B. Walker

Library of Congress Cataloging-in-Publication Data

Halpern, Alvin M.
Schaum's outline of theory and problems of beginning physics II: waves, electromagnetism, optics, and modern physics/Alvin Halpern, Erich Erlbach.
p. cm. — (Schaum's outline series)
Includes index.
ISBN 0-07-025707-8
1. Physics. I. Erlbach, Erich. II. Title.
QC23.H213 1998
530—dc21

98-24936
CIP

McGraw-Hill
A Division of The ***McGraw-Hill*** *Companies*

This book is dedicated to

Edith Erlbach, beloved wife of Erich Erlbach

and

to the memory of Gilda and Bernard Halpern,
beloved parents of Alvin Halpern

Preface

Beginning Physics II: Waves, Electromagnetism, Optics and Modern Physics is intended to help students who are taking, or are preparing to take, the second half of a first year College Physics course that is quantitative in nature and focuses on problem solving. From a topical point of view the book picks up where the first volume, *Beginning Physics I: Mechanics and Heat* leaves off. Combined with volume I it covers all the usual topics in a full year course sequence. Nonetheless, *Beginning Physics II* stands alone as a second semester follow on textbook to any first semester text, or as a descriptive and problem solving supplement to any second semester text. As with *Beginning Physics I*, this book is specifically designed to allow students with relatively weak training in mathematics and science problem solving to quickly gain quantitative reasoning skills as well as confidence in addressing the subject of physics. A background in High School algebra and the rudiments of trigonometry is assumed, as well as completion of a first course covering the standard topics in mechanics and heat. The second chapter of the book contains a mathematical review of powers and logarithms for those not familiar or comfortable with those mathematical topics. The book is written in a "user friendly" style so that those who were initially terrified of physics and struggled to succeed in a first semester course can gain mastery of the second semester subject matter as well. While the book created a "coaxing" ambiance all the way through, the material is not "watered down". Instead, the text and problems seek to raise the level of students' abilities to the point where they can handle sophisticated concepts and sophisticated problems, in the framework of a rigorous noncalculus-based course.

In particular, *Beginning Physics II* is structured to be useful to pre-professional (premedical, predental, etc.) students, engineering students and science majors taking a second semester physics course. It also is suitable for liberal arts majors who are required to satisfy a rigorous science requirement, and choose a year of physics. The book covers the material in a typical second semester of a two semester physics course sequence.

Beginning Physics II is also an excellent support book for engineering and science students taking a calculus-based physics course. The major stumbling block for students in such a course is not calculus but rather the same weak background in problem solving skills that faces many students taking non-calculus based courses. Indeed, most of the physics problems found in the calculus based course are of the same type, and not much more sophisticated than those in a rigorous non-calculus course. This book will thus help engineering and science students to raise their quantitative reasoning skill levels, and apply them to physics, so that they can more easily handle a calculus-based course.

ALVIN HALPERN
ERICH ERLBACH

To the Student

The Preface gives a brief description of the subject matter level, the philosophy and approach, and the intended audience for this book. Here we wish to give the student brief advice on how to use the book. *Beginning Physics II* consists of an interweaving of text with solved problems that is intended to give you the opportunity to learn through exploration and example. The most effective way to gain mastery of the subject is to go through each problem as if it were an integral part of the text (which it is). The last section in each chapter, called *Problems for Review and Mind Searching*, gives additional worked out problems that both review and extend the material in the book. It would be a good idea to try to solve these problems on your own before looking at the solutions, just to get a sense of where you are in mastery of the material. Finally, there are supplementary problems at the end of the chapter which given only numerical answers. You should try to do as many of these as possible, since problem solving is the ultimate test of your knowledge in physics. If you follow this regime faithfully you will not only master the subject but you will sense the stretching of your intellectual capacity and the development of a new dimension in your ability. Good luck.

Contents

Chapter 1

Wave Motion

1.1 PROPAGATION OF A DISTURBANCE IN A MEDIUM

In our study of mechanics we considered solids and fluids that were at rest or in overall motion. In thermodynamics we started to explore the internal behavior of large systems, but for the most part addressed equilibrium states where there is a well defined pressure and temperature of our system. In our study of transfer of heat (see, e.g., Schaum's Beginning Physics I, Chap. 17), we discussed the transfer of thermal energy within a medium, from a "hot region" to a "cold region". In the case of convection this transfer took place by the actual movement of physical matter, the more energetic molecules (hot gas or liquid), from one location to another; in the case of conduction, it took place by means of transference of thermal energy from one layer of molecules to an adjacent layer and then on to the next layer, and so on, without the displacement of the physical matter itself over macroscopic distances. In the present chapter we will discuss the transfer, not of thermal energy, but rather of mechanical energy, through a solid, liquid or gas, by means of wave motion—also a process in which the physical matter itself does not move over significant distances beyond their initial positions, while the energy can be transferred over large distances. The transferred energy can carry information, so that wave motion allows the transfer of information over large distances as well.

Of course, one way to communicate information over distance is to actually transfer matter from one location to another, such as throwing stones in coded sequences (e.g., three stones followed by two stones, etc.). This means of communication is very limited and cumbersome and requires a great amount of energy since large objects have to be given significant kinetic energy to have them move. Instead, we can take advantage of the inter-molecular forces in matter to transfer energy (and information) from molecular layer to molecular layer and region to region, without the conveyance of matter itself. It is the study of this process that constitutes the subject of wave motion.

Propagation of a Pulse Wave Through a Medium

Consider a student holding one end of a very long cord under tension S, with the far end attached to a wall. If the student suddenly snaps her hand upward and back down, while keeping the cord under tension, a pulse, something like that shown in Fig. 1-1(*a*) will appear to rapidly travel along the cord away from the student. If the amplitude of the pulse (its maximum vertical displacement) is not large compared to its length, the pulse will travel at constant speed, v, until it reaches the tied end of the cord. (We will discuss what happens when it hits the end later in the chapter). In general, the shape of the pulse remains the same as it travels [Fig. 1-1(*b*)], and its size diminishes only slightly (due to thermal losses) as it propagates along the cord. By rapidly shaking her hand in different ways, the student can have pulses of different shapes [e.g., Fig. 1-1(*c*)] travelling down the cord. As long as the tension, S, in the cord is the same for each such snap, and the amplitudes are not large, the speed of all the pulses in the cord will be the same no matter what their shapes [Fig. 1-1(*d*)].

Problem 1.1.

(*a*) For the cases of Fig. 1-1, in what direction are the molecules of the cord moving as the pulse passes by?

(*b*) If actual molecules of cord are not travelling with the pulse, what is?

(*c*) What qualitative explanation can you give for this phenomenon?

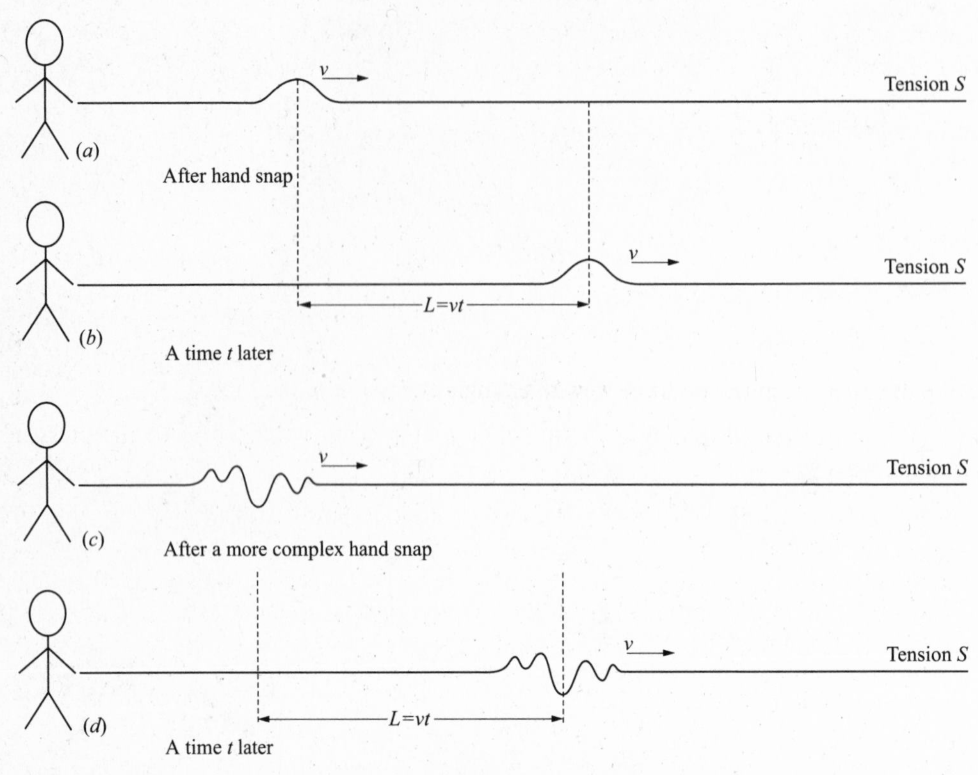

Fig. 1-1

Solution

(*a*) We can understand the motion of the cord molecules as the pulse approaches a point in the cord and passes by. First the molecules at a given horizontal point on the cord move upward, until the maximum of the pulse passes the point, at which time the molecules are at the maximum vertical displacement (the amplitude); then the molecules move back down until they return to their normal position as the pulse passes by. Thus the molecules move perpendicular to the direction in which the pulse moves.

(*b*) The shape of the pulse travels as one set of molecules after another go through the vertical motion described in part (*a*). The pulse carries energy—the vertical kinetic energy of the moving molecules, and the associated potential energy due to momentary stretching of the cord, in the pulse region.

(*c*) As the tension in the cord is increased forces between adjacent molecules get stronger, resisting the effort to pull the cord apart. When the student snaps the end of the cord upward the adjacent molecules are forced upward as well, and so are the next set of molecules and the next set, and so on. All the molecules in the cord don't move upward at the same instant, however, because it takes some time for each succeeding set of molecules to feel the resultant force caused by the slight motion of the prior set away from them. While the successive groups of molecules are being pulled upward, the student snaps her hand back down, so the earlier molecules are reversing direction and moving back down. The net effect is that successive sets of molecules down the length of the cord start moving upward while further back other sets are feeling the pull back down. This process causes the pulse to, in effect, reproduce itself over and over again down the cord.

The pulse in the cord is an example of a **transverse** wave, where molecules move to and fro at right angles to the direction of propagation of the wave. Another type of wave, in which the molecules actually move to and fro along the direction of the propagation of the wave is called a **longitudinal** wave. Consider a long straight pipe with air in it at some pressure P, and a closely fitting piston at one end. Suppose a student suddenly pushes the piston in and pulls it back out. Here the molecules of air first move forward along the tube and then back to their original positions, while the wave pulse travels in a parallel direction [Fig. 1-2]. (Since it is hard to visually display the longitudinal pulse, we indicate its location by showing a shaded area in the figure, darker meaning greater displacement). This air pulse is a primitive example of a sound wave. Longitudinal sound waves also occur in liquids and solids, as one experiences by hearing sound under water or by putting ones ear to a railroad track.

Problem 1.2.

(*a*) Drawing analogy from the transverse wave in Problem 1.1(*a*), describe the pulse that you would expect occurs when the student jerks the piston in and out, in the piston-tube arrangement described above in the text.

(*b*) Describe the pulse from the point of view of changing pressure in the tube.

(*c*) What in the transverse wave of Fig. 1-1(*a*) behaves in a manner analogous to the pressure in the longitudinal wave?

Solution

(*a*) Aside from the different nature of the intermolecular forces in the two cases (solid vs. gas), the similarities are considerable. Just as the molecules in the cord first communicate upward motion and then downward motion, the air molecules in the tube first communicate motion away from the piston and then motion toward the piston. The maximum displacement of the molecules away from their normal, or equilibrium, positions represents the amplitude of the pulse. A reasonable speculation is that the longitudinal pulse travels with some definite velocity (characteristic of the air) along the tube, and maintains its shape, with some diminution in amplitude due to thermal losses.

Note. This is in fact what actually does happen.

(*b*) When the piston is first pushed in it compresses the air between the piston and the layer of air in the tube not yet moving, so there is a small increase in pressure, ΔP, above the ambient pressure of the air, P. This increase drops rapidly to zero as the compression reverts to normal density as the air molecules further along move over. As the piston is pulled back to its original position a rarefaction occurs as molecules rush back against the piston but molecules further along the tube have not yet had time to respond, so there is a small decrease in pressure, ΔP, that again disappears as the molecules further on come back to re-establish normal density.

(*c*) The displacement of the transverse pulse of Fig. 1-1(*a*) is always positive (as is the displacement of the longitudinal wave in part (*a*) above), while the "pressure wave", ΔP, described in (*b*) above, first goes positive, drops back through zero to become negative, and then returns to zero. One quantity in the

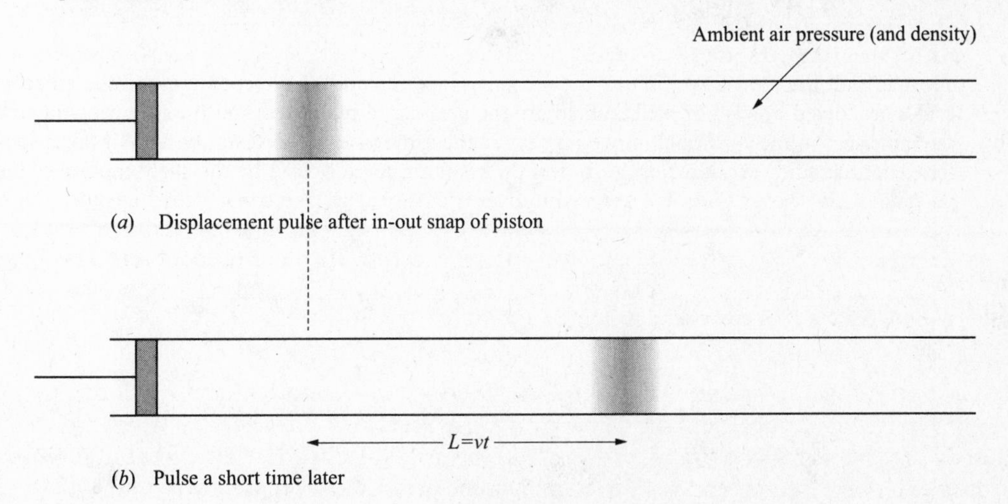

(*a*) Displacement pulse after in-out snap of piston

(*b*) Pulse a short time later

Fig. 1-2

transverse wave that behaves analogously is the vertical velocity of the molecules of the cord. This transverse velocity (not to be confused with the velocity of propagation of the pulse) is first positive (upward), then becomes zero at maximum amplitude, and then turns negative (downward), becoming zero again after the pulse passes by. ΔP behaves exactly the same way. Indeed, from this analogy, we can surmise that the change in pressure is zero where the air molecules are at maximum displacement from their equilibrium position, just as the velocity is zero when the cord molecules are at maximum displacement.

These results can be illustrated by examining two graphs representing either the transverse pulse in the cord or equally well the longitudinal pulse in the tube. In Fig. 1-3(*a*) we show a graph representing at a given instant of time, and on some arbitrary scale, the vertical displacement from equilibrium of the molecules of the transverse pulse in the cord of Problem 1.1(*a*). Figure 1-3(*b*) then represents, at the same instant of time, and with the same horizontal scale but arbitrary vertical scale, a graph of the vertical (transverse) velocities of the corresponding points along the cord. The displacements and velocities of the various points in the cord in these "snapshot" graphs also shows the "real time" behavior that any one point in the cord would have as the pulse passed by.

Equally well, Fig. 1-3(*a*) can represent, at a given instant of time, and on some arbitrary scale, a vertical plot of the longitudinal displacement of gas molecules from their normal (or equilibrium) positions, with the horizontal representing the various equilibrium (if no pulse were passing) positions along the tube. For this case, Fig. 1-3(*b*) can then represent, at the same instant, and on the same horizontal but arbitrary vertical scale, the changes of pressure (ΔP) at corresponding points along the tube. Again, as with the cord, we note that for our pulse moving through the tube to the right, the graphs of the various points in this snapshot of the pulse also represent the behavior at a given point along the tube, as the pulse passes by in real time. We see that ΔP is positive at the front (right-most) end of the pulse, first increasing and then decreasing to zero (normal pressure) as the maximum vertical longitudinal displacement passes by, then turning negative, first increasingly negative and then decreasingly negative,

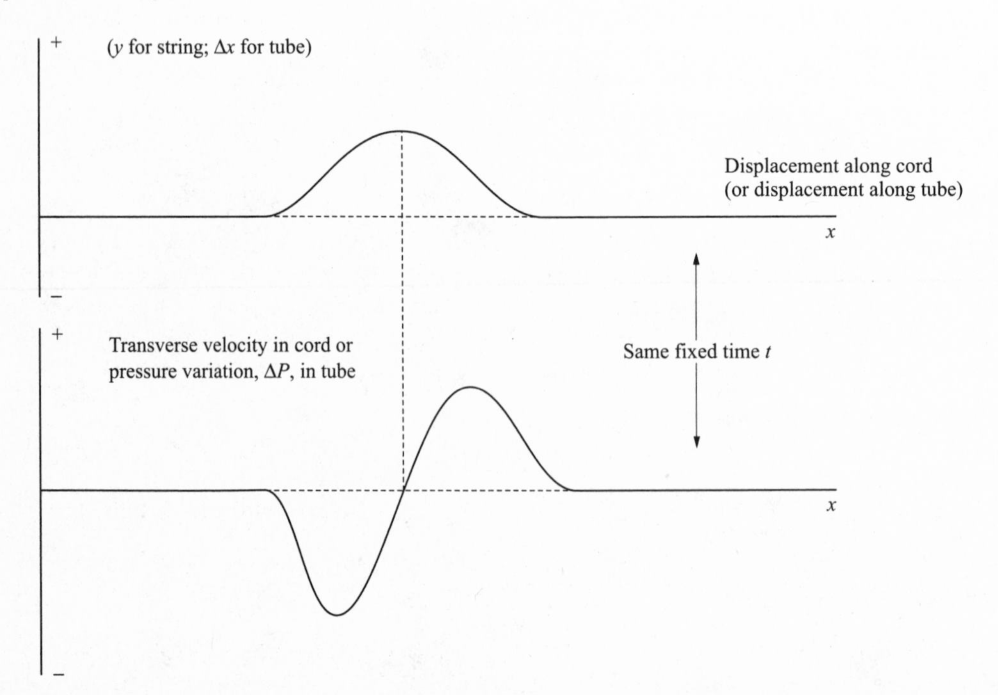

Fig. 1-3

until reaching zero (normal pressure) as the pulse completely passes the given point. The same graph can also represent the longitudinal velocities of the moving air molecules, and we see that the velocities of the air molecules at various points along the tube behave like ΔP at those points.

Velocity of Propagation of Waves

Using the laws of mechanics, it is possible to derive the actual velocities of propagation of waves such as transverse waves in a cord or rope and longitudinal waves in a gas, liquid or solid. We will not do that here (but we will do one case in a problem later on). Instead, we will use qualitative arguments to show the reasonableness of the expressions for the velocities. Consider first the case of transverse waves in a cord. What are the factors that would affect the velocity of propagation? First we note that the more quickly a molecule responds to the change in position of an adjacent molecule, the faster the velocity of propagation would be. The factor in a cord that impacts the most on this property is how taut the cord is, or how much tension, S, it is under. The greater the tension the stronger the intermolecular forces, and the more quickly each molecule will move in response to the motion of the other. Thus increasing S will increase the velocity of propagation, v_p. On the other hand, the more massive the cord is, the harder it will be for it to change its shape, or to move up and down, because of inertia. The important characteristic, however, is not the mass of the cord as a whole, which depends on how long it is, but rather on a more intrinsic property such as the mass per unit length: μ. Then, increased μ means decreased v_p. The simplest formula that has these characteristics would be $v_p = S/\mu$. A quick check of units shows that $S/\mu = N \cdot \text{m/kg} = (\text{kg} \cdot \text{m/s}^2)\text{m/kg} = \text{m}^2/\text{s}^2$. By taking the square root we get units of velocity so we can guess:
For transverse waves in a cord

$$v_p = (S/\mu)^{1/2}. \qquad (1.1)$$

As it turns out, this is the correct result. (Our qualitative argument allows the possibility of a dimensionless multiplication factor in Eq. (*1.1*), such as 2, $\sqrt{2}$ or π, but in a rigorous derivation it turns out there are none!)

Similarly, in obtaining the propagation velocity of sound in a solid, consider a bar of length, L, and cross-sectional area, A. The strength of the intermolecular forces are measured by the intrinsic stiffness, or resistance to stretching, of the bar, a property which does not depend on the particular length or cross-section of our sample. We have already come across a quantity which measures such intrinsic stiffness independent of L and A: the Young's modulus of the material, Y, defined as the stress/strain (see, e.g., Beginning Physics I, Chap. 11.1), and which has the dimensions of pressure. Thus the larger Y, the larger v_p for the bar. As in the case of the cord, there is an inertial factor that impedes rapid response to a sudden compression, and the obvious intrinsic one for the bar is the mass/volume, or density, ρ. (Note that the mass per unit length would not work for the bar because it depends on A, and we have already eliminated dependence on A in the stiffness factor). Again, we try stiffness/inertia $= Y/\rho$, but this has the dimensions of $(\text{N/m}^2)/(\text{kg/m}^3) = \text{m}^2/\text{s}^2$. This is the same as the dimensions of S/μ for transverse waves in a cord, so we know we have to take the square root to get velocity:

For longitudinal waves in a solid

$$v_p = (Y/\rho)^{1/2} \qquad (1.2)$$

For a fluid the bulk modulus, $B =$ (change in pressure)/(fractional change in volume) $= |\Delta p/(\Delta V/V)|$ replaces Y as the stiffness factor, yielding:

For longitudinal waves in a fluid

$$v_p = (B/\rho)^{1/2} \qquad (1.3)$$

As with Eq. (*1.1*), for transverse waves in a cord, these last two equations turn out to be the correct results, without any additional numerical coefficients, for longitudinal waves in a solid or fluid.

Problem 1.3.

(*a*) Calculate the velocity of a pulse in a rope of mass/length $\mu = 3.0$ kg/m when the tension is 25 N.

(*b*) A transverse wave in a cord of length $L = 3.0$ m and mass $M = 12.0$ g is travelling at 6000 cm/s. Find the tension in the cord.

Solution

(*a*) From Eq. (*1.1*) we have:

$$v_p = (S/\mu)^{1/2} = [(25 \text{ N})/(3.0 \text{ kg/m})]^{1/2} = 2.89 \text{ m/s}.$$

(*b*) Again using Eq. (*1.1*) we have:

$$S = \mu v_p{}^2 = (M/L)v_p{}^2 = [(0.012 \text{ kg})/(3.0 \text{ m})](60 \text{ m/s})^2 = 14.4 \text{ N}.$$

Problem 1.4.

(*a*) If the speed of sound in water is 1450 m/s, find the bulk modulus of water.

(*b*) A brass rod has a Young's modulus of $91 \cdot 10^9$ Pa and a density of 8600 kg/m^3. Find the velocity of sound in the rod.

Solution

(*a*) Recalling that the density of water is 1000 kg/m^3, and using Eq. (*1.3*), we have:

$$B = (1450 \text{ m/s})^2(1000 \text{ kg/m}^3) = 2.1 \times 10^9 \text{ Pa}.$$

(*b*) From Eq. (*1.2*):

$$v_p = [(91 \times 10^9 \text{ Pa})/(8600 \text{ kg/m}^3)]^{1/2} = 3253 \text{ m/s}.$$

Problem 1.5. Consider a steel cable of diameter $D = 2.0$ mm, and under a tension of $S = 15$ kN. (For steel, $Y = 1.96 \cdot 10^{11}$ Pa, $\rho = 7860$ kg/m^3).

(*a*) Find the speed of transverse waves in the cable.

(*b*) Compare the answer to part (*a*) with the speed of sound in the cable.

Solution

(*a*) We need the mass/length, $\mu = \rho A$, where A is the cross-sectional area of the cable, $A = \pi D^2/4$. From the data for the cable:

$$\mu = (7860 \text{ kg/m}^3)(3.14)(0.0020 \text{ m})^2/4 = 0.0247 \text{ kg/m}.$$

Substituting into Eq. (*1.1*) we get: $v_p = [(15 \cdot 10^3 \text{ N})/(0.0247 \text{ kg/m})]^{1/2} = 779$ m/s.

(*b*) The speed of longitudinal (sound) waves is given by Eq. (*1.2*), which yields:

$$v_p = [(1.96 \cdot 10^{11} \text{ Pa})/(7860 \text{ kg/m}^3)]^{1/2} = 4990 \text{ m/s},$$

which is 6.41 times as fast as the transverse wave.

Problem 1.6.

(*a*) Assume the cable in Problem 1.5 is 1000 m long, and is tapped at one end, setting up both a transverse and longitudinal pulse. Find the time delay between the two pulses arriving at the other end.

(*b*) What would the tension in the cable have to be for the two pulses to arrive together?

Solution

(*a*) We find t_1 and t_2, the respective times for the transverse and longitudinal pulses to reach the end:

$$t_1 = (1000\text{ m})/(779\text{ m/s}) = 1.28\text{ s};\qquad t_2 = (1000\text{ m})/(4990\text{ m/s}) = 0.20\text{ s}.$$

$$\Delta t = t_1 - t_2 = 1.08\text{ s}.$$

(*b*) Here the speed of the two pulses must be the same so, as noted in Problem 1.5(*b*), the new transverse speed must be 6.41 times faster than before. Since the linear density, μ, does not change significantly, we see from Eq. (*1.1*) that the tension must increase by a factor of $6.41^2 = 41.1$. Thus, the new tension is

$$S' = 41.1(15\text{ kN}) = 617\text{ kN}.$$

1.2 CONTINUOUS TRAVELLING WAVES

Sinusoidal Waves

We now re-consider the case of the student giving a single snap to the end of a long cord (Fig. 1-1). Suppose, instead, she moves the end of the cord up and down with **simple harmonic motion** (SHM), of amplitude A and frequency $f = \omega/2\pi$, about the equilibrium (horizontal) position of the cord. We choose the vertical (y) axis to be coincident with the end of the cord being moved by the student, and the x axis

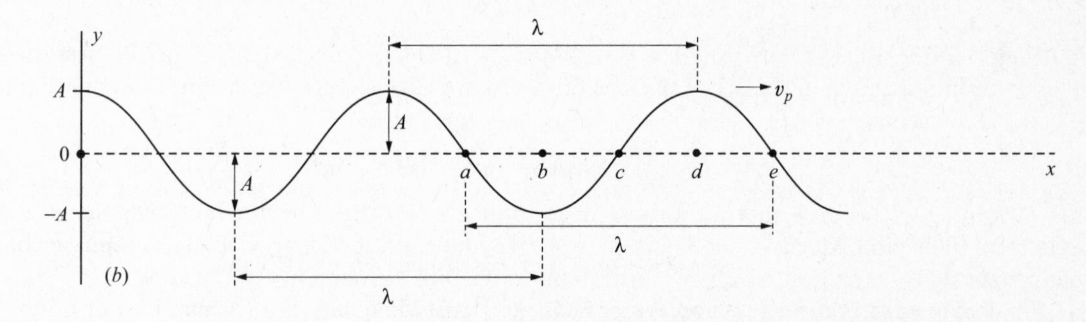

Travelling wave when pt. x=0 is oscillating vertically with SHM. Snapshot at $t=t_0$

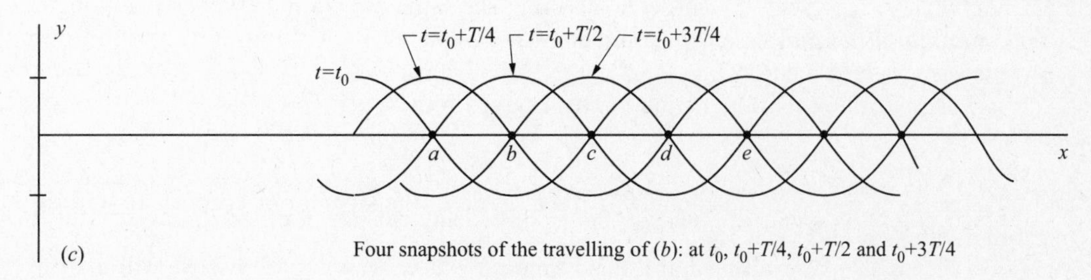

Four snapshots of the travelling of (*b*): at t_0, $t_0+T/4$, $t_0+T/2$ and $t_0+3T/4$

Fig. 1-4

to be along the undisturbed cord, as shown in Fig. 1-4(*a*). Let $y_0(t)$ represent the vertical position of the point on the cord corresponding to the student's end ($x = 0$) at any time t. Then, assuming $y_0 = 0$ (and moving upward) at $t = 0$ we have: $y_0 = A \sin(\omega t)$ for the simple harmonic motion of the end of the cord.

> ***Note.*** Recall that in general for SHM, $y = A \cos(\omega t + \theta_0)$, where θ_0 is an arbitrary constant that defines where in the cycle we are at $t = 0$. Choosing $\theta_0 = 0$ corresponds to being at maximum positive displacement at $t = 0$, while choosing $\theta_0 = 3\pi/2$ gives us our present result.

Every change in position of the cord at the student's end is propagated to the right with the velocity of propagation, v_p. This means that at any horizontal point x along the cord the molecules of cord will mimic the same up and down motion as the student initiated at the end point ($x = 0$), and with the same amplitude, A (if we ignore thermal losses). Let us call $y_x(t)$ the vertical position of the cord at a definite horizontal position, x, along the cord, at any time t. $y_x(t)$ will mimic what y_0 was at an earlier time t':

$$y_x(t) = y_0(t') \tag{1.4}$$

where $(t - t')$ is the time interval it takes for the signal to go from the end ($x = 0$) to the point x of interest. Since the signal travels at speed v_p we must have: $x = v_p(t - t')$, or $(t - t') = x/v_p \Rightarrow t' = t - x/v_p$. Finally, recalling our expression for $y_0(t)$, and using our expression for t' in Eq. (*1.4*), we get:

$$y_x(t) = A \sin[\omega(t - x/v_p)] \tag{1.5}$$

Note that Eq. (*1.5*) gives us the vertical displacement of any point x along the cord, at any time t. It thus gives us a complete description of the wave motion in the cord. As will be seen below, this represents a travelling wave moving to the right in the cord. This result presumes, of course, that the cord is very long and we don't have to concern ourselves with what happens at the other end. Eq. (*1.5*) can be reexpressed by noting that $\omega(t - x/v_p) = \omega t - (\omega/v_p)x$. We define the **propagation constant** for the wave, k as: $k = \omega/v_p$, or:

$$v_p = \omega/k \tag{1.6}$$

Recalling that the dimensions of ω are s^{-1} (with the usual convention that the dimensionless quantity, ωt, is to be in radians for purposes of the sine function), we have for the dimensions of k: m^{-1}. In terms of k, Eq. (*1.5*) becomes:

$$y_x(t) = A \sin(\omega t - kx) \tag{1.7}$$

Eqs. (*1.5*) and (*1.7*) indicate that for any fixed position x along the cord, the cord exhibits SHM of the same amplitude and frequency with the term in the sine function involving x acting as a phase constant that merely shifts the time at which the vertical motion passes a given point in the cycle.

Eqs. (*1.5*) and (*1.7*) can equally well represent the longitudinal waves in a long bar, or a long tube filled with liquid or gas. In that case $y_x(t)$ represents the longitudinal displacement of the molecules from their equilibrium position at each equilibrium position x along the bar or tube. Note that y_x for a longitudinal wave represents a displacement along the same direction as the x axis. Nonetheless, the to and fro motion of the molecules are completely analogous to the up and down motion of molecules in our transverse wave in a cord.

It is worth recalling that the period of SHM is given by:

$$T = 1/f = 2\pi/\omega \tag{1.8}$$

and represents the time for one complete vertical cycle of the SHM in our cord (or to and fro motion for our longitudinal waves).

Eq. (*1.7*) can also be examined at a fixed time t for all x. In what follows we will use the example of the transverse wave in the cord, since it is easier to visualize. For any fixed t, Eq. (*1.7*) represents a snapshot in time of the shape of the cord. Clearly for fixed t this represents a sinusoidal wave in the variable x. The spatial periodicity of this wave, i.e. the length along the x axis that one moves to go through one complete cycle of the wave, is called the **wavelength**: λ. Since a sine wave repeats when its argument (angle) varies through 2π, we see that for fixed t in Eq. (*1.7*), the sine wave will repeat when $x \to (x + \lambda)$ with $k\lambda = 2\pi$. Rearranging, we get:

$$\lambda = 2\pi/k \tag{1.9}$$

which is the spatial analogue of Eq. (*1.8*) for the period. The dimensions of λ are meters, as expected. A snapshot of the cord (at a moment t when $y_0 = A$) is shown in Fig. 1-4(*b*).

Problem 1.7. A student holds one end of a long cord under tension $S = 10$ N, and shakes it up and down with SHM of frequency $f = 5.0$ Hz and amplitude 3.0 cm. The velocity of propagation of a wave in the cord is given as $v_p = 10$ m/s.

(*a*) Find the period, T, the angular frequency, ω, and wavelength, λ, of the wave.

(*b*) Find the maximum vertical displacement of any point on the cord.

(*c*) Find the maximum vertical velocity and vertical acceleration of any point on the cord.

Solution

(*a*) $T = 1/f = 0.20$ s; $\omega = 2\pi f = 6.28(5.0 \text{ Hz}) = 31.4$ rad/s. To get λ we use Eqs. (*1.6*) and (*1.9*): $k = \omega/v_p = (31.4 \text{ s}^{-1})/(10 \text{ m/s}) = 3.14 \text{ m}^{-1}$; $\lambda = (2\pi)/k = 2.0$ m.

(*b*) Assuming no losses, the amplitude, A, is the same everywhere along the cord, so $A = 3.0$ cm.

(*c*) Noting that all the points on the cord exercise SHM of the same frequency and amplitude, and recalling the expressions for maximum velocity and acceleration (Beginning Physics I, Chap. 12, Eqs. 12.10b,c) we have: $v_{max} = \omega A = (31.4 \text{ s}^{-1})(3.0 \text{ cm}) = 0.942$ m/s; $\alpha_{max} = \omega^2 A = (31.4 \text{ s}^{-1})^2(3.0 \text{ cm}) = 29.6$ m/s.

Problem 1.8.

(*a*) Re-express the travelling wave equation, Eq. (*1.7*) in terms of the period T and the wavelength, λ.

(*b*) Find an expression for the velocity of propagation, v_p, in terms of the wavelength, λ, and frequency, f.

Solution

(*a*) Recalling that $\omega = 2\pi f = 2\pi/T$, and that $\lambda = 2\pi/k$, Eq. (*1.9*), we have, substituting into Eq. (*1.7*):

$$y = A \sin (2\pi t/T - 2\pi x/\lambda) = A \sin [2\pi(t/T - x/\lambda)] \tag{i}$$

(*b*) From Eq. (*1.6*) we have: $v_p = \omega/k = 2\pi f/(2\pi/\lambda)$, or:

$$v_p = \lambda f \tag{ii}$$

Eq. (*ii*) of Problem (*1.8*) is a very general result for all travelling sinusoidal waves and can be illustrated intuitively by examining the travelling wave in Fig. 1-4(*b*). Consider the cord at point e in Fig. 1-4(*b*). At the instant shown (time $t = 0$) $y_e = 0$. As the wave moves to the right a quarter of a wavelength the crest originally at point d is now above point e, so the cord at point e has moved to its maximum positive position which is $\frac{1}{4}$ of the period of SHM, T. When the wave moves another $\frac{1}{4}$ wavelength the position originally at point c arrives at point e, so the cord at point e is now back at equilibrium, corresponding to another $\frac{1}{4}$ period. After moving another $\frac{1}{4}$ wavelength the wave originally at point b is over point e, so the cord at point e is now at its negative maximum, corresponding to another $\frac{1}{4}$ period. Finally, when the last quarter wavelength has moved over, the wave originally at point a is now over point e, and the cord at point e is now back to the equilibrium, completing the final

$\frac{1}{4}$ of the SHM period. Clearly, then, the wave has moved a distance λ to the right in the time of one SHM period, T. So, speed = distance/time, or:

$$v_p = \lambda/T = \lambda f \tag{1.10}$$

Of course, we have been assuming that Eq. (*1.5*), or equivalently, Eq. (*1.7*), represents a travelling wave moving to the right with velocity v_p. In the next problem we demonstrate that this actually follows from the wave equation itself.

Problem 1.9. Show by direct mathematical analysis that Eq. (*1.7*) is a travelling wave to the right with velocity: $v_p = \omega/k = \lambda f$.

Solution

Consider the wave shown in Fig. 1-4(*b*), which represents a snapshot at time, t, of a cord with a wave obeying Eq. (*1.7*). We consider an arbitrary point, x, along the cord corresponding to a particular position on the wave form, and ask what is the change in position, Δx, along the cord of the chosen vertical point on the wave form in a new snapshot of the cord taken a short time, Δt, later.

Since a given vertical position corresponds to a definite "angle" or phase of the sine wave, we have from Eq. (*1.7*), Δx and Δt obey: $[\omega t - kx] = [\omega(t + \Delta t) - k(x + \Delta x)]$. Canceling like terms we get:

$$\omega \Delta t - k\Delta x = 0 \Rightarrow \Delta x/\Delta t = \omega/k \tag{i}$$

Since Δx represents the distance the chosen point on the wave form moves in a time Δt, we have $\Delta x/\Delta t$ represents the speed of the chosen point on the wave form. Furthermore, since ω/k is a positive constant, all points on the wave form move at the same speed (as expected), and in the positive x (to the right) direction. This speed is just the velocity of propagation, so $v_p = \Delta x/\Delta t$ or, $v_p = \omega/k = \lambda f$, the desired result.

Problem 1.10.

(*a*) Consider the situation in Problem 1.7. If the student shakes the cord at a frequency of 10 Hz, all else being the same, what is the new wavelength of the travelling wave?

(*b*) Again assuming the situation of Problem 1.7, but this time the tension in the cord is increased to 40 N, all else being the same. What is the new wavelength?

(*c*) What is the wavelength if the changes of parts (*a*) and (*b*) both take place?

(*d*) Do any of the changes in parts (*a*), (*b*), (*c*) affect the transverse velocity of the wave in the cord? How?

Solution

(*a*) The velocity of propagation remains fixed if the tension, S, and mass per unit length, μ, remain the same. Therefore, if we use primes to indicate the new frequency and velocity we must have: $v_p = \lambda f = \lambda' f'$. For our case $f' = 10$ Hz so, from Problem 1.7, $v_p = 10$ m/s $= \lambda' f' = \lambda'$ (10 Hz) $\Rightarrow \lambda' = 1.0$ m. (Or, starting from the situation in Problem 1.7, $f = 5$ Hz and $\lambda = 2.0$ m for fixed v_p, if the frequency doubles the wavelength must halve, giving $\lambda' = 1.0$ m.)

(*b*) From Eq. (*1.1*), v_p increases as the square root of the tension, S. Here the tension has doubled from the value in Problem 1.7, so the new velocity of propagation is $v_p' = (\sqrt{2})v_p = 1.414(10 \text{ m/s}) = 14.1$ m/s. Since the frequency has remained the same we have:

$$v_p' = \lambda' f' \Rightarrow 14.1 \text{ m/s} = \lambda'(5.0 \text{ Hz}) \Rightarrow \lambda' = 2.82 \text{ m}.$$

(*c*) Combining the changes in (*a*) and (*b*), we have:

$$v_p' = \lambda' f' \Rightarrow 14.1 \text{ m/s} = \lambda'(10 \text{ Hz}) \Rightarrow \lambda' = 1.41 \text{ m}.$$

(*d*) As can be seen in Problems 1.7, the transverse velocity and acceleration are determined by ω and A. In none of parts (*a*), (*b*), or (*c*) is A affected. In part (*b*) $\omega = 2\pi f$ is not changed either, so no change in transverse velocity and acceleration takes place. In parts (*a*) and (*c*) the frequency has doubled, so ω

doubles as well. Then, the maximum transverse velocity, v_{max} doubles to 1.88 m/s, and the maximum transverse acceleration quadruples to 118 m/s^2.

Problem 1.11. Using the analysis of Problem 1.9, find an expression for a travelling sinusoidal wave of wavelength λ and period T, travelling along a string to the left (along the negative x axis).

Solution

As usual we define $k = 2\pi/\lambda$ and $\omega = 2\pi/T$ for our wave travelling to the left. From Eq. (*i*) of Problem 1.9 we see that if the phase of our sine wave had a plus instead of minus sign, [i.e., was $\omega t + kx$], then our analysis of the motion of the wave motion would lead to: $\Delta x/\Delta t = -\omega/k$. This corresponds to a negative velocity: $v_p = -\omega/k$. The wave equation itself is then:

$$y_x(t) = A \sin(\omega t + kx) \qquad (i)$$

This wave clearly has the same period of vertical motion at any fixed point on the string, and the same wavelength, as a wave travelling to the right [Eq. (*1.7*)] with the same A, k, and ω.

Problem 1.12. Two very long parallel rails, one made of brass and one made of steel, are laid across the bottom of a river, as shown in Fig. 1-5. They are attached at one end to a vibrating plate, as shown, that executes SHM of period $T = 0.20$ ms, and amplitude $A = 19$ μm. Using the speeds of sound (velocities of propagation) given in Problem 1.4 for water and brass, and in Problem 1.5 for steel:

(*a*) Find the wavelengths of the travelling waves set up in each rail and in the water.

(*b*) Compare the maximum longitudinal displacement of molecules in each rail and in water to the corresponding wavelengths.

(*c*) Compare the maximum longitudinal velocity of the vibrating molecules in each rail and in water to the corresponding velocities of propagation.

Solution

(*a*) For each material, $v_p = \lambda f$, with $f = 1/T = 5000$ s^{-1}. For steel [from Problem 1.5(*b*)], $v_{p,s} = 4990$ m/s, so $\lambda_s = (4990 \text{ m/s})/(5000 \text{ s}^{-1}) = 0.998$ m. For brass [from Problem 1.4(*b*)], $v_{p,b} = 3253$ m/s, so $\lambda_b = 3253/5000 = 0.651$ m. For water (from Problem 1.4(*a*))$v_{p,\omega} = 1450$ m/s, so $\lambda_w = 1450/5000 = 0.290$ m.

(*b*) For all cases, assuming no losses, the amplitude is 19 μm $= 1.9 \cdot 10^{-5}$ m, which is more than a factor of 10^4 smaller than the wavelengths for all three cases.

(*c*) For each case the maximum SHM velocity is $v_{max} = \omega A = 2\pi f A$, which yields: $v_{max} = 6.28(5000$ s$^{-1})(1.9 \cdot 10^{-5}) = 0.596$ m/s. Again, these are very small compared to the propagation velocity in each material.

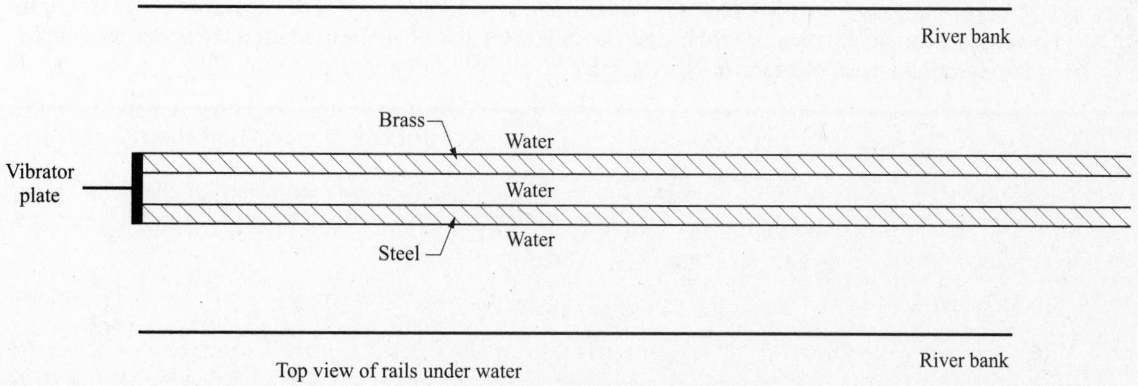

Fig. 1-5

Problem 1.13. Write the specific equation describing the travelling longitudinal wave in the steel rail of Problem 1.12. Assume x is measured from the vibrator end of the rail.

Solution

The general equation is given by Eq. (*1.7*). For steel $\omega = 2\pi f = 6.28(5000\ \text{s}^{-1}) = 31{,}400\ \text{s}^{-1}$; $k = 2\pi/\lambda = 6.28/(0.998\ \text{m}) = 6.29\ \text{m}^{-1}$; $A = 1.9 \cdot 10^{-5}$ m. Substituting into Eq. (*1.9*) we get:

$$y_x(t) = (1.9 \times 10^{-5}\ \text{m}) \sin [(31{,}400\ \text{s}^{-1})t - (6.29\ \text{m}^{-1})x] \qquad (i)$$

This could also be obtained by substitution of appropriate quantities into Eq. (*1.5*) or Eq. (*i*) of Problem 1.8.

Problem 1.14. The equation of a transverse wave in a cord is given by:

$$y_x(t) = (2.0\ \text{cm}) \sin [2\pi/(0.040\ \text{s}) + 2\pi x/(0.50\ \text{m})] \qquad (i)$$

(*a*) Find the amplitude, wavelength and frequency of the wave.

(*b*) Find the magnitude and direction of the velocity of propagation, v_p.

(*c*) Find the maximum transverse velocity and acceleration of the wave.

Solution

(*a*) We could compare Eq. (*i*) with Eq. (*1.7*), to get ω and k, but Eq. (*i*) is given in a way that is more easily translated using Eq. (*i*) of Problem (1.8). There a comparison shows:

$$A = 2.0\ \text{cm}; \qquad T = 0.040\ \text{s} \Rightarrow f = 1/T = 25\ \text{s}^{-1}; \qquad \lambda = 0.50\ \text{m}.$$

(*b*) In magnitude, $v_p = \lambda f = (0.50\ \text{m})(25\ \text{s}^{-1}) = 12.5$ m/s; the direction is along the negative x axis, because of the plus sign in the argument of the sine function (see Problem 1.11).

(*c*) $v_{max} = \omega A = 2\pi f A = 6.28(25\ \text{s}^{-1})(2.0\ \text{cm}) = 3.14$ m/s; $\alpha_{max} = \omega^2 A = \omega v_{max} = 6.28(25\ \text{s}^{-1})(3.14\ \text{m/s}) = 493\ \text{m/s}^2$.

Energy and Power in a Travelling Sinusoidal Wave

When a wave travels in a medium it carries energy. To calculate the energy in a given wave, and the rate at which energy transfers (power) from one point to another in the medium, we require a detailed knowledge of the wave and the medium in which it travels. For the case of a transverse sinusoidal wave travelling in a cord, or a longitudinal sinusoidal wave travelling in a rail or tube, it is not hard to calculate the energy per unit length and the power transfer across any point or cross-section. Consider the case of the wave in a cord of linear density μ. As the wave travels, all the molecules in any length L of the cord are executing SHM of amplitude A and angular frequency ω, although they are all out of phase with each other. The total energy of SHM equals the maximum kinetic energy which, for a particle of mass m, is just: $\frac{1}{2}mv_{max}^2$, where v_{max} is the maximum transverse velocity, $v_{max} = \omega A$. Since all the particles have the same maximum velocity, and the mass in a length L is μL, we have for the energy, E_L, in a length L of cord: $E_L = \frac{1}{2}\mu L\omega^2 A^2$. Dividing by L to get the energy per unit length, $\varepsilon = E_L/L$, we have:

$$\varepsilon = \tfrac{1}{2}\mu\omega^2 A^2 \qquad (1.11)$$

To find the power, or energy per unit time passing a point in the cord, we just note that the wave travels at speed v_p, so that in time t a length $v_p t$ of wave passes any point. The total energy passing a point in time t is then $\varepsilon v_p t$. Dividing by t to get the power, P, we have:

$$P = \varepsilon v_p = \tfrac{1}{2}\mu\omega^2 A^2 v_p \qquad (1.12)$$

Problem 1.15. Assume that the travelling transverse wave of Problem 1.14 is in a cord with $\mu = 0.060$ kg/m.

(*a*) Find the energy per unit length in the wave.

(*b*) Find the power transferred across any point as the wave passes by.

Solution

(*a*) From Problem 1.14 we have $A = 2.0$ cm and $\omega = 2\pi f = 6.28(25 \text{ Hz}) = 157 \text{ s}^{-1}$. Applying Eq. (*1.11*) we have:

$$\varepsilon = \tfrac{1}{2}(0.060 \text{ kg/m})(157 \text{ s}^{-1})^2(0.020 \text{ m})^2 = 0.296 \text{ J/m}.$$

(*b*) Noting that $P = \varepsilon v_p$, and from Problem 1.14 that $v_p = 12.5$ m/s, we get:

$$P = (0.296 \text{ J/m})(12.5 \text{ m/s}) = 3.70 \text{ W}.$$

Problem 1.16.

(*a*) How are Eqs. (*1.11*) and (*1.12*) modified for the case of a longitudinal sinusoidal wave in a rail or tube?

(*b*) Find the energy/length and power of the longitudinal sinusoidal wave in the steel rail of Problem 1.12, if the cross-sectional area is 20 cm^2. The density of steel is 7860 kg/m^3.

Solution

(*a*) From the derivation in the text, we see that the mass per unit length is needed for both ε and P, irrespective of whether the waves are transverse or longitudinal. For our rail or tube filled with fluid, the usual quantity given is the mass/volume or density, ρ. If the cross-sectional area of the rail or tube is labelled C_A, we have: $\mu = \rho C_A$, and Eqs. (*1.11*) and (*1.12*) are still valid as written.

(*b*) From Problem 1.12 we have for the steel rail: $v_p = 4990$ m/s, $\omega = 2\pi f = 2\pi(5000 \text{ Hz}) = 31{,}400 \text{ s}^{-1}$, and $A = 1.9 \cdot 10^{-5}$ m. Noting that $\mu = \rho C_A = (7860 \text{ kg/m}^3)(2.0 \cdot 10^{-3} \text{ m}^2) = 15.7$ kg/m, and substituting into Eq. (*1.11*), we get: $\varepsilon = \tfrac{1}{2}(15.7 \text{ kg/m})(31{,}400 \text{ s}^{-1})^2(1.9 \cdot 10^{-5} \text{ m})^2 = 2.79$ J/m. Similarly, $P = \varepsilon v_p = (2.79 \text{ J/m})(4990 \text{ m/s}) = 13{,}900$ W.

It is important to note that the equations for travelling waves, Eqs. (*1.5*) or (*1.7*) describe ideal sinusoidal waves that are travelling forever (all times t) and extend from $x = -\infty$ to $x = +\infty$. Real sinusoidal waves are typically finite in length, from several to hundreds of wavelengths long, and are called "wave trains". Thus, if the student starts her SHM motion of one end of the cord at some instant of time $t = 0$, and stops at some later time, t_f, Eqs. (*1.5*) and (*1.7*) do not exactly describe the cord at all times t and at all positions x. Still, for long wave trains, these equations do describe the wave motion accurately during those times and at those positions where the wave is passing by.

1.3 REFLECTION AND TRANSMISSION AT A BOUNDARY

Reflection and Transmission of a Pulse

Until now we have assumed our cords, rails, etc., were very long so we did not have to deal with what happened when our wave hit the other end. In this section we consider what happens at such an end. Consider the long cord, under tension S, of Fig. 1-1, with the single pulse travelling to the right. Assume that the other end is tightly tied to a strong post. Figures 1-6(*a*) to (*d*) shows the cord at different times before and after the pulse hits the tie-down point. It is found that the pulse is *reflected* back, turned upside down, but with the same shape and moving at the same speed, now to the left. The amplitude will also be the same except for the thermal energy losses along the cord and at the end. There always will be some losses but we ignore them here for simplicity. This reversal of the pulse can be understood by applying the laws of mechanics to the end of the cord, but the mathematics is too complicated for presentation here. We can, however, give a qualitative explanation.

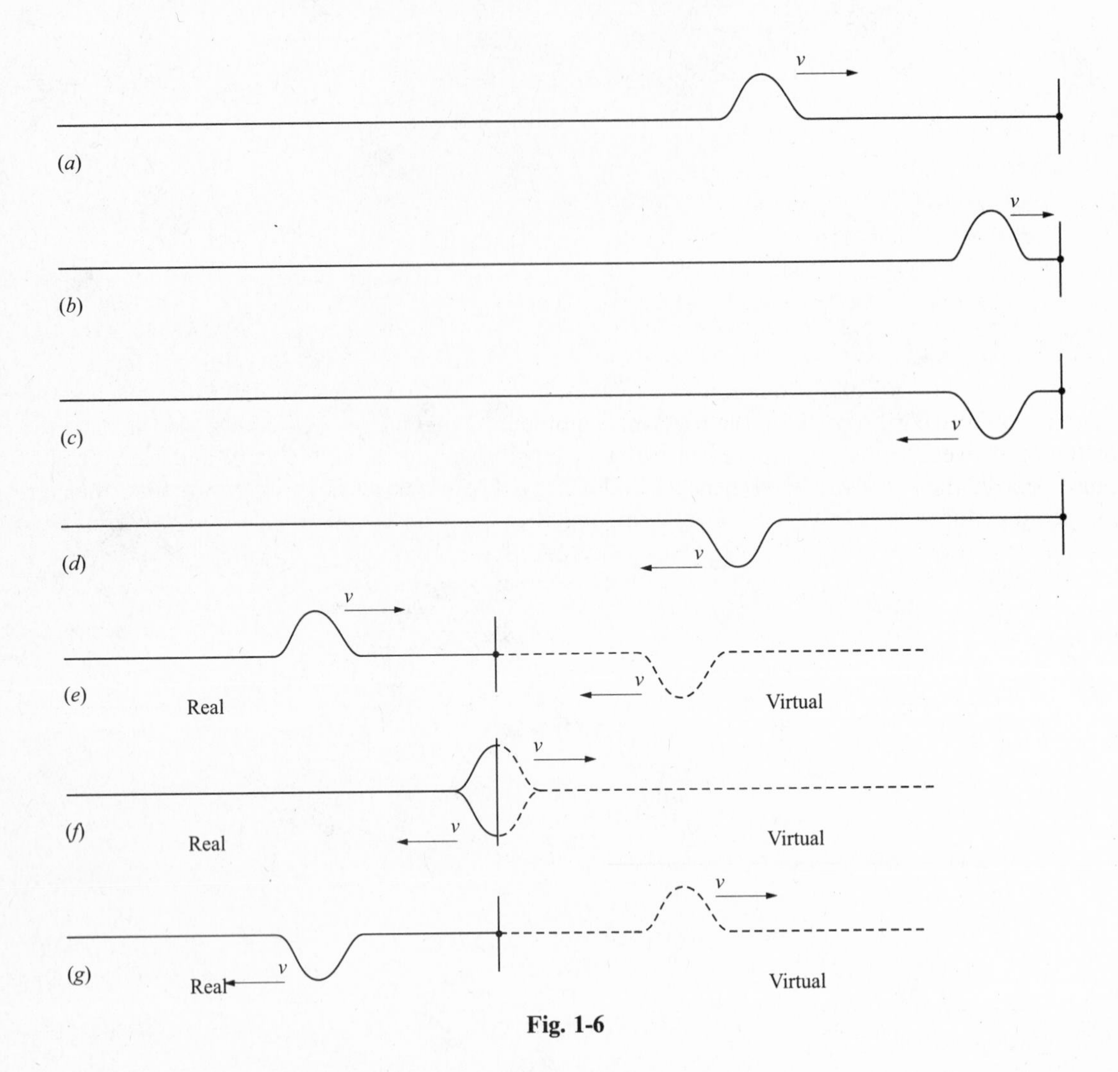

Fig. 1-6

As the wave travels along the cord the molecules communicate their transverse motion, and associated momentum and energy, to the next layer along the cord. In this way when the pulse passes a portion of the cord, that portion returns to rest while the next portion goes through its paces. As the pulse reaches the end of the cord it can't transfer its upward momentum because that point is tied down. Instead, the cord near the end first gets stretched slightly upward, like a stiff spring, as the first half of the pulse reaches it and like a stiff spring almost instantaneously snaps down in response, sending the molecules in the opposite transverse direction. As the second half of the pulse arrives the cord near the end is stretched downward. Again almost instantaneously springing the molecules back up. In effect the cord near the end mimics the original up–down snap of the student who originated the pulse at the other end, but this time it's a down–up snap so the pulse is upside down, as shown. The newly created pulse then travels back along the cord with the same characteristic velocity of propagation, v_p.

There is a nice way of visualizing the reflection process. We think of the tied down end of the cord as being a mirror, with the reflection of the cord and the pulse appearing to the right of the "mirror" point (dotted). Since this is merely a reflection and not real it is called the **virtual** pulse. This virtual pulse differs from a visual reflection of an object in a flat glass mirror only by its being upside down. In every other way it has the same properties as the visual image: it is as far to the right of the "mirror" point as the actual pulse is to the left, has the same shape, and is travelling to the left with the same speed as the real pulse is travelling to the right. The real pulse and virtual pulse reach the mirror point at the same time. We then imagine that what happens next is that the real pulse continues on past the mirror as a virtual pulse, while what was originally the virtual pulse emerges to the left of the "mirror" point as the new real pulse. For the short time while the real and virtual pulses are passing the "mirror" point, parts of both are real and have equal and opposite displacements at the "mirror"point. The effect is that they cancel each other out at that point so that, as necessary, the end of the cord doesn't move. This process is depicted in Fig. 1-6(*e*) to (*g*). This model actually gives an accurate representation of what actually happens to the cord upon reflection.

If the far end of the cord were not tied down, but instead ended with a light frictionless loop around a greased pole, we would again get a reflected pulse, but this time it would not be upside down. This case is shown in Fig. 1-7(*a*) to (*d*). Again the mathematical demonstration of this phenomena is beyond the scope of the book but a qualitative explanation can be given. Here, as the pulse reaches the end there is no more cord to pick up the transverse momentum and energy of the pulse, so this time the end of the cord overextends upward before being whipped back down, as the front and back ends of the pulse deposit their transverse momentum and energy. The net effect is an up–down snap that directly mimics the student's original up–down snap, and the right-side up pulse travels to the left, as shown. Again, we can use our "mirror" point approach to consider the reflection process. Here, however, the virtual pulse is right side up, just as a visual image in a flat mirror would be, and the overlap of the

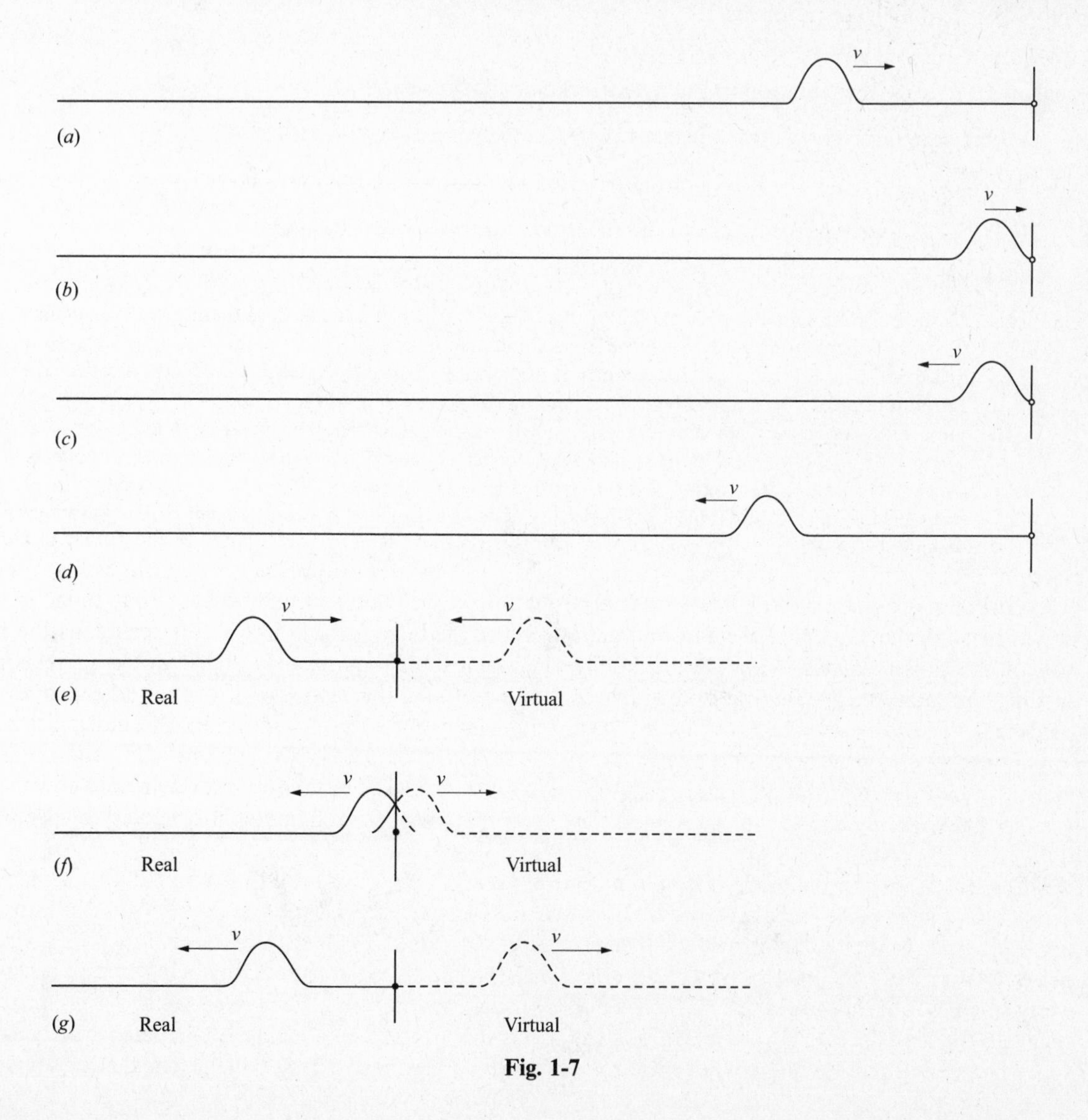

Fig. 1-7

pulses as they pass the end point reinforce rather than cancel each other, leading to an exceptionally large amplitude at the end point, as expected. The situation in every other way is the same as for the tied down cord, and is depicted in Fig. 1-7(*e*) to (*g*).

In each of the two cases discussed, the pulse reflects off the barrier at the far end. In the first case we say the reflection is "180° out of phase". This terminology originates not from the case of reflection of the single pulse, but rather from the case of reflection of a travelling sinusoidal wave as in Fig. 1-4(*b*). At the tied down end the upside-down reflection for the pulse would be equivalent to a half wave-length, or 180°, shift upon reflection in the travelling wave. The second case, with the "free" end, is a reflection that is "in phase", since the sinusoidal wave just reflects back without a flip-over.

The two cases are the extreme examples of possible boundary conditions. In one case the end is rigidly held down by the molecules of the bar to the right of it, so it cannot move at all; in the other case the end has no molecules to the right of it that exert any up–down constraints of any kind. A more general case is somewhere in-between these two extremes. Consider the case of two long ropes, A and B, of linear densities μ_a and μ_b, respectively, attached as shown in Fig. 1-8(*a*), with the combination held under tension S. A pulse is shown travelling to the right through the first rope. We can ask what happens when the pulse hits the interface? We would expect that part of the pulse will reflect off the interface back along rope A and that part will be *transmitted* to the right along rope B. This behavior is explored in the following problems.

Problem 1.17. For the situation in Fig. 1-8(*a*), assume that $\mu_a < \mu_b$.

(*a*) Describe qualitatively what happens to the pulse after it hits the interface.

(*b*) Describe qualitatively the height of the reflected and transmitted pulses.

(*c*) Describe qualitatively the speeds of the reflected and transmitted pulses.

Solution

(*a*) As the pulse hits the interface the molecules of the first rope suddenly find themselves conveying their transverse momentum and energy to a more massive material. This is somewhat like the case of the tied down barrier, discussed above but not as extreme. As a consequence we will get a reflected pulse 180° out of phase. This time, however, the molecules to the right of the interface, in rope B, will pick up some of the transverse momentum and energy of the molecules in rope A, just as if someone had snapped that end of rope B up and down, and part of the pulse will be transmitted to rope B, and continue moving to the right. The transmitted pulse is in phase, since it is a direct response to the transverse motion of the molecules in rope A. The reflected and transmitted pulses are shown (not to scale) in Fig. 1-8(*b*).

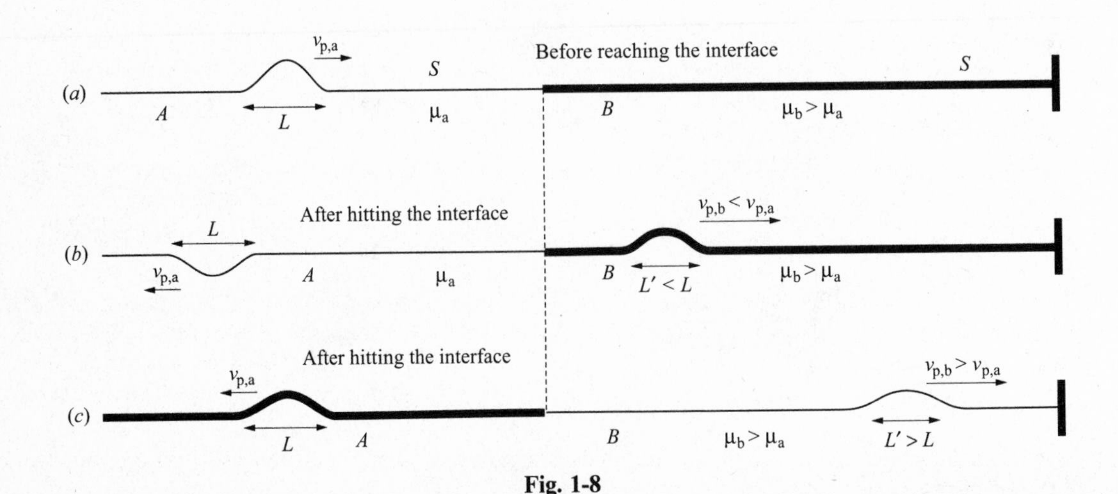

Fig. 1-8

(*b*) Since the total available energy comes from the original pulse, and is now shared between the reflected and transmitted pulses, those two pulses will have diminished energy which will most visibly manifest itself in reduced amplitude of each pulse. Determining the exact distribution of energy in the two pulses is beyond the scope of this book.

(*c*) Once the pulses leave the interface they must travel with the characteristic velocities of waves in the respective ropes. The reflected wave will travel to the left with the same magnitude velocity as the incoming pulse, $v_{p,a} = (S/\mu_a)^{1/2}$. The transmitted wave will travel to the right with the velocity $v_{p,b} = (S/\mu_b)^{1/2}$. Since rope B is more massive than rope A, we have $v_{p,b} < v_{p,a}$.

Problem 1.18. Suppose in the previous problem rope A were more massive than rope B ($\mu_a > \mu_b$). How would the answers to parts (*a*) to (*c*) change? Describe the length of the reflected and transmitted pulses,

Solution

(*a*) The new situation is depicted in Fig. 1-8(*c*). The only change in our answer to part (*a*) is that the reflected wave will be upright, or in phase. Here the molecules in rope B are more easily pushed up and down than those of rope A, and the conditions more closely correspond to the cord with the "free" end described earlier in the text.

(*b*) Energy reasoning is the same, but amplitude of transmitted pulse might be larger.

(*c*) The answer is basically the same, except that $v_{p,b} > v_{p,a}$.

The initial and reflected pulses have the same length. The transmitted pulse is longer because the speed in rope B is larger and the front of the pulse moves further before the back hits the interface.

Reflection and Transmission of a Sinusoidal Travelling Wave

We now extend our discussion to travelling waves that reach an interface.

Problem 1.19. Assume that a travelling sinusoidal wave of amplitude $A = 0.40$ cm, frequency $f = 40$ Hz and wavelength $\lambda = 0.50$ m is moving to the right in rope A of Fig. 1-8(*a*). Rope B has a linear density twice that of rope A. Assume that we have a finite wave train many wavelengths long, but still small in length compared to the length of the ropes, and that it has not yet reached the interface. The common tension in the ropes is $S = 200$ N.

(*a*) Find the velocity of propagation, $v_{p,a}$ in rope A.

(*b*) Find the linear density, μ_a, of rope A.

(*c*) Find the velocity of propagation of a wave in rope B.

Solution

(*a*) $v_{p,a} = \lambda f = (0.50 \text{ m})(40 \text{ Hz}) = 20$ m/s.

(*b*) From Eq. (*1.1*): $v_{p,a}^2 = S/\mu_a \Rightarrow \mu_a = (200 \text{ N})/(20 \text{ m/s})^2 = 0.50$ kg/m.

(*c*) From the information given, $\mu_b = 2\mu_a = 1.00 \text{ kg/m} \Rightarrow v_{p,b} = (S/\mu)^{1/2} = [(200 \text{ N})/(1.00 \text{ kg/m})]^{1/2} = 14.1$ m/s. [Or, equivalently, $\mu_b = 2\mu_a \Rightarrow v_{p,b} = v_{p,a}/\sqrt{2} = (20 \text{ m/s})/1.414 = 14.1$ m/s].

Problem 1.20. When the wave train of Problem 1.19 hits the interface, part of the wave will be reflected and part will be transmitted through to rope B. Here we address only the transmitted wave.

(*a*) What is the frequency and wavelength of the transmitted wave.

(*b*) Assuming that half the energy of the incoming wave transmits and half reflects, find the amplitude of the transmitted wave. [*Hint*: See Problem 1.15, and Eq. (*1.12*).]

Solution

(*a*) The frequency will be the same in the transmitted as in the initial wave: $f_b = f_a$. This follows from the fact that the stimulating SHM comes from the incoming wave and the interface must move up and down at a common rate. Thus, $f_b = 40$ Hz. We can determine the wavelength from the requirement that: $v_{\rm p,\,b} = f_b \lambda_b$. Using Problem 1.19(*c*), we have: $\lambda_b = (14.1 \text{ m/s})/(40 \text{ Hz}) = 0.35$ m, a shorter wavelength than in rope A.

(*b*) The transmitted power must be half the power of the incoming wave, since half the incoming energy is reflected back. From Eq. (*1.12*), we have for the incoming wave: $P_I = \frac{1}{2}\mu_a \omega^2 A_I^2 v_{\rm p,\,a}$. For the transmitted wave we have: $P_T = \frac{1}{2}\mu_b \omega^2 A_T^2 v_{\rm p,\,b}$ where, ω is common to both waves, $\mu_b = 2\mu_a$, and from Problem 1.19(*c*), $v_{\rm p,\,b} = v_{\rm p,\,a}/\sqrt{2}$. We then have: $P_T = \frac{1}{2}P_I \Rightarrow \frac{1}{2}\mu_b \omega^2 A_T^2 v_{\rm p,\,b} = \frac{1}{4}\mu_a \omega^2 A_I^2 v_{\rm p,\,a}$. Canceling ω^2 on both sides, and noting the relationships of the velocities and the linear densities, we have:

$$2A_T{}^2/\sqrt{2} = \tfrac{1}{2}A_I^2.$$

Substituting $A_I = 0.40$ cm, we get: $A_T = 0.238$ cm.

The statement (Problem 1.20(*a*)) that the transmitted frequency is the same as the incoming frequency can be rigorously demonstrated, and is a very general statement about waves moving across a boundary or interface. Whatever changes occur as a wave moves across a boundary from one medium to another, the frequency stays the same. Since the velocities of propagation typically change from medium to medium, fixed frequency implies the wavelengths must change in accordance with Eq. (*1.10*).

1.4 SUPERPOSITION AND INTERFERENCE

We now address the question of what happens when two waves pass the same point on a cord (or in any medium) at the same time. For all materials through which waves travel, as long as the amplitudes of the waves are small, we have what is known as the *principle of superposition*, which can be expressed generally as follows:

> The actual vector displacement of molecules from their equilibrium position, at any given location in a medium, at any instant of time, when more than one wave is travelling through that medium, is just the vector sum of the displacements that each wave would separately have caused at that same location at that same instant of time.

For a sinusoidal wave travelling along a cord and its reflection from an interface, the displacements are in the same transverse y direction. Similarly, for sound waves in a long rail, the direct and reflected longitudinal displacements are again in the same longitudinal x direction. In a large body of water, however, one can imagine two or more waves, travelling in different directions, passing a single point. In that case the displacements can be in quite different directions. Even in a cord, if the cord is along the x axis, one could conceivably have one transverse wave travelling to the right with a displacement in the y

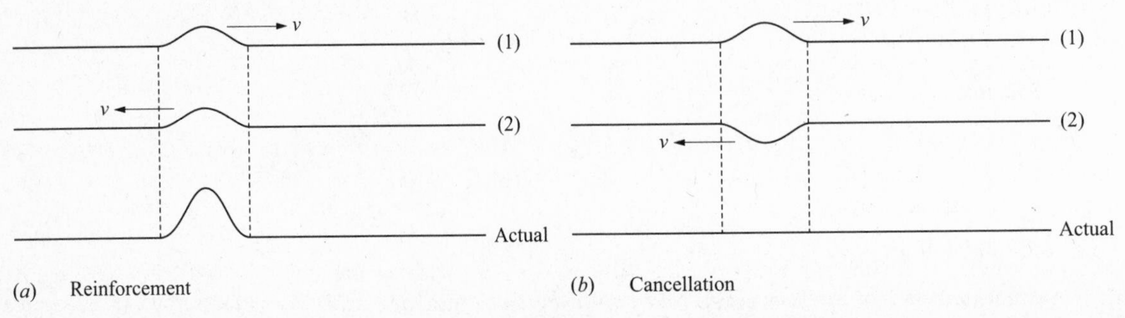

Fig. 1-9

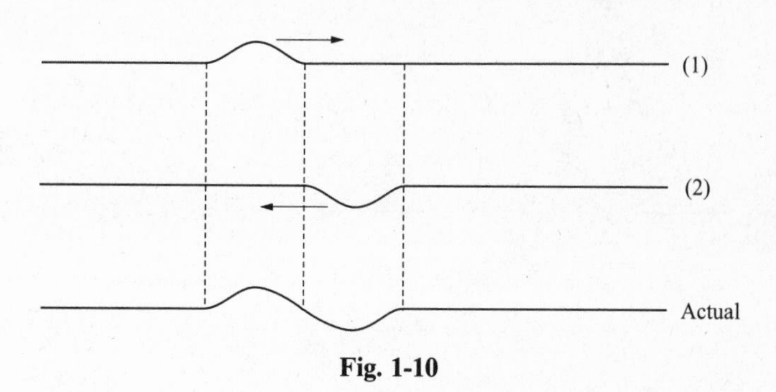

Fig. 1-10

direction, and another wave travelling to the left with a displacement in the z direction. The actual displacement of the cord would then be the vector sum of the two displacements. Figures 1-9 and 1-10 show a variety of situations demonstrating the principle of superposition applied to two transverse waves in the y direction passing each other in a cord. In each case there are three pictures; the first two showing the individual waves and the last the combined (actual) wave at that instant.

When two waves pass the same point in a medium they are said to interfere. If they correspond to long wave trains having the same wavelength, then certain regular patterns can appear, such as points that never move and points that move maximally. Such patterns are called *interference patterns.* In Fig. 1-11(*a*) to (*d*) we consider the case of two transverse sinusoidal waves of the same amplitude and wavelength travelling in opposite directions along a fixed portion of a cord. Each sub-figure has three pictures representing each wave separately and then the combined wave. The sub-figures are $\frac{1}{4}$ of a period apart, corresponding to each wave having moved $\frac{1}{4}$ wavelength. The relative positions of the two waves therefore move $\frac{1}{2}$ wavelength apart from sub-figure to sub-figure. An examination of the actual "superimposed" wave for each of the four cases reveals some interesting features. First, there are some points on the cord that seem not to move at all as the waves pass each other (points a, b, c, d, e) while other points midway between them move up and down with double the amplitude of either wave (points α, β, γ, δ). The actual wave motion of the cord is therefore not a travelling wave, since in a travelling wave every point on the cord moves up and down in succession. The wave caused by the interference of these two travelling waves is therefore called a *standing wave.* It has the same frequency since the points α, β, γ, δ move from positive maximum to negative maximum in half a period, as with a point on either travelling wave alone. Also, the distance between successive positive peaks (or successive negative peaks) is exactly one wavelength. The points that don't move are called *nodes* and the points that move maximally are called *anti-nodes.* The result shown can be derived mathematically by considering the equations of the two waves and adding them, as shown in the following problem.

Problem 1.21. Two long sinusoidal wave trains of the same amplitude and frequency are travelling in opposite directions in a medium (either transverse waves in a cord, or longitudinal waves in a rail or tube filled with fluid). Using the law of superposition find a mathematical expression for the resultant wave form. [*Hint*: $\sin(A \pm B) = \sin A \cos B \pm \sin B \cos A$]

Solution

Letting y_{x+} and y_{x-} represent the travelling waves along the positive and negative x axis, respectively we have:

$$y_{x+}(t) = A \sin(\omega t - kx) \quad \text{and} \quad y_{x-}(t) = A \sin(\omega t + kx + \theta_0)$$

where θ_0 is a constant phase which is included to account for the waves having been set up (initial conditions) so that different parts of the two waves happen to cross each other at a particular location x at a given time t. The choice of θ_0 will only affect the absolute positions of the nodes and anti-nodes, but not

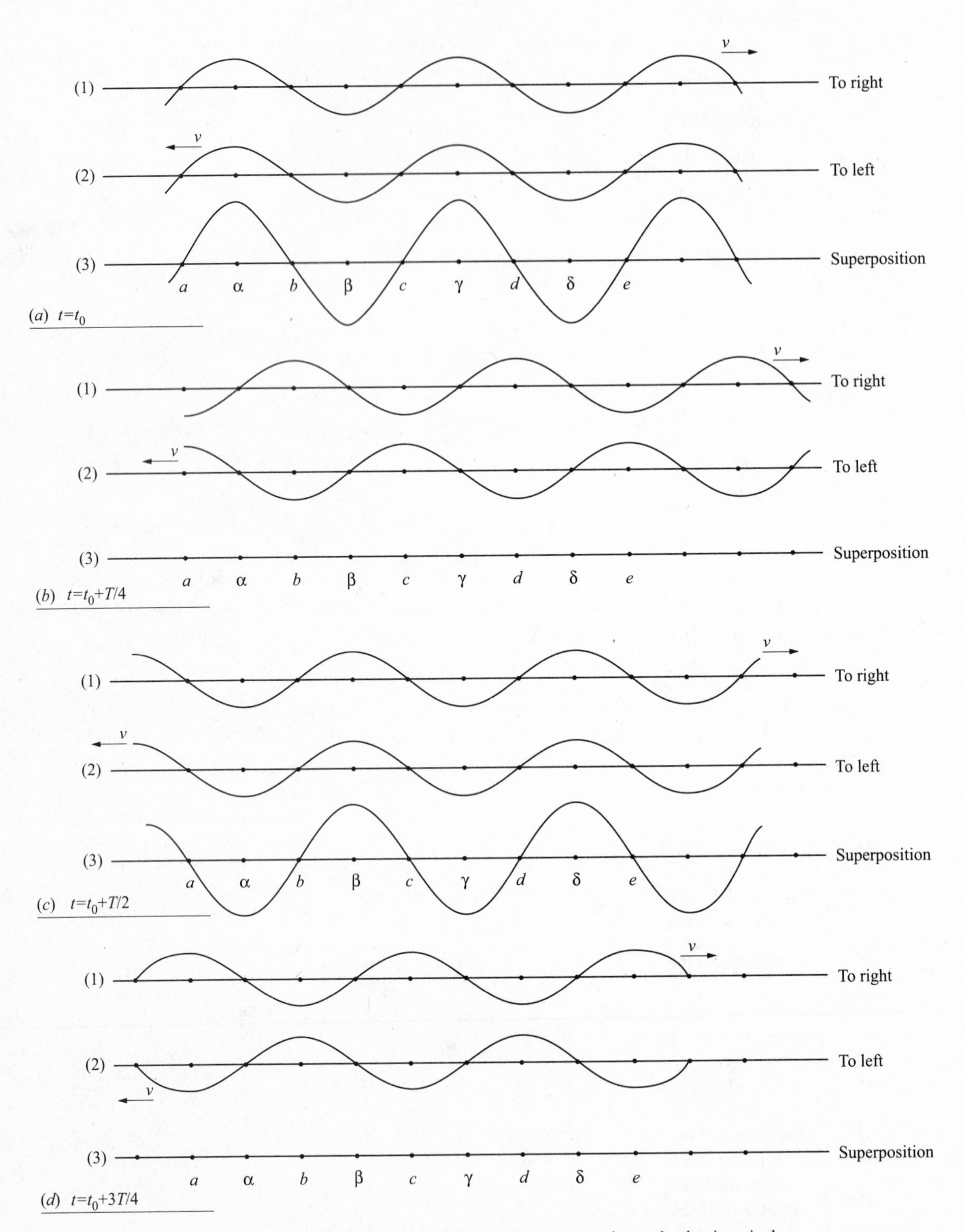

Two sinusoidal travelling waves of same frequency passing each other in a single cord. For (a)–(d): (1) shows wave travelling to the right; (2) shows wave travelling to the left; (3) shows the actual shape of cord at that instant. Unperturbed cord with reference points shown in each diagram for reference.

Fig. 1-11

any other characteristics of the resulting wave, so we choose $\theta_0 = 0$ for simplicity. By the law of superposition,

$$y_x(t) = y_{x+}(t) + y_{x-}(t) = A \sin(\omega t - kx) + A \sin(\omega t + kx) \qquad (i)$$

Using the trigonometric identity supplied in the hint, we have:

$$y_x(t) = A[\sin \omega t \cos kx - \sin kx \cos \omega t] + A[\sin \omega t \cos kx + \sin kx \cos \omega t] \qquad (ii)$$

Combining terms, we see that the second terms in each bracket cancel out to yield:

$$y_x(t) = 2A \sin \omega t \cos kx \qquad (iii)$$

Problem 1.22.

(*a*) Show that the result of Problem 1.22, Eq. (*iii*), is a standing wave and find the location and separation of the nodes.

(*b*) Find the location of the anti-nodes, and the amplitude of wave motion at those points.

(*c*) Interpret the behavior of the standing wave between any two nodes, and between those two nodes and the adjacent two nodes.

Solution

(*a*) This is clearly a standing wave since at each x for which $\cos kx = 0$ we have an immobile point, or node, for all times t. Recalling that $kx = 2\pi x/\lambda$, and that the cosine then vanishes at values of x for which: $2\pi x/\lambda = \pi/2, 3\pi/2, 5\pi/2, \ldots$, we have nodes at: $x = \lambda/4, 3\lambda/4, 5\lambda/4, \ldots$ The distance between successive nodes is thus $\lambda/2$.

(*b*) Similarly, the anti-nodes correspond to values of x for which we have the maximum possible wave amplitudes. Eq. (*iii*) of Problem 1.21 implies that for any given point x, the molecules oscillate in SHM of angular frequency ω, and of amplitude:

$$2A \cos kx = 2A \cos(2\pi x/\lambda) \qquad (i)$$

The cosine has alternating maximum values of ± 1, and at the corresponding points, x, the molecules oscillate with SHM of amplitude $2A$. The $\cos(2\pi x/\lambda) = \pm 1$ points occur at $2\pi x/\lambda = 0, \pi, 2\pi, 3\pi, \ldots$, with positive values at even multiples of π and negative values at odd multiples of π. The corresponding values of x are: $0, \lambda, 2\lambda, 3\lambda, \ldots$ and $\lambda/2, 3\lambda/2, 5\lambda/2$, respectively. The separation between adjacent anti-nodes, without reference to whether the cosine is $\pm$, is $\lambda/2$ (the same as the separation of adjacent nodes). Furthermore, comparing to part (*a*) the anti-nodes are midway between the nodes, and from node to next anti-node is a distance of $\lambda/4$.

(*c*) At the anti-nodes the equation of SHM are, from Eq. (*i*):

$$y_{\text{anti}}(t) = \pm 2A \sin \omega t \qquad (ii)$$

The only difference between the oscillations when the cosine is $+1$ as opposed to -1, is that they are 180° out of phase. As can be seen from Eq. (*ii*): when $y = +2A$ at one anti-node, $y = -2A$, at the next anti-node, and so on. All the points between two adjacent nodes oscillate "in phase" with each other—that is they reach their positive maxima, given by Eq. (*i*), at the same time. The points between the *next* two nodes, also oscillate in phase with each other, but exactly 180° out of phase with the points between the first two nodes. The overall shape and behavior of the waves is illustrated in Fig. 1-12, which depicts the "envelope" of the wave as each point varies between the maximum positive and negative transverse positions corresponding to the various positions along the cord. The vertical arrows at the anti-nodes are intended to depict the relative direction of transverse motion of points between adjacent pairs of nodes.

Resonance and Resonant Standing Waves

Many physical systems, when stimulated can be made to vibrate or oscillate with definite frequencies. Examples are a simple pendulum of a particular length, a mass at the end of a spring, a tuning

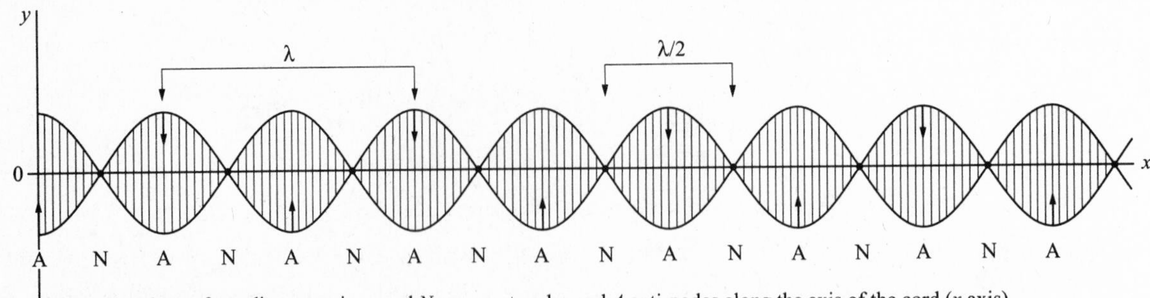

Envelope of standing wave in a cord N represent nodes and A anti-nodes along the axis of the cord (x axis). The two sine wave outlines correspond to the two maximum distortions of the cord from equilibrium and occur one-half period ($T/2$) apart.

Fig. 1-12

fork, and a children's swing in the park. In each of these cases there is a single "natural" frequency associated with the system. In other cases, such as a violin or guitar string, an organ pipe, or more complex structures such as a bridge or a building, many "natural" frequencies of vibration can occur. Such frequencies, which are characteristic of the particular system or structure, are called the *resonant frequencies* of the system. If one stimulates a system by shaking it with SHM of arbitrary frequency and low amplitude, the system will respond by vibrating at the stimulating frequency. The amplitude of the system's responsive vibration to such stimuli will generally be quite small. However, if one stimulates the system at one of its resonant frequencies, one can stimulate huge amplitude oscillations, sometimes to the point of destroying the structure. This is true because, when stimulating a system at a resonant frequency it is extremely easy to transfer energy to the system. A simple example of such a transfer is a mother pushing a child on a swing while talking to a friend. If the mother times her pushes to always coincide with the moment the swing is moving down and away from her, the force she exerts does positive work on each push and therefore transfers energy to the swing, increasing its amplitude. If on the other hand, she is distracted by conversation and pushes slightly off frequency, she will sometimes push the swing while it is still moving toward her, hence doing negative work so that the swing loses energy. If there are as many times when the swing loses energy as when it gains energy, the amplitude of the swing will not build up. The same is true of stimulating any system by vibrating it at a given frequency. If the frequency is a resonant frequency, each stimulating vibration will reinforce the previous one and pump energy into the system. If the stimulating vibration has a frequency even a little off the resonant frequency, over time there will be as many vibrations that lose energy to the system as gain energy to it, as in the case of the mother pushing the swing. This is the reason that army troops are ordered to "break step" while marching across a bridge; if the troop's "in-step" march is at the same frequency as one of the resonant frequencies of the bridge, the bridge could start to vibrate with ever increasing amplitude and actually break apart as a consequence. In the following problems we will determine the resonant frequencies of some simple systems.

Problem 1.23. Consider a long cord of length L, mass/length μ, under tension S, tied rigidly to a post at one end with the other end tied to a variable frequency metal vibrator, which vibrates with SHM of amplitude less than 0.05 mm, as shown in Fig. 1-13(a). At certain frequencies it is found that standing waves of very large anti-node amplitudes (possibly several centimeters or larger) appear in the cord. These are called resonant standing waves. Find the only frequencies for which such resonant standing waves can occur.

Solution

In Problem 1.22 we saw that standing waves have their nodes separated by $\lambda/2$, with anti-nodes midway between. For our cord under tension S, if resonant standing waves were to occur both ends of the cord would have to be nodes. This is true even at the vibrator end because the maximum transverse

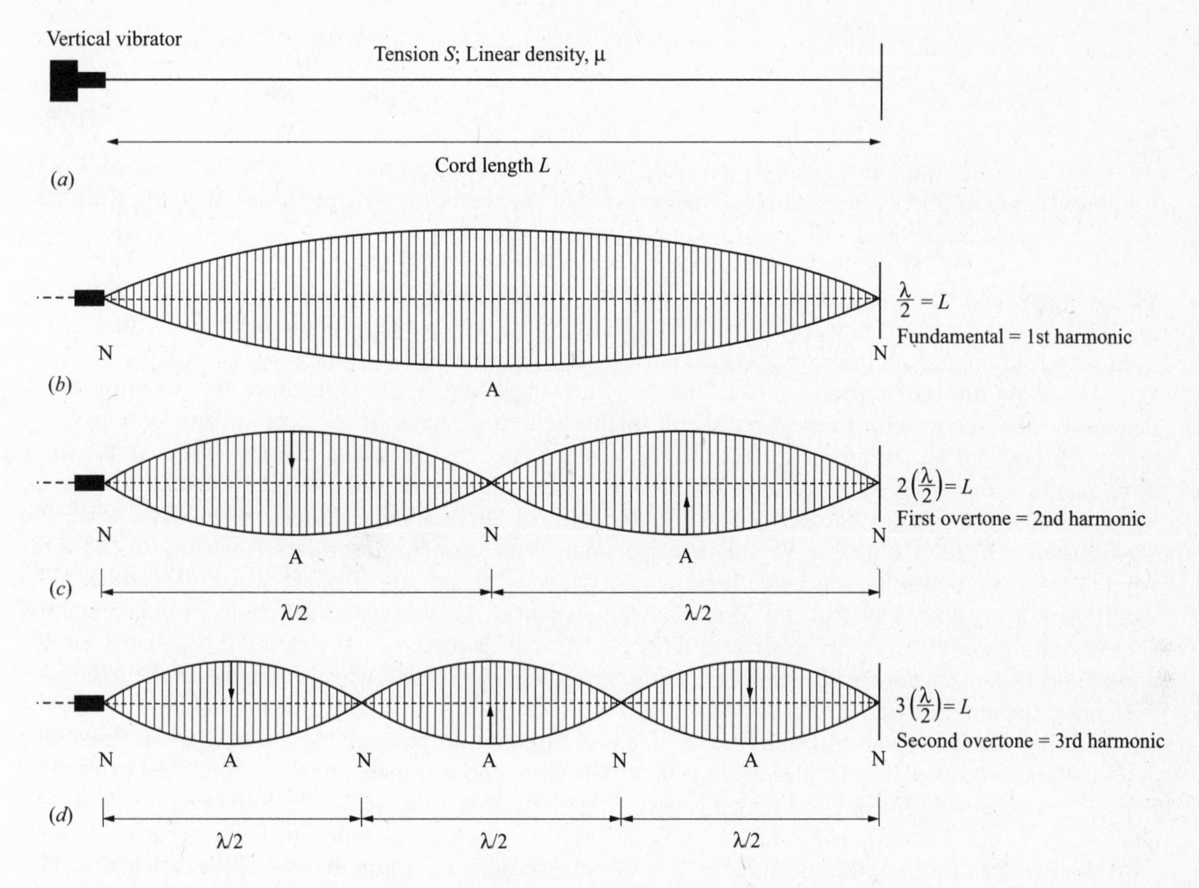

Envelopes of resonant standing waves in a cord. (Equilibrium position of cord shown as dashed line.)

Fig. 1-13

displacement of the vibrator is negligible ($<1\%$) of the amplitude of the resonant standing wave. On this basis, we must have a whole number, n, of half wavelengths fit into the length L, or, $n\lambda/2 = L$. The allowed wavelengths, therefore, are:

$$\lambda_n = 2L/n \qquad n = 1, 2, 3, \ldots \tag{i}$$

Since frequency and wavelength are related to the speed of waves in the cord: $v_p = \lambda f$, we have for the allowed frequencies:

$$f_n = v_p/\lambda_n = n(v_p/2L) \tag{ii}$$

From Problem 1.23, we note that the lowest possible frequency, called the *fundamental*, has the value, recalling Eq. (*1.1*):

$$f_F = f_1 = v_p/2L = (S/\mu)^{1/2}/2L \tag{1.13a}$$

In terms of the fundamental the other allowed frequencies become:

$$f_n = nf_F \qquad n = 1, 2, 3, \ldots \tag{1.13b}$$

For completeness, we repeat Eq. (*i*) of Problem 1.23 for the corresponding wavelengths:

$$\lambda_n = 2L/n \qquad n = 1, 2, 3, \ldots \tag{1.13c}$$

The integer multiples of the fundamental frequency are called *harmonics*. Thus, f_n is the nth harmonic, with the fundamental itself being the first harmonic. Another commonly used terminology is to call the successive frequencies, above the fundamental, *overtones*. Thus, for our vibrating cord, the first overtone is the second harmonic, the second overtone is the third harmonic, etc. Recalling that $v_p = (S/\mu)^{1/2}$ we can re-express Eq. (*1.13b*) as:

$$f_n = n[(S/\mu)^{1/2}/2L] \qquad (1.13d)$$

In Problem 1.23 we found the only possible resonant frequencies that could exist in the cord. Using the ideas of reflection of waves, and of resonance discussed above, we now show that these all are, indeed, resonant frequencies of the cord. Consider again the cord of Fig. 1-13(*a*). As the vibrator executes its low amplitude oscillations it sends a continuous travelling wave down the cord. This wave reflects off the tied down end, travels back toward the vibrator with the same frequency and wavelength and then reflects a second time, again travelling to the right. The process then repeats many times with multiple reflections before the amplitude dies down. The even reflections (2nd, 4th, · · ·) are travelling to the right while the odd reflections (1st, 3rd, · · ·) are travelling to the left. Since the vibrator keeps generating new waves which reflect back and forth, the actual shape of the cord at any instant is the sum of the overall superposition of the waves travelling to the right and the left. The overall wave travelling to the right is itself the sum of the newly generated wave and all the even reflected waves, while the overall wave travelling to the left is the sum of all the odd reflected waves. In general, the overall waves in either direction would be quite small. For example, the waves travelling to the right (original plus the multiple even reflections) would typically all be out of phase with one another. The superposition of these waves at any point on the string, at a given instant of time would, therefore, involve adding about as many positive as negative vertical displacements, yielding a net displacement, that would be almost negligible. The same would be true of the overall wave travelling to the left, being the sum of the multiple odd reflections.

In order for the overall travelling waves in either direction to be large the multiple waves travelling in that direction have to all be travelling in phase with one another—crest to crest and trough to trough. The condition for this to happen is easy to see. Consider the even reflections; each wave travels a distance of $2L$ to undergo a double reflection. For the original wave and all the subsequent double reflections to be in phase, the distance $2L$ has to correspond to a whole number of wavelengths. This would ensure that the crests of successive double reflections appear at the same place at the same time. (Note that there is a change of a half wavelength at each reflection, but since there are an even number of reflections the overall shift is a whole number of wavelengths and does not change the relative phases.) A similar argument holds for the odd reflected waves, which travel to the left; again these involve double reflections and we again must have a whole number of wavelengths fitting into $2L$. Thus, for either case—travelling to the right or to the left—the condition for in-phase reinforcement of reflected waves occurs at wavelengths that obey: $n\lambda_n = 2L$, where n is a positive integer, or equivalently: $\lambda_n = 2L/n$. This is the same result we obtained in Problem 1.23. Under these conditions we have, on net, a giant travelling wave to the right and a giant travelling wave to the left at essentially the same amplitude and wavelength, yielding a giant, or resonant, standing wave. Thus, the results of Problem 1.23 do, in fact, represent the actual requirements for resonant standing waves. The envelope of the first three resonant standing waves for a cord of length L, such as that of Fig. 1-13(*a*), are shown in Fig. 1-13(*b*) to (*d*) (as in Fig. 1-12, the vertical arrows indicate the relative direction of the transverse motion of molecules in the cord between successive pairs of nodes).

Problem 1.24. A rope of length $L = 0.60$ m, and mass $m = 160$ g is under tension $S = 200$ N. Assume that both ends are nodes, as in Problem 1.23.

(*a*) Find the three longest resonant wavelengths for the rope.

(*b*) Find the fundamental frequency and first two overtones for resonant standing waves in the rope.

(*c*) How would the results to part (*a*) and (*b*) change if the tension in the rope were 800 N?

Solution

(*a*) The wavelengths obey $\lambda_n = 2L/n = (1.20\text{ m})/n \Rightarrow$ longest wavelength $= \lambda_1 = 1.20$ m; next longest $= \lambda_2 = 0.60$ m; third longest $= \lambda_3 = 0.40$ m.

(*b*) From Eqs. (*1.13*) we see that we need the speed of propagation of the wave in the rope: $v_p = (S/\mu)^{1/2}$. Noting that $\mu = m/L = (0.16\text{ kg})/(0.60\text{ m}) = 0.267\text{ kg/m}$, we get: $v_p = [(200\text{ N})/(0.267\text{ kg/m})]^{1/2} = 27.4\text{ m/s}$. Then, from Eqs. (*1.13*) we get: for our fundamental (1st harmonic), $f_1 = (27.4\text{ m/s})/(1.20\text{ m}) = 22.9\text{ Hz}$; for our first overtone (2nd harmonic), $f_2 = 2f_1 = 45.8\text{ Hz}$; for our second overtone (3rd harmonic), $f_3 = 3f_1 = 68.7\text{ Hz}$. (The same results could be obtained directly from $f_n = v_p/\lambda_n$.)

(*c*) If the tension, S, quadruples the velocity, v_p, doubles. Since the resonant wavelengths depend only on the length of the cord, they remain the same as in part (*a*). The corresponding frequencies, however, are proportional to v_p, and therefore all double as well.

Problem 1.25. Suppose that we have the exact situation of Problem 1.24, except that now the far end of the cord is not tied down, but has a frictionless loop free to slide on a vertical bar, as in Fig. 1-7.

Extending the kind of reasoning employed in Problem 1.23, what would the allowed resonant wavelengths and frequencies be?

Solution

Since the far end is free to snap up and down without the constraining effect of any addition cord to the right, it should be an anti-node for any resonant standing waves that appear. The left end must again be a node. The standing wave envelope of the longest wavelength that fits this condition is shown in Fig. 1-14(*a*). Recalling that the distance from an anti-node to the next node is $\lambda/4$, we have: $\lambda_1 = 4L$. The next possible situations are shown in Figs. 1-14(*b*) and (*c*) from which we deduce: $\lambda_3 = 4L/3$, and $\lambda_5 = 4L/5$, where we label the successive wavelengths by the odd integer denominators. From this it is not hard to deduce that in general:

$$\lambda_n = 4L/n \qquad n = 1, 3, 5, \ldots \tag{i}$$

and the allowed frequencies are:

$$f_n = v_p/\lambda_n = n[v_p/4L] \qquad n = 1, 3, 5, \ldots. \tag{ii}$$

Problem 1.26.

(*a*) What is the relationship between the harmonics and overtones for the situation of Problem 1.25?

(*b*) If in Problem 1.24 the far end of the rope were looped to slide freely without friction on a vertical greased bar, what would the change be for the answer to Problem 1.24(*a*)?

(*c*) What would the corresponding change be for the answer to Problem 1.24(*b*)?

Solution

(*a*) From Problem 1.25, Eq. (*ii*), the fundamental frequency is $f_1 = v_p/4L$. This is the lowest frequency for our "free end" cord, and therefore the first harmonic. The first overtone, which is the next allowed frequency, is now $f_3 = 3f_1$, which, by virtue of being three times the fundamental, is by definition the third harmonic. Similarly, the third overtone is the fifth harmonic, and so on. For this case we see that only the odd harmonics are allowed frequencies.

(*b*) The three longest wavelengths now obey Eq. (*i*) of Problem 1.25, and are therefore: $\lambda_1 = 2.40$ m; $\lambda_3 = 0.80$ m; $\lambda_5 = 0.48$ m.

(*c*) The fundamental frequency is now: $v_p/4L = (27.4\text{ m/s})/(2.40\text{ m}) = 11.4\text{ Hz}$. Note that this is half the fundamental with the cord tied down. The first and second overtones are the third and fifth harmonics, respectively: $f_3 = 3(11.4\text{ Hz}) = 34.2\text{ Hz}$; $f_5 = 5(11.4\text{ Hz}) = 57.0\text{ Hz}$.

In our discussion of resonant standing waves in a cord, we assumed there was an SHM vibrator stimulating the waves at one end. It is also possible to stimulate standing waves with a non-SHM stimulus, such as bowing (as with violin strings), plucking (as with guitar strings), or hammering (as with piano wires). In these cases each stimulating disturbance can be shown to be equivalent to a combination of many SHM standing waves over a broad range of frequencies. As might be expected, only the

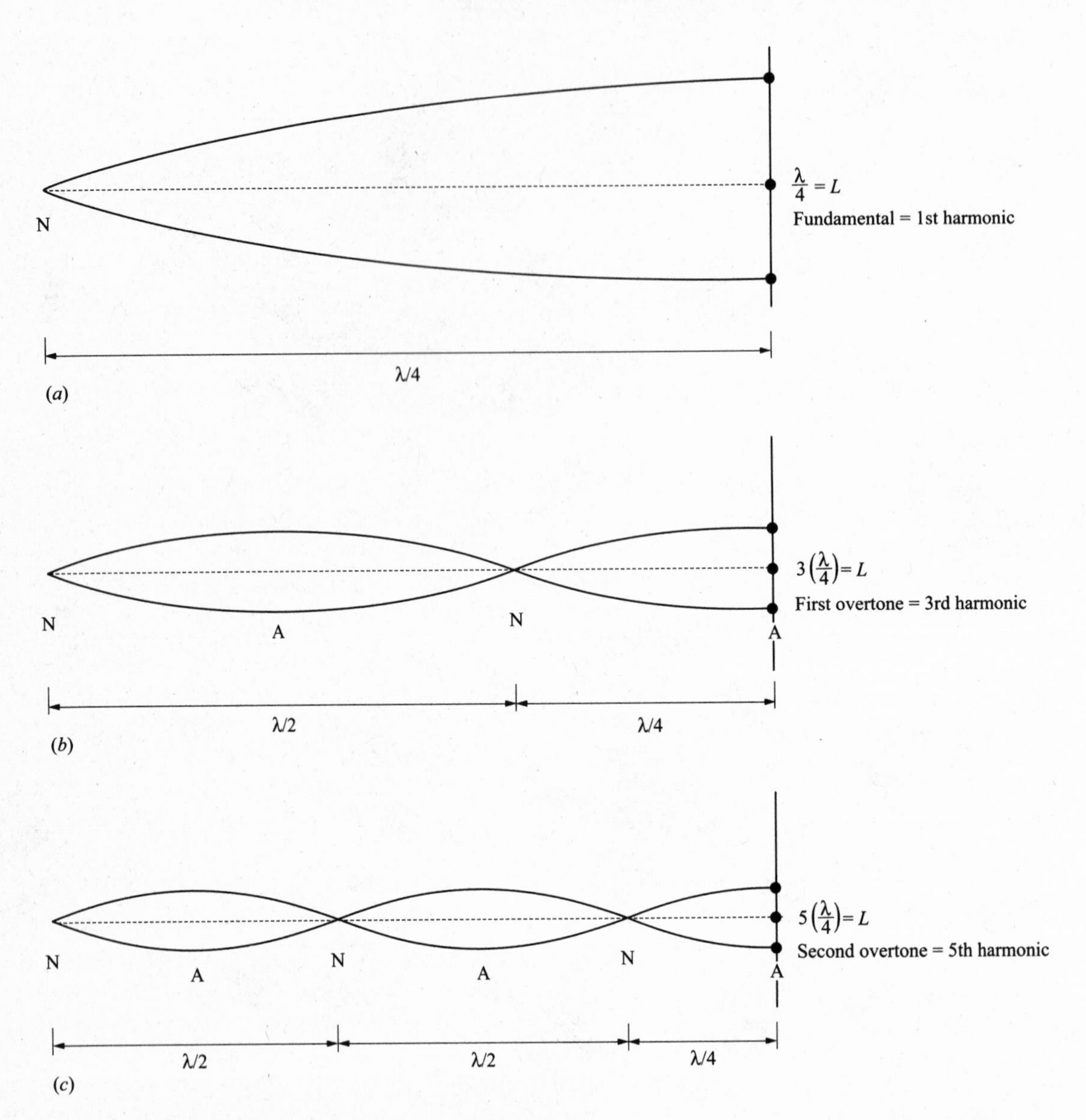

Envelopes of resonant standing waves in a cord - one end an anti-node

Fig. 1-14

resonant frequencies can absorb this energy efficiently, so the net effect is simultaneous creation of the allowed resonant standing waves with different amplitudes. Typically, only the first several allowed harmonics receive much energy. The exact distribution of energy among the harmonics depends on the details of the stimulating disturbance, and of the medium in which the disturbance takes place. If the stimulus is very short lived, the standing waves last only a short time as well, since their energy rapidly transfers itself to surrounding materials such as the air, and/or dissipates into thermal energy. The distinctive sound of different musical instruments, even when sounding the same note, is a function of the different amplitudes of the harmonics that are generated. This will be discussed further in the next chapter.

The results we obtained for resonant transverse standing waves, have their counterpart in longitudinal waves, and we will briefly explore this case. For simplicity, we will consider the case of sound waves in the air in an organ pipe. Consider the organ pipe of length L, as shown in Figs. 1-15(a) and (b). In both cases one end of the pipe has an opening to the atmosphere (with a reed to help outside vibrations

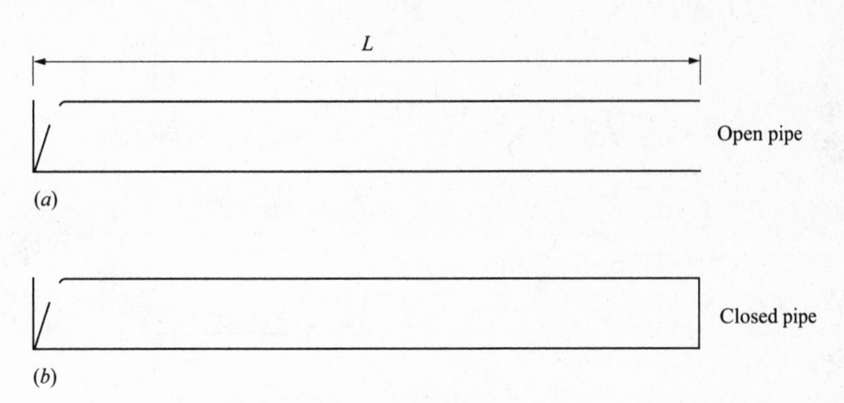

Fig. 1-15

enter the pipe). In Fig. 1-15(*a*) the far end of the pipe is open, and is therefore called an *open* organ pipe; in the case of Fig. 1-15(*b*), the far end of the pipe is closed and is therefore called a *closed* organ pipe. Following the reasoning of Problems 1.23 and 1.25, we would expect ends that are open to the atmosphere to be anti-nodes for the to and fro motion of the air molecules, while a sealed end would be a node for such motion. This turns out to be true for nodes at a closed end, but the anti-nodes at the open ends are actually slightly beyond the edges of the pipe, the extent depending on the cross-sectional area and other geometrical properties of the ends of the pipe. We will ignore this "end" effect in our simple analysis. For the case of the pipe open on both ends, the first few allowed standing waves are shown in Figs. 1-16(*a*) to (*c*). The wave envelope is shown just below the pipe, as a transverse representation of the to and fro motion of the molecules from their equilibrium positions at various points along the pipe. Since the distance between successive anti-nodes is always a half wavelength, we see that a whole number of half-wavelengths fit into the length, L, of the pipe, or the allowed wavelengths are:

$$\lambda_n = 2L/n \qquad n = 1, 2, 3, \ldots . \tag{1.14a}$$

The allowed frequencies are then given by:

$$f_n = v_p/\lambda_n = n(v_p/2n) \qquad n = 1, 2, 3, \ldots \tag{1.14b}$$

where v_p here represents the speed of sound in air. For the case of the pipe closed at one end, we have a node at that end, and an anti-node at the other end. This is similar to the case of transverse waves in a cord tied down at one end, and the other end free to move up and down, as discussed above. The first few allowed wavelengths for this case are shown in Figs. 1-17(*a*) to (*c*). As can be seen, a whole number of half-wavelengths plus a quarter-wavelength or, equivalently, an odd number of quarter-wavelengths, must fit into L, so that:

$$\lambda_n = 4L/n \qquad n = 1, 3, 5, \ldots \tag{1.15a}$$

and the allowed frequencies are then:

$$f_n = v_p/\lambda_n = n(v_p/4L) \qquad n = 1, 3, 5, \ldots . \tag{1.15b}$$

Note that these are the same formulas as Problem 1.25, Eqs. (*i*), (*ii*), as expected.

Problem 1.27.

(*a*) In earlier discussions we showed that for travelling longitudinal waves in a fluid, the pressure variation, ΔP, was maximum when the displacement of the molecules was zero, and vice versa. Show that for standing waves the pressure anti-nodes occur at displacement nodes, and vice versa.

(*b*) Given the results of part (*a*) show that our intuitive presumption that displacement anti-nodes in an organ pipe occur at the ends of the pipe open to the atmosphere makes sense.

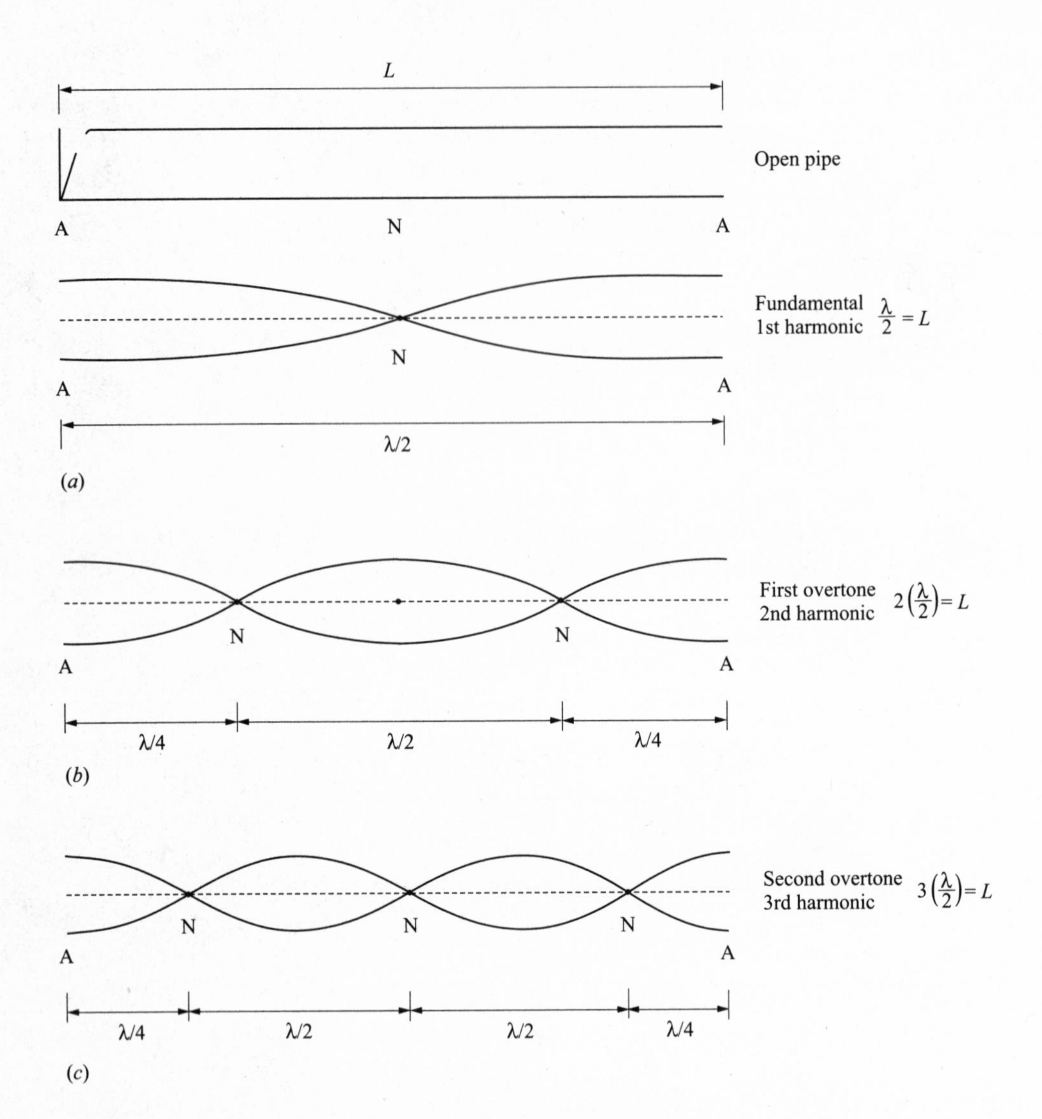

Resonant standing waves in an open pipe. Vertical represents air displacement along pipe from equilibrium position.

Fig. 1-16

Solution

(*a*) Consider a displacement node for an organ pipe, such as the middle one depicted in Fig. 1-16(*a*). On either side of the node the air molecules are moving in opposite directions. When these molecules are rushing toward the node that create a maximum condensation, and a corresponding increase in pressure; when they are racing away from the node they create a maximum rarefaction, and a corresponding decrease in pressure. In effect, at the displacement nodes the air is squeezed together and pulled apart just like an accordion. No other points in the standing wave have such a drastic variation in pressure as do the displacement nodes, so these are pressure anti-nodes. On the other hand, on either side of a displacement anti-node, such as the middle one in Fig. 1-16(*b*), the air molecules are moving to and fro in the same direction with the same amplitudes, so no condensation or rarifications occur and ΔP remains zero. Thus the displacement anti-nodes are pressure nodes.

(*b*) At the ends of the pipe that are open to the atmosphere the exposure to the equilibrium pressure of the air outside the pipes doesn't allow for the type of pressure variation, ΔP found in the large resonant waves in the pipe, so these locations are in effect nodes for ΔP. Since such nodes are anti-nodes for displacement, we have our result.

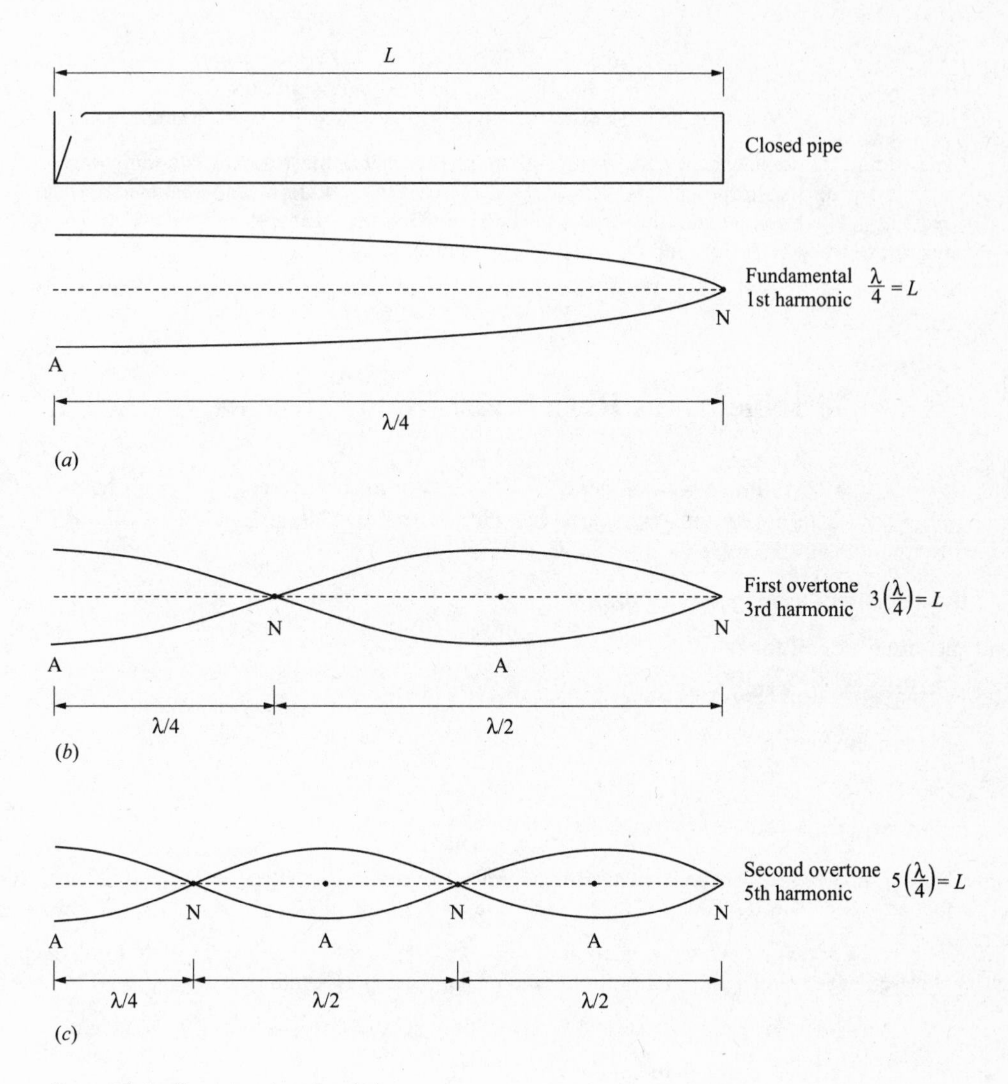

Resonant standing waves in a closed pipe.

Fig. 1-17

Problem 1.28.

(*a*) An open organ pipe is of length $L = 6.0$ ft. Find the wavelength of the fundamental and first two overtones.

(*b*) Find the fundamental frequency and those of the first two overtones. [*Hint*: Assume the speed of sound in air is 1100 ft/s].

(*c*) If one end of the pipe is now closed, what changes does one have in the wavelengths and frequencies of parts (*a*) and (*b*)?

(*d*) Explain qualitatively why the tone of a trumpet drops when a mute is placed over the horn end?

Solution

(*a*) From Eq. (*1.14a*), we have: $\lambda_n = 2(6.0 \text{ ft})/n \Rightarrow \lambda_1 = 12.0$ ft; $\lambda_2 = 6.0$ ft; $\lambda_3 = 4.0$ ft.

(*b*) From Eq. (*1.14b*), we have: $f_n = v_p/\lambda_n = (1100 \text{ ft/s})/\lambda_n \Rightarrow$ Fundamental $= f_1 = 91.7$ Hz; first overtone $= f_2 =$ 2nd harmonic $= 183$ Hz; second overtone $= f_3 =$ 3rd harmonic $= 275$ Hz.

(c) From Eq. (1.15a), we have: $\lambda_n = 4(6.0 \text{ ft})/n$, odd n only $\Rightarrow \lambda_1 = 24.0$ ft; $\lambda_3 = 8.0$ ft; $\lambda_5 = 4.8$ ft. Similarly, for the frequencies, $f_n = v_p/\lambda_n = (1100 \text{ ft/s})/\lambda_n$, odd n only $\Rightarrow$ Fundamental $= f_1 = 45.8$ Hz; first overtone $= f_3 =$ 3rd harmonic $= 138$ Hz; second overtone $= f_5 =$ 5th harmonic $= 229$ Hz.

(d) The trumpet is somewhat similar to an organ pipe in generating resonant standing waves. With the mute covering the horn we have moved to a closed pipe situation, and the fundamental drops in frequency. The resonant standing waves in the trumpet is what generates the sound wave that reaches our ear, so we hear the lowered frequency of the fundamental.

Problems for Review and Mind Stretching

Problem 1.29. A SHM travelling wave of period of $T = 3.0$ s and wavelength $\lambda = 30$ m moves to the right in a long cord. The maximum transverse velocity in the cord is $v_{max} = 2.5$ cm/s, and the power transmitted by the wave is 0.60 W.

(a) Find the velocity of propagation of the wave, v_p.

(b) Find the amplitude of the wave.

(c) Find the mass per unit length of the cord, μ.

(d) Find the tension in the cord, S.

Solution

(a) From Eq. (1.10), $v_p = \lambda f = \lambda/T = (30 \text{ m})/(3.0 \text{ s}) = 10$ m/s.

(b) We can find the amplitude from the maximum transverse velocity: $v_{max} = \omega A = 0.025$ m/s. Recalling that $\omega = 2\pi/T = 6.28/(3.0 \text{ s}) = 2.09 \text{ s}^{-1}$, we have: $(0.025 \text{ m/s}) = (2.09 \text{ s}^{-1})A \Rightarrow A = 1.20$ cm.

(c) We cannot obtain μ from the expression for v_p, Eq. (1.1), unless we know S, which is not given. We do, however, have enough information to obtain μ from the expression for the power, Eq. (1.12):

$$P = \tfrac{1}{2}\mu\omega^2 A^2 v_p \Rightarrow (0.60 \text{ W}) = \tfrac{1}{2}\mu(2.09 \text{ s}^{-1})^2(0.0120 \text{ m})^2(10 \text{ m/s}) \Rightarrow \mu = 191 \text{ g/m}.$$

(d) Now that we have μ we can obtain S from Eq. (1.1):

$$v_p = (S/\mu)^{1/2} \Rightarrow S = (0.191 \text{ kg/m})(10 \text{ m/s})^2 = 19.1 \text{ N}.$$

Problem 1.30. When a musical instrument plays, it is almost invariably the fundamental frequency that dominates, and with which the human ear identifies the musical note. For a stringed instrument, such as a piano or base fiddle, the fundamental is given by Eq. (1.13a): $f_F = (S/\mu)^{1/2}/2L$.

(a) For the case of a piano the "strings" are wire wound and of different lengths and thicknesses. Explain, using Eq. (1.13a), what this accomplishes and explain how a piano tuner "fine tunes" the piano.

(b) If one bows a base fiddle string one can double the frequency by pressing a certain point on the string while bowing; where is that point?

Solution

(a) From Eq. (1.13a) we see that there are three variables that affect the fundamental frequency: the length and mass/length of the string, each of which decreases the frequency when it increases, and the tension which increase the frequency when it increases. The basic construction of the piano has strings of different lengths and thicknesses intended to generate different tones. The short thin strings have the highest frequencies while the longer thicker strings generate the lower notes. Without increased thickness of the long strings, the piano strings would have to be much longer to get the lowest frequencies.

In grand pianos the strings are on a horizontal frame, called the sound board, and the shape of the piano reflects the lengths of the strings; in an upright piano the strings are vertical and the differing lengths are hidden inside the piano. Once the lengths and thicknesses of the strings are set, the only variable is the tension, and a piano tuner can fine tune the frequencies by adjusting the tensions of the strings.

(*b*) The fundamental has a node at each end of the string and an anti-node at the midpoint. By pressing ones finger at the midpoint one forces that point to be a node, so that only the first harmonic and above can be heard. Since the first harmonic is half the wavelength it is double the frequency.

Problem 1.31. A long tube has a close fitting piston initially all the way into the tube, and a tuning fork is vibrating just in front of the tube, as shown in Fig. 1–18(*a*). As the piston is slowly pulled back (to the left in the figure) the sound gets very loud as it passes the 13 cm, 41 cm, and 69 cm marks as measured from the right end of the tube.

(*a*) What is the next point at which the loudness will rise?

(*b*) Assuming that the speed of sound is 350 m/s, find the frequency of the tuning fork.

(*c*) Assuming the tuning fork is replaced by one with double the frequency, what would the first three loud points have been as the piston was pulled out?

Solution

(*a*) The tube acts like a closed organ pipe with regard to resonant standing waves. The tuning fork has a definite frequency, and an associated definite wavelength. As the piston is pulled back, the first possible resonance occurs when a node is at the piston face and an anti-node is at, or just outside, the open end of the pipe—with the distance between node and anti-node corresponding to a quarter-wavelength (Fig. 1–18(*b*)). The second possible resonance should occur when the piston is moved sufficiently far

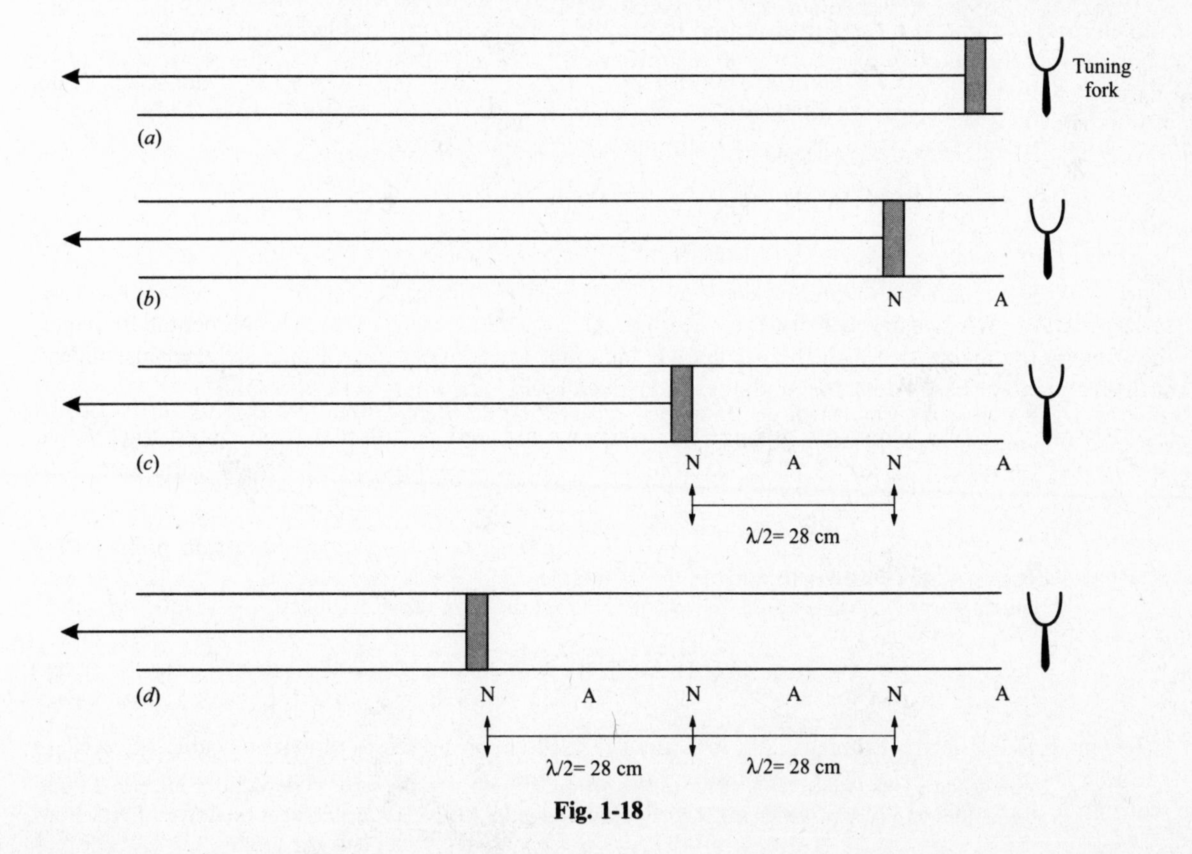

Fig. 1-18

back so that a second node occurs between the piston and the open end of the pipe. Since the wavelength is fixed, one node remains at its original location, and the second node must be at the new position of the piston (Fig. 1-18(*c*)). The distance between the nodes is precisely a half-wavelength and from the given data corresponds to:

$$\lambda/2 = (41 \text{ cm}) - (13 \text{ cm}) = 28 \text{ cm} \Rightarrow \lambda = 56 \text{ cm}. \qquad (i)$$

The next resonance should occur when the piston is pulled sufficiently far back that a third node occurs, the other two nodes maintaining at their prior locations (Fig. 1-18(*d*)). Again, we have from the given data:

$$\lambda/2 = (69 \text{ cm}) - (41 \text{ cm}) = 28 \text{ cm} \Rightarrow \lambda = 56 \text{ cm} \qquad (ii)$$

This confirms our previous result, and makes clear that the next resonance will occur when the piston is pulled another 28 cm further in, or at the 97 cm mark. Note that the first data point of 13 cm is 1 cm short of a quarter wavelength, indicating that the anti-node is actually 1 cm outside the mouth of the pipe.

(*b*) From Eqs. (*i*) and (*ii*), we have $\lambda = 56$ cm, so:

$$f = v_p/\lambda = (350 \text{ m/s})/(0.56 \text{ m}) = 625 \text{ Hz}.$$

(*c*) If the frequency doubled to 1250 Hz, the wavelength would halve to 28 cm. Then a quarter-wavelength would be 7.0 cm. Assuming the anti-node is still 1 cm outside the open end of the pipe, the first resonance occurs when the piston is at the 6 cm mark. The next two positions of the piston for resonance are half-wavelengths further in, at:

$$(6 \text{ cm}) + (14 \text{ cm}) = 20 \text{ cm}, \qquad \text{and at } (20 \text{ cm}) + (14 \text{ cm}) = 34 \text{ cm}.$$

Problem 1.32. A cord of length $L = 3.0$ m and total mass $m = 400$ g, is connected at one end to a vibrator, and at the other end to a very long and massive steel cable under tension $S = 400$ N. When the vibrator is turned on the cord is found to rapidly develop a large, stable, transverse standing wave consisting of five equal sections. The cable is observed to have a transverse travelling SHM wave to the right at 15 m/s.

(*a*) Find the wavelength and frequency of the standing wave in the cord.

(*b*) Find the wavelength and frequency of the travelling wave in the cable.

Solution

(*a*) Since the cord vibrates in five sections, and each section has a node at each end so that it is a half-wavelength long, we have:

$$\lambda/2 = L/5 = (3.0 \text{ m})/5 = 0.60 \text{ m} \Rightarrow \lambda = 1.20 \text{ m}.$$

Knowing the wavelength, the frequency can now be determined from a knowledge of the speed of propagation $v_{p,c}$ in the cord. We have:

$$v_{p,c} = (S/\mu)^{1/2} = (SL/m)^{1/2} = [(400 \text{ N})(3.0 \text{ m})/(0.40 \text{ kg})]^{1/2} = 54.8 \text{ m/s}.$$

$$f = v_{p,c}/\lambda = (54.8 \text{ m/s})/(1.20 \text{ m}) = 45.6 \text{ Hz}.$$

(*b*) The frequency must be the same as the driving frequency of the cord which is $f = 45.6$ Hz. The wavelength is now determined from the velocity of propagation in the steel cable, $v_{p,s} = 15$ m/s:

$$\lambda_s = v_{p,s}/f = (15 \text{ m/s})/(45.6 \text{ Hz}) = 0.329 \text{ m}.$$

Problem 1.33. Refer to Problem 1.32. Assuming that there are no losses of cord or cable wave energy to other forms (e.g., thermal, to surrounding air, etc.), find the power delivered by the vibrator to the cord if the amplitude of the travelling wave in the cable is 0.60 mm.

Solution

Since no wave energy is lost to other forms, all the energy the vibrator pours into the cord stays as wave energy in the cord or the cable. Since the standing waves in the cord quickly become stable their total energy content is fixed; therefore, the power in the cord from the vibrator must equal the power transferred by the cord to the cable. This in turn equals the power in the cable wave, which is determined from Eq. (1.12): $P = \frac{1}{2}\mu\omega^2 A^2 v_p$. We have all the required information to obtain P except for the linear density of the steel cable. This can be determined by rearranging the formula for $v_{p,s}$ to obtain:

$$\mu = S/v_{p,s}^2 = (400\ \text{N})/(15\ \text{m/s})^2 = 1.78\ \text{kg/m}.$$

$$P = \tfrac{1}{2}(1.78\ \text{kg/m})[6.28 \times (45.6\ \text{Hz})]^2(0.60 \cdot 10^{-3}\ \text{m})^2(15\ \text{m/s}) = 0.394\ \text{W}.$$

Thus the vibrator transmits 0.394 W of power to the cord.

Supplementary Problems

Problem 1.34.

(*a*) Find the velocity of propagation of a transverse wave in a long rubber hose of mass/length 0.23 kg/m when the tension is 88 N.

(*b*) Find the wavelength of a travelling wave in the hose if the frequency is 9.0 Hz.

(*c*) If the tension in the hose doubled, all else being the same, what would be the new answer to part b?

Ans. (*a*) 19.6 m/s; (*b*) 2.18 m; (*c*) 3.08 m.

Problem 1.35.

(*a*) Compare the speed of sound in a tube filled with mercury and in a steel rail. [*Date*: $Y_s = 196$ GPa; $B_m = 26$ GPa; $\rho_s = 7.86\ \text{g/cm}^3$; $\rho_m = 13.6\ \text{g/cm}^3$].

(*b*) If the mercury and the steel rail are exposed to the same SHM vibration, and the wavelength of the wave in mercury is 30 cm, what is the frequency of the vibration, and what is the wavelength in the steel?

Ans. (*a*) $v_{p,s} = 4990$ m/s, $v_{p,m} = 1380$ m/s; (*b*) 4600 Hz, 1.08 m.

Problem 1.36.

(*a*) Assume that the mercury tube and the steel rail of Problem 1.35 are of equal length and are lined up alongside each other. A longitudinal pulse is stimulated in each at the same end at the same time. When the pulse in the steel reaches the other end the pulse in the mercury is 20 m behind. Find the length of the tube or rail.

(*b*) Assume the rail is now placed end to end with the tube of mercury, and the steel just makes a tight fit into the end of the tube so it is in direct contact with the mercury at that end. A pulse starts at the other end of the mercury and reflects and transmits at the interface. Find the time lag time for the reflected pulse to reach the front end after the transmitted pulse reaches the far end of the rail.

(*c*) Is the reflected pulse inverted or upright?

Ans. (*a*) 27.6 m; (*b*) 0.0145 s; (*c*) inverted.

Problem 1.37. A transverse wave in a long rope is given by: $y = (0.44\ \text{cm}) \sin(90t + 15x)$ where x and t are in meters and seconds, respectively.

(*a*) Find the amplitude, wavelength and frequency of the wave.

(*b*) Is the wave travelling in the positive or negative x direction?

(*c*) What is the velocity of propagation of the wave?

Ans. (*a*) 0.44 cm, 0.419 m, 14.3 Hz; (*b*) negative; (*c*) 6.00 m/s.

Problem 1.38.

(*a*) Referring to the wave in Problem 1.37, find the maximum transverse velocity and acceleration of a point on the rope.

(*b*) Assuming the tension in the rope is 80 N, find the mass/length of the rope.

(*c*) Find the energy per unit length associated with the wave, and the power transmitted by the wave across a given point in the rope.

Ans. (*a*) 0.395 m/s, 35.5 m/s^2; (*b*) 2.22 kg/m; (*c*) 0.173 J, 1.04 W.

Problem 1.39. A longitudinal SHM travelling wave passes through a long tube filled with water. The maximum displacement of water molecules from their equilibrium positions is 2.0 mm, and their maximum velocity is ± 3.0 m/s. When the water is replaced by ethyl alcohol a wave of the same frequency has a wavelength three quarters as long. [*Data*: Speed of sound in water is 1400 m/s; specific gravity of ethyl alcohol is 0.810].

(*a*) Find the frequency and wavelength of the wave in water.

(*b*) Find the speed of sound in ethyl alcohol.

(*c*) Find the bulk moduli of water and ethyl alcohol, respectively.

Ans. (*a*) 239 Hz, 5.86 m; (*b*) 1050 m/s; (*c*) $1.96 \cdot 10^9$ Pa; $0.893 \cdot 10^9$ Pa.

Problem 1.40. Water waves near the shore give the appearance of being travelling transverse waves. Actually they are more closely described by a combination of transverse and longitudinal waves. This becomes apparent when one watches a small object floating in the water—it not only bobs up and down but also moves to and fro. The frequencies and wavelengths of these two simultaneous, but mutually perpendicular, waves are the same but their amplitudes, of course, need not be the same.

(*a*) If the observed crest to crest distance of the travelling water waves is 3.0 m and the speed of the waves toward the shore is 2.0 m/s, find the frequency of the waves.

(*b*) If, for a floating object, the crest to trough distance is 0.80 m, and the to and fro motion has a maximum separation of 0.6 m, find the amplitudes of the transverse and longitudinal waves.

(*c*) If the transverse and longitudinal waves are in phase (e.g., they both reach their maximum displacements at the same time) what is the actual maximum displacement of a floating object from its equilibrium position, and what is the actual path of the object?

(*d*) If the waves were out of phase by 90° (e.g., transverse maximum when longitudinal is at equilibrium), qualitatively describe the path of the floating object.

Ans. (*a*) 0.667 Hz; (*b*) 0.4 m, 0.3 m; (*c*) 0.5 m, straight line at 53° to horizontal; (*d*) elliptical path.

Problem 1.41. Consider the "virtual reflection" model used in the text to describe the behavior of a reflected wave at the end of a long cord that is either tied down or looped without friction. Draw to scale three wavelengths of a SHM wave nearing the end of the cord and the corresponding virtual reflection of those wavelengths beyond the end of the cord, for:

(*a*) the case of the end being tied down.

(*b*) The case of the end free to move up and down without friction.

[*Hint*: When the real and virtual waves meet at the end of the cord the superposition of the two must correspond to the actual behavior of that point on the cord.]

Problem 1.42. A cord of $\mu = 30$ g/m is tightly tied between two strong posts a distance 3.0 m apart, so that it is under a tension of 100 N. The cord is plucked so that resonant standing waves are set up.

(*a*) Find the three longest wavelengths that are allowed in the cord.

(*b*) Find the corresponding frequencies.

(*c*) The tension is now adjusted so that the new fundamental frequency corresponds to that of the second overtone in part (*b*); find the new tension.

Ans. (*a*) 6.0 m, 3.0 m, 2.0 m; (*b*) 9.62 Hz, 19.2 Hz, 28.9 Hz; (*c*) 900 N.

Problem 1.43. A cord of $\mu = 30$ g/m has a vibrator at one end with frequency $f = 120$ Hz, and the other end is connected to hanging weights over a frictionless pulley, as shown in Fig. 1-19. The distance between the end of the cord in contact with the vibrator and the point of contact of the cord with the pulley is $L = 1.2$ m. Both these points can be considered nodes for resonant standing waves. A mass of $M = 8.0$ kg is initially hanging from the cord, and this mass is slowly increased in tiny increments.

(*a*) What are the five longest wavelengths for which resonant standing waves can possibly occur?

(*b*) Find a formula relating the hanging mass M (in kg) to the wavelength λ (in m) for the specific cord and frequency given.

(*c*) As M rises above 8.0 kg, find all the masses for which a resonant standing wave appears in the cord, and give the corresponding number of sections in which the cord vibrates.

Ans. (*a*) 2.4 m, 1.2 m, 0.8 m, 0.6 m, 0.48 m; (*b*) $M = (\mu f^2/g)\lambda^2 = 44.1\lambda^2$; (*c*) 10.2 kg (5), 15.9 kg (4), 28.2 kg (3), 63.5 kg (2), 254 kg (1).

Problem 1.44. An open organ pipe has a 1st overtone frequency of 90 Hz, and a third harmonic wavelength of 2.56 m.

(*a*) Find the length of the organ pipe.

(*b*) Find the speed of sound in the air.

Ans. (*a*) 3.84 m; (b) 346 m/s.

Problem 1.45. Suppose the organ pipe of Problem 1.44 is closed at one end, and the same air conditions prevail.

(*a*) Find the three longest resonant wavelengths.

(*b*) Find the frequency of the fifth overtone.

Ans. (*a*) 15.36 m, 5.12 m, 3.07 m; (b) 248 Hz.

Problem 1.46. A brass rod is firmly clamped at the center and when tapped at one end resonates longitudinally with a fundamental of 2000 Hz.

(*a*) What characterizes the resonant standing waves at the rod ends?

(*b*) What next higher frequency would you expect to be resonating under the conditions described, and why?

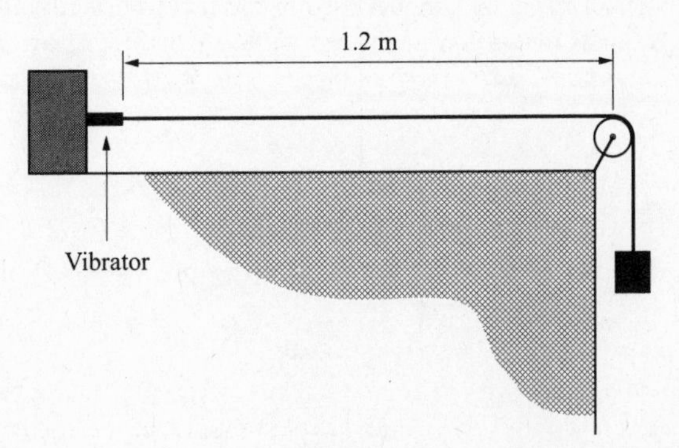

Fig. 1-19

(*c*) What would the fundamental be if the rod were clamped at one end?

(*d*) What would the next higher frequency be for the case of part (*c*)?

Ans. (*a*) anti-nodes; (*b*) 6000 Hz, center point must be a node; (*c*) 1000 Hz, (*d*) 3000 Hz.

Problem 1.47. A closed organ pipe is measured to be one meter longer than an adjacent open organ pipe. It is found that the second overtone of the open pipe is the same frequency as the third overtone of the closed pipe.

(*a*) Find the length of the open pipe.

(*b*) Find the difference in the wavelengths of the fundamentals of the two pipes.

Ans. (*a*) 6.0 m; (*b*) 16 m.

Problem 1.48. A variable frequency vibrator is held over a 3.0 m high open cylinder, as shown in Fig. 1-20. The frequency starts out at 20 Hz and is slowly increased. A substantial rise in loudness first occurs at 28.0 Hz. The speed of sound in the air in the cylinder is 340 m/s.

(*a*) Assuming the data given is correct to four decimal places, find the wavelength of the fundamental.

(*b*) How far outside the opening of the cylinder does the anti-node of the fundamental occur?

(*c*) As the frequency continues to rise when will the next rise in loudness occur?

Ans. (*a*) 12.14 m; (*b*) 0.035 m; (*c*) 84 Hz.

Problem 1.49. For the same cylinder as in Problem 1.48 the vibrator is replaced by a vibrating tuning fork of frequency 100 Hz. Water is slowly poured into the mouth of the cylinder and it is observed that at certain specific levels a strong increase in loudness occur. [*Hint*: Assume anti-nodes always occurs 0.03 m above the top of the cylinder.]

(*a*) What is the wavelength corresponding to this frequency?

(*b*) How high up in the cylinder does the water rise when the first loudness occurs?

(*c*) What are the water heights for additional loud points?

Ans. (*a*) 3.4 m; (*b*) 0.48 m; (*c*) Only one other water height, at 2.18 m.

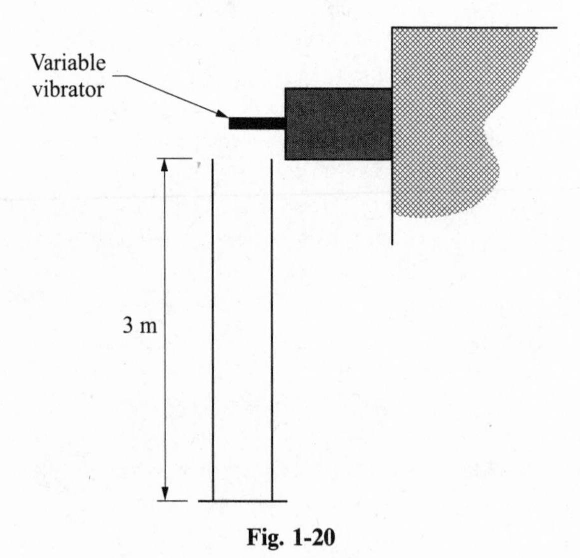

Fig. 1-20

Chapter 2

Sound

2.1 MATHEMATICAL ADDENDUM—EXPONENTIAL AND LOGARITHMIC FUNCTIONS

While exponentials and logarithms are not major mathematical tools in a second semester physics course, there are enough references to them to warrant a brief mathematical review here. We start with the **exponential function**:

$$y = A^x \tag{2.1}$$

where A is an arbitrary number.

Integer Powers of a Number

When the exponent, x, is a positive integer we have the usual powers of A: $A^1 = A$; $A^2 = A \cdot A$; $A^3 = A \cdot A \cdot A$, etc. By convention, $A^0 = 1$. For negative integers we define $A^{-1} = 1/A$; $A^{-2} = 1/A^2$ and in general: $A^{-n} = 1/A^n = (1/A)^n$. Clearly, by examination of a few simple examples, we have for any two non-negative integers: $A^n \cdot A^m = A^{(n+m)}$; $A^n \cdot A^{-m} = A^{n-m}$. These two can be combined into the single statement:

$$A^n \cdot A^m = A^{(n+m)} \tag{2.2}$$

where now n and m are any positive, negative or zero integers.

Problem 2.1. Show that for any two non-negative integers, n and m:

(*a*) $(A^n)^m = (A^m)^n = A^{(n \cdot m)}$

(*b*) $A^n \cdot A^{-m} = A^{-n} \cdot A^m = A^{-(n \cdot m)}$

Solution

(*a*) This can be shown by generalizable example. Consider the case $n = 2$ and $m = 3$. Then: $(A^2)^3 = A^2 \cdot A^2 \cdot A^2 = A^6$; Similarly, $(A^3)^2 = A^3 \cdot A^3 = A^6$. Since $n \cdot m = 6$ we have our result. This reasoning works for all positive integers, n and m. If n or m is zero, the result follows from the definition.

(*b*) Again by example, consider $n = 2$ and $m = 3$. Then: $(A^2)^{-3} = (1/A^2) \cdot (1/A^2) \cdot (1/A^2) = 1/A^6 = A^{-6}$, and so on. Again, the reasoning works for all positive integers and zero.

The results of Problem 2.1 can be summarized as:

$$(A^n)^m = A^{(n \cdot m)} \tag{2.3}$$

for any positive, negative or zero integers, n and m.

Fractional Powers of a Number

We now turn to fractional powers. By definition, $A^{1/2} = \sqrt{A}$; $A^{1/3} = \sqrt[3]{A}$ and, in general, $A^{1/n}$ = the nth root of A. This means that: $(A^{1/n})^n = A$. As for integers, it is understood that: $A^{-1/n} = 1/A^{1/n} = (1/A)^{1/n}$. The nth roots are defined for all positive numbers A, but for n even this is not true for negative A (e.g., no number times itself equals a negative number, so square roots, $n = 2$, are not defined). We will assume that A is positive when dealing with fractional powers. We now ask the question what do we mean by: $A^{n/m}$, where n is any integer and m is a non-zero integer. An obvious definition is:

$$A^{n/m} = (A^n)^{1/m} \tag{2.4}$$

For this to make sense, we must have that if $n'/m' = n/m$, $A^{n'/m'} = A^{n/m}$, or, $(A^{n'})^{1/m'} (A^n)^{1/m}$. This is true, and we illustrate it for the simple example $n' = 2n$ and $m' = 2m$. We must show that $(A^{2n})^{1/2m} = (A^n)^{1/m}$. Let $B = A^n$, and $C = A^{2n} = B^2$. Since $C = B^2$, the 2mth root of C is, indeed, the mth root of B, and we have our result. For Eq. (*2.4*) to be a useful definition, it must also be shown that $(A^{1/m})^n = (A^n)^{1/m}$.

Problem 2.2. Show that $(A^{1/m})^n = (A^n)^{1/m} = A^{n/m}$ for any integer n, and non zero integer m.

Solution

Let $A^{1/m} = B$. We then have to show that:

$$B^n = (A^n)^{1/m} \tag{i}$$

Noting from the definition that $B^m = A$, we substitute B^m for A in the right side of Eq. (*i*), to get:

$$(A^n)^{1/m} = [(B^m)^n]^{1/m} = (B^{m \cdot n})^{1/m} = B^{(m \cdot n/m)} = B^n \tag{ii}$$

which is just the result, Eq. (*i*), that we needed to prove.

It can be demonstrated that Eqs. (*2.2*) and (*2.3*) are valid for any two positive or negative fractional powers, a and b:

$$A^a \cdot A^b = A^{(a+b)} \tag{2.5a}$$

$$(A^a)^b = A^{a \cdot b} \tag{2.5b}$$

where we have generalized the definition of negative powers to include fractions:

$$1/A^a = A^{-a} \tag{2.5c}$$

General Powers of a Number and their Properties

Powers of A can be defined not only for all proper and improper fractions as we have done, but more generally for all real numbers, including the myriad of numbers such as $\sqrt{2}$ and π, which are infinite non-repeating decimals, and correspond to points on a line (such as the x axis of a graph), but cannot be expressed as a fraction. For all such powers, Eqs. (*2.5*) hold. Returning, to our exponential function, Eq. (*2.1*): $y = A^x$, we see that for positive A, x can be any number on the real line, from $-\infty$ to ∞, and that y takes on positive values which depend on A and x. There are two values of A that are most often used in dealing with powers. One is $A = 10$, which is particularly useful since for historical reasons numbers are most often expressed in "base 10 ", i.e., using the **decimal** system. (As most students now know from computer science courses, one can express numbers in the **binary** (base 2) system, and in fact any positive integer can be used as the base for the integer system). For the powers of 10 our exponential function becomes: $y = 10^x$. Every time x increases by a unit ($x \rightarrow x + 1$), y increases by a multiplicative factor of $10^1 = 10$. It is this rapid increase in y with increasing x that characterizes an exponential function. Of course, if x is negative, every decrease in x by one unit causes y to decrease by a factor of 10, so that exponential decreases are drastic as well. Figure 2-1(*a*) and (*b*) shows graphs of the exponential function $y = 10^x$.

The Logarithmic Function

As x increases continuously from $-\infty$ to ∞, y is always positive and increases continuously from 0 to ∞, passing through $y = 1$ when $x = 0$. Thus, for any positive y there is a unique number x for which

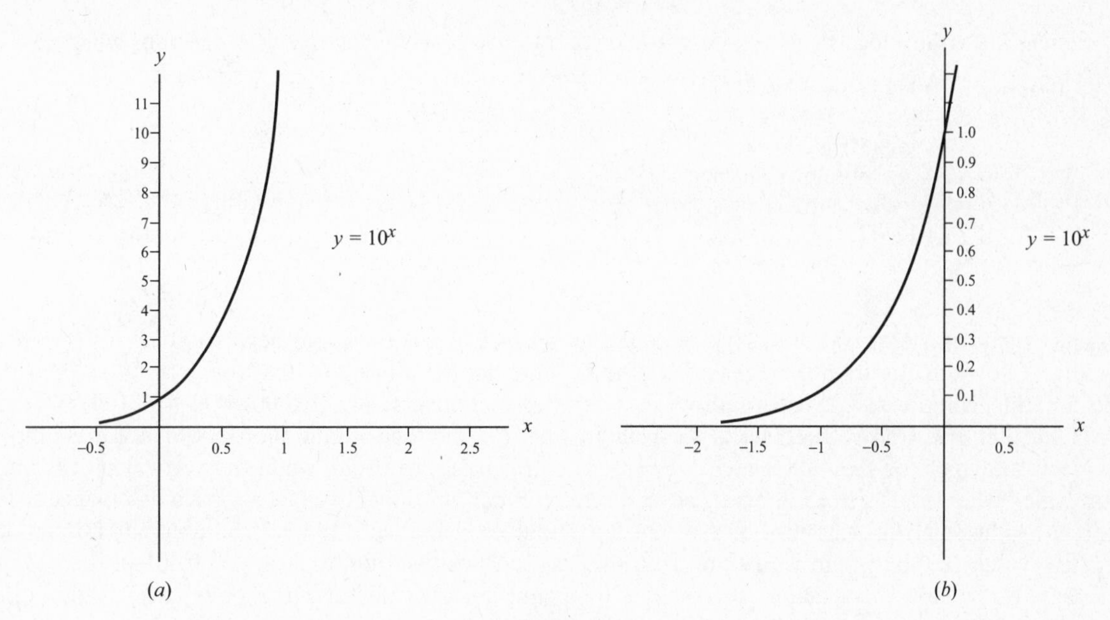

Fig. 2-1

$y = 10^x$ holds. The **logarithmic function** "to the base 10" is then defined as: $\log_{10} y = x$. As such, the base 10 logarithm is the inverse function to $y = 10^x$:

$$y = 10^x \Leftrightarrow \log_{10}(y) = x \tag{2.6}$$

For example: $\log_{10}(1) = \log_{10}(10^0) = 0$; $\log_{10}(10) = \log_{10}(10^1) = 1$; $\log_{10}(100) = \log_{10}(10^2) = 2$; $\log_{10}(1000) = \log_{10}(10^3) = 3$. Often it is understood that "log" stands for $\log_{10}$, and the subscript is omitted. The logarithmic function can be shown to obey the following rules:

$$\log(A \cdot B) = \log(A) + \log(B) \tag{2.7a}$$

$$\log(A^z) = z \cdot \log(A) \tag{2.7b}$$

for all positive A and B, and all z. It follows from Eqs. (*2.7a*) and (*2.7b*) that:

$$\log(A/B) = \log(A) - \log(B) \tag{2.7c}$$

Problem 2.3.

(*a*) Show that for any two positive numbers, A and B, $\log(A \cdot B) = \log(A) + \log(B)$.

(*b*) Show that $\log(A^z) = z \cdot \log(A)$ for all positive A and any z.

(*c*) Show that $\log(A/B) = \log(A) - \log(B)$

(*d*) Using the results of parts (*a*) and (*b*), find the following in terms of log (2) and/or log (3): log (8), log (18), log (27), log (80), $\log(\frac{1}{2})$, $\log(\frac{1}{8})$, $\log(\frac{3}{8})$.

Solution

(*a*) Since A and B are positive we know we can find real numbers a and b such that: $A = 10^a$ and $B = 10^b$. Then, $\log(A) = a$, and $\log(B) = b$. Next, we have: $\log(A \cdot B) = \log(10^a \cdot 10^b) = \log(10^{(a+b)}) = \mathrm{a} + \mathrm{b} = \log(A) + \log(B)$, which is the desired result.

(*b*) Again, let $A = 10^a$, so that $a = \log(A)$. Then, $A^z = (10^a)^z = 10^{a \cdot z} \Rightarrow \log(A^z) = \log(10^{a \cdot z}) = a \cdot z = z \log(A)$, which is the desired result.

(*c*) $\log(A/B) = \log(A \cdot B^{-1}) = \log(A) + \log(B^{-1}) = \log(A) - 1 \cdot \log(B) = \log(A) - \log(B)$.

(*d*) $\log(8) = \log(2^3) = 3\log 2$;
$\log(18) = \log(2 \cdot 9) = \log(2) + \log(3^2) = \log 2 + 2\log 3$;
$\log(27) = \log(3^3) = 3\log 3$;
$\log(80) = \log(8) + \log(10) = \log 8 + 1$.
$\log(\frac{1}{2}) = \log(2^{-1}) = -1 \cdot \log 2 = -\log 2$,
$\log(\frac{1}{8}) = \log(2^{-3}) = -3 \cdot \log 2$
$\log(\frac{3}{8}) = \log 3 - \log(8) = \log 3 - 3\log 2$.

Figure 2-2 shows a plot of $x = \log(y)$. Note that x is positive for $y > 1$ and negative for $y < 1$, properties that follow directly from the behavior of the exponential function, $y = 10^x$. Notice that the logarithmic function compresses huge variations into smaller increments: $\log(10{,}000) = 4$, $\log(100{,}000) = 5$, $\log(1{,}000{,}000) = 6$, and so on. This is the principle behind the logarithmic plot in which, for example, the horizontal axis for the independent variable is equi-spaced, as usual, while the vertical plot is on a "log scale" where equi-spaced intervals correspond to equal multiplicative factors, such as powers of 10. Figure 2.3 shows a log scale plot of $y = 10^x$. Note that on a log scale the vertical axis has no zero or negative values, so the origin is just an arbitrarily chosen positive number, e.g, $y = 0.001$ in the case of Fig. 2-3. Equal upward spacings correspond to a multiplicative factor (10 in our case) while equal downward spacings correspond to decreases by the same factor.

Natural Exponential and Logarithm

There is a number with special mathematical properties, called the **naperian base**, e, that is particularly useful for exponentials. The number e, like $\sqrt{2}$ and π, is a real number that cannot be expressed as a fraction. Its approximate value is: e = 2.7183. The exponential function using e is: $y = e^x$, sometimes called the "natural" exponential function, and is often written as: $y = \exp(x)$. The inverse function, called the **natural logarithm**, is $x = \log_e(y)$. $\log_e$ is often given the shorthand notation "ln" so that $\log_e(y)$ is expressed as: $\ln(y)$. The counterpart of Eq. (*2.6*) is then:

$$y = e^x \equiv \exp(x) \Leftrightarrow x = \ln(y) \tag{2.8}$$

The basic rules for exponentials, Eqs. (*2.5*), hold for exp (x), as do the logarithmic rules, Eqs. (*2.7*), with "log" replaced by "ln" in those expressions.

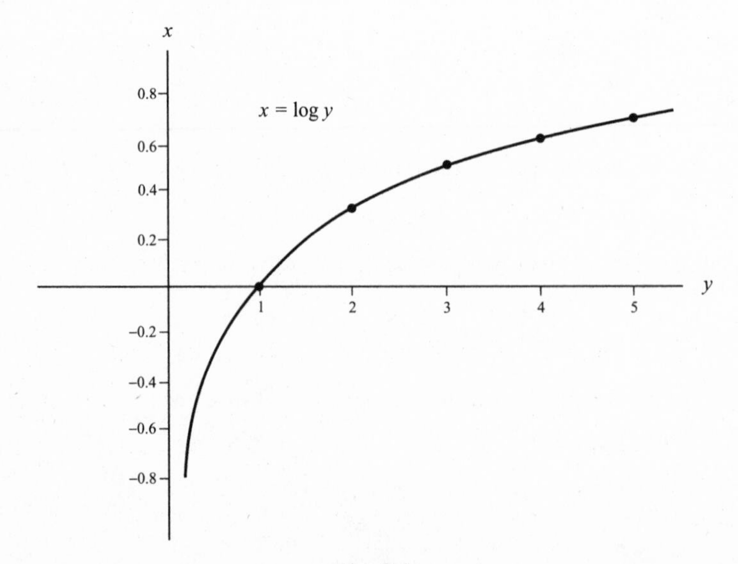

Fig. 2-2

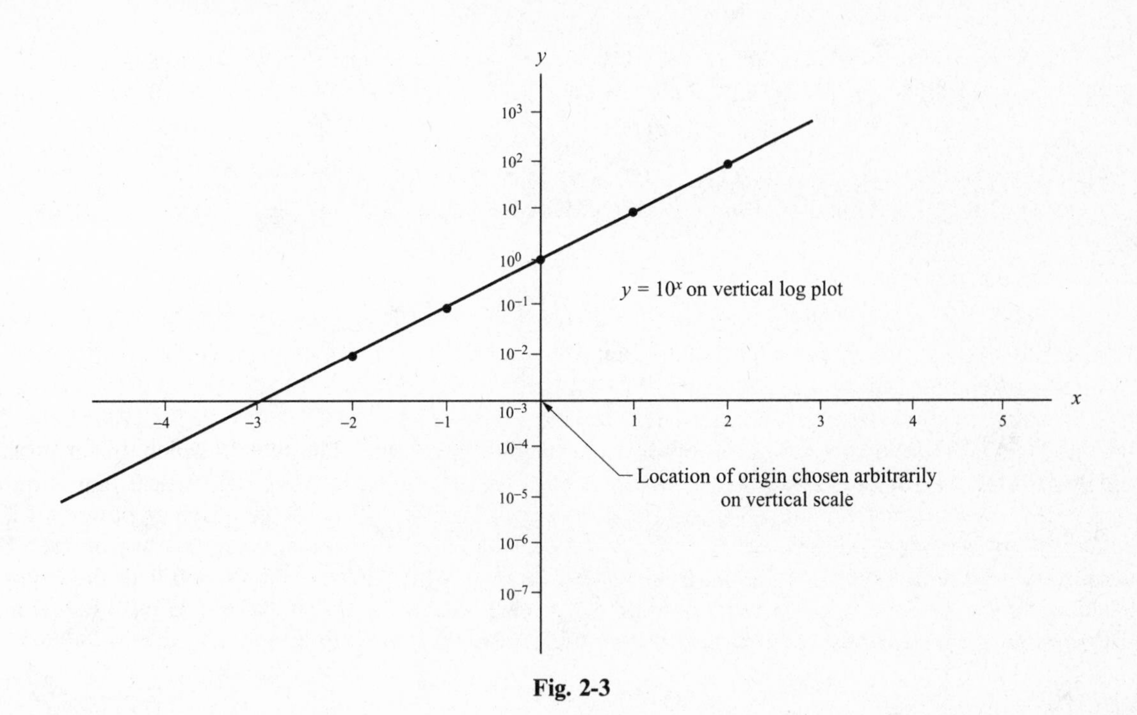

Fig. 2-3

Problem 2.4.

(*a*) Find the values of exp (x) for $x = 0, 1, 2, -1, \frac{1}{2}, \frac{3}{2}$ (keep values to four place accuracy).

(*b*) Find the values of ln (z) for $z = 1, 10, 10^2, \frac{1}{2}$.

Solution

(*a*) These results can be obtained from our approximate value of e given above, or more easily using the "e^x" function on a good calculator. $\exp(0) = e^0 = 1$(exact); $\exp(1) = e^1 = e = 2.718$; $\exp(2) = e^2 = 7.389$; $\exp(-1) = 1/e = 0.3679$; $\exp(\frac{1}{2}) = \sqrt{e} = 1.649$; $\exp(\frac{3}{2}) = \exp(1) \cdot \exp(\frac{1}{2}) = (2.718)(1.649) = 4.482$.

(*b*) $\ln(1) = 0$; from ln tables or calculator: $\ln(10) = 2.303$; $\ln(100) = 2 \ln(10) = 4.605$; $\ln(\frac{1}{2}) = -\ln(2) = -0.6931$.

Problem 2.5.

(*a*) Show that the natural and decimal logarithms are related by:

$$\ln(A) = \ln(10) \cdot \log(A) \qquad (i)$$

(*b*) Use the results of part (*a*) to find ln (8) and ln (18) in terms of log 2 and log 3.

Solution

(*a*) Let $A = 10^x$, and $10 = e^a$. Then: $A = (e^a)^x$. Using the rules for logarithms:

$$\ln(A) = \ln[(e^a)^x] = x \cdot \ln(e^a) = x \cdot a \qquad (ii)$$

We also have, however:

$$\log(A) = \log(10^x) = x \qquad (iii)$$

and

$$\ln(10) = \ln(e^a) = a \qquad (iv)$$

Substituting (*iii*) and (*iv*) into the right side of (*ii*) we get our result: $\ln(A) = \ln(10) \cdot \log(A)$.

(*b*) $\ln(8) = 3 \cdot \ln(2) = 3 \cdot \ln(10) \cdot \log 2$. Using the value of $\ln(10)$ from Problem 2.4 we have: $\ln(8) = 6.908 \cdot \log 2$. Similarly, $\ln(27) = 3 \cdot \ln(3) = 3 \cdot \ln(10) \cdot \log(3) = 6.908 \cdot \log(3)$.

2.2 PROPAGATION OF SOUND—VELOCITY, WAVE-FRONTS, REFLECTION, REFRACTION, DIFFRACTION AND INTERFERENCE

Sound Velocity in Air

In Chap. 1 we discussed the propagation of waves in different media. In the case of fluids we saw that the general formula for the velocity of propagation, v_p, of longitudinal sound waves, was:

$$v_p = (B/\rho)^{1/2} \tag{2.9}$$

where B is the **bulk modulus** and ρ the density of the fluid. The bulk modulus is given by (see, e.g. Beginning Physics I, Chap. 11, Sec. 3)

$$B = \text{stress/strain} = -\Delta P/(\Delta V/V) \tag{2.10}$$

where ΔP is the increase in hydrostatic pressure on the fluid and $\Delta V/V$ is the consequent fractional change in volume. Since the volume decreases as the pressure increases, the minus sign assures that B is positive. To calculate the bulk modulus for air we can reasonably assume that the ideal gas law holds:

$$PV = nRT \tag{2.11}$$

Eq. (*2.11*) gives us a relationship between P and V, and would allow us to calculate the actual value of B if we knew that T was constant during the compression (or rarefaction) of gas. We know from the study of heat and thermodynamics that a compression of a gas is generally accompanied by a temperature rise unless there is adequate time for heat to flow from the compressed gas to the surroundings so that a common temperature with the surroundings is maintained during the compression period, and the process is isothermal. In the case of longitudinal waves, the compressions and rarefactions at a given location are typically very rapid, so that there is no time for a significant amount of heat to flow to or from the surrounding air as the local air goes through its accordion-like paces. Indeed, under such conditions the process is adiabatic rather than isothermal. Thus, instead of Eq. (*2.11*), we can use the relationship between P and V for an ideal gas undergoing an adiabatic process (see, e.g., Beginning Physics I, Problem 18.9):

$$PV^{\gamma} = \text{constant} \tag{2.12}$$

where γ is the ratio of the molar heat capacity at constant pressure to that at constant volume: $\gamma = c_{\text{mol, p}}/c_{\text{mol.v}}$. Eq. (*2.12*) implies that a change in P must be accompanied by a corresponding change in V, so we should have a relationship between ΔP and ΔV and hence an expression for B. Using the calculus it can be shown that for small ΔP:

$$\Delta P/\Delta V = -\gamma P/V \tag{2.13}$$

Substituting into Eq. (*2.10*) we get:

$$B_{\text{adiabatic}} = \gamma P \tag{2.14}$$

From this, and the ideal gas law, it can be shown that:

$$v_p = [\gamma RT/M]^{1/2} \tag{2.15}$$

where M is the molecular mass of the gas. We derive this result in the following problem.

Problem 2.6.

(*a*) Using Eqs. (*2.9*), (*2.11*) and (*2.14*), find an expression for the speed of sound in an ideal gas in terms of the temperature, T, and pressure P, of the gas.

Solution

Substituting Eq. (*2.14*) into Eq. (*2.9*), we get:

$$v_p = [\gamma P/\rho]^{1/2} \qquad (i)$$

The ideal gas law, Eq. (*2.11*), can be re-expressed in terms of the density, ρ, and the molecular mass, M, by recalling that: $n = m/M$, where n = number of moles in our gas, m = mass of the gas. The molecular mass, M, represents the mass per mole of the gas (see, e.g., Beg. Phys. I, Problem 16.7).

Then:

$$PV = nRT = mRT/M \qquad (ii)$$

Dividing both sides by V, and noting that $\rho = m/V$, we get:

$$P = \rho RT/M \qquad (iii)$$

Substituting (*iii*) into (*i*) we get: $v_p = [\gamma RT/M]^{1/2}$, which is the desired result, Eq. (*2.15*).

Problem 2.7. Recall that γ for monatomic and diatomic gases are approximately $5/3 = 1.67$, and $7/5 = 1.40$, respectively. Also, the gas constant $R = 8.31$ J/mol · K, and atmospheric pressure is $P_A = 1.013 \cdot 10^5$ Pa.

(*a*) Calculate the velocities of sound in hydrogen and helium at $P = P_A$ and $T = 300$ K (27 °C).

(*b*) Calculate the velocity of sound in air at P_A at: $T = 273$ K, $T = 300$ K, and $T = 373$ K. Assume $M_{air} = 29.0$ kg/kmol.

Solution

(*a*) Using Eq. (*2.15*), and noting that helium is monatomic, with molecular mass 4.0 kg/km, we have:

$$v_p = [1.67(8310 \text{ J/kmol} \cdot \text{K})(300 \text{ K})/(4.0 \text{ kg/kmol})]^{1/2} = 1020 \text{ m/s}.$$

Similarly, for hydrogen, which is diatomic and has molecular mass 2.0 kg/kmol, we have:

$$v_p = [1.40(8310 \text{ J/kmol} \cdot \text{K})(300 \text{ K})/(2.0 \text{ kg/kmol})]^{1/2} = 1321 \text{ m/s}.$$

(*b*) Here, the dominant gases are oxygen and nitrogen, which are diatomic, so:

$$v_p = [1.40(8310 \text{ J/kmol} \cdot \text{K})\text{T}/(29.0 \text{ kg/kmol})]^{1/2} = 20\text{T}^{1/2} \text{ m/s}.$$

Substituting our temperatures, we get:

$$T = 273 \text{ K} \Rightarrow v_p = 330 \text{ m/s}; \qquad T = 300 \text{ K} \Rightarrow v_p = 346 \text{ m/s}; \qquad T = 373 \text{ K} \Rightarrow v_p = 386 \text{ m/s}.$$

It is interesting to note that the formula for the speed of sound in a gas, Eq. (*2.15*), is very similar to the equation for the root-mean-square velocity of the gas molecules themselves (see, e.g., Beg. Phys. I, Problem 16.12): $v_{rms} = (3RT/M)^{1/2}$. Both the velocity of sound and the mean velocity of the molecules decrease with molecular mass and increase with temperature, and v_p is slightly less than v_{rms} for the same gas and temperature.

Waves in Two and Three Dimensions

In Chap. 1 all of the waves we considered were constrained to propagate in one dimension, such as transverse waves in a cord or sound waves in a rail or tube. Waves in bulk material such as air, tend to spread out in all available directions. A two-dimensional analogue of this is the ripple effect when a stone is tossed into the still water of a pond. The disturbance of the water surface at the point of contact

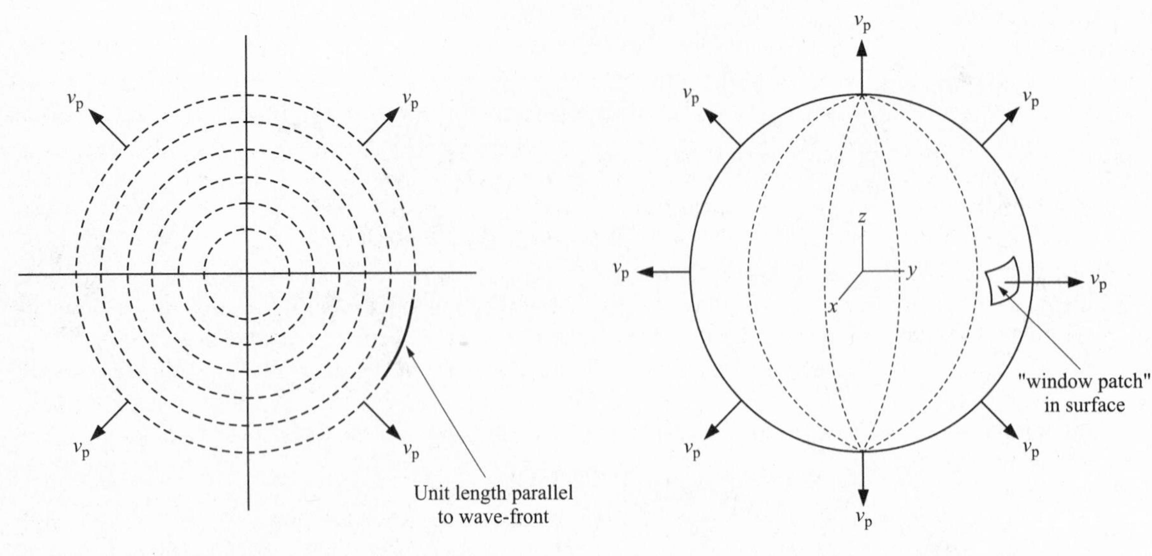

(*a*) Concentric circular ripples

(*b*) Expanding spherical wave-front (snapshot)

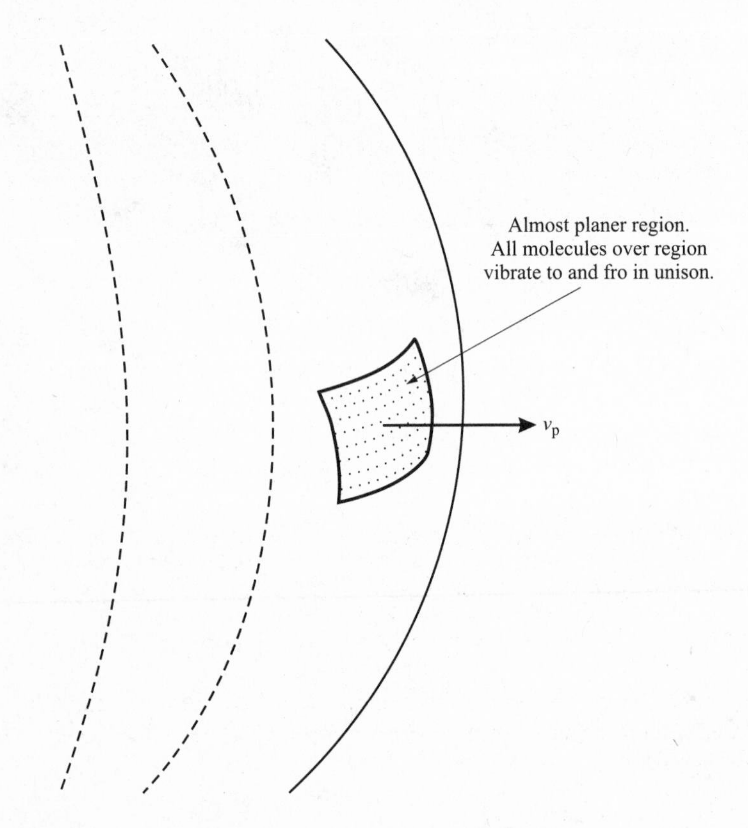

(*c*) Magnification of small "window patch" on large spherical wave-front

Fig. 2-4

sends circular ripples moving out at the appropriate wave propagation speed for this system. The fact that the ripples are circular tells us that the disturbance is traveling at equal speed in all directions. This demonstrates that the speed and direction of the disturbance over the water surface has nothing to do with the speed and direction of the stone. Once a disturbance in any material is created, the wave propagation of the disturbance is *characteristic of the material through which the wave moves.* If we draw an imaginary line through the crest (or trough) of one of the ripples at a given instant of time, we are looking at the same phase of the disturbance at all different locations on the water surface. Such a line is called a **wave-front**. If we plot this wave-front at many different instants of time we get a clear picture of how the disturbance moves through the water surface. For our water surface the motion of the wave-front mimics our own observation of the motion of the ripple. If a given point in our pond were disturbed with a constant frequency vibrator, the wave would consist of a continuous train of circular ripples and corresponding wave-fronts, with crests and troughs spaced equally from one another. Figure 2-4(*a*) shows a pictorial display of the wave-fronts, with the origin as the location of the vibrator.

This analysis can be extended to sound waves in three dimensions. Consider a medium such as air at rest; a disturbance (such as caused by a snap of the fingers) or a continuous simple harmonic vibration (such as caused by a vibrating tuning fork) will have wave-fronts that travel in all directions with equal speed, just as the ripples in the water. In the case of the longitudinal sound waves in the still air at constant temperature, the disturbance propagates in three dimensions with constant velocity, so the wave-fronts now take the form of spherical shells—at least until they hit some boundary. A representation of the wave-front of a spherical wave in three dimensions is shown in Fig. 2-4(*b*) where, again, the origin is at the source of the disturbance.

Note. The direction of propagation of the wave at any location is perpendicular to the wave front at that location.

Energy and Power in Waves in Two and Three Dimensions

In general, it is complicated and beyond the scope of this book to quantitatively describe wave motion in two or three dimensions. Nonetheless, there are a number of characteristics that can be described fairly easily. For example, for our water ripples, the energy of the wave in any small wave-front region, and the associated power transmitted through a unit length parallel to the wave-front, see Fig. 2-4(*a*), are being diluted as the circular wave-front expands to larger circumference. Since the circumference of a ripple increases in proportion to its growing radius R, the power per unit wave-front length must decrease as $1/R$.

A similar analysis can be made for our SHM sound waves in three dimensions. The energy and power of the wave, per unit area perpendicular to the direction of propagation of the wave [see, e.g., the small "window" in Fig. 2-4(*b*)] now fall off as $1/R^2$. The power per unit area perpendicular to the direction of propagation is called the intensity, I, and is given by:

$$I = P/A \tag{2.16}$$

where P is the power transmitted through a "window" concentric to the wave-front and of cross-sectional area A, as shown schematically in Fig. 2-4(*c*).

Problem 2.8. A spherical sound wave emanates from a small whistle suspended from a ceiling of a very large room, emitting a single frequency simple harmonic wave.

(*a*) If the power generated by the whistle is 0.0020 W, find the intensity of the spherical wave 1.0 m, 2.0 m, and 3.0 m from the source. [*Hint*: Recall that the surface area of a sphere of radius r is $4\pi r^2$].

(*b*) Find the power passing through an imaginary circular window of area 12.0 cm^2, which is facing (parallel to) the wave fronts and at a distance of 3.0 m from the source.

Solution

(*a*) All the power must pass through any imaginary concentric spherical shell, and by symmetry will flow out with equal intensity in all directions. For a spherical shell of radius r, the intensity would thus be: $I = P/A = P/(4\pi r^2)$. Substituting the value of P and the various r values into this relationship, we get:

$$r = 1 \text{ m}, I = 0.159 \text{ mW/m}^2; \qquad r = 2 \text{ m}, I = 0.0398 \text{ mW/m}^2; \qquad r = 3 \text{ m}, I = 0.0177 \text{ mW/m}^2.$$

(*b*) Since I represents power per unit area passing perpendicular to the imaginary window, we must have:

$$P_A = IA = (0.0177 \cdot 10^{-3} \text{ W})(12.0 \cdot 10^{-4} \text{ m}^2) = 0.212 \text{ mW}.$$

Problem 2.9.

(*a*) Assuming a simple harmonic disturbance in the water of a pond, how would you expect the amplitude of any given ripple to change with radius R as the ripple expands out? Ignore thermal losses.

(*b*) Repeat for the sound wave in Problem 2.8; If the wave amplitude were 0.20 mm at $R = 1.0$ m, what is the amplitude at $R = 3.0$ m?

Solution

(*a*) We recall from Chap. 1 that the power of a wave of a given frequency and velocity of propagation is proportional to the square of the amplitude. Since the power falls off as $1/R$ for the case of our circular ripples, the amplitude of the wave decreases as $1/\sqrt{R}$.

(*b*) In this case the power falls off as $1/R^2$, so the wave amplitude falls off as $1/R$. If the amplitude were 0.20 mm at $R = 1.0$ m, then it would be 1/3rd that amount, or 0.0667 mm, at $R = 3.0$ m.

Plane Waves

Another interesting result for our water ripples and our spherical sound waves is that at large distances from the source, a small portion of the circular (or spherical) wave-front looks almost like a straight line (or flat plane) at right angles to the direction of motion of the wave. For a spherical sound wave whose source is a long way off, the wave-front appears to be a planar surface perpendicular to the direction of motion of the wave, as long as we are observing a portion of the wave-front whose dimensions are small compared with the distance to the source. Thus, for example, the imaginary "window" shown in Fig. 2-4(*c*) is almost planar if the dimensions are small compared to the distance from the source of the wave. A wave moving through space in which the wave-front is planar is called a **plane wave**, and is characterized by the fact that every point on the planar wave-front is in phase at the same time. Thus, the air molecules are all vibrating in and out along the direction of motion of the wave (longitudinal) in lock-step at all points on the wave-front in the window of Fig. 2-4(*c*).

Since all the points on a plane wave act in unison, the wave equation for such a wave is exactly the same as for our longitudinal waves in a long tube. Indeed, if x is along the direction of wave propagation, then the wave-front is parallel to the (y, z) plane. Under those circumstances our SHM sound wave is described by:

$$d_{y,z}(x, t) = A \sin (2\pi t/T - 2\pi x/\lambda) \tag{2.17}$$

where $d_{y,z}$ represents the displacement of air molecules at any point (y, z) on the wave-front, and at a distance x (measured from some convenient point) along the direction of wave propagation, at any time t. A, T and λ are the amplitude, period and wavelength of the wave, and all three are constant for all y and z in our plane wave region. Thus $d_{y,z}$ does not depend on y or z in our plane wave region. We note that in reality the amplitude A does decrease with increasing x but only slightly if we limit ourselves to changes in x that are small compared to the distance from the source of the wave-front.

Problem 2.10. Consider the spherical wave in Problem 20.8(*b*).

(*a*) Could the portion of the wave passing through the imaginary window be considered a plane wave?

(*b*) When the portion of the wave-front passing through the imaginary window, which is 3.0 m from the source, moves an additional 4.0 cm, how much larger is the area it will occupy?

(*c*) What will be the intensity of the wave of Problem 20.8(*b*) when it moves an additional 4.0 cm, as in part (*b*), above?

Solution

(*a*) The imaginary window is like a circular patch of radius less than 2 cm on a spherical surface of radius 3.0 m. It therefore is, indeed, almost flat and a plane wave would be an excellent approximation to the part of the spherical wave passing through it.

(*b*) The new area would correspond to the equivalent window on a concentric spherical shell of radius 3.04 m. Since the areas go as the square of the radius, the ratio of the window areas would be (letting a_1 and a_2 represent the areas at 3.00 m and 3.04 m, respectively):

$$a_2/a_1 = r_2{}^2/r_1{}^2 = (3.04/3.00)^2 = 1.027 \Rightarrow a_2 = 1.027a_1 = 12.3\ \text{cm}^2, \qquad \text{or a 2.7\% increase.}$$

(*c*) Since the portion of the wave passing through the first window is precisely the portion of the wave passing through the second window, the new intensity will just be: $I_2 = P_{A1}/A_2$, where P_{A1}, the power through the first window, was already calculated in Problem 2.8(*b*). Substituting in numbers we get:

$$I_2 = (0.212 \cdot 10^{-3}\ \text{W})/(12.3 \cdot 10^{-4}\ \text{m}^2) = 0.172\ \text{W/m}^2.$$

Note. This is only slightly less than $I_1 = 0.177\ \text{W/m}^2$, as calculated in Problem 2.8(*a*). Indeed, I_2 could have been calculated by noting that if the area has gone up by 2.7% the intensity must go down by 2.7% so: $I_2 = I_1/1.027 = 0.172\ \text{W/m}^2$.

Problem 2.11.

(*a*) Find an expression for the intensity of a sinusoidal planar sound wave traveling through air in terms of the density of air ρ, the angular frequency ω, the amplitude A, and the velocity of propagation v_p, of the wave.

(*b*) A sinusoidal planar sound wave travels through air at atmospheric pressure and a temperature of $T = 300$ K. The intensity of the wave is $5.0 \cdot 10^{-3}\ \text{W/m}^2$. Find the amplitude of the wave if the frequency is 2000 Hz. (Assume the mean molecular mass of air is M = 29 kg/kmol and $\gamma = 1.40$.)

Solution

(*a*) From Eq. (*1.12*) and Problem 1.16 we have for the power P of the wave passing through a cross-sectional area C_A perpendicular to the direction of propagation:

$$P = \tfrac{1}{2}\rho C_A \omega^2 A^2 v_p \qquad (i)$$

From the definition, $I = P/C_A$, and dividing we get:

$$I = \tfrac{1}{2}\rho\omega^2 A^2 v_p \qquad (ii)$$

(*b*) Recalling (Problem 2.6) that gas pressure is $p = \rho RT/M$, we have for air at atmospheric pressure:

$$1.013 \cdot 10^5\ \text{Pa} = \rho(8314\ \text{J/kmol})(300\ \text{K})/(28.8\ \text{kg/kmol}), \qquad \text{and} \qquad \rho = 1.17\ \text{kg/m}^3.$$

Similarly (Problem 2.6), $v_p = [\gamma p/\rho]^{1/2} = [1.40(1.013 \cdot 10^5\ \text{Pa})/(1.17\ \text{kg/m}^3)]^{1/2} = 348$ m/s. Substituting into (*ii*) above, we get:

$$5 \cdot 10^{-3}\ \text{W/m}^2 = \tfrac{1}{2}(1.17\ \text{kg/m}^3)(6.28)^2(2000\ \text{Hz})^2(348\ \text{m/s})A^2 \Rightarrow A = 3.94 \cdot 10^{-7}\ \text{m}.$$

By examining a region of space where a sound wave can be approximated by a plane wave (or the corresponding two-dimensional region on the surface of a lake where a ripple wave can be approximated by a "straight line" wave-front) one can gain interesting insight into many wave phenomena. These wave phenomena are very similar to those associated with light waves, which we will be studying later on. They include: reflection, refraction, interference and diffraction. These are examined in the following sections.

Reflection and Refraction of Sound

When sound wave-fronts hit a barrier, such as the floor or a wall for the case of the whistle in Problem 2.9, or the side of a mountain or canyon wall for the case of a person making a noise in the great outdoors, part of the wave reflects and part is transmitted into the barrier. The part that is transmitted can penetrate deeply into the barrier material, or it can quickly lose amplitude with wave energy converting to thermal energy (absorption). The rate of absorption depends on such factors as the composition of the barrier, its elasticity and the frequency of the wave. The part of the wave that is reflected has diminished amplitude but the same frequency and velocity as the original wave, and hence the same wavelength. The echo we hear in a canyon is a consequence of such a reflection, and the time elapsed between emission of a sound and the echo we hear can be used to roughly measure the speed of sound in the air if the distances are known, or the distances if the speed of sound is known.

Problem 2.12.

(*a*) A man standing 3360 ft from a high cliff hits a tree stump with an axe, and hears the faint echo 6.4 s later. What is the velocity of sound in the air that day?

(*b*) A child standing with his parents somewhere between the two walls of a wide canyon shouts "hello". They hear two loud echoes, which one parent times with a stop watch. The first echo arrived after an interval of 1.2 s, while the second arrived 1.8 s later. How wide is the canyon? Assume the same speed of sound as in part (*a*).

Solution

(*a*) The sound created by the axe hitting the stump first travels the distance, x, to the cliff, where it is reflected and makes the return trip of the same distance to the man. We must then have that: $2x = v_p t$, where t is the elapsed time for the echo and v_p is the velocity of sound in air. Substituting the given values for x and t in the equation, we have:

$$v_p = 2(3360 \text{ ft})/(6.4 \text{ s}) = 1050 \text{ ft/s}.$$

(*b*) The situation is shown schematically in Fig. 2-5. The family is clearly not midway between the two walls because the echoes took different times. Letting x and y be the respective distances to the near and far walls, we have:

$$2x = v_p t_1 = (1050 \text{ ft/s})(1.2 \text{ s}) = 1260 \text{ ft} \Rightarrow x = 630 \text{ ft}.$$

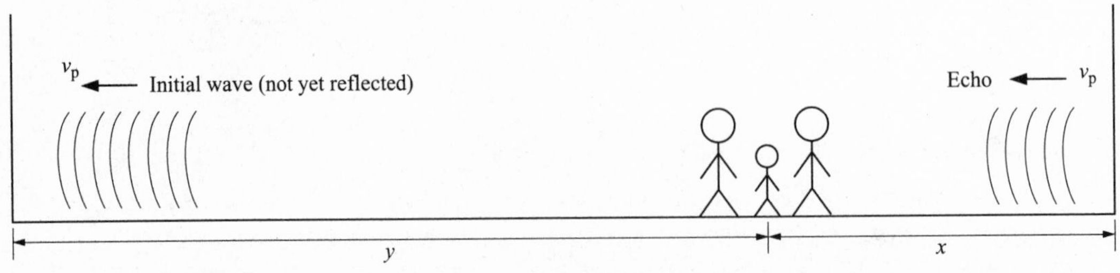

Fig. 2-5

Similarly, we have:

$$2y = v_p t_2 = (1050 \text{ ft/s})(1.8 \text{ s} + 1.2 \text{ s}) = 3150 \text{ ft} \Rightarrow y = 1575 \text{ ft}.$$

Then the distance between the walls is

$$d = x + y = 2205 \text{ ft}.$$

The reflection of sound is of great importance in modern high frequency detection devices. Sonar is used by submarines to find and map out objects at various distances from the sub. The time between emission and detection of reflected sound pulses is measured, and from a knowledge of the speed of sound the distance to the object is determined. Ultrasound is used in medical imaging by detecting changes in tissue density in the body through examination of reflected and transmitted waves.

In our discussion of the velocity of sound in air, we concluded that the velocity is temperature dependent, as shown in Eq. (*2.15*). In general, the layers of air above the ground are not at a constant temperature. Depending on circumstances, e.g. time of day or night, specific atmospheric conditions, etc., the layers of air near the earth's surface can be either colder or warmer than the layers above. Consider the portion of a sound wave emitted from the horn of a ship at sea. Part of the wave initially travels parallel to the surface of the sea, and to an observer at some distance from the ship the wave-fronts can be approximated by those of a plane wave traveling from the ship to the observer. Indeed, if the air is at a uniform temperature, a cross section of the wave-fronts in some local region would look something like that in Fig. 2.6(*a*). Suppose the wave passes a region where the temperature is higher at sea level and drops with increasing altitude. Eq. (*2.15*) indicates that the propagation velocity would be highest at sea level and decreasing upward. Then the bottom of a given wave-front would move faster than a point higher up and the wave-front would start to bend as shown in Fig. 2-6(*b*). Since the direction of propagation of a wave is perpendicular to the wave-front, the wave velocity would start to develop an upward component and would therefore not carry as far along the sea surface. On the other hand, if the layers of air at the water surface were colder than those above, the speed of the wave-front near the surface would be less than that above, and we would get the effect shown in Fig. 2-6(*c*). Here the net effect is that more of the wave-front from higher levels is pushed down toward the surface, ensuring that a substantial amount of wave energy would travel along the surface, and thus be audible a long way off. In general, when a wave travels through a medium of varying densities (for example, layers of air at different temperatures) the velocity of different parts of the wave-front are different, and the direction of propagation of the wave changes as a consequence. This is called **refraction**, and will be discussed in greater detail in our discussion of light waves.

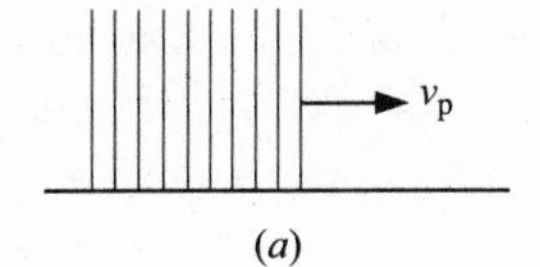

(*a*)

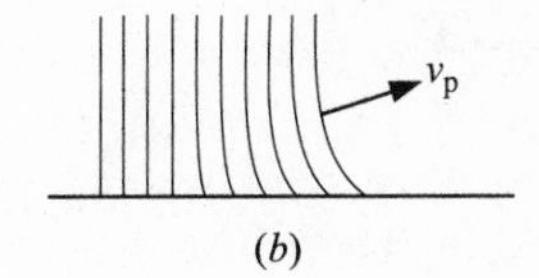

(*b*)

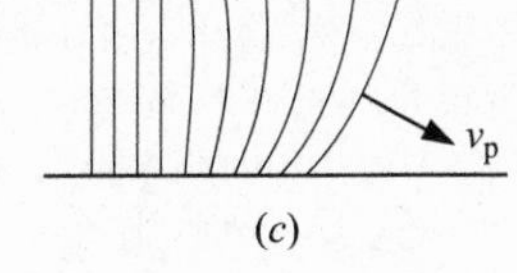

(*c*)

Fig. 2-6

Interference and Diffraction

We can now examine interference and diffraction of sound waves. Interference was already encountered in Chap. 1, and was used to demonstrate the formation of standing waves in a cord or a tube. **Interference** is the effect of having more than one wave passing a given point, and the possibility that the two waves will reinforce or weaken each other as a consequence of the phase difference between the waves. For the case of continuous waves, such as sinusoidal traveling waves, the ability of two waves to completely cancel each other out at a given point over an extended time period (e.g. a node for standing waves) requires three things to be true:

1. The frequencies of the two waves are the same.
2. The amplitudes of the two waves are the same.
3. The wave vibrations are in the same direction in space.

The most important of these requirements is the first. If the frequencies are not identical (or almost so) then at any point in space the relative phases of the two waves would rapidly be changing due to the different frequencies and the positive or negative interference would average to zero over even short time intervals. Our second requirement is less important because even if the amplitudes are somewhat different and one does not get complete cancellation, the interference effect might still be quite significant. The third requirement does not come up in the context of sound waves in a tube, since the vibrations of all waves are constrained to be in the same direction. This is no longer true for waves traveling through space. Nonetheless, if the direction of vibrations of two waves passing a point make a relatively small angle with each other, one could still get substantial cancellation of the two waves—assuming of course that requirements 1 and 2 is satisfied.

2.3 HUMAN PERCEPTION OF SOUND

Intensity Scale of Sound Waves

The human ear responds to the intensity of the sound waves hitting it with the perception of loudness. While the sense of loudness is a physiological and psychological response of human beings and varies somewhat from person to person (and therefore is not exactly the same as sound intensity), it is true that the human ear can perceive an exceptionally broad range of sound intensities. To describe that range it is useful to create a logarithmic scale called the **decibel scale** (db), which gives a quantitative measure to "loudness", which we label n, and define as:

$$n = 10 \log (I/I_0) \tag{2.18}$$

where I is the intensity of sound and I_0 is a fixed reference intensity taken to approximate the lowest level of sound audible to a human being: $I_0 = 1.0 \cdot 10^{-12}$ W/m². It turns out that an intensity of about $I = 1.0$ W/m² represents the highest intensity to which the ear can respond without feeling pain. Substituting this intensity into Eq. (*2.18*) we obtain $n = 10 \log (1.0/1.0 \cdot 10^{-12}) = 10 \log (10^{12}) = 120$ db, so that the threshold of pain is 120 db. Note that because of the logarithmic scale each factor of 10 increase in intensity corresponds to an addition of 10 db. Thus a thousand-fold increase in noise corresponds to a thirty decibel increase in loudness level.

Problem 2.13. A powerful firecracker is tossed in the air and explodes 5 m from a person walking nearby. The peak sound power generated by the explosion is 16 W.

(*a*) What is the intensity of sound that enters the persons ear?

(*b*) To how many decibels does this correspond?

(*c*) At what distance r from the explosion would the person's ear have to be if the sound was at the threshold of pain?

Solution

(*a*) We assume the energy disperses in a spherically symmetric shell away from the burst site, so

$$I = P/4\pi r^2 \Rightarrow I = (16\ \text{W})/[12.56 \cdot (5.0\ \text{m})^2] = 5.09 \cdot 10^{-2}\ \text{W/m}^2.$$

(*b*) $n = 10 \log [(5.09 \cdot 10^{-2}\ \text{W/m}^2)/(1.0 \cdot 10^{-12}\ \text{W/m}^2)] = 107$ db.

(c) Here $n = 120$, so: $12 = \log(I/I_0) \rightarrow I = I_0 \cdot 10^{12} = 1.0\ \text{W/m}^2$. Then: $16\ \text{W} = (1.0\ \text{W/m}^2)(12.56)r^2$, or $r = 1.13$ m.

Problem 2.14. A symphonic passage produces a sound level at a person's ear in the auditorium of 60 db while a person speaking in the next row produces a sound level of 40 db at the same ear. What is the ratio of the intensities of the two sounds?

Solution

$$n_1 - n_2 = 10\log(I_1/I_0) - 10\log(I_2/I_0) = 10\log(I_1/I_2) \Rightarrow 20\ \text{db} = 10\log(I_1/I_2), \quad \text{or: } I_1/I_2 = 10^2.$$

While human perception of loudness will approximately follow the decibel scale it does not exactly do so. Indeed, loudness perception is dependent on a variety of factors specific both to individuals in a species and the species as a whole. An important factor for human (and other) species is the frequency of the sound. The human ear is most sensitive to frequencies in the range 1000–6000 Hz and people with the most acute hearing are able to detect sounds at I_0 intensity, or 0 db, at those frequencies. As the frequency drops below 1000 Hz this threshold of hearing rapidly rises to higher and higher intensities requiring about 30 db at 100 Hz and 100 db at 20 Hz—about the lowest frequency that human beings can hear. For frequencies higher than 6000 Hz the threshold intensity rises relatively slowly (about 20 db) as the frequency reaches toward 12,000–15,000 Hz, and then rises more rapidly to about 100 db at 20,000 Hz—about the highest frequency that human beings can hear. Thus, a 30 decibel sound at 40 Hz will be inaudible, while the same level sound at 1000 Hz will sound quite loud. As it turns out the threshold of pain is about 120 decibels at all frequencies from 20 to 20,000 Hz. It should be noted that few people can hear the full frequency range from 20 to 20,000 Hz, and most cannot hear even intermediate frequencies at the lowest threshold intensities.

Reverberation Time

When a sound is emitted in a closed environment such as a room or auditorium it takes a certain amount of time for the intensity of the sound to dissipate. This is because the sound reflects off the walls and the people and objects in the room, and dies down only as a consequence of the absorption of some of the energy by each object at each reflection. The reverberation time is defined as the time it takes for the intensity of a given steady sound to drop 60 db (or six orders of magnitude in intensity) from the time the sound source is shut off. Reverberation times are important because if they are too long successive sounds run into one another and can make it difficult to make out speech (too much echo). For music the quality of the performance is negatively impacted if the reverberation time is too long or too short, the latter case corresponding to a thin or dry effect. Reverberation times depend on the total acoustic energy pervading the room, the surface areas of the absorbing materials and their absorption coefficients. The absorption coefficient of a surface is defined as the fraction of sound energy that is absorbed at each reflection. Thus, an open window has an absorption coefficient of 1 since all the energy passes out of it and none reflects back in. Heavy curtains have a coefficient of about 0.5, and acoustic ceiling tiles have a coefficient of about 0.6. Wood, glass, plaster, brick, cement, etc. have coefficients that range from 0.02 to 0.05. A formula that gives good estimates of the **reverberation time** was developed by Sabine, a leading acoustic architect, and is given by:

$$t_r = 0.16V/A \tag{2.19}$$

where t_r is the reverberation time (s), V is the volume of the room (m^3) and A is called the absorbing power of the room. The absorbing power A is just the sum of the products of the areas of all the absorbing surfaces (m^2) and their respective absorption coefficients.

Problem 2.15.

(*a*) Find the reverberation time for an empty auditorium 15 m wide by 20 m long by 10 m high. Assume that the ceilings are acoustic tile, the side walls are covered with heavy drapes, and that the floor and the back and front walls are concrete. Assume the following absorption coefficients: ceiling tiles, 0.6; drapes, 0.5; concrete, 0.02.

(*b*) How would the answer to part (*a*) change if the auditorium were filled with 50 people, each with an absorbing power of 0.4. Assume no change in absorption of the floor.

Solution

(*a*) We use Sabine's formula, Eq. (*2.19*). The volume of the room is $15 \times 20 \times 10 = 3000 \text{ m}^3$, so $V = 3000$. To get the absorbing power we multiply absorption coefficients by areas:

$$A = 0.6(300) + 0.5(200 + 200) + 0.02(300 + 150 + 150) = 392.$$

Then:
$$t_r = 0.16(3000)/392 = 1.22 \text{ s}.$$

(*b*) The only change from part (*a*) is that the absorbing power A is increased by the contribution of the people: $A = 392 + 0.4(50) = 412$. Then: $t_r = 0.16(3000)/412 = 1.17$ s.

Problem 2.16. If it were desirable to raise the reverberation time of the auditorium in part (*b*) of Problem 2.15 to 1.70 s, How many m^2 of drapes would need to be removed if the walls behind them were concrete?

Solution

For t_r to be 1.70 s, we must have an absorbing power A given by:

$$1.70 = 0.16(3000)/A \Rightarrow A = 282.$$

If x is the number of m^2 of drape that need to be removed, exposing a like amount of concrete, we have, recalling the absorbing power of Problem 2.15(*b*):

$$412 - 0.5x + 0.02x = 282 \Rightarrow 130 = 0.48x \Rightarrow x = 271.$$

Thus, 271 m^2 of the original 400 m^2 of drapes must be removed.

Quality and Pitch

In addition to loudness, humans can distinguish other sound factors related to frequencies and combinations of frequencies of sound. When a note on a musical instrument is played, the fundamental is typically accompanied by various overtones (harmonics, i.e., integer multiples of the fundamental) with differing intensity relative to that of the fundamental. The intensities of the harmonics will vary from instrument to instrument. The sound of harmonics is pleasing to the ear, and while the note is identified by the listener with the fundamental frequency, the same note from different instruments will sound differently as a consequence of the different harmonic content. The time evolution of the note also contributes to the different sounds. These different sound recognitions by the human ear are called the *quality* of the note. The **pitch** of a note is the human perception of the note as "high" or "low" and is closely related to the frequency but is not identical to it. The pitch involves human subjective sense of the sound. While a higher frequency will be perceived as a higher pitch, the same frequency will be perceived as having slightly different pitches when the intensity is changed: higher intensity yields lower pitch. Another difference between frequency and pitch is the perception of simultaneous multiple frequencies. As noted above, when the human ear hears a fundamental and harmonics it perceives the pitch as that of the fundamental.

With regard to musical notes, it is found that certain combinations of notes have particularly pleasing sounds. The frequency of such notes are found to be close whole number ratio to each other. In particular two notes an octave apart are in the ratio of 1 to 2, and such notes are labeled with the same letter. Thus, middle C on the piano, which corresponds to 264 Hz and the C an octave above (C') is 528 Hz while the C an octave below is 132 Hz. Similarly, the notes C, E, G form what is called a major triad in "the key of C", having frequencies in the ratio of 4 to 5 to 6. Their actual frequencies are then: C = 264 Hz, E = 330 Hz, G = 396 Hz. Similarly, F, A, C' and G, B, D' form major triads in the keys of F and G, with actual frequencies: F = 352 Hz, A = 440 Hz, C' = 528 Hz, G = 396 Hz, B = 495 Hz, D' = 594 Hz. All the main piano notes of the C octave can be determined from these triads, recalling that C = $\frac{1}{2}$ C' and D = $\frac{1}{2}$D'. The octave then has seven notes: C, D, E, F, G, A, B. If one starts with D and tries to make a major triad in "the key of D" new notes would be necessary. In general practice five new notes are added to the piano octave in part to address this problem: C#, D#, F#, G#, and A#. The new scale is then: C, C#, D, D#, E, F, F#, G, G#, A, A#, B. In the **diatonic scale** in the key of C, the original seven notes have the same frequencies as given above. A quick check shows that for the original seven, any two adjacent notes are either in the ratio of 9/8 or 10/9 or 16/15. The intervals between adjacent notes that have either of the first two ratios are called whole-note intervals, while those with the last ratio are called half-note intervals. The new notes are placed between those that have the 9/8 or 10/9 ratios, so that all adjacent notes are approximately half-note intervals. Even with the added notes, if one tried to have the major triad in all keys many more notes would be necessary. To avoid this the **equally tempered scale** was created, in which all the twelve notes of the octave are tuned so that the ratio of any two adjacent notes are the same. Since there are twelve notes in the octave the adjacent notes must be in the ratio of the twelfth root of 2, $(2)^{1/12} = 1.05946$, so that C' = 2C, D' = 2D, etc. By agreement the note A is taken as 440 Hz, and all the other notes are then determined. In this scale the notes have slightly different frequencies than in the diatonic scale. The advantage is that for this choice every note has a major triad, while the disadvantage is that the ratios are not exactly 4 to 5 to 6 for any key. In the key of C for example, the new frequencies are: C = 261.6 Hz, E = 329.6 Hz, G = 392.0 Hz, so the ratios are 3.97 to 5 to 5.95. Since the ear finds it more pleasing to have the ratios: 4 to 5 to 6, a piano tuned in the diatonic scale will sound better than the even tempered scale in the keys of C, F and G, but would sound worse in some other keys such as D, E, and A, etc.

2.4. OTHER SOUND WAVE PHENOMENA

Beats

In discussing interference of waves we noted that it was necessary to have the same frequency if one was to have interference effects observable. Nonetheless, if we have two frequencies that differ only by a few Hz we can indeed detect "interference" effects that oscillate in time slowly enough to be easily detectable. Consider two sound waves of equal amplitude A, and slightly different frequencies, f_1 and f_2, traveling along the x axis. At a given point in space the actual disturbance of the air molecules from their equilibrium positions can be expressed as: $x = A\cos(2\pi f_1 t) + A\cos(2\pi f_2 t + \phi)$, where ϕ is the relative phase of the two waves at some arbitrary instant of time, t_0. Since the frequencies are different this relative phase is of no significance since the relative phases of the waves continually change as time goes on. We therefore simplify the mathematics by setting $\phi = 0$. Then: $x = A[\cos 2\pi f_1 t) + \cos 2\pi f_2 t)]$. Using the trigonometric identity: $\cos\theta_1 + \cos\theta_2 = 2\cdot\cos[(\theta_1 - \theta_2)/2]\cdot\cos[(\theta_1 + \theta_2)/2]$, we get:

$$x = 2A\cdot\cos[2\pi t(f_1 - f_2)/2]\cdot\cos[2\pi t(f_1 + f_2)/2] \tag{2.20}$$

We let $f = (f_1 + f_2)/2$, and $\Delta f = (f_1 - f_2)$. Then f is the average of the two frequencies and is midway between them, while Δf is the difference of the two frequencies. Since the frequencies are very close, the last cosine term on the right of Eq. (*2.20*) approximates the oscillation of either original wave, while the other cosine term represents a very slow oscillation at frequency $\Delta f/2$. For example, if $f_1 = 440$ Hz (middle A on a properly tuned keyboard) and $f_2 = 437$ Hz (e.g., middle A on an out of tune keyboard), $\Delta f/2$ would equal 1.5 Hz. Then, in Eq. (*2.20*) the expression: $2A\cdot\cos[2\pi t(f_1 - f_2)/2]$ can be thought of as a slowly varying amplitude for the "average" oscillation at $f = 441.5$ Hz. This variable amplitude reaches two maximal values: $2A$ and $(-2A)$ in each complete cycle. Each will correspond to a maximal loudness in the sound, called a **beat**. Since there are two such beats in each cycle of the $\Delta f/2$ Hz oscillation, the number of beats per second is: $\Delta f = (f_1 - f_2)$ Hz. In other words, the number of beats per second is just the difference of the two frequencies. Note that we have been assuming that $f_1 > f_2$ so that Δf is positive. However, it doesn't matter which is larger since $\cos(\theta) = \cos(-\theta)$. Thus, in our analysis we can more generally use $\Delta f = |f_1 - f_2|$. Because beats are most clearly audible as the frequencies are closest they are an excellent vehicle for tuning an instrument against a known standard frequency such as that of a tuning fork.

Problem 2.17.

(*a*) A piano tuner is testing middle A on the piano against a standard tuning fork with the exact frequency of 440 Hz. She hears four beats per second, and starts to decrease the tension in the piano cord. The beats increase to five per second. What is the frequency of the cord before and after her adjustment?

(*b*) What must the piano tuner do next to correctly tune the piano?

Solution

(*a*) The original frequency of the piano cord differed from the 440 Hz by 4 Hz, and was therefore either 444 Hz or 436 Hz. To choose between these two we note that reducing tension in the cord drops the frequency. Since by decreasing tension the number of beats increased, she must have started with the A cord at lower frequency than the tuning fork and it got lower still. It therefore was at 436 Hz to start, and dropped to 435 Hz after the adjustment.

(*b*) She has to increase the tension slowly and listen to a decrease in the number of beats against the tuning fork. When the beats are no longer audible the cord is properly tuned.

The Doppler Shift

We now turn to a phenomenon we all recognize from every day life. When an ambulance or police car with its siren screaming approaches you it appears to have one pitch, but when it passes by and moves away from you the pitch seems to drop noticeably. Clearly the mechanical siren did not change, so what did? This change in pitch is an example of what is called the **Doppler shift** and is caused by motion of the source of a sound wave through the air (as in the example of the siren) or by the motion of the listener through the air. In the case of the source moving, the wave-fronts of successive crests of the sound wave are bunched up in the direction of motion through the air, while they are more separated in the direction opposite to the motion. In either case the crests all move through the air with the characteristic propagation velocity of the medium, v_p. Thus, if the source is moving toward (away from) the listener, the listener would detect shorter (longer) wavelengths or higher (lower) frequencies.

The moving source situation is depicted in Fig. 2-7. We consider a source emitting a sinusoidal wave with frequency $f_s = 1/T$. If the source is stationary (relative to the air) as in Fig. 2-7(*a*), two successive positive crests are emitted a time T (one period) apart, say at times t_0 and $t_0 + T$. The successive crests move off as spherical wave-fronts traveling at velocity v_p, and at some later time would appear as depicted in the figure. The distance between the crests in any radial direction, including to the right and left in the figure, is just the wavelength, which is given as $\lambda_s = v_p T = v_p/f_s$. If, on the other hand, the source were moving to the right with some velocity v_s (typically much smaller than v_p) relative to the air, then the second crest emitted at time $t_0 + T$ will be emitted from a location to the right of the one emitted at time t_0. The distance between the two points of emission would be $x = v_s T$, as shown in Fig. 2-7(*b*). Once emitted, both wave fronts travel relative to the medium with the characteristic velocity v_p, and again spread out as spherical shells, but the shells are of course no longer concentric. It is easy to see that the distance between the crests now depends on the direction in which one is interested. If a listener is off to the right (source traveling directly toward listener) the effective wavelength (crest to crest distance) will be: $\lambda_{eff} = v_p T - v_s T$ since the second crest is closer to the first by the distance x the

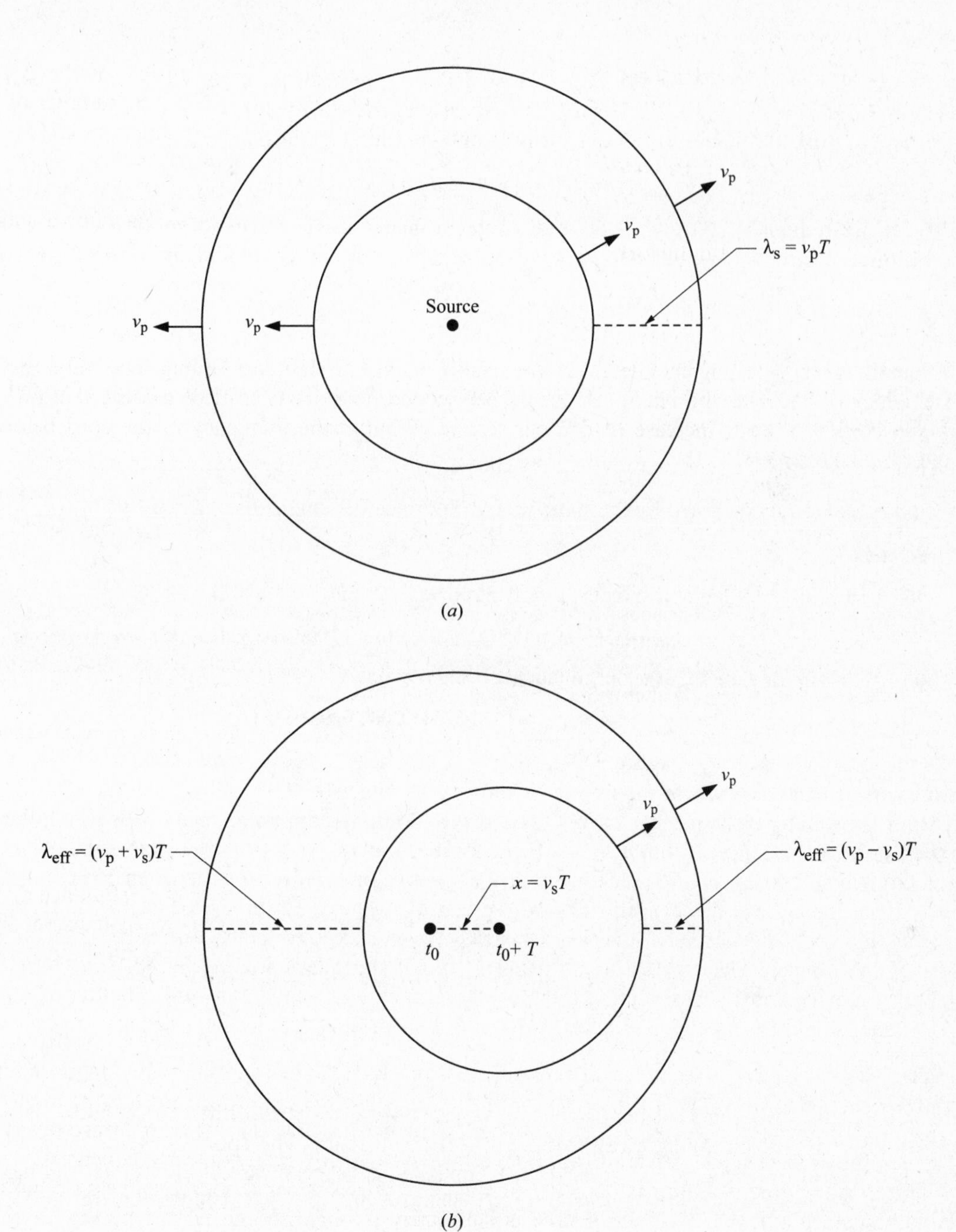

Fig. 2-7

source moved before emitting the second crest. Similarly, if the listener were off to the left (source moving away from listener) the effective wavelength would be $\lambda_{eff} = v_pT + v_s T$, since the second crest was emitted a distance x to the right of the first crest. Combining these two cases, and redefining v_s so that it is positive (negative) when moving away from (toward) the listener, we have:

$$\lambda_{eff} = (v_p + v_s)T = (v_p + v_s)/f_s \tag{2.21}$$

where f_s is the stationary source frequency and v_s is negative for the source moving toward the listener. It should be noted that as long as the source keeps moving at constant velocity the previous discussion will hold for all succeeding crests and the forward and backward traveling waves will indeed have wavelength λ_{eff}. Since the waves still travel with velocity v_p the effective frequency is just: $f_{eff} = v_p/\lambda_{eff}$, or $\lambda_{eff} = v_p/f_{eff}$. Substituting for λ_{eff} in our formula and shifting terms around we get finally:

$$f_{eff} = [v_p/(v_p + v_s)]f_s \tag{2.22}$$

where f_{eff} is the frequency heard by a stationary listener facing along the line of motion of the source, f_s is the stationary source frequency and v_s is positive (negative) for source moving away from (toward) the listener.

Problem 2.18. A fire engine has a siren with a frequency of 1000 Hz. The engine is hurtling down the street at 25 m/s in the direction of a pedestrian standing on the curb. Assume the speed of sound in air is 350 m/s, and that there is no wind blowing.

(*a*) What is the frequency of the siren heard by the pedestrian as the engine approaches?

(*b*) What is the frequency heard by the same pedestrian once the engine has passed by?

Solution

(*a*) Using Eq. (*2.22*) and noting that $v_p = 350$ m/s, $v_s = -25$ m/s, $f_s = 1000$ Hz, we get:

$$f_{eff} = [(350 \text{ m/s})/(350 \text{ m/s} - 25 \text{ m/s})](1000 \text{ Hz}) = 1077 \text{ Hz}.$$

(*b*) We again use Eq. (*2.22*), the only difference being that now $v_s = 25$ m/s. Then:

$$f_{eff} = [350/375] \cdot 2000 = 933 \text{ Hz}.$$

We now consider the case where the listener is moving at some speed, v_L, relative to the air, toward or away from the source. If the listener moves toward the source the apparent speed with which the crests pass the listener is no longer v_p but $v_p + v_L$. If the listener moves away from the source the corresponding velocity would be $v_p - v_L$. We redefine v_L to be positive or negative for the listener moving toward or away from the source, respectively, so that we can always express the speed of the crests past the listener as: $v_p + v_L$. The wavelength is the distance between successive crests and is not affected by the motion of the listener. The wavelength is either λ_s (for a stationary source) or λ_{eff} [as given by Eq. (*2.21*)] for a moving source. Considering the more general case of a moving source, the frequency heard by the listener would be: $f_L = (v_p + v_L)/\lambda_{eff}$. Substituting from Eq. (*2.21*) for λ_{eff}, we get:

$$f_L = [(v_p + v_L)/(v_p + v_s)]f_s \tag{2.23}$$

where v_L is positive (negative) for the listener moving toward (away from) the source, and v_s is positive (negative) for the source moving away from (toward) the listener. The special case of the source not moving is obtained by setting $v_s = 0$ in Eq. (*2.23*). Similarly, the special case of the listener not moving is obtained by setting $v_L = 0$, reproducing Eq. (*2.22*). The use of Eq. (*2.23*) is illustrated in the following problems.

Problem 2.19. Consider the case of Problem 2.18, except that now the listener is driving a car initially moving toward the fire engine with a speed of 15 m/s.

(*a*) Find the frequency heard by driver before passing the fire engine.

(*b*) Find the frequency heard by the driver after passing the fire engine.

Solution

(*a*) We use Eq. (*2.23*) with $v_s = -25$ m/s (toward listener) and $v_L = 15$ m/s (toward source), and again $f_s = 1000$ Hz and $v_p = 350$ m/s. Then,

$$f_L = [(350 \text{ m/s} + 15 \text{ m/s})/(350 \text{ m/s} - 25 \text{ m/s})](1000 \text{ Hz}) = 1123 \text{ Hz}.$$

(*b*) Here the only difference in Eq. (*2.23*) is that both v_s and v_L change signs (away from listener and away from source, respectively):

$$f_L = [(350 - 15)/(350 + 25)](1000\text{ Hz}) = 893\text{ Hz}.$$

Problem 2.20. Suppose in Problem 2.19 the automobile were moving at the same speed of 15 m/s but this time in the same direction as the fire engine. All else being the same:

(*a*) Find the frequency heard by the driver before the fire engine overtakes the automobile.

(*b*) Find the frequency heard by the driver after the fire engine overtakes the automobile.

(*c*) Suppose after the fire engine passes the automobile, the automobile speeds up to match the speed of the engine. What would be the frequency heard by the driver?

Solution

(*a*) Here the fire engine is traveling toward the listener who is traveling at a slower speed in the same direction, so, v_s is negative (toward listener) while v_L is negative (away from source), so we have from Eq. (*2.23*):

$$f_L = [(350\text{ m/s} - 15\text{ m/s})/(350\text{ m/s} - 25\text{ m/s})](1000\text{ Hz}) = 1031\text{ Hz}.$$

(*b*) Here the fire engine has passed the listener who is now following the fire engine at the slower speed of the automobile. Now v_s is positive (away from listener), while v_L is also positive (toward source), so we get:

$$f_L = [(350 + 15)/(350 + 25)](1000\text{ Hz}) = 973\text{ Hz}.$$

(*c*) We again apply Eq. (*2.23*), with the same sign conventions for v_s and v_L as in part (*b*). The only difference is that v_l is now 25 m/s. Then:

$$f_L = [(350 + 25)/(350 + 25)](1000\text{ Hz}) = 1000\text{ Hz},$$

the actual frequency of the source.

Note that the answer to Problem 2.20(*c*) is a general result: If the source and listener are both moving in the same direction with the same speed, the listener hears the actual frequency of the source. The Doppler shift occurs in any medium in which waves travel and a comparable phenomenon occurs with light waves, although the formulas for light are somewhat different.

Shock Waves

In the Doppler shift we assumed that the velocity of the source (or listener) is less than the velocity of propagation of the wave through the medium. There are circumstances where that is not the case, such as the travel of a **supersonic** (faster than the speed of sound) jet aircraft (SST). When supersonic motion occurs a compressional wave, due to the object cutting through the air, is emitted by the traveling body and forms what is called a shock wave. The shock wave moves at a specific angle relative to the direction of motion of the object through the air, and can sometimes be of sufficient intensity to cause a loud booming sound, as in the case of an SST. To understand this phenomenon we consider an object moving to the right at supersonic speed v through the air. As the object passes any point the disturbance of the air at that point expands out in a spherical ripple. Since the object travels faster than sound it is always beyond the shell of any previous ripple. This is shown in Fig. 2-8. The object is shown at its location at time t_3 while the ripples from earlier times t_1, and t_2 are also shown. We draw a tangent line to the emitted ripples to get the wave-front of the shock wave, which makes an angle θ with the direction of motion. If R is the radius of the ripple starting at time t_1, after a time $(t_3 - t_1)$ has elapsed, and x is the distance the object has moved in that time interval, we can see from the figure that

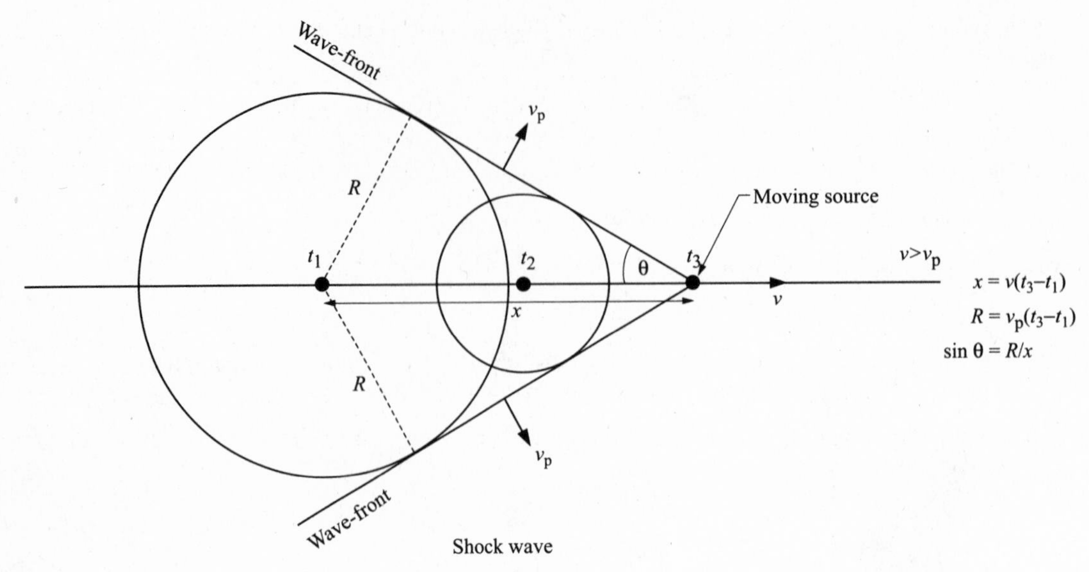

Fig. 2-8

R/x is the ratio of the opposite side to the hypotenuse of a right triangle with angle θ as shown. Then:

$$R/x = v_p/v = \sin\theta \tag{2.24}$$

The direction of propagation of the shock wave is perpendicular to the wave-front and makes an angle $(90° - \theta)$ to the direction of motion of the object. Shock waves accompany speeding bullets as well as SSTs, and an example in a medium other than air is the bow wave of a speed boat in water.

Problem 2.21.

(*a*) A supersonic airliner creates a shock wave whose wave-front makes an angle of 65° with its direction of motion through the air. Find the speed of the airliner, assuming the speed of sound in air is $v_p = 350$ m/s.

(*b*) A bullet moves through the air at 2000 m/s. Find the angle the shock wave makes with the direction of motion of the bullet.

Solution

(*a*) We have $v_p/v = \sin\theta \Rightarrow$

$$v = v_p/\sin\theta = (350 \text{ m/s})/\sin(65°) = 386 \text{ m/s (or 869 mph)}$$

(*b*) $v_p/v = \sin\theta \Rightarrow \sin\theta = 350/2000 = 0.175 \Rightarrow \theta = 10.1°$.

Problems for Review and Mind Stretching

Problem 2.22. Find the expression for the velocity of propagation of a sound wave [replacing Eq. (*2.15*)] if the compression were isothermal rather than adiabatic.

Solution

From the ideal gas law, PV is constant for constant T, so: $(P + \Delta P)(V + \Delta V) = PV$, which results in: $PV + P\Delta V + V\Delta P + \Delta P\Delta V = PV$. Canceling like terms on both sides and noting that for infinitesimal changes in P and V the term $\Delta P \Delta V$ is negligible compared to the others, we get: $P\Delta V = -V\Delta P$ or: $\Delta P/\Delta V = -P/V$. Then, from Eq. (*2.10*) the bulk modulus becomes: $B = P$. Comparing to Eq. (*2.14*) and recalling Eq. (*2.9*), Eq. (*2.15*) changes to: $v_p = [RT/M]^{1/2}$.

Problem 2.23. In the equally tempered scale

(*a*) find the frequencies of the following notes: middle G, B, D, and D′ (one octave above D) and

(*b*) What are the ratios in the major triad G to B to D′? [express as $x:5:y$]

Solution

(*a*) Recalling that middle A = 440 Hz, we have: $G = A/[(2)^{1/12}]^2 = 440/(1.059{,}46)^2 = 392$ Hz; $B = A \cdot [(2)^{1/12}]^2 = 493.9$ Hz; $D = G/[(2)^{1/12}]^5 = 392/1.334{,}84 = 293.7$ Hz: $D' = 2D = 587.4$.

(*b*) $x/5 = G/B = 392/493.9 \Rightarrow x = 3.97$; $5/y = B/D' = 493.9/587.4 = 0.7937 \Rightarrow y = 5.95$. Thus, G to B to D = 3.97 to 5 to 4.95.

Note. This is the same as any other major triad ratio in the equally tempered scale since all the adjacent notes are related by the same multiplicative factor.

Problem 2.24. Two pianos are tuned to different scales. One is the equally tempered scale and the other is the diatonic scale in the key of C.

(*a*) What is the beat frequency when the note G is struck on both pianos?

(*b*) What is the beat frequency when the note G′ is struck on both pianos?

Solution

(*a*) From the text, G on the diatonic scale is 396 Hz, while from Problem 2.23, G on the tempered scale is 392 Hz. Since the beat frequency is just the frequency difference between the two notes, we have: beat frequency = 4 Hz.

(*b*) Since in either scale G′ = 2G for an octave shift, the difference in frequencies is also doubled, so the beat frequency is now 8 Hz.

Problem 2.25. Consider Problem 2.19, with the fire engine traveling down the street at 25 m/s toward the car which is traveling toward the engine at 15 m/s. As before, the engine's 1000 Hz siren is blaring. Suppose a wind of 10 m/s were blowing in the direction from the engine to the car, all else being the same.

(*a*) What is the frequency of the siren heard by the listener in the car as the engine approaches?

(*b*) What is the frequency of the siren heard by the listener once the engine has passed by?

(*c*) How would part (*a*) change if the wind velocity was 30 m/s (gale force wind).

Solution

(*a*) We still can use Eq. (*2.23*) if we recall that the velocities v_s and v_L in that equation represent the velocities of source and listener, respectively, relative to the medium in which the sound travels, namely the air. In Problem 2.19 the air was assumed to be still so the velocities relative to the ground and those relative to the air were the same. This is no longer the case when the mass of air is moving at 10 m/s from fire engine toward car. In our present case we must consider the velocities as seen from the frame of reference at rest relative to the air. In that frame of reference the fire engines velocity relative to the air is still pointed toward the listener, so, with our sign convention: $v_s = -(25\text{ m/s} - 10\text{ m/s}) = -15$ m/s. The listener's velocity relative to the air is still toward the source, so: $v_L = (15\text{ m/s} + 10\text{ m/s}) = 25$ m/s. v_p, the velocity of propagation in sound waves in air is, of course unchanged, so

$v_p = 350$ m/s. Substituting into Eq. (*2.23*) we get:
$f_L = [(350 \text{ m/s} + 25 \text{ m/s})/(350 \text{ m/s} - 15 \text{ m/s})](1000 \text{ Hz}) = 1119$ Hz.

(*b*) The only difference in Eq. (*2.23*) is that both v_s and v_L change signs (away from listener, away from source), so:

$$f_L = [(350 - 25)/(350 + 15)](1000 \text{ Hz}) = 890 \text{ Hz}.$$

(*c*) The only difference is that the fire engine would appear to be going "backward" relative to the air, so v_s would be positive. This automatically comes out of the equation: $v_s = -(25 \text{ m/s} - 30 \text{ m/s}) = +5$ m/s. v_L is still toward the engine and so: $v_L = (15 \text{ m/s} + 30 \text{ m/s}) = 45$ m/s. Then: $f_L = [(350 + 45)/(350 + 5)(1000 \text{ Hz}) = 1113$ Hz.

Problem 2.26. A jeep travels in a canyon at a speed of 10 m/s perpendicular to the parallel cliff walls that form the canyon boundary. The jeep blows its 200 Hz horn as it passes the midpoint between the cliffs. An observer at rest on the canyon floor has an instrument which joins the two echoes into a single wave signal and re-emits their combined sound. What is the beat frequency heard from the instrument?

Solution

The two cliffs detect the same frequencies that a listener at rest near each cliff would hear. When the sound reflects off these cliffs it reflects the same frequency that hit the cliffs. For cliff 1, in front of the jeep, we have a Doppler shift in frequency given by Eq. (*2.23*) with $v_L = 0$ and $v_s = -10$ m/s. Then, the reflected frequency for an observer at rest in the canyon is: $f_{1L} = (350 \text{ m/s})/(350 \text{ m/s} - 10 \text{ m/s})(200 \text{ Hz}) = 206$ Hz. For the other cliff, behind the jeep we again use Eq. (*2.23*) with $v_L = 0$, but now $v_s = 10$ m/s, so the reflected frequency is: $f_{2L} = 350/(350 + 10)(200 \text{ Hz}) = 194$ Hz. The beat frequency heard is therefore $206 - 194 = 12$ Hz.

Supplementary Problems

Problem 2.27. Using the fact that $(6.5)^{1/2} = 2.5495$ and $(6.5)^3 = 274.625$:

(*a*) Find the value of $(6.5)^{1/4}$.

(*b*) Find the value of $(6.5)^{3.5}$.

(*c*) Find the value of $(6.5)^{3/2}$.

(*d*) Find the value of $(6.5)^{3.25}$.

Ans. (*a*) 1.5967; (*b*) 700.16; (*c*) 16.572; (*d*) 438.49

Problem 2.28. Reduce the following expressions to most simplified terms:

(*a*) $\log [x^3 \cdot y^{-1/2}/z^n]$; (*b*) $\ln [x^{y+z}/\exp (yz)]$.

Ans. (*a*) $3 \log x - (\frac{1}{2})\log y - n \log z$; (*b*) $y \ln x + z \ln x - yz$

Problem 2.29. Suppose one defines a logarithmic function to an arbitrary base a: $\log_a (x)$, where a is a positive number.

(*a*) Find an expression for $\log_a (x)$ in terms of $\log (x)$ [*Hint*: See Problem 2.5].

(*b*) Find the value of $\log_{100} (2)$.

Ans. (*a*) $\log_a (x) = \log_a (10) \cdot \log (x)$; (*b*) $(\frac{1}{2}) \log (2)$

Problem 2.30. The velocity of sound in CO_2 at 300 K is found to be 270 m/s. Find the ratio of specific heats, γ.

Ans. 1.29

Problem 2.31. A spherical sinusoidal sound wave has an intensity of $I = 0.0850\ \text{W/m}^2$ at a distance of $r = 2.0$ m from the source.

(*a*) Find the intensity of the wave at $r = 3.0$ m and $r = 4.0$ m.

(*b*) Find the total power transmitted by the spherical wave at $r = 2.0$ m, $r = 3.0$ m and $r = 4.0$ m.

Ans. (*a*) 0.0378 W/m^2, 0.0213 W/m^2; (*b*) 4.27 W for all three cases.

Problem 2.32. If the amplitude of the wave in Problem 2.31 is $A = 0.450$ mm at $r = 1.0$ m, find the amplitude at $r = 2.0$ m, 3.0 m, and 4.0 m.

Ans. 0.225 mm, 0.150 mm, 0.113 mm

Problem 2.33. Redo Problem 2.32 for the case of circular sinusoidal ripples in on the surface of a pond.

Ans. 0.318 mm, 0.260 mm, 0.225 mm

Problem 2.34. A sinusoidal plane-wave traveling through the air in the x direction has an intensity $I = 0.0700$ W/m^2 and an amplitude $A = 0.0330$ mm. The density of air is 1.17 kg/m^3 and the velocity of propagation is $v_p = 350$ m/s.

(*a*) Find the frequency of the wave [*Hint*: See Problem 2.11].

(*b*) Find the energy passing through a 3.0 cm by 4.0 cm rectangular window parallel to the ($y \cdot z$) plane in a 15 s time interval.

Ans. (*a*) 89.2 Hz; (*b*) $1.26 \cdot 10^{-3}$ J

Problem 2.35.

(*a*) A woman faces a cliff and wishes to know how far away it is. She calls out and hears her echo 4.0 s later. If the speed of sound is 350 m/s how far is she from the cliff?

(*b*) A surface ship uses sonar waves (high frequency sound waves) emitted below the water line, to locate submarines or other submerged objects. Testing the sonar using an object at a known distance of 4000 m, the time interval between emission of the signal and its return is found to be 5.52 s. What is the speed of sound in seawater?

Ans. (*a*) 700 m; (*b*) 1450 m/s

Problem 2.36. A person at a distance of 1200 m from an explosion hears an 80 db report. How close would the person have been to the explosion if the report were just at the threshold of pain?

Ans. 12.0 m.

Problem 2.37. A point source emits three distinct frequencies of 100 Hz, 1000 Hz and 10,000 Hz, each with the same power level. A student whose threshold of hearing is 0 db at 1000 Hz can just barely make out the 1000 Hz signal at a distance of 60 m from the source, but cannot hear the other tones at that distance. As the student moves closer to the source she first detects the 10,000 Hz signal at 2.0 m from the source, and first detects the 100 Hz signal at 25 cm from the source.

(*a*) What is the student's threshold of hearing at 10,000 Hz?

(*b*) What is the student's threshold of hearing at 100 Hz?

Ans. (*a*) 29.5 db; (*b*) 47.6 db

Problem 2.38. A classroom has dimensions $h = 5.0$ m, $w = 10$ m, $l = 7.0$ m, and the reverberation time of the classroom when empty is 1.80 s.

(*a*) Find the absorbing power of the room.

(*b*) If a class of 20 students with 1 professor comes into the room, find the new reverberation time of the room. [Assume each person contributes an additional absorbing power of 0.4.]

Ans. (*a*) 31.1; (*b*) 1.42 s

Problem 2.39. For the classroom of Problem 2.38(*a*), assume the absorption coefficient for the walls is 0.05 and the absorbing power of the floor and furniture is 8.0. What is the absorption coefficient of the ceiling?

Ans. 0.21

Problem 2.40. A specially constructed room has ceiling, floor and one pair of opposite walls acting as near-perfect sound reflectors (absorption coefficient zero). The remaining two opposite walls, of areas 30 m^2 each and separated by a distance of 6.0 m, have absorption coefficients of 0.50. Assume that a single pulse of sound is emitted from one of these walls toward the other at 80 db, and assume that the pulse reflects back and forth between the two walls as a plane-wave.

(*a*) Do a direct calculation of the reverberation time for this case (i.e., the time for the sound level to drop to 20 db). Assume the speed of sound is 350 m/s.

(*b*) What is the result one gets from Sabine's formula [Eq. (*2.19*)]?

(*c*) Explain the discrepancy?

Ans. (*a*) 0.34 s; (*b*) 0.96 s; (*c*) plane-wave model assumes shortest possible time between absorptions; realistically, much of the sound energy will bounce one or more times from the perfect reflecting surfaces for each reflection from the absorbing surfaces, so a longer time is involved

Problem 2.41.

(*a*) Show that in the diatonic scale in the key of C, the seven notes C, D, E, F, G, A, B indeed satisfy the condition that the ratios of frequencies of adjacent notes are either 9/8 or 10/9 or 16/15.

(*b*) If the same seven notes are tuned to the equally tempered scale, what are the deviations of the frequencies above (or below) the diatonic scale values, in Hz and in percent of frequency?

Ans. (*a*) D/C = 9/8, E/D = 10/9, F/E = 16/15; G/F = 9/8, A/G = 10/9, B/A = 9/8, C′/B = 16/15; (*b*) C: (2.4 or 0.9%), D: (3.3 or 1.1%), E: (0.4 or 0.12%), F: (2.8 or 0.8%), G: (4.0 or 1.0%), A: 0, B: (1.1 or 0.2%)

Problem 2.42. Using the results of Problem 2.41(*b*):

(*a*) How many beats would one hear if D on the diatonic and even tempered scales were played simultaneously?

(*b*) If one wished to tune the even tempered D note to the diatonic scale, would one increase or decrease the tension in the piano string?

Ans. (*a*) 3.3; (*b*) increase

Problem 2.43. When tuning fork A is struck at the same time as tuning fork B the beat frequency is 3 Hz. When tuning fork B is struck at the same time as tuning fork C the beat frequency is 5 Hz.

(*a*) When tuning fork A is struck at the same time as tuning fork C what beat frequency is expected?

(*b*) If tuning fork C has the lowest of the three frequencies, at 300 Hz, what are the possible frequencies of tuning forks A and B?

Ans. (*a*) 8 Hz or 2 Hz; (*b*) B is 305 Hz and A is either 308 Hz or 302 Hz.

Problem 2.44. An ambulance with siren blowing travels at 20 m/s toward a stationary observer who hears a frequency of 1272.7 Hz. Assume the speed of sound in air is 350 m/s.

(*a*) What is the actual frequency of the siren?

(*b*) What is the frequency heard by the observer once the ambulance has passed?

Ans. (*a*) 1200 Hz; (*b*) 1135 Hz

Problem 2.45. An observer travels toward a stationary whistle of frequency 1200 Hz at a speed of 20 m/s. Assume the speed of sound in air is 350 m/s.

(*a*) What is the frequency heard by the observer?

(*b*) What is the frequency heard by the observer after passing the whistle?

Ans. (*a*) 1268.6 m/s; (*b*) 1131 Hz

Problem 2.46. An observer travels north at 10 m/s and sees an ambulance traveling south at 10 m/s with siren blowing. The actual frequency of the siren is 1200 Hz. Assume the speed of sound in air is 350 m/s.

(*a*) What is the frequency heard by the observer?

(*b*) What is the frequency heard by the observer after passing the ambulance?

Ans. (*a*) 1270.6 Hz; (*b*) 1133 Hz

Problem 2.47. An automobile travels along a road parallel to railroad tracks at 60 ft/s. A train coming from behind is traveling at 90 ft/s and blows its whistle. After the train passes the automobile it blows its whistle again. A passenger in the car notes that the drop in frequency in the sound of the whistle from before passing the automobile to after is 80 Hz. Assume the speed of sound in air is 1100 ft/s.

(*a*) Find the actual frequency of the whistle.

(*b*) What are the frequencies heard by the passenger before and after the train passes?

Ans. (*a*) 1457 Hz; (*b*) 1500 Hz (before), 1420 Hz (after)

Problem 2.48. A supersonic airplane is heading due south and shock waves are observed to propagate in the directions 70° east of south and 70° west of south. The speed of sound in air is 350 m/s.

(*a*) What is the angle made by the shock wave-fronts with the direction of travel of the airplane?

(*b*) What is the speed of the airplane?

Ans. (*a*) 20°; (*b*) 1023 m/s

Problem 2.49. A bullet travels at five times the speed of sound.

(*a*) What is the angle that the shock wave-front makes with the direction of travel of the bullet?

(*b*) If the angle were half that value how many times faster than sound is the speed of the bullet?

Ans. (*a*) 11.5°; (*b*) ten

Chapter 3

Coulomb's Law and Electric Fields

3.1 INTRODUCTION

In previous chapters, we learned about the laws of mechanics, including how to determine the type of motion that occurred under the application of various kinds of force. The forces that we discussed were tensions in wires, normal and friction forces at surfaces, elastic forces of springs and the force of gravitation. Of these forces, only the gravitational force is "fundamental" in the sense that it does not arise as an application of a more basic underlying force. The other forces are all manifestations of two more basic and related forces between particles, known collectively as the electromagnetic interaction. In this chapter, we begin our investigation of the forces and applications of the electromagnetic interaction. In a later chapter, we will discuss two additional fundamental interactions, both of which are associated with the nucleus of atoms, the "strong interaction" and the "weak interaction". To the best of our current knowledge of the laws of nature, these four interactions (gravitation, electromagnetic, strong and weak) are all that are needed to understand the phenomena of nature of which we are aware. Even these four, however, are not totally distinct from each other, but appear instead, to be manifestations of a deeper and more encompassing interaction. It is the hope of many physicists that sometime in the future we will be able to unify all of these interactions under one basic "Theory of Everything" (TOE).

The electromagnetic interaction is comprised of two related forces that we will discuss separately, electric forces and magnetic forces. All interactions between atoms and molecules are actually different aspects of electromagnetic interactions, but they arise from complicated applications of the fundamental laws and require the ideas of quantum mechanics before they can be properly understood. Thus the forces of friction, tension, etc. are the subtle results of electromagnetic forces. Before we can consider how electromagnetism leads to those forces we must first discuss the basic ideas of each force. In the next three chapters we will discuss the basic concepts of electricity, and the subsequent chapters will do the same for magnetism.

3.2 ELECTRIC CHARGES

All material consists of particles that have a property called **electric charge**. These particles are either the nuclei of atoms, which are positively charged, or the electrons that surround the nuclei, which are negatively charged. All charged particles exert a force on each other called the **electric force**. Usually one cannot detect the charge easily because atoms, and even molecules (made up of two or more atoms), have an equal amount of positive and of negative charge in their natural state, leaving the material neutral. The charged nature of materials is evident when one succeeds in removing some charge of one polarity (sign) thus leaving the material charged with the other polarity. The simplest way to accomplish this is to rub together two materials (many different types of material can be used), such as amber with fur, or a plastic rod with a plastic sheet. In the process, one material becomes positively charged, while the other receives an equal amount of negative charge. This is an example of the **Law of conservation of charge** which requires that the total amount of charge remains unchanged. If one starts with uncharged materials the initial charge present is zero. Then the total charge after it has been separated must still add to zero, requiring that there be equal amounts of positive and negative charge present.

When one separates the charge in this (or in any other) manner one can explore the force between the charges. The exact law for the magnitude and direction of this force will be discussed in the next section. Here we will discuss the force qualitatively. It is found that the force is one of attraction between charges of opposite polarity and of repulsion between charges of like polarity. Furthermore, the magnitude of the force decreases as the charges move further apart. Thus two positive (or two negative) charges repel each other while positive charges attract negative charges and vice versa. Once one has a sample of positive charge, one can determine whether a different charge is negative or positive by seeing whether it is attracted or repelled by the positive charge. It is by convention that the charge on the amber rod that has been rubbed with fur is considered negative. This convention results in the fact that nuclei are positively charged while electrons carry negative charge. Furthermore, all electrons, being identical particles, carry the same amount of negative charge. Nuclei of different atoms have different positive charge, but one of the building blocks of nuclei, the proton, has a definite positive charge which is equal and opposite to that of the electron. (The other building block of nuclei, the neutron, is electrically neutral, i.e. has zero charge.) Atoms are neutral because the number of electrons surrounding the nucleus equals the number of protons in the nucleus. Charge is measured in units of **Coulomb** (C), which is defined by an experiment in magnetism. Using this unit, the magnitude of the charge of an electron, labeled e, is 1.60×10^{-19}C.

In many materials, called **conductors**, there are some charges, usually electrons in the outer reaches of the atoms, which are free to move in the material. If the conducting material is uncharged, then the electrons are uniformly distributed in the material, with each electron being attracted to a fixed, positively charged, nucleus. If other charges are inserted in the conductor, then the free charges move in response to the electrical forces that occur. Since it is the electrons that move, a piece of conductor can be given a negative charge by adding some electrons from elsewhere, or a positive charge by removing some electrons to another location.

Problem 3.1.

(*a*) If a conductor is charged with an amount of charge, Q, show that this charge is located on the surface.

(*b*) If a charge Q approaches a neutral conductor (but doesn't touch it), show that the conductor will be attracted by the charge Q.

Solution

(*a*) If the charge Q is negative it will consist of electrons that are free to move since we are considering the case of a conductor. These "extra" electrons will on net repel each other, and therefore move as far apart as possible. When the electrons reach the surface they can move no further and they will therefore all be found distributed on the surface, leaving the interior neutral. Note that a similar effect occurs if the charge Q is positive. If the charge is positive it really means that some negatively charged electrons have been removed from the material, leaving a net positive charge on the material. Some of the remaining free electrons will move themselves to the interior of the material, attracted by the positive charge there, leaving the surface charged positively. (While it is less obvious than for the case of Q negative, in fact the electrons shift so that all of the net positive charge Q is on the surface, and again the interior remains neutral.)

(*b*) This is best understood in two steps. First, when a charge Q approaches a neutral conductor it will exert a force on the charges within the conductor. If Q is negative the free electrons, having the same sign as Q, will be repelled, while if Q is positive the free electrons will be attracted. In either case, the free charges will move so that that part of the conductor near Q will contain an excess of charge of sign opposite to that of Q, while that part of the conductor far from Q will contain an equal charge of the same sign as Q. Given this fact, we reach step two in our reasoning. The charge near Q will cause the conductor to be attracted to Q while the charge far from Q will cause the conductor to be repelled from Q. The force of attraction will be larger than the force of repulsion since the charges that are attracted to Q are nearer to Q than the equal charges that are repelled, and the magnitude of the force is larger if the charges are nearer each other. Therefore, the net force will be one of attraction.

Note. Although it is the free electrons that actually move in the conductor, the same result would be true if positive charges were free to move as well. Indeed, it is even more convenient to think of either type of charge moving—so we can speak of "positive charges repelled" even if what really happens is that negative charges are attracted. We will often use this common language.

Problem 3.2. Two conducting spheres are touching each other on an insulated table. A rod that is positively charged approaches the two spheres from the left, as in Fig. 3-1(*a*).

(*a*) What is the charge distribution on the spheres while they are in contact with each other in the presence of the rod?

(*b*) While the charged rod remains in place one separates the spheres. Will they be charged, and if so, with what polarity?

(*c*) While the charged rod remains in place and the spheres are touching, one "grounds" the far end of the second sphere, as in Fig. 3-1(*b*). This means that one provides a path by which charges can move to a large uncharged reservoir, such as the ground. After removing the grounding path, and then the charged rod, the spheres are separated. Will they be charged, and if so with what polarity?

Solution

(*a*) In the presence of the rod, positive charge will be repelled to the far side of the touching spheres and negative charge will accumulate on the near side, as shown in Fig. 3-1(*a*). This is because electrons can move from one sphere to the other while they are in contact. Thus the nearer sphere will have an excess of negative charge and the sphere that is further away will have an excess of positive charge.

(*b*) If one now separates the two spheres they will retain their positive (further sphere) and negative (nearer sphere) charge. This is called charging by induction, since the rod induces the charges on the materials without actually transferring charge to them.

(*c*) When the spheres are touching in the presence of the charged rod, negative charges are induced on the near sphere, and positive charges are repelled to the far sphere and from there to the ground. (Remember the note after Problem 3.1!) After one removes the grounding path, the touching spheres will contain a net amount of negative charge, most of which will be on the sphere near the rod. When the rod is removed, this negative charge will redistribute itself evenly over the surface of both spheres. If the spheres are then separated, each sphere will be negatively charged. This is an alternative way of charging by induction.

Problem 3.3. Repeat Problem 3.2 if the charged rod touches the spheres.

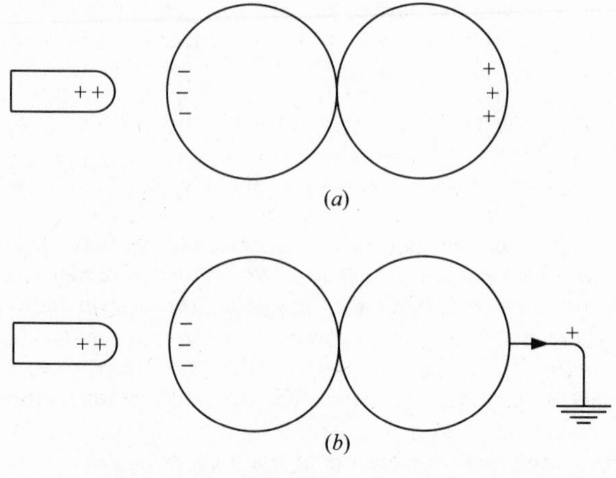

Fig. 3-1

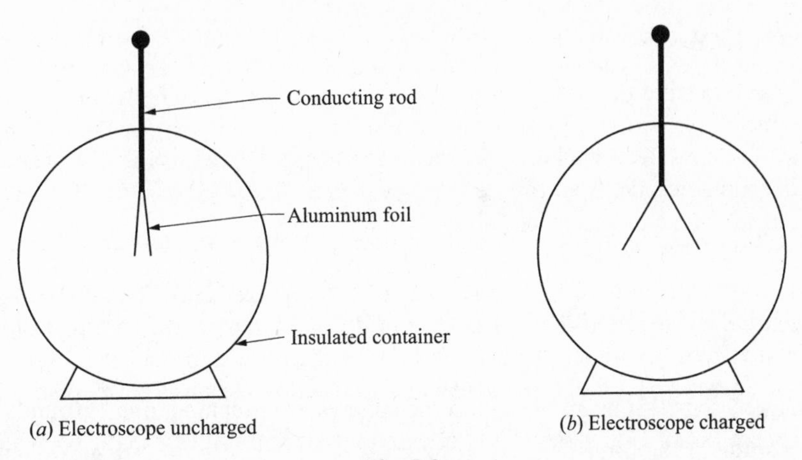

(*a*) Electroscope uncharged

(*b*) Electroscope charged

Fig. 3-2

Solution

(*a*) When the rod touches the spheres the positive charge redistributes itself over the surfaces of the rod and the spheres. Thus all three (rod and two spheres) become positively charged.

(*b*) If one separates the spheres (either before or after removing the rod) each sphere will retain its positive charge.

(*c*) When the rod touches the spheres that are grounded all the charge moves as far away as possible, which means that it moves to the ground. None of the objects will now have a charge. When one removes the ground and then the rod and then separates the spheres, no transfer of charge occurs, and the spheres will be uncharged.

Problem 3.4. An electroscope is built by attaching two pieces of aluminum foil to a conducting rod and mounting the assembly in an insulated container, as in Fig. 3-2(*a*).

(*a*) If the rod and foil is initially uncharged, what happens to the aluminum foil if one approaches the top of the rod with a positive charge? With a negative charge?

(*b*) If the rod and foil initially contains some positive charge, what happens to the foil if one approaches the top of the rod with a positive charge? With a negative charge?

Solution

(*a*) Since there is initially no charge on the rod and foil, the two foil pieces will not exert any electrical forces on each other and they will both hang down vertically. If positive charge approaches the top of the rod, then negative charge will be attracted to the top of the rod (or equivalently, positive charge will appear to be repelled down the rod to the two foil pieces). Each of the foil pieces will become positively charged and they will repel each other. They will come to static equilibrium with the force of the earth's gravity by moving apart as in Fig. 3-2(*b*). The same will be true for an approaching negative charge except that each foil piece will now be negatively charged.

(*b*) If the rod and foils are initially charged with positive charge, then this positive charge will be distributed over all the surfaces, including the foils. Each foil will therefore be positively charged, and the two foils will initially move apart as in Fig. 3-2(*b*). If one approaches the top of the rod with a positive charge, then more positive charge will be pushed to the foils, the electrical force will increase and the foils will move further apart. If one approaches the top of the rod with negative charge, some of the positive charge on the foils will be attracted to the top, and the charge on the foils will be reduced. Then the electrical force decreases, and the foils will move closer together. We see that this is a sensitive means of determining the sign of the charge on some materials, without transferring any of the charge from that material.

3.3 COULOMB'S LAW

We are now ready to state the quantitative law which gives the force between two charged particles (**Coulomb's law**). Since force is a vector, we must give both the magnitude and the direction of the force. Suppose charge q_1 is located at a distance r to the left of charge q_2, as in Fig. 3-3. The force on q_2 due to the charge q_1 has a magnitude given by:

$$|F| = |kq_1q_2/r^2|, \qquad \text{where } k = 1/4\pi\varepsilon_0 = 9.99 \times 10^9 \text{ N} \cdot \text{m}^2/\text{C}^2 \tag{3.1}$$

and where k and ε_0 are constants related as shown for later convenience. The direction of the force is along the line joining the charges. If the charges are of the same sign, then the force is one of repulsion, i.e. it is directed *away* from q_1, while if the charges are of opposite sign then the force is one of attraction, i.e. it is directed *toward* q_1. Of course there will also be a force exerted by q_2 on q_1, which will have the same magnitude as the force on q_2 but will be in the opposite direction. We see that we can calculate the magnitude of the force neglecting the signs of the charges, and then use our knowledge of the signs to determine the direction of the force. This is because a minus sign only reverses the direction of a vector.

The formula for the magnitude of the force is identical in form to that for the gravitational force between two masses. That formula is given in Beginning Physics I, Chap. 5, Eq. (*5.1*), as $F = Gm_1m_2/r^2$. For the electrical force we use k instead of G and we use charges instead of masses. There is a major difference due to the fact that masses are always positive and that the gravitational force is always attractive. For the electrical force we can have both positive and negative charges resulting in either attractive or repulsive forces.

Problem 3.5. An electron is at a distance of 0.50×10^{-10} m from a proton. [*Data*: $e = 1.6 \times 10^{-19}$ C; $m_e = 9.1 \times 10^{-31}$ kg; $m_p = 1.67 \times 10^{-27}$ kg; $G = 6.67 \times 10^{-11}$ N $\cdot$ m^2/kg^2]. What is the electrical force exerted on the electron by the proton and what is the gravitational force between these particles? Compare the two forces.

Solution

Using Eq. (*3.1*), $|F_e| = ke^2/r^2 = (9.0 \times 10^9 \text{ N} \cdot \text{m}^2/\text{C}^2)(1.6 \times 10^{-19} \text{C})^2/0.50 \times 10^{-10} \text{m})^2 = 9.2 \times 10^{-8}$ N. Using the equation for the gravitational force we get that $F_g = Gm_e m_p/r^2 = (6.67 \times 10^{-11} \text{ N} \cdot \text{m}^2/\text{kg}^2)(9.1 \times 10^{-31} \text{ kg})(1.67 \times 10^{-27} \text{ kg})/(0.50 \times 10^{-10} \text{ m})^2 = 4.05 \times 10^{-47}$ N. This is 4.4×10^{-40} times smaller than the electrical force. From this we can deduce that, on an atomic and molecular scale, gravitational forces are negligible compared to electrical forces and can almost invariably be neglected.

Problem 3.6. Two charges are located on the x axis with $q_1 = 2.3 \times 10^{-8}$ C at the origin, and with $q_2 = -5.6 \times 10^{-8}$ C at $x = 1.30$ m (see Fig. 3-4). Find the force exerted by these two charges on a third charge of $q = 3.3 \times 10^{-8}$ C, which is located at: (*a*) $x = 0.24$ m on the x axis and (*b*) $x = 1.55$ m on the x axis.

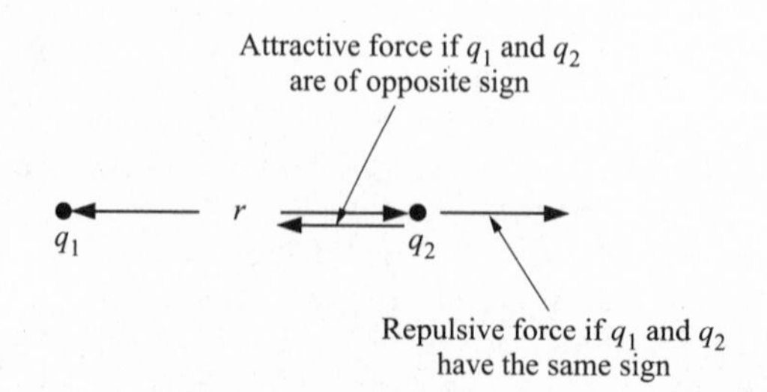

Fig. 3-3

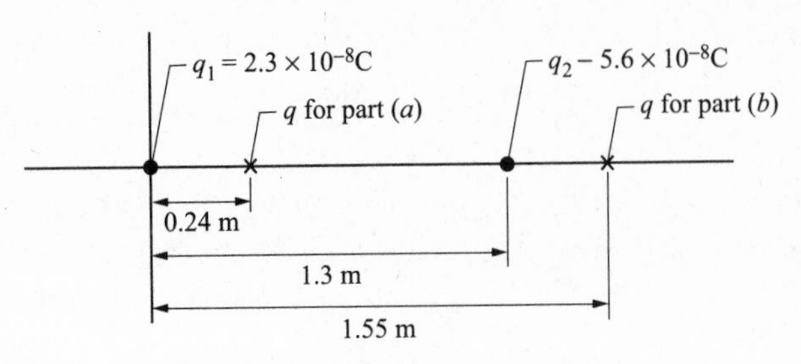

Fig. 3-4

Solution

(a) We must calculate the force exerted by each charge on q, both in magnitude and in direction and then add the two forces to get the total force on q. Now, $F_{1q} = kqq_1/r_{1q}^2 = 9.0 \times 10^9(2.3 \times 10^{-8})(3.3 \times 10^{-8})/0.24^2 = 1.19 \times 10^{-4}$ N. Since both charges are positive, the force on q is repulsive and points in the $+x$ direction. Similarly, $F_{2q} = kqq_2/r_{2q}^2 = (9 \times 10^9)(5.6 \times 10^{-8})(3.3 \times 10^{-8})/(1.30 - 0.24)^2 = 0.15 \times 10^{-4}$ N. Since the charges are of opposite sign, the force on q is one of attraction. Since q_2 is further out on the x axis than q, the force is again in the positive x direction. The total force is then $(1.19 + 0.15) \times 10^{-4}$ N $= 1.34 \times 10^{-4}$ N in the $+x$ direction.

(b) Using the same procedure as in part (a) we get $F_{1q} = (9.0 \times 10^9)(2.3 \times 10^{-8})(3.3 \times 10^{-8})/(1.55)^2 = 2.84 \times 10^{-6}$ N in the $+x$ direction, and $F_{2q} = 9.0 \times 10^9(5.6 \times 10^{-8})(3.3 \times 10^{-8})/(1.55 - 1.3)^2 = 2.66 \times 10^{-4}$ N. This force is again one of attraction. However, since q is to the right of q_2 the force on q is to the left, in the $-x$ direction. The total force is therefore $(2.66 - 0.03) \times 10^{-4}$ N in the $-x$ direction.

Problem 3.7. Solve Problem 3.6 for the case that q is on the y axis at $y = 1.03$ m (see Fig. 3-5).

Solution

Again we calculate the two forces exerted on q. $F_{1q} = kqq_1/r_{1q}^2 = 9.0 \times 10^9(2.3 \times 10^{-8})(3.3 \times 10^{-8})/(1.03)^2 = 6.44 \times 10^{-6}$ N. Since the force is one of repulsion the direction is in the $+y$ direction. To calculate $F_{2q} = kqq_2/r_{2q}^2$, we must first find r_{2q}. From the figure we see that $r_{2q} = (1.03^2 + 1.3^2)^{1/2} = 1.66$ m. Then $F_{2q} = 9 \times 10^9(5.6 \times 10^{-8})(3.3 \times 10^{-8})/(1.66)^2 = 6.05 \times 10^{-6}$ N. This is a force of attraction, so it is directed along the line joining q with q_2 and pointing toward q_2 as in Fig. 3-5. To get $\mathbf{F} = \mathbf{F}_{1q} + \mathbf{F}_{2q}$, we use components to add the vectors. $F_x = 0 + F_{2q} \cos\phi = (6.05 \times 10^{-6})(1.3/1.66) = 4.74 \times 10^{-6}$. $F_y = F_{1q} - F_{2q} \sin\phi = 6.44 \times 10^{-6} - (6.05 \times 10^{-6})(1.03/1.66) = 2.69 \times 10^{-6}$ N. Then $|F| = (F_x^2 + F_y^2)^{1/2} = 5.45 \times 10^{-6}$ N. Since both components of $\mathbf{F}$ are positive, it points at some angle θ above the x axis in the first quadrant, where $\tan\theta = F_y/F_x = (2.69/4.74) = 0.568 \rightarrow \theta = 29.6°$.

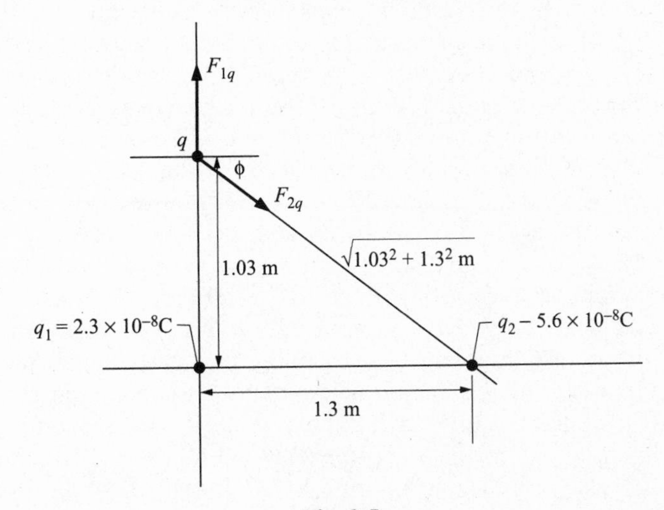

Fig. 3-5

Problem 3.8. A particle of charge $Q = 4.1 \times 10^{-6}$ C is placed at the origin. If one wants to exert a force of 6.3×10^{-6} N on the particle in the $+x$ direction with a positive charge of 1.6×10^{-7} C where must the charge be placed?

Solution

Since $F = kqQ/r^2$, we must have $F = 6.3 \times 10^{-6}$ N. Therefore $6.3 \times 10^{-6} = (9 \times 10^9)(1.6 \times 10^{-7})(4.1 \times 10^{-6})/r^2 \rightarrow r^2 = 937\ \text{m}^2 \rightarrow r = 30.6$ m. Therefore, the charge must be placed at a distance of 30.6 m from the charge Q. To get the direction of the force in $+x$, the charge must be along the x axis. Furthermore, since both charges are positive, the force on Q is repulsive, and if F points in the positive x direction we require the other charge to be on the negative x axis (to the left of Q). The charge must therefore be at $x = -30.6$ m.

3.4 THE ELECTRIC FIELD—EFFECT

The electrical force between two charges and the gravitational force between two masses are cases in which a force seemingly exists between objects that are not in contact with each other. This is called "action at a distance". All common forces, such as friction, tension, normal, elastic, pushes, etc. cannot exist unless the bodies are in contact. It is much easier to understand the idea of "contact forces" than to understand how forces can exist without contact, as is seemingly the case for electrical and gravitational forces. For instance, if a mass moves and therefore its distance to another mass changes, does the force change instantaneously, or must there be a time lag to enable the second mass to realize that the first mass has moved. In order to avoid such problems it would be convenient if the electrical and gravitational forces could be viewed in an alternative fashion which would eliminate the need to think in terms of "action at a distance". The alternative view, that accomplishes this objective, is to introduce the concept of a field. As this new concept developed over the years it became more than just a convenience, so that the modern view considers fields to be the required approach to the understanding of physical phenomena.

We will develop the concept of a field in the case of the electrical force, but it is equally applicable to the gravitational force. In Chap. 6, when we discuss magnetic forces, we will automatically use the concept of a field in developing that subject. The idea is as follows. Consider two charges, q and Q, separated by a distance r. Assume q is fixed at the origin of our coordinate system. A force is exerted on Q, and we have viewed this force as being exerted by the distant q. Instead of this, we will think of that force as being exerted by a new entity, an "**electric field**", which exists at every point in space, including the location of Q. In general, the electric field has both magnitude and direction, both of which can vary from point to point in space, as well as change from moment to moment. The electric field is thus a vector field, usually represented by the symbol **E**, which can vary with x, y, z, t. As will be seen from the defining equation, Eq. (*3.2a*) below, the electric field for our simple case of a charge q at the origin will have magnitude kq/r^2 for every point in space a distance r from the origin, and will have a direction radially away from (or toward) the origin if q is positive (or negative). It is the vector field, **E** at the position of Q that we can now consider to exert the force on Q just like a contact force. Most important the field that Q finds itself in does not depend on the value of Q at all. We no longer think of action at a distance, since the electric field exists at the location of Q. If Q moves to a different position then the field at this new position exerts a force on Q. Only a field at the same location as Q can exert a force on Q. We do not have to know how the field was established at the position of Q in order to know the force that the field exerts on Q, we only need to know its vector value. It is a separate question, that we will also have to discuss, as to how the field gets established, but once the field is established we know that the field at the position of Q can exert a force on Q. It is clear that if we had other charges at different locations in addition to q, the electric field would be quite complicated. Furthermore, there may be other means of establishing an electric field as well. In this section we will concentrate on determining the *effect* of a field, **E**, on a charge Q that is at the position of **E**, and in the next section we will discuss the *source* of electric fields. We will only consider fields that stay constant in time, unless otherwise stated.

Suppose one has an electric field **E** at a certain position in space. If we now place a charge Q at that point the electric field will exert a force on the charge, given by:

$$\mathbf{F} = Q\mathbf{E} \tag{3.2a}$$

This vector equation gives the magnitude of **F** as:

$$|\mathbf{F}| = |Q\mathbf{E}| \tag{3.2b}$$

The direction of **F** is the same as the direction of **E** if Q is positive, and opposite to **E** if Q is negative. The units for E are N/C (in the next chapter, after we define the potential in units of volts, we will show that an alternative unit for **E** is V/m). From the definition, Eq. (*3.2*), we see that the **electric field** can be thought of as the force per unit charge exerted on any charge at a point in space (see Problem 3.9).

Problem 3.9. A charge of 2.3×10^{-4} C is in an electric field and feels a force of 0.34 N in the $-x$ direction. What is the electric field at that point?

Solution

Using Eq. (*3.2b*), we get for the magnitude of E, $|E| = |F/Q|$. Therefore, $|E| = (0.34 \text{ N})/(2.3 \times 10^{-4} \text{ C}) = 1.48 \times 10^3$ N/C. Since the charge Q is positive, the electric field and the force are in the same direction, so E points in $-x$. We see from this that the field can be measured by placing a "test charge" at a point in space, and measuring the force on that charge. The field is then the force per unit charge. This is often used as the definition of E.

Problem 3.10. An electric field exists in space. At the origin, the field is 735 N/C and points in the $+x$ direction. At $x = 3$ m along the x axis, the field is 404 N/C and points in the $-y$ direction.

(*a*) What force is exerted on a charge of 0.018 C when the charge is at the origin? When the charge is at $x = 3$?

(*b*) What force is exerted on a charge of -0.032 C when the charge is at the origin? When the charge is at $x = 3$?

Solution

(*a*) Using Eq. (*3.2b*), the magnitude of the force at the origin is $F = QE = (0.018 \text{ C})(735 \text{ N/C}) = 13.2$ N. Since the charge Q is positive, the force is in the same direction as E, in the $+x$ direction. At $x = 3$, the magnitude of the force is $(0.018 \text{ C})(404 \text{ N/C}) = 7.27$ N, and the direction is $-y$ (the direction of **E**).

(*b*) Using the same reasoning, we get the force at the origin to have a magnitude of $|F| = (0.032 \text{ C})(735 \text{ N/C}) = 23.5$ N. The direction is $-x$, since the charge is negative. At $x = 3$, the force has a magnitude of $(0.032 \text{ C})(404 \text{ N/C}) = 12.9$ N. The direction of the force is $+y$, opposite to **E**, since the charge is negative.

Problem 3.11. A uniform electric field exists in space, pointing in the $+x$ direction. The field has a magnitude of 546 N/C. At time $t = 0$, a charge of 1.6×10^{-19} C is located at the origin.

(*a*) If the charge is positive and is initially at rest, describe, qualitatively, the motion that occurs.

(*b*) If the charge is negative and is initially moving to the right with a velocity of 2.3×10^4 m/s, describe qualitatively the motion that occurs.

(*c*) For the case of part (*b*), where is the particle at $t = 1.8 \times 10^{-10}$ s, if it has a mass of 9.1×10^{-31} kg?

Solution

(*a*) The force on the charge is $546 \text{ N/C}(1.6 \times 10^{-19} \text{ C}) = 8.74 \times 10^{-17}$ N. The direction is to the right on this positive charge. Since this force is constant, the acceleration is also constant. Therefore, we are dealing with motion under constant acceleration. If the initial velocity is zero, then the position of the particle will be given by $x = (\frac{1}{2})at^2$, where $a = F/m = qE/m$.

(*b*) For a negative charge, the force will be in the $-x$ direction, and will still equal 8.74×10^{-17} N. Again, we have motion under constant acceleration. Here the initial velocity was to the right and the acceleration is to the left. Therefore the particle will slow down, stop, and then move to the left with increasing speed.

(*c*) The acceleration is $a = F/m = (-8.74 \times 10^{-17}\ \text{N})/(9.1 \times 10^{-31}\ \text{kg}) = 9.60 \times 10^{13}\ \text{m/s}^2$. Recalling that for constant acceleration $x = v_0 t + (\frac{1}{2})at^2$, and using $v_0 = 2.3 \times 10^4$ m/s and $a = -9.60 \times 10^{13}\ \text{m/s}^2$, then at $t = 1.8 \times 10^{-10}$ s, $x = 2.3 \times 10^4(1.8 \times 10^{-10}) + (\frac{1}{2})(-9.60 \times 10^{13})(1.8 \times 10^{-10})^2 = 2.58 \times 10^{-6}$ m.

Problem 3.12. A uniform electric field exists in space, pointing in the $-y$ direction. The field has a magnitude of 546 N/C. At time $t = 0$, a negative charge of 1.6×10^{-19} C is located at the origin, and is moving in the positive x direction with a speed of 2.3×10^4 m/s. The charge has a mass of 9.1×10^{-31} kg. Where is the particle at $t = 1.8 \times 10^{-8}$ s?

Solution

The force on the charge is 546 N/C$(1.6 \times 10^{-19}\ \text{C}) = 8.74 \times 10^{-17}$ N. The direction is opposite to E for this negative charge, i.e. in the $+y$ direction. Since the force is constant, the acceleration is also constant, and equals $F/m = (8.74 \times 10^{-17}\ \text{N})/(9.1 \times 10^{-31}\ \text{kg}) = 9.6 \times 10^{13}\ \text{m/s}^2$ in $+y$. Therefore, we are dealing with two-dimensional motion under constant acceleration. The particle will move in a parabola, as was discussed in the chapters on mechanics. We calculate the x and y motions separately. In x we have constant velocity, since there is no acceleration in that direction. Thus, at $t = 1.8 \times 10^{-10}$ s, $x = (2.3 \times 10^4\ \text{m/s})(1.8 \times 10^{-10}\ \text{s}) = 4.14 \times 10^{-6}$ m. For y we use the equation for constant acceleration, with zero initial velocity, $y = (\frac{1}{2})at^2 = (\frac{1}{2})(9.6 \times 10^{13})(1.8 \times 10^{-10})^2 = 1.56 \times 10^{-6}$ m. Thus, the particle will be located at (4.14 μm, 1.56 μm) at t = 1.8×10^{-10} s.

3.5 THE ELECTRIC FIELD—SOURCE

We have already stated that one source of an electric field is a charge, q. The charge establishes an electric field everywhere in space and we need a formula for the magnitude and direction of this field at any given point. Consider the charge q in Fig. 3-6. We desire the field at point P, which is at a distance r from q in the direction of the vector **r**. The magnitude of **E** is given by:

$$\mathbf{E} = k|q|/\mathbf{r}^2 = (1/4\pi\varepsilon_0)|q|/\mathbf{r}^2 \tag{3.3}$$

and the direction is along the line joining q and the point P. If the charge q is positive, the direction is *away* from q, and for q negative, the direction is *toward* q (see Fig. 3-6). At any point is space the electric field created by a charge points away from a positive charge, and toward a negative charge.

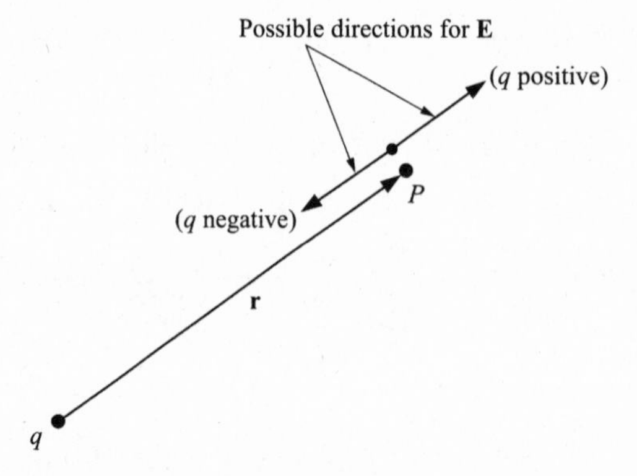

Fig. 3-6

Problem 3.13. A charge of 2.5×10^{-8} C is located at the origin.

(*a*) What electric field (magnitude and direction) does this charge produce at a point on the x axis at $x = 0.87$ m?

(*b*) Repeat for a point on the y axis at $y = -0.79$ m.

(*c*) How would your answer change if the charge q were -2.5×10^{-8} C?

Solution

(*a*) The magnitude is given by Eq. (*3.3*), $E = kq/r^2 = (9.0 \times 10^9)(2.5 \times 10^{-3})/(0.87)^2 = 2.97 \times 10^2$ N/C. The direction is along the line joining q and P, which in this case is the x axis. Since the charge q is positive the direction is away from the charge. At P, which is to the right of q, this direction is the positive x direction.

(*b*) The magnitude is $E = (9.0 \times 10^9)(2.5 \times 10^{-8})/(0.79)^2 = 360$ N/C. The direction is along the line joining q and P which is the y axis. Since q is positive E points away from q which means the negative y direction, since P is below q.

(*c*) If q is negative this does not affect the magnitude of E. The direction, however, is reversed. Thus, in (*a*) the field points toward q, or in the negative x direction, and in (*b*) the field points in the positive y direction.

Problem 3.14. Consider a charge q at the origin and a charge Q at a distance r from q. Use the electric field approach to calculate the force on Q by the field established by q, and show that the result is the same as Coulomb's Law for all possible polarities of q and Q.

Solution

Let us take the line joining the charges to be the x axis. The charge q establishes a field at the position of Q with a magnitude equal to $k|q|/r^2$. This field exerts a force on Q of magnitude $|QE| = k|qQ|/r^2$, which is the same as for Coulomb's Law.

To get the direction let us take each case separately. Suppose q is positive. Then the field at Q is in the x direction and away from q. If Q is also positive, the force is in the same direction as E, which is away from q, or repulsive, as required by Coulomb's Law. If Q is negative the force is opposite to **E**, which is toward q, or attractive, as required by Coulomb's Law. Now, suppose q is negative. Then the electric field is in the x direction, but toward q. If Q is positive, then the force is in the direction of **E**, and points toward q, as required by Coulomb's Law for charges of opposite sign. If Q is negative, the force is opposite to **E**, and thus away from q again as required by Coulomb's Law. Thus, as required, both approaches give the same result.

Problem 3.15. A charge of 1.25×10^{-7} C is at a distance of 0.38 m from a second charge of -5.3×10^{-7} C, as in Fig. 3-7.

(*a*) What electric field do these charges produce at a point, P_1, on the line joining the charges and midway between the charges?

(*b*) What electric field do these charges produce at a point, P_2, on the line joining the charges and at a distance of 0.28 m past the second charge?

Solution

(*a*) From Eq. (*3.2*), and the fact that forces add vectorially, we know the same is true of the fields set up by different charges. The field **E**, will be the vector sum of the fields produced by each charge. Now, $|\mathbf{E}_1| = k|q_1|/r_1^2 = 9.0 \times 10^9(1.25 \times 10^{-7}\ \text{C})/(0.38/2\ \text{m})^2 = 3.12 \times 10^4$ N/C. The direction is away from q_1 along the line joining the charges (which is toward q_2 since the point is between the charges). Also, $|\mathbf{E}_2| = k|q_2|/r_2^2 = (9.0 \times 10^9)(5.3 \times 10^{-7}\ \text{C})/(0.19\ \text{m})^2 = 1.32 \times 10^5$ N/C. The direction is toward q_2, which is the same direction as $\mathbf{E}_1$. Adding these two fields gives $\mathbf{E} = 1.63 \times 10^5$ N/C, since the two fields are in the same direction.

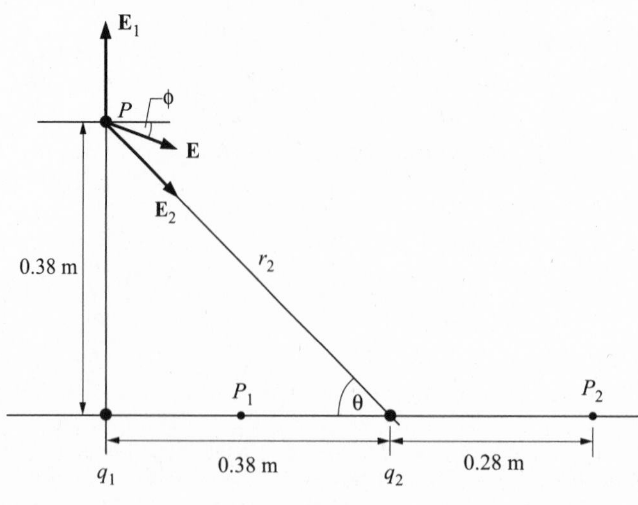

Fig. 3-7

(*b*) Again, the field **E**, will be the vector sum of the fields produced by each charge. Now, $|\mathbf{E}_1| = k|q_1|/r_1^2 = 9.0 \times 10^9(1.25 \times 10^{-7}\ \text{C})/(0.38\ \text{m} + 0.28\ \text{m})^2 = 2.58 \times 10^3$ N/C. The direction is away from q_1 along the line joining the charges (which is also away from q_2 since the point is past the second charge). Also, $|\mathbf{E}_2| = k|q_2|/r_2^2 = (9.0 \times 10^9)(5.3 \times 10^{-7}\ \text{C})/(0.28\ \text{m})^2 = 6.08 \times 10^4$ N/C. The direction is toward q_2, which is the direction opposite to $\mathbf{E}_1$. Adding these two fields vectorially gives $\mathbf{E} = (6.08 - 0.26) \times 10^4 = 5.82 \times 10^4$ N/C, since the two fields are in the opposite direction. The direction of the total field is toward the charges along the line joining the charges.

Problem 3.16. Consider the same charges as in Problem 3.15. Calculate the electric field at a point, *P*, which is 0.38 m from the first charge along a line through q_1 perpendicular to the line joining the charges, as in Fig. 3-7.

Solution

The field **E** is again the sum of $\mathbf{E}_1$ and $\mathbf{E}_2$. First we calculate $\mathbf{E}_1$ whose magnitude is $k|q_1|/r_1^2 = (9.0 \times 10^9)(1.25 \times 10^{-7}\ \text{C})/(0.38\ \text{m})^2 = 7.79 \times 10^3$ N/C. The direction is away from q_1, or in the positive y direction as in the figure. Similarly, $|\mathbf{E}_2| = (9.0 \times 10^9)(5.3 \times 10^{-7})/(0.38^2 + 0.38^2)\ \text{m}^2 = 1.65 \times 10^4$ N/C. The direction is along the line joining P and q_2, and toward q_2, since q_2 is negative. This is shown in the figure. To get **E**, we must add these two vectors together. Thus $E_x = E_2 \cos\theta = 1.65 \times 10^4(\cos 45) = 1.167 \times 10^4$. $E_y = E_1 - E_2 \sin\theta = (7.79 \times 10^3 - 1.65 \times 10^4 \sin 45) = -3.88 \times 10^3$. Therefore, $|\mathbf{E}| = [(1.167 \times 10^4)^2 + (3.88 \times 10^3)^2]^{1/2} = 12.30 \times 10^3$ N/C. This field is at an angle ϕ with the x axis given by $\tan\phi = -3.88/11.67 = -0.332$, or $\phi = -18.4°$, as in the figure.

Problem 3.17. Two positive charges are located a distance of 0.58 m apart on the x axis. At the origin, the charge is 5.5×10^{-8} C, and at $x = 0.58$ the charge is 3.3×10^{-8} C. At what point is the electric field equal to zero?

Solution

The electric field is the vector sum of the fields from the two charges. For this vector sum to be equal to zero, the two fields must be equal in magnitude and opposite in direction. The only points where the fields are along the same line are points on the x axis. Therefore the point where the field is zero must be on the x axis. There are three regions on the x axis: to the left of the origin, between the charges and to the right of the second charge. In the region to the left of the origin the field from each charge is to the left, away from each (positive) charge. To the right of the second charge each field is to the right, again away from each charge. Therefore the fields cannot add to zero. Only in the region between the charges are the two fields in opposite directions, with the field from the charge at the origin to the right and the field from the second charge to the left. Therefore the point of zero field must lie between the charges. The exact location is determined by the condition that the magnitudes must be equal. If the field is zero at x, then we must have that $kq_1/x^2 = kq_2/(0.58 - x)^2$. Then $q_1(0.58 - x)^2 = q_2 x^2$, or taking square roots, $(0.58 - x) = \pm(q_2/q_1)^{1/2}x$. This means that $0.58 - x = \pm 0.77x$. The two solutions are: $x_1 = 0.33$ and $x_2 = 2.5$. Only the first solution has the point between the charges and therefore E is zero at $x = 0.33$ m.

Problem 3.18. In Problem 3.17, suppose the second charge was negative, -3.3×10^{-8} C. Where would the field be zero?

Solution

As in Problem 3.17 the point would have to be on the x axis in order that the two fields are parallel. However, the point cannot be between the charges, since, in that region, both fields point in the $+x$ direction and cannot add to zero. Therefore the point must be either to the left of the origin or to the right of the second charge. In order for the two fields to have equal magnitude, the point must be nearer to the smaller charge than to the larger charge. In this problem that means that the point is to the right of the second charge. Suppose the point is at position x, which must be greater than 0.58 m. Then the equality of the magnitudes gives: $kq_1/x^2 = k|q_2|/(x - 0.58)^2$, or $|q_1/q_2|(x - 0.58)^2 = x^2$. Taking square roots, we get: $(5.5/3.3)^{1/2}(x - 0.58) = \pm x = 1.29(x - 0.58)$, giving $x_1 = 2.58$ m and $x_2 = 0.33$ m. Only $x_1 > 0.58$, so the point of zero field must be $x = 2.58$ m.

Problem 3.19. A charge of 2.8×10^{-8} C is located at the origin.

(*a*) At what point on the x axis must one place an equal positive charge so that the field at $x = 0.53$ m is 550 N/C in the positive x direction?

(*b*) At what point on the x axis must one place an equal negative charge so that the field at $x = 0.53$ m is 550 N/C in the positive x direction?

(*c*) At what point on the x axis must one place an equal positive charge so that the field at $x = 0.53$ m is 550 N/C in the negative x direction?

Solution

(*a*) The field at $x = 0.53$ is the sum of the fields from the two charges. The field from the first charge is $E_1 = kq_1/r_1{}^2 = (9.0 \times 10^9)(2.8 \times 10^{-8})/(0.53)^2 = 897$ N/C, and is in the $+x$ direction. Therefore, the field from q_2 must equal $(897 - 550)$ N/C $= 347$ N/C, and be directed in the $-x$ direction. The second, positive charge must therefore be to the right of the point, i.e. at $x > 0.53$ m. Then $kq_2/(x - 0.53)^2 = 347$, $(x - 0.53)^2 = (9.0 \times 10^9)(2.8 \times 10^{-8})/347 = 0.726$, $(x - 0.53) = \pm 0.85$, and $x_1 = 1.38$ m $x_2 = -0.28$ m. The only acceptable solution is $x = 1.38$ m.

(*b*) Again, the field from the second charge must equal 347 N/C to the left. Since q_2 is negative, the second charge must be to the left of the point, i.e. $x < 0.53$ m. Then, $k|q_2|/(0.53 - x)^2 = 347$, giving $(0.53 - x) = \pm 0.85$, and $x_1 = -0.28$, $x_2 = 1.38$. Now, the only acceptable solution is $x = -0.28$ m, which is to the left of the point.

(*c*) Using the same analysis as in (*a*), the field from q_1 is 897 N/C in the $+x$ direction. In order to get $E = 550$ N/C in the $-x$ direction, we need $E_2 = (897 + 550)$ in the $-x$ direction. Then the second positive charge must be to the right of the point at 0.53 m. Then $kq_1/(x - 0.53)^2 = 1447$, $(x - 0.53) = \pm 0.42$, and $x_1 = 0.95$, $x_2 = 0.11$ m. The correct answer is $x = 0.95$ m since we require $x > 0.53$.

Problem 3.20. Suppose we have a ring of radius r, as in Fig. 3-8. Positive charge is uniformly distributed along the ring, with a linear charge density of λ C/m throughout its length.

(*a*) Calculate the electric field produced by the charge on the ring at the center of the ring (point P_1 in the figure).

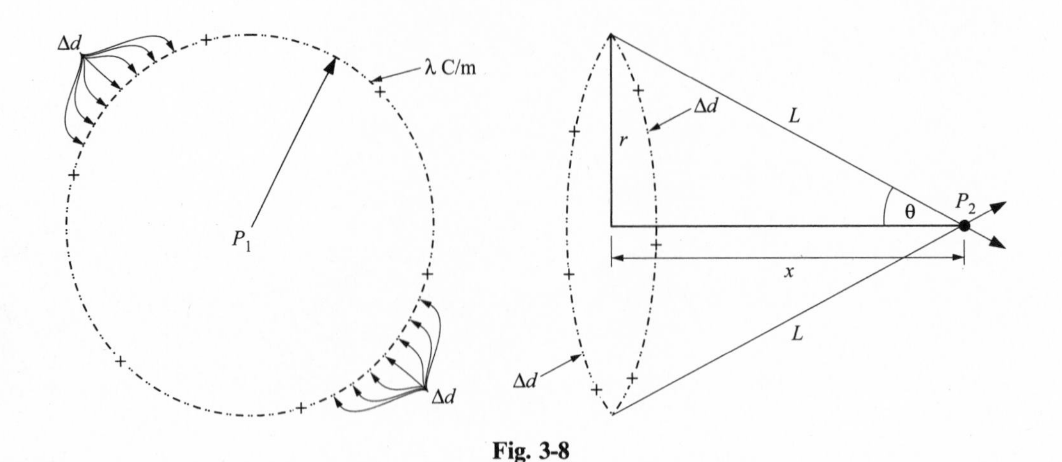

Fig. 3-8

(*b*) Repeat for a point on the axis of the ring at a distance x from the plane of the ring (point P_2 in Fig. 3-8).

(*c*) If $r = 0.15$ m and $\lambda = 6.3 \times 10^{-8}$ C/m, what field is produced at the point $x = 0.20$ m?

Solution

(*a*) Consider any small segment of the ring, of length Δd, which contains an amount of charge $\lambda \Delta d$. It will produce a field at the center of the ring pointing away from the point, and of magnitude $k(\lambda \Delta d)/r^2$. The same region of the ring on the other end of the diagonal will produce a field of the same magnitude but in the opposite direction. Thus the field produced by these regions will cancel. This is true for all opposite parts of the ring so that the total field produced by all the charges on the ring will add to zero. Thus the field at the center of the ring will be zero.

(*b*) Consider the two small segments at the top and at the bottom of the ring, each of length Δd. Each segment has a charge of $\lambda \Delta d$, and contributes a field ΔE at P_2 of magnitude $\Delta E = k\lambda \Delta d/(r^2 + x^2)$, since the distance L from the charge to the point is $(r^2 + x^2)^{1/2}$. The direction of each field contribution is away from the charge, and therefore in the direction shown in the figure. When one adds these two contributions the components perpendicular to the axis of the ring will cancel, leaving only the components parallel to the axis. This will also be true for all other opposite pair segments of the ring and the net field which will be produced is one parallel to the axis and equal to the sum of the parallel components of the field due to all segments Δd. Each region of length Δd contributes a parallel component of $[k\lambda \Delta d/(r^2 + x^2)] \cos \theta = k\lambda \Delta d x/(r^2 + x^2)^{3/2}$, where we have used the fact that $\cos \theta = x/(r^2 + x^2)^{1/2}$. Thus, each region of length Δd contributes the same parallel component of field at point P_2 in the positive x direction, and the total field is obtained by adding all the Δd contributions. Noting the

$$\sum_{\text{loop}} \Delta d = 2\pi r,$$

we have

$$E = k\lambda(2\pi r)x/(r^2 + x^2)^{3/2} \text{ and points along the axis of the ring} \qquad (3.4)$$

(*c*) Substituting in Eq. (*3.4*), we get $E = (9.0 \times 10^9)(6.3 \times 10^{-8}\text{ C/m})(2\pi)(0.15\text{ m})\ (0.20\text{ m})/(0.15^2 + 0.20^2)^{3/2} = 683$ N/C.

Problem 3.21. Consider a sphere of radius R, which is uniformly charged throughout its volume. Show, by symmetry arguments, that at a point, P, outside the sphere, at a distance r from the center of the sphere (*a*) the field must point in the radial direction away from the center of the sphere (see Fig. 3-9) and (*b*) the magnitude of the field depends only on r and not on any other coordinate of the point P.

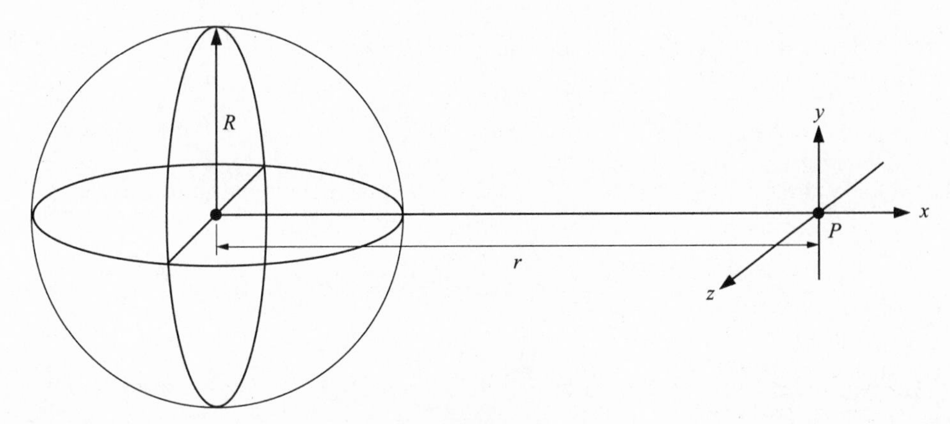

Fig. 3-9

Solution

(*a*) At point P we draw a coordinate system x, y, z as shown. At this point the $\pm y$ directions are indistinguishable from each other since the sphere appears identical from both directions. Thus if there is a field in one of those directions there should also be a field in the other, and therefore there cannot be a field in either direction. The same symmetry argument is applicable to the $\pm z$ directions. The only possible direction for the field is therefore the x direction. Since the field must point away from positive charge, the field must be radially away from the center of the sphere.

(*b*) All points at the same distance r from the center of the sphere are equivalent since the sphere appears identical to each point. Thus they must all have the same magnitude for the electric field, and the field cannot vary with any coordinate other than r.

In fact we will show in the next section that the magnitude of the field outside the sphere is given by $E = kQ/r^2$, where Q is the total charge on the sphere. This is the same field that would be produced by a point charge Q located at the center of the sphere. This result is the identical to the case of the gravitational force (Beginning Physics I, Section 5.3, Eq. (5.2)).

In order to obtain the fields produced by other charge distributions we can make use of calculus to add the contributions from all the charges in the distribution. For instance, for a solid disk, with a surface charge distribution of σ, we could divide the disk into rings and add the field of each ring at a point along the axis (see Fig. 3-10). The field points along the positive x axis, since the contribution of each ring is in that direction. The field along the symmetry axis of a ring is given by Eq. (3.4). The radius

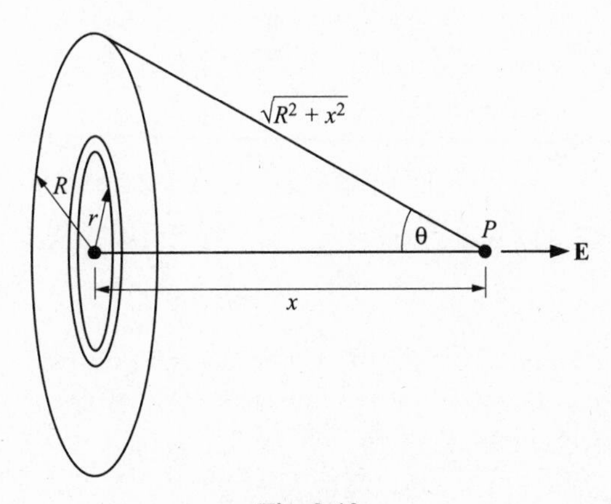

Fig. 3-10

r varies from ring to ring while x stays fixed. Using calculus to add all the contributions results in a field of

$$E = 2\pi k\sigma[1 - x/(R^2 + x^2)^{1/2}] = 2\pi k\sigma[1 - \cos\theta] \tag{3.5}$$

due to the entire disk.

Similarly, for a uniformly charged rod, of length $2L$ and linear charge density λ, the field at a distance of x from the center of the rod (see Fig. 3-11) is given by

$$E = 2kL\lambda[1/x(L^2 + x^2)^{1/2}] = kQ[1/x(L^2 + x^2)^{1/2}] = kQ\cos\theta/x^2, \tag{3.6}$$

where $Q = 2L\lambda$ is the total charge on the rod.

Problem 3.22. Consider a solid disk of radius R, which is uniformly charged throughout its area with a charge density σ C/m^2.

(*a*) Show that if the point, P, is at a distance x from the center of the disk, with $x \gg R$, the field approaches the field of a point charge Q.

(*b*) Show that as the disk becomes very large ($R \gg x$), the field approaches the value $E = \sigma/2\varepsilon_0$.

Solution

(*a*) The field is given by Eq. (*3.5*) $E = 2\pi k\sigma[1 - x/(R^2 + x^2)^{1/2}] = 2\pi k\sigma[1 - 1/(R^2/x^2 + 1)^{1/2}]$. As $x \gg R$, $R^2/x^2 \to 0$, and the second term in the parenthesis approaches 1. We must use the binomial expansion for this term to see how it differs from 1, and $[1/(1 + R^2/x^2)^{1/2}] \to 1 - (1/2)R^2/x^2$. Then $E \to 2\pi k\sigma(1/2)(R^2/x^2) = kQ/x^2$, where $Q = \sigma\pi R^2$ = total charge on disk, and x is the distance from the disk. This is the same as the field produced by a point charge Q at a distance of x from the charge. The same result can be obtained more easily by examining the field due to a ring of charge for $x \gg r$. From Eq. (*3.4*) we ignore r in the denominator to get $E = k\lambda(2\pi r)x/x^3 = kq/x^2$ where $q = 2\pi r\lambda$ is the total charge on the ring. Since the disk can be thought of as a series of concentric rings, and x is the same for each ring, all the rings contribute kq/x^2 when $x \gg R$, and the sum is just kQ/x^2 where Q is the total charge on the disk.

(*b*) We again use the formula for the field of a disk, $E = 2\pi k\sigma[1 - x/(R^2 + x^2)^{1/2}]$, but now we let $R \gg x$. Then the second term in the parenthesis approaches zero, and

$$E \to 2\pi k\sigma = \sigma/2\varepsilon_0, \tag{3.7}$$

where we recall the definition, $k = 1/(4\pi\varepsilon_0)$

The direction of the field is away from the large plane disk for a positive charge distribution. If the charge distribution is negative, the field would point toward the disk. Thus, along the symmetry axis of our uniform disk the field is perpendicular to the disk and does not depend on x, *for* x $\ll$ R.

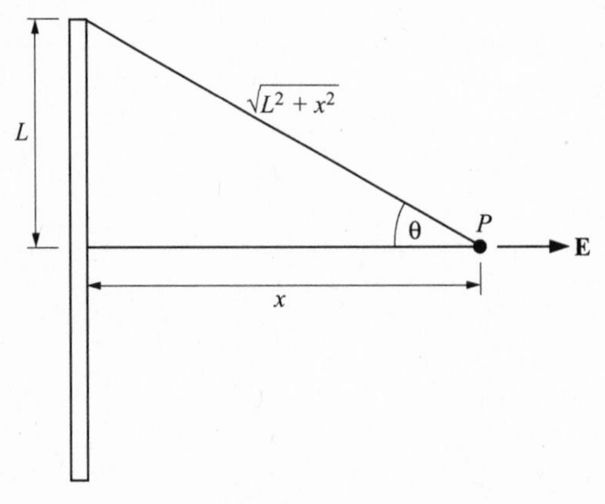

Fig. 3-11

This second result can be extended to the field produced by any large uniform planar charge distribution along any line perpendicular to the plane. If one is near enough to the charge distribution, i.e. if $x \ll R$, where x is the distance to the plane and R is the closest distance of the line to the edge of the plane (see Fig. 3-12), then the field is given by $E = 2\pi k\sigma$. Note that the planar charge distribution does not have to be disk shaped, as long as one is "far" from the edge (i.e. $x \ll R$).

Problem 3.23. Two parallel plates, of area A, are separated by a distance d, as in Fig. 3-13. The distance d is much smaller than the linear dimension of the area. One plate has a positive charge density of σ, while the other has an equal negative charge density $-\sigma$.

(*a*) What is the electric field at a point, P_1, between the plates which is not near the edge of the plates?

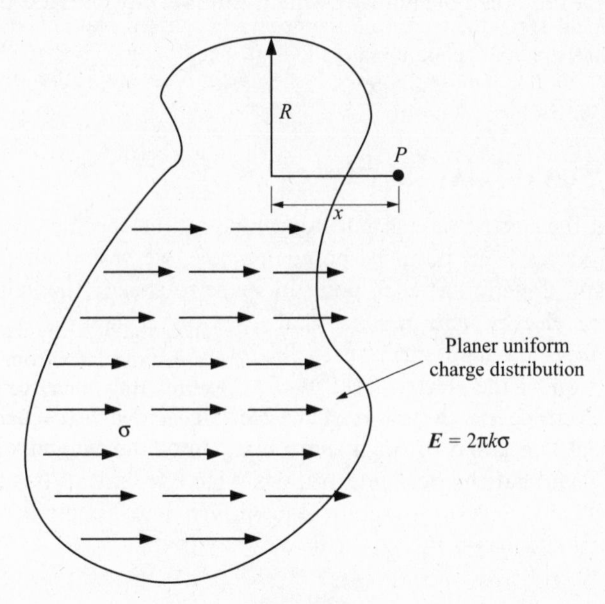

Fig. 3-12

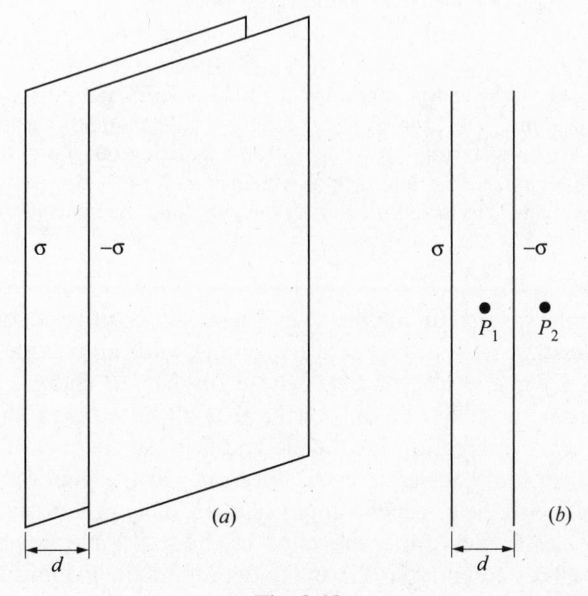

Fig. 3-13

(*b*) What is the electric field at point P_2 outside the negatively charged plate but much nearer to the plate than to the edge?

Solution

(*a*) The field close to the surface of the plate is given by Eq. (*3.7*) $E = \sigma/2\varepsilon_0$, as long as we are far from the edge of the plate. Each plate thus produces a field of the same magnitude. The field from the positively charged plate points away from the plate, and the field from the negatively charged plate points toward the plate. At point P_1, both fields point toward the right, and the total field is therefore, $E = \sigma/2\varepsilon_0 + \sigma/2\varepsilon_0 = \sigma/\varepsilon_0$. The field is the same everywhere in the region between the plates as long as one is not near the edges of the plates. This arrangement of charge is called a parallel plate capacitor, and is especially useful because it produces this uniform field.

(*b*) Since the distance between the plates is very small the approximation $E = \sigma/2\varepsilon_0$ for the magnitude of the field still holds. Here, the field from the positively charged plate points to the right, while the field from the negatively charged plate points to the left. The total field is therefore zero.

3.6 THE ELECTRIC FIELD—GAUSS' LAW

We have learned that the electric field is a vector which can be specified by giving its magnitude and its direction. If the field varies from point to point in space, we would have to draw a vector of the appropriate magnitude and direction at each point in space to specify the field. There is another useful approach for picturing the electric field in space that gives one a new insight into the electric field. In this approach, we trace lines through space in such a way that as the line passes through a point it always aims in the direction of the electric field at that point. Such lines are called **electric field lines.** Thus one can always determine the direction of the electric field at a point in space by drawing the tangent to the electric field line going through that point. Since the tangent to any line allows for two possible directions, it is clear that the field lines we draw have to have a positive sense associated with them (just like the straight line axis on a graph). We will see that this picture of electric field lines will also permit us to determine the magnitude of the field at any point.

Problem 3.24. Consider a point charge q located at the origin. Draw the electric field lines produced by this charge. How do these lines change if the charge is negative?

Solution

The field produced by the charge q has a magnitude of kq/r^2 and points away from the charge at any point in space. The field lines must therefore be along the radii emanating from the charge. At any point in space the field lines are lines starting at the charge and radiating out along the radii. The direction of the lines is away from the charge. The lines are shown in Fig. 3-14. If the charge is negative, then the only change that has to be made is to have the arrows pointing along the lines *toward* the charge instead of away from the charge.

We see in these simple cases that all the lines begin on positive charges and end on negative charges. There are no lines that start or end at other points, and no lines that close on themselves like circles do. This is the case for any electric field whose origin is a charge. If one has a collection of charges producing an electric field it will still be true that all lines begin and end on charges. At any point in space there is only one direction for the electric field, so there is only one line going through each point. It cannot happen that lines cross each other since at the point of crossing there would then be two directions for the electric field, which cannot happen. This insight will often allow us to draw a qualitative picture of the electric field for a collection of charges. First, however, we must develop one more concept which we will use to understand how to determine the magnitude of the electric field from these field lines.

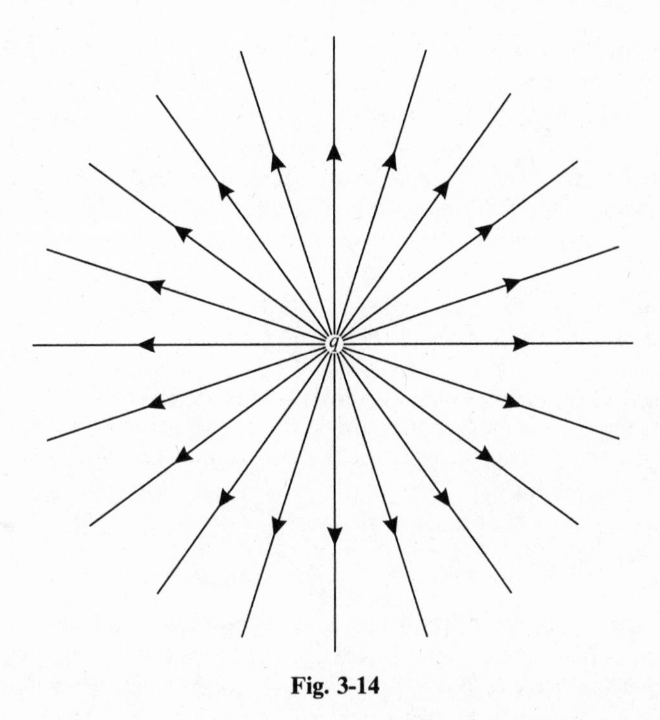

Fig. 3-14

Problem 3.25. Consider a point charge q located at the origin. Assume that we draw a large number, N, of representative electric field lines emanating from the charge, uniformly distributed over space (e.g. see Fig. 3-14).

(*a*) How many electric field lines pass through the closed sphere surrounding the charge at radius r?

(*b*) What is the number of electric field lines per unit area that passes through the closed sphere at radius r?

(*c*) If one equates the number of field lines drawn per area to the magnitude of the electric field at this radius, what would N (the total number of field lines drawn) have to be, in terms of the charge q?

Solution

(*a*) All the lines that are emitted from the charge pass through this closed area. This is because the lines are radial straight lines that do not stop anywhere within the sphere, and the sphere encompasses them all. Thus N lines pass through the closed sphere at any radius r.

(*b*) At a radius r, the closed sphere has a surface area of $4\pi r^2$. Since the lines have been drawn uniformly over the sphere the number of lines per unit area of the sphere is the same everywhere and equals $N/4\pi r^2$.

(*c*) The field produced by the charge q has a magnitude of $(1/4\pi\varepsilon_0)q/r^2$. Equating this to the number of lines per unit area, we get $N/4\pi r^2 = q/4\pi\varepsilon_0 r^2$, or $N = q/\varepsilon_0$.

Choosing N, the representative number of field lines drawn from the charge q, to equal q/ε_0 is particularly useful because we can now deduce the magnitude of the electric field at any point P at any distance R from the charge in terms of the lines/area at that point. The lines/area at a point P is defined as the number of lines passing through a small "window" centered on the point P and facing perpendicular to the field lines, divided by the small area of the window. For the case of a point on the sphere of radius r in Problem 3.25, we saw the lines/area is just $N/4\pi\varepsilon_0 r = q/4\pi\varepsilon_0 r$, which indeed is the correct electric field (as it must be since that is how we defined N in the first place). However, if we choose any point P at any distance R from q, we can draw a sphere of radius R through P centered on q, and again conclude that the lines/area is just $N/4\pi R^2 = q/4\pi\varepsilon_0 R^2$ which is the correct field at point P. It is not surprising that this works since the number of lines N, once chosen, is constant, and the number passing through a unit area decreases as $1/R^2$ with distance R from the point charge, q (since the surface area of a sphere goes up as R^2). Since the electric field also falls as $1/R^2$, lines/area and field are proportional, and the proper choice of N makes them equal everywhere. If instead of a single charge we had two or more charges, the field lines would be much more complicated looking (curved rather than straight, some starting at a positive charge and ending on a negative charge with others wandering off to infinity, etc.). Nonetheless, it is still true that if one chooses the lines/area at one point to equal $|E|$ then upon following the lines near that point to any other location, the lines/area at the new location will still equal $|E|$ at that location. This property is at the heart of Gauss's law, which we will now demonstrate. The lines/area (as defined above) when chosen to equal $|E|$, is called the **flux density**, and the number of lines passing through any given area is called the **flux** through that area.

> ***Note.*** Thinking in terms of lines/area is a very useful pictorial device for understanding the behavior of the electric field, but ultimately all the results we obtain are expressible directly in terms of the electric field E and do not depend on the artifact of lines that are drawn through space.

In what follows, we will always assume that lines/area are chosen to equal $|E|$.

Having defined flux as the number of lines passing through an arbitrary area, we now examine how to calculate such a flux when the area is not perpendicular to the lines. We will also define a positive or negative sense to the flux depending on whether the electric field points "outward" through the area or "inward" through the area, as defined below. In Fig. 3-15(a) we draw an electric field which is not perpendicular to a small area. We designate the area by a vector **A** whose direction is perpendicular to

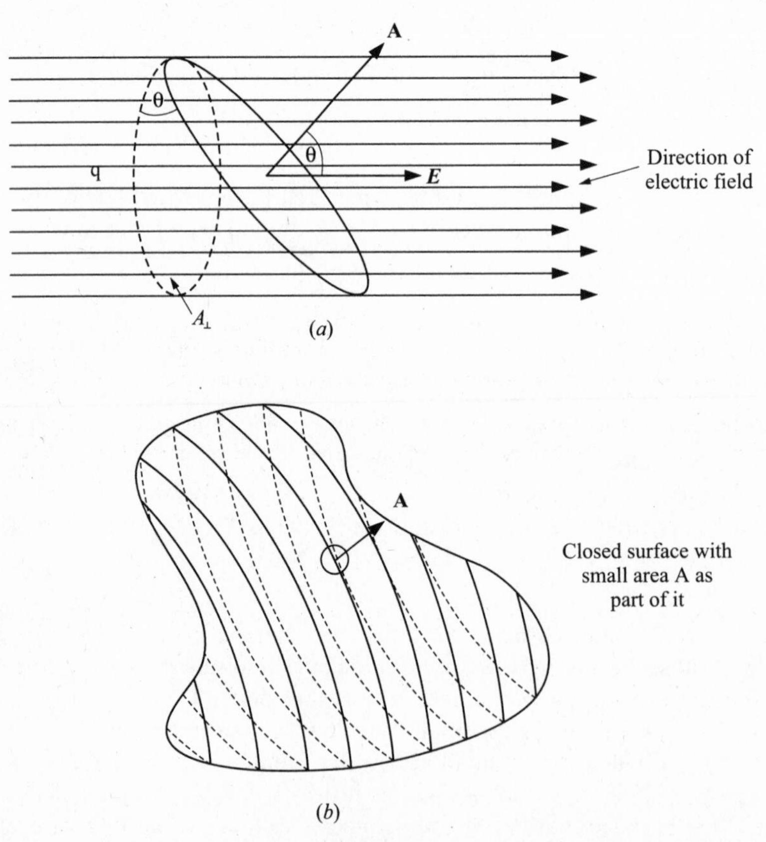

Fig. 3-15

the area and whose magnitude is equal to the area. The direction of the arrow on the vector, which indicates the positive direction for the area, is arbitrary. If the small area is part of a closed surface (e.g., a spherical shell, a cylindrical shell, a rectangular shell, an egg-shaped shell, etc.) then the convention is that the little surface areas making up the closed surface are positive outward from the enclosure. In Fig. 3-15(*b*) we show the area **A** as part of such a closed surface. For the case of Fig. 3-15(*a*) the field lines clearly pass through the area **A** in the positive sense, so the flux is positive. If the field lines pointed in the opposite direction, i.e. to the left in Fig. 3-15(*a*), then the flux would be negative. To actually calculate the flux through the small area A, we proceed as follows. If we cast the shadow of the area A on a plane perpendicular to the field lines, we would get the dashed area $A_\perp$ shown in the figure. As can be seen, the number of field lines passing through A and $A_\perp$ are identical. Since $A_\perp$ is perpendicular to the field lines, the magnitude of flux F passing through $A_\perp$ is just (lines/area) × $A_\perp$ or $F = EA_\perp$, where E is the magnitude of the electric field at **A**. Note that the angle θ between **A** and **E** is the same as the angle between A and $A_\perp$. It is not hard to see that the shadow area and original area are related by $A_\perp = A \cos\theta$, so $F = EA \cos\theta$. If $\theta > 90°$ then $\cos\theta$ goes negative, and this corresponds to the field lines passing through **A** in the negative sense, so the formula holds quite generally:

$$F = EA \cos\theta \tag{3.8}$$

Noting that $E \cos\theta$ is just the component of **E** parallel to **A**, we see that the flux through **A** is just the component of **E** along **A** times the magnitude of A: $F = E_\parallel A = E(\cos\theta)A$. (This is an equivalent definition of flux that needs no reference to "field lines per unit area" or other intuitive constructs). Eq. (*3.8*) is a general result for the electric flux, i.e. for the number of lines passing through a *small* area, since the **E** vector does not vary over the small area. It can be extended to a large area as well, giving flux $= E(\cos\theta)A$, provided that neither E nor θ varies over the area. If there is a variation of E or θ, then one has to divide the area into many small parts ΔA [as could be done with the surface of Fig. 3-15(*b*)], evaluate the flux for each part, and add all the flux together. Then flux $= \sum [E(\cos\theta)\Delta A]$. For a closed surface, if this sum is positive then net flux leaves the surface, i.e. more lines leave the surface than enter the surface. If the sum is negative, then on net more lines have entered the surface from outside than leave. Recalling Problem 3.25(*c*), we see that the flux through the spherical surface due to charge q is just the total number of lines, $N = q/\varepsilon_0$.

Problem 3.26. Consider a closed surface of arbitrary shape, as in Fig. 3-16. Suppose a single charge Q_1 is located at some point within the surface, and a second charge Q_2 is located outside the surface.

(*a*) What is the total flux passing through the surface due to the charge Q_1?

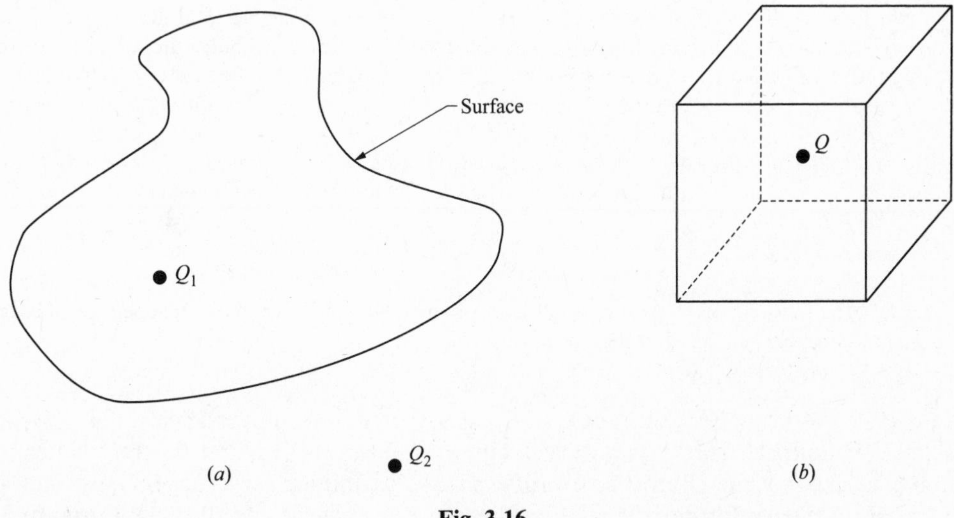

Fig. 3-16

(*b*) What is the total flux passing through the surface due to Q_2?

(*c*) If a charge Q is located at the center of a cube, Fig. 3-16(*b*), how much flux is passing through one side of the cube?

Solution

(*a*) All the lines leaving from (or converging on) the charge Q_1 pass through this closed area. This is because the lines that start or end on the charge Q_1 do not stop anywhere within the surface, and the sphere is closed. From earlier work we know that Q_1 has associated with it Q_1/ε_0 lines, and the flux through the closed surface is therefore Q_1/ε_0. If the charge is positive the field lines will point away from the charge passing from the inside to the outside of the closed surface, and using our area convention the flux will be positive. If the charge is negative the field lines point inward and the flux will be negative.

(*b*) Charge Q_2 emits electric field lines. These lines are straight lines emanating out from Q_2 in all directions. Some of these lines will not pass through any part of the surface of the closed surface and will therefore contribute nothing to the flux passing through the closed surface. Any line that enters the volume through one part of the surface must continue onward and eventually exit from the surface. The net number of lines entering the surface is therefore zero since any line that enters also exits. Thus any charge, such as Q_2 that is outside the closed surface contributes no net flux over the entire closed surface.

(*c*) The six sides of the cube form a closed surface. Therefore the total flux through all six sides must equal Q/ε_0. By symmetry, each side of the cube is equivalent, and an equal amount of flux passes through each side. Therefore, the flux passing through each side is $Q/6\varepsilon_0$.

Problem 3.27. Consider a closed surface of arbitrary shape, as in Fig. 3-16. Suppose there are several charges with combined net charge Q_{in} located at points within the surface, and several other charges with combined net charge Q_{out} located outside the surface. All these charges contribute to the total electric field and associated field lines. Find the flux through the closed surface due to the total electric field.

Solution

The electric field $\mathbf{E}_T$ at any point in space is just the vector sum of the electric fields due to all the charges inside and outside the surface. We also know that the flux through any small area $\mathbf{A}$ is just the component of $\mathbf{E}_T$ parallel to $\mathbf{A}$ times A. Since the component of the sum of vectors is just the sum of the components, we have $E_{T\parallel} = \sum E_\parallel$ (individual charges) and flux through $\mathbf{A}$ is $\sum$ (fluxes of individual charges through A). Since this is true for every small area $\mathbf{A}$, we conclude that the total flux = sum of fluxes due to individual charges for our closed surface. The outside charges do not contribute anything to the net flux passing through this closed area, as we showed in the previous problem. Each inside charge q_i contributes a flux q_i/ε_0 where q_i can be positive or negative. Since $\sum q_i = Q_{in}$ the flux through the closed surface will equal Q_{in}/ε_0.

Recalling that the total flux due to $\mathbf{E}_T$ is just $\sum [E_T(\cos\theta)\Delta A]$, where the sum is over the entire closed surface, the result of the last problem leads to a powerful result known as **Gauss' law:**

For any closed surface,

$$\sum [E(\cos\theta)\Delta A] = Q_{in}/\varepsilon_0 \tag{3.9}$$

where E is the total electric field due to all charges in the universe (we have dropped the subscript T) and Q_{in} is the total charge enclosed in the surface.

From our development of Gauss' law it is clear that it is a universal law that is true even in situations that are not particularly symmetrical. However, for cases of symmetry the law turns out to be very useful in calculating the electric field produced by distributions of charge. Gauss' law is derived directly from the particular form of the relationship between a charge and the field it produces. In fact it is an alternative formulation of this relationship. Testing the validity of Gauss' law is equivalent to testing the form of Coulomb's law which we used to define the electric field.

The practical use of Gauss' law in cases of symmetry involves choosing a particular imaginary closed surface (called a **Gaussian surface**) through which either the flux is zero or the field contributing to the flux is constant. In that case it is easy to evaluate the sum leading to the total flux. If part of the surface is chosen so that the field lines are parallel to it (i.e. $\mathbf{E} \perp \mathbf{A}$) then that part of the surface has zero flux passing through it. If part of the surface can be chosen as a planar region where the field is constant and perpendicular to the plane, then the flux will just equal the product of E and the area, EA. We will illustrate these cases in the following problems.

Note. While we typically associate charges with discrete particles such as electrons and nuclei, from a macroscopic point of view net charges can often be thought of as distributed smoothly through space with a charge density (charge per unit volume) ρ, which can vary from location to location or be constant (uniform density) as the case may be.

Problem 3.28. Consider a uniformly charged sphere of radius R. The charge per unit volume is ρ. Calculate the field produced by this sphere at a point outside the sphere, i.e. for $r > R$.

Solution

Because of the symmetry we know that the magnitude of the field will be the same at all points at the same distance from the center of the sphere and the direction will be radially outward (or inward) from the center of the sphere. If we choose a closed concentric sphere of radius r as our Gaussian surface (see Fig. 3-17), we know that the field will be the same at every point on the sphere. Furthermore we also know from symmetry (see, e.g. Problem 3.21) that the field will be along a radius drawn from the center of the sphere. This direction is perpendicular to the surface of the sphere at every point so that E is parallel to $\mathbf{A}$, and $\cos\theta = 1$ in Eq. (*3.9*). Then the sum needed to calculate the flux through the closed sphere will just equal E multiplied by the surface area of the sphere. Since the surface area of a sphere is $4\pi r^2$, the total flux will equal $E(4\pi r^2)$. By Gauss' law this total flux must equal the total charge within the Gaussian surface. Since the charged sphere is completely within the Gaussian surface, Q_{in} will equal the total charge Q on the charged sphere. This total charge is just ρ times the volume of the charged sphere, or $Q_{\text{in}} = \rho(4/3)\pi R^3$. Carrying through the mathematics, using Eq. (*3.9*), $\sum [E(\cos\theta)\Delta A] = Q_{\text{in}}/\varepsilon_0$, we get

$$E(4\pi r^2) = Q/\varepsilon_0 = \rho(4/3)\pi R^3/\varepsilon_0, \qquad \text{and} \qquad E = \rho R^3/(3\varepsilon_0 r^2)$$

This is just the result we quoted in Problem 3.21. We will use Gauss' law to calculate the field inside the sphere ($r < R$) in a later problem.

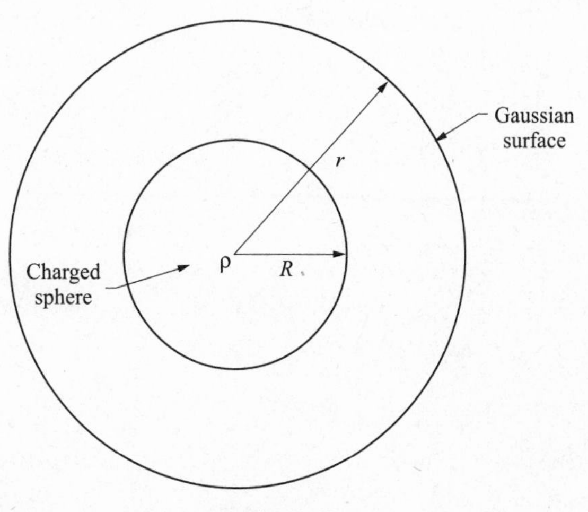

Fig. 3-17

Problem 3.29. A charge Q is placed at the center of a conducting spherical shell of inner radius R_1 and outer radius R_2 (see Fig. 3-18). The sphere has no net charge.

(*a*) Use Gauss' law to find the electric field in the hollow spherical region of the sphere, $r < R_1$.

(*b*) Show, using your knowledge of conductors, that for a conductor in equilibrium (no moving charges) the electric field in the interior of the conductor is zero.

(*c*) Use Gauss' law to calculate the charge collected on the inner surface of the sphere (at R_1).

(*d*) Calculate the field outside the sphere, i.e. for $r > R_2$.

(*e*) If the sphere had a net charge of Q', what change, if any, would there be to the answers in parts (*a*), (*c*) and (*d*)?

Solution

(*a*) Because of the symmetry we know that E will be the same at all points at the same distance from the center of the sphere and will point radially. If we choose a closed spherical surface of radius r as our Gaussian surface (see Fig. 3-18), we then know that the magnitude of the field will be the same at every point on this surface, and its direction is perpendicular to the surface at every point so that E is parallel to **A**. To get the field within the sphere we draw this Gaussian surface at $r < R_1$. Then the sum [Eq. (*3.9*)] needed to calculate the flux through the closed sphere will just equal E multiplied by the surface area of the sphere. Again noting that the surface area of a sphere is $4\pi r^2$, the total flux will equal $E(4\pi r^2)$. By Gauss' Law this total flux must equal the total charge within the Gaussian surface. Since the only charge within the sphere is Q, we have $\sum [E(\cos\theta)\Delta A] = Q/\varepsilon_0$, $\rightarrow E(4\pi r^2) = Q/\varepsilon_0$, and $E = Q/(4\pi\varepsilon_0 r^2) = kQ/r^2$. This is just the result we would get for a free charge Q.

(*b*) Since conductors are filled with freely moving charges, if an electrical field existed in the interior, charges would be pushed or pulled and hence be moving. In equilibrium, the charges must arrange themselves on the surface of the conductor so that the net field (due to *all* the charges everywhere) is zero throughout the interior of the conductor.

(*c*) For this part we draw a Gaussian surface as a concentric spherical shell within the conductor, at r such that $R_1 < r < R_2$. At every point on this surface the field is zero, since it is in the conducting region. Thus the total flux through the surface is zero. From Gauss' law this means that the total charge within the sphere is zero. If we look inside the Gaussian sphere we note that there is a charge Q at its center and some other possible charge on the inner surface of the conducting sphere. For the total charge to be zero requires that the charge on the inner surface of the conducting sphere be equal to

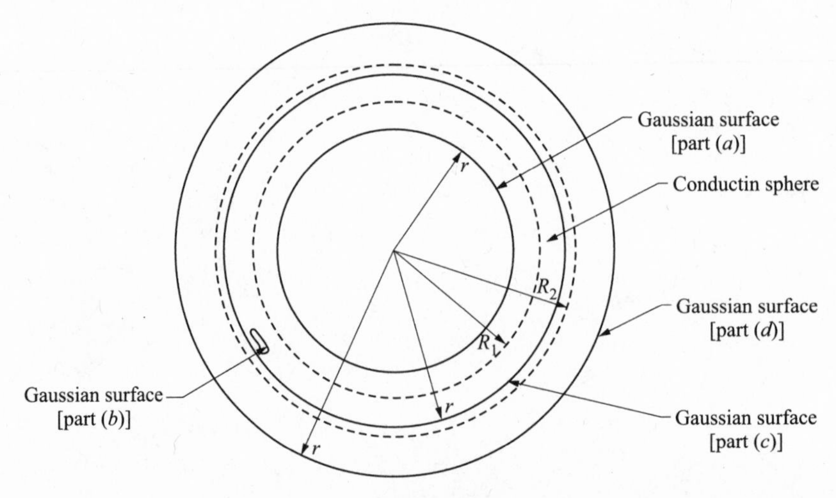

Fig. 3-18

$-Q$. Note that this means that the outer surface of the sphere has a charge of $+Q$, since the net charge on the conducting sphere was given as zero.

(*d*) For this part we draw the Gaussian surface as a sphere with a radius, r, greater than R_2. The total flux through the Gaussian surface will again equal $E(4\pi r^2)$. The total charge within the sphere is the charge at the center plus the charge on the conducting sphere. There is no net charge on the conducting sphere (although there is $-Q$ on its inner surface and $+Q$ on its outer surface). Thus the total charge within the Gaussian surface is Q. Using Gauss' law gives $E(4\pi r^2) = Q/\varepsilon_0$, or $E = Q/4\pi\varepsilon_0 r^2 = kQ/r^2$.

(*e*) All of the charge Q' must appear on the surface of the conductor, and the arguments of part (*b*) still hold. Thus nothing is changed about parts (*a*) or (*c*), since they depend only on the central charge and the fact that the field inside the conductor vanishes. Part (*c*), however, does tell us that since $(-Q)$ appears on the inner surface of the conductor, $(Q' + Q)$ must appear on the outer surface. In part (*d*) the total charge within the Gaussian surface is changed to $(Q + (-Q) + Q + Q') = (Q + Q')$, since the conducting sphere now has a net charge Q'. Then the field outside the conducting sphere is $E = k(Q + Q')/r^2$.

Problem 3.30. Consider a uniformly charged sphere of radius R. The charge per unit volume is ρ.

(*a*) Calculate the field produced by this sphere at a point inside the sphere ($r < R$).

(*b*) If the charged sphere has a radius of 0.33 m, and carries a total charge of 8.6×10^{-6} C, calculate the field at $r = 0.10$ m and at $r = 0.40$ m.

Solution

(*a*) Because of the symmetry we know that E will be the same at all points at the same distance from the center of the sphere. If we choose a closed sphere of radius r as our "Gaussian surface" (see Fig. 3-19), we know that E will be the same at every point on the sphere. Furthermore we also know from symmetry that the field will be along a radius drawn from the center of the sphere. This direction is perpendicular to the sphere at every point so that E is parallel to **A**. To get the field within the sphere we draw this Gaussian surface at $r < R$. As in previous problems with this symmetry, the flux through the closed sphere will just equal E multiplied by the surface area of the sphere, or $F = E(4\pi r^2)$. By Gauss' law this total flux must equal the total charge within the Gaussian surface. This corresponds to a sphere of radius $r < R$. This charge q will equal ρ times the volume of the Gaussian sphere,

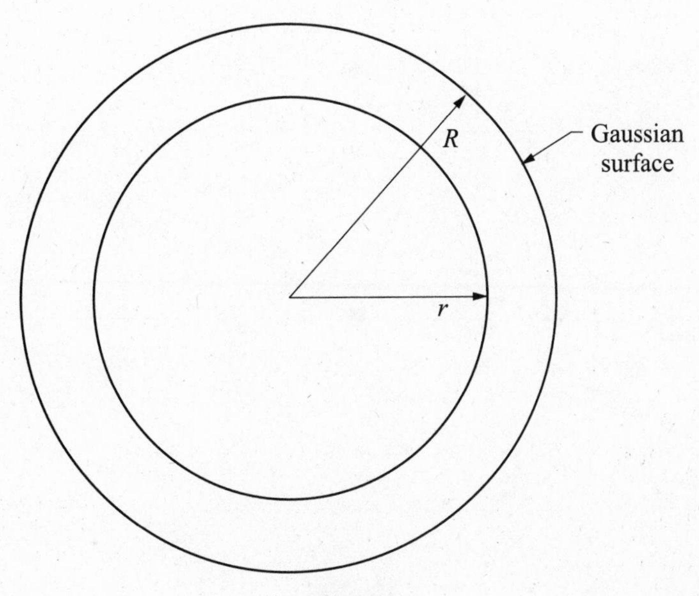

Fig. 3-19

$q = \rho(4\pi r^3/3)$. Eq. (3.9) then yields: $\sum [E(\cos\theta)\Delta A] = q/\varepsilon_0 \rightarrow E(4\pi r^2) = q/\varepsilon_0 = \rho(4\pi r^3/3)/\varepsilon_0$ and $E = \rho r/3\varepsilon_0$. The total charge on the charged sphere is $Q = \rho(4\pi R^3/3)$. Thus $E = kQr/R^3$ (where we recall $k = 1/4\pi\varepsilon_0$). The field therefore increases linearly from zero (at $r = 0$) until it reaches the edge of the sphere (at $r = R$). Then the field decreases as $1/r^2$, as we saw in problem 3.28.

(*b*) If the radius of the charged sphere is 0.33 m and the total charge is 8.6×10^{-6} C, then $\rho = (8.6 \times 10^{-6}\ \text{C})/[(4\pi/3)(0.33\ \text{m})^3 = 5.72 \times 10^{-5}\ \text{C/m}^3$. At $r = 0.10$ m, which is inside the sphere, the field will therefore be $E = \rho r/3\varepsilon_0 = 5.72 \times 10^{-5}(0.10)/3(8.85 \times 10^{-12}) = 2.16 \times 10^5$ N/C. At $r = 0.40$ m, which is outside the sphere, we use the formula developed in Problem 3.28, $E = kQ/r^2 = 9.0 \times 10^9\ (8.6 \times 10^{-6})/(0.40)^2 = 4.8 \times 10^5$ N/C.

Problem 3.31. Consider a long wire carrying a uniform charge per unit length of λ.

(*a*) Calculate the field produced by this wire at a distance of r from the axis of the wire.

(*b*) If the charged wire has a charge per unit length of 5.6×10^{-6} C/m, calculate the field at $r = 0.10$ m.

Solution

(*a*) Because of the symmetry we know that the field will be the same at all points at the same perpendicular distance r from the axis of the wire. We also know from symmetry that the field cannot have a component along the direction parallel to the wire since there is no difference between the direction to the right or the left. Similarly, the field cannot have a component that circulates around the wire since there is no difference between the two directions of circulation. Therefore, for any point, the direction of the field must be along the line radiating out perpendicularly from the axis of the wire to that point. This direction is shown for several points in Fig. 3-20. We choose as our Gaussian surface a cylinder of radius r and length L, with axis along the wire as in the figure. The closed surface is thus made up of two flat disks (end faces) and a cylindrical surface. On each disk the field is parallel to the surface and therefore $\mathbf{E} \perp \mathbf{A}$ ($\cos\theta = 0$), and the flux through any point is zero. On the outer surface the field is perpendicular to the surface at every point so that $\mathbf{E}$ is parallel to $\mathbf{A}$. Then the sum needed to calculate the flux through the outer surface will just equal E multiplied by the cylindrical surface area. That surface area is $2\pi rL$, and the total flux through the closed surface will equal $E(2\pi rL)$. By Gauss' law this total flux must equal the total charge within the Gaussian surface. The charge within the surface

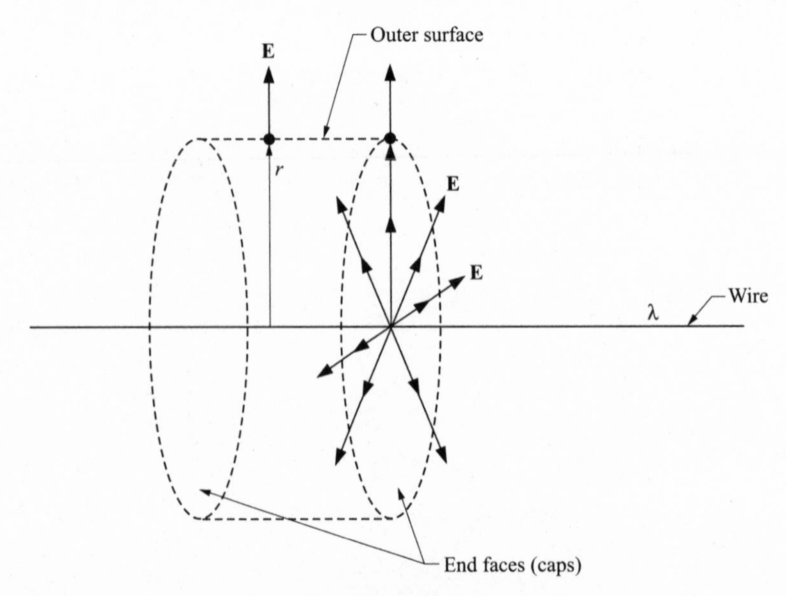

Fig. 3-20

will equal the charge on the wire that is within the volume of the Gaussian surface, i.e. within a cylinder of length L. This charge Q will equal λL. We now use Eq. (3.9), $\sum [E(\cos\theta)\Delta A] = Q/\varepsilon_0$, to get $E(2\pi rL) = Q/\varepsilon_0 = \lambda L/\varepsilon_0$ and $E = \lambda/2\pi\varepsilon_0 r$.

(*b*) Substituting in the equation for λ and r, we get: $E = (5.6 \times 10^{-6}\ \text{C/m})/2\pi(8.85 \times 10^{-12})(0.10\ \text{m}) = 1.01 \times 10^6\ \text{N/C}$.

Problem 3.32. Consider a large planar plate carrying a uniform charge per unit area of σ, as in Fig. 3-21.

(*a*) Calculate the field produced by this plate at a distance of x from the plate.

(*b*) If the charged plate has a charge per unit area of $2.1 \times 10^{-6}\ \text{C/m}^2$, calculate the field at a distance $x = 0.10$ m.

Solution

(*a*) Because of the symmetry we know that the field will be the same at all points at the same distance x from the plate as long as we are far from the edge of the plate. We also know from symmetry that the field cannot be along the direction parallel to the plate since there is no difference between any parallel direction. Therefore, the field must be along the direction perpendicular to the plane. We choose as our Gaussian surface the "pillbox" cylinder of base area A and length $2x$ perpendicular to the plate, as in the figure. The closed surface has two flat caps of area A, one on each side of the plate, and a cylindrical surface. On the cylindrical surface the field is parallel to the surface and therefore $\mathbf{E} \perp \mathbf{A}$ ($\cos\theta = 0$), and the flux through this surface is zero. On the caps the field is perpendicular to the surface at every point so that $\mathbf{E}$ is parallel to $\mathbf{A}$ ($\cos\theta = 1$). The field is directed away from the positive charge on the plate, and is therefore to the left on the left plate and to the right on the right plate. In each case the field will be in the same direction as $\mathbf{A}$, since we always choose $\mathbf{A}$ to point from the inside to the outside on the Gaussian surface. The flux through each cap will therefore equal EA, and the total flux through the closed surface will equal $2EA$. By Gauss' law this total flux must equal the total charge within the Gaussian surface. The charge within the surface will equal the charge on the plate that is within the volume of the Gaussian surface, i.e. within the cylinder of base area A. This charge Q will equal σA. We now use Eq. (3.9), $\sum [E(\cos\theta)\Delta A] = Q/\varepsilon_0$, to get $E(2A) = Q/\varepsilon_0 = \sigma A/\varepsilon_0$ and $E = \sigma/2\varepsilon_0$. We see that the field is constant, independent of x as long as we are far from the edge of the plate.

(*b*) Substituting in the equation for σ, we get: $E = (2.1 \times 10^{-6}\ \text{C/m}^2)/2(8.85 \times 10^{-12}) = 1.19 \times 10^5\ \text{N/C}$.

The examples in these problems illustrate the power of this technique in cases of appropriate symmetry. More examples will be discussed in the supplementary problems.

Problem 3.33. Consider a solid conducting object that has no net charge. Outside of the object there are charges and there are electric fields produced as a result of the presence of these charges.

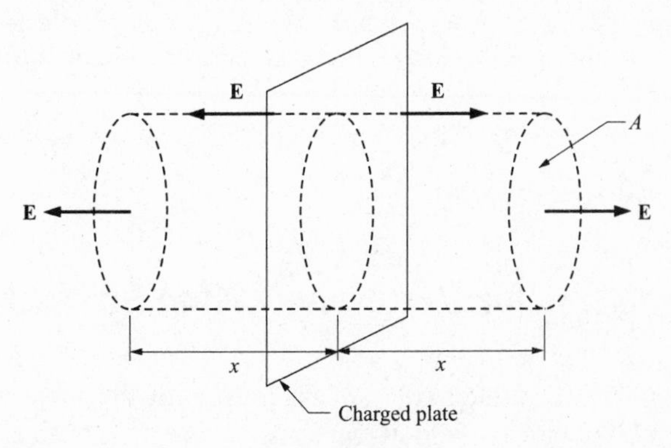

Fig. 3-21

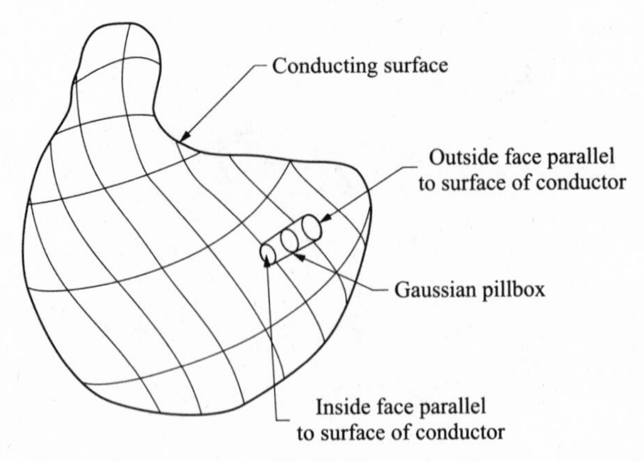

Fig. 3-22

(*a*) What can one say about the charge density in the conductor?

(*b*) What must the direction of the electric field be just outside the conductor?

(*c*) Use Gauss' law to find an expression for the electric field at any point just outside the conductor.

Solution

(*a*) Since no charges can accumulate in the interior of the conductor, all charges appear on the surface, with some surface distribution σ. Indeed σ adjusts itself from point to point on the surface to ensure the field inside the conductor vanishes, as required for equilibrium [see Problem 3.29(*b*)].

(*b*) While the field vanishes in the interior of the conductor, just up to the surface, fields can exist at the surface and beyond. At the surface, however, the fields cannot have a component parallel to the surface, otherwise the surface charges would feel a force and be in motion. Thus, in equilibrium, the field at the surface (and just beyond) must be perpendicular to the surface.

(*c*) We draw a tiny Gaussian "pillbox' about the point of interest on the surface of the conductor. The pillbox is drawn with the cylindrical part perpendicular to the conductor surface (which for small enough region is almost planar), and with one end face of the pillbox in the conductor and the other end just outside, as in Fig. 3-22. From part (*b*) the field just outside the surface is perpendicular to the end face of the pillbox so no flux passes through the cylindrical portion and the flux through the end face is just EA, where E is the field at the surface and A the area of the end face of the pillbox. From part (*a*) the field inside the conductor vanishes so no flux passes through the portion of the pillbox in the conductor. The *total* flux through the pillbox is thus EA. The charge enclosed by the pillbox is just σA, where σ is the surface charge density at (and near) the point P. From Gauss' law we have $EA = \sigma A/\varepsilon_0 \rightarrow E = \sigma/\varepsilon_0$. Clearly for σ positive the field points away from the conductor and for σ negative it points toward the conductor. Remember that this result is true *just* outside the conductor. As we move away from the conductor the field changes both in magnitude and in direction.

Problems for Review and Mind Stretching

Problem 3.34. Two particles with equal charges, q, are located at the corners of an equilateral triangle of side r, as in Fig. 3-23. Find the electric field at the third corner if (*a*) both charges are negative and (*b*) q_1 is positive and q_2 is negative.

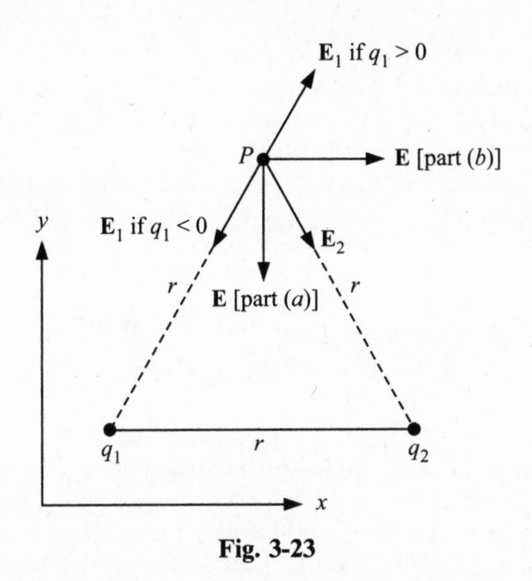

Fig. 3-23

Solution

(*a*) The magnitude of the field produced by each charge is $|E| = kq/r^2$. The direction of E_1 is toward q_1 and of $\mathbf{E}_2$ is toward q_2, as shown. The sum of these two fields is in $-y$, and equals $2[(kq/r^2)\cos 30°] = kq\sqrt{3}/r^2$.

(*b*) Again, the magnitude of the field produced by each charge is $|E| = kq/r^2$. The direction of $\mathbf{E}_1$ is away from q_1 and of $\mathbf{E}_2$ is toward q_2, as shown. The sum of these two fields is in $+x$, and equals $2[(kq/r^2)\sin 30°] = kq/r^2$.

Problem 3.35. Two square parallel plates are charged with a surface charge density of 4.7×10^{-6} C/m^2, with the top plate positive and the bottom plate negative. The plates are 6.2 mm apart, and have a side of 4.8 m, as shown in Fig. 3-24. An electron enters the region from the left side, at the midpoint between the plates and moving parallel to them. It is deflected by the field within the plates so that it just misses the edge of the plate as it emerges out the other side. Assume that the field is uniform everywhere within the plates and zero outside of the plates.

(*a*) What is the electric field within the plates?

(*b*) What is the acceleration of the electron?

(*c*) What was the initial velocity of the electron?

Solution

(*a*) The magnitude of the field produced by parallel plate capacitors was calculated in Problem 3.23 as $|E| = \sigma/\varepsilon_0 = (4.7 \times 10^{-6}\ \text{C/m}^2)/(8.85 \times 10^{-12}) = 5.31 \times 10^5$ N/C. The direction is from the positive charge to the negative charge, which is in $-y$.

(*b*) The acceleration of the electron is $F/m = eE/m = (1.6 \times 10^{-19}\ \text{C})(5.31 \times 10^5\ \text{N/C})/(9.1 \times 10^{-31}\ \text{kg}) = 9.34 \times 10^{16}\ \text{m/s}^2$. Since the electron has a negative charge the direction of the force (and acceleration) is opposite to that of the electric field and is in $+y$.

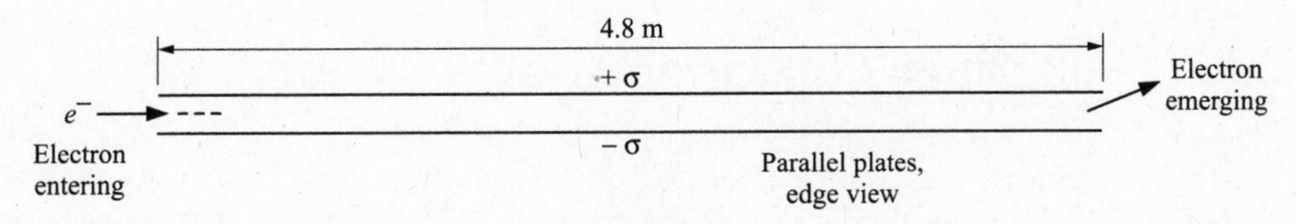

Fig. 3-24

(c) In the y direction the initial velocity is zero and the distance traveled before exiting the plates is 3.1 mm $= 3.1 \times 10^{-3}$ m. For constant acceleration, we can calculate the time needed from $y = (1/2)at^2$, or $t = (2y/a)^{1/2} = [(2)(3.1 \times 10^{-3})/(9.33 \times 10^{16})]^{1/2} = 2.58 \times 10^{-10}$ s. In the x direction, the electron travels at constant speed for a distance of 4.8 m during this same time. Then $x = vt$, and $v = x/t = (4.8)/(2.58 \times 10^{-10}) = 1.86 \times 10^{10}$ m/s. Since the electron is accelerated upward the electron will exit near the upper plate.

Problem 3.36. Four uniformly charged bars are arranged in a square of side 0.76 m, as in Fig. 3-25(a). Two adjacent sides have a positive charge of 4.9×10^{-7} C and the other two sides have the same negative charge. Calculate the field at the center of the square.

Solution

The field at the center is the sum of the fields produced by each of the four sides. The field from each side can be calculated from Eq. (3.6), $E = kQ \cos\theta/x^2$, where θ and x for a typical side are shown in Fig. 3-25(b). For each of the sides, $x = 0.38$ m and $\theta = 45°$. Therefore each side produces a field at the center whose magnitude is $|E| = (9.0 \times 10^9)(4.9 \times 10^{-7})(\cos 45°)/0.38)^2 = 2.16 \times 10^4$. The direction of the field of each side is toward a negatively charged side and away from a positively charged side. Thus the fields from the right and from the left sides are both to the right at P, and those two, when added together produce a field of 4.32×10^4 N/C at P. Similarly, the fields from the top and bottom sides are both downward and add to 4.32×10^4 downward. Adding these two fields together results in a field of $4.32 \times 10^4\sqrt{2} = 6.11 \times 10^4$ N/C pointed towards the lower right corner.

Problem 3.37. What electric field is produced by a long straight wire charged with a linear charge density of λ C/m?

Solution

The field is the field produced by a rod of length L as we let the length L approach infinity. The field from a rod is given by Eq. (3.6), $E = 2k\lambda L/[x(L^2 + x^2)^{1/2}]$, where L and x are shown in Fig. 3-25(b). As $L \to \infty$, the numerator and denominator both approach infinity. We divide both numerator and denominator by L in order to be able to evaluate the limit, and get $E = 2k\lambda/[x(1 + (x/L)^2)^{1/2}]$. Now, as $L \to \infty$, $E \to 2k\lambda/x$, which is the field of a long wire. The field points away from the wire for a positive λ, and toward the wire for a negative λ. This is the same result as in Problem 3.31, obtained using Gauss' law.

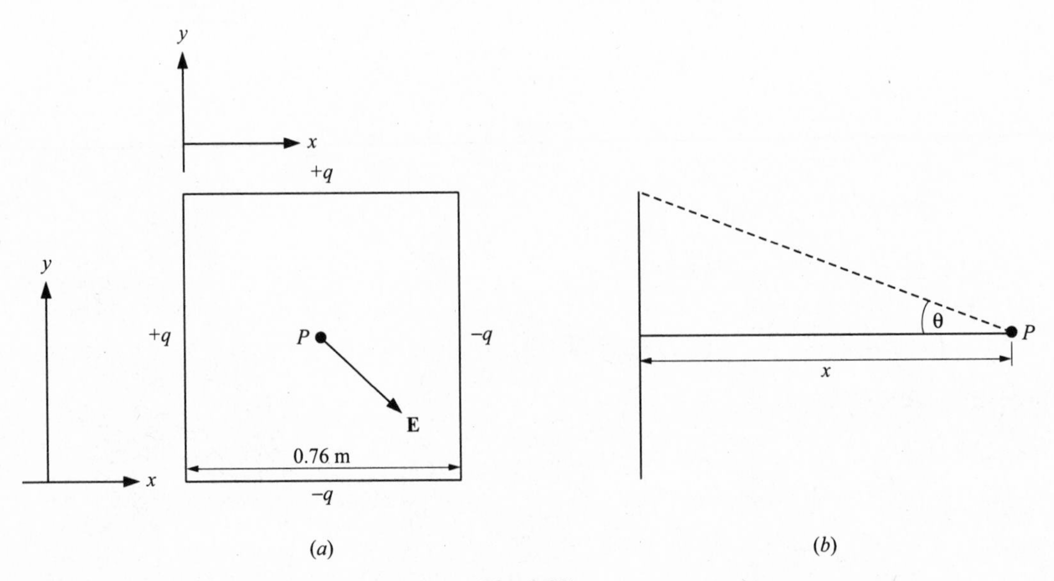

Fig. 3-25

Problem 3.38. Two huge parallel plates are separated by a distance of 1.3 m. The left plate is charged positively, with 5.4×10^{-6} C/m^2, and the right plate is charged negatively with 7.9×10^{-6} C/m^2; see Fig. 3-26(*a*). Assume that the plates are large enough that the field is unaffected by the edges of the plate.

(*a*) What is the electric field at a point to the left of the left plate?

(*b*) What is the field at a point between the plates?

(*c*) What is the field at a point to the right of the right plate?

Solution

(*a*) The field at any point is the sum of the fields produced by each of the plates. For a large plate the field of each plate is $\sigma/2\varepsilon_0$ everywhere, and pointing away from a positively charged plate and toward a negatively charged plate (Problem 3.32). Thus the field from the left plate equals $E_l = (5.4 \times 10^{-6})/2(8.85 \times 10^{-12}) = 3.05 \times 10^5$ N/C and from the right plate the field is $E_r = (7.9 \times 10^{-6})/2(8.85 \times 10^{-12}) = 4.46 \times 10^5$ N/C. In the region to the left of the left plate E_l is to the left and E_r is to the right. The sum of these fields is $E = 1.41 \times 10^5$ N/C to the right.

(*b*) In the region between the plates both E_l and E_r are to the right, so the field is $E = 7.51 \times 10^5$ N/C to the right.

(*c*) In the region to the right of the right plate E_l is to the right and E_r is to the left. The total field is therefore 1.41×10^5 N/C to the left.

Problem 3.39. In Problem 3.38, a large uncharged conducting plate is placed between the charged plates, see Fig. 3-26(*b*).

(*a*) Show that the electric field is unchanged everywhere except within the conducting plate.

(*b*) What is the surface charge density on the two surfaces of the conducting plate?

Solution

(*a*) The field at any point is the sum of the fields produced by the charges on each of the plates. On the conducting plate there will be charge on each of the two surfaces, but one will be positively charged

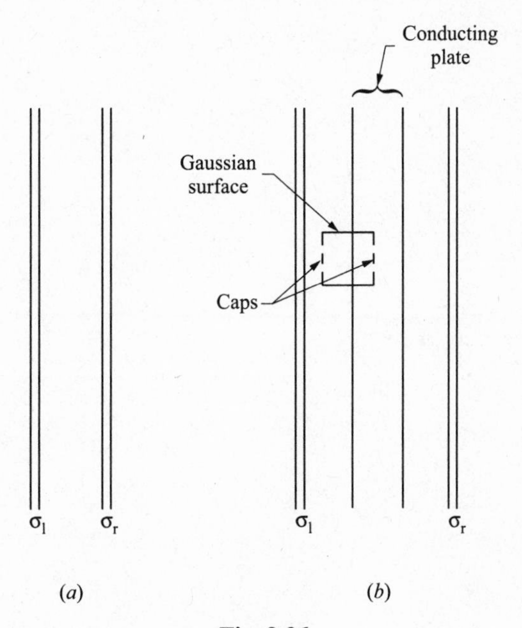

Fig. 3-26

and the other one will be negatively charged. The charges will be equal in magnitude since the plate has no net charge. Therefore the conducting plate will contribute no net field anywhere except within the conducting plate since each surface contributes an equal and opposite field. Thus the field is the same as it would be everywhere except within the conducting plate.

(*b*) In the region within the conducting plate the field must be zero. Let us draw a Gaussian surface as a cylinder with one cap within the conducting plate and the other cap between the left plate and the conducting plate, as in the figure. The only part of this surface that contributes to the flux is this last cap. On the inner cap the field is zero and on the outside surface of the cylinder the field is parallel to the surface. The field at the outer cap is 7.51×10^5 N/C to the right as in Problem 3.38(*b*). The area vector points out of the Gaussian surface which means to the left. Thus the flux through this surface is -7.51×10^5 A, which is also the total flux. By Gauss' law, this must equal the total charge within the Gaussian surface divided by ε_0. This charge is $\sigma A/\varepsilon_0$. Then $\sigma A = -6.64 \times 10^{-6}$ A, and $\sigma = -6.64 \times 10^{-6}\text{C/m}^2$ on the left side of the conducting plate, and $\sigma = +6.64 \times 10^{-6}\text{C/m}^2$ on the right side of the conducting plate. Note that this is just the special application of the general result relating the surface charge on a conductor to the field just outside (Problem 3.33)

Problem 3.40.

(*a*) Two equal charges, one positive and one negative are separated by a distance *d*. Sketch the electric field lines that result from this.

(*b*) Repeat the above if both charges are positive.

Solution

(*a*) Field lines originate on positive charges and terminate on negative charges. If there were only one charge present the lines would be straight lines along the radii from the charge, directed away from the positive (or toward the negative) charge. If both charges are present the lines from each cannot cross each other as we showed earlier. They must either bend out of each other's way or connect to each other. In the case of the two equal but oppositely charged sources they easily connect to each other as is seen in Fig. 3-27(*a*).

(*b*) For two charges of the same polarity the lines cannot combine since they are both directed away from the charges. Thus they bend out of each other's way as in Fig. 3-27(*b*).

Problem 3.41. A long cylinder of radius *R* is uniformly charged with a charge density ρ, as in Fig. 3-28.

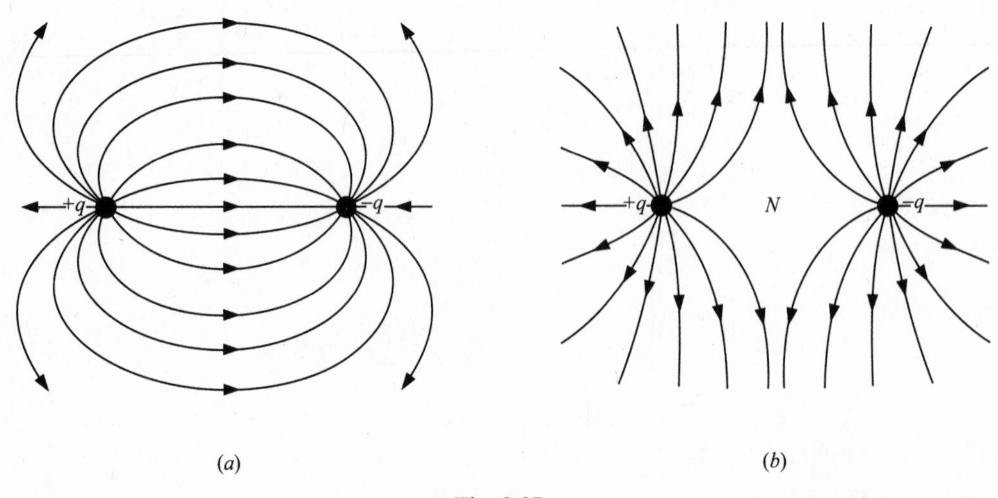

Fig. 3-27

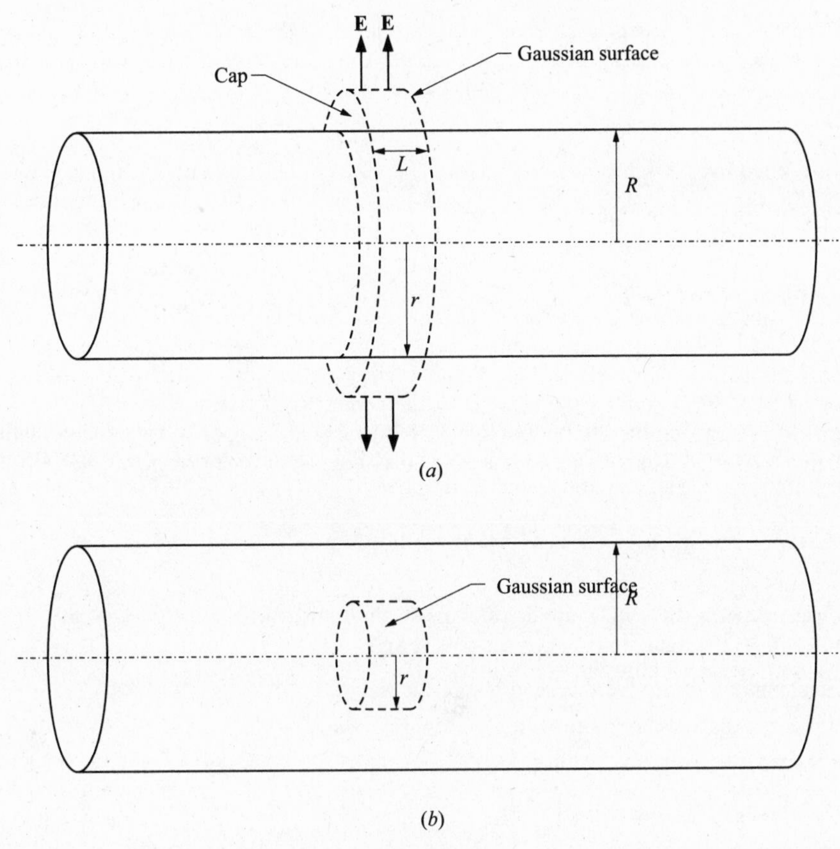

Fig. 3-28

(*a*) Use Gauss' law to calculate the electric field at a point outside the cylinder at a distance r from the axis of the cylinder, i.e. for $r > R$.

(*b*) Repeat the above for a point within the cylinder, i.e. for $r < R$.

Solution

(*a*) We draw a Gaussian surface in the form of a cylinder of radius r and length L, as in Fig. 3-28(*a*). As in Problem 3.31, the field must always be in the radial direction perpendicular to the axis of the cylinder. Therefore the field is parallel to the surface on the two caps of the cylinder and those two surfaces do not contribute to the flux entering or leaving the surface. On the outer surface of the cylinder, the field is perpendicular to the surface (A is parallel to **E**), and the flux equals $EA = E(2\pi rL)$. By Gauss' Law, this equals Q_{in}/ε_0, and $Q_{in} = \rho(\pi R^2 L)$. Then $E = \rho R^2/2\varepsilon_0 r$.

(*b*) In this case we draw a Gaussian surface as a cylinder with a radius r within the cylinder as in Fig. 3-27(*b*). Again, the flux through the closed surface will equal $E(2\pi rL)$, and must equal Q_{in}/ε_0. In this case $Q_{in} = \rho\pi r^2 L$, and therefore, $E = \rho r/2\varepsilon_0$.

Supplementary Problems

Problem 3.42. A force of 1.6 N acts to the right on a charge of 3.8×10^{-6} C at the origin, due to another charge at $x = 0.058$ m. What is the other charge?

Ans. -1.57×10^{-7} C

Problem 3.43. Two charges, q_1 and q_2 exert forces on a charge of 6.9×10^{-5} C which is located at the origin. The charge q_1 is -9.8×10^{-4} C and is located at $x = 0.88$ m. The total force on the charge is 603 N in the $+x$ direction.

(*a*) What force is exerted by q_1?

(*b*) What force is exerted by q_2?

Ans. (*a*) 786 N, in $+x$ direction; (*b*) 183 N in $-x$

Problem 3.44. In Problem 3.43 the charge q_2 is located at $x = 0.085$ m. What is q_2?

Ans. $+2.12 \times 10^{-6}$ C

Problem 3.45. An electric field at the origin has a magnitude of 845 N/C, and is directed at an angle of $+65°$ with the negative x axis. A charge of magnitude 0.34 C is located at the origin. What are the x and y components of the force on the charge if the charge is: (*a*) positive and (*b*) negative?

Ans. (*a*) $F_x = -121$ N, $F_y = 260$ N; (*b*) $F_x = 121$ N, $F_y = -260$ N

Problem 3.46. A charge of -0.061 C is initially moving to the right with a velocity of 54 m/s. It is moving in an electric field of 888 N/C to the right, and has a mass of 0.72 kg.

(*a*) What is the maximum distance the charge moves to the right?

(*b*) What velocity does it have when it returns to the origin?

(*c*) What velocity does it have at $x = 5.75$ m?

Ans. (*a*) 19.4 m (*b*) -54 m/s; (*c*) ± 45.3 m/s

Problem 3.47. A proton near the surface of the earth is in equilibrium under the force of gravity and the force of an electric field. What electric field is required? [$m_p = 1.67 \times 10^{-27}$ kg]

Ans. 1.02×10^{-7} N/C, upwards

Problem 3.48. A charge of -1.97×10^{-5} C produces an electric field of 740 N/C at the origin, in the $+x$ direction. Where is the charge located?

Ans. $x = +15.5$ m

Problem 3.49. Two charges, one at the origin and one at $x = 1.6$ m produce a field of 1.59×10^3 N/C at $x = 9.11$ m. The charge at the origin has a charge of 1.11×10^{-5} C. How much charge does the second charge have?

Ans. 2.44×10^{-6} C

Problem 3.50. Two charges, one at the origin and one at $x = 1.6$ m produce a field 785 N/C at $x = 9.11$ m. The charge at the origin has a charge of 1.11×10^{-5} C. How much charge does the second charge have?

Ans. -2.62×10^{-6} C

Problem 3.51. Three charges are at the corners of a square of side 2.0 m, as in Fig. 3-29. The charges are $q_1 = 5.0 \times 10^{-6}$ C, $q_2 = 3.0 \times 10^{-6}$ C, and $q_3 = -6.0 \times 10^{-6}$ C. Find the field at the fourth corner, produced by: (*a*) q_1, (*b*) q_2, (*c*) q_3 and (d) all the charges together.

Ans. (*a*) 5.62×10^3 N/C at 45° as shown; (*b*) 6.75×10^3 N/C in $+y$; (*c*) 1.35×10^4 N/C in $-x$; (*d*) 1.43×10^4 N/C at angle of $+48°$ from $-x$ axis.

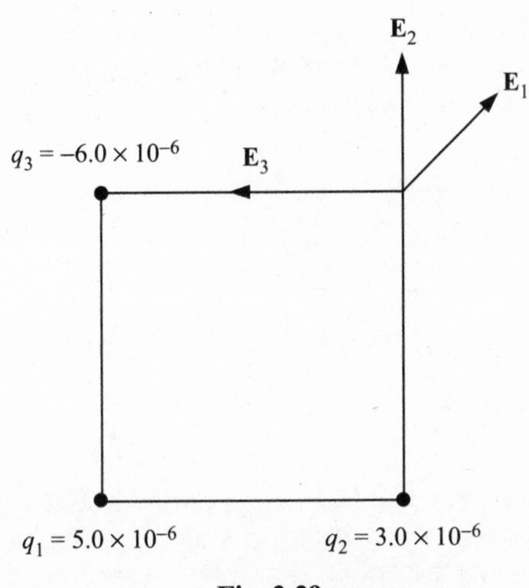

Fig. 3-29

Problem 3.52. Two positive charges on the x-axis produce a total field of zero at the origin. One charge q_1 has twice the charge of the other, i.e. $q_1 = 2q_2$. What is the ratio of the coordinates of the charges, x_1/x_2?

Ans. $-\sqrt{2}$

Problem 3.53. Two positive charges of 2.3×10^{-6} C are placed at diagonally opposite corners of a square of side 3.6 m.

(*a*) What field is produced by these two charges at the third corner?

(*b*) Where should one place a third equal charge in order to get a field of zero at the origin?

Ans. (*a*) 2.26×10^3 N/C along the other diagonal; (*b*) 3.03 m from the third corner along the diagonal [direction of the field of part (*a*)] in the direction away from the square.

Problem 3.54. Two equal and opposite charges $\pm q$, are located, respectively at $\pm d/2$ on the x axis (they are separated by the small distance d). Assume that the distance d is small compared with any other distances in the problem, and that $qd = p$.

(*a*) What is the electric field at a large distance r from the charges along the positive y axis?

(*b*) What is the electric field at a large distance r from the charges along the positive x axis?

Ans. (*a*) kp/r^3 in $-x$ direction; (*b*) $2kp/r^3$ in $+x$ direction

Problem 3.55. A ring of radius 0.96 m has a charge of -6.7×10^{-7} C uniformly distributed along its circumference. What field does it produce at a point on its axis at a distance of 1.35 m from the plane of the ring?

Ans. 1.79×10^3 N/C pointing toward the ring

Problem 3.56. Two long parallel wires, each containing the same positive charge of 3.8×10^{-6} C/m, are a distance of 1.25 m apart. What electric field do they produce at a point located at a distance of 0.75 m from one wire, if that point is: (*a*) between the wires and (*b*) further away from the other wire?

Ans. (*a*) 4.56×10^4 N/C, toward the further wire; (*b*) 1.25×10^5 N/C, away from both wires

Problem 3.57. Two long parallel wires, each containing a charge of 3.8×10^{-6} C/m but of opposite sign, are a distance of 1.25 m apart. What electric field do they produce at a point located at a distance of 0.75 m from the positive wire, if that point is: (*a*) between the wires and (*b*) further away from the other wire?

Ans. (*a*) 2.28×10^5 N/C, toward the negative wire; (*b*) 5.70×10^4 N/C, away from both wires

Problem 3.58. Two long parallel wires are a distance of 1.25 m apart. One wire has a positive charge of 3.8×10^{-6} C/m. What linear charge density must the other wire have if the field is to be zero at a point located at a distance of 0.75 m from the first wire, if that point is: (*a*) between the wires and (*b*) further away from the other wire?

Ans. (*a*) $+2.53 \times 10^{-6}$ C/m; (*b*) -1.01×10^{-5} C/m

Problem 3.59. Three large parallel dielectric plates have surface charge densities of 1.3×10^{-8} C/m^2, -3.5×10^{-8} C/m^2 and 2.9×10^{-8} C/m^2. What electric field is produced at a point: (*a*) to the left of all the plates, (*b*) between the first and second plates, (*c*) between the second and third plates and (*d*) to the right of all the plates?

Ans. (*a*) 395 N/C to the left; (*b*) 1.07×10^3 N/C to the right; (*c*) 2.88×10^3 N/C to the left; (*d*) 395 N/C to the right

Problem 3.60. A large dielectric plate has a surface charge density of 1.30×10^{-6} C/m^2 and is parallel to a long wire that has a linear charge density of -9.8×10^{-7} C/m (see Fig. 3-30). The plate and wire are separated by a distance of 0.035 m. What magnitude electric field is produced at a point: (*a*) P_1, at a distance of 0.020 m from the wire along the line perpendicular to the plate and (*b*) P_2, at a distance of 0.40 m from the wire along a line parallel to the plate?

Ans. (*a*) 9.55×10^5 N/C to the right; (*b*) 8.57×10^4 N/C

Problem 3.61. A charge of 2.3×10^{-7} C is at the center of a tetrahedron (see Fig. 3-31). What is the electric flux through one of the four sides?

Ans. $6.50 \times 10^3 \text{C} \cdot \text{m}^2$

Problem 3.62. A spherical conducting sphere has a radius of 2.3×10^{-3} m and carries a uniform charge of 3.5×10^{-8} C. It is surrounded by a concentric hollow conducting sphere of inner radius 0.055 m and outer radius 0.075 m.

(*a*) What is the electric field at $r = 2.1 \times 10^{-3}$ m?

(*b*) What is the electric field at $r = 4.1 \times 10^{-2}$ m?

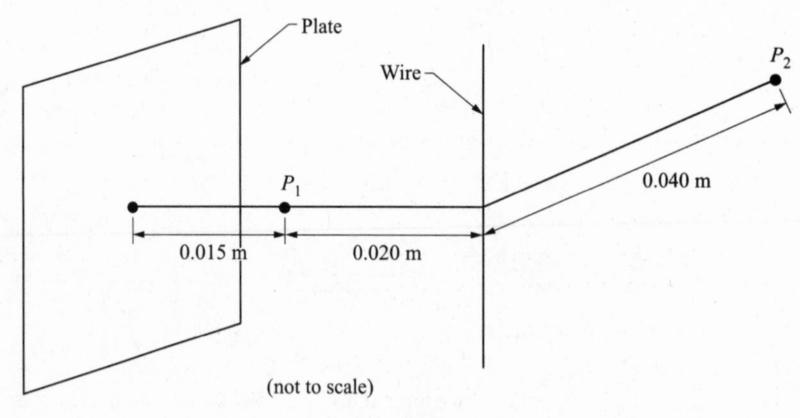

Fig. 3-30

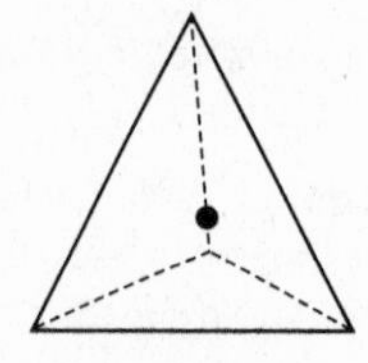

Fig. 3-31

(*c*) What is the electric field at $r = 6.1 \times 10^{-2}$ m?

(*d*) What is the electric field at $r = 8.1 \times 10^{-2}$ m?

(*e*) What is the charge on the inner surface of the hollow sphere?

Ans. (*a*) 0; (*b*) 1.87×10^5 N/C; (*c*) 0; (*d*) 4.80×10^4 N/C; (*e*) -3.5×10^{-8} C

Problem 3.63. A charge Q is surrounded by a concentric hollow conducting sphere of inner radius 0.055 m and outer radius 0.075 m. At large r the field is measured to be $E = 345/r^2$ N/C.

(*a*) What is the charge Q?

(*b*) What is the surface charge density on the inner surface and on the outer surface of the hollow sphere?

Ans. (*a*) 3.83×10^{-8} C; (*b*) -1.01×10^{-6} C/m^2, 5.42×10^{-7} C/m^2

Problem 3.64. A sphere of radius 0.055 m is uniformly charged at a density of 7.3×10^{-6} C/m^3.

(*a*) What is the total charge Q on the sphere?

(*b*) What is the field at $r = 0.044$ m?

(*c*) What is the field at $r = 0.066$ m?

Ans. (*a*) 5.09×10^{-9} C; (*b*) 1.21×10^4 N/C; (*c*) 1.05×10^4 N/C

Problem 3.65. A sphere of radius 0.055 m is uniformly charged and produces a field of 2.5×10^4 N/C at a distance of 1 m from the center of the sphere.

(*a*) What is the total charge Q on the sphere?

(*b*) What is the charge density in the sphere?

(*c*) What is the field at $r = 0.026$ m?

Ans. (*a*) 2.78×10^{-6} C; (*b*) 3.99×10^{-3} C/m^3; (*c*) 3.90×10^6 N/C

Problem 3.66. An electric field is pointing in the $+x$ direction, and is uniform with a magnitude of 467 N/C. What is the flux through a planar area of 7.3×10^{-2} m^2 if the orientation of this area is: (*a*) in the yz plane; (*b*) in the xy plane and (*c*) at an angle of 18° from the yz plane (see, e.g. Fig. 3-32)?

Ans. (*a*) 34.1; (*b*) 0; (*c*) 32.4

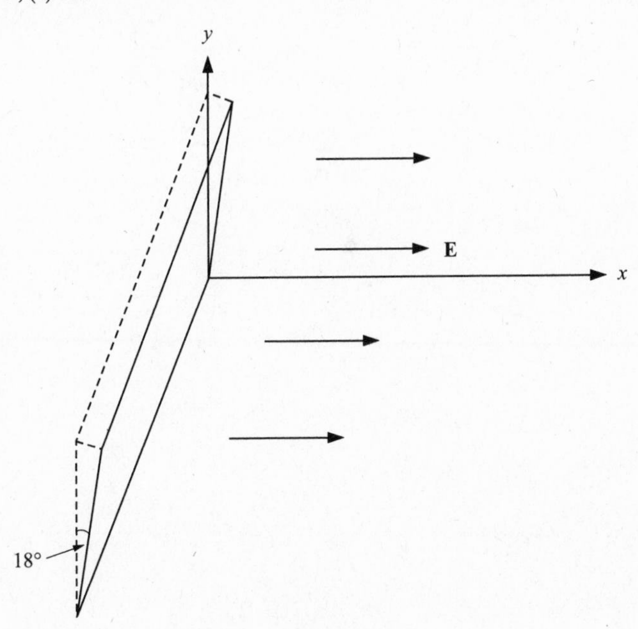

Fig. 3-32

Problem 3.67. A long conducting wire carries a charge of $\lambda = 6.8 \times 10^{-10}$ C/m, and has a radius of 0.45 cm. It is surrounded by a long conducting hollow cylinder of inner radius 0.65 cm and outer radius 0.67 cm, that carries a charge of $-\lambda$.

(*a*) What is the electric field within the long wire?

(*b*) What is the field between the wire and the cylinder, at $r = 0.50$ cm?

(*c*) What is the field outside the cylinder?

Ans. (*a*) 0; (*b*) 2.45×10^3 N/C; (*c*) 0

Problem 3.68. A long conducting wire carries a charge of $\lambda = 6.8 \times 10^{-10}$ C/m, and has a radius of 0.45 cm. It is surrounded by a long conducting hollow cylinder of inner radius 0.65 cm and outer radius 0.67 cm, that carries a charge of -4.5×10^{-10} C/m.

(*a*) What is the electric field within the long wire?

(*b*) What is the field between the wire and the cylinder, at $r = 0.50$ cm?

(*c*) What is the field outside the cylinder, at $r = 0.88$ cm?

Ans. (*a*) 0; (*b*) 2.45×10^3 N/C; (*c*) 470 N/C

Chapter 4

Electric Potential and Capacitance

4.1 POTENTIAL ENERGY AND POTENTIAL

In the previous chapter we learned about the force due to the electrical interaction and the electric field concept used to describe that force. The interaction is very similar to the interaction of masses with each other described by the gravitational interaction. Forces in general, as we learned in Chap. 6 of Beginning Physics I, Sec. 6.3, are able to do work, and the work that they do can be transformed into kinetic energy. For forces that are "conservative" the work done can be expressed in terms of a change in potential energy associated with those forces. In the case of the gravitational force due to the Earth, for example, the potential energy is given by $U_p = \text{mgh}$ near the surface of the earth (where the force of gravity is a constant) and, more generally, $U_p = -GmM/r$ for greater distances r from the center of the earth. When some forces are conservative and others are not, the work–energy theorem can be expressed as total work (non-conservative) equals the total change in kinetic energy plus the total change in potential energy (due to all conservative forces). We now consider the electrical force. Is this force also conservative, and, if so, what is its potential energy?

Problem 4.1. By analogy to the force of gravitation (*a*) show that the electric force is conservative and (*b*) derive the formula for the potential energy of two charges, q and Q, separated by a distance r.

Solution

(*a*) The force of gravity is given in magnitude by $F_g = GmM/r^2$, and is a force of attraction along the line joining the masses. The electrical force between charges q and Q is given in magnitude by $F_e = kqQ/r^2$, and is a force along the line joining the charges. This force is attractive for charges of opposite sign and negative for charges of the same sign. When this force is attractive it is identical to the force of gravity if one interchanges charges for masses and the constant k for G. Therefore, it is clearly also conservative just as the force of gravity is conservative. If the force is between charges of the same sign, so that the force is repulsive, the work done by the force is the same as would be done by the same charges if they were of opposite sign, except that the work is the negative of that done by the attractive force. Since the attractive force is conservative, the work however depends only on the starting and ending points and not on what happened in between. This will also be true of the repulsive force which is therefore also conservative. Therefore the electric force is conservative, and work can be written in the form of a change in potential energy.

(*b*) By analogy with the force of gravity the potential energy can be written down immediately by substituting k for G and $-qQ$ for mM. We need the minus sign because for two positive charges the work is of the opposite sign to that two positive masses. The potential energy of two charges q and Q separated by a distance r is then given by:

$$U_p = kqQ/r = (1/4\pi\varepsilon_0)qQ/r \tag{4.1}$$

A quick examination of signs shows that this equation works for arbitrary sign charges.

This formula can be used to calculate the potential energy for arbitrary sets of charges. This follows because energy is a scalar, and the total potential energy is determined by adding together, algebraically, the potential energy between pairs of charges.

We note that in Eq. (*4.1*) the zero of potential energy has been chosen when $r \to \infty$. If the charges are of the same sign then the potential energy increases as the charges approach each other. This follows because an external force must do positive work in forcing the charges closer together against their mutual repulsion. When such charges are left to themselves they try to move to regions of lower potential energy. This corresponds to the fact that the repulsive electrical force now does positive work by moving the charges further apart, thus causing a decrease in their potential energy. If the charges are of opposite sign then the potential energy becomes more negative (decreases) as the charges approach each other, and less negative (increases) as they are forced further apart. If left to themselves, these charges would move closer, seeking regions of lower potential energy.

If we fix the position of one charge, Q, and allow the second charge, q, to move, then the potential energy will vary with the position of the second particle. One could say that the system changes its potential energy and that this change in potential energy depends on the change in the position of the second charge. We could associate a specific potential energy with each point in space in a manner similar to associating an electric field to each point in space. From Eq. (*4.1*) we note that this potential energy is proportional to the moving charge. The potential energy per unit charge, U_p/q, then depends only on the position of the moving charge, as well as on the magnitude and sign of the stationary charge. Similarly, if one had many stationary charges, the potential energy of the entire system changes as the moving charge goes from one point to another, and is proportional to this moving charge. Again, the potential energy per charge depends only on the position of the moving charge and on the characteristics of the stationary charges. We can view this as a situation in which the stationary charges provide each point in space with a scalar value, called the potential, V, such that the potential energy of the system will equal qV if the moving charge is at that point in space. (We ignore here the potential energy between the fixed charges, which remains unchanged as the charge q moves.) The unit for potential V is the **volt** (V), which is the same as J/C. As the charge moves there will be a change in **potential energy**, ΔU_p which will equal q times the change in the potential at each point. In summary:

$$U_p = qV, \tag{4.2a}$$

and

$$\Delta U_p = q\Delta V \tag{4.2b}$$

The quantity ΔV is the "**potential difference**" between the two points, and depends on the stationary charges Q_i that produce this potential at all points in space. It is independent of the characteristics of the moving charge, q, whose potential energy changes. The potential is related to the potential energy in the same manner that the electric field is related to the electric force. Whenever an electric field is produced by some set of charges, Q_i, it acts as the source of the force distribution in space; it also can be thought of as the source of the potential distribution in space. If one places another charge, q, at some position in space, the electric field will exert a force of $\mathbf{F} = q\mathbf{E}$ on the charge, and the system will have a potential energy of $U_p = qV$, where $\mathbf{E}$ and V are the field and the potential at that point. The work done by the force $\mathbf{F} = q\mathbf{E}$ in moving the charge q from one location to another is just $-\Delta U_p = -q\Delta V$, from the usual relationship between work and potential energy. Clearly $\mathbf{E}$ and ΔV are related in exactly the same way that $\mathbf{F}$ and ΔU_p are related. This is discussed in greater detail in Sect. 4.3. One can change $\mathbf{E}$ and V by changing the source charges, Q_i and their position.

Problem 4.2. Two charges, $Q_1 = 3.3 \times 10^{-6}$ C and $Q_2 = -5.1 \times 10^{-6}$ C are located at the origin and at $x = 0.36$ cm, respectively. A third charge, $q = 9.3 \times 10^{-7}$ C, is moved from far away ($r = \infty$) to a point on the y axis, $y = 0.48$ cm.

(*a*) What is the potential energy between q and Q_1 at this point?

(*b*) What is the potential energy between q and Q_2 at this point?

(*c*) What is the change in potential energy of the system as one moves q from far away to this point?

(*d*) What is the potential difference between the point at ∞ and this point?

Solution

(*a*) The potential energy between any two charges is kqQ/r. Thus the potential energy between q and Q_1 is $U_p = (9.0 \times 10^9)(9.3 \times 10^{-7}\ \text{C})(3.3 \times 10^6\ \text{C})/0.48 \times 10^{-2}\ \text{m} = 5.75$ J.

(*b*) The distance between q and Q_2 is $(0.36^2 + 0.48^2)^{1/2}$ cm = 0.60 cm. Thus the potential energy between q and Q_2 is $U_p = (9.0 \times 10^9)(9.3 \times 10^{-7}\ \text{C})(-6.3 \times 10^{-6}\ \text{C})/0.60 \times 10^{-2}\ \text{m} = -8.79$ J.

(*c*) When q is far away the potential energy between q and each of the charges Q is zero. There is potential energy of the system between Q_1 and Q_2, but that potential energy does not change as one moves q from point to point. As one moves the charge q to the final point the potential energy changes because of the interaction between q and the Q. The final potential energy is $U_p = 5.75\ \text{J} - 8.79\ \text{J} = -3.04$ J. Therefore $\Delta U_p = -3.04 - 0 = -3.04$ J.

(*d*) Since $\Delta V = \Delta U_p/q$, the potential difference is $\Delta V = -3.26 \times 10^6$ V.

4.2 POTENTIAL OF CHARGE DISTRIBUTIONS

The previous problem illustrated how to calculate the potential energy in the case of two fixed point charges and a moving charge, and then how to use that potential energy to obtain the potential. We can clearly use this procedure to calculate the potential produced by any number of point charges at all points in space. We can thus calculate the potential produced by a collection of particles or by a distribution of charge.

Problem 4.3. Calculate the potential produced by a point charge Q located at the origin at a point distant from the charge by r.

Solution

Our method is to calculate the potential energy, U_p, at the desired point if one places a "test charge" q at that point. Then the potential will equal U_p/q. Using Eq. (*4.1*), we get $U_p = kqQ/r$, and then:

$$V = kQ/r = (1/4\pi\varepsilon_0)Q/r \tag{4.3a}$$

This is the potential produced by a single charge Q at a point that is distant from the charge by r. If we have a collection of charges, Q_i, then the potential will equal:

$$V = k \sum Q_i/r_i = (1/4\pi\varepsilon_0) \sum Q_i/r_i \tag{4.3b}$$

Problem 4.4. A charge of 1.75×10^{-6} C is placed at the origin. Another charge of -8.6×10^{-7} C is placed at $x = 0.75$ m.

(*a*) What is the potential at a point halfway between the charges?

(*b*) What is the electric field at that point?

(*c*) If an electron is placed at that point, what force acts on it, and how much potential energy does it have?

Solution

(*a*) The potential equals $k \sum Q_i/r_i$. Thus $V = (9.0 \times 10^9)[(1.75 \times 10^{-6}\ \text{C}/0.375\ \text{m}) + (-8.6 \times 10^{-7}\ \text{C}/0.375\ \text{m})] = 2.14 \times 10^4$ V. Since V is a scalar we were able to add the values algebraically.

(*b*) To calculate the electric field we must calculate the magnitude and direction of the fields produced by each source and then add them vectorially. Thus $E = \mathbf{E}_1 + \mathbf{E}_2$. Now $|\mathbf{E}_1| = kQ_1/r^2 = (9.0 \times 10^9)(1.75 \times 10^{-6}\ \text{C})/0.375^2 = 1.12 \times 10^5$ N/C. Since Q_1 is positive this field is directed along $+x$. Similarly, $|\mathbf{E}_2| = (9.0 \times 10^9)(8.6 \times 10^{-7}\ \text{C})/0.375^2 = 5.50 \times 10^4$ N/C. Since Q_2 is negative, the field points toward Q_2 which is also in the $+x$ direction. Then the total field is 1.67×10^5 N/C in $+x$.

(*c*) An electron has a charge of -1.6×10^{-19} C. Therefore the force on it is $\mathbf{F} = q\mathbf{E} = (1.6 \times 10^{-19}\ \text{C})(1.67 \times 10^5\ \text{N/C}) = 2.67 \times 10^{-14}$ N. The direction is opposite to $\mathbf{E}$ since q is negative, so $\mathbf{F}$ is in $-x$. The potential energy is $qV = (-1.6 \times 10^{-19}\ \text{C})(2.14 \times 10^4\ \text{V}) = -3.42 \times 10^{-15}$ J.

Problem 4.5. Refer to the two fixed charges of Problem 4.4. At what two points on the x axis is the potential zero?

Solution

If the point of zero potential is between the charges, and the distance from the origin to the point is x, then the first charge is at a distance of x and the second charge is at a distance $(0.75 - x)$ from the point. The total field is $k[Q_1/x + Q_2/(0.75 - x)] = 0$. Q_1 is positive and Q_2 is negative. Substituting for the charges, we get: $(1.75 \times 10^{-6}/x) = 8.6 \times 10^{-7}/(0.75 - x)$. Then $(0.75 - x) = 0.49x$, $1.49x = 0.75$, $x = 0.50$ m. If the point of zero potential is not between the charges, and the distance from the origin to the point of zero potential is x, then the first charge is at a distance of x and the second charge is at a distance $(x - 0.75)$ from the point. (Recall that in Eq. (*4.3a*), r is always positive.) The total field is $k[Q_1/x + Q_2/(x - 0.75)] = 0$. Again, Q_1 is positive and Q_2 is negative. Substituting values for the charges, we get: $(1.75 \times 10^{-6}/x) = 8.6 \times 10^{-7}/(x - 0.75)$. Then $(x - 0.75) = 0.49x$, $0.51x = 0.75$, $x = 1.47$ m. A quick check for finite points on the negative x axis shows that the potential cannot vanish there. Of course, the potential also vanishes at $x \to \pm\infty$.

Problem 4.6. Four equal charges of 5.7×10^{-7} C are placed on the corners of a square whose side has a length of 0.77 m.

(*a*) What is the electric field at the center of the square?

(*b*) What is the electric potential at the center of the square?

(*c*) If one brought a charge of 6.8×10^{-7} C from rest at ∞ to the center of the square, what is the change in the potential energy of the system?

(*d*) How much work must be done by an outside force to bring in this charge?

Solution

(*a*) All the charges produce fields of the same magnitude at the center, since they have the same charge and are equidistant from the center. The charges at opposite corners produce fields that are in opposite directions, thus canceling each other. The total field at the center is therefore zero.

(*b*) The potential at the center is the sum of the contribution from each of the four charges. Each charge produces the same potential, kq/r, where r is the distance from the corner to the center. Thus $r = 0.77/\sqrt{2} = 0.544$ m. The total potential is therefore $V = 4(9.0 \times 10^9)(5.7 \times 10^{-7}\ \text{C})/0.544 = 3.77 \times 10^4$ V. We see that the potential can be non-zero even at a point where the electric field is zero.

(*c*) The change in the potential is the difference between the potential at the center of the square and the potential at ∞. Thus $\Delta V = 3.77 \times 10^4 - 0 = 3.77 \times 10^4$ V. The change in potential energy is $q\Delta V = (6.8 \times 10^{-7}\ \text{C})(3.77 \times 10^4\ \text{V}) = 0.026$ J. Thus the system gained 0.026 J of energy. (This makes sense since all the charges are positive so potential energy increases as the fifth charge is brought closer.)

(*d*) The work done by outside (non-conservative) forces equals the change in the total mechanical energy of the system. Since there is no change in kinetic energy, the outside work will equal the change in the potential energy, $W_{\text{outside}} = 0.026$ J.

Problem 4.7. A total charge of 5.4×10^{-6} C is uniformly distributed along a ring of radius 0.89 m.

(*a*) What is the potential at the center of the ring?

(*b*) What is the potential at a point on the axis of the ring at a distance of 0.98 m from the plane of the ring?

Solution

(*a*) All the charge is located at a distance of $r = 0.89$ m from the center of the ring. Each part of the charge therefore contributes the same scalar potential at the center, and the total potential is $kQ/r = (9.0 \times 10^9)(5.4 \times 10^{-6}\ \text{C})/(0.89\ \text{m}) = 5.46 \times 10^4$ V.

(*b*) Now all the charge is located at a distance of $(r^2 + x^2)^{1/2} = (0.89^2 + 0.98^2)^{1/2} = 1.32$ m, and the potential is $(9.0 \times 10^9)(5.4 \times 10^{-6}\ \text{C})/(1.32) = 3.68 \times 10^4$ V.

Note how easy it is to calculate the potential in Problem 4.7 in comparison with finding the electric field in a comparable problem in Chap. 3. This, of course, is a consequence of the potential being a scalar while the field is a vector.

4.3 THE ELECTRIC FIELD—POTENTIAL RELATIONSHIP

We know that the electric field is the force per charge and the potential is the potential energy per charge. The force and the potential energy are related by the work–energy theorem, and therefore the electric field and the potential must be related in the same manner. We would like to develop that relationship in more detail at this time. It is useful to do this by considering an opposing force to the electric force.

When an outside force (non-electric) **F**, is exerted on a charge in an electric field, and is adjusted to *always* be equal and opposite to the electric force, then the positive (negative) work done by that force in moving the charge from one location to another will equal the increase (decrease) in the electric potential energy of the charge. If no work is done by this outside force either because the force is zero (hence there is no electric field) or the force is perpendicular to the direction in which the charge moves, then there will not be any change in the electric potential energy of the charge. Therefore there is a change in potential energy (and a corresponding change in potential) only if there is a component of the electric field in the direction of motion. If one moves perpendicular to **E** [along $\Delta d_\perp$ in Fig. 4-1(*a*)], there is no change in V. If one moves in the direction of **E** [along $\Delta d_\parallel$ in Fig. 4-1(*a*)], then, for constant **E**, the change in potential energy is $|\mathbf{F}|d = -q|\mathbf{E}|\Delta d_\parallel$, and the change in potential will equal $\Delta V = -|\mathbf{E}|\Delta d$. If the field is at an angle of θ with the direction of motion (Δd in Fig. 4-1), then the change in potential will equal $\Delta V = -|\mathbf{E}|\Delta d\cos\theta$. If the field is not constant, then one must divide the path into small segments over which the field can be considered to be a constant and add the contribution from each segment. Thus, in general;

$$\Delta V = -\sum|\mathbf{E}|\cos\theta\,\Delta d, \tag{4.4}$$

where the sum is evaluated along the path of the particle [see Fig. 4-1(*b*)]. We have already learned that for a conservative force the result of this calculation depends only on the beginning and ending points, so we can choose any path between those points that we want in evaluating the sum. This relationship can be used to calculate ΔV between any two points if the field **E** is known along a path joining those points. Eq. (*4.4*) also shows that an equivalent unit for E is V/m.

Problem 4.8. Two parallel plates carry a surface charge density of $\pm\sigma$, respectively, and are separated by a small distance d. Assume that the size of the plates is always large compared with the distance to the plates.

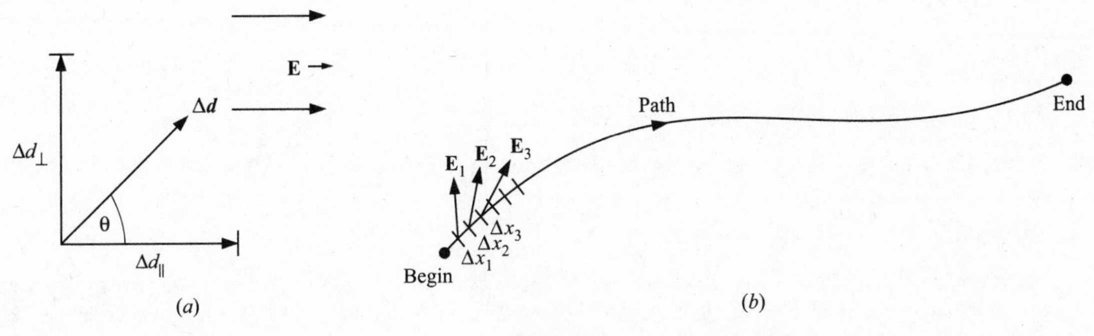

Fig. 4-1

(*a*) What is the electric field in the region between the plates?

(*b*) What is the potential difference between a point on one plate and a point on the other plate (e.g. points P_1 and P_2 in Fig. 4-2)?

(*c*) Which plate, the positive or the negative plate, is at the higher potential?

Solution

(*a*) We learned in the previous chapter that the field between the plates points from the positive plate to the negative plate, and has a constant magnitude of σ/ε_0.

(*b*) Since the field is constant and pointing along the direction perpendicular to the plates, we choose our path in two parts starting at the point P_1 as shown in Fig. 4-2. Along path 1 we move parallel to the field to the second plate, and along path 2 we move along the second plate, perpendicular to E, until the final point. Along path 2 there is no ΔV since we are moving perpendicular to **E**. Along path 1, $|\Delta V| = |E|d = \sigma d/\varepsilon_0$. Thus the potential difference is, in magnitude, equal to $\sigma d/\varepsilon_0$.

(*c*) Along path 1 the field is in the same direction as the displacement. Therefore, from Eq. (*4.4*), $\Delta V = V_2 - V_1 = -\sigma d/\varepsilon_0$, and the potential decreases as we move from the positive plate (P_1) to the negative plate (P_2), and the positive plate is at the higher potential, V_1. This illustrates the fact that the potential always decreases as we move along the direction in which the field points. Since the field points away from positive charge and towards negative charge, the potential decreases as we move away from positive or toward negative charge.

Problem 4.9. An isolated conducting sphere is charged with a total charge, Q, of 6.0×10^{-8} C, and has a radius of 1.35 m.

(*a*) What is the field inside the sphere, and what is the field outside the sphere?

(*b*) What is the potential at a distance r from the sphere, if r is outside the sphere?

(*c*) What is the potential at the surface of the sphere?

(*d*) What is the potential at a point r within the sphere?

(*e*) If instead of a conducting sphere we had a thin uniform spherical shell of charge, again with no other charges nearby, how would the answers to (a)–(d) change?

Solution

(*a*) We learned in Chap. 3 that the field inside a conductor is zero, and that the field outside an isolated conducting sphere, where the surface charge is uniformly distributed, is the same as if all the charge were concentrated at a point at the center of the sphere. Therefore the field is kQ/r^2 for $r > R$, and zero for $r < R$.

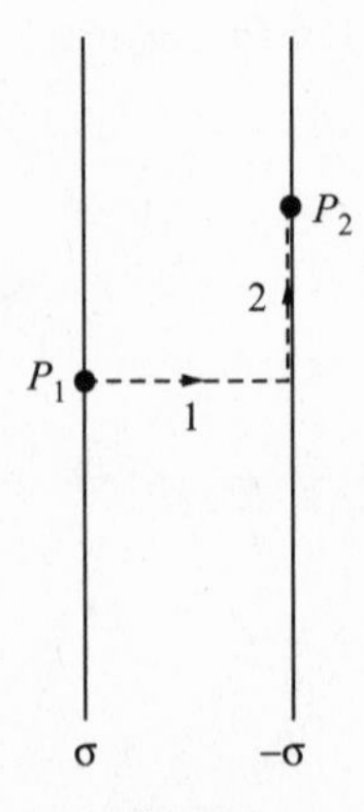

Fig. 4-2

(*b*) The field outside the sphere is identical to that of a point charge located at the center of the sphere. The sum to be evaluated [Eq. (*4.4*)] for the case of the sphere is therefore just the result for a point charge, as long as we remain outside the sphere. Therefore the difference in potential between a point at $r > R$ and a point at ∞ is $\Delta V = kQ/r$. Since the potential at ∞ is chosen to be zero, $V = kQ/r$.

(*c*) At the surface $r = R$. Thus $V_{\text{surface}} = (9.0 \times 10^9)(6.0 \times 10^{-8}\ \text{C})/1.35\ \text{m} = 400\ \text{V}$.

(*d*) The field inside the sphere is zero. Therefore if one moves from any point inside to any other point inside the sphere there will be no change in potential. The potential is the same everywhere within the sphere. At the surface the potential is 400 V, so the potential remains at 400 V for any other point $r < R$.

(*e*) By Gauss' law (choosing concentric spherical surfaces of radius $r < R$) since no charge is enclosed within the shell, the electric field will still be zero. The field outside could again be that of a point charge at the center so part (*a*) is unchanged. Similarly, the results of parts (*b*), (*c*) and (*d*) will be unchanged.

Problem 4.10. A charge Q_1 of 5.5×10^{-7} C is at the center of a conducting spherical shell that has an inner radius of 0.87 m and an outer radius of 0.97 m (see Fig. 4-3). The conducting sphere has a total charge of -2.3×10^{-7} C.

(*a*) How much charge Q_2 is there on the inner surface of the conducting sphere, and how much charge Q_3 is there on the outside surface?

(*b*) By adding the contributions from all charges, calculate the potential at a point at a distance of 1.05 m from the center.

(*c*) By adding the contributions from all charges, calculate the potential at a point at a distance of 0.95 m from the center.

(*d*) By adding the contributions from all charges, calculate the potential at a point at a distance of 0.45 m from the center.

Solution

(*a*) We know that in static equilibrium (no charges in motion) the electric field within the conducting shell is zero as it must be within any conductor. We draw a Gaussian surface at a radius within the conductor, and note that the flux through that surface is zero, since the field is zero. Therefore the total charge inside that surface must be zero. The only charges inside the surface are on the inner surface of the shell and at the center. Therefore the charge on the inner surface must be $Q_2 = -Q_1 = -5.5 \times 10^{-7}$

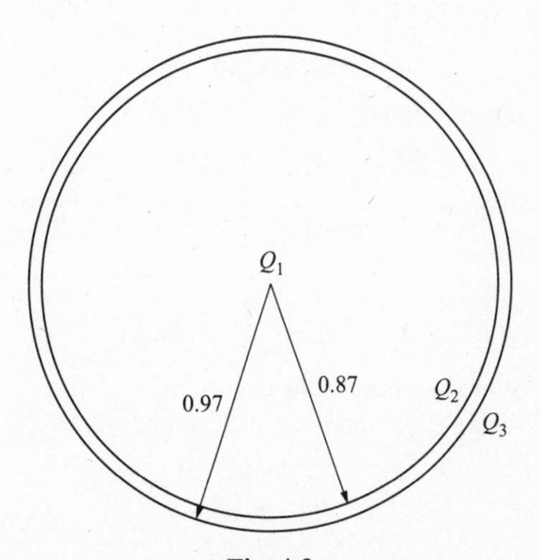

Fig. 4-3

C. The total charge on the sphere is given as -2.3×10^{-7} C which must equal $Q_2 + Q_3 = -2.3 \times 10^{-7} = Q_3 + (-5.5 \times 10^{-7})$, giving $Q_3 = 3.2 \times 10^{-7}$ C.

(*b*) We showed in the previous problems that the potential of a point charge is $V = kQ/r$. We also showed that the potential due to a uniform spherical surface charge distribution at a radius R is equal to $V_{\text{outside}} = kQ/r$ if $r > R$, and $V_{\text{inside}} = kQ/R$ if $r < R$. In our problem there are three charge distributions: a point charge at the center, a surface charge at $R = 0.87$ m and another surface charge at $R = 0.97$ m. If $r = 1.05$ m then we are seeking the potential outside each charge distribution. The total potential is then $V = V_1 + V_2 + V_3 = kQ_1/r + kQ_2/r + kQ_3/r = (9.0 \times 10^9)[(5.5 - 5.5 + 3.2) \times 10^{-7}\ \text{m}]/(1.05\ \text{m}) = 2.74 \times 10^3$ V.

(*c*) At $r = 0.95$ m, we are outside of charges Q_1 and Q_2, but within charge Q_3. Therefore $V_3 = kQ_3/R_3 = (9.0 \times 10^9)(3.2 \times 10^{-7}\ \text{C})/0.97\ \text{m} = 2.97 \times 10^3$ V. Furthermore, $V_1 + V_2 = k(Q_1 + Q_2)/r = (9.0 \times 10^9)(5.5 - 5.5) \times 10^{-7}/0.95 = 0$. Thus $V = 2.97 \times 10^3$ V.

Note. We could also have derived this result from the fact that E is zero within the conducting sphere, and therefore the potential within the sphere is the same as it is on the outer (or inner) surface. On the outer surface the potential, from part (*a*) is $k(3.2 \times 10^{-7})/0.97$, which is the same as we found.

(*d*) At $r = 0.45$ m, we are outside of the point charge but inside the two surface charges. The potential from the point charge is $kQ_1/r = (9.0 \times 10^9)(5.5 \times 10^{-7}\ \text{C})/0.45\ \text{m} = 1.1 \times 10^4$ V. The potential from the surface charges is $V_2 + V_3 = k(Q_2/R_2 + Q_3/R_3) = (9.0 \times 10^9)[(-5.5 \times 10^{-7}/0.87) + (3.2 \times 10^{-7}/0.97)] = -2.72 \times 10^3$ V. The total potential is then $1.1 \times 10^4 - 2.72 \times 10^3 = 8.29 \times 10^3$ V.

We have seen in the previous problems how to calculate the potential if the electric field is constant, or if the electric field is produced by a point charge, or if the electric field is produced by a spherical surface distribution. For other cases, one must use one of two methods to evaluate the potential difference between two points: (1) calculate the electric field everywhere along a path and then use the sum in Eq. (*4.4*) to calculate the difference in potential, or (2) use the charge distribution to calculate the potential at every point using Eq. (*4.3b*) and then calculate the difference between the potential at the points. We summarize some results from using such methods, together with the results we have already obtained.

For a point charge,

$$V = (1/4\pi\varepsilon_0)Q/r \tag{4.3a}$$

For a collection of charges,

$$V = (1/4\pi\varepsilon_0) \sum Q_i/r_i \tag{4.3b}$$

For a spherical surface charge at radius R;

$$V = (1/4\pi\varepsilon_0)Q/r \qquad \text{for} \qquad r > R \tag{4.5a}$$

and

$$V = (1/4\pi\varepsilon_0)Q/R \qquad \text{for} \qquad r < R \tag{4.5b}$$

For a long wire,

$$\Delta V = V_2 - V_1 = -(\lambda/2\pi\varepsilon_0) \ln (r_2/r_1) \tag{4.6}$$

for r_1 and r_2 any two perpendicular distances from the wire.

For a long cylinder of length L with symmetric surface charge on the cylindrical portion at radius R ($r, R \ll L$);

$$V = -(\lambda/2\pi\varepsilon_0) \ln (r/R') \qquad \text{for} \qquad r > R \tag{4.7a}$$

$$V = -(\lambda/2\pi\varepsilon_0) \ln (R/R') \qquad \text{for} \qquad r < R \tag{4.7b}$$

where R' is an arbitrary distance. It is often useful to set $V = 0$ at the radius of the cylinder, which is equivalent to setting $R' = R$.

For a large, uniformly charged infinitesimally thin plate of surface charge density σ,

$$\Delta V = V_2 - V_1 = -\sigma(|x_2| - |x_1|)/2\varepsilon_0 \tag{4.8}$$

where $|x_2|$ and $|x_1|$ are perpendicular distances on either side of the plate, and $|x_1|, |x_2| \ll L$, where L is the distance to the edge of the plate.

Problem 4.11. A coaxial cable (see Fig. 4-4) consists of a long, conducting wire, of radius R_1 with a linear charge density of λ, and a long conducting coaxial cylindrical shell, with an inner radius R_2 and an outer radius R_3, and with a symmetric linear charge density of $-\lambda$. We assume the length to be much greater than any of the radial distances of interest.

(*a*) What is the potential due to the cable at a point at a radial distance from the axis r, such that $r > R_3$?

(*b*) What is the potential at a point within the outer cylindrical shell, at $R_2 < r < R_3$?

(*c*) What is the potential at a point between the wire and the cylinder at $R_1 < r < R_2$?

(*d*) What is the potential at a point within the wire, at $r < R_1$?

Solution

(*a*) We use Eq. (*4.7a*) for each of the three surface charges since the point in question is outside both cylindrical distributions. Then $V = 0$, since the total enclosed linear charge density is $\lambda - \lambda = 0$.

(*b*) We note that the charge on the outer cylinder is all on the inner surface. This is because the field within the conductor is zero, and therefore, from **Gauss' law** the total charge within a Gaussian surface must be zero. Then the charge on the inner surface must cancel the charge on the wire, and equal $-\lambda$. Therefore the point within the cylinder is also outside all the charge distributions, and the result is the same as in (a), i.e. $V = 0$.

(*c*) In this case the point in question is outside of the wire but within the surface distribution on the outer cylinder. Using Eq. (*4.7a*) for the wire and Eq. (*4.7b*) for the cylinder we have for the potential: $V = V_1 + V_2 = (-\lambda/2\pi\varepsilon_0)\ln(r/R') - (-\lambda/2\pi\varepsilon_0)\ln(R_2/R') = (-\lambda/2\pi\varepsilon_0)\ln(r/R_2)$ (where we recall $\ln(A/B) = \ln A - \ln B$).

(*d*) Since we are now within the inner conducting cylinder where the field is zero, the potential must equal its value at the surface. Thus, $V = (-\lambda/2\pi\varepsilon_0)\ln(R_1/R_2)$.

Note. One could also get this result by adding the contributions of the two surface charge distributions. Then $V = V_1 + V_2 = (-\lambda/2\pi\varepsilon_0)\ln(R_1/R') - (-\pi/2\pi\varepsilon_0)\ln(R_2/R') = (-\lambda/2\pi\varepsilon_0)\ln(R_1/R_2)$.

Problem 4.12. Two large thin parallel plates are a distance D apart, and have surface charge densities

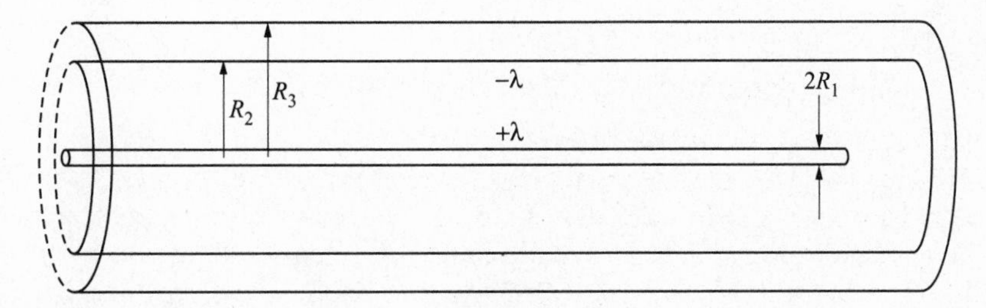

Fig. 4-4

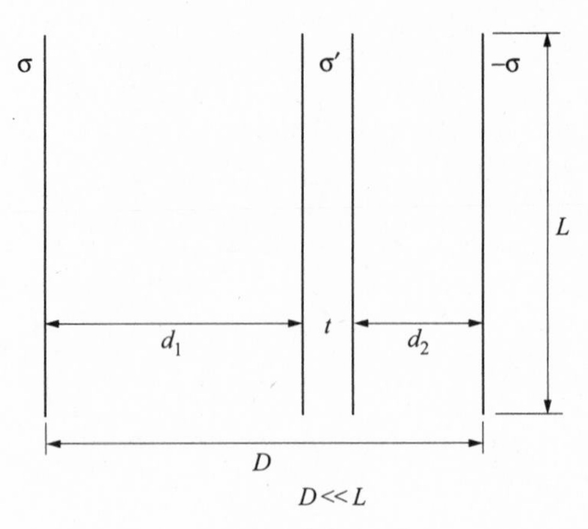

Fig. 4-5

of $\pm\sigma$, as in Fig. 4-5. A large conducting plate, of thickness t, is placed with one side at a distance of d_1 from the positive plate, as in the figure. The conducting plate has a charge density of σ'.

(*a*) What is the surface charge distribution on the two sides of the conducting plate?

(*b*) What is the difference in potential between the positive plate and the conducting plate?

(*c*) What is the difference in potential between the positive and the negative plates?

Solution

(*a*) The field within the conducting plate must be zero, as it is within any conductor. Each charge distribution produces a field of $\sigma/2\varepsilon_0$ pointing away from positive and toward negative charge. The field within the conductor has four contributions: (1) from the positive plate with charge distribution σ, (2) from the negative plate with charge distribution $-\sigma$, (3) from the side of the conducting plate near the positive charge with a charge distribution labeled σ_1, and (4) from the other side of the conducting plate with a charge distribution $\sigma_2 = (\sigma' - \sigma_1)$. The fields produced are: $E = E_1 + E_2 + E_3 + E_4 = (1/2\varepsilon_0)[\sigma + \sigma + \sigma_1 - (\sigma' - \sigma_1)] = (1/2\varepsilon_0)(2\sigma + 2\sigma_1 - \sigma') = 0$. Thus, $\sigma_1 = (\sigma'/2) - \sigma$. On the other side of the plate the charge distribution is then $\sigma_2 = (\sigma'/2) + \sigma$. (As a check we add $\sigma_1 + \sigma_2$ to get σ'.)

(*b*) To obtain the difference of potential between two points we calculate the field in the region between the points and, for a constant field perpendicular to the plates use the fact that $\Delta V = -E\Delta x$, where ΔV is the final-minus-initial potential as we move through Δx. In the region between the positive plate and the conducting plate, the field is $E = [\sigma - (\sigma'/2)]/\varepsilon_0$ to the right. We get this result either by adding the field from all four distributions or by using Gauss' law. By adding the contributions we get $E = (1/2\varepsilon_0)[\sigma - (\sigma'/2 - \sigma) - (\sigma'/2 + \sigma) - (-\sigma)] = [\sigma - (\sigma'/2)]/\varepsilon_0$. This field is to the right if the number is positive. Then the difference of potential between the positive plate and the conducting plate is given by $\Delta V = V_c - V_+ = -[\sigma - (\sigma'/2)]d_1/\varepsilon_0$, or $V_+ - V_c = [\sigma - (\sigma'/2)]/\varepsilon_0$.

(*c*) Using the same procedure we obtain the field between the conducting plate and the negative plate to be $E = [\sigma + (\sigma'/2)]/\varepsilon_0$. Then the difference of potential between the conducting plate and the negative plate is given by $\Delta V = V_- - V_c = -[\sigma + (\sigma'/2)]d_2/\varepsilon_0$. The difference of potential between the positive and the negative plates is therefore: $V_+ - V_- = (V_+ - V_c) + (V_c - V_-) = [\sigma - (\sigma'/2)]d_1/\varepsilon_0 + [\sigma + (\sigma'/2)]d_2/\varepsilon_0 = (1/\varepsilon_0)[\sigma(d_1 + d_2) + (\sigma'/2)(d_2 - d_1)]$.

4.4 EQUIPOTENTIALS

In our discussion so far we have learned how to use information about the electric field to obtain the potential difference between two points. We now shift our attention to the reverse process, obtaining the electric field from a knowledge of the potential. At every point there is an electric field pointing in some direction. If we move to a different point along that direction, then the potential will change.

However, if we move to a different point perpendicular to that direction, the potential will not change. Thus, for example, for the uniform field between large parallel plates, for every plane perpendicular to **E**, the potential remains the same at every point in the plane. Even for non-uniform fields, if we continue moving from point to point, always in a direction perpendicular to the electric field at that point, we will sweep out a surface with all points on that surface at the same potential. This surface is called the "**equipotential surface**". This idea can be used to obtain the direction of the electric field at any point if we know the potential everywhere in the region. We do this by sweeping out the various equipotential surfaces, and noting that the electric field lines are perpendicular to those surfaces. Once we have the direction of the electric field we can easily obtain its magnitude. We move a distance Δd in the direction of the electric field, between nearby equipotential surfaces and note the difference in potential. We know that along the direction of the electric field $\Delta V = -E\Delta d$, giving $E = -\Delta V/\Delta d$. The minus sign means that E is positive in the direction that ΔV is negative, i.e. **E** points from high to low potential. Thus, a knowledge of how V varies in a region around a point allows us to obtain the magnitude and the direction of the electric field at that point.

Problem 4.13. The potential produced by a point charge is $V = kQ/r$. Use this information to: (*a*) determine the shape of the equipotential surfaces, (*b*) determine the direction of the electric field at any point and (*c*) determine the actual value of the electric field at any point.

Solution

(*a*) The potential at a point at a distance r from the charge is given as $V = kQ/r$. All other points at the same distance r from the charge have the same potential. Therefore the equipotential surface consists of all points equidistant from the source at a distance r. This is the surface of a sphere of radius r. The equipotential surfaces are therefore concentric spherical surfaces.

(*b*) The direction of the electric field is perpendicular to the equipotential surfaces. That direction, for spheres, is in the direction of the radius. Thus the electric field must point along a radius. We know that it points from high to low potential. If Q is positive, then the potential decreases as r increases. Therefore the field points in the direction away from the charge, as we expected. For a negative charge the potential becomes less negative as r increases, which means that V increases as r increases. Then E points toward smaller r, or toward the center.

(*c*) The magnitude and direction of E along a radius is given by $|E| = \Delta V/\Delta d$, if Δd is along the direction of the field. Here $\Delta d = \Delta r$. If we move along a radius from r_1 to r_2, the difference in potential is $\Delta V = V_2 - V_1 = kQ(1/r_2 - 1/r_1) = kQ(r_1 - r_2)/r_1 r_2$. For very small $\Delta r = r_2 - r_1$ we can set $r_1 = r_2 = r$ in the denominator to get $\Delta V = -kQ\Delta r/r^2$. Then $E = -\Delta V/\Delta r = kQ\Delta r/r^2\Delta r = kQ/r^2$, as expected.

Problem 4.14. Two large parallel plates carry charge distributions of $\pm\sigma$. The positive plate is at $x = 0$, and the negative plate is at $x = d$, where x is measured perpendicular to the plates. The potential at any point can be shown to be given by $V = V_0(1 - x/d)$ when $0 < x < d$, i.e. between the plates, and where V_0 and 0 are the potentials at the positive and negative plates, respectively.

(*a*) What are the equipotential surfaces?

(*b*) What is the direction of the electric field at a point located at a distance x from the positive plate?

(*c*) What is the magnitude of the electric field at this point?

Solution

(*a*) The potential at a point at a distance x from the positive plate is given as $V = V_0(1 - x/d)$. All other points at the same distance x from the plate have the same potential. Therefore the equipotential surface consists of all points equidistant from the plate at a distance x. This surface is a plane parallel to the plates. The equipotential surfaces are therefore planes parallel to the plates.

(*b*) The direction of the electric field is perpendicular to the equipotential surfaces. That direction, for a plane parallel to the y–z plane, is in the direction of x. Thus the electric field must point along x. We know that it points from high to low potential. The potential decreases from V_0 to zero as one increases x from zero to d. Therefore the field is in the $+x$ direction.

(*c*) The magnitude of E is given by $|E| = \Delta V/\Delta x$, if Δx is along the direction of the field. If we move along the field from x_1 to x_2, the difference in potential is $\Delta V = V_2 - V_1 = V_0[(1 - x_2/d) - (1 - x_1/d)] = V_0(x_1 - x_2)/d = -V_0\,\Delta x/d$. Then $|E| = V_0\,\Delta x/d\Delta x = V_0/d$, as expected.

Problem 4.15. The electric field lines for a particular situation are shown in Fig. 4-6(*a*). Along the curved field line *OACD* the electric potential decreases linearly by 4.0 V every 3.0 m. At point *A* the potential, V_A, is 40 V.

(*a*) On the figure, draw the direction of the electric field at *A*.

(*b*) Calculate the magnitude of the electric field at *A*.

(*c*) Calculate the potential, V_C, at point *C*, which is 3.0 m from *A*.

(*d*) Calculate the potential, V_B, at point *B* which is 0.010 m along a line perpendicular to the field line through *A*.

Solution

(*a*) The field is tangent to the electric field line at any point. It points from high to low potential. Since the potential is decreasing as one moves along the line toward *C*, the field points in that direction. The direction is shown in Fig. 4-6(*b*).

(*b*) The magnitude of the field is equal to $\Delta V/\Delta x$ if one moves along the direction of E. When moving from *A* to *C* one is indeed moving in the direction of *E*, and $\Delta V/\Delta x = 4.0\text{ V}/3.0\text{ m} = |E| = 1.33\text{ V/m}$. Ordinarily this would be the average magnitude of *E* over the 3.0 m distance, but because the potential decreases linearly it is the actual magnitude at any point along the line.

(*c*) We can obtain V_C from $\Delta V = V_C - V_A = -E\Delta x = -(1.33\text{ V/m})(3.0\text{ m}) = -4.0\text{ V}$. Then $V_C = 40 - 4.0 = 36\text{ V}$.

(*d*) Point *B* is along a direction perpendicular to the electric field. Therefore the potential does not change as one moves from *A* to *B*. Thus $V_B = V_A = 40\text{ V}$.

The result that we have obtained for calculating the electric field from a knowledge of the potential everywhere can be written in a different form. If one moves a small distance Δx in the x direction from a given point, and the electric field makes an angle θ with the x axis at that point, then the change in potential in that direction, $\Delta V_x = -E\cos\theta\,\Delta x = -E_x\Delta x$. Thus $E_x = -\Delta V_x/\Delta x$, where ΔV_x is the

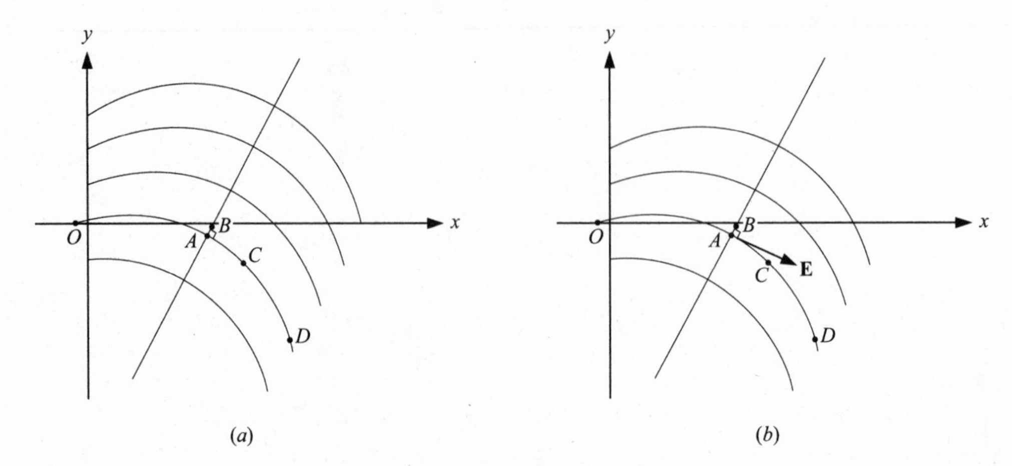

Fig. 4-6

change in V as one moves in the x direction. Similarly, $E_y = -\Delta V_y/\Delta y$, and $E_z = -\Delta V_z/\Delta z$. If we have the three components of the vector **E**, then we have all the information needed to characterize **E** at that point. The vector, whose components are determined by calculating the rate of change of V in each direction ($\Delta V_x/\Delta x$, $\Delta V_y/\Delta y$, $\Delta V_z/\Delta z$), is called, in mathematical terminology, the **gradient** of V, and written as ∇V. Then our expression relating the electric field to the potential at every point in space can formally be expressed as $\mathbf{E} = -\nabla V$. As you may have guessed this is a calculus relationship and allows one to carry out sophisticated analyses beyond the scope of this book.

Problem 4.16. Fig. 4-7 shows the value of the electric potential at various points in the x–y plane. The potential at the origin is 75 V. At points along the x and y axes, at a distance of 0.65 m from the origin, the potentials are as shown.

(*a*) Calculate the x and y components of the electric field at the origin. Assume the potential varies linearly with distance in both the x and y directions.

(*b*) What is the magnitude and direction of the electric field at the origin?

(*c*) What can one say about the electric field at other points near the origin?

Solution

(*a*) To get E_x we must calculate $E_x = -\Delta V_x/\Delta x = -(65 - 75)\text{V}/0.65 \text{ m} = 15.4 \text{ V/m}$. Similarly, $E_y = -\Delta V_y/\Delta y = -(80 - 75)\text{V}/0.65 \text{ m} = -7.7 \text{ V/m}$. Thus the field has components in $+x$ and in $-y$ of 15.4 V/m and 7.7 V/m, respectively.

(*b*) $E = (E_x{}^2 + E_y{}^2)^{1/2} = 17.2$ V/m. If θ is the angle of E below the positive x axis, we have $\tan\theta = |E_y/E_x| = 0.50 \rightarrow \theta = 26.6°$.

(*c*) Since the potential varies linearly in the region from -0.65 m to $+0.65$ m in both the x and y directions, both E_x and E_y will be constant in that region. Thus **E** will be uniform for all points near the origin.

Problem 4.17.

(*a*) Show that the surface of a conductor (in static equilibrium) is always an equipotential surface irrespective of the charge on the surface or of nearby charges.

(*b*) Show that a hollow region inside a conductor that has no charges in it has no electric field in it as well.

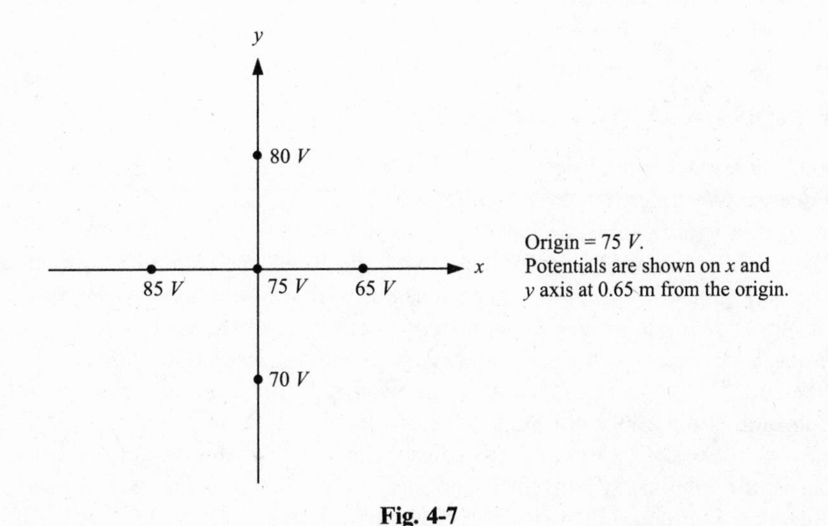

Fig. 4-7

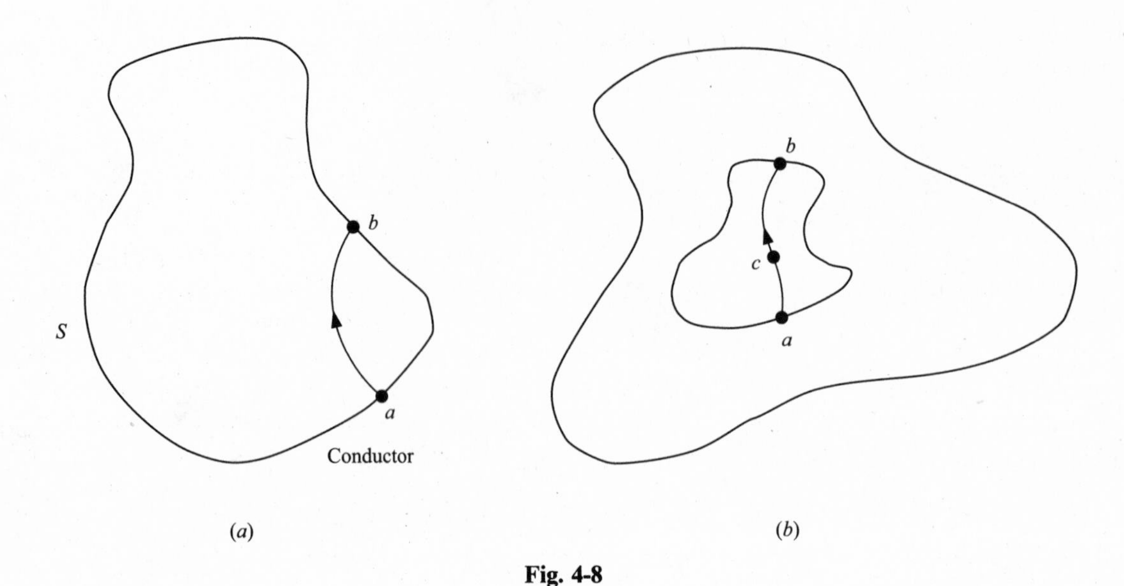

Fig. 4-8

Solution

(*a*) Consider the conductor shown in Fig. 4-8 with surface *S*, and consider two points on the surface, *a* and *b*. We can use Eq. (*4.4*) along any path leading from *a* to *b* to obtain $\Delta V = V_b - V_a$, including the path shown through the conductor. For the path chosen, which is wholly in the conductor, E is zero everywhere along the path. Therefore, $\Delta V = 0 \rightarrow V_b = V_a$. Since this is true for all points *a* and *b* on the surface, the surface must be an equipotential. (Indeed, the whole conductor is an equipotential, by the same argument.)

(*b*) Consider the hollow in the conductor shown in Fig. 4-8(*b*). Suppose there were an electric field at any point *c* in the hollow. If we trace the electric field line through point *c* it would have to start at some point *a* on the inner surface and end at some other point *b*. This is because the electric field lines always start and end on charges or go off to infinity. Since the electric field points in the same direction everywhere on the field line from *a* to *b*, applying Eq. (*4.4*) to the path along the field line, $\cos\theta$ is always equal to one and the sum must be a positive (non-zero) value. Therefore, $V_b - V_a \neq 0$ and the surface cannot be an equipotential. Since we have just shown in part (*a*) that it must be an equipotential, our hypothesis that an electric field existed at point *c* cannot be true. Since point *c* was chosen arbitrarily, we must have $\mathbf{E} = 0$ at all points in the hollow. (This implies that the hollow is also an equipotential region with the same value as the conductor.) This result is no longer true if a charge were placed in the hollow region.

4.5 ENERGY CONSERVATION

The potential energy associated with the electrical force can be used in the same manner as any other potential energy. We note that the potential energy of any charge is given by qV, and the change in potential energy that is used in most energy related problems is $\Delta U_p = q\Delta V$. A positive charge gains energy as it moves to a region of higher potential (ΔV positive) and, unless restricted by other forces, will tend to move to regions of lower potential. A negative charge, such as an electron, will lose energy as it moves to a higher potential (q negative and ΔV positive), and therefore tends to move to a region of higher potential. When an electron moves through a difference of potential of one volt it gains or loses $e(1) = 1.6 \times 10^{-19}$ J of energy. This amount of energy is called an **electron-volt**, or eV. If the electron moves through a difference of potential of x volts, the electron gains or loses x electron-volts of energy. This is a very convenient unit of energy to use whenever one discusses the motion of an electron, or other particle with a similar charge, since the energy the particle gains (loses) in eV is numerically equal to the difference of potential in volts through which it moves.

Problem 4.18. An electron moves from the positive to the negative terminal of a 9 V battery. How much potential energy did it gain or lose? Did it gain or did it lose potential energy?

Solution

The change in potential energy was 9 eV, since the electron moved through a difference of potential of 9 volts. This corresponds to (9 eV)(1.6×10^{-19} J/eV) = 1.44×10^{-18} J. Since the charge on the electron is negative, and the change in potential was also negative, the electron gained potential energy. This is in accordance with our discussion that negative charges tend to move to higher potentials in order to lose potential energy, and they gain potential energy in moving to lower potentials.

Problem 4.19. We want to produce protons with a kinetic energy of 4.3×10^{-15} J. Through what difference of potential should we accelerate them in order to obtain that kinetic energy, assuming that they start from rest and that there are no other forces present?

Solution

Since only the electric force is present, and the electric force is conservative, we can use conservation of energy in this problem. If we start with a stationary proton, then the proton has no initial kinetic energy. The increase in kinetic energy must equal the decrease in potential energy. Thus the positively charged proton must move through a difference in potential that will result in the loss of 4.3×10^{-15} J. This means that it must move through ΔV such that $q\Delta V = -4.3 \times 10^{-15}$ J, or $\Delta V = (-4.3 \times 10^{-15}$ J)/1.6×10^{-19} C $= -2.69 \times 10^4$ V. Alternatively, we could have converted 4.3×10^{-15} J into eV by dividing by 1.6×10^{-19} J/eV, obtaining 2.69×10^4 eV. Then we can say that a proton must have fallen through a decrease of 2.69×10^4 V to lose that amount of potential energy.

Problem 4.20. A proton is moving directly toward a fixed nucleus containing 23 protons. The speed of the proton when it is at a distance of 5.8×10^{-9} m from the nucleus is 2.4×10^6 m/s. The proton has a charge of 1.6×10^{-19} C and a mass of 1.67×10^{-27} kg.

(*a*) What was its kinetic and potential energy at this initial distance?

(*b*) At what distance from the nucleus does the proton stop, i.e. what is the distance of nearest approach? (Assume the nucleus remains stationary.)

Solution

(*a*) The kinetic energy of the proton is $(1/2)m_p v^2 = (0.5)(1.67 \times 10^{-27}\ \text{kg})(2.4 \times 10^6\ \text{m/s})^2 = 4.81 \times 10^{-15}$ J. The potential energy is $U_p = kqQ/r = (9.0 \times 10^9)(1.6 \times 10^{-19}\ \text{C})(23 \times 1.6 \times 10^{-19}\ \text{C})/(5.8 \times 10^{-9}\ \text{m}) = 9.14 \times 10^{-19}$ J. The total energy is therefore nearly all kinetic energy and equals 4.81×10^{-15} J.

(*b*) By conservation of energy, the total energy must be the same as the proton moves toward the nucleus. At the point of nearest approach, the kinetic energy is zero, since $v = 0$. Therefore, the potential energy must equal the original energy. Thus, $kqQ/r = 4.81 \times 10^{-15}\ \text{J} = (9.0 \times 10^9)(1.6 \times 10^{-19}\ \text{C})(23 \times 1.6 \times 10^{-19}\ \text{C})/r = 5.30 \times 10^{-27}/r$. Then $r = 1.10 \times 10^{-12}$ m.

Problem 4.21. Four charged particles are placed at the corners of a square of side 0.39 m. The particles have charges of 2.3 μC, -5.6 μC, 7.9 μC and -1.3 μC as in Fig. 4-9.

(*a*) How much work was done by outside forces to place those particles in their positions if they were originally very far away?

(*b*) If an electron starts with no velocity very far away, what velocity does it have when it reaches the center of the square? ($m_e = 9.1 \times 10^{-31}$ kg)

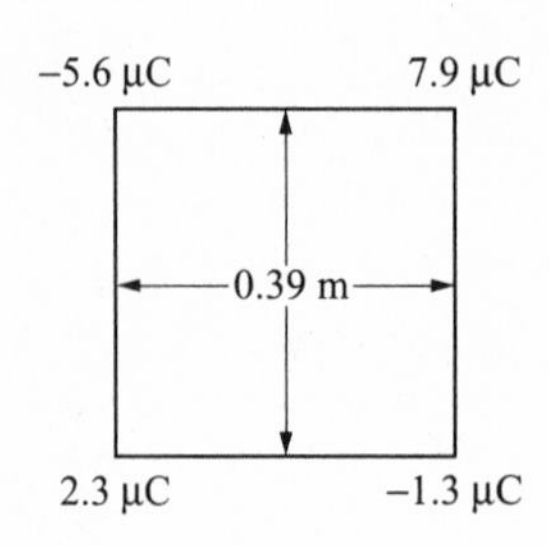

Fig. 4-9

Solution

(*a*) We will assemble the particles one at a time. To place the first particle (2.3 μC) in place requires no work ($W_1 = 0$) since there are no forces present as yet. To place the next particle (-5.6 μC) in place the outside work W_2 must be equal to the change in potential energy. This equals $W_2 = kQ_1Q_2/r_{12} = (9.0 \times 10^9)(2.3 \times 10^{-6}\ \text{C})(-5.6 \times 10^{-6}\ \text{C})/0.39\ \text{m} = -0.30$ J. To place the next particle we must again supply the added potential energy. This additional potential energy is due to the interaction with both of the particles already in place. Thus $W_3 = kQ_3(Q_1/r_{13} + Q_2/r_{23}) = (9.0 \times 10^9)(7.9 \times 10^{-6})[(2.3 \times 10^{-6}/0.39\sqrt{2}) + (-5.6 \times 10^{-6}/0.39)] = -0.72$ J. Similarly, to add the fourth particle requires work of $W_4 = kQ_4(Q_1/r_{14} + Q_2/r_{24} + Q_3/r_{34}) = (9.0 \times 10^9)(-1.3 \times 10^{-6})[(2.3 \times 10^{-6}/0.39) + (-5.6 \times 10^{-6}/0.39\sqrt{2}) + (7.9 \times 10^{-6}/0.39)] = -0.19$ J. The total work is therefore $W_{\text{total}} = W_1 + W_2 + W_3 + W_4 = -0.30 - 0.72 - 0.19 = -1.21$ J.

(*b*) With all the four particles in place, the potential at the center is $V = V_1 + V_2 + V_3 + V_4 = k(Q_1 + Q_2 + Q_3 + Q_4)/r = (9.0 \times 10^9)(2.3 - 5.6 + 7.9 - 1.3) \times 10^{-6}/0.195\sqrt{2} = 1.08 \times 10^5$ V. At a large distance, the potential is zero. Therefore the electron loses potential energy equal to 1.08×10^5 eV. This is converted into kinetic energy. Then, $(1/2)mv^2 = (1.08 \times 10^5\ \text{eV})(1.6 \times 10^{-19}\ \text{J/eV}) = 1.73 \times 10^{-14}$ J. The mass of an electron is 9.1×10^{-31} kg, so $v^2 = 2(1.73 \times 10^{-14})/9.1 \times 10^{-31} = 3.80 \times 10^{16}$, and $v = 1.9 \times 10^8$ m/s.

Problem 4.22. Two large, thin parallel plates, of length L, are perpendicular to the x axis and carry charge distributions of $\pm\sigma$ (as in Fig. 4-10). The positive plate is at $x = 0$, and the negative plate is at $x = d$. The potential at any point is given as $V = V_0(1 - x/d)$ for $0 < x < d$, i.e. between the plates. An electron starts at the bottom, halfway between the plates, with an upward speed of v_0. The electron just passes the end of the plate at the top. Assume that the field is uniform throughout the region between the plates, and the potential is as given above. Give your answers in terms of L, d, v_0, σ and e (where e, as always, is the magnitude of the electron charge).

(*a*) How much kinetic energy, ΔK, did the electron gain until it leaves the region between the plates?

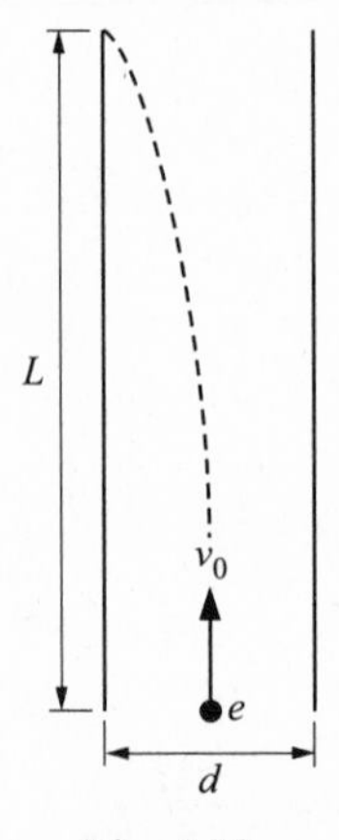

Fig. 4-10

(*b*) What is the x component of the velocity of the electron?

(*c*) How much time does it take for the electron to move through the plates?

Solution

(*a*) We will use conservation of energy to solve this part of the problem. The gain in kinetic energy ΔK must equal the loss in potential energy. This loss is equal to $e\Delta V = e(V_f - V_i) = e(V_0 - V_0 d/2) = eV_0 d/2$. Thus the gain in kinetic energy is $eV_0 d/2$. Recalling that the potential difference across the plates is just $(V_0 - 0) = Ed = \sigma d/\varepsilon_0$, we have finally $\Delta K = e\sigma d^2/2\varepsilon_0$.

(*b*) The gain in kinetic energy is $K_f - K_i = (\frac{1}{2})m(v_f^2 - v_i^2) = (\frac{1}{2})m(v_{fx}^2 + v_{fy}^2 - v_{iy}^2)$ where we recall $v_{ix} = 0$. Now, $v_y = v_0$ does not change, so $\Delta K = (\frac{1}{2})mv_{fx}^2$ and using our results in (*a*) we get: $v_{fx} = [(e/m)\sigma d^2/\varepsilon_0]^{1/2}$.

(*c*) Since v_y does not change, the time to move a distance of L in y is $t = L/v_0$.

Note. If we wanted we could solve for V_0, since we must also have $v_{fx} = at$ where acceleration $a = |(e/m)E| = (e/m)\sigma/\varepsilon_0$, and we can solve for t and insert in $t = L/v_0$.

4.6 CAPACITANCE

We have seen that positive work is required by an outside force to separate opposite charges that were initially together. For instance, we may have two metal surfaces which were initially uncharged, and then remove negative charge from one surface and place this charge on the other surface. The first surface that lost negative charge becomes positively charged, and the other surface gains the same negative charge. The more charge that we transfer the harder it becomes to transfer the next unit of charge because of the Coulomb forces between the charges, and the more work we have to do to transfer additional charge. This work is manifested in the resultant potential energy of the final distribution of charge.

When a given distribution of charge is reached, we wish to be able to calculate the potential everywhere in space. This will allow us to determine the energy necessary to bring another charge from one location to another. We know that each conductor surface will be an equipotential surface once charges have reached their equilibrium positions. Therefore each surface has its own potential and potential differences exist between the various surfaces. For a particular pair of conductors we label this potential difference ΔV. Since we can always set our zero of potential at our will, we can take one of the surfaces to have zero potential and the other to have a potential V which will equal ΔV. Therefore we will call the potential difference between the two surfaces V.

Let us consider the case of two isolated conductors (labeled 1 and 2) with charge $+Q$ on one and $-Q$ on the other, and a potential difference V between them. Depending on the shape of the conductors and their positions relative to each other, the charges on the conducting surfaces will distribute themselves with some definite (but not necessarily uniform) charge distribution, σ_1 and σ_2. In general, σ_1 and σ_2 will vary from point to point on the respective surfaces. In principle, the potential and electric field everywhere outside and on the conductors, can be determined by dividing the surfaces into tiny segments and calculating the potential (or electric field) at any point by adding the contributions of all the electric charges in all the tiny segments. It is not hard to see that if we doubled (or halved, or tripled) the electrical charges in *all* segments on both surfaces we would not disturb the equilibrium on those surfaces, and furthermore the potential and electric field everywhere would also double (or halve, or triple) as a consequence. This is equivalent to saying that if we doubled the total charges (Q and $-Q$) on both isolated conductors (and waited for equilibrium to return), the potential V between them would double (as would the surface charge distributions σ_1 and σ_2, everywhere on the surfaces). From this we conclude that V is proportional to Q, as long as the geometry stays the same. Thus, if for example we transfer charge between one conductor and the other, V would increase in proportion to the increases in $\pm Q$ on the surfaces. We can therefore write $V = (1/C)Q$, where $1/C$ is the constant of proportionality, or equivalently, $Q = CV$, and the constant C is called the **capacitance** of the system. This constant

C depends on the geometry of the conductors, their size, shape and position, but it does not depend on the charge on the plates. For any particular geometry we can calculate its capacitance by assuming a certain charge and calculating the resultant V. Then $C = Q/V$, and for any other Q this ratio remains the same. The unit for capacitance is the **farad** (F). A capacitance of one farad is very large, and more common capacitances are μF (10^{-6} F) or pF (10^{-9} F). If we build a unit containing two conductors with relatively large surfaces close to each other (but not touching) we call this object a **capacitor** whose capacitance is C. The name derives from the fact that C represents the capacity of the two conductors to store charge on their surfaces per unit potential difference (per volt) between them. A large capacitance means that the capacitor holds a lot of charge per volt, while a small capacitance means that only a small amount of charge is held per volt. We will first discuss the calculation of capacitances for several specific geometries, and the use of these results. Then we will discuss the energy needed to charge a capacitor and the interpretations of these results. The most common capacitor geometry is that of two close parallel, conducting plates.

Problem 4.23. A "parallel plate capacitor" consists of two parallel plates, of area A, separated by a small distance d and carrying charges of $\pm Q$ (as in Fig. 4-11). Assume that the field is uniform throughout the region between the plates.

(*a*) What is the field between the plates?

(*b*) What is the potential difference between the plates?

(*c*) What is the capacitance of this parallel plate capacitor?

Solution

(*a*) The field was calculated in Problem 3.23, and equals $E = \sigma/\varepsilon_0$. Ignoring edge effects, the surface charge, σ, is uniformly distributed and $\sigma = Q/A$, giving $E = Q/\varepsilon_0 A$. This is a uniform field pointing from the positive to the negative plate.

(*b*) As shown in Problem 4.8(*b*), the potential difference between the plates is just $V = Ed = \sigma d/\varepsilon_0 = Qd/\varepsilon_0 A$. The positive plate is at the higher potential.

(*c*) Using the results of (*b*), we get $C = Q/V = Q/(Qd/\varepsilon_0 A) = \varepsilon_0 A/d$.

Problem 4.23 shows that the capacity of a parallel plate capacitor can be written as

$$C = \varepsilon_0 A/d \text{ (parallel plate capacitor)} \tag{4.9}$$

Note. The capacitance (ability to hold, or store, charge per volt) increases in proportion to the cross-sectional area of the plates, A. Thus doubling the area doubles C. The capacitance also varies in inverse proportion to the separation distance, d. Thus halving d doubles C as well.

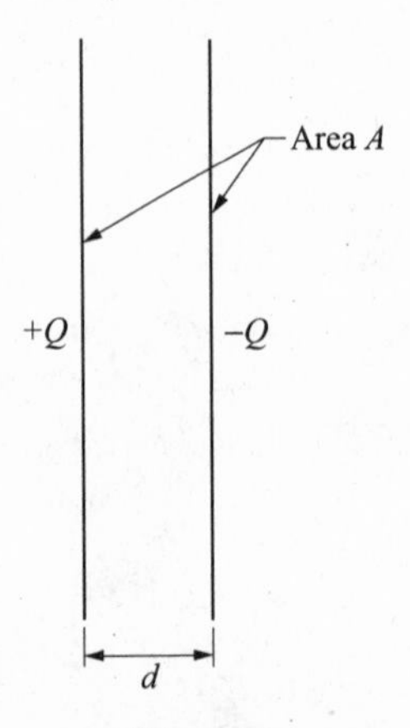

Fig. 4-11

Problem 4.24. A parallel plate capacitor has a capacitance of 2.5 μF and an area of 156 m^2.

(*a*) What is the distance between the plates?

(*b*) If one applies a voltage of 75 V to the capacitor how much charge is collected on each plate?

(*c*) How much work is needed to move an additional charge of 1.8×10^{-8} C from the negative to the positive plate?

Solution

(*a*) The capacitance is given by $C = \varepsilon_0 A/d = 2.5 \times 10^{-6} = (8.85 \times 10^{-12})(156 \text{ m}^2)/d$. Thus $d = 5.52 \times 10^{-4}$ m.

(*b*) The charge is given by $Q = CV = (2.5 \times 10^{-6} \text{ F})(75 \text{ V}) = 1.88 \times 10^{-4}$ C.

(*c*) Since the charge we are moving is small compared to the charge already there the potential will hardly change as we move the charge. Therefore the work needed, which is just the increase in potential energy, will be given by $\Delta QV = (1.8 \times 10^{-8} \text{ C})(75 \text{ V}) = 1.35 \times 10^{-6}$ J.

Problem 4.25. A parallel plate capacitor is built from plates with areas of 888 m^2 each and a separation of 1.6×10^{-4} m. The maximum electric field that can exist in the capacitor before the air ionizes causing sparking is 3.1×10^6 V/m.

(*a*) What is the capacitance of this capacitor?

(*b*) What is the maximum voltage that can be applied to this capacitor?

Solution

(*a*) The capacitance is given by $C = \varepsilon_0 A/d = (8.85 \times 10^{-12})(888 \text{ m}^2)/(1.6 \times 10^{-4} \text{ m}) = 4.91 \times 10^{-5}$ F.

(*b*) The maximum electric field that the capacitor can stand before electrical breakdown is 3.1×10^6 V/m. The electric field is equal to $Q/\varepsilon_0 A = CV/\varepsilon_0 A = 3.1 \times 10^6$. Thus $V = (8.85 \times 10^{-12})$ (888 m^2) $(3.1 \times 10^6 \text{ V/m})/4.91 \times 10^{-5} \text{ F} = 496$ V. This could have been derived more simply using the relationship that $V = dE$ for a uniform field, giving $V = (3.1 \times 10^6 \text{ V/m})(1.6 \times 10^{-4} \text{ m}) = 496$ V.

Problem 4.26. A capacitor consists of two thin concentric hollow metal spherical shells of radii r_1 and r_2 ($r_1 < r_2$) with charges Q and $-Q$, respectively

(*a*) What is the capacitance of this capacitor?

(*b*) Show that all the charges reside on the outer surface of the inner shell and the inner surface of the outer shell.

Solution

(*a*) The potential produced by a uniform spherical shell of charge Q was calculated earlier and given by Eqs. (*4.5*): $V = (1/4\pi\varepsilon_0)Q/r$ for $r > R$ and $V = (1/4\pi\varepsilon_0)Q/R$ for $r < R$. On the outer surface of the outer spherical shell the potential is zero, since we are outside of each shell and the potential is therefore $V = V_1 + V_2 = kQ/r + k(-Q)/r = 0$, $r \geq r_2$. On the outer surface of the inner shell the potential from sphere two is still $-kQ/r_2$ but the potential from the first sphere is kQ/r_1. Thus $V = kQ(1/r_1 - 1/r_2)$, which is also the potential difference between the shells (since the potential at the second shell is zero). Then $C = Q/V = 4\pi\varepsilon_0/(1/r_1 - 1/r_2)$.

(*b*) Since the potential is constant everywhere in the outer shell and beyond (actually zero) the electric field is zero everywhere in this region. Since $E = \sigma/\varepsilon_0$ just outside a conducting surface, we have $\sigma = 0$ on the outside of the outer sphere, and all the charge, $-Q$, resides on the inside surface. Similarly, in the hollow region within the inner shell the potential is constant [(Eq. (*4.5*)] and the electric field again vanishes. Thus σ/ε_0 on the inner surface vanishes as well, and the entire charge Q resides on the outer surface of the inner shell.

Problem 4.27. The two shells of Problem 4.26 have radii of 1.6 m and 1.8 m.

(*a*) What is the capacitance of this arrangement?

(*b*) How much voltage must be applied across the shells to store a charge of 3.7×10^{-8} C on the shells?

Solution

(*a*) The capacitance was derived in the previous problem and equals $C = 4\pi\varepsilon_0/(1/r_1 - 1/r_2) = 4\pi(8.85 \times 10^{-12})/[1/1.6 \text{ m} - 1/1.8 \text{ m}) = 1.60 \times 10^{-9} \text{ F} = 1.6 \text{ nF}$.

(*b*) The charge is given by $Q = CV$, so $V = Q/C = (3.7 \times 10^{-8} \text{ C})/(1.6 \times 10^{-9} \text{ F}) = 23.1$ V.

4.7 COMBINATION OF CAPACITORS

Capacitors have many applications in electrical circuits, both using constant sources of voltage such as batteries (Chap. 3), and using time varying sources of voltage (Chap. 9) such as supplied by the electric utility. Often one uses combinations of capacitors and we inquire into the result of making such combinations. There are two basic different ways in which one can combine capacitors. The two are called series and parallel combinations. We will see later that the same types of combinations can be applied to resistors as well. In what follows we will assume that the pair of close conductors constituting a capacitor is sufficiently far from the conductors making up the next capacitor, that we do not have to worry about "cross-capacitance" between the two capacitors. In addition, all connections between capacitors are made with conducting wire, and the conductors and wires so connected must all be at the same potential when we have equilibrium. For visual simplicity we will carry out our discussion in the context of parallel plate capacitors.

First we discuss what is called the parallel connection of capacitors. Here one side of all the capacitors are kept at a common potential by being connected to each other by a conducting wire, while the other sides of all the capacitors are kept at a (different) common potential by connection to a second conducting wire. This is illustrated in Fig. 4-12. Here the two sides of C_1 (the symbol for a capacitor is —||—) are connected to points *a* and *b* by conducting wires and so are the two sides of capacitor C_2. If one has three capacitors in series one would connect C_3 between the same two points. The left sides of the capacitors are thus at a common potential, and the right sides are at a different common potential. The potential difference across each capacitor is the same, since in each case it will equal $V_a - V_b$. This

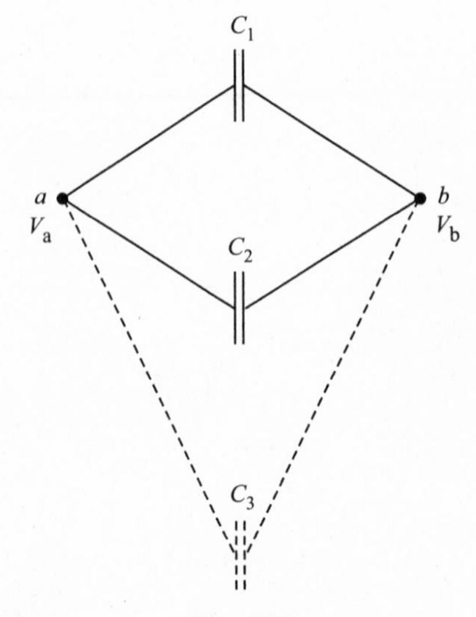

Fig. 4-12

is the defining characteristic of all parallel circuits: each branch has the same potential difference or voltage. We will use the next problem to develop the properties of a parallel circuit.

Problem 4.28. Consider the two capacitors in Fig. 4-12, connected between a difference of potential, $V = V_a - V_b$.

(*a*) What is the charge on the plates of each capacitor?

(*b*) What is the total charge collected on the equipotential surfaces connected to points *a* and *b*?

(*c*) If one replaced the two capacitors with a single capacitor, collecting the same charge between the two points, what capacitance would it have? (This is called the "equivalent" capacitor.)

(*d*) If $C_1 = 2.3\ \mu\text{F}$ and $C_2 = 5.7\ \mu\text{F}$, what is the equivalent capacitance of the combination?

Solution

(*a*) $Q_1 = C_1 V$ and $Q_2 = C_2 V$; i.e., if $V_a > V_b$, Q_1 and Q_2 will appear on the left plates of C_1 and C_2, respectively, while $-Q_1$ and $-Q_2$ will appear on the right plates of C_1 and C_2.

(*b*) The total charge is just the sum of Q_1 and Q_2 on side *a* and $-(Q_1 + Q_2)$ on side *b*.

(*c*) The equivalent capacitance would have to be charged to $(Q_1 + Q_2)$ when the potential difference across it is V. Thus, $C_{eq}V = Q_1 + Q_2 = C_1V + C_2V = (C_1 + C_2)V$. Dividing out by V we get:

$$C_{eq} = C_1 + C_2 \text{ (parallel combination)} \tag{4.10a}$$

(*d*) Using the given values for C_1 and C_2 we get $C_{eq} = (2.3 + 5.7)\ \mu\text{F} = 8.0\ \mu\text{F}$.

If capacitor C_3 were also connected as shown in Fig. 4-12 the same reasoning as in Problem 4.28 would lead to $C_{eq} = C_1 + C_2 + C_3$. In general, for any number of parallel capacitors,

$$C_{eq} = \sum C_i \tag{4.10b}$$

The other possible way to combine two capacitors is in series. Consider the two capacitors in Fig. 4-13. Here one plate of the first capacitor is connected to point *a* and the second plate is connected to the first plate of the next capacitor through point *c*. The second plate of the second capacitor is connected to point *b*. If there are more capacitors in series then the second is connected to the third and so on until the last is connected to point *b*. Now the potential across C_1 need not be the same as is the potential V_2 across C_2, since $V_1 = V_a - V_c$, and $V_2 = V_c - V_b$ and points *a* and *b* are not connected. Indeed the total voltage between *a* and *b* is $V = V_1 + V_2$. If we examine the figure more closely, we note if the first plate of C_1 accumulates charge $+Q_1$ (inserted or removed through point *a*), then the second plate of C_1 will have a charge of $(-Q_1)$. This follows because if it did not, the electric field immediately outside the plate would not vanish, and charges would flow in the wire (through point *c*) until the field vanished. This would occur when the charge is $-Q_1$. From where did this $-Q_1$ charge come? It must have come from the first plate of the second capacitor. In that case the second capacitor has the same charge as the first, $+Q_1$ on its first plate. Using the same reasoning as for the first capacitor, we conclude that the second capacitor will have charge $-Q_1$ on its second plate (where we presume that point *b* is connected to other parts of the circuit to or from which charges can flow). We are now ready

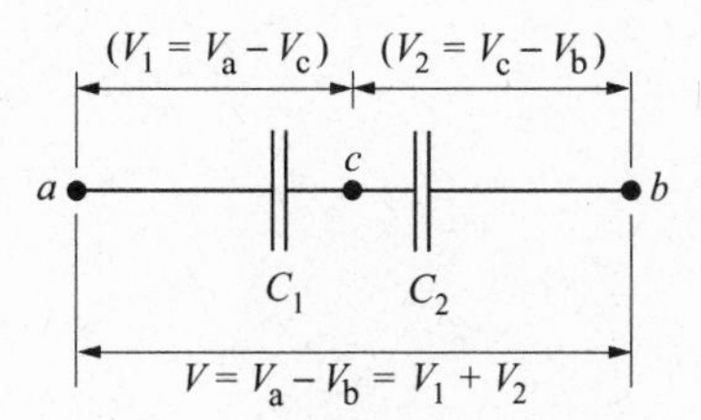

Fig. 4-13

to calculate the equivalent capacitance that we could use to replace C_1 and C_2.

Problem 4.29. Consider the two capacitors in series in Fig. 4-13. Calculate the equivalent capacitance.

Solution

We have just shown that each capacitor contains the same charge which we call Q. This is the charge which is supplied by the source of potential between a and b, and is the charge that will be on the equivalent capacitor that we can use to replace the combination of C_1 and C_2. Now $V_1 = Q/C_1$ and $V_2 = Q/C_2$. Then $V = V_1 + V_2 = Q(1/C_1 + 1/C_2) = Q/C_{eq}$. Thus

$$1/C_{eq} = 1/C_1 + 1/C_2 \text{ (series combination)} \tag{4.11a}$$

The same reasoning as used in Problem 4.29 can be used to generalize to any number of series capacitors:

$$1/C_{eq} = \sum (1/C_i) \tag{4.11b}$$

Often we have situations in which a number of capacitors are used in a circuit, some in series and some in parallel. In many cases we can combine the results of Eqs. (*4.10*) and (*4.11*) to obtain an overall equivalent capacitance.

Problem 4.30. Consider the combination of capacitors shown in Fig. 4-14(*a*). Here $C_1 = 2.5\ \mu F$, $C_2 = 3.5\ \mu F$, $C_3 = 5.6\ \mu F$ and $C_4 = 1.3\ \mu F$.

(*a*) What is the equivalent capacitance of C_2 and C_3?

(*b*) What is the equivalent capacitance between points a and b?

(*c*) If a voltage of 10.5 V were provided between points a and b, what charge would accumulate on the equivalent capacitance?

(*d*) For case (*c*), what charge accumulates on capacitor C_1? On capacitor C_4?

(*e*) What charge accumulates on capacitor C_2? On capacitor C_3?

Solution

(*a*) Capacitors C_2 and C_3 are in parallel (points c and d play the role of points a and b of Fig. 4-12). They can therefore be replaced with an equivalent capacitance of $C_{eq} = C_2 + C_3 = (3.5 + 5.6)\ \mu F = 9.1\ \mu F$ [see Fig. 4-14(*b*)].

(*b*) If we replace C_2 and C_3 with an equivalent capacitance $C_{eq} = 9.1\ \mu F$, we then have three capacitors in series. Using Eq. (*4.11b*), we get the final equivalent capacitance to be $1/C_{f,eq} = 1/C_1 + 1/C_{eq} + 1/C_4 = 1/2.5 + 1/9.1 + 1/1.3 = 1.28$, and $C_{f,eq} = 0.78\ \mu F$ [see Fig. 4-14(*c*)].

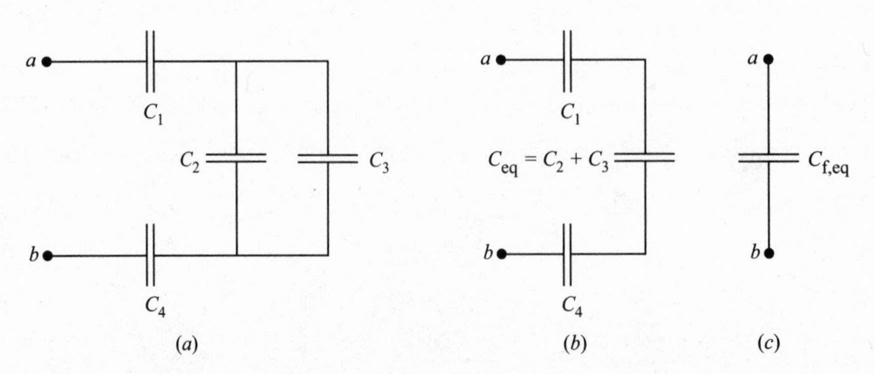

Fig. 4-14

(*c*) The voltage across $C_{f,\,eq}$ equals 10.5 V. Then the charge on the equivalent capacitor is $Q = C_{f,\,eq}V = 8.21 \times 10^{-6}$ C.

(*d*) In a series circuit, the charge on each capacitor is the same and is equal to the charge on the equivalent capacitor. Thus the charge on both C_1 and on C_4 is 8.21×10^{-6} C.

(*e*) The total charge on the two parallel capacitors C_2 and C_3 is the charge on C_{eq} which equals 8.21×10^{-6} C. This charge is distributed between C_2 and C_3. To get the individual charge Q_2 or Q_3 we need the voltage across each capacitor. We know that, for a parallel combination, the voltage across each capacitor is the same and is equal to the voltage across the equivalent capacitor. We can easily calculate the voltage across the equivalent capacitor $V' = Q/C_{eq} = (8.21 \times 10^{-6}\ \text{C})/(9.1\ \mu\text{F}) = 0.90$ V. Then $Q_2 = C_2V' = (3.5 \times 10^{-6}\ \text{F})(0.90) = 3.16 \times 10^{-6}$ C and $Q_3 = C_3V' = (5.6 \times 10^{-6}\ \text{F})(0.90) = 5.05 \times 10^{-6}$ C. Note that $Q_2 + Q_3 = 8.21 \times 10^{-6}$ C, as required.

4.8 ENERGY OF CAPACITORS

As stated previously, whenever we charge a capacitor we must do work to bring more positive charge to the plate that was already positively charged, and similarly to the negative plate. This work is converted into potential energy of the capacitor, which can be viewed as the energy stored by the charges that have been separated. As we will see, we can also take an alternative viewpoint that the effect of separating the charges is to produce an electric field in space, and that the accumulated energy is stored in these electric fields.

If a capacitor is charged to a difference of potential V, then the work by an outside force that is needed to transfer an additional small charge ΔQ from the negative to the positive plate is $(\Delta Q)V = Q(\Delta Q)/C$. Using arguments similar to those used to calculate the potential energy of the spring (Beginning Physics I, Problem 6.8), we can show that the work needed to accumulate a charge of Q on the capacitor is $W = (\frac{1}{2})Q^2/C$. Then the energy stored in a capacitor can be written as

$$U_p = (\tfrac{1}{2})Q^2/C = (\tfrac{1}{2})CV^2 = (\tfrac{1}{2})QV \qquad (4.12)$$

Problem 4.31. Derive the expression for the electrical potential energy stored in a capacitor C with charge Q [Eq. (*4.12*)]

Solution

We know that when the capacitor is charged to some value q_i the potential is given by $V_i = q_i/C$. The work necessary to bring the next increment of charge, Δq, across [so that the new plate charge will be $(q_i + \Delta q)$ and $-(q_i + \Delta q)$], is given by: $\Delta W_i = V_i\Delta q$. In Fig. 4-15 we show a plot of potential difference vs. charge for our capacitor, as well as the increment from q_i to $q_i + \Delta q$. Clearly, ΔW_i is just the area under the V vs. q curve between the adjacent dotted vertical lines. The total work done in bringing charges across, starting from $q = 0$ to $q = Q$ is just the triangular area under the V vs. q curve between the origin and $q = Q$. This is just: $W = (\frac{1}{2})QV = (\frac{1}{2})Q^2/C = (\frac{1}{2})CV^2$ as indicated in Eq. (*4.10*).

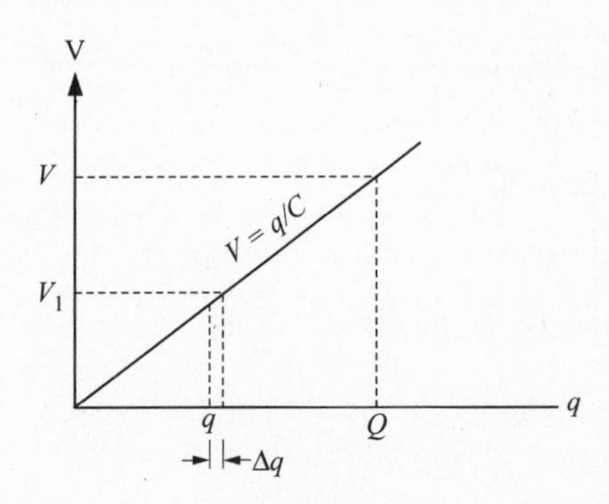

Fig. 4-15

Problem 4.32. A capacitor with $C = 82.3\ \mu F$ is charged to a voltage of 110 V.

(*a*) How much charge is accumulated on the capacitor?

(*b*) How much potential energy is stored in the capacitor?

(*c*) If the voltage on the capacitor is to be increased to 150 V, what additional work will have to be done?

(*d*) If the capacitor is discharged from 150 V to 75 V, how much work can be done by the electric field on the moving charges?

Solution

(*a*) The charge on the capacitor is $Q = CV = (82.3\ \mu F)(110\ V) = 9.05 \times 10^{-3}$ C.

(*b*) The potential energy is given by Eq. (*4.12*) as $U_p = (\frac{1}{2})CV^2 = (\frac{1}{2})(82.3 \times 10^{-6}\ F)(110\ V)^2 = 0.50$ J. (Alternatively, $U_p = (\frac{1}{2})Q^2/C = (\frac{1}{2})(9.05 \times 10^{-3}\ C)^2/(82.3\ \mu F) = 0.50$ J).

(*c*) The final potential energy is $(1/2)(82.3 \times 10^{-6})(150)^2 = 0.93$ J. The additional work is $W = \Delta U_p = U_{pf} - U_{pi} = 0.93 - 0.50 = 0.43$ J.

(*d*) When the electric field does positive work the electric potential energy decreases by a like amount. Thus $W = -\Delta U_p = U_{pi} - U_{pf} = 0.93\ J - (\frac{1}{2})(82.3 \times 10^{-6})(75)^2\ J = 0.70$ J.

Problem 4.33. Consider the combination of capacitors used in Problem 4.30, with the voltage of 10.5 V between points *a* and *b* (Fig. 4-14).

(*a*) What is the total potential energy stored in the combination?

(*b*) What is the energy stored on each of the capacitors?

Solution

(*a*) We showed that the equivalent capacitance of the combination between points *a* and *b* is 0.78 μF. Then the total energy stored is $(\frac{1}{2})C_{f,\,eq}V^2 = (\frac{1}{2})(0.78\ \mu F)(10.5)^2 = 4.3 \times 10^{-5}$ J.

(*b*) For each capacitor we can use either $U_p = (\frac{1}{2})CV^2$ or $U_p = (\frac{1}{2})Q^2/C$. On C_1 and C_4 we know that the charge is 8.21×10^{-6} C, so the energies are $U_{p1} = (\frac{1}{2})(8.21 \times 10^{-6})^2/2.5\ \mu F = 1.35 \times 10^{-5}$ J, and $U_{p4} = (\frac{1}{2})(8.21 \times 10^{-6})^2/1.3\ \mu F = 2.59 \times 10^{-5}$ J. For C_2 and C_3 we know that $V' = 0.90$ V. Thus $U_{p2} = (\frac{1}{2})3.5\ \mu F)(0.90)^2 = 1.4 \times 10^{-6}$ J, and $U_{p3} = (\frac{1}{2})(5.6\ \mu F)(0.90)^2 = 2.3 \times 10^{-6}$ J. The total energy is then $(1.35 + 2.59 + 0.10 + 0.23) \times 10^{-5}\ J = 4.3 \times 10^{-5}$ J, as we found in part (*a*).

The energy that is stored in a capacitor can be viewed as the energy stored by the charge that has been separated. As a result of separating these charges, electric fields are established in space. We can therefore, alternatively, view the work done in separating the charges as the work required to produce these electric fields. The energy stored would then be viewed as the energy stored in these electric fields. We will illustrate this view by using a parallel plate capacitor as an example, but the result we derive will be valid for all situations in which electric fields are established.

Problem 4.34. Consider a parallel plate capacitor whose plates have an area of *A* and are separated by a distance *d*. As shown previously the capacitance is given as $C = \varepsilon_0 A/d$. A difference of potential *V* is established between the plates.

(*a*) Derive an expression for the energy stored in the capacitor in terms of the dimensions of the capacitor and the (constant) electric field within the capacitor.

(*b*) Derive an expression for the "**energy density**" (the energy per unit volume) within the capacitor.

Solution

(*a*) We know that the electric field within a parallel plate capacitor is $E = V/d$, and that the energy stored is $U_p = (\frac{1}{2})CV^2 = (\frac{1}{2})(\varepsilon_0 A/d)(Ed)^2 = (\frac{1}{2})(\varepsilon_0 E^2)(Ad)$.

(*b*) The volume within the capacitor is Ad. In this volume the electric field is given by the formula we used (again ignoring slight edge effects). Outside of this volume, the electric field is essentially zero. Thus the energy density is $U_{pd} = (\frac{1}{2})\varepsilon_0 E^2$. This is a general expression for the energy density (we will modify this slightly in the next section)

$$U_{pd} = (\tfrac{1}{2})\varepsilon_0 E^2 \tag{4.13}$$

Problem 4.35. A parallel plate capacitor has a capacitance of 2.6 μF. The plates are separated by a distance of 0.63 mm.

(*a*) If a voltage of 34 V is applied to the plates of the capacitor, calculate the energy stored in the capacitor.

(*b*) Calculate the electric field within the capacitor.

(*c*) Calculate the energy density within the capacitor.

(*d*) Use the results of parts (*a*) and (*b*) to obtain the area A of the capacitor plates

(*e*) Calculate the energy stored in a cylindrical volume of base area $A' = 0.36$ m^2 extending from one plate to the other within the capacitor.

Solution

(*a*) The energy stored is $(\frac{1}{2})CV^2 = (\frac{1}{2})(2.6\ \mu\text{F})(34\ \text{V})^2 = 1.50 \times 10^{-3}$ J.

(*b*) The electric field within the capacitor is $E = V/d = (34\ \text{V})/(0.63 \times 10^{-3}\ \text{m}) = 5.40 \times 10^4$ V/m.

(*c*) The energy density is given by $U_{pd} = (\frac{1}{2})(\varepsilon_0 E^2) = (\frac{1}{2})(8.85 \times 10^{-12})(5.40 \times 10^4)^2 = 1.29 \times 10^{-2}$ J/m^3.

(*d*) $U_p = U_{pd}(Ad) \rightarrow 1.5 \times 10^{-3}\ \text{J} = (1.29 \times 10^{-2}\ \text{J/m}^2)(0.63 \times 10^{-3}\ \text{m})A \rightarrow A = 185\ \text{m}^2$.

(*e*) The volume of the cylinder is $Ad = (0.36\ \text{m}^2)(0.63 \times 10^{-3}\ \text{m}) = 2.27 \times 10^{-4}$ m^3. The energy stored in that volume is the energy density times the volume $= 1.29 \times 10^{-2}(2.27 \times 10^{-4}) = 2.93 \times 10^{-6}$ J.

4.9 DIELECTRICS

So far we have discussed only cases in which charges establish electric fields and potentials in empty space or on conductors. If the region includes other, non-conducting materials, even when the materials are not charged (neutral), the atoms and molecules within that material may alter the fields that are otherwise produced. We have already seen that when neutral conductors are placed near free charges, the free charges in the conductors redistribute themselves on the surface and thereby produce fields of their own which must be added to the fields of the original charges. Unlike conductors other neutral materials do not have free charges and we must consider what mechanism might cause electrical effects to arise.

Normal **insulating materials** consist of atoms and molecules that are composed of positively charged nuclei and negatively charged electrons that are tightly bound together with no loose outer electrons that are free to roam. In the presence of an electric field the positive and negative charges in the atoms and molecules are pulled in opposite directions. As a result, the atoms and molecules will become somewhat "**polarized**" with the positive and negative charges becoming slightly separated from their equilibrium positions. This separation is expected to be approximately proportional to the magnitude of the electric field as long as the field is not too large. The (slightly) separated charges will produce their own electric field which must be added to the field established by the original charges. In general this can lead to many complications, and we will consider only a special case in which the effect can be easily understood.

Consider a parallel plate capacitor which is filled with some insulating material. We call this material a "**dielectric**" since, as we will show, it will produce its own electric field in a direction opposite to the original field. If we place a surface charge distribution of $\pm\sigma$ on the plates of the capacitor, this charge will produce an electric field of σ/ε_0 within the capacitor. The field will point from the positive to the negative plate. This field will cause a polarization of the material such that each atom will have its positive charge move closer to the negative plate (see Fig. 4-16). We will then have tiny "**dipoles**" throughout the material with positive charge to the left and negative charge to the right. In the interior of the dielectric the material remains uncharged since the shifting of negative charge slightly to the right from one parallel layer will be compensated by negative charge shifting into that layer from the next layer to the left. Only at the surfaces, next to the plates, will charge accumulate. On the left surface in Fig. 4-16 the electrons that shift to the right are not compensated for and a net positive charge appears; on the right surface negative charges moving from the layer just to the left of the surface accumulate on the surface, and cannot be compensated for by electrons moving further to the right. Since the bulk of the dielectric remains neutral, the net "polarization" charges on the two surfaces of the dielectric are equal and opposite. Thus the dielectric develops a surface charge next to each of the plates which is of opposite sign to the original charge on the plates. This is equivalent to an additional charge added to the plates which produces its own electric field in a direction opposite to the original field. The total field within the dielectric will therefore be reduced in this region. If the polarization is proportional to the field, then the new total field will be proportional to the field that would be produced in the absence of the dielectric material. We can then write that $E = E_0/\kappa$, where E is the total field in the presence of the dielectric, E_0 is the field that would be present without the dielectric and κ is the "**dielectric constant**" of the material. These dielectric constants vary from material to material, and some common examples are given in Table 4-1.

With this electric field the potential difference between the plates is $V = Ed = E_0 d/\kappa = \sigma d/\kappa\varepsilon_0 = Qd/A\kappa\varepsilon_0$, where σ and Q represent the free charge density and free total charge on the capacitor plates. Recalling that the capacitance without the dielectric is $C_0 = \varepsilon_0 A/d$, we have $V = Q/\kappa C_0 = Q/C$, where C is the true capacitance in the presence of the dielectric. Thus $C = \kappa C_0 = \kappa\varepsilon_0 A/d = \varepsilon A/d$, where $\varepsilon = \kappa\varepsilon_0$ is called the "**permittivity**" of the material. (correspondingly, ε_0 is called the **permittivity of free space.**) Since κ is always greater than 1, the addition of a dielectric within a capacitor increases the capacitance by the factor κ.

The energy stored in the capacitor is still given by $U_p = (\frac{1}{2})CV^2$, but both C and V are modified for a particular free charge Q on the plates. More charge is needed on each plate to produce the same

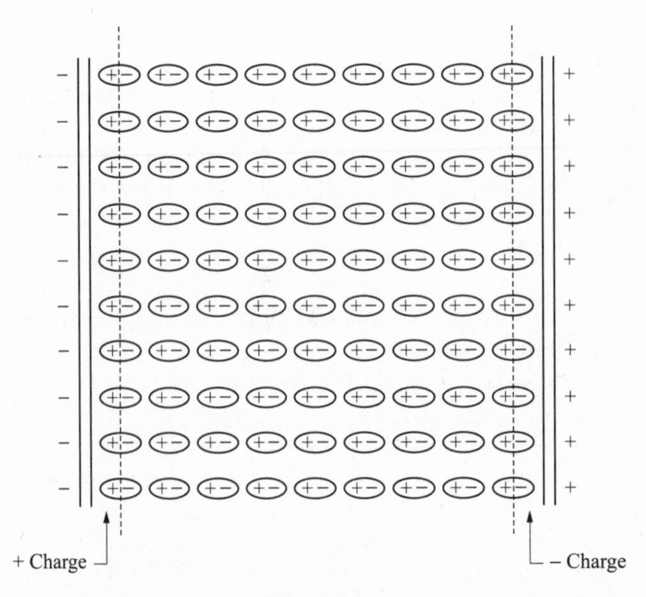

Fig. 4-16

Table 4.1. Dielectric Constants in Common Materials

Material	Dielectric constant
Vacuum	1
Air	1.0005
Teflon	2.1
Paper	3.3
Mica	3–6
Glass	5–10
Water	80.4

potential difference. Correspondingly, the energy density within the dielectric is modified from its value in vacuum, and is given by $U_{pd} = (\frac{1}{2})\varepsilon E^2$.

Problem 4.36. A parallel plate capacitor has plates with an area of 71 m^2. The plates are separated by a distance of 0.63 mm and the capacitor is filled with a dielectric of dielectric constant $\kappa = 2.6$. A voltage of 34 V is applied to the plates of the capacitor. Calculate (*a*) the capacitance of the capacitor, (*b*) the electric field within the capacitor, (*c*) the energy density within the capacitor, (*d*) the surface charge and charge density on the plates of the capacitor (the free charges) and (*e*) the surface charge and charge density on the dielectric layer near the plates.

Solution

(*a*) The capacitance is $C = \kappa\varepsilon_0 A/d = (2.6)(8.85 \times 10^{-12})(71\text{ m}^2)/(0.63 \times 10^{-3}\text{ m}) = 2.6\ \mu\text{F}$.

(*b*) The electric field within the capacitor is $E = V/d = (34\text{ V})/(0.63 \times 10^{-3}\text{ m}) = 5.40 \times 10^4\text{ V/m}$.

(*c*) The energy density is given by $U_{pd} = (\frac{1}{2})(\varepsilon E^2) = (\frac{1}{2})(2.6 \times 8.85 \times 10^{-12})(5.40 \times 10^4)^2 = 3.35 \times 10^{-2}$ J/m^3.

(*d*) The charge on the plates is $Q = CV = 2.6\ \mu\text{F}(34\text{ V}) = 8.84 \times 10^{-5}$ C. The charge density is $\sigma = Q/A = 1.25\ \mu\text{C/m}^2$.

(*e*) The electric field within the capacitor is produced by two parallel charge distributions, that on the plates and that on the surface of the dielectric. Since the two distributions are of the opposite sign, the field produced is $E = (\sigma - \sigma_d)/\varepsilon_0 = (Q - Q_d)/A\varepsilon_0$. Now from part (*b*) $E = V/d = 5.4 \times 10^4$ V/m and $(\sigma - \sigma_0)/(8.85 \times 10^{-12}) = 5.4 \times 10^4 \rightarrow \sigma - \sigma_0 = 4.78 \times 10^{-7}\text{ C/m}^2$. Recalling σ from part (*d*) we have $\sigma_d = 1.25\ \mu\text{C/m}^2 - 0.48\ \mu\text{C/m}^2 = 0.77\ \mu\text{C/m}^2$. The total surface charge on the dielectric is then $Q_d = \sigma_d A = (0.77 \times 10^{-6})(71) = 5.47 \times 10^{-5}$ C.

Problem 4.37. A potential difference of 25 V is maintained across the plates of a parallel plate capacitor. The plates have an area of 43 m^2 and are separated by 1.56 mm.

(*a*) What is the capacitance of the capacitor if it is filled with air?

(*b*) How much energy is stored in this capacitor?

(*c*) What is the energy stored in the capacitor if it is filled with a dielectric of dielectric constant $\kappa = 1.9$ and the potential is held fixed?

(*d*) How much work is done when the dielectric is inserted between the plates?

(*e*) How much charge is on the plates with and without the dielectric?

Solution

(*a*) The capacitance is $C = \varepsilon_0 A/d = (8.85 \times 10^{-12})(43\ \text{m}^2)/(1.56 \times 10^{-3}\ \text{m}) = 0.244\ \mu\text{F}$.

(*b*) The energy stored $= (\frac{1}{2})CV^2 = (\frac{1}{2})(0.244\ \mu\text{F})(25\ \text{V})^2 = 7.62 \times 10^{-5}$ J.

(*c*) The energy stored is changed because the capacitance is increased to $\kappa C_0 = 1.9(0.244\ \mu\text{F}) = 0.464\ \mu\text{F}$. Then the energy stored is 1.44×10^{-4} J.

(*d*) The work done is the change in the energy stored, which equals $(1.44 - 0.76) \times 10^{-4}\ \text{J} = 6.8 \times 10^{-5}$ J. This work is done in the process of increasing the charge on the plates, as the dielectric is inserted, to keep the voltage across the capacitor fixed.

(*e*) The charge in each case equals $Q = CV$. For air, $Q = (0.244\ \mu\text{F})(25\ \text{V}) = 6.1 \times 10^{-6}$ C. For the dielectric, $Q = (0.464\ \mu\text{F})(25\ \text{V}) = 1.16 \times 10^{-5}$ C.

Problems for Review and Mind Stretching

Problem 4.38. A square, of side 0.38 m, has a charge $Q_1 = 7.6 \times 10^{-8}$ C at each of three corners, and a charge $Q_2 = -5.3 \times 10^{-8}$ C at the fourth corner, as in Fig. 4-17.

(*a*) What electric field is produced at the center of the square?

(*b*) What potential is produced at the center of the square?

(*c*) How much work must be done by an outside force to just remove Q_2 to a very large distance $(\rightarrow \infty)$?

Solution

(*a*) The magnitude of the field produced by each charge is $|E| = kQ/r^2$. The directions of E from the Q_1 at the two opposite corners are opposite and therefore cancel out. The direction of $\mathbf{E}_1$ is toward q_2, and has a magnitude of $|E_1| = (9.0 \times 10^9)(7.6 \times 10^{-8}\ \text{C})/(0.38/\sqrt{2}\ \text{m})^2 = 9.47 \times 10^3$ V/m. The direction of $\mathbf{E}_2$ is also toward q_2 since Q_2 is negative, and has a magnitude of $|E_2| = (9.0 \times 10^9)(5.3 \times 10^{-8}\ \text{C})/(0.38/\sqrt{2}\ \text{m})^2 = 6.61 \times 10^3$ V/m. The sum of these two fields is toward Q_2, and equals $(9.47 + 6.61) \times 10^3\ \text{V/m} = 1.61 \times 10^4$ V/m. This is the total field at the center of the square.

(*b*) The potential at the center is the scalar sum of the potential due to each charge. It therefore equals $V = 3V_1 + V_2 = k(3Q_1 + Q_2)/r = (9.0 \times 10^9)(3 \times 7.6 - 5.3) \times 10^{-8}\ \text{C}/(0.38/\sqrt{2}\ \text{m}) = 5.86 \times 10^3$ V.

(*c*) To calculate the work needed to remove Q_2 far away, we must calculate the change in potential energy between the case of Q_2 at infinity and at its present position. The change that occurs is that the potential energy between Q_2 and the three other charges becomes zero at ∞, while the potential energy between the fixed three charges does not change. When Q_2 is at its present position its potential energy equals the sum of kQ_1Q_2/r_{12} for each of the three charges. Two of the charges are at a distance of 0.38 m from Q_2, and the third charge is at a distance of $0.38\sqrt{2}$ m from Q_2. Thus $U_{\text{pi}} = (9.0 \times 10^9)(7.6 \times 10^{-8}\ \text{C})(-5.3 \times 10^{-8}\ \text{C})(2/0.38\ \text{m} + 1/0.38\sqrt{2}\ \text{m}) = -2.58 \times 10^{-4}$ J. The change in potential energy, which is the work that is needed, is 2.58×10^{-4} J.

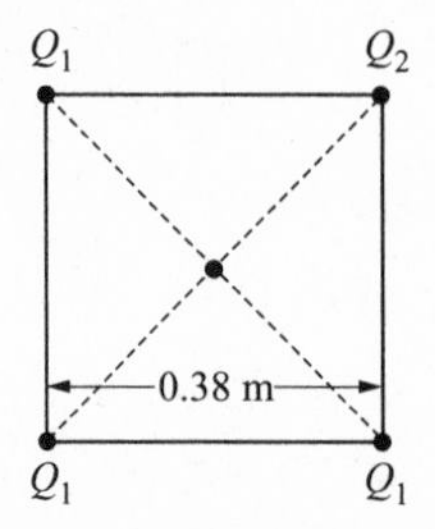

Fig. 4-17

Problem 4.39. A dipole consists of a positive charge q at $x = d/2$ and a negative charge $-q$ at $x = -d/2$ (as in Fig. 4-18). The dipole "moment", p, is defined as $p = qd$, where d is the distance between the charges.

(*a*) What is the potential produced by this dipole at a point on the x axis far from the dipole, i.e. at $x \gg d$?

(*b*) What is the potential produced by this dipole at a point on the y axis?

(*c*) What is the potential produced by this dipole at a point (x,y) far from the dipole, i.e. $r = (x^2 + y^2)^{1/2} \gg d$?

Solution

(*a*) The potential is the sum of the potential from the two charges. Thus $V = kq/(x - d/2) - kq/(x + d/2)$. Combining by using the common denominator gives, $V = kq[(x + d/2) - (x - d/2)]/[(x + d/2)(x - d/2)] = kqd/(x^2 - d^2/4) \approx k(qd)/x^2 = kp/x^2$, since $x \gg d$. In the numerator we were unable to neglect d compared to x, because the x canceled upon subtraction and we are left with d as a multiplicative factor, but in the denominator the x^2 term clearly dominates.

(*b*) In this case the potential is $V = kq/[y^2 + (d/2)^2]^{1/2} - kq/[y^2 + (-d/2)^2]^{1/2} = 0$.

(*c*) The distance from the charges to the point (x,y) is $[(x - d/2)^2 + y^2]^{1/2}$ and $[(x + d/2)^2 + y^2]^{1/2}$ for the positive and negative charges, respectively. For $r \gg d$, each of these is approximately equal to $r = (x^2 + y^2)^{1/2}$, and we can use this approximation whenever we are not subtracting the two distances from each other. We can write $V = kq[1/[(x - d/2)^2 + y^2]^{1/2} - 1/[(x + d/2)^2 + y^2]^{1/2}]$. Combining using a common denominator we get $V = [kq/r^2]\{[(x + d/2)^2 + y^2]^{1/2} - [(x - d/2)^2 + y^2]^{1/2}\}$, where we have used the approximation that $[(x \pm d/2)^2 + y^2]^{1/2} \approx (x^2 + y^2)^{1/2} = r$ in the denominator. Now, $[(x + d/2)^2 + y^2]^{1/2} = [x^2 + dx + d^2/4 + y^2]^{1/2} \approx [r^2 + dx]^{1/2} \approx r(1 + dx/2r^2)$. Similarly, $[(x - d/2)^2 + y^2]^{1/2} = [x^2 - dx + d^2/4 + y^2]^{1/2} \approx [r^2 - dx]^{1/2} \approx r(1 - dx/2r^2)$. Then $V \approx (kq/r^2)[(r + dx/2r) - (r - dx/2r)] = kqdx/r^3 = kp \cos\theta/r^2$. This result gives us the correct answer for part (*a*) when $\theta = 0$ and for part (*b*) when $\theta = 90°$.

Problem 4.40. A charge of $Q_1 = 4.35 \times 10^{-8}$ C is at the center of a conducting spherical shell of inner radius $r_1 = 0.93$ m and outer radius $r_2 = 1.07$ m. The shell itself has a charge of $Q' = -7.55 \times 10^{-8}$ C.

(*a*) What charge Q_1 is on the inner surface of the sphere and what charge Q_2 is on the outer surface?

(*b*) What is the potential at $r = 1.55$ m?

(*c*) What is the potential at $r = 1.00$ m?

(*d*) What is the potential at $r = 0.67$ m?

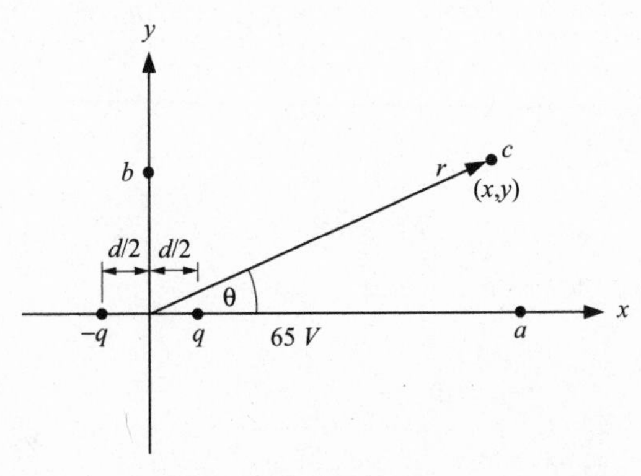

Fig. 4-18

Solution

(*a*) The charge on the inner surface must equal $Q_1 = -Q = -4.35 \times 10^{-8}$ C in order that the field is zero within the conducting sphere. Then the charge on the outer surface must equal $Q_2 = -3.20 \times 10^{-8}$ C so that the total charge on the shell is $Q' = -7.55 \times 10^{-8}$ C.

(*b*) The potential at any point is the sum of the potential produced by the three charges: Q, Q_1 and Q_2. The potential from Q is kQ/r, and the potential from the charges on the surfaces is given by Eq. (*4.5*): (a) $V = (1/4\pi\varepsilon_0)Q/r$ for $r > R$ and (*b*) $V = (1/4\pi\varepsilon_0)Q/R$ for $r < R$. At $r = 1.55$ m we are outside of all the charge distributions, and the total potential is $V = k(Q + Q_1 + Q_2)/r = (9.0 \times 10^9)(4.35 - 4.35 - 3.20) \times 10^{-8}$ C/1.55 $= -186$ V.

(*c*) At $r = 1.00$ m, we are outside of Q and Q_1, but inside Q_2. Then $V = k(Q + Q_1)/r + kQ_2/r_2 = 0 + (9.0 \times 10^9)(-3.20 \times 10^{-8}\text{ C})/1.07 = -269$ V.

(*d*) At $r = 0.67$ m, we are inside Q_1 and Q_2, and $V = kQ/r + k(Q_1/r_1 + Q_2/r_2) = (9.0 \times 10^9)(4.35 \times 10^{-8}\text{ C})/0.67\text{ m} + (9.0 \times 10^9)(-4.35 \times 10^{-8}\text{ C}/0.93\text{ m} - 3.20 \times 10^{-8}\text{ C}/1.07\text{ m}) = -106$ V.

Problem 4.41. The capacitance of two concentric spherical shells was calculated in Problem 4.26 as $C = 4\pi\varepsilon_0/(1/r_1 - 1/r_2)$. Show that as $r_1 \to r_2$, the capacitance approaches $\varepsilon_0 A/d$, where A is the surface area of the sphere and $d = r_2 - r_1$.

Solution

The capacitance can be written as $C = 4\pi\varepsilon_0 r_1 r_2/(r_2 - r_1)$. As $r_1 \to r_2$, $C \to 4\pi\varepsilon_0 r^2/d = \varepsilon_0 A/d$. This is just the formula for a parallel plate capacitor of area A separated by d. Thus the two spherical surfaces behave like two parallel surfaces separated by d.

Problem 4.42. A coaxial cable consists of an inner conducting cylinder of radius r_1 and a coaxial conducting cylindrical shell of inner radius r_2. Calculate the capacitance between the inner and outer cylinders for one meter of this cable.

Solution

We assume that the inner cylinder has a charge of $+Q$ and the outer cylinder has a charge of $-Q$. To calculate the potential difference between the cylinders we make use of the formulas given for charged cylinders in Eq. (*4.7*) for a long cylinder with surface charge at R: (*a*) $V = -(\lambda/2\pi\varepsilon_0)\ln(r/R')$ for $r > R$; (*b*) $V = -(\lambda/2\pi\varepsilon_0)\ln(R/R')$ for $r < R$. Here $\lambda = Q/L$, and R' is an arbitrary radius, usually taken as R. The potential at r_2 will then equal $V = V_1 + V_2 = 0$, since we get opposite contributions from the two surface charges using Eq. (*4.7a*) for both. At r_1, we must use Eq. (*4.7b*) for V_2, since we are now at $r < r_2$. Then $V = -(\lambda/2\pi\varepsilon_0)\ln(r_1/R') - (-\lambda/2\pi\varepsilon_0)\ln(r_2/R') = (\lambda/2\pi\varepsilon_0)\ln(r_2/r_1) = (Q/2\pi\varepsilon_0 L)\ln(r_2/r_1)$. The capacitance per unit length $C/L = Q/VL = 2\pi\varepsilon_0/\ln(r_2/r_1)$.

Problem 22.43. Several capacitors are connected as in Fig. 4-19(*a*). The capacitors have capacitance of: $C_1 = C_6 = 2.5\ \mu$F, $C_2 = C_3 = C_4 = 1.5\ \mu$F, $C_5 = 3.5\ \mu$F. The charge on C_3 is $Q_3 = 5.3 \times 10^{-6}$ C.

(*a*) What is the equivalent capacitance between points a and f?

(*b*) What is the difference of potential between points c and d?

(*c*) What is the difference of potential between points b and e?

(*d*) What is the difference of potential between points a and f?

(*e*) What is the charge on each capacitor?

Solution

(*a*) We first calculate the equivalent capacitance of the three capacitors that are in series, C_2, C_3 and C_4. This is given by $1/C_{eq} = 1/C_2 + 1/C_3 + 1/C_4 = 3(1/1.5\ \mu\text{F})$, or $C_{eq} = 0.50\ \mu$F. The circuit can then be

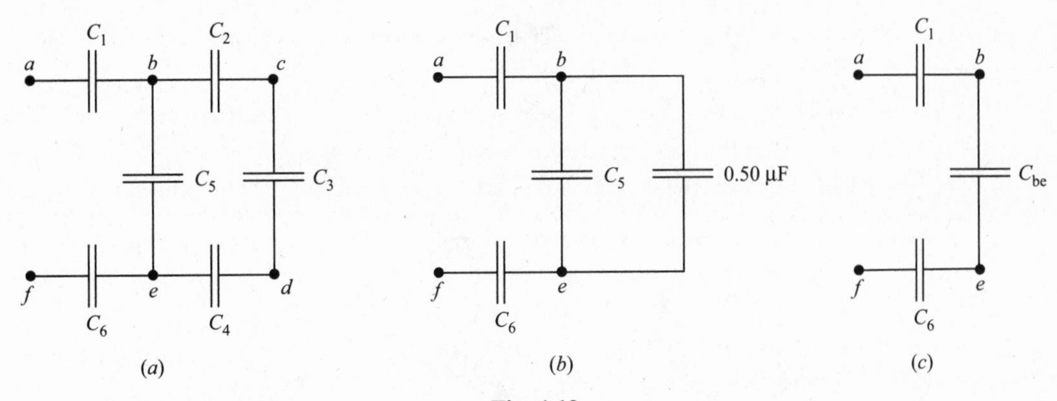

Fig. 4-19

redrawn as in Fig. 4-19(*b*). We then combine this capacitor and the parallel capacitor C_5 with an equivalent capacitor C_{be} of $C_{be} = (3.5 + 0.5)\ \mu F$, as in Fig. 4-19(*c*). Finally, we combine the three series capacitors in this figure to get the equivalent capacitance between *a* and *f*, $C_{af} = 1/C_1 + 1/C_{be} + 1/C_6$, giving $C_{af} = 0.952\ \mu F$.

(*b*) The potential difference between the points *c* and *d* is the potential across $C_3 = Q_3/C_3 = (5.3 \times 10^{-6}\ C)/(1.5 \times 10^{-6}\ F) = 3.53\ V$.

(*c*) The potential difference between the points *b* and *e* is the potential across each of the parallel capacitors Fig. 4-19(*b*). The charge on the 0.50 μF capacitor is the same as on each of the three series capacitors, C_2, C_3 and C_4, which is 5.3×10^{-6} C. Thus $V_{be} = (5.3 \times 10^{-6}\ C)/(0.50 \times 10^{-6}\ F) = 10.6\ V$.

(*d*) The potential difference V_{af} will equal Q/C_{af} where Q is the common charge on each of the three series capacitors in Fig. 4-19(*c*). The charge on C_{be} can be calculated as $C_{be}V_{be} = (4.0 \times 10^{-6}\ F)(10.6\ V) = 4.24 \times 10^{-5}\ C$. Then $V_{af} = (4.24 \times 10^{-5}\ C)/(0.952 \times 10^{-6}\ F) = 44.5\ V$.

(*e*) In part (*d*) we already used the fact that $Q_1 = Q_6 = Q_{be} = 4.24 \times 10^{-5}$ C [Fig. 4-19(*c*)]. From Fig. 4-19(*a*) we see that $Q_2 = Q_3 = Q_4 = 5.3 \times 10^{-6}$ C. From Fig. 4-19(*b*) we see that $Q_5 = C_5 V_{be} = (3.5 \times 10^{-6}\ F)(10.6\ V) = 3.71 \times 10^{-5}\ C$.

Problem 4.44. In a certain region of space the equipotential surfaces are the surfaces of concentric spheres. The potential is given as $V = -V_0 r/r_0$, where $V_0 = 38$ V, is the potential at $r_0 = 0.35$ m and r is the distance from the center of the concentric spheres.

(*a*) What is the direction of the electric field at a distance r from the center of the spheres?

(*b*) What is the magnitude of the field at this value of r?

(*c*) If a particle with a charge of 6.1×10^{-7} C and mass 9.3×10^{-15} kg has a speed of 3.8×10^5 m/s at $r = 0.35$ m, what is the speed of this particle when it reaches $r = 2.8$ m?

Solution

(*a*) The electric field lines are always perpendicular to the equipotential surface and point from high to low potential. The direction that is perpendicular to the surface of concentric spheres is the radial direction. Therefore the field points along a radius. Since the potential decreases as r increases (it becomes more negative), the field points away from the center (outward) along the radius.

(*b*) The magnitude of the electric field is given by $|E| = |\Delta V/\Delta d|$ when Δd is along the direction of the field lines. To get $|E|$ we calculate V at r and at $(r + \Delta r)$ and subtract to get ΔV. This gives us $|\Delta V| = (V_0/r_0)[(r + \Delta r) - r] = V_0 \Delta r/r_0$. Thus $|E| = \Delta V/\Delta r = V_0/r_0$, and the magnitude of E is constant throughout the region.

(*c*) We use conservation of energy in this part. This requires that the sum of the potential and kinetic energy be the same at both points. The potential energy is $U = qV$ and the kinetic energy is $K = (\frac{1}{2})mv^2$. Initially $K = (\frac{1}{2})(9.3 \times 10^{-15}\ \text{kg})(3.8 \times 10^5\ \text{m/s})^2 = 6.71 \times 10^{-4}$ J, and $U_p = q(-V_0) = (6.1 \times 10^{-7}\ \text{C})(-38\ \text{V}) = -2.32 \times 10^{-5}$ J. At $r = 2.8$ m, $U_p = q(-V_0 r/r_0) = (6.1 \times 10^{-7}\ \text{C})(-38 \times 2.8/0.35) = -1.85 \times 10^{-4}$ J. Then adding kinetic and potential energies, $6.71 \times 10^{-4} - 2.32 \times 10^{-5} = -1.85 \times 10^{-4}\ \text{J} + K$ giving $K = 8.33 \times 10^{-4}$ J. Then $v_f = 4.23 \times 10^5$ m/s.

> ***Note.*** Newton's 2nd law could be easily used to get this result only if the initial velocity were along a radius. Our result is quite general.

Problem 4.45. A charge Q produces an electric field of magnitude $|E| = kQ/r^2$. How much energy is stored by this electric field in a spherical shell at radius r and thickness Δr, where $\Delta r \ll r$?

Solution

Within this shell the electric field can be considered constant since r hardly varies. The energy density is given by $U_{pd} = (\frac{1}{2})\varepsilon_0 E^2 = (\frac{1}{2})\varepsilon_0[(1/4\pi\varepsilon_0)Q/r^2]^2$. For a thin shell the volume is equal to the surface area of the shell times the thickness of the shell, or volume $= 4\pi r^2 \Delta r$. The energy stored equals $U_{pd} \times$ volume $= Q^2 \Delta r/8\pi\varepsilon_0 r^2$.

Problem 4.46. A parallel plate capacitor C is given a charge Q with air between the plates. The capacitor is then isolated so that no charge can be added or removed from the plates. Then a dielectric, of dielectric constant κ, is inserted between the plates, filling $\frac{1}{3}$ of the volume (see Fig. 4-20).

(*a*) What is the potential difference between the plates when there is air between the plates?

(*b*) What is the potential difference between the plates when the dielectric material is between the plates?

(*c*) What is the capacitance of the plates when the dielectric material is between the plates?

Solution

(*a*) The potential difference is $V = Q/C$.

(*b*) The electric field is now produced by the charges on the plates and also by the polarization surface charges on the dielectric material. The charge on the dielectric material does not produce any field in the region outside of the dielectric since the two surfaces are oppositely charged and they add to zero outside the material. Within the material (as discussed in the text for the case of dielectric filling the

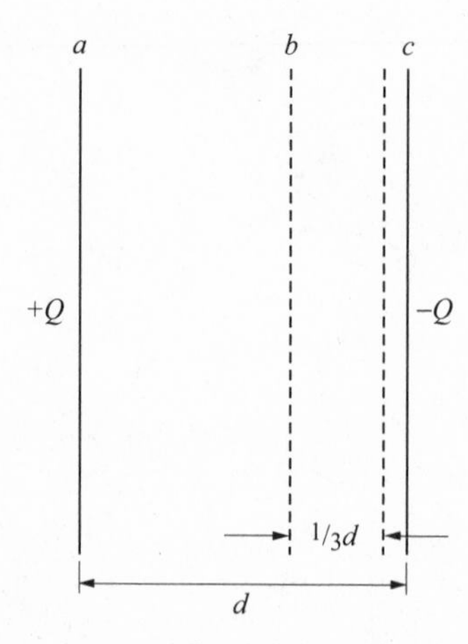

Fig. 4-20

entire space) the electric field will be reduced to E_0/κ, where $E_0 = Q/\varepsilon_0 A$, the field for the dielectric free capacitor. If we now move along a line from the positive plate to the negative plate, the potential difference from a to b is $E_0(2d/3)$, and the potential difference from b to c is $(E_0/\kappa)(d/3)$. Then $V = E_0 d(2/3 + 1/3\kappa) = (Qd/\varepsilon_0 A)(2/3 + 1/3\kappa) = Q(2/3 + 1/3\kappa)/C$.

(*c*) The new capacitance is $C' = Q/V = C/(2/3 + 1/3\kappa)$.

Supplementary Problems

Problem 4.47. A charge of 6.8×10^{-7} C is at a distance of 0.96 m from a second charge. The potential energy of the combination is -3.8×10^{-3} J. What is the charge on the other charge?

Ans. -6.0×10^{-7} C

Problem 4.48. Three charges are at the corners of an equilateral triangle of side 2.5 cm. The charges have charge of 5.3×10^{-8} C, -6.9×10^{-8} C and -9.9×10^{8} C. What is the total potential energy of the combination?

Ans. -7.5×10^{-4} J

Problem 4.49. Two charges of $q = 5.6 \times 10^{-7}$ C are located on the x axis at $x = \pm 0.76$ m.

(*a*) What is the potential at $x = 1.52$ m on the x axis?

(*b*) What is the potential at $x = -1.52$ m an the x axis?

(*c*) What is the potential at $y = 1.52$ m on the y axis?

(*d*) What is the potential at the origin, $x = y = 0$?

Ans. (*a*) 8.84×10^3 V; (*b*) 8.84×10^3 V; (*c*) 5.93×10^3 V; (*d*) 1.33×10^4 V

Problem 4.50. A charge of 5.3×10^{-7} C is located at the origin and a second charge of -4.5×10^{-7} C is on the x axis at $x = 2.1$ m. At what two points on the x axis is the potential equal to 500 V? (Refer to Problem 4.5 for a similar problem.)

Ans. $x = 1.073$ m and $x = -4.15$ m

Problem 4.51. A charge of 4.5×10^{-7} C is at $x = -0.19$ m and a charge of -5.3×10^{-7} C is at $x = +0.19$ m. At what point or points on the y axis is the potential equal to -500 V?

Ans. $y = \pm 1.43$ m

Problem 4.52. A ring of uniformly distributed charge has a radius of 1.81 m and contains a total charge of 6.5×10^{-7} C.

(*a*) At what distance from the plane of the ring is the potential equal to 1100 V along the axis of the ring?

(*b*) How much work must be done to move a charge of 3.8×10^{-7} C from this point to the center of the ring?

Ans. (*a*) 5.0 m; (*b*) 8.10×10^{-4} J

Problem 4.53. A large plane sheet has a surface charge density of 3.7×10^{-8} C/m^2. Point a is at a distance of 2.1 cm to the left of the sheet, point b is 1.1 cm to the left, point c is 1.1 cm to the right and point d is 2.1 cm to the right of the sheet.

(*a*) What is the potential difference between points a and b, $V_a - V_b$?

(*b*) What is the potential difference between points b and c, $V_b - V_c$?

(*c*) What is the potential difference between points c and d, $V_c - V_d$?

Ans. (*a*) -20.9 V; (*b*) 0; (*c*) 20.9 V

Problem 4.54. Two large parallel plane sheets are uniformly charged and separated by 5.6 cm. The sheet on the left has a surface charge density of 3.7×10^{-8} C/m^2 and the one on the right has a surface charge density of -1.3×10^{-8} C/m^2. Point a is between the sheets at a distance of 1.2 cm from the left sheet, point b is between the sheets at a distance of 1.2 cm from the right sheet and point c is to the right of both sheets at a distance of 1.2 cm from the right sheet.

(*a*) What is the potential difference between points a and b, $V_b - V_a$?

(*b*) What is the potential difference between points c and b, $V_c - V_b$?

Ans. (*a*) 90.6 V; (*b*) 50.2 V

Problem 4.55. A charge of 6.2×10^{-7} C is at the center of a charged conducting spherical shell of inner radius 0.86 m and outer radius 0.91 m. At a distance of 1.00 m from the charge, the potential is 4.92×10^3 V.

(*a*) What charge is on the sphere?

(*b*) What is the potential on the surface of the sphere?

(*c*) What is the potential at a point within the sphere at a distance of 0.50 m from the central charge?

Ans. (*a*) -7.33×10^{-8} C; (*b*) 5.41×10^3 V; (*c*) 1.01×10^4 V

Problem 4.56. A charge Q_1 is at the center of a charged conducting spherical shell of inner radius 0.54 m and outer radius 0.77 m that has a charge Q_2. At a point 0.40 m from the central charge, the potential is 985 V and on the sphere the potential is 880 V.

(*a*) What is the charge Q_1?

(*b*) What is the charge Q_2?

Ans. (*a*) 1.80×10^{-8} C; (*b*) 5.72×10^{-8} C

Problem 4.57. A long straight wire has a uniform charge of 6.3×10^{-9} C/m. What is the difference of potential between a point a which is 0.62 m to the left of the wire and a point b that is 0.13 m to the right of the wire, i.e. what is $V_a - V_b$?

Ans. -177 V

Problem 4.58. Two long wires are parallel to each other, separated by a distance of 0.43 m, and have uniform charges of 1.9×10^{-9} C/m and -7.3×10^{-9} C/m, respectively. Point a is midway between the wires and point b is 0.20 m from the negatively charged wire (and 0.63 m from the positively charged wire). What is the difference of potential $V_a - V_b$?

Ans. 46.3 V

Problem 4.59. Two long wires are each uniformly charged, with one along the x axis and the other along the y axis. The one along the x axis has a charge of 1.9×10^{-9} C/m, and the one along the y axis has a charge of 2.5×10^{-9} C/m. Point a is at (0.15, 0.15), point b is at (0.45, 0.15), point c is at (0.15, 0.45) and point d is at (0.45, 0.45).

(*a*) What is the potential difference $V_b - V_a$?

(*b*) What is the potential difference $V_c - V_a$?

(*c*) What is the potential difference $V_d - V_a$?

Ans. (*a*) 49.4 V; (*b*) 37.6 V; (*c*) 87.0 V

Problem 4.60. A long straight line carries a uniform charge of 6.6×10^{-9} C/m. A long conducting cylindrical shell, carrying a charge of -4.8×10^{-9} C/m is coaxial with the line and has an inner radius of 0.25 m and an outer radius of 0.27 m. Use $R = 0.25$ m for calculating the potential.

(*a*) What is the linear charge density on the inner and on the outer surface of the cylinder?

(*b*) What is the potential at $r = 0.36$ m?

(*c*) What is the potential at $r = 0.27$ m, the surface of the cylinder?

(*d*) What is the potential at $r = 0.15$ m?

Ans. (*a*) -6.6×10^{-9} C/m and 1.8×10^{-9} C/m; (*b*) -11.8 V; (*c*) -2.5 V; (*d*) 58.2 V

Problem 4.61. A long wire has a uniform positive charge distribution along its length.

(*a*) What are the equipotential surfaces for this wire?

(*b*) In which direction does the electric field point?

Ans. (*a*) cylindrical surfaces coaxial with the wire; (*b*) radially outward

Problem 4.62. A long straight wire carries a charge of 4.9×10^{-7} C/m. A short segment of insulating wire, of length 0.077 m, is parallel to the long wire, and carries a total charge of 6.8×10^{-6} C. How much work is needed to move this short wire from a distance of 5.3 m to 3.1 m from the long wire?

Ans. 3.22×10^{-2} J

Problem 4.63. A dipole is at the origin, oriented along the x axis. The dipole moment is 6.7×10^{-9} C · m, with the positive charge on the positive x side. Two charges of $\pm 5.0 \times 10^{-6}$ C are separated by a distance of 0.39 m and placed along the x axis with the positive charge nearer the dipole at a distance of 2.10 m. Refer to Problem 4.39 for the potentials.

(*a*) What is the potential energy of the charges in this position?

(*b*) If the charges are rotated by 90° and shifted so that the charges are now both at $x = 2.10$ m and $y = \pm 0.195$ m, what is the potential at this position?

(*c*) How much work by an outside force was done to turn the charges?

Ans. (*a*) 1.97×10^{-5} J; (*b*) 0; (*c*) -1.97×10^{-6} J

Problem 4.64. A certain charge distribution gives a potential of $V = -A/r^4$, where A is a positive constant and r is the distance from the origin.

(*a*) What are the equipotential surfaces for this potential?

(*b*) In which direction does the electric field point?

(*c*) What is the magnitude of the electric field? (*Hint*: See Problem 4.13)

Ans. (*a*) spherical surfaces centered on the origin; (*b*) radially in; (*c*) $4/r^5$

Problem 4.65. A proton has a speed of 6.0×10^6 m/s. The mass of a proton is 1.67×10^{-24} kg, and the charge is the same as on an electron (except that it is positive).

(*a*) What is the kinetic energy of the proton in Joules and in eV?

(*b*) If all the kinetic energy was gained by falling through a difference of potential, what difference in potential is required?

Ans. (*a*) 3.01×10^{-14} J $= 1.88 \times 10^5$ eV; (b) 188 keV

Problem 4.66. An electron is moving with constant speed in a circle around a proton. The centripetal force is supplied by the electrical force between the proton and the electron. The radius of the orbit is $r = 0.53 \times 10^{-10}$

(*a*) What is the potential energy of the system in eV?

(*b*) Use the equation relating the (mass) $\times$ (centripetal acceleration) to the electrical force to deduce the kinetic energy of the electron in eV directly from the result of (*a*).

(*c*) What is the total energy of the system in eV?

(*d*) How much energy is needed to ionize the system, i.e. to remove the electron to a position at rest at infinity (total energy equal to zero)?

Ans. (*a*) -27.2 eV; (*b*) 13.6 eV; (*c*) -13.6 eV; (*d*) 13.6 eV

Problem 4.67. A particle, of mass 1.8×10^{-27} kg and charge 1.6×10^{-19} C is fixed to the origin. Another charge, of mass 9.1×10^{-31} kg and charge -1.6×10^{-19} C is initially at a distance of 9.3×10^{-10} m from the origin and moving directly away from the origin with a speed of 5.14×10^5 m/s. At what distance from the origin does this second particle stop and reverse its direction?

Ans. 1.8×10^{-9} m

Problem 4.68. A capacitor is built out of two closely spaced concentric spherical shells separated by a distance of 0.83 mm. The capacitance is 25 nF. What is the radius of the shells? (Refer to Problem 4.41.)

Ans. 0.43 m

Problem 4.69. A certain capacitor has an electric field of 2.85×10^5 V/m when 120 V are across the capacitor.

(*a*) What is the distance between the plates?

(*b*) If the area of the plates is 33 m^2, what is the capacitance of the capacitor?

(*c*) What is the energy in the capacitor when the voltage across the capacitor is 120 V?

(*d*) What is the electrical energy density in the capacitor at this voltage?

Ans. (*a*) 0.42 mm; (*b*) 0.69 μF; (*c*) 5.0×10^{-3} J; (*d*) 0.359 J/m^3

Problem 4.70. Four capacitors are connected in series and a voltage of 12 V is connected across the circuit. The capacitances are 1.3 μF, 2.5 μF, 6.8 μF and 0.92 μF.

(*a*) What is the equivalent capacitance of the circuit?

(*b*) What is the voltage across each capacitor?

(*c*) What is the total energy stored in the system?

Ans. (*a*) 0.416 μF; (*b*) 3.84 V, 2.00 V, 0.73 V, 5.42 V; (*c*) 3.0×10^{-5} J

Problem 4.71. Four capacitors are connected in parallel and a voltage of 12 V is connected across the circuit. The capacitances are 1.3 μF, 2.5 μF, 6.8 μF and 0.92 μF.

(*a*) What is the equivalent capacitance of the circuit?

(*b*) What is the charge stored on each capacitor?

(*c*) What is the total energy stored in the system?

Ans. (*a*) 11.5 μF; (*b*) 15.6 μC, 30 μC, 82 μC, 11 μC; (*c*) 8.28×10^{-4} J

Problem 4.72. Four capacitors are connected as in Fig. 4-21 and a voltage of 12 V is connected across the circuit. The capacitances are 1.3 μF, 2.5 μF, 6.8 μF and 0.92 μF.

(*a*) What is the equivalent capacitance of the circuit?

(*b*) What is the charge stored on each capacitor?

(*c*) What is the total energy stored in the system?

Ans. (*a*) 2.55 μF; (*b*) 10.5 μC, 20.1 μC, 26.9 μC, 3.6 μC; (*c*) 1.84×10^{-4} J

Problem 4.73. A capacitor filled with air has a capacitance of 25 μF. What capacitance would the capacitor have if it were filled with paper?

Ans. 82.5 μF

Problem 4.74. An air filled capacitor has a capacitance of 25 μF. If 1/4 of its volume were filled with paper, what capacitance would it have? (See Problem 4.46.)

Ans. 30.3 μF

Problem 4.75. An air filled capacitor has a capacitance of 25 μF, and a constant voltage of 18 V is across the capacitor.

(*a*) How much charge is stored on this capacitor?

(*b*) If the capacitor were filled with paper, and the voltage remained the same, how much charge would be stored on the capacitor?

(*c*) How much energy is stored in the system in each case?

Ans. (*a*) 4.5×10^{-4} C; (*b*) 1.49×10^{-3} C; (*c*) 4.05×10^{-3} J, 1.34×10^{-2} J

Problem 4.76. A capacitor has an area of 91 m^2 and the plates are separated by 0.86 mm. We want the capacitor to have a capacitance of 25 μF. What must be the dielectric constant of the material filling the capacitor to give this capacitance?

Ans. 26.7

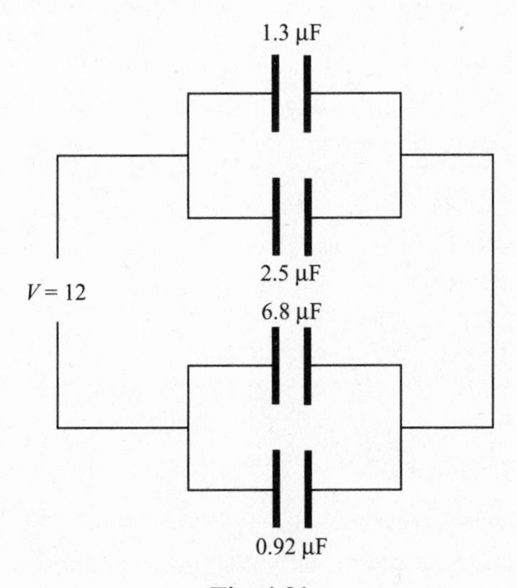

Fig. 4-21

Chapter 5

Simple Electric Circuits

5.1 CURRENT, RESISTANCE, OHM'S LAW

We have learned previously that an electric field exerts a force on charged particles. If there is an electric field in a material that has free charges (those not held tightly to the nuclei) then those charges will be induced to move. In the previous chapters we noted that when charges are placed on a conductor the charges will quickly move to the surface in such a way that the field will be reduced to zero within the conductor, thus achieving electrical equilibrium. Indeed, if there were an electric field in the conductor, the "free" charges would keep moving, so the vanishing of the electric field is a requirement for equilibrium. Furthermore, as we saw, a consequence of the vanishing of the electric field in the conductor is that the entire conductor is an equipotential region. Suppose, however, we never allow equilibrium to be reached. For example, suppose we insert charge on one side of the conductor and let charge escape from the other side in a continuous way so that the conductor can never reach equilibrium. In this case we can maintain an electric field in the conductor and charges will be continuously moving due to the force caused by the electric field. Also, there will now be a potential difference between the two ends of the conductor, with the electric field pointing from high to low potential, as we learned previously. As the charges are pushed by the electric field from one end to the other, positive work is done, and the potential difference represents the electrical energy lost per unit charge that completes the trip. To maintain the steady flow of charges therefore requires an external source of energy that in effect takes charges leaving one end and brings them back to the other end, thus replenishing the lost electrical energy. The external source, in effect, maintains a net positive—negative charge separation between the two ends of the conductor, which of course is what causes the potential difference (and associated electric field) to be maintained.

There are many external sources of energy that can maintain the charge separation in a steady way. A **battery**, which we will discuss later in this chapter uses chemical means to separate charges. An **electric generator** uses magnetic fields to generate electric potentials in a manner to be discussed in a later chapter. The important point is that we can produce *differences of potential* which can be sustained between two points, and which will cause electric fields to exist within conductors. The free charges in the conductor will then continuously move through the potential difference (voltage) and the external source of energy will have to continually supply charges to replenish those that have moved away and maintain the voltage. The energy per unit charge supplied by the external source in maintaining the voltage is called the EMF ("**electromotive force**"—although it is not a force, but the name has stuck for historical reasons), and it is the EMF that replenishes the electrical energy lost as the charges flow within the conductor. In a steady state situation the external energy supplied per unit charge returned to the front end of the conductor exactly equals the electrical energy per unit charge expended in moving a unit charge through the conductor (from front to back). Hence the EMF equals the voltage across the conductor.

Suppose that we have a wire, with a uniform cross-sectional area A and a length L, as in Fig. 5-1 and that there is an EMF that maintains a difference of potential $V_1 - V_2$ across the ends of the wire. If V_1 is greater than V_2, there will be an electric field within the wire pointing from V_1 to V_2, i.e. to the right in the figure. This will exert a force on the charges in the conductor which is to the right for positive charges and to the left for negative charges. Consequently, if there are free positive charges they will move to the right and if there are free negative charges they will move to the left. In either case the electric force does positive work. In general, if we move positive charges to the right then the right side will tend to become positively charged and the left side will tend to become negatively charged. Similarly, if we move negative charges to the left, the right side will still tend to become positively charged and the left side negatively charged. We therefore see that positive charge moving to the right has the

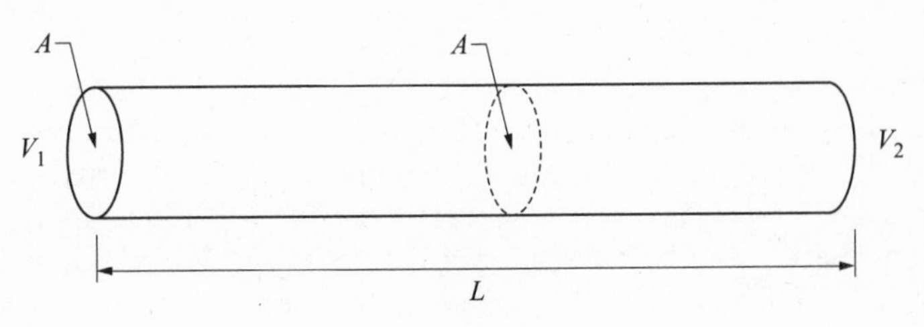

Fig. 5-1

same effect as negative charge moving to the left. In both cases we are in effect transporting positive charges to the right. By convention it is usual to talk *as if* positive charge is moving, even if in fact it is negative charge that is moving in the opposite direction. If the moving "positive" charge is removed when it reaches the end (by the source of EMF) and returned to the front, then there will be a steady flow of positive charge from the high to the low potential side of the wire. The amount of charge that flows through the wire per second is called the **current**. The symbol we use for current is I, and the unit is **ampere** (one ampere is one coulomb/s). By our convention, the direction of the current is the direction of flow of positive charge. As noted above, this means that the current always flows from high to low potential irrespective of whether the charges that are really moving are positive or negative. Of course, for positive charges the current is actually in the same direction as the charges that are moving while for negative charges the current is flowing in the direction opposite to the direction in which the charges are moving. In conductors we always have current carried by electrons that are negatively charged, but there are materials in which the flow contains positive charges or both positive and negative charges. Mathematically, the current is defined as

$$I = \Delta q/\Delta t \tag{5-1}$$

where Δq is the effective positive charge passing a cross-section of the conducting wire in the time Δt, and the direction of I is the direction of flow of positive charge.

It can easily be shown that the current flowing into one end of a conducting wire (say the left end in Fig. 5-1) must be the same as the current flowing out the right side. If the current flows were even slightly imbalanced there would be a build up of charges and associated electric field that would rapidly return the currents to equality. This turns out to be true even if the wire is quite long, and even when we are dealing with time varying currents (a later chapter) unless those time variations are extremely rapid. By similar reasoning we see that the current is the same at every position along the wire, even if the cross-sectional area of the wire is different at different locations.

Problem 5.1. A current consists of electrons that are moving to the left. The magnitude of the current is 3.4 A.

(*a*) What is the direction of the current?

(*b*) If the charge on each electron is 1.60×10^{-19} C, how many electrons are passing through the area of the wire per second?

Solution

(*a*) Since the electrons are negatively charged the current flows in the direction opposite to the electrons. Therefore, the current flows to the right.

(*b*) The amount of charge transported per second equals the current, and $I = 3.4\ \text{A} = 3.4\ \text{C/s}$. If there are n electrons transported per second, there would be ne coulombs transported per second. Thus $I = ne = n(1.60 \times 10^{-19}\ \text{C}) = 3.4\ \text{A}$, and $n = 2.13 \times 10^{19}$ electrons/s.

Since the potential difference across a wire is just the work done by the electric field in the wire in moving a charge from one end to another, we expect that increasing the potential difference will cause a corresponding increase in the electric field. The electric field, in turn, exerts the force on the free charges in the wire (the electrons) that causes the current to flow. The larger the electric field the larger the current we should get. Thus, increasing the potential difference across a wire will increase the current. For a given potential difference, we would expect that the amount of current would depend on the material the wire is made of, as well as on such factors as the cross-sectional area and length of the wire. For some materials we expect the current to be small since they may be resistant to the flow of current. We define a quantity called the **resistance** of the wire, R, as the ratio of voltage across the wire to the current flowing through the wire, $R = V/I$, or $V = IR$. The wire itself is called a **resistor**. The unit for R is V/A which we call an **ohm** (Ω). It is not necessarily true that R will be a constant, independent of I and V, and there are many cases where R does vary. However, for most ordinary conducting materials and for ordinary currents, R is very nearly a constant. We call such materials ohmic materials, because they satisfy **Ohm's law** that $R = V/I$ is a constant, or in other words, the current is directly proportional to the voltage, with R as the constant of proportionality:

$$V = IR \tag{5-2}$$

Unless stated to the contrary we will discuss only ohmic materials in the remainder of this chapter. The symbol that we use for a resistor is ⩘.

Problem 5.2. A certain resistor has a current of 1.8 A when a potential difference of 120 V exists across the resistor.

(*a*) What is the resistance of the resistor?

(*b*) If the potential difference is only 50 V, what current will flow in the resistor?

Solution

(*a*) The resistance is $R = V/I = (120\text{ V})/(1.8\text{ A}) = 66.7\ \Omega$.

(*b*) The current is $I = V/R = (50\text{ V})/(66.7\ \Omega) = 0.75\text{ A}$.

As stated earlier, we expect the resistance to depend on the dimensions of the conductor as well as on the material from which it is made. For instance, if we double the cross-sectional area A of the wire, we would expect to double the current since this is equivalent to having two adjacent wires of the same material and length with the same voltage across them, each of cross-sectional area A, and hence each carries the same current and the total current is the sum of the two currents. Thus, from Ohm's law, the resistance halves when we double the area and R is inversely proportional to the area. If we double the length, d, of the wire (keeping the same potential difference) then we can think of the wire as made up of two identical segments with the same current flowing. Then the potential difference across each segment would be equal, and have one half of the potential difference across the whole wire. From Ohm's law half the voltage with the same current means half the resistance. Thus the whole wire has double the resistance as does each half, and the resistance is directly proportional to the length. Thus $R \propto (d/A)$, or

$$R = \rho d/A, \tag{5-3}$$

where the constant of proportionality, ρ, depends on the material being used and has the dimensions of $\Omega \cdot \text{m}$. This quantity ρ is called the **resistivity** of the material. For ohmic materials, ρ is a constant. Materials that conduct electricity very easily have low resistivities and materials that resist the flow of current have high resistivities. There is a tremendous range of values for resistivities between good conductors and good insulators. In fact, there are materials that have zero resistivity at low enough temperatures, and these materials are called superconductors. In Table 5.1, we list the of resistivity of some materials.

While Ohm's law has been expressed as a relationship between potential difference and current, underlying this is the relationship between the electric field in a wire and the consequent rate of flow of the charges. For a uniform wire of length d and cross-section A under potential difference V, the electric

Table 5.1 Values of Resistivity of Materials

Material	Resistivity ($\Omega \cdot m$)
Metals:	
Silver	1.47×10^{-8}
Copper	1.72×10^{-8}
Gold	2.44×10^{-8}
Aluminum	2.63×10^{-8}
Tungsten	5.51×10^{-8}
Steel	20×10^{-8}
Lead	22×10^{-8}
Mercury	95×10^{-8}
Semiconductors:	
Pure carbon	3.5×10^{-5}
Pure germanium	0.60
Pure silicon	2300
Insulators:	
Amber	5×10^{14}
Mica	10^{11}–10^{15}
Teflon	10^{16}
Quartz	7.5×10^{17}

field along the wire is constant and we have $V = Ed$. Then from Ohm's law and Eq. (5.3) we have $Ed = I\rho d/A$, or $E = \rho(I/A)$. We define the current density J as the current/unit cross-section area so:

$$J = I/A, \tag{5-4}$$

and we get:

$$E = \rho J \tag{5-5}$$

which is actually a more fundamental statement of Ohm's law in terms of the electric field, and does not depend on the dimensions of the wire but only on the nature of the material of which it is made.

Problem 5.3. A resistor is made from pure carbon, and has a length of 0.21 m. The resistance of the resistor is 25 Ω.

(*a*) What is the cross-sectional area of the resistor?

(*b*) If the voltage across the resistor is 100 V, find the current density.

(*c*) Find the electric field.

Solution

(*a*) The resistivity of carbon is $3.5 \times 10^{-5}\ \Omega \cdot m$. Therefore, using $R = \rho d/A = 25\ \Omega = (3.5 \times 10^{-5}\ \Omega \cdot m)(0.21\ m)/A$, we get $A = 2.94 \times 10^{-7}\ m^2 = 2.94 \times 10^{-1}\ mm^2$.

(*b*) From Ohm's law $V = IR \rightarrow I = (100\ V)/(25\ \Omega) = 4\ A$. Using the results of part (*a*), $J = (4A)/(2.94 \times 10^{-7}\ m^2) = 1.36 \times 10^7\ A/m^2$.

(*c*) $Ed = V \rightarrow E = (100\ V)/(0.21\ m) = 476\ V/m$; or, using Eq. (5.5), $E = \rho J = (3.5 \times 10^{-5}\ \Omega \cdot m)(1.36 \times 10^7\ A/m^2) = 476\ \Omega \cdot A/m = 476\ V/m$, as before.

Problem 5.4.

(*a*) A resistor, of length L and area A has a resistance R. If the same volume of material is doubled in length, what is the new resistance?

(*b*) If the same voltage is applied to both resistors, how are the currents related? How are the current densities related? How are the electric fields related?

Solution

(*a*) The volume is AL and remains the same. Therefore, for our new resistor, $A'L' = AL = A'(2L)$, and $A' = A/2$. Then $R' = \rho L'/A' = \rho(2L)/(A/2) = 4\rho L/A = 4R$. Thus the resistance is quadrupled.

(*b*) Since the resistance quadrupled, the currents are related by $I' = I/4$. Furthermore $Ed = E'd' = E'(2d) \rightarrow E' = E/2$, so the electric field is halved. Lastly, $J' = I'/A' = (I/4)/(A/2) = I/2A = J/2$.

The mechanism by which resistance is produced in a material can be thought of as follows. When the free electrons are accelerated by the force of the electric field their velocity increases in the direction of the force. If there were no other interactions, then the velocity would continue to increase until the electron leaves the wire. This, of course, is not what happens in an ohmic material. (Such situations can occur in a vacuum, but even in vacuum tubes, which were the first means for providing the "non-linear" elements of modern technology, charges were slowed down by accumulated "space charge" near the electrodes.) In solid materials the electrons collide with the atoms of the material and with any impurities in the material and thereby lose energy and momentum to the material. Under normal conditions, this results in electrons that, on average, acquire a constant "drift" velocity in the direction of the force which they maintain throughout their travel in the material. This drift velocity, v_D, increases linearly with the electric field, and hence with the voltage across the resistor. We can calculate the current in terms of this drift velocity if we know the density of free electrons in the material.

Problem 5.5. At a certain voltage across a resistor the electrons develop a drift velocity, v_D, in the direction opposite to the electric field. There are n free electrons/m^3 in the material, and the cross-sectional area of the resistor is A.

(*a*) Find an expression for the current produced by these electrons in terms of n, e, v_D and A.

(*b*) Find an expression for the resistivity of this material in terms of n, e, v_D and the electric field E.

Solution

(*a*) The current is defined as the amount of positive charge passing through the area A per second. The amount of positive charge will be equal to the number of electrons passing through the area, times the charge on an electron, except that it will be in the direction opposite to the direction of motion of the electron, since the electron has negative charge. During a time Δt, the number of electrons that pass through a given cross-sectional area will be all those that are near enough to the area to reach it in the time Δt. Since the electrons travel a distance $v_D\Delta t$ in this time, the number of electrons passing through A are all the free electrons in an imaginary cylinder of length $v_D\Delta t$ and area A. Since n is the number of electrons per unit volume and the volume of the cylinder is $Av_D\Delta t$, we have that the number passing A in time Δt equals $nAv_D\Delta t$, and the charge passing through the area is $nev_D\Delta t$. Thus the current is $I = \Delta q/\Delta t = neAv_D$, and the direction of the current is in the direction of the electric field.

(*b*) The resistance of the resistor is $R = V/I = Ed/I = Ed/neAv_D$, where we have used the result of part (*a*). From Eq. (*5.3*) we also have $R = \rho d/A$ and equating the two expressions we get $\rho = E/nev_D$. If the drift velocity is proportional to the electric field, then this ratio, $v_D/E = \mu$ (the mobility of the electrons) is constant and $\rho = 1/ne\mu$, or $\sigma = 1/\rho = ne\mu$, where σ is called the **conductivity** of the material.

5.2 RESISTORS IN COMBINATION

As in the case of capacitors we often build circuits with a combination of resistors. There are two basic ways to connect two resistors, in series and in parallel. We will discuss each case separately and then discuss combinations of many resistors.

If resistors are connected in series, as in Fig. 5-2, it is clear that each resistor carries the same current as the other resistors. This is because any charge entering one resistor in series will exit from that resistor and then must enter the next resistor. Since the voltage across a resistor equals IR and the resistors do not necessarily have the same resistance, there will be different voltages across each resistor. In the next problem we calculate the equivalent resistance of several resistors connected in series.

Problem 5.6. Two resistors R_1 and R_2 are connected in series as in Fig. 5-2(*a*). What is the equivalent resistance of these resistors?

Solution

The voltage across the combination is $V = V_a - V_c = (V_a - V_b) + (V_b - V_c) = V_1 + V_2$, where V_1 is the voltage across R_1 and V_2 is the voltage across R_2. We also know that the current is the same in the two resistors, and we call this current I. The equivalent resistance will be $V/I = R = (V_1 + V_2)/I = V_1/I_1 + V_2/I_2 = R_1 + R_2$ or $R = R_1 + R_2$. It is not hard to see that this can be extended to any number of resistors in series: $R_1, R_2, R_3, \ldots$ and we have:

$$R_{eq} = R_1 + R_2 + R_3 + \ldots = \sum R_i \text{ (resistors in series)} \tag{5-3}$$

It makes sense that the equivalent resistance is greater than any individual resistor since the current has to overcome the resistance of each resistor before the current reaches the end of the series. In the case of parallel resistors, shown in Fig. 5-2(*b*), the opposite is true. In this case the current that comes from the source of potential can go through one of two paths, and the equivalent resistance should be reduced because there are two paths that are available. This is shown in the next problem.

Problem 5.7. Two resistors, R_1 and R_2 are connected in parallel as in Fig. 5-2(*b*). Calculate the equivalent resistance of these resistors.

Note. We assume that the wires leading up to the resistors themselves have negligible resistance.

Solution

In this case, both resistors are connected between points a and b, so that the voltage across each is $V_a - V_b = V = V_1 = V_2$. This is true for all parallel circuits—the voltage is the same across all the elements. Since the current through each individual resistor is $I_i = V/R_i$ and the resistances can be different, the current in each resistor can be different. The current flowing out of point a divides into two paths, with some of the current flowing along path one through R_1 and the remainder along path two through R_2.

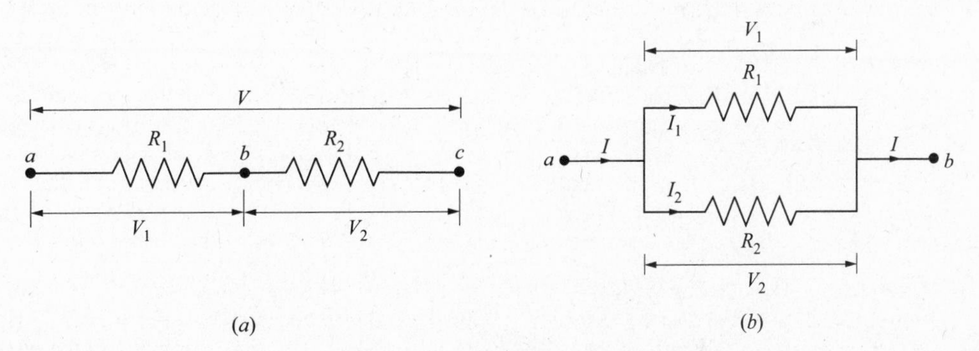

Fig. 5-2

Therefore, $I_1 + I_2 = I$, and noting $I = V/R_{eq}$, $I_1 = V/R_1$ and $I_2 = V/R_2$, we get (dividing out the common V):

$$1/R_{eq} = 1/R_1 + 1/R_2, \tag{5-4a}$$

If we have more than two resistors in parallel a similar analysis leads to:

$$1/R_{eq} = 1/R_1 + 1/R_2 + 1/R_3 + \dots = \sum 1/R_i \tag{5-4b}$$

There are major practical differences between connecting resistors in series and in parallel in a circuit. In a series connection, the current is the same for all the elements. If the current is reduced to zero in one of the resistors because the resistor "burns out" (i.e., no longer allows charges through—in effect becoming a resistor with an infinite resistance), then the current will become zero in all the resistors. Therefore one avoids connecting light bulbs or other devices in series, because if one of them burns out they will all go out. For the same reason, a fuse *is* connected in series. A **fuse** is a device which has very low resistance, and is made of material that will melt (i.e. burn out) when the current gets too high. The fuse burns out before other wires or resistors burn out or get so hot that nearby objects catch fire. When the fuse burns out all current ceases in the series circuit. For **parallel connections**, the current flowing through any parallel path i is V/R_i, and, for a fixed potential difference, this current is not affected by adding or subtracting other parallel resistors. The current flowing into the parallel circuit increases as additional resistors are connected between the same points, and the additional current is just the current flowing through that added resistor. The equivalent resistance will decrease because more current will flow for the same voltage.

It is important to note that Figs. 5-2(*a*) and 5-2(*b*) represent only parts of what is called a circuit. For example, in Fig. 5-2(*a*) the current leaving point c presumably travels through other wires and through a source of EMF which drives the current through these additional wires connected back to point a. Fig. 5-3(*a*) and 5-3(*b*) show very simple closed circuits for the cases of Fig. 5-2(*a*) and (*b*). Since in these simple cases no other resistors appear in the circuit the EMF equals the voltage across both resistors in each case. The voltage represents the electrical energy lost/unit charge moving through the resistors [from $a \to b \to c$ in 5-3(*a*) and from $a \to b$ in 5-3(*b*)] while the EMF is the *non* electrical energy/unit charge delivered by the source of EMF (as charges pass through) to replace the lost electrical energy. As long as the source of EMF keeps supplying energy we have a "steady state" situation in which the current will keep flowing. EMF is discussed in more detail in the next section.

Problem 5.8. Consider the series portion of a circuit shown in Fig. 5-4. The current in the circuit flows from a to b and is 2.3 A.

(*a*) What is the equivalent resistance?

(*b*) What is the voltage across the entire circuit? Which point, a or b, is at the higher potential?

(*c*) What is the voltage across each resistor?

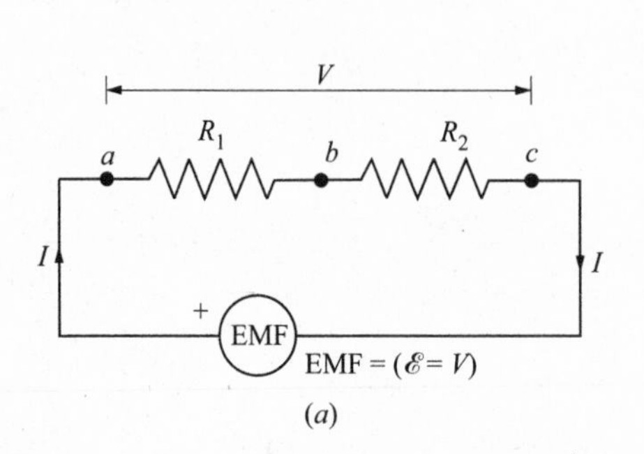

(*a*)

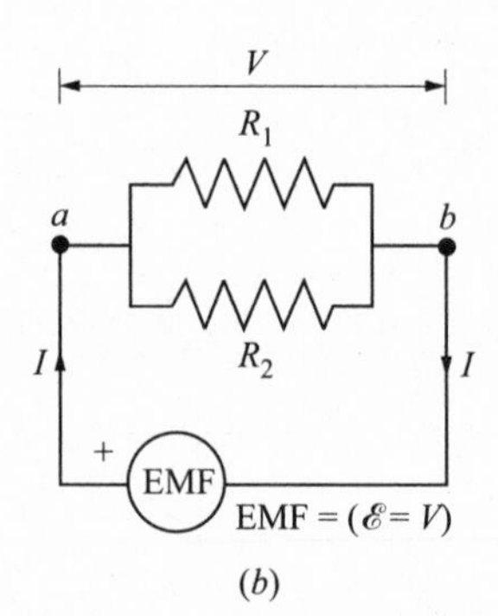

(*b*)

Fig. 5-3

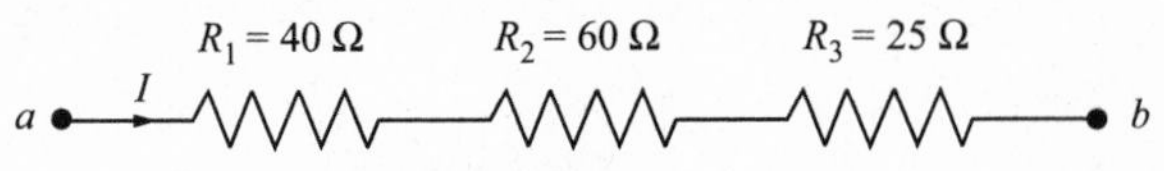

$R_1 = 40\ \Omega$ $R_2 = 60\ \Omega$ $R_3 = 25\ \Omega$

a I b

Fig. 5-4

Solution

(*a*) The equivalent resistance is the sum of all the resistances, or $R_{eq} = 40 + 60 + 25 = 125\ \Omega$.

(*b*) The voltage across the entire circuit is $V_{total} = IR_{eq} = (2.3\ \text{A})(125\ \Omega) = 288$ V. Since the current flows from a to b, and the electric field does positive work in pushing charges through the resistors, energy is lost as the charges move through. Thus the potential at a is higher (by 288 V) than the potential at b.

(*c*) The voltage across each resistor is IR_i. Thus $V_1 = (2.3\ \text{A})(40\ \Omega) = 92$ V, $V_2 = (2.3\ \text{A})(60\ \Omega) = 138$ V and $V_3 = (2.3\ \text{A})(25\ \Omega) = 58$ V.

> ***Note.*** Adding the voltages gives $92 + 138 + 58 = 288$ V, which is the voltage we calculated in part(*b*).

Problem 5.9. Three resistors are connected in parallel, as in Fig. 5-5. The potential difference between a and b is 75 V.

(*a*) What is the equivalent resistance of this circuit?

(*b*) What is the current flowing from point a?

(*c*) What is the current in each resistor?

Solution

(*a*) The equivalent resistance is given by $1/R_{eq} = \Sigma\,(1/R_i) = 1/40 + 1/60 + 1/25 = 0.817$, or $R_{eq} = 12.2\ \Omega$.

(*b*) The total current is $I_{tot} = V/R_{eq}$. Thus $I_{tot} = 6.13$ A.

(*c*) The current in each resistor is $I_i = V/R_i$. Thus $I_1 = (75\ \text{V})/(40\ \Omega) = 1.88$ A, $I_2 = (75\ \text{V})/(60\ \Omega) = 1.25$ A, $I_3 = (75\ \text{V})/(25\ \Omega) = 3.0$ A. [The total current is $1.88 + 1.25 + 3.0 = 6.13$ A, as in part(b).]

Problem 5.10. Consider the combination of resistors shown in Fig. 5-6(*a*). The voltage drop from a to b is 82 V.

(*a*) Calculate the equivalent resistance of the circuit.

(*b*) How much current is flowing through R_1?

(*c*) How much current is flowing through R_5?

(*d*) How much current is flowing through R_2?

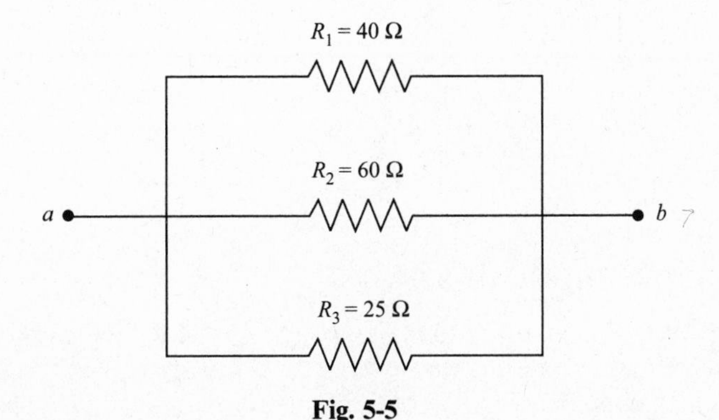

Fig. 5-5

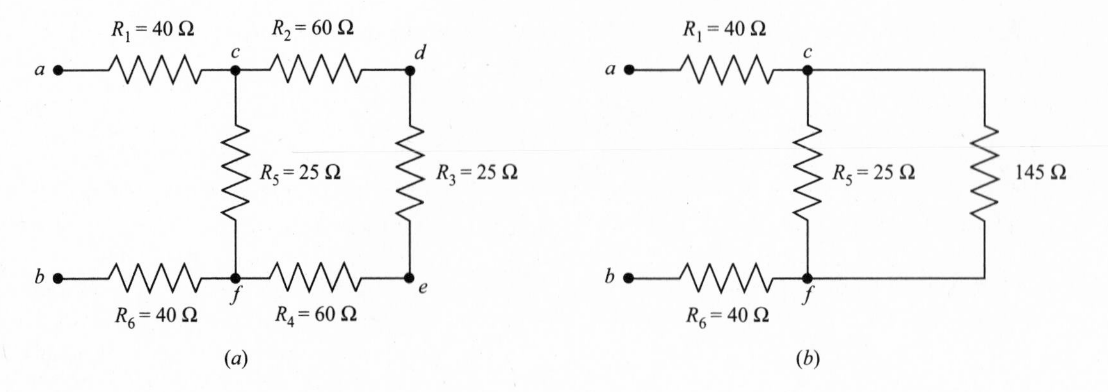

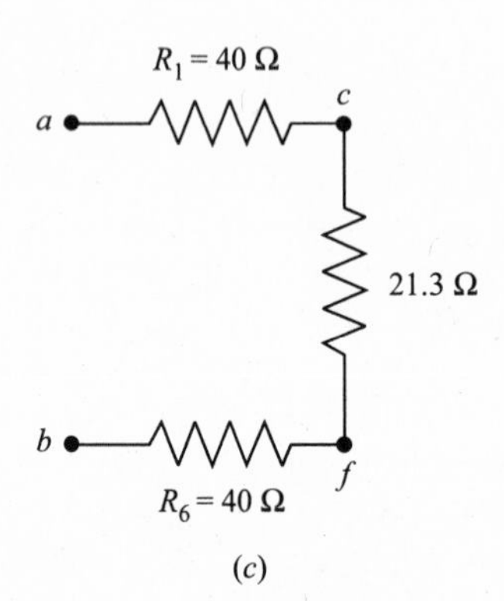

Fig. 5-6

(*e*) What is the voltage across R_4?

Solution

(*a*) We must combine resistors part by part. First we note that R_2, R_3 and R_4 are in series. We can therefore replace them with an equivalent resistance of $60 + 25 + 60 = 145\ \Omega$, as in Fig. 5-6(*b*). This resistance is now in parallel with R_5, and we can combine these two to get $1/R = 1/25 + 1/145 = 0.0469$, and $R = 21.3\ \Omega$, as in Fig. 5-6(*c*). Now, this resistance is in series with R_1 and R_6, giving a final equivalent resistance of $R_{eq} = 40 + 21.3 + 40 = 101.3\ \Omega$.

(*b*) The current flowing through R_1 is the total current flowing from *a* to *b* [e.g., Fig. 5-6(*c*)]. Thus $I_1 = I = V/R_{eq} = (82\ \text{V})/(101.3\ \Omega) = 0.809\ \text{A}$.

(*c*) The current flowing through R_5 equals the voltage between *c* and *f* divided by R_5. The voltage between *c* and *f* can be most easily calculated using Fig. 5-6(*c*). Here the current in each part of the series circuit is 0.809 A. Therefore $V_{cf} = (0.809\ \text{A})(21.3\ \Omega) = 17.2\ \text{V}$. Then $I_5 = (17.2\ \text{V})/(25\ \Omega) = 0.690\ \text{A}$.

(*d*) The current flowing through R_2 is the same as the current flowing through each resistor in the series R_2, R_3 and R_4, which is the current in the equivalent 145 Ω resistor in Fig. 5-6(*b*). That current is $V_{cf}/145 = I_2 = (17.2\ \text{V})/(145\ \Omega) = 0.119\ \text{A}$.

(*e*) The voltage across R_4 equals $I_4 R_4$. Now $I_4 = I_2 = 0.119\ \text{A}$. Therefore $V_4 = (0.119\ \text{A})(60\ \Omega) = 7.12\ \text{V}$.

5.3 EMF AND ELECTROCHEMICAL SYSTEMS

We mentioned previously that, in order to maintain a potential difference between two points in the presence of a current, there must be a non-electrical source of energy replenishing the energy lost by the charges moving through that potential difference. The energy supplied/unit charge by this source is called the EMF, whether the means for providing the EMF is chemical, magnetic, mechanical or any other process. The symbol that we use for an EMF is $\mathscr{E}$.

Let us examine in more detail the mechanism of an EMF. Basically any source of EMF is a device in which positive and negative charges are separated. The two ends of such a device are called *terminals*. On one terminal positive charge will accumulate and on the other terminal negative charge will accumulate. The positively charged terminal is called the **anode** and the negatively charged terminal is the **cathode**. This separated charge will establish an electric field within the source of EMF that points from the anode to the cathode. This field, in turn, exerts a force on positive charges within the device tending to push them back through the device toward the cathode, and a force on negative charges tending to push them back through the device to the anode. In order to keep the positive charges on the positive terminal and the negative charges on the negative terminal, the device—the source of EMF-exerts a non-electrical force that opposes the electrical force and continues to push positive charges toward the anode and negative charges to the cathode through the device. The flow of positive charges through the device is analogous to pushing water upward through a vertical pipe. As water moves upward to the top (the "anode" side) the force of gravity tries to push the water back to the bottom (the "cathode" side). To keep the water moving upward, a "non-gravity" force, such as that provided by a pump, must push upward against gravity. In the case of our source of EMF, this "other" force may derive from chemical interactions as in a battery, from magnetic forces as in an electric generator or from some other mechanical source. Just as in the water case, where the gravitational potential energy of the water is increased as the water is forced upward, this other force causes the charges to flow to the anode, increasing the electrical potential energy and hence causing a potential difference between the anode and the cathode. Assuming no thermal losses as the charges move through the device to the anode, we must have, by conservation of energy, that the potential difference is equal to the EMF of the source. If, outside the device, the anode is connected back to the cathode, then a current will flow through this "external" part of the circuit from the anode to the cathode. An example of such a circuit would be connecting points a and c in Fig. 5-2(a) to the anode and cathode respectively of the device. This is shown in Fig. 5-3(a). The charges lose electrical energy moving from the high to the low terminals through the external circuit, and are then forced by the non-electrical force back to the anode through the EMF device thus replenishing their electrical energy. (The analog for our water system would be water flowing back downward through a set of other pipes and then returning through the vertical pipe with the pump.) The potential difference established between the terminals when no current flows (because no wire has been connected between the terminals) is called the "**open circuit EMF**".

If we connect a single wire of resistance R between the terminals, then current will be made to flow through that resistance by the voltage, V, that is established across the terminals. If this wire were the only resistance in the circuit, then the current that flows would equal $I = V/R = \text{EMF}/R$. In practice, however, there is also some heat loss within the source of EMF due to the thermal agitation of molecules as the charges flow within the source. In this case it is no longer true that $\text{EMF} = V$, because some non-electrical energy is lost in the form of thermal energy. To a good approximation, this loss is proportional to the current, so $\text{EMF} = V - Ir$, where r is the proportionality constant. As can be seen; r has the same dimensions as resistance and is called the "**internal resistance**", R_{int}, of the source, and treated like any other resistance. If no current flows through the source, then the potential difference across the source is just equal to the open circuit EMF of the source, since there is no voltage drop due to the current across the internal resistance. However, if current flows in the external circuit, the same current will also flow in the source and a voltage drop of $V_{\text{int}} = IR_{\text{int}}$ will occur across the internal resistance R_{int}. The voltage applied to the rest of the circuit will then equal $\mathscr{E} - IR_{\text{int}}$. A common source of EMF is a battery, which uses chemical forces and hence chemical energy to force the current through the battery from cathode to anode.

Problem 5.11. A battery has an open circuit EMF of 2.0 V and an internal resistance of 0.94 Ω. A resistance of 22 Ω is connected between the terminals of the battery as in Fig. 5-7.

(a) What is the current flowing through the resistance?

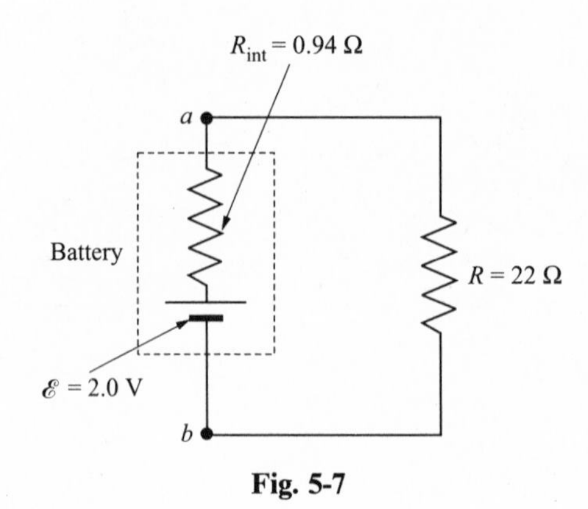

Fig. 5-7

(*b*) What is the potential difference across the terminals of the battery when the resistance is connected?

Solution

(*a*) The EMF produced by the source will cause current to flow through the series circuit of R and R_{int}. Thus the current will be $(2.0\ \text{V})/(22 + 0.94\ \Omega) = 0.087$ A.

(*b*) The potential difference across the terminals of the battery will equal the EMF minus the voltage drop across the internal resistance. Thus $V = V_a - V_b = 2.0 - (0.94)(0.087) = 1.92$ V.

Problem 5.12. A resistance, R, is connected to a battery and the current that flows is measured. When $R = 40\ \Omega$, the current is 0.240 A, and when $R = 60\ \Omega$, the current is 0.162 A. What is the EMF of the battery and what is its internal resistance?

Solution

When $R = 40\ \Omega$, the current will equal $I_1 = 0.240\ \text{A} = \mathscr{E}/(40 + R_{int})$, or $\mathscr{E} = (0.240\ \text{A})(40 + R_{int}\ \Omega)$. Similarly, when $R = 60\ \Omega$, $\mathscr{E} = (0.162\ \text{A})(60 + R_{int})\ \Omega$. These are two equations in the two unknowns, $\mathscr{E}$ and R_{int}, which can be easily solved, since they both give the $\mathscr{E}$ in terms of R_{int}. Thus $0.162(60 + R_{int}) = 0.240(40 + R_{int}) \rightarrow R_{int}(0.240 - 0.162) = 9.72 - 9.60 = 0.12$, $R_{int} = 1.54\ \Omega$. Then, $\mathscr{E} = 0.162(60 + 1.54) = 9.97$ V.

Within a source of EMF the charges are made to flow in a direction opposite to the electric force. This means that work must be done in separating the charges within the source. This work must come from some source of energy: chemical, mechanical, nuclear, solar, etc. As noted, in the case of a battery the source is chemical, and the energy stored in the battery is reduced whenever the battery supplies current to an external circuit. In due time that energy is exhausted and the battery must either be replaced or "recharged". In **recharging** a battery, energy must be delivered to the chemicals within the battery and be stored in the form of chemical energy of the molecules of the medium. To add energy requires that current must flow within the battery from the positive to the negative terminal, which is opposite to the direction in which it flows when the battery is **discharging**. To accomplish this one uses a different source of EMF, such as a generator, and applies a voltage across the terminals of the battery from this external source which will try to force current to flow in the desired direction. If the EMF of the external source is greater than the EMF of the battery, then current will flow in the direction determined by the external source. In that case the battery will receive energy and, if the battery is of the type that can be recharged, that energy will be stored in the battery.

Problem 5.13. A battery with EMF 6.0 V and internal resistance 1.6 Ω, is being recharged from a generator with an EMF of 8.2 V and internal resistance 2.1 Ω, as in Fig. 5-8.

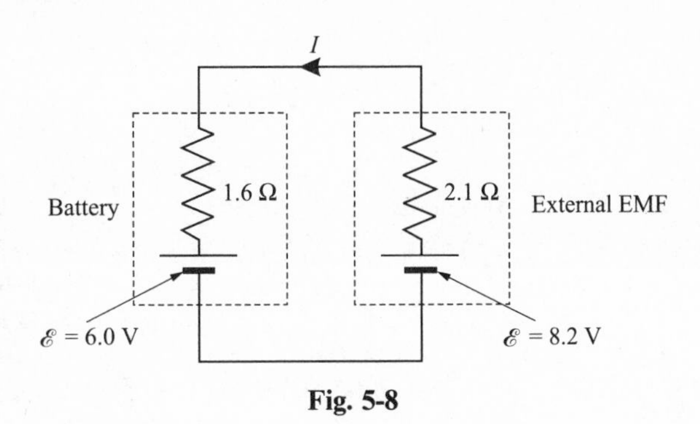

Fig. 5-8

(*a*) How much current flows in the circuit?

(*b*) How long does it take for 15,000 C of charge to be transferred to the battery?

(*c*) How much work is done during this time?

Solution

(*a*) The external EMF attempts to send current from its positive terminal in the direction of I in the figure. The battery attempts to send current in the opposite direction. Since the external EMF is greater than the EMF of the battery, the current will flow in the direction shown. The total EMF that is responsible for the current flow is the difference between the two EMFs, and equals $(8.2 - 6.0)\text{V} = 2.2$ V. The total resistance in the circuit is $(1.6 + 2.1)\Omega = 3.7\ \Omega$. The current is therefore $I = 2.2\ \text{V}/3.7\ \Omega = 0.595$ A.

(*b*) The total charge that flows is $q = It = (0.595\ \text{A})t = 15{,}000$ C. Therefore, $t = (15{,}000\ \text{C})/(0.595\ \text{A}) = 25,$ 210 s = 7.0 h.

(*c*) The work done in moving a charge q through an EMF, $\mathcal{E}$, is $q\mathcal{E}$. Thus the work done in moving 15,000 C through the battery EMF of 6.0 V is (15,000 C)(6.0 V) = 90,000 J = 90 kJ.

Note. The work supplied by the external source was (15,000 C)(8.2 V) = 123 kJ. The difference in these amounts was dissipated in the two internal resistors. The amount of heat dissipated in a resistor can easily be calculated as will be discussed in a future section.

5.4 ELECTRIC MEASUREMENT

To measure currents and voltages in circuits we need an instrument that is sensitive to the flow of current. Such an instrument is a **galvanometer** which detects the effect of current in a magnetic field that we will discuss in a later chapter. The deflection of the needle of the galvanometer depends on the current flowing through the galvanometer. Both **ammeters** (that measure currents) and **voltmeters** (that measure voltages) make use of the galvanometer as the basic measuring tool.

Ammeters

To measure the current in a circuit, it is obvious that one must place the measuring instrument in series within the circuit so that the same current flows in the meter as in the circuit. It would seem that the current read on the meter will then equal the current in the circuit. There is, however, a slight complication. Consider the circuit of Fig. 5-9. Here a source of EMF causes a current I to flow through a resistor R. The current will be $I = V/R$. If we wish to measure this current, we would have to place an ammeter in series with the resistor, R. The ammeter itself, however, has resistance, R_A. Therefore, if we place the ammeter in the circuit in series with R, the current will change to $I = V/(R + R_A)$, which is what the ammeter will measure. (We can correct for this if we know R and R_A, but we would prefer to be able to take the measurement at face value.) In order to minimize the effect of the ammeter on the

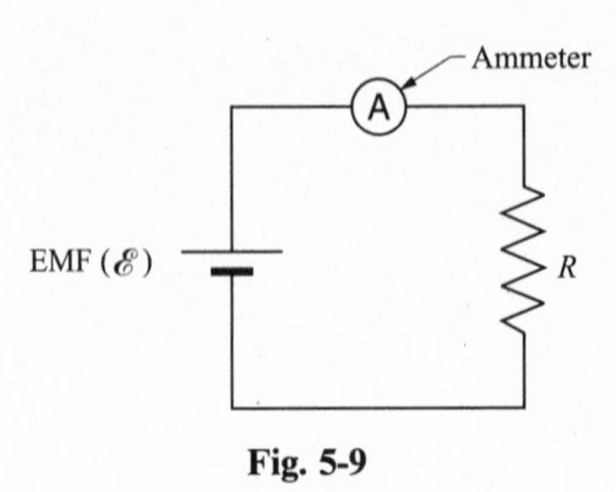

Fig. 5-9

current we must build our meters to have a very small resistance compared with the resistance R in the circuit we are measuring. Thus ammeters must always have small resistances to be accurate in their measurements.

Problem 5.14. In the circuit of Fig. 5-9, the EMF is 2.5 V, the resistance R is 25 Ω, and the ammeter has a resistance of 0.32 Ω.

(*a*) What current flows in the circuit if the ammeter is not present?

(*b*) What current does the ammeter measure when placed in the circuit?

(*c*) In terms of R and R_A, what correction must be applied to the measurement to get the current in the absence of the meter?

Solution

(*a*) The current is $I = V/R = (2.5\ \text{V})/(25\ \Omega) = 0.10$ A.

(*b*) The current is $I_A = V/(R + R_A) = (2.5\ \text{V})/(25 + 0.32)\Omega = 0.0987$ A.

(*c*) Using the relationship that $I = V/R$ and $I_A = V/(R + R_A)$ and dividing, we get $I/I_A = (R + R_A)/R$, or $I = I_A(R + R_A)/R$. Substituting in our values of R and R_A and the measured value from part (*b*), we indeed get the result of part (*a*). Try it.

To construct an ammeter from a current sensitive galvanometer one must know the current, I_{max}, at which the galvanometer obtains full-scale deflection. If one uses this galvanometer itself in series in the circuit, then the maximum current that can be measured is I_{max}. In order to use this galvanometer to measure larger currents one places this galvanometer in parallel with another, smaller resistance, so that only a small fraction of the current flows through the galvanometer. The maximum deflection on the galvanometer will then still occur when I_{max} flows through the galvanometer, but this occurs when the current flowing through the parallel circuit is much greater than I_{max}. This low parallel resistor also ensures that the overall resistance of the ammeter is also small. The following problem illustrates this phenomenon.

Problem 5.15. Consider the galvanometer in Fig. 5-10, which can be connected to be in parallel with various resistors by the switch S. Current enters on the left and flows through the galvanometer and any one of the parallel resistors that is connected by the switch. The galvanometer has a resistance $R_G = 3.2$ Ω, and has its maximum deflection at a current of 2.0×10^{-2} A.

(*a*) What resistance R_A is required to build an ammeter whose maximum deflection occurs at a current of 0.20 A? What is the resistance of this ammeter?

(*b*) What resistance R_B is required to build an ammeter whose maximum deflection occurs at a current of 2.0 A? What is the resistance of this ammeter?

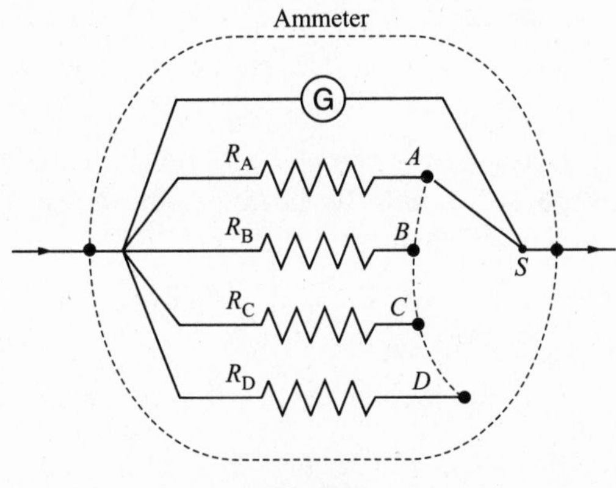

Fig. 5-10

(c) What resistance R_C is required to build an ammeter whose maximum deflection occurs at a current of 20 A? What is the resistance of this ammeter?

Solution

(a) To build an ammeter with a maximum deflection when a current of 0.20 A enters the ammeter, the current through the galvanometer must be 0.020 A at this point. Then the current through the parallel resistor R_A must be 0.18 A so that the total current in the parallel circuit is 0.20 A. The voltage across the parallel paths is the same for both the elements, i.e. for R_G and for R_A, and therefore $V_G = V_A = I_G R_G = I_A R_A = 0.020(3.2) = 0.18 R_A$, giving $R_A = 0.36\ \Omega$. The resistance of this parallel circuit is $1/R = 1/3.2 + 1/0.36$, $R = 0.32\ \Omega$.

(b) Using the same procedure as in part (a), but with $I_B = 2.0$ A, we get $R_B = (0.020\ \text{A})(3.2\ \Omega)/(1.98\ \text{A}) = 0.032\ \Omega$. The resistance of the parallel circuit is then $1/R = 1/3.2 + 1/0.032$, giving $R_B = 0.0317\ \Omega$.

(c) Using the same procedure as in part (a), but with $I_C = 20$ A, we get $R_C = (0.020\ \text{A})(3.2\ \Omega)/\ (20\ \text{A}) = 0.0032\ \Omega$. The resistance of the parallel circuit is then $1/R = 1/3.2 + 1/0.0032$, giving $R_B = 0.0032\ \Omega$.

Voltmeters

We now turn our attention to the question of constructing a voltmeter from a current sensitive galvanometer. We realize that for an instrument to measure the voltage between two points, for instance the voltage across a resistor R, we must connect the instrument between those two points. This means that a voltmeter must be connected in parallel with the circuit element whose voltage we seek. This is shown in Fig. 5-11. The voltmeter will then read the same voltage as exists across R, i.e. $V_A - V_B$. Since the galvanometer we are using to construct a voltmeter is sensitive to current, we are really measuring the current flowing through the voltmeter. If the voltmeter consists of a resistor R' in series with the galvanometer, then the current in the galvanometer will equal $I = V/(R' + R_G) = V/R_V$. Thus $V =$

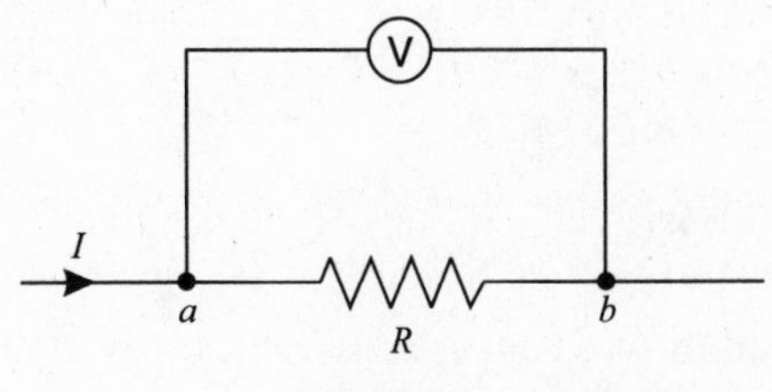

Fig. 5-11

IR_V, where R_V is the equivalent resistance of the voltmeter. A measurement of I is equivalent to a measurement of V, since the two are proportional. This is how voltmeters are generally constructed.

Problem 5.16. A voltmeter is constructed to operate in various voltage ranges by using a switch to select the resistance that is placed in series with the galvanometer. In Fig. 5-12, the galvanometer has a resistance of 3.2 Ω and has maximum deflection at a current of 0.020 A.

(*a*) What resistance, R_A is needed so that the voltmeter will have a maximum deflection at a voltage of 10 V?

(*b*) What resistance, R_B is needed so that the voltmeter will have a maximum deflection at a voltage of 100 V?

(*c*) What resistance, R_C is needed so that the voltmeter will have a maximum deflection at a voltage of 10^3 V?

Solution

(*a*) The current through the galvanometer must be 0.02 A when the voltage across the circuit is 10 V. Since $V = I(R_A + R_G)$, $R_A + R_G = (10\text{ V})/(0.02\text{ A}) = 500\ \Omega = 3.2 + R_A$ and $R_A = 496.8\ \Omega$

(*b*) For this range, the current through the galvanometer must be 0.02 A when the voltage across the circuit is 100 V. Since $V = I(R_B + R_G)$, $R_B + R_G = (100\text{ V})/(0.02\text{ A}) = 5000\ \Omega = 3.2 + R_B$, and $R_B = 4997\ \Omega$.

(*c*) For this range, the current through the galvanometer must be 0.02 A when the voltage across the circuit is 10^3 V. Since $V = I(R_C + R_G)$, $R_C + R_G = (10^3\text{ V})/(0.02\text{ A}) = 50{,}000\ \Omega = 3.2 + R_C$ and $R_C = 49{,}997\ \Omega \approx 50{,}000\ \Omega$.

In Fig. 5-11, the voltage across the resistor had a certain value V before we attached the voltmeter. Usually, the resistor is part of a larger circuit which supplies a certain current to the resistor in the circuit. If we attach the voltmeter, we are inserting a parallel path for the current, and some of the current will flow through the voltmeter instead of the resistor. Then the voltage across the resistor will be reduced as a result of this diminished flow through R. Thus, the insertion of the voltmeter can change the voltage that we are trying to measure. In order to minimize this change, we require that very little current be diverted through the voltmeter. This can be accomplished by making the resistance of the voltmeter very large compared to R. If this is not the case, one has to correct the reading to account for the effect of the voltmeter.

Problem 5.17. In Fig. 5-11, current of 0.020 A enters point a from the left. Assume this current remains the same whether or not the voltmeter is in the circuit. The resistance $R = 25\ \Omega$, and $R_V = 2500\ \Omega$.

(*a*) Without the voltmeter in the circuit, what is the voltage across the resistor, R?

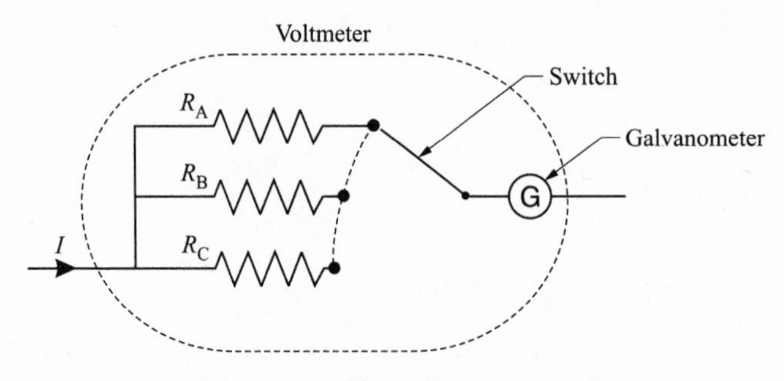

Fig. 5-12

(*b*) With the voltmeter in the circuit, what is the voltage across the circuit?

(*c*) In terms of R and R_V, how would we correct a reading on the voltmeter to give us the voltage in the absence of the voltmeter?

Solution

(*a*) The voltage across R is $I_R R = (0.02\ \text{A})(25\ \Omega) = 0.50\ \text{V}$.

(*b*) Now the voltage across R is the voltage across the parallel circuit of R and R_V. Thus $R_{eq} = (1/R + 1/R_V)^{-1} = 24.75\ \Omega$. Then $V = (0.02\ \text{A})(24.75\ \Omega) = 0.495\ \text{V}$.

(*c*) With the voltmeter in place, the voltage reading is $V' = I[(RR_V)/(R + R_V)]$, since $R_{eq} = (RR_V)/(R + R_V)$. Without the voltmeter, the voltage is $V = IR$. Then $V/V' = (R + R_V)/R_V \rightarrow V = V'(R + R_V)/R_V$. [Check to see if this is correct by substituting in for part (*b*) to get the result of part (*a*).]

Now suppose we have a resistor in a circuit and we want to measure both the current through the resistor and the voltage across the resistor, in order to determine its resistance which is V/I. We know that we should place the ammeter in series with the resistor so it has the same current as the resistor. Similarly we know that we should place the voltmeter in parallel with the resistor so that it will measure the same voltage as is across the resistor. We cannot simultaneously do both of these things, as can be seen from Fig. 5-13. Here we have connected the ammeter A in series with the resistor R. The current flowing through the resistor from a to b also flows through the ammeter from b to c. Now we try to connect the voltmeter across R. One side is obviously attached to a, but we have a problem with the other side. If we connect it to point b then it will indeed be across R, and measure the voltage across R. However, the ammeter will then no longer be in series with R, and will measure the current flowing both through R and through the voltmeter. If we connect the end of the voltmeter to point c, then the ammeter will remain in series with R and measure only the current through R, but the voltmeter will be across the series combination of R and the ammeter and measure the voltage across both. There is no way to avoid this if one uses this circuit, and we must make the appropriate corrections if we know R_A and R_V.

Problem 5.18. In the circuit of Fig. 5-13, the meters have the following resistances: $R_A = 3.2\ \Omega$, $R_V = 2500\ \Omega$. These meters are used to measure the resistance of the resistor, by simultaneously measuring I and V.

(*a*) If no corrections are made, what is the resistance R in terms of the measured I and V?

(*b*) If the voltmeter is connected to point b, what is the corrected value of R in terms of the measured I and V, and the resistances of the meters?

(*c*) If the voltmeter is connected to point c, what is the corrected value of R in terms of the measured I and V, and the resistances of the meters?

(*d*) If the readings on the meters for case (*b*) are $I = 0.21$ A and $V = 10.3$ V, what is the resistance R? What would the uncorrected value of R be?

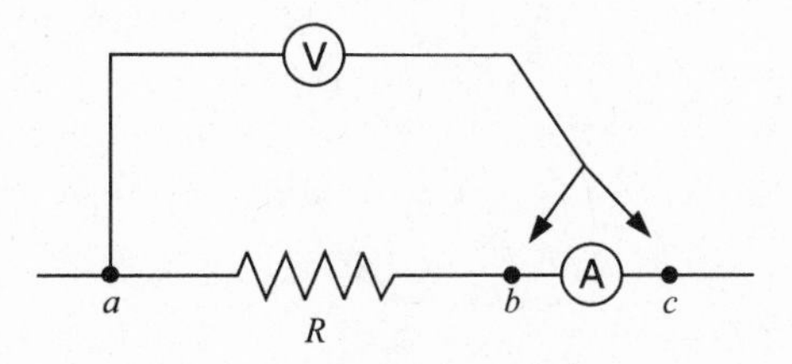

Fig. 5-13

(*e*) If for case (*c*) the current was adjusted so the ammeter read the same value $I = 0.21$ A, what would the reading on the voltmeter be? What would the uncorrected value of R then be?

Solution

(*a*) $R = V/I$.

(*b*) If the voltmeter is connected to *b*, then the voltage on the voltmeter is the voltage across R. The current in the ammeter, however, is the current in the parallel circuit of R and R_V. Thus $I = V/R_{eq} = V(1/R + 1/R_V)$. Thus $1/R = I/V - 1/R_V$, which we can use to get R.

(*c*) Now the current read on the ammeter is the current in R. The voltage on the voltmeter is the voltage across the series circuit of R and R_A. Thus $V = I(R + R_A)$, and $R = (V/I) - R_A$.

(*d*) For case (*a*), $1/R = (0.21 \text{ A})/(10.3 \text{ V}) - 1/2500\ \Omega = 0.0200$, $R = 50.0\ \Omega$. From part (*a*) the uncorrected value would be $R = V/I = 10.3/0.21 = 49\ \Omega$.

(*e*) We have $V = I(R + R_A) = (0.21 \text{ A})(50\ \Omega + 3.2\ \Omega) = 11.2$ V; the uncorrected value of R would be $R = V/I = (11.2 \text{ V})/(0.21 \text{ A}) = 53\ \Omega$. We see clearly that the measuring instruments can have an effect on the measurements and we must be careful to check that they do not give us incorrect results.

It is clear that the ideal way to measure a resistance is to use meters that do not draw any current when they are in the circuit. This would be a case of a **null measurement** where the result depends on adjusting a dial until the meter reads zero. An example of a null measurement is the equal arm balance used to measure weights. Here one adjusts the position of the known weight until there is no deflection of the arm, and determines the unknown weight from the position of the known weight. The corresponding instrument that is used to measure resistance using a null method is the **Wheatstone bridge**. This can be used to measure an unknown resistance by adjusting known resistances until the current in a galvanometer is zero. The circuit for the Wheatstone bridge is shown in Fig. 5-14. Here, the unknown resistor is X, and the other (known) resistors M, N and P are adjusted so that no current flows through the galvanometer G when the EMF is applied to the circuit. This means that no current flows through G, between points *b* and *c*, when both switches are closed. For no current to flow in the galvanometer, there must be no voltage difference between points *b* and *c*. Therefore placing the galvanometer between those points does not disturb the circuit, and the currents and voltages that existed before connecting G are maintained. Then no adjustments are necessary for the resistance of the galvanometer. The operation of the bridge is the subject of the next problem.

Problem 5.19. In the Wheatstone bridge of Fig. 5-14, the switch S_1 is closed and the EMF is applied to the circuit. If the current in the galvanometer is zero when S_2 is closed, and the voltage across points *a* and *b* is V, what is:

(*a*) the current in the resistor N? in the resistor M?

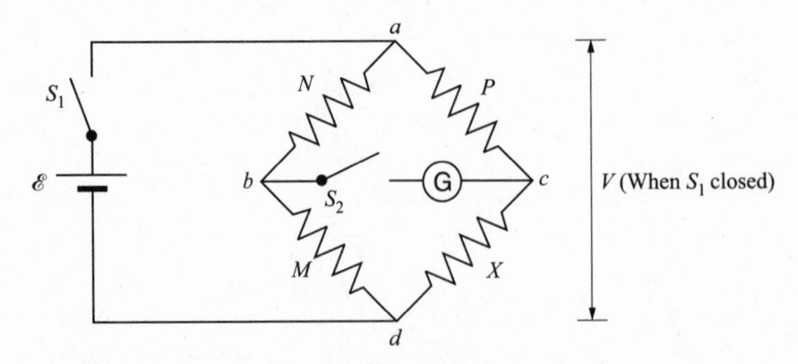

Fig. 5-14

(*b*) the current in the resistor P? in the resistor X?

(*c*) the resistance X in terms of the resistances M, N and P?

Solution

(*a*) There is no current in G. Therefore, all the current that flows through N also flows through M, and the two can be considered to be in series. Then the current in N and in M is $I_1 = V/(N + M)$.

(*b*) Using the same analysis as in (*a*), the resistors P and X are in series and the current through both is $I_2 = V/(P + X)$.

(*c*) The difference in potential between point b and point d is $MI_1 = MV/(M + N)$. Similarly, the difference in potential between point c and point d is $XI_2 = XV/(P + X)$. But these differences are equal since there is no potential difference between b and c. Thus $MV/(M + N) = XV/(P + X)$. Cross multiplying we get: $X(M + N) = M(P + X)$, or $XN = MP$, $X = MP/N$. In using the bridge one has known adjustable resistors M, P and N, and adjusts one or more of them until the bridge is "balanced" i.e. until there is no current in the galvanometer.

5.5 ELECTRIC POWER

We mentioned previously that the work that is done within the source of EMF is available to the external circuit. We can easily calculate where this energy is dissipated, either in the form of work or heat. When a positively, charged particle moves from the anode to the cathode (or a negatively charged particle moves from the cathode to the anode), the electrical energy that the particle loses equals qV, where q is the charge on the particle and V is the difference of potential through which the particle moves. The rate at which the energy is lost, the **power**, equals $P = \Delta(qV)/\Delta t = V(\Delta q/\Delta t) = VI$. This power is available for work (turning a motor) or for heat (in an electric heater or light bulb). This derivation of the power available was not dependent on the circuit containing only ohmic elements. It is universally true. In fact it is true even for time varying currents and voltage at any instant of time. Thus,

$$P = IV \tag{5-5}$$

If the circuit contains only ohmic resistors, we can easily express the power loss in the resistors. If we have a potential difference V across a resistance R, then $V = IR$ and

$$P = IV = I(IR) = I^2R = V^2/R \tag{5-6}$$

In these formulas, the current I is the current through the resistor R, and the voltage V is the voltage across the resistor R. For parallel or series circuits the power will differ from resistor to resistor, but for each resistor the power will equal I^2R or V^2/R using its own current and voltage. The total power will equal the sum of the power of each element, which will be the same as $I^2R_{eq} = V^2/R_{eq}$, with I and V the total current and voltage for the circuit. This power is dissipated in the resistors as heat. We can understand the dissipation process as one in which the particles transfer energy to the material of the resistor via the collisions that produced the resistance in the first place. This energy that is transferred to the material heats up the material. If the resistor gets sufficiently hot, the material will emit visible light, as in the filament of an incandescent light bulb.

Problem 5.20. In the circuit segment of Fig. 5-2(*b*), the voltage across the circuit is 55V. If $R_1 = 25\ \Omega$ and $R_2 = 35\ \Omega$, calculate (*a*) the current in each resistor; (*b*) the power dissipated in each resistor and (*c*) the power dissipated in the equivalent resistance. Compare the answer to this with the sum of the answers to (*b*).

Solution

(*a*) The voltage across each resistor is 55 V. Thus the current for each resistor is $I = V/R$. Then $I_1 = (55\ \text{V})/(25\ \Omega) = 2.2$ A, and $I_2 = (55\ \text{V})/(35\ \Omega) = 1.57$ A.

(*b*) Since the voltage across each resistor is the same, the power in each resistor can be calculated using $P = V^2/R$. Thus $P_1 = (55\text{ V})^2/(25\ \Omega) = 121$ W, and $P_2 = (55\text{ V})^2/(35\ \Omega) = 86.4$ W. Alternatively, we could have used $P = I^2R$, using the current appropriate to each resistor. Then $P_1 = (2.2\text{ A})^2(25\ \Omega) = 121$ W, and $P_2 = (1.57\text{ A})^2(35\ \Omega) = 86.4$ W.

(*c*) The equivalent resistance is $R_{eq} = (25)(35)/(25 + 35) = 14.6\ \Omega$. The total power is therefore $P_{tot} = (55)^2/14.6 = 207.4$ W. This equals the sum of $P_1 + P_2 = 121 + 86.4$.

Problem 5.21. In the circuit segment of Fig. 5-4, the current entering the circuit is 2.1 A. Calculate (*a*) the voltage across each resistor, (*b*) the power dissipated in each resistor and (*c*) the power dissipated in the equivalent resistance. Compare the answer to this with the sum of the answers to (*b*).

Solution

(*a*) The voltage across each resistor is $V = IR = (2.1\text{ A})R$, since the current is the same through each resistor. Thus $V_1 = (2.1\text{ A})(40\ \Omega) = 84$ V, $V_2 = (2.1\text{ A})(60\ \Omega) = 126$ V, $V_3 = (2.1\text{ A})(25\ \Omega) = 52.5$ V.

(*b*) The power dissipated in each resistor can be calculated using I^2R or V^2/R for each resistor. Thus $P_1 = (2.1\text{ A})^2(40\ \Omega) = 176.4$ W (or $P_1 = 84^2/40 = 176.4$), $P_2 = (2.1\text{ A})^2(60\ \Omega) = 264.6$ W, $P_3 = (2.1\text{ A})^2(25\ \Omega) = 110.25$ W.

(*c*) The equivalent resistance is $R_{eq} = (40 + 60 + 25) = 125\ \Omega$. The total power is therefore $P_{tot} = (2.1\text{ A})^2(125) = 551.25$ W. This equals the sum of $P_1 + P_2 + P_3 = 176.4 + 264.6 + 110.25 = 551.25$ W.

Problem 5.22. Refer to Problem 5.10, in which we solved for the equivalent resistance of the circuit in Fig. 5-6, as well as the current in each resistor. Use the results of that problem to calculate (*a*) the total power developed in the circuit and (*b*) the power developed in each resistor.

Solution

(*a*) The equivalent resistance of the circuit was calculated to be $R_{eq} = 101.3\ \Omega$. Therefore, the total power developed in the circuit is $V_{tot}^2/R_{eq} = (82\text{ V})^2/(101.3\ \Omega) = 66.4$ W

(*b*) The current in each resistor is:

$$I_1 = I_6 = 0.809\text{ A}$$

$$I_2 = I_3 = I_4 = 0.119\text{ A}$$

$$I_5 = 0.690\text{ A}$$

Then:

$$P_1 = (0.809\text{ A})^2(40\ \Omega) = 26.6\text{ W}$$

$$P_2 = (0.119\text{ A})^2(60\ \Omega) = 0.85\text{ W}$$

$$P_3 = (0.119)^2(25) = 0.35\text{ W}$$

$$P_4 = (0.119)^2(60) = 0.85\text{ W}$$

$$P_5 = (0.690)^2(25) = 11.9\text{ W}$$

$$P_6 = (0.809)^2(40) = 26.2\text{ W}$$

The total power, calculated by adding the power in each resistor equals $P_{tot} = 66.4$ W, as in part (*a*).

Problem 5.23. A light bulb is rated at 60 W, 120 V. Assume that this bulb is an ohmic resistance.

(*a*) What is the resistance of the bulb?

(*b*) If one applies a voltage of 75 V to the bulb what power is developed in the bulb?

(*c*) If three bulbs, rated at 25 W, 40 W and 60 W, are connected in series and 120 V is imposed across the entire circuit, how much power is developed in each bulb?

Solution

(*a*) The rating of the bulb means that, if 120 V is across the bulb, then 60 W is developed in the bulb. Since $P = V^2/R$, $R = V^2/P = (120\ \text{V})^2/(60\ \text{W}) = 240\ \Omega$.

(*b*) The power is $V^2/R = (75\ \text{V})^2/(240\ \Omega) = 23.4$ W. It is clear that this bulb will be very dim since only 23.4 W is developed, rather than the rated 60 W.

(*c*) The three bulbs are in series and therefore they have the same current, but not the same voltage. In order to calculate the power, we need to calculate the resistance of each bulb, as well as the common current. To calculate each individual resistance, we use the rating of the bulb, which states that, if 120 V is across the resistor, it will develop the rated wattage. Thus, the rated wattage will equal $(120\ \text{V})^2/R$. Then, $R_1 = (120)^2/25 = 576\ \Omega$, $R_2 = (120)^2/40 = 360\ \Omega$, $R_3 = 240\ \Omega$. To calculate the current we use the fact that $R_{eq} = (576 + 360 + 240) = 1176\ \Omega$, and calculate $I = V_{tot}/R_{eq} = 0.102$ A. Then, $P_1 = I^2R_1 = (0.102\ \text{A})^2(576\ \Omega) = 6.00$ W, $P_2 = (0.102)^2(360) = 3.75$ W, $P_3 = (0.102)^2(240) = 2.50$ W. The total power developed is 12.25 W. This could have been calculated using $P = IV = (0.102\ \text{A})(120\ \text{V}) = 12.24$ W.

Problem 5.24. An air conditioner is rated at 1.0 kW, 120 V.

(*a*) How much current does it draw?

(*b*) If the air conditioner is run for 3.0 h, how much energy is used?

Solution

(*a*) The power is $P = IV = 1.0 \times 10^3\ \text{W} = I(120\ \text{V})$, so $I = 8.33$ A.

(*b*) The power is the energy used per unit time, $P = \Delta E/\Delta t$. If the power is constant, then $\Delta E = P\Delta t = (1000\ \text{W})(3.0\ \text{h} \times 3600\ \text{s/h}) = 1.08 \times 10^7$ J. This could also be calculated in mixed units by noting that the power developed was 1.0 kW and was used for three hours, thus consuming 3 kW-h of energy, where one kW-h (**kilowatt-hour**) is the energy consumed for one hour at a rate of one kW. This is actually the unit of energy used by the electric utilities in billing their customers for electrical energy. The conversion to Joules is 1 kW-h = 3.6×10^6 J.

Problems for Review and Mind Stretching

Problem 5.25. A solenoid consists of wire wound around a cylinder of radius 0.36 m. If the wire is made of tungsten, of radius $r = 0.34$ cm, and has 2500 turns, what is the resistance of the solenoid?

Solution

The resistance is given by $R = \rho L/A$. For tungsten, $\rho = 5.51 \times 10^{-8}\ \Omega$-m. The area is $A = \pi r^2 = \pi(0.0034\ \text{m})^2 = 3.63 \times 10^{-5}\ \text{m}^2$, and $L = (2500\ \text{turns})(2\pi)(0.36\ \text{m}) = 5.65 \times 10^3$ m. Thus $R = (5.51 \times 10^{-8}\ \Omega \cdot \text{m})(5.65 \times 10^3\ \text{m})/(3.63 \times 10^{-5}\ \text{m}^2) = 8.6\ \Omega$.

Problem 5.26. In the circuit of Fig. 5-15, the resistors have resistances of: $R_1 = 25\ \Omega$, $R_2 = 45\ \Omega$, $R_3 = 150\ \Omega$, $R_4 = 78\ \Omega$, $R_5 = 18\ \Omega$, $R_6 = 55\ \Omega$. The current through R_1 is $I_1 = 0.98$ A.

(*a*) What is the equivalent resistance of the circuit?

(*b*) What are the currents in each resistor?

(*c*) What is the EMF of the battery? (Assume zero internal resistance.)

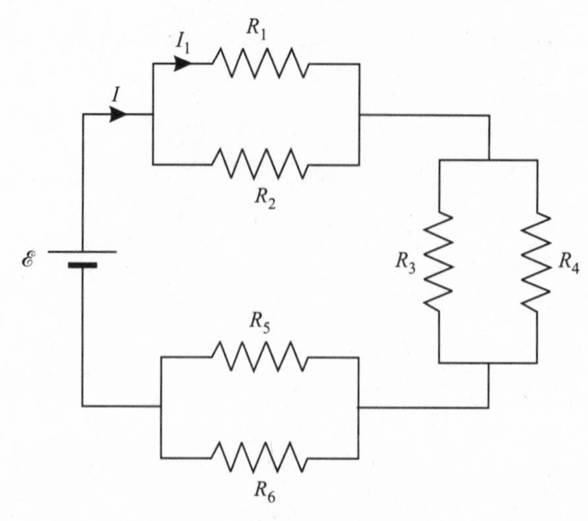

Fig. 5-15

Solution

(*a*) R_1 and R_2 are in parallel and we can combine them as a single resistance with $R = (R_1R_2)/(R_1 + R_2) = (25\ \Omega)(45\ \Omega)/(25 + 45)\Omega = 16.1\ \Omega$. Similarly, R_3 and R_4 are in parallel with each other and can be combined as a single resistor with $R = (150\ \Omega)(78\ \Omega)/(150 + 78)\Omega = 51.3\ \Omega$. The two resistors, R_5 and R_6 can be similarly combined into $R = (55\ \Omega)(18\ \Omega)/(55 + 18)\Omega = 13.6\ \Omega$. These three resistances are now in series, yielding a final equivalent resistance of $R_{eq} = (16.1 + 51.3 + 13.6)\Omega = 81\ \Omega$.

(*b*) Since $I_1 = 0.98$ A, $V_1 = (0.98\text{ A})(25\ \Omega) = 24.5$ V. Since R_2 is parallel to R_1, it has the same voltage, and then $I_2 = (24.5\text{ V})/(45\ \Omega) = 0.54$ A. The current through the parallel combination (and the current from the battery and through each of the other parallel combinations) is therefore $I = 0.98 + 0.54 = 1.52$ A. The voltage across R_3 and across R_4 is therefore $V_3 = V_4 = (1.52\text{ A})(51.3\ \Omega) = 78.0$ V. Then $I_3 = (78.0\text{ V})/(150\ \Omega) = 0.52$ A, and $I_4 = (78.0\text{ V})/(78\ \Omega) = 1.00$ A. Similarly, the voltage across R_5 and across R_6 is $V_5 = V_6 = (1.52\text{ A})(13.6\ \Omega) = 20.7$ V. Thus, $I_5 = (20.7\text{ V})/18\ \Omega) = 1.15$ A and $I_6 = (20.7\text{ V})/(55\ \Omega) = 0.37$ A.

(*c*) The EMF equals $I_{tot}R_{eq} = 123$ V.(This could also have been calculated as the sum of the voltages across the three parallel circuits, $V = 24.5 + 78.0 + 20.7 = 123$ V.)

Problem 5.27. For the same circuit as in Problem 5.26, calculate (*a*) the total power consumed and (*b*) the power in each of the resistors.

Solution

(*a*) The total power consumed is $I^2R_{eq} = (1.52\text{ A})^2(81\ \Omega) = 187$ W. Alternatively $P = IV = (1.52\text{ A})(123\text{ V}) = 187$ W (or $P = V^2/R_{eq} = (123)^2/81 = 187$ W).

(*b*) For each resistor we can use either $I_i^2R_i$ or I_iV_i or $V_i^2R_i$. We will use different equations for the various resistors in order to demonstrate the use of each equation.

$$P_1 = I_1{}^2R_1 = (0.98\text{ A})^2(25\ \Omega) = 24.0\text{ W}$$

$$P_2 = I_2\,V_2 = (0.54\text{ A})(24.5\text{ V}) = 13.2\text{ W}$$

$$P_3 = V_3{}^2/R_3 = (78.0\text{ V})^2/(150\ \Omega) = 40.6\text{ W}$$

$$P_4 = I_4{}^2R_4 = (1.00\text{ A})^2(78\ \Omega) = 78.0\text{ W}$$

$$P_5 = I_5\,V_5 = (1.15\ A)(20.7\text{ V}) = 23.8\text{ W}$$

$$P_6 = V_6{}^2/R_6 = (20.7\text{ V})^2/(55\ \Omega) = 7.8\text{ W}$$

If we add these together, we get $P_{tot} = 187$ W, as in (*a*).

Problem 5.28. A battery has an EMF of $\mathscr{E} = 26$ V, and is connected to a series circuit of two resistors, $R_1 = 35\ \Omega$ and $R_2 = 17\ \Omega$.

(*a*) What is the current in R_1 and the voltage across R_1?

(*b*) If one tries to measure the current by placing an ammeter with an internal resistance of 2.3 Ω in the circuit, what is the current read on the ammeter?

(*c*) If one tries to measure the voltage across R_1 by placing a voltmeter which has an internal resistance of 150 Ω across R_1, what voltage is read on the voltmeter?

(*d*) If one tries to calculate the resistance of R_1 by using the measured current and voltage, what resistance would be calculated? By what percentage does this differ from the actual R_1?

Solution

(*a*) The equivalent resistance of the circuit is $(35 + 17)\Omega = 52\ \Omega$. Thus the current is $I = (26\text{ V})/(52\ \Omega) = 0.50$ A. The voltage across R_1 is $V_1 = (35\ \Omega)(0.50\text{ A}) = 17.5$ V.

(*b*) If one places an ammeter in series with R_1 and R_2, the total resistance is now $(R_1 + R_2 + R_A) = 54.3\ \Omega$. The current read on the ammeter is therefore $I = (26\text{ V})/(54.3\ \Omega) = 0.48$ A.

(*c*) If one places a voltmeter across R_1, then that voltmeter will be in parallel with R_1. The equivalent resistance of this parallel circuit will be $R = (35\ \Omega)(150\ \Omega)/(35 + 150)\Omega = 28.4\ \Omega$. This is in series with R_2, and the equivalent resistance of the whole circuit is now $(17 + 28.4) = 45.4\ \Omega$. The current is therefore $(26\text{ V})/(45.4\ \Omega) = 0.57$ A. This is the current flowing through the equivalent parallel resistance also. The voltage read on the voltmeter is therefore $V = (0.57\text{ A})(28.4\ \Omega) = 16.3$ V.

(*d*) If one uses the measured values as the current through the resistor and the voltage across the resistor, one would obtain that $R = V/I = (16.3\text{ V})/(0.48\text{ A}) = 34.0\ \Omega$. This is a bit different from the actual value of R_1 which is 35 Ω. The percentage difference is $(35 - 34)/35 = 0.029 = 2.9\%$.

Problem 5.29. In the circuit shown in Fig. 5-16, the resistances are $R_1 = 45\ \Omega$, $R_2 = 58\ \Omega$, $R_3 = 99\ \Omega$, $R_4 = 103\ \Omega$ and $R_5 = 66\ \Omega$. The power dissipated in the circuit is 185 W.

(*a*) What is the current in the circuit?

(*b*) What is the voltage across the circuit?

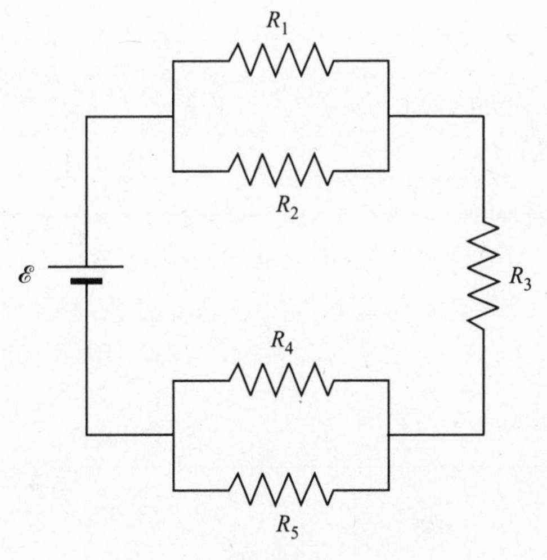

Fig. 5-16

Solution

(*a*) We first calculate the equivalent resistance of the circuit. R_1 and R_2 are parallel, and can be replaced by a resistance of $(45\ \Omega)(58\ \Omega)/(45 + 58)\Omega = 25.3\ \Omega$. Similarly, R_4 and R_5 can be replaced with a resistance of $(66)(103)/(66 + 103) = 40.2\ \Omega$. The two are in series with R_3 and the equivalent resistance of the entire circuit is $R_{eq} = (25.3 + 99 + 40.2) = 164.5\ \Omega$. The power equals $P = I^2 R_{eq} = 185$ W, and therefore $I^2 = (185\ \text{W})/164.5\ \Omega) = 1.12$, and $I = 1.06$ A.

(*b*) The voltage is $V = IR_{eq} = (1.06\ \text{A})(164.5\ \Omega) = 174$ V. (Alternatively, $P = V^2/R_{eq}$ and $V^2 = PR_{eq} = 185(164.5) = 3.04 \times 10^4$, and $V = 174$ V.)

Supplementary Problems

Problem 5.30. A certain wire has a diameter of 0.55 mm, and contains 8.5×10^{28} free electrons/m^3. If the wire has a current of 3.1 A, what is the drift velocity of the electrons?

Ans. 9.5×10^{-4} m/s

Problem 5.31. A long wire carries a current of 3.6 A when 12 V are placed across the wire. The wire has a length of 25 m and its cross-section has a radius of 0.30 mm.

(*a*) What is the resistivity of the material of the wire?

(*b*) What is the current density in the wire?

(*c*) How much charge passes through a cross-section of the wire in 3.3 minutes?

Ans. (*a*) $3.8 \times 10^{-8}\ \Omega \cdot \text{m}$; (*b*) $1.27 \times 10^7\ \text{A/m}^2$; (*c*) 713 C

Problem 5.32. A wire, with a square cross-section of side 0.21 mm, is made of copper. How long must the wire be to provide a resistance of 0.35 Ω?

Ans. 0.90 m

Problem 5.33. In the circuit shown in Fig. 5-17, calculate (*a*) the current in the 10 Ω resistor; (*b*) the voltage across the 2.0 Ω resistor and (*c*) the power dissipated in the 1.0 Ω resistor.

Ans. (*a*) 1.0 A; (*b*) 1.33 V; (*c*) 0.44 W

Problem 5.34. In the series circuit unit shown in Fig. 5-18, R_1 is a "60 W" bulb, R_2 is a "40 W" bulb and R_3 is a "100 W" bulb. A current flows such that the voltage across R_2 is 120 V. Calculate (*a*) the current in R_1; (*b*) the power dissipated in R_3 and (*c*) the voltage between the points *a* and *b*.

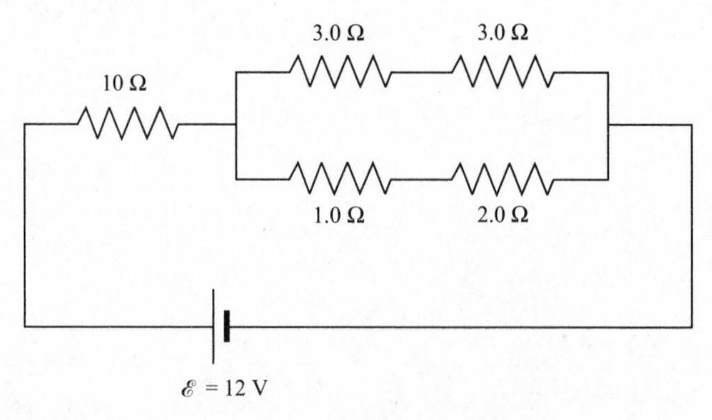

Fig. 5-17

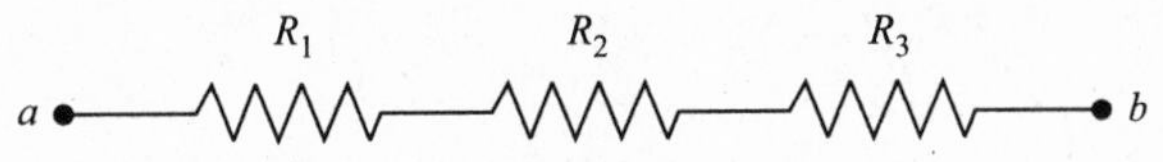

R_1 R_2 R_3

a b

Fig. 5-18

Ans. (*a*) 0.33 A; (*b*) 16 W; (*c*) 248 V

Problem 5.35. Two equal resistors are connected in parallel.

(*a*) Case 1: When a difference of potential of 120 V is connected across the resistors, the total power dissipated is 240 W. What is the resistance of each resistor?

(*b*) Case 2: When each resistor carries a current of 2.0 A, the total power dissipated is 480 W. What is the equivalent resistance of the parallel circuit?

Ans. (*a*) 120 Ω; (*b*) 30 Ω

Problem 5.36. In the circuit unit shown in Fig. 5-19, the current entering the circuit is 3.0 A, in the direction shown. The potential at point *a* is zero. Calculate (*a*) the potential difference across R_3; (*b*) the current in R_1; (*c*) the potential of point *c* and (*d*) the total power dissipated in the circuit.

Ans. (*a*) 300 V; (*b*) 2.0 A; (*c*) −440 V; (*d*) 1320 W

Problem 5.37. In the circuit unit shown in Fig. 5-20, a difference of potential of 180 V exists between points *a* and *b*, with $V_a < V_b$.

(*a*) In which direction does the current flow in resistor R_1, and in which direction do the electrons move in R_1?

(*b*) What is the current in R_2?

(*c*) What power is dissipated in R_3?

(*d*) How much potential energy does each electron lose when it passes through R_2?

Ans. (*a*) from *b* to *a*, from *a* to *b*; (*b*) 1.13 A; (*c*) 114 W; (*d*) 1.26×10^{-17} J

Problem 5.38. In the circuit shown if Fig. 5-21, the circuit elements have the following values: $R_1 = 3.0\ \Omega$, $R_3 = 2.5\ \Omega$, $R_4 = 1.5\ \Omega$, $C_1 = 1.0\ \mu F$, $C_2 = 4.0\ \mu F$, $C_3 = 2.0\ \mu F$. The battery produces an EMF of 12 V. After the capacitors have been fully charged, no current flows through the capacitors, but a steady current flows through the resistors. The voltage between points *b* and *c* is 8.0 V.

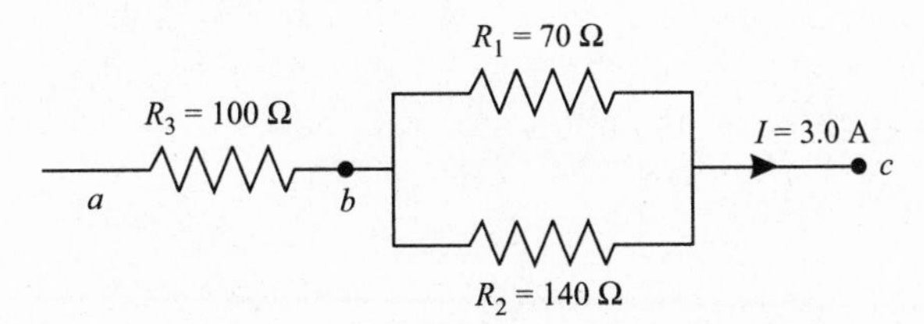

Fig. 5-19

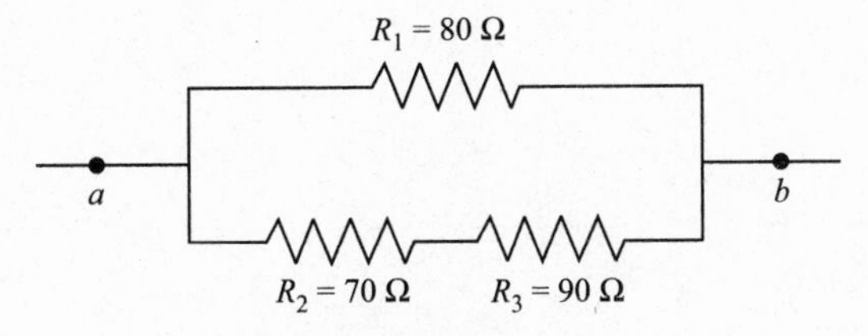

Fig. 5-20

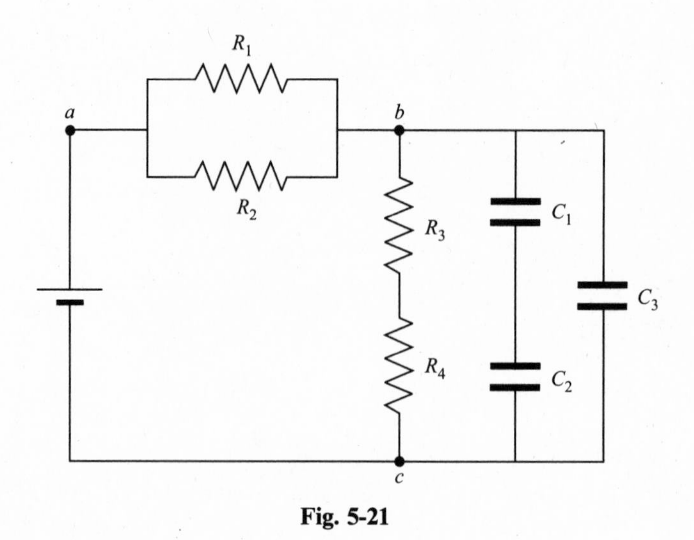

Fig. 5-21

(a) What is the voltage between points a and b?

(b) What is the current in R_3?

(c) What is the resistance of R_2?

(d) What is the charge stored on C_3?

(e) What is the charge stored on C_1?

Ans. (a) 4.0 V; (b) 2.0 A; (c) 6.0 Ω; (d) 1.6×10^{-5} C; (e) 6.4×10^{-6} C

Problem 5.39. In the Wheatstone bridge discussed in Problem 5.19, the fixed, known resistors have values: $M = 500.1\ \Omega$, $N = 333.4\ \Omega$. If the bridge is balanced when P is adjusted to $1.386 \times 10^3\ \Omega$, what is the resistance of the unknown resistor?

Ans. $2.079 \times 10^3\ \Omega$

Problem 5.40. A battery $\mathcal{E}_1$ is used to charge a second battery $\mathcal{E}_2$, as in Fig. 5-22. $\mathcal{E}_1 = 12.2$ V, $\mathcal{E}_2 = 9.0$ V and its internal resistance is 0.96 Ω. The variable resistor, R, is adjusted to make the current equal to 0.35 A.

(a) What is the resistance of R?

(b) What is the potential difference between the terminals of $\mathcal{E}_2$ while it is being charged?

(c) How long must one charge the battery in order to deliver a total of 6500 J of energy to the battery? What fraction of that will be stored as chemical energy in the battery?

Ans. (a) 8.18 Ω; (b) 9.34 V; (c) 1.98×10^3 s = 0.552 h; 96.3%

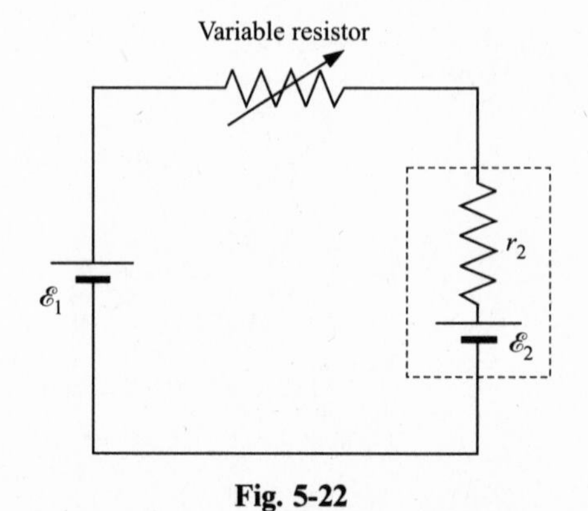

Fig. 5-22

Problem 5.41. An ammeter is constructed using a galvanometer that has a resistance of 15.1 Ω and a maximum deflection at a current of 0.100 A. The circuit used to enable measurements at ranges of 1.00 A, 10.0 A and 100 A is shown in Fig. 5-23. One of the terminals of the ammeter is point *a*, and the other terminal is chosen for the range needed. What are the resistance values needed for R_1, R_2 and R_3 ?

Ans. $R_1 = 0.0168\ \Omega$, $R_2 = 0.151\ \Omega$, $R_3 = 1.51\ \Omega$

Problem 5.42. A voltmeter is constructed using a galvanometer that has a resistance of 15.1 Ω and a maximum deflection at a current of 1.21×10^{-3} A. The circuit used to enable measurements at ranges of 1.00 V, 3.0 V and 10 V is shown in Fig. 5-24. One of the terminals of the voltmeter is point *a*, and the other terminal is chosen for the range needed. What are the resistance values needed for R_1, R_2 and R_3 ?

Ans. $R_1 = 811\ \Omega$, $R_2 = 1653\ \Omega$, $R_3 = 5785\ \Omega$

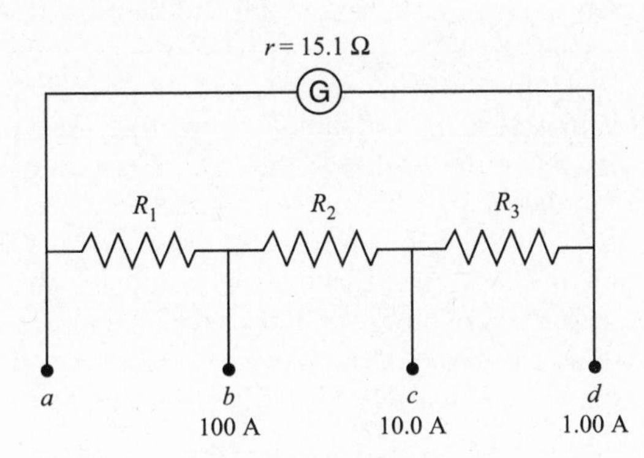

Fig. 5-23

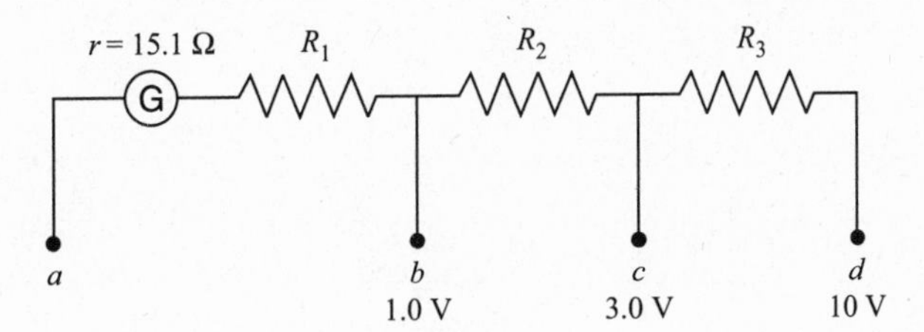

Fig. 5-24

Chapter 6

Magnetism—Effect of the Field

6.1 INTRODUCTION

In previous chapters, we learned about forces exerted by one mass on another mass (gravitational force) and by one charge on another charge (electrical force). Experimentally we find that there is also a force exerted by one *moving* charge on another *moving* charge (in addition to the electrical force). This force is the **magnetic force**. The most common occurrence of this force is when two magnets attract (or repel) each other, but this attraction (or repulsion) is due to subtle properties of the materials, which we will leave to a later chapter.

In discussing the magnetic force, we will use the concept of a **magnetic field**, for which we use the symbol **B**. The magnetic field is a vector, and is the link between the two moving charges that interact with each other. One of the charges is the *source* of the field, and this field, in turn, has the *effect* of exerting a force on the second moving charge. Thus, the magnetic field has two aspects: (1) its effect—to exert a force on a moving charge and (2) its source—the origin of the field, which can be another moving charge, or possibly there may be another means of producing the field. These two aspects are totally independent of each other and therefore we will discuss each in a separate chapter.

The unit for a magnetic field is a **tesla** (T) in our system. A more common unit which is widely used in practice is the **gauss** (G). One gauss equals 10^{-4} tesla. The strength of the magnetic field near the surface of the earth is approximately one gauss.

Note that in general a magnetic field can vary from point to point in space, and can also change from moment to moment. For the present we will assume that the magnetic field remains constant in both space (uniform magnetic field) and time.

6.2 FORCE ON A MOVING CHARGE

In this chapter the *effect* of the magnetic field, **B**, will be discussed. This means that we ask ourselves the following question. Given a magnetic field produced by some means, which is not necessarily of concern to us, what is the force, **F**, that this field, **B**, exerts on a charge, q, moving with a velocity, **v**?

Note. We have to find a vector, **F**, that results from some interaction of a scalar, q, and two vectors, **v** and **B**. This is depicted in Fig. 6-1.

We seek to know the magnitude and direction of the as yet unknown force, **F**, that the magnetic field **B** exerts on the charge q moving with velocity **v** when the angle between the vectors **v** and **B** (when

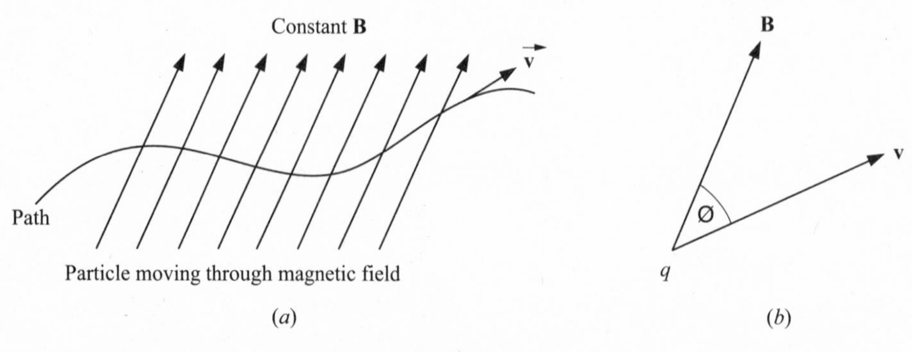

Fig. 6-1

their two tails are touching) is ϕ. We do this in two steps. First we discuss the magnitude, and then we discuss the direction of the force.

Magnitude of the Force

The formula for the magnitude of the force is:

$$|F| = |qvB \sin \phi| \tag{6.1}$$

We have used absolute value signs, since the magnitude is always positive. The sign of q does not affect the *magnitude* of the force. It will, however affect the *direction* of the force. Note that the force is zero when the angle ϕ = zero or 180°, i.e. when the velocity and the magnetic field are along the same line. Also, the largest force occurs when $\sin \phi$ is ± 1, i.e. when the velocity is perpendicular to the magnetic field (see Fig. 6-2).

Problem 6.1.

(*a*) A charge of 2×10^{-6} C is moving with a velocity of 3×10^4 m/s at an angle of 30° with a magnetic field of 0.68 T. What is the magnitude of the force exerted on the charge?

(*b*) What is the magnitude of the force if the charge were -2×10^{-6} C?

(*c*) What is the magnitude of the force if the angle ϕ were 150°?

Solution

(*a*) Substituting $q = 2 \times 10^{-6}$ C, $v = 3 \times 10^4$ m/s, $\phi = 30°$ and B = 0.68 T into Eq. (*6.1*), we get $|F| = (2 \times 10^{-6})(3 \times 10^4)(\sin 30°)(0.68) = 0.0204$ N.

(*b*) Since only the absolute value of each variable enters, the answer is the same as for part (*a*).

(*c*) Since $|\sin 150°| = \sin 30°$, the answer is still the same.

Problem 6.2. A charge of 3×10^{-5} C is at the origin in Fig. 6.3. There is a uniform magnetic field of 0.85 T pointing in the positive x direction. Calculate the magnitude of the force exerted on the charge if it is moving with a velocity of 2×10^5 in the direction (*a*) from A to B; (*b*) from A to E; (*c*) from D to A; (*d*) from A to F; and (*e*) from A to H.

Solution

In all five cases, $qvB = (3 \times 10^{-5})(2 \times 10^5)(0.85) = 5.1$ N. The difference between each case is the value of $\sin \phi$. Thus, the solution for each case is

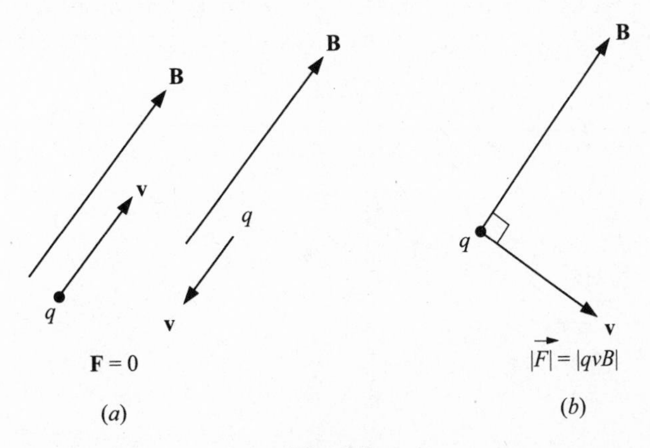

Fig. 6-2

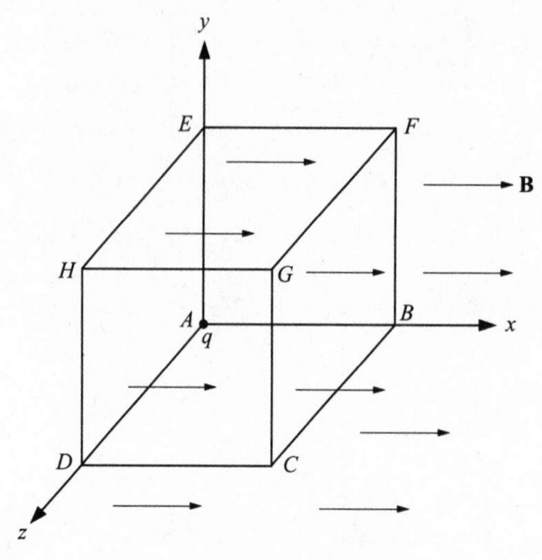

Fig. 6-3

(a) $\sin \phi = 0$ and therefore $|F| = 0$ N.

(b) $\sin \phi = 1$ and therefore $|F| = 5.1$ N.

(c) $\sin \phi = 1$ and therefore $|F| = 5.1$ N.

(d) ϕ is 45° so $\sin \phi = 0.707$ and therefore $|F| = 3.61$ N.

(e) ϕ is 90°, $\sin \phi = 1$ and therefore $|F| = 5.1$ N.

Direction of the Force

The direction of the force is perpendicular to both **v** and **B**, and it is therefore necessary to consider the problem in three dimensions. The solution is done in two steps. First one determines the line along which the force acts and then one determines the proper direction along that line.

The two vectors **v** and **B** can be considered as forming a plane. In Fig. 6-4, we draw the plane formed by a combination of vectors **v** and **B**. The direction that is perpendicular to this plane we call the normal to the plane. You can picture placing the palm of your right-hand in this plane containing

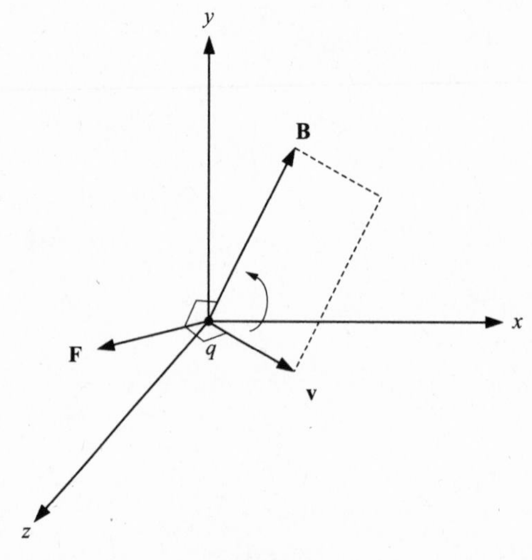

Fig. 6-4

both **v** and **B**. The normal is the direction perpendicular to your palm. On Fig. 6-4, we also draw this normal direction for this combination of **v** and **B**. This normal direction is the line along which the force vector lies. This direction is perpendicular to both **v** and **B**. We now have to choose between the two possible directions along this line. This is done by using the "**right-hand rule**". There are many different ways of applying a right-hand rule, and if you already know a particular method, you should continue to use that method. Here is one method that you can use. For the case of a positive charge, when the vectors **v** and **B** are tail to tail, curl the fingers of your right-hand in the direction from **v** to **B** with your thumb perpendicular to the other fingers. Your thumb then points in the direction of the force. If the charge is negative, then the direction is reversed.

> ***Note.*** Both the magnitude and the direction of the magnetic force are completely analogous to the magnitude and direction of the vector torque discussed in Chap. 10, Section 10.3 when we replace **r** and **F** in that chapter by (q**v**) and **B**.

Problem 6.3. Determine the direction of the force in Problems 6.1(*a*), (*b*) and (*c*).

Solution

(*a*) The orientation of v and **B** is shown in Fig. 6-5 (*a*). Both **v** and **B** are in the plane of the paper. The perpendicular to the paper is the line going in and out of the paper. Curling the fingers of our right-hand from **v** to **B**, we see the perpendicular thumb points out of the paper. Thus the direction of the force is *out* of the paper. We use a dot, reminding us of the point of an arrow, to indicate that the force is out of the paper.

(*b*) The only change from (*a*) is that the sign of the charge is negative. Therefore, we reverse the direction of **F**, and it is now *into* the paper. We use a cross, reminding us of the cross hair at the back of an arrow, to indicate that the force is into the paper.

(*c*) Suppose that the directions of **v** and **B** are as shown in Fig. 6-5 (*b*). Rotating our fingers through the 150° angle from **v** to **B**, the thumb points into the paper. Thus, the direction of the force is *into* the paper.

Problem 6.4. Determine the direction of the force in Problem 6.2.

Solution

(*a*) Since the magnitude of the force is zero, there is obviously no direction needed.

(*b*) **v** is in the y direction, and **B** is in the x direction. The plane formed by these vectors is the x–y plane. The normal to this plane is the z direction (either $+z$ or $-z$). We use the right-hand rule to choose

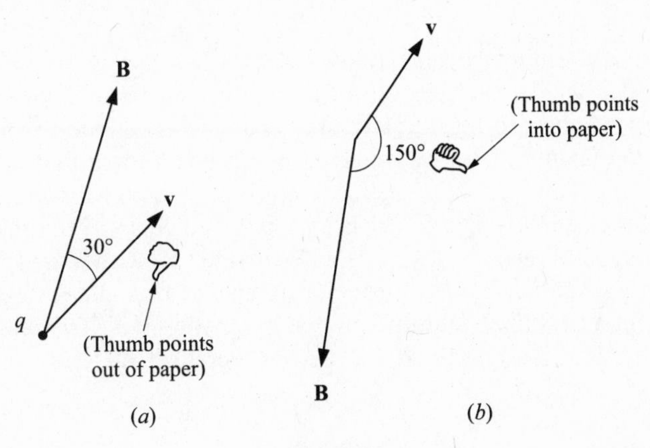

Fig. 6-5

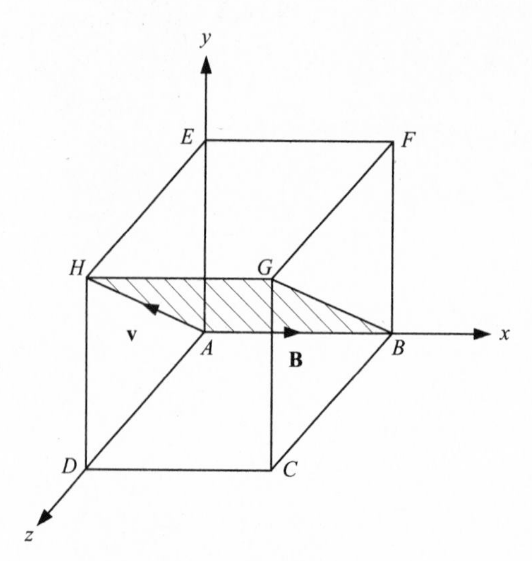

Fig. 6-6

between $+z$ and $-z$. Rotating from the positive y direction (the direction of **v**) to the positive x direction (the direction of **B**) our thumb points in the $-z$ direction. This is the direction of **F**.

(*c*) **v** is in the $-z$ direction, and **B** is in the x direction. The plane formed by these vectors is the x–z plane. The normal to this plane is the y direction (either $+y$ or $-y$). We use the right-hand rule to choose between $+y$ and $-y$. Rotating our fingers from the negative z direction (the direction of **v**) to the positive x direction (the direction of **B**) our hand now pushes in the $-y$ direction. This is the direction of **F**.

(*d*) **v** is in the x–y plane, at an angle of 45° with the positive x axis, and **B** is in the x direction. The plane formed by these vectors is the x–y plane. The normal to this plane is the z direction (either $+z$ or $-z$). We use the right-hand rule to choose between $+z$ and $-z$. Rotating from **v** to the positive x direction (the direction of **B**) our thumb points into the paper, which is the $-z$ direction. This is the direction of **F**.

(*e*) **v** is in the y–z plane, at an angle of 45° with the positive z axis, and **B** is in the x direction. The plane formed by these vectors is shown in Fig. 6-6 (plane *ABGH*). The normal to this plane is parallel to the direction *DE* (or *ED*). We use the right-hand rule to choose between *DE* and *ED*. Curling our fingers from the direction of **v** toward the positive x direction (the direction of **B**) our hand now pushes in the *DE* direction. This is the direction of **F**.

6.3 APPLICATIONS

If the magnetic force is the only force exerted on a moving charged particle, then the particle will move with constant speed. This is because the force is always perpendicular to the direction of the motion, and a force perpendicular to the velocity only changes the direction and not the magnitude of the motion. To change the magnitude of the velocity, one needs a force that is parallel to the velocity, which the magnetic force does not provide. The force, and thus the acceleration, is also perpendicular to **B**. Suppose that, in addition, the magnetic field is also perpendicular to the initial velocity. The entire motion (both **v** and **a**) are now in the plane perpendicular to **B**, with **v** and **a** each having constant magnitude. This is exactly what is needed to produce **circular motion** at constant speed. The centripetal force needed for the circular motion is supplied by the magnetic force. The magnitude of the magnetic force must equal the centripetal force required and we can therefore say that

$$qvB = mv^2/R \tag{6.2}$$

or

$$R = mv/qB \tag{6.3}$$

This is a formula for the radius of the circle traversed by the particle of mass, m, charge, q, moving with a velocity, v, in a perpendicular magnetic field, B. There are many applications of this relationship. The circular motion that a magnetic field can create is useful in many diverse areas. We will discuss some of these applications below.

Applications of Circular Motion

If one has a charged particle of unknown sign, one can use the circular motion created by a magnetic field to determine the sign of the charge. Suppose one has a charged particle moving upward, as in Fig. 6-7, in a magnetic field that is directed into the paper. The resultant circular path of the particle could be either path 1 or path 2, depending on whether the force is in the direction of $\mathbf{F}_1$ or $\mathbf{F}_2$. If the charge is positive, then the right-hand rule shows that the force is in the direction of $\mathbf{F}_1$ and the particle moves along path 1. On the other hand, if the charge is negative, the force is reversed and points in the direction of $\mathbf{F}_2$ causing the particle to move along path 2. Thus, by making the particle go in a circle in a perpendicular magnetic field, one can determine the sign of the charge.

Problem 6.5. A particle, with a charge equal to that of an electron is moving in a circle of radius 5 cm in a magnetic field of 0.2 T. What momentum does this particle have?

Solution

Using Eq. (*6.3*), $R = mv/qB$, one gets that the momentum, p, which is mv equals $p = mv = qBR$. Thus, $p = 1.60 \times 10^{-19}\,(0.2)(0.05) = 1.60 \times 10^{-21}$.

Problem 6.6. Two negatively charged particles, each with a charge equal to that of an electron are moving in a circle with the same velocity. The circular motion is due to a perpendicular magnetic field of 0.2 T. One of the particles is a charged atom of carbon with approximately 12 times the mass of a hydrogen atom, and the other is a charged atom of unknown mass. The radius of the circular path of the carbon atom is R_c, while that of the unknown atom is R_u.

(*a*) Show that one can get the unknown mass by measuring the ratio of the two radii.

(*b*) If $R_u = 1.33R_c$ what is the unknown atom?

Solution

(*a*) $R_u = m_u v/qB$, and $R_c = m_c v/qB$

Thus $R_u/R_c = m_u/m_c$

(*b*) $m_u = 1.33m_c = 16$ hydrogen masses, so the unknown atom is oxygen.

This problem illustrates the principle behind the operation of a mass spectrometer. In practice, a mass spectrometer typically does not have the particles moving with the same speed. Instead, each particle gains its speed by being accelerated from near rest through the same difference of potential, V. This is illustrated in Problem 6.7.

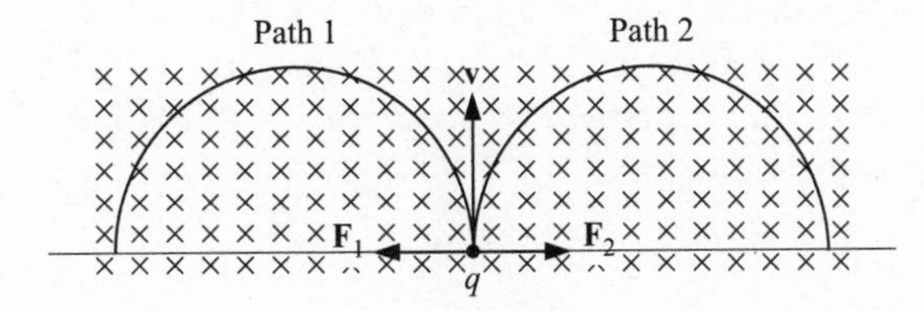

Fig. 6-7

Problem 6.7. The two charged particles in the previous problem are accelerated through a difference of potential, V, and then they travel in a perpendicular magnetic field, B.

(*a*) Derive an expression for the ratio of the two radii of their circular paths. Figure 6-8(*a*) illustrates the geometry.

(b) In this case, what would R_u/R_c be?

Solution

(*a*) At a the charges have no velocity. As they travel to point b, they lose potential energy equal to eV since b is at a higher potential and the charges are negative. This lost potential energy is converted to kinetic energy, so that at b the particles have a kinetic energy of

$$KE = (\tfrac{1}{2})mv^2 = eV, \qquad \text{or} \qquad v^2 = 2eV/m$$

$$v = \sqrt{2eV/m}$$

Each particle enters the magnetic field region with the velocity corresponding to its mass, and is then turned into a circular path with the appropriate radius. Thus, from Eq. (*6.3*) and with $q = e$

$$R^2 = m^2v^2/q^2B^2 = m^2(2eV/m)/e^2B^2 = 2V(m/e)/B^2$$

$$R^2B^2 = 2Vm/e \qquad \text{and} \qquad m = eR^2B^2/2V$$

The ratio of the masses is therefore

$$m_u/m_c = (R_u/R_c)^2.$$

(*b*) Again, by measuring the ratio of the radii, one can get the ratio of the masses. Since we are assuming that the unknown mass is oxygen, $R_u/R_c = \sqrt{4/3} = 1.15$.

There are many other uses to which one can put the ability of the magnetic field to produce circular motion. For instance, nearly all particle accelerators use magnetic fields to make particles return to an area where they are accelerated. There is often only one small region where particles are given an increase in speed, as a result of a parallel electric field, and the circular motion created by the magnetic

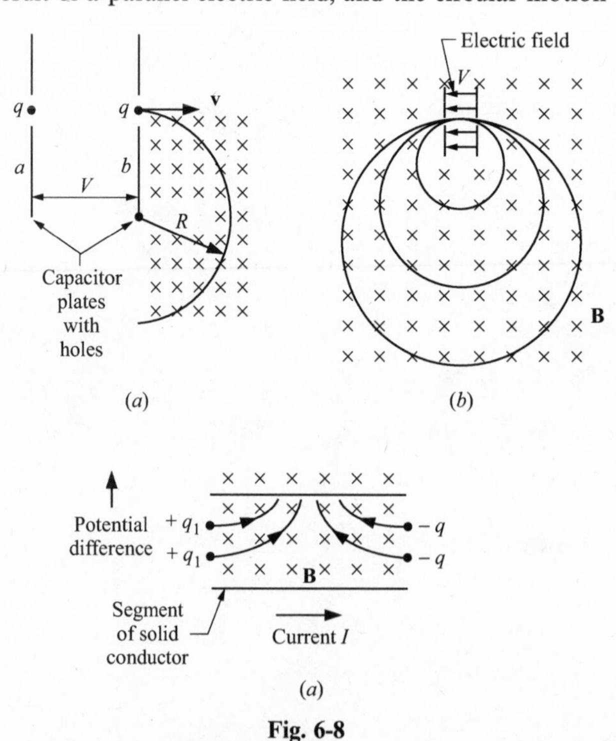

Fig. 6-8

field causes the particles to return to this region regularly and receive additional kinetic energy, as shown in Fig. 6-8(*b*).

Another application is in the "**Hall Effect**", which is used to determine the sign of the charges that produce currents in various solid conductors. Here, one deflects the moving charges in an electric current in the direction of the magnetic force. If a current is flowing to the right, this could be the result of positive charges moving to the right, or of negative charges moving to the left. For a perpendicular magnetic field, for instance a field going into the paper, both the positive and the negative charges are deflected upward, since a negative charge moving in a direction opposite to a positive charge has the same direction for the force. This is illustrated in Fig. 6-8(*c*). Thus, the magnetic field will deflect the conducting charges upward and one can detect the sign of the charge by determining whether positive or negative charge has gathered at the top. If positive charge is at the top, and negative at the bottom, then the top surface will be at a higher potential than the bottom, as we learned from a the case of a parallel plate capacitor. The opposite potential difference would result if negative charges gathered at the bottom. By this method, it has been determined that in some materials, called **semiconductors**, there are cases of positive as well as negative charge conductors.

Another interesting phenomenon occurs if the magnetic field is not perpendicular to the velocity vector. In that case one can resolve the velocity vector into one component which is parallel to the magnetic field and another component that is perpendicular to the magnetic field (see Fig. 6-9). The parallel component will be unaffected by the magnetic field since there is no force produced by B on a parallel velocity. The perpendicular component, however will be deflected into a circular path. circling the direction of the magnetic field. The resultant motion will be a spiral around the field direction (see the figure). There are many cases in which this actually happens, such as when particles in the "solar wind" meet the magnetic field in the earth's atmosphere.

Velocity Selector

In the previous problems there were cases of only a magnetic force, and some cases of both electrical and magnetic forces acting on the particles. But in the region in which the electrical force was active (when the particles were accelerated by the difference of potential), there was no magnetic force, and in the region of the magnetic force (the circular motion) there was no electrical force. By using a combination of both electric and magnetic fields, we can produce a mechanism to separate out particles of a particular velocity. This is known as a **velocity selector**, and is shown in Fig. 6-10, and described in the following problem.

Problem 6.8. The particle, of charge q, is moving with velocity, $\mathbf{v}$, in a region between two charged parallel plates a distance d apart, that produce a uniform electric field, $\mathbf{E}$. A uniform magnetic field, $\mathbf{B}$, pointing into the paper, also exists in this region.

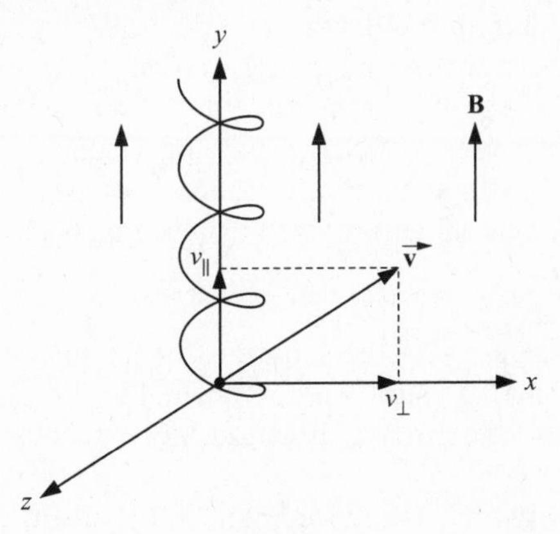

Fig. 6-9

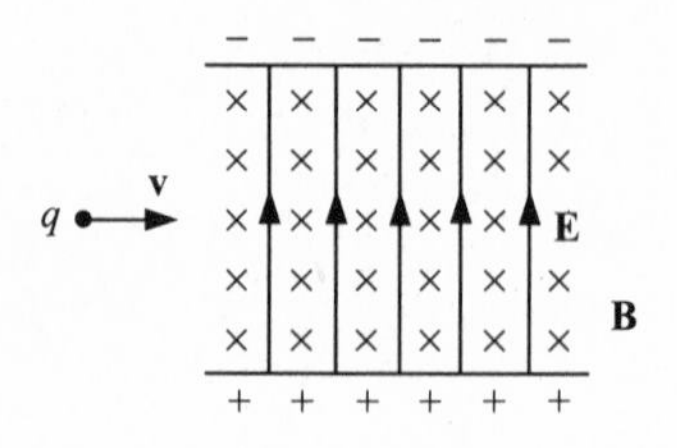

Fig. 6-10

(*a*) For what velocity, **v**, will the electrical and magnetic forces be equal and opposite, thus canceling each other's effect?

(*b*) Find the velocity when $V = 100$ volts, $d = 2.0$ cm and $B = 0.5$ T.

Solution

(*a*) The electrical force, F_E, will equal $q\mathbf{E}$ and point in the direction of **E** (for a positive charge). The magnetic force, $\mathbf{F}_B$, will equal qvB, and point in the direction opposite to $\mathbf{F}_E$. Thus, the two will cancel if their magnitudes are equal. This is also true for a negative charge, since the direction of both forces changes. Therefore there will be no net force if $qE = qvB$, or $E = vB$.

> ***Note.*** For a velocity of $v = E/B$ there is no force to deflect the particle, and it will travel in a straight line. If the particle has a bigger velocity than this, then the magnetic force will exceed the electrical force, and the particle will be deflected in the direction of the magnetic force. Similarly, if the velocity is smaller than this velocity, then the magnetic force will be less than the electrical force and the particle will be deflected in the direction of the electrical force. Thus only particles with this particular velocity will be undeflected, and they can be easily selected out from the rest. We can choose the velocity we want by varying E, simply by changing the potential difference across the two plates, which is producing the electric field.

(*b*) Remembering that the electric field produced by two parallel plates is V/d, where d is the distance between the plates, we have

$$v = E/B = V/dB = 100 \text{ V}/(2 \times 10^{-2} \text{ m})(0.5 \text{ T}) = 10^4 \text{ m/s}.$$

6.4 MAGNETIC FORCE ON A CURRENT IN A WIRE

Whenever current flows in a wire, one has charge that is moving. If a segment of the wire is in a magnetic field, then the magnetic field will exert a force on that segment of the wire. To obtain this force one has to determine how to adjust the formula for a single moving charge to accommodate a current in a wire. The answer to this is that all one has to do is to substitute IL for qv in the equation for the force. Here, I is the current flowing in the wire, and **L** is a vector whose magnitude is the length of the segment of the wire, and the direction of **L** is the direction of the current. We can see this intuitively by noting that in a small time Δt the amount of charge passing a point in the wire is $q = I\Delta t$. If this charge moves with an average velocity v, it will cover a distance $L = v\Delta t$. Thus $q/I = L/v$ or $qv = IL$. Therefore, the magnitude of the force on a segment is given by (see Fig. 6-11)

$$|F| = |ILB \sin \phi| \qquad (6.4)$$

The direction of the force is calculated by the same procedure that was used for a single charge. The force is perpendicular to both **L** and **B**, and therefore normal to the plane containing those vectors. We use the right-hand rule to choose the correct direction along this normal, where the direction of the current replaces the direction of **v**.

The force calculated in this manner is the force on that segment of wire, of length L, carrying the current I. Each segment of the wire is affected separately by the magnetic field, and we can separately

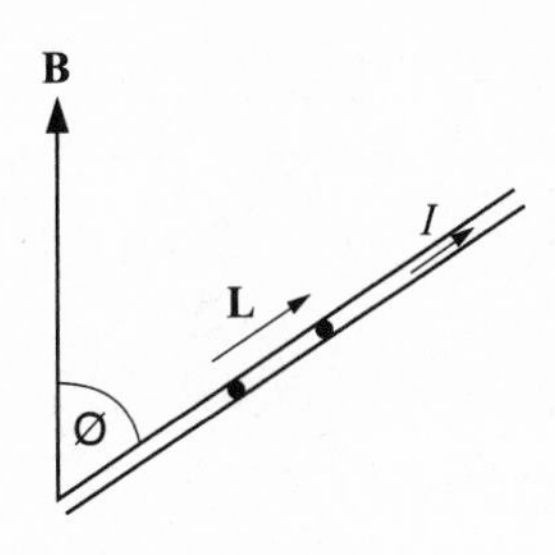

Fig. 6-11

calculate the magnitude and direction of the force on each segment. To get the total force on the wire we would then add together, vectorially, the forces on each segment. This is illustrated by the next problem.

Problem 6.9. Consider a wire *abcd*, in the shape shown in Fig. 6-12 which is in a magnetic field **B** pointing out of the paper. The current is 1.5 A, the magnetic field is 0.3 T and the lengths are $L_1 = 0.5$ m and $L_2 = L_3 = 0.8$ m.

(*a*) Calculate the force (magnitude and direction) acting on segment *ab*.

(*b*) Calculate the force (magnitude and direction) acting on segment *bc*.

(*c*) Calculate the force (magnitude and direction) acting on segment *cd*.

(*d*) Calculate the force (magnitude and direction) acting on the wire *abcd*.

Solution

(*a*) Using Eq. (*6.4*), the magnitude of the force is $|F_1| = IL_1B$ since $\phi = 90°$. The direction is perpendicular to $\mathbf{L}_1$ (to $+y$) and to **B** (out of the paper) and thus in the x direction, either $+x$ or $-x$. Using our right-hand rule, (we rotate our fingers from $\mathbf{L}_1$ to **B**) and our thumb then points in the $+x$ direction, which is therefore the direction of $\mathbf{F}_1$. For the magnitude we get $F_1 = (1.5 \text{ A})(0.5 \text{ m})(0.3 \text{ T}) = 0.225$ N.

(*b*) Applying the same formula to segment *bc*, we get the magnitude of the force to be $|F_2| = IL_2 B$. The direction of the force is perpendicular to $\mathbf{L}_2$ (to $+x$) and to **B**, and thus in the y direction. To choose between $\pm y$, we use our right-hand rule and find that $\mathbf{F}_2$ is in the $-y$ direction. For the magnitude we get $F_2 = (1.5 \text{ A})(0.5 \text{ m})(0.3 \text{ T}) = 0.360$ N.

(*c*) Applying the same formula to segment *cd*, we get the magnitude of the force to be $|F_3| = IL_3 B$. The direction of the force is perpendicular to $\mathbf{L}_3$ (to $-y$) and to **B**, and thus in the x direction. To choose between $\pm x$, we use our right-hand rule and find that $\mathbf{F}_3$ is in the $-x$ direction. The magnitude of F_3 is the same as F_1.

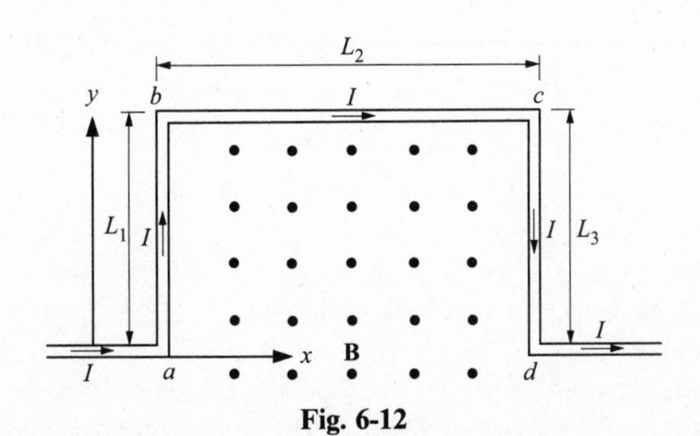

Fig. 6-12

(*d*) The force on *abcd* is the vector sum of the forces on the three segments. Since F_1 and F_3 are in opposite directions and of equal magnitude, they cancel each other when added together. Thus

$$\mathbf{F}_{\text{total}} = \mathbf{F}_1 + \mathbf{F}_2 + \mathbf{F}_3 = \mathbf{F}_2 = 0.3 \text{ N in the } -y \text{ direction.}$$

Problem 6.10. Consider a cube, with a side of 0.5 m, as shown in Fig. 6-13. A current of 2 A is flowing along the x direction. Calculate the force (magnitude and direction) on the segment *ab* when the magnetic field of 0.3 T points in (*a*) the x direction; (*b*) the $-y$ direction. (*c*) the direction from *a* to *h*; and (*d*) the direction from *a* to *c*.

Solution

(*a*) Using Eq. (*6.4*), the magnitude of the force is $|F_1| = 0$ since $\phi = 0$.

(*b*) Now the angle $\phi = 90°$, and therefore $|F| = ILB = 2(0.5)(0.3) = 0.3$ N. To get the direction, we note that L and **B** are in the x–y plane and the normal to that plane is the z direction. To choose between $\pm z$, we apply the right-hand rule, curling our fingers from **L** to **B**. Our perpendicular thumb then faces the $-z$ direction, which is the direction of **F**.

(*c*) The angle between **L** and **B** is still 90° (not 45°) since *ah* is in the y–z plane, which is perpendicular to the direction of **L** (the x direction). Therefore $|\mathbf{F}| = 0.3$ N. Getting the direction is somewhat harder. The plane of **L** and **B** is now *abgh* whose normal is along the diagonal *ed* (or *de*). Using the right-hand rule the thumb faces the direction *ed*, which is the direction of **F**.

(*d*) The angle between **L** (*ab*) and **B** (*ac*) is now 45°. Thus $|F| = 2(0.5)(0.3) \sin 45 = 0.212$ N. The plane of **L** and **B** is now the x–z plane whose perpendicular is the y direction. The right-hand rule selects between $\pm y$. Rotating our fingers from **L** to **B**, our thumb faces in the $-y$ direction which is therefore the direction of **F**.

Problem 6.11. Consider an equilateral triangle, with a side of 0.5 m, as shown in Fig. 6-14. A current of 2 A is flowing around the triangle in the direction shown. Calculate the force (magnitude and direction) on each segment of the triangle, and on the whole triangle when a magnetic field of 0.3 T points in the direction *ab*.

Solution

Along *ab*, the force is zero, since *ab* is along the direction of **B**.

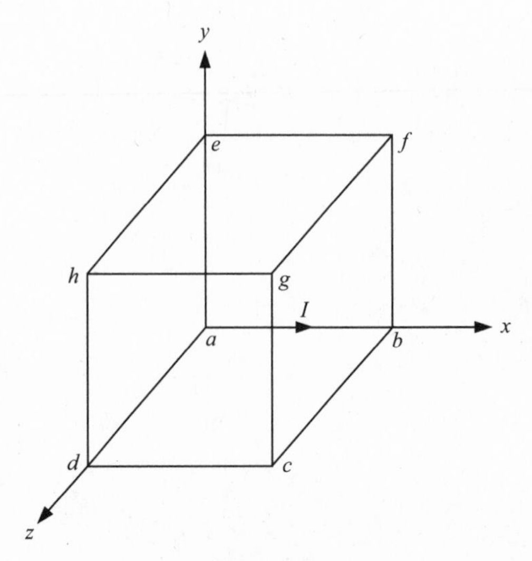

Fig. 6-13

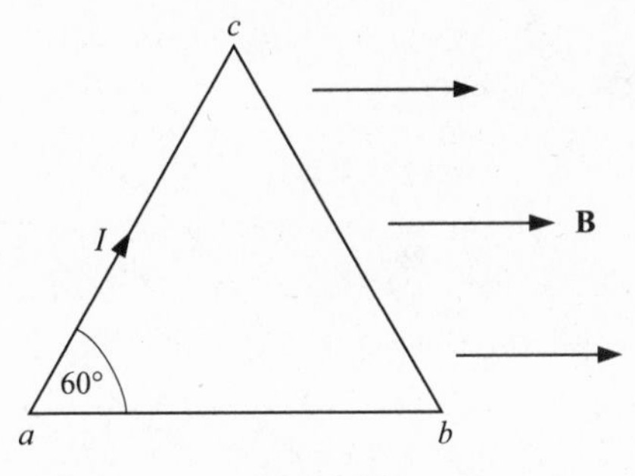

Fig. 6-14

Along ac, the force is $|F| = 0.5\ (2)\ (0.3) \sin 60° = 0.26$ N. The plane containing **L** and **B** is the plane of the paper whose normal is in or out of the paper. The right-hand rule gives the correct direction as into the paper.

Again, along cb, the force is $|F| = 0.5\ (2)\ (0.3) \sin 60° = 0.26$ N. The plane containing **L** and **B** is the plane of the paper whose normal is in or out of the paper. The right-hand rule gives the correct direction as out of the paper.

Adding these forces vectorially gives $\mathbf{F} = 0$.

Problem 6.12. Consider a circular metal disc, which is free to rotate about its center. The bottom of the disc is in contact with a pool of liquid mercury as seen in Fig. 6-15. A battery is connected between the center of the disc and the pool of mercury, so that a current flows vertically downward from the center to the mercury. If a magnetic field is established in the direction out of the paper, what motion, if any, will the disc undergo?

Solution

Because we have current flowing downward in the bottom of the disc, we have a case of current in a magnetic field. The magnetic force will be perpendicular to **L** (which is downward) and to **B** (which is out of the paper) and will therefore point in the horizontal direction. The right-hand rule determines that the direction is to the left. Since the disc is not free to move from its position (it is only free to rotate), the only possible motion is a rotation about its center. On the lower half of the disc there is a force to the left, and there is no force on the top of the disc. The disc will therefore rotate in the clockwise direction. This is an example of a very crude **electromagnetic motor**.

6.5 MAGNETIC TORQUE ON A CURRENT IN A LOOP

We have seen in Problem 6.11, that the magnetic field did not exert a net force on a triangle in which current was flowing around the perimeter. This is an example of current flowing around a closed loop, where the net magnetic force will always be zero. However, even with no net force it is possible

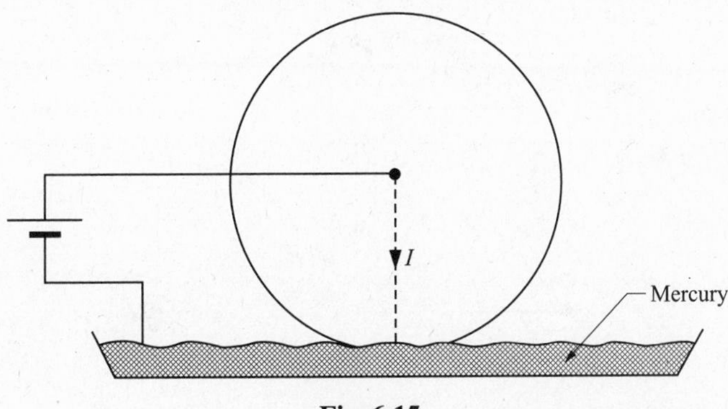

Fig. 6-15

that there will be a tendency for the coil to rotate, and this tendency is determined by the torque (or moment) which these forces exert on the coil. Let us examine the case of a rectangular coil, $PQRS$, whose sides are of lengths a and b, and which carries a current I. This coil is in a magnetic field which is constant throughout the area of the coil. In Fig. 6-16(a) the coil is pictured in three dimensions, while Fig. 6-16(b) shows the same coil projected on the x–y plane. The magnetic field is in the x direction.

Problem 6.13. Find an expression for the torque, Γ, on coil $PGRS$ in Fig. 6-16. A current, I, flows, as shown. Discuss the rotation of the coil.

Solution

We will first calculate the forces acting on each of the four sides of the coil, in order to determine the net force (which we know should be zero) and the torque which may be exerted. On each of sides PS and QR, the magnitude of the force is $IbB \sin \phi$ [Fig. 6-16(b)]. The direction of the two forces are opposite to each other, since the current flows in the opposite direction for the two sides. For one of the sides the force is in the $+z$ direction [out of the paper in Fig. 6-16(b)], and for the other side the force is in the opposite, or $-z$ direction. The line of action of these two forces is clearly the same and they exert no net torque on the coil. Thus any net torque will have to come from sides PQ and SR. The force exerted on each of these sides is $IbB \sin 90°$ and the directions are opposite to each other. The net force will therefore be zero, as we expected. However, the line of action of these two forces will not be identical and there will usually be a net torque. Let us calculate the direction of the forces and their line of action, and from this information we will then be able to calculate the torque.

On side PQ the force will be perpendicular to PQ (the z direction) and to $\mathbf{B}$ (the x direction). The force is therefore in the $\pm y$ direction. Since the current in the coil is flowing in the direction $PQRS$, then using the right-hand rule gives a force in the $-y$ direction. For this same current direction, the force on side SR will be in the $+y$ direction. This is depicted in Fig. 6-17. The line of action for the force on PQ is the y axis, while the line of action for the force on side SR is the line parallel to the y axis, but at a distance of $a \sin \phi$ from that axis. If we take the torque about the origin, only the force on SR will contribute and the torque will be $\Gamma = F a \sin \phi = IbB\, a \sin \phi = IAB \sin \phi$, where $A = ab =$ the area of the coil. (Actually, the torque will be the same about any axis because the two forces producing the torque are equal in magnitude and opposite in direction, thus forming a couple; see Ibid., 9.2, p. 234). This torque will try to rotate the coil about the z axis in the counter-clockwise direction, until the plane of the coil is parallel to the y axis. At this point, the angle ϕ is zero, and the torque is zero. Thus the coil will try to line up with its plane perpendicular to the magnetic field. If the current in the coil had been in the opposite direction, from Q to P, then the direction of all the forces would have been reversed, and the torque would have been in the direction to rotate the coil clockwise around the y axis. Again, when the plane of the coil is perpendicular to the magnetic field the torque will be zero.

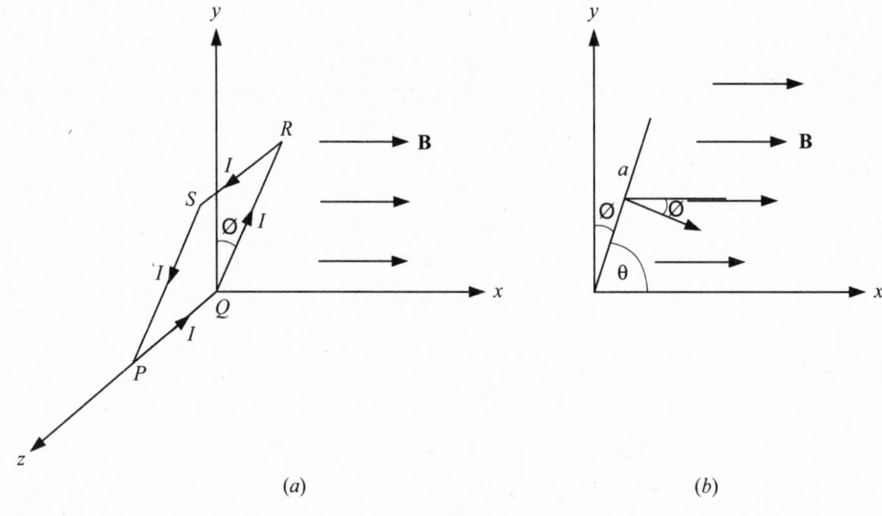

Fig. 6-16

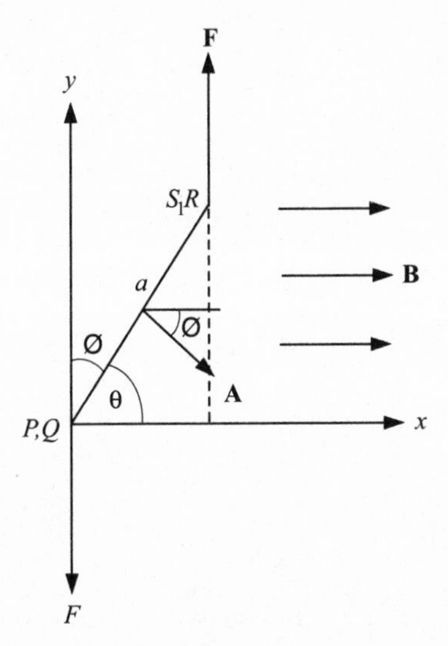

Fig. 6-17

It is useful to define a vector area for the coil, **A**, whose magnitude is $A = ab$, and whose direction is perpendicular to the plane of the coil. The $\pm$direction of **A** is determined by the right-hand rule. Curl the fingers of your right-hand around the coil in the direction of the current. Your thumb then points in the positive **A** direction. Thus ϕ is the angle between **A** and **B**, as can be seen in Fig. 6-17. We see that in general the torque is given by

$$\Gamma = IAB \sin \phi \tag{6.5}$$

where Γ tends to rotate the coil in the same direction as rotating the vector **A** through ϕ to **B**. When **A** is parallel to **B**, $\phi = 0$ and the torque is zero.

We define a new vector, **M**, the magnetic dipole moment of the coil, whose magnitude is IA and whose direction is the same as **A** (with the convention we defined earlier). If the coil consists of several turns, then each turn has a magnetic moment IA, and the entire coil has a magnetic moment NIA, where N is the number of turns in the coil. The torque will turn the coil in the direction of making **M** point in the direction of **B**.

For the case shown in Fig. 6-16, where the current flows from P to Q, the torque will rotate the coil in the counter-clockwise direction, until the plane of the coil is parallel to the y axis. At that point, the torque is zero. If the coil rotates past the y axis, then the torque will again try to align **M** with **B**, and the rotation will now be clockwise. Thus the torque will always be rotating the coil back to the equilibrium position, and we see that the coil is in *stable* equilibrium, when **M** is parallel to **B**. If the current were in the opposite direction, then **M** would point in the opposite direction (see Fig. 6-18). If the coil then starts in position (1) in the figure the torque would be clockwise, trying to rotate the coil further away from the y axis. If the coil starts on the other side of the y axis (position 2 in the figure), the torque would be counter-clockwise, again rotating the coil away from the y axis. Of course, when the coil is precisely lined up with the y axis ($\phi = 180°$), the torque is zero, and the coil is in equilibrium, but the equilibrium is unstable because any move away from the y axis will result in the coil continuing to rotate even more, rather than returning to the equilibrium position. After the coil has rotated 180°, the coil will have its vector **M** pointing in the direction of **B**, and the coil will be in stable equilibrium.

Although this result was derived for the special case of a rectangle, the result is valid for any coil shape, with the moment of the coil equaling $M = NIA$, and the torque on the coil equaling $MB \sin \phi$, with the usual counter-clockwise, clockwise conventions.

It should be clear that this phenomenon of a torque on a coil can be used to build a motor, which will continuously rotate in the magnetic field. Such motors are built by constructing a coil from many

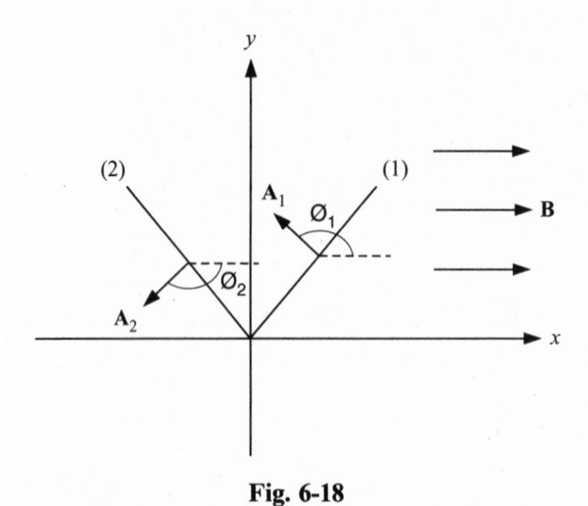

Fig. 6-18

turns (to increase **M** and thereby, the torque), and suspending the coil on an axis in a constant magnetic field. The direction of the current in the coil is chosen to make the coil rotate in one particular direction, for instance clockwise. When the coil passes the y axis the direction of the torque would normally reverse, making the coil turn counter-clockwise. In order to prevent this from happening, we arrange to have the current direction reverse as the coil passes through the y axis, thus maintaining a clockwise torque. This is accomplished by the split in the rings where the current enters from the source of EMF (see Fig. 6-19).

Problem 6.14. Consider a circular ring carrying a current of 2 A. The plane of the ring is at an angle of 60° to the yz plane, as shown in Fig. 6-20. The ring has a radius of 1.5 m, and is in a uniform magnetic field of 0.3 T pointing in the positive x direction. What torque is exerted on the coil?

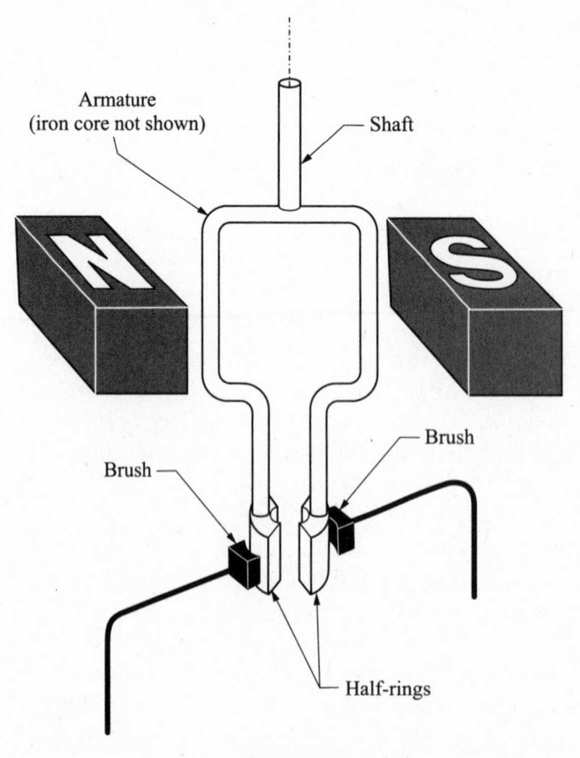

Fig. 6-19

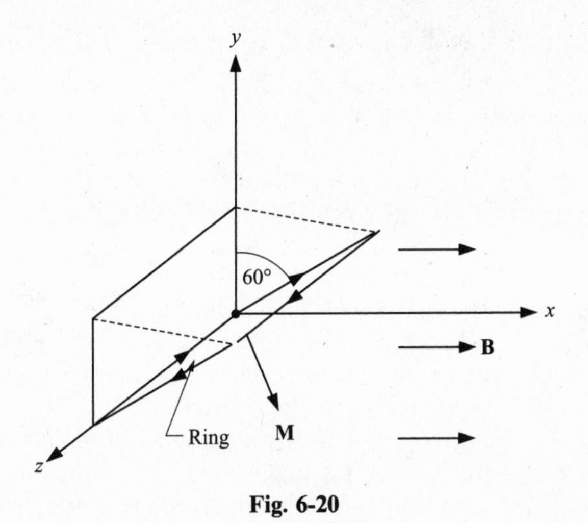

Fig. 6-20

Solution

The magnitude of the torque is $IAB \sin \phi = 1.5\ (\pi r^2)\ (0.3) \sin 60°$ where 60° is the angle between **M** and **B**. Thus the torque will equal 2.75 N · m. For the direction of current shown in the figure, the vector **M** points below the xz plane. Since the torque tries to align **M** with **B**, it will try to rotate the plane upward toward the yz plane.

Another example of a torque exerted by a magnetic field is a charged particle which is spinning. Consider the case of a charged sphere spinning on its axis (see Fig. 6-21). Every part of the sphere is moving in a circle about the axis, and we therefore have charges going in concentric circles which make a current. This is like many different coils, all with planes perpendicular to the axis of rotation, or with area vectors parallel to the axis. For positive charge, the current is in the same direction as the velocity of the charge, and for the rotation in the figure, the area vector is vertically upward. For a negative charge, the current is opposite to the velocity, and the area vector would be vertically downward. The magnetic moment vector is in the same direction as the area vector, upward for positive charge and

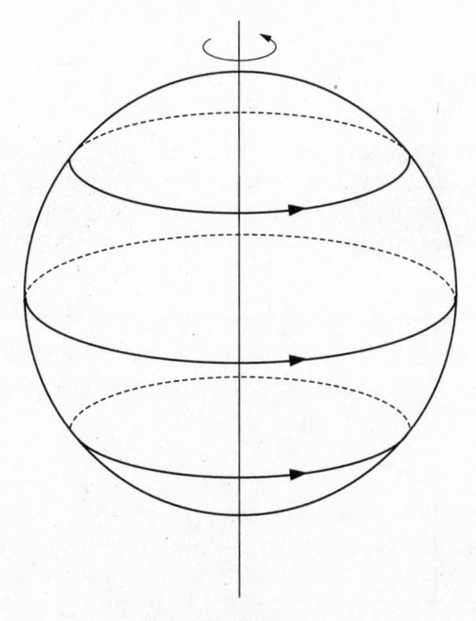

Fig. 6-21

downward for negative charge. In a magnetic field there will be a torque on this spinning sphere which tries to align the moment vector in the direction of **B**. Since spinning particles are like tiny spinning spheres, a charged spinning particle tries to rotate so that its spin axis is along the magnetic field. The direction of rotation about this magnetic field will be interchanged for positive and for negative particles. This is part of the meaning when one talks about electrons with spin "up" or spin "down". In both cases the spin axis aligns with a magnetic field, either parallel or anti parallel to **B**.

The last example of magnetic moments in a magnetic field that we will discuss is that of a compass needle in a magnetic field. We will find out later that a bar magnet consists of many charged particles producing a circulating current about the axis of the magnet. The particles produce this current either because they are spinning in unison about parallel axes to the axis of the magnet, or because they are moving in orbit-like paths circulating the magnet axis. Thus a magnet is actually similar to a coil with a magnetic moment parallel to its axis. Magnets are often described by "poles" at each end. The direction of the moment of the magnet is from the south to the north pole of the magnet. The north pole is the end that tries to align itself facing north, when the magnet is free to rotate about a vertical axis through its center. Such a magnet is called a **compass**. This alignment is due to the magnet being affected by the Earth's magnetic field. In any magnetic field, such as the intrinsic field of the earth, the magnet rotates, with the north pole of the magnet lining up parallel to the magnetic field. Since the Earth's magnetic field points approximately due north, this use of the magnetic needle (compass) has been an ancient method of determining the northerly direction.

Problems for Review and Mind Stretching

Problem 6.15. An electron is in an upward, vertical magnetic field of 0.8 T. What horizontal velocity must the electron have (magnitude and direction) for the magnetic force to be 1.6×10^{-13} N to the east?

Solution

The magnitude of the force is $|F| = |qvB \sin \phi|$. Since B is vertical and **v** is horizontal, $\phi = 90°$ and $\sin \phi = 1$. Thus,

$$1.6 \times 10^{-13} = (1.6 \times 10^{-19})v(0.8)(1) \qquad (i)$$

Therefore,

$$v = 1.25 \times 10^{6} \text{ m/s} \qquad (ii)$$

Since **F** is perpendicular to **v**, and they both are horizontal, **v** must be in the north–south direction. Suppose **v** is north. Then rotating **v** upward toward **B** would give east as the direction of **F** for a positive charge (right-hand rule). However, an electron has a negative charge, so the force on an electron moving north is to the west, which is not the direction we seek. Thus, the velocity must be south.

Problem 6.16. A particle with a charge of 2×10^{-9} C is moving horizontally toward the east at point a, as shown in Fig. 6-22. The particle has a mass of 5×10^{-15} kg and is moving with a velocity of 4×10^{4} m/s. We want to make this charge move in a circle through point b, which is 1.0 m south of a. What magnetic field (magnitude and direction) is required?

Solution

Since the circle is in the horizontal plane, the magnetic field must be vertical (either in or out of the paper). At point a, the (positively charged) particle is moving to the east, and the centripetal force needed to make the particle move in the desired circle must be to the south. Using the right-hand rule, we find that if the field is out of the paper, the force would be to the south, while if the field is into the paper the force is to the north. Thus, the direction of the field must be out of the paper.

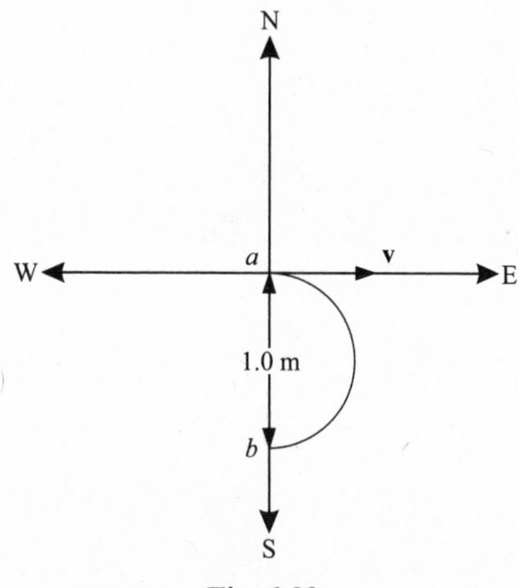

Fig. 6-22

The radius of the circle must be half of the distance between a and b, or

$$R = 0.5 \text{ m} \tag{i}$$

Using

$$R = mv/qB \tag{ii}$$

$$B = mv/qR = (5 \times 10^{-15})(4 \times 10^4)/(2 \times 10^{-9})(0.5) = 0.2 \text{ T} \tag{iii}$$

Problem 6.17. A metal rod of length 0.025 m, is free to roll along a railing as in Fig. 6-23. There is a uniform magnetic field of 0.03 T in the entire region, pointing into the plane of the railing, as shown. A current of 20 A is flowing through the railing and rod, as a result of the battery and resistor shown in the figure.

(*a*) What force (magnitude and direction) is exerted on the rod?

(*b*) If the polarity of the battery is changed, what change, if any, occurs to the force?

Solution

(*a*) The current is flowing from a to b in the rod. The magnitude of the force is:

$$|F| = ILB \sin \phi = (20)(0.025)(0.3) \sin 90° = 0.15 \text{ N} \tag{i}$$

The direction of the force is perpendicular to L (which points from a to b) and to **B** (which points into the paper), and is therefore along the direction of the railing. The right-hand rule (curl the fingers of the right-hand from **L** to **B** and then **F** is in the direction of the thumb) gives the direction of **F** to the right.

(*b*) Changing the polarity of the battery reverses the direction of the current (and therefore **L**). The force is then reversed and is to the left.

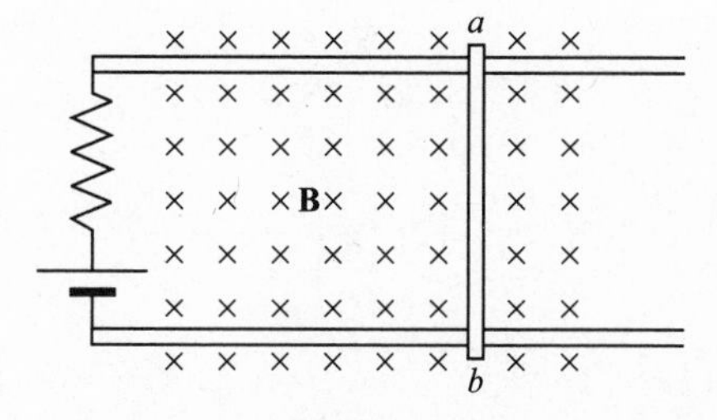

Fig. 6-23

Problem 6.18. A coil with 100 turns is free to rotate about an axis AA' as in Fig. 6-24. There is a magnetic field of 0.05 T in the plane of the coil, as shown in the figure. The coil has an area of 0.07 m^2, and a current, I, of 2×10^{-3} A is made to flow in the coil in the direction shown.

(*a*) Calculate the torque produced by the magnetic field on the coil while in this position.

(*b*) Calculate the torque produced by the magnetic field on the coil if the coil rotates through an angle θ from this initial position.

Solution

(*a*) The torque produced by the magnetic field is given by

$$|\Gamma| = MB \sin \phi, \text{ where } \phi \text{ is the angle between } \mathbf{M} \text{ and } \mathbf{B} \tag{i}$$

$$M = NIA = (100)(2 \times 10^{-3})(0.07) = 0.014 \tag{ii}$$

and the direction is perpendicular to the plane of the coil. Using our convention (curl the fingers of the right-hand in the direction of I, and the thumb points in the direction of M), we get that **M** is into the paper. Thus

$$\Gamma_B = (0.014)(0.05)(1) = 7 \times 10^{-4} \text{ N} \cdot \text{m} \tag{iii}$$

The torque tends to rotate the coil in the direction you get by rotating **M** to **B**.

(*b*) If the plane of the coil rotates by θ then the angle ϕ becomes $(90 \pm \theta)$. The torque, Γ_B' is now

$$\Gamma_B' = (0.014)(0.05) \sin (90 \pm \theta) = 7 \times 10^{-4} \cos \theta \tag{iv}$$

Problem 6.19. In Problem (6.18) a wire along axis AA' can produce a restoring torque on the coil, given by $|\Gamma_R| = 7 \times 10^{-4}\, \theta$, where θ is the angle, measured in radians, through which the coil rotates from the original position. The coil will be in equilibrium if the restoring torque is equal to the magnetic torque in magnitude and opposite in direction.

(*a*) Calculate the restoring torque produced by the wire on the coil at angles of 5°, 10° 15°, 20°, 25°, and 30°.

(*b*) Calculate the current needed in the wire so that the magnetic torque equals, in magnitude, the restoring torque at each of those angles.

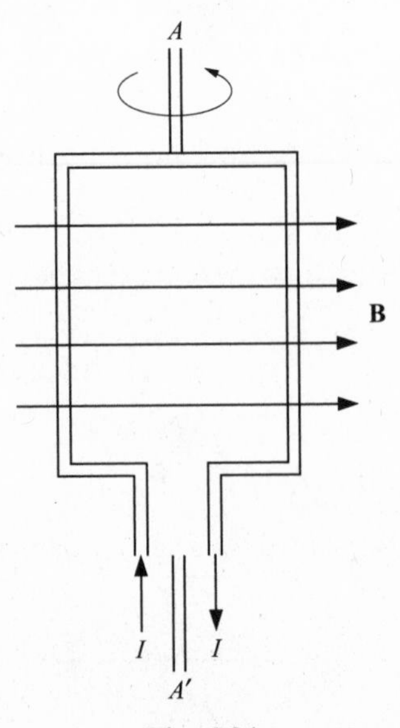

Fig. 6-24

(*c*) Plot a graph of the angle of equilibrium vs. current, using the data calculated in (*a*) and (*b*).

Solution

(*a*) $|\Gamma_R| = 7 \times 10^{-4}\ \theta$, where the angle θ is in radians (*i*)

Thus:

θ (degree)	θ (radians)	Γ_R
5	0.087	0.61×10^{-4}
10	0.175	1.22×10^{-4}
15	0.262	1.83×10^{-4}
20	0.349	2.44×10^{-4}
25	0.436	3.05×10^{-4}
30	0.524	3.67×10^{-4}

(*b*) From the previous problem,

$$|\Gamma_B| = MB \sin\phi = NIA \sin(90° \pm \theta) = 7I \cos\theta \quad (ii)$$

If $$|\Gamma_B| = |\Gamma_R|,\ 7I\cos\theta = 7 \times 10^{-4}\ \theta \quad (iii)$$

$$I = 10^{-5}\ \theta/\cos\theta \quad (iv)$$

Thus

θ (degree)	θ (radians)	I
5	0.087	0.87×10^{-6}
10	0.175	1.78×10^{-6}
15	0.262	2.71×10^{-6}
20	0.349	3.71×10^{-6}
25	0.436	4.81×10^{-6}
30	0.524	6.05×10^{-6}

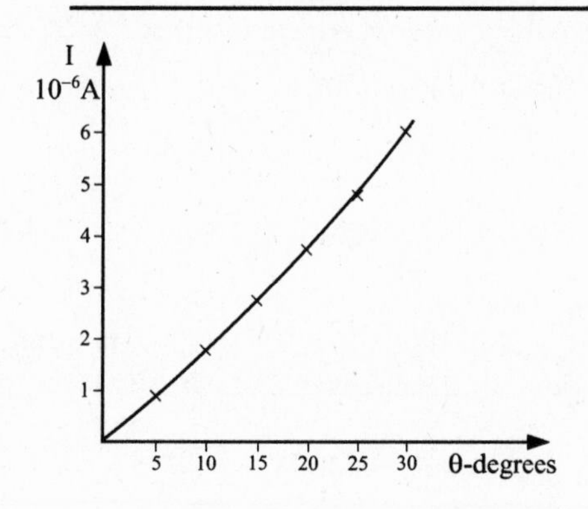

Supplementary Problems

Problem 6.20. A magnetic field of 0.3 T is in the x-direction. Calculate the force (magnitude and direction) on a charge of 3×10^{-5} C moving with a velocity of 3×10^6 m/s in the direction shown in Fig. 6-25.

Ans. (*a*) 27 N, into paper; (*b*) 0; (*c*) 13.5 N, into paper; (*d*) 13.5 N, into paper

Problem 6.21. A magnetic field of 0.3 T is out of the paper. Calculate the force (magnitude and direction) on a charge of 3×10^{-5} C moving with a velocity of 3×10^6 m/s in the direction shown in Fig. 6-25.

Ans. (*a*) 27 N, in $+x$ direction; (*b*) 27 N, in $-y$ direction; (*c*) 27 N, 60° below $+x$ axis; (*d*) 27 N, 60° above $+x$ axis

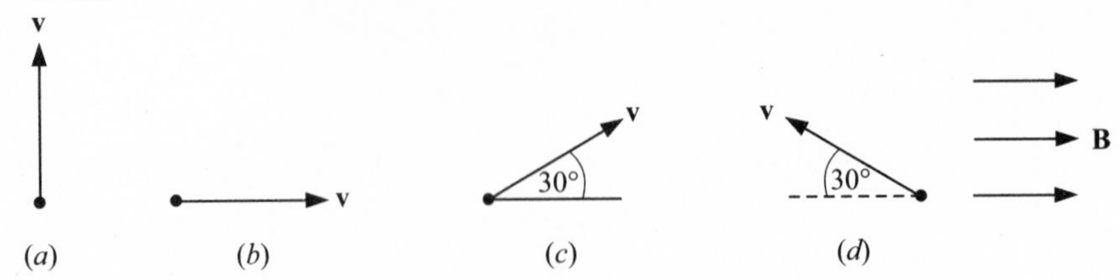

Fig. 6-25

Problem 6.22. A particle with charge -1.6×10^{-19} C is moving horizontally in the air above the earth with a speed of 7×10^4 m/s. It has a mass of 1.67×10^{-27} kg. What magnetic field (magnitude and direction) is needed so that the magnetic force cancels out the gravitational force of the earth?

Ans. 1.5×10^{-12} T, horizontal and perpendicular to **v**. (This illustrates that magnetic forces, for normal values of B, are very much larger than gravitational forces.)

Problem 6.23. A vertical magnetic field of 0.3 T causes a charged particle, moving with a velocity of 3×10^5 m/s, to move in a circle of radius 0.01 m. What is the charge to mass ratio (q/m) of this particle?

Ans. 10^8 C/kg

Problem 6.24. A magnetic field, coming out of the paper, causes a charged particle to move around a circle in a clockwise direction. Is the particle positively or negatively charged?

Ans. positively charged

Problem 6.25. A magnetic field, of 0.6 T, going into the paper, causes a charged particle to move in a horizontal circle. The particle has mass 1.67×10^{-27} kg and charge 1.6×10^{-19} C.

(*a*) How long does it take for the particle to go once around the circle?

(*b*) How many times per second does the particle go around the circle?

Ans. (*a*) 1.33×10^{-7} s; (*b*) 7.4×10^6 Hz

Problem 6.26. Two isotopes have masses of 9.87×10^{-26} kg and 9.97×10^{-26} kg, and each have a charge of 1.6×10^{-19} C. They each move in a circle with a velocity of 5×10^6 m/s in a magnetic field of 0.5 T. After going through a semi-circle, they strike a screen perpendicular to their velocity. How far apart are they on this screen?

Ans. 0.125 m

Problem 6.27. Particles move through a velocity selector, in which the fields are $E = 10^4$ V/m and $B = 0.5$ T. Outside of the velocity selector, there is only the magnetic field of 0.5 T. The particles have a charge of 1.6×10^{-19} C, and a mass of 6.7×10^{-27} kg.

(*a*) What is the velocity of those particles that leave the velocity selector?

(*b*) What is the radius of the circle in which the particles move after they leave the velocity selector?

(*c*) If the electric field is doubled, what is the radius of the circle in which the particles now move?

Ans. (*a*) 2×10^4 m/s; (*b*) 1.68×10^{-3} m; (*c*) 3.35×10^{-3} m

Problem 6.28. One wants to build a velocity selector to select particles with a speed of 6×10^6 m/s. A magnetic field of 0.3 T is available.

(*a*) What electric field is needed?

(*b*) If the electric field is produced by parallel plates, spaced 2 mm apart, what voltage must be applied?

(*c*) If one wants to select particles with half of this velocity, what voltage is needed?

Ans. (*a*) 1.8×10^6 V/m; (*b*) 3.6×10^3 V; (*c*) 1.8×10^3 V

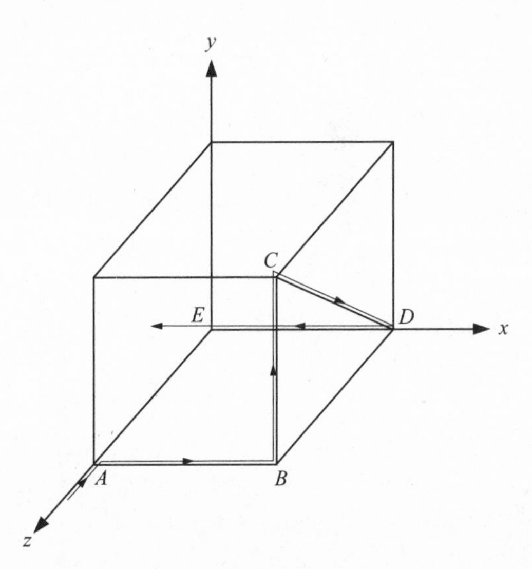

Fig. 6-26

Problem 6.29. Consider a cube, with a side of 0.5 m, as shown in Fig. 6-26. A magnetic field of 0.3 T points in the y direction. A current of 2 A flows along the direction $ABCDE$. Calculate the force (magnitude and direction) on (*a*) segment AB; (*b*) segment BC; (*c*) segment CD; (*d*) segment DE; and (*e*) the entire path $ABCDE$.

Ans. (*a*) 0.3 N, $+z$ direction; (*b*) 0; (*c*) 0.3 N, $+x$ direction; (*d*) 0.3 N, $-z$ direction; (*e*) 0.3 N, $+x$ direction

Problem 6.30. Consider a cube, with a side of 0.5 m, as shown in Fig. 6-26. A magnetic field of 0.3 T points in the $-x$ direction. A current of 2 A flows along the direction $ABCDE$. Calculate the force (magnitude and direction) on (*a*) segment AB; (*b*) segment BC; (*c*) segment CD; (*d*) segment DE; and (*e*) the entire path $ABCDE$.

Ans. (*a*) 0; (*b*) 0.3 N, $+z$ direction; (*c*) 0.42 N, in y–z plane 45° above the $-z$ direction; (*d*) 0;
(*e*) $F_x = 0.3$ N, $F_y = -0.3$ N, $F_z = 0$

Problem 6.31. Consider a cube, with a side of 0.5 m, as shown in Fig. 6-26. A magnetic field of 0.3 T points in the $-z$ direction. A current of 2 A flows along the direction $ABCDE$. Calculate the force (magnitude and direction) on (*a*) segment AB; (*b*) segment BC; (*c*) segment CD; (*d*) segment DE; and (*e*) the entire path $ABCDE$.

Ans. (*a*) 0.3 N, $+y$ direction; (*b*) 0.3 N, $-x$ direction; (*c*) 0.3 N, $+x$ direction; (*d*) 0.3 N, $-y$ direction; (*e*) 0

Problem 6.32. A square plate of length 0.5 m, with a mass of 0.03 kg, is hinged and free to rotate about the z axis. Current flows along three edges in the direction shown in Fig. 6-27(*a*), and there is a magnetic field of 0.6 T in the $-x$ direction. For this current, the plate is in equilibrium at an angle, θ, of 30°.

(*a*) What is the direction of the net magnetic force on the plate?

(*b*) By taking torques about the z axis, determine the current flowing along the edges, assuming that the center of gravity of the plate is at its center.

Ans. (*a*) $+y$ direction; (*b*) 0.5 A

Problem 6.33. A square plate of length 0.5 m, with a mass of 0.03 kg, is hinged and free to rotate about the z axis. Current flows along three edges in the direction shown in Fig. 6-27(*a*), and there is a magnetic field of 0.6 T in the $+y$ direction. For this current, the plate is in equilibrium at an angle, θ, of 30°.

(*a*) What is the direction of the net magnetic force on the plate?

(*b*) By taking torques about the z axis, determine the current flowing along the edges, assuming that the center of gravity of the plate is at its center.

Ans. (*a*) $+x$ direction; (*b*) 0.25 A

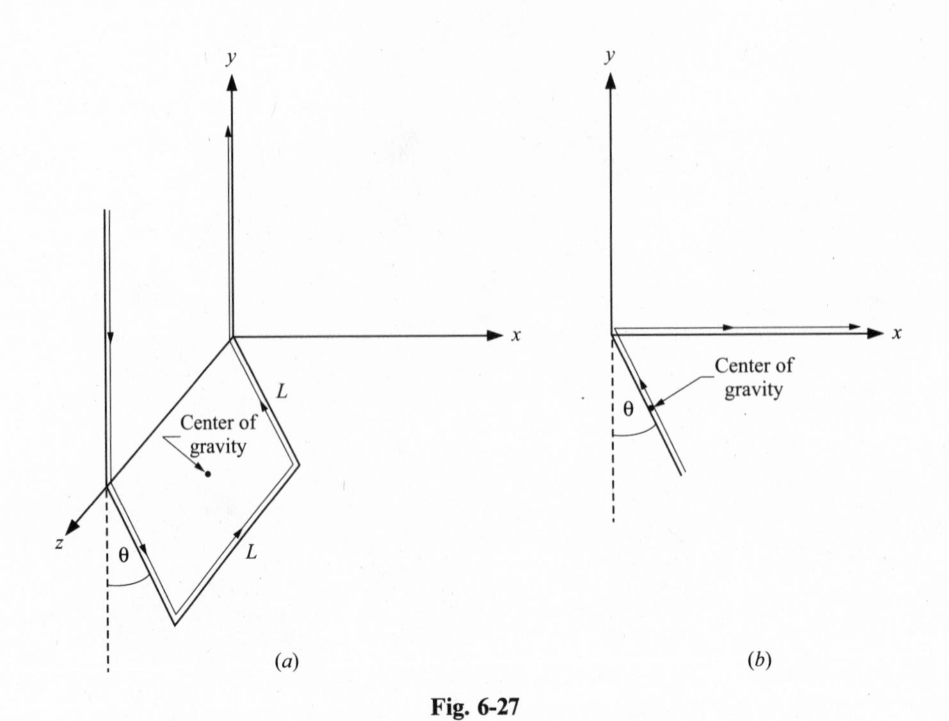

Fig. 6-27

Problem 6.34. An electron is orbiting about a proton with a speed of 3×10^7 m/s in a circle of radius 1.5×10^{-10} m (see Fig. 6-28).

(*a*) What current is moving in the circle? Is it clockwise or counter-clockwise?

(*b*) What is the magnetic moment due to this current?

(*c*) Is the magnetic moment in or out of the paper?

Ans. (*a*) 0.0051 A, clockwise; (*b*) 3.6×10^{-22} A · m^2; (*c*) into

Problem 6.35. A current of 2 A flows along the edges of the rectangle *ABCD* in Fig. 6-29. The sides of the rectangle are 0.06 m and 0.10 m, respectively. What torque is exerted on the rectangle by a magnetic field of 1.1 T, if the magnetic field points (*a*) in the *x* direction?; (*b*) in the *y* direction?; (*c*) in the *z* direction?; and (*d*) in the direction from *D* to *B*?

Ans. (*a*) 0; (*b*) 0.0132 N · m, $+z$ direction; (*c*) 0.0132 N · m, $-y$ direction; (*d*) 0.0132 N · m, direction from *A* to *C*

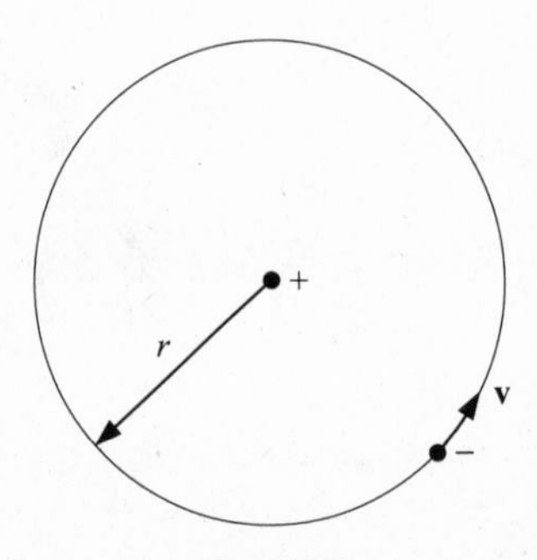

Fig. 6-28

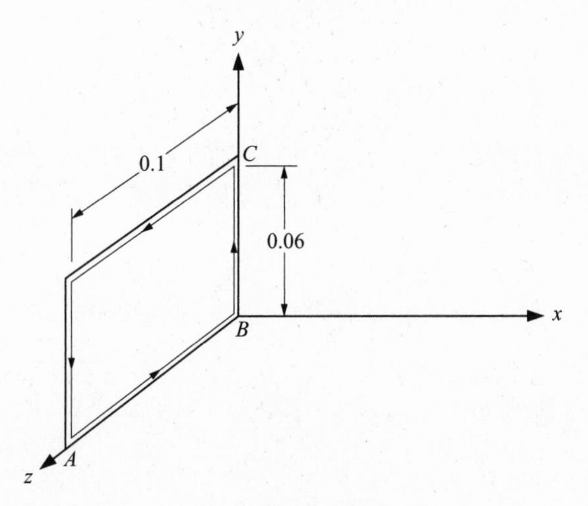

Fig. 6-29

Problem 6.36. A rectangular gate, of sides 2 m × 1.5 m, carries 1500 turns of wire with a current of 0.2 A along its edge. One wants the gate to swing open, with a torque of 90 N · m. What magnetic field is needed?

Ans. 0.1 T

Problem 6.37. A circular coil, of 2000 turns, and area 0.15 m^2, carries a current of 0.3 A. It is in the earth's magnetic field of 1.6×10^{-5} T, which we will assume is directed due north [see Fig. 6-30(*a*)].

(*a*) If the coil is in stable equilibrium in the x–y plane, does the current flow clockwise or counter-clockwise in the coil?

(*b*) If one turns the coil so that the plane of the coil makes an angle, $\theta = 30°$ with the x-axis, as in Fig. 6-30(*b*), what torque is exerted on the coil?

Ans. (*a*) clockwise; (*b*) 1.87×10^{-7} N · m

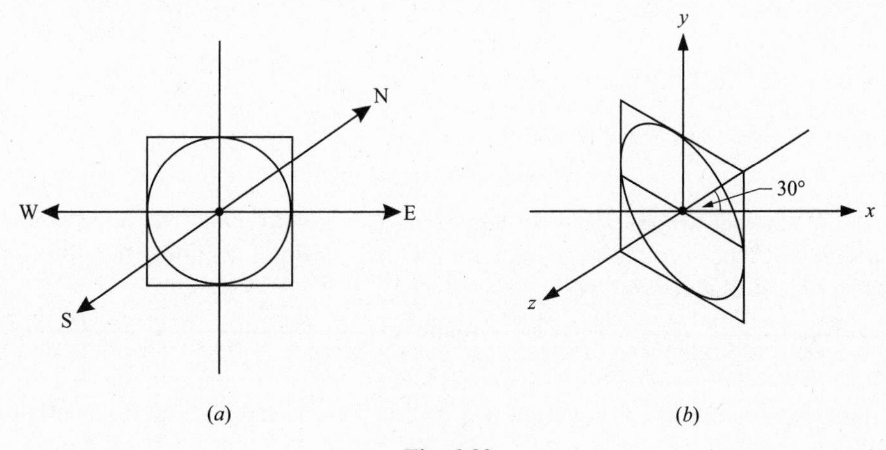

Fig. 6-30

Chapter 7

Magnetism—Source of the Field

7.1 INTRODUCTION

In the previous chapter, we learned about the effect of a magnetic field on a moving charge (or on a current carrying wire). In this chapter, we will discuss the origin of the magnetic field. We recall that the source of a gravitational field, which exerts a force on one mass, is another mass, and the source of an electric field, which exerts a force on one charge, is another charge. We will therefore not be surprised to find that one source of a magnetic field, which exerts a force on a moving charge, is another moving charge. Indeed, the basic origin of a magnetic field is a moving charge or an equivalent current in a wire. In a later chapter, we will learn that there is another basic source for a magnetic field, namely an electric field that varies with time. In this chapter we will develop the concepts and equations needed to understand the magnetic fields produced by moving charges.

7.2 FIELD PRODUCED BY A MOVING CHARGE

To obtain the magnetic field produced by a charge, q, moving with velocity $\mathbf{v}$, at a point located at a displacement $\mathbf{r}$ from the charge, we need a mathematical expression for the field in terms of q, $\mathbf{v}$ and $\mathbf{r}$. The geometry is shown in Fig. 7-1. In this figure, a charge, q, located at point a is moving with velocity $\mathbf{v}$, as shown. We seek the magnitude and direction of the magnetic field at point b, displaced from point a by the vector $\mathbf{r}$. Thus, the point b is at a distance r from point a along a line that makes an angle ϕ with the velocity $\mathbf{v}$. As was the case with the force exerted by the magnetic field, we are looking for a vector (in this case $\mathbf{B}$) which is formed from some combination of two vectors (in this case $\mathbf{v}$ and $\mathbf{r}$) and a scalar q. And once again we will discuss separately the magnitude and the direction of this vector $\mathbf{B}$. The results we express here were determined from a wide array of experimental studies of magnetic fields and their sources.

Magnitude of the Field

The formula for the magnitude of the field is:

$$|B| = (\mu_0/4\pi)|qv \sin \phi/r^2| \tag{7.1}$$

where $(\mu_0/4\pi)$ is a constant which, for our system of units, is equal to 10^{-7} T · m/A. (The 4π is included for later convenience.) This formula, together with the prescription for finding the direction of the field, is known as the **Law of Biot and Savart**.

We have used absolute value signs, since the magnitude is always positive. The magnitude of the field $\mathbf{B}$ does not depend on the sign of the charge nor on the sign of $\sin \phi$ (which in any case is positive between 0° and 180°). The direction of the field will, however, be dependent on the sign of q.

This formula tells us that the field is zero if ϕ is zero. This occurs if the point b lies along the line of $\mathbf{v}$, i.e. if the present path of the charge would carry it through the point b. In order for a magnetic field

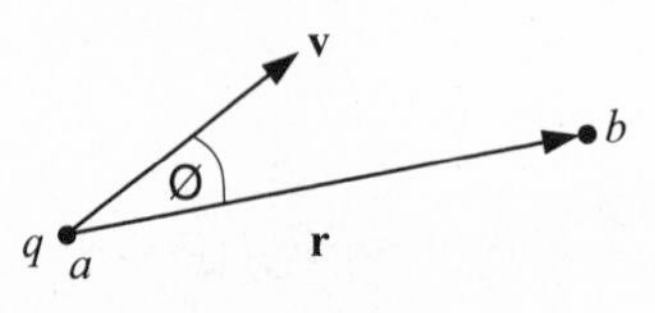

Fig. 7-1

to be produced at a point b, that point must lie at some non-zero distance from the extended line of $\mathbf{v}$. The largest magnetic field is produced when ϕ is 90°. This occurs when the point b is located along the line perpendicular to $\mathbf{v}$ at point a.

The magnitude of $\mathbf{B}$ decreases as $1/r^2$ with the distance from point a. This is reminiscent of the dependence of $\mathbf{g}$ and of $\mathbf{E}$ on the distance from their respective sources. As expected, the field increases with both q and v. Thus the field gets bigger for charges which move fast and for those that have a lot of charge. The field decreases as one goes to points that are further away from the charge and for those at smaller angles to the line along which the charge moves.

Problem 7.1.

(*a*) A charge of 2×10^{-6} C is moving with a velocity of 3×10^4 m/s when passing point a in Fig. 7-1. What is the magnitude of the field at point b if that point is at a distance of 2×10^{-3} m from point a at an angle ϕ of 30°?

(*b*) What is the magnitude of the field if the charge were -2×10^{-6} coulomb?

(*c*) What is the magnitude of the field if the angle ϕ were 150°?

Solution

(*a*) Substituting $q = 2 \times 10^{-6}$ coulomb, $v = 3 \times 10^4$, $\phi = 30°$ and $r = 2 \times 10^{-3}$ into Eq. (*7.1*), we get $|B| = (10^{-7})(2 \times 10^{-6})(3 \times 10^4)(\sin 30°)/(2 \times 10^{-3})^2 = 7.5 \times 10^{-4}$ T.

(*b*) Since only the absolute value of each variable enters, the answer is the same as for part (*a*).

(*c*) Since $\sin 150° = \sin 30°$, the answer is still the same.

Problem 7.2. A charge of 3.0×10^{-5} C is at the origin in Fig. 7-2, moving in the positive x direction with velocity 2.0×10^6 m/s. The length of each side of the cube is 2.0×10^{-3} m. Calculate the magnitude of the field at (*a*) point B; (*b*) point E; (*c*) point H; (*d*) point C; and (*e*) point F.

Solution

In all five cases, $(\mu_0/4\pi)\, qv = (10^{-7})(3.0 \times 10^{-5})(2.0 \times 10^5) = 6.0 \times 10^{-7}$. The difference between each case is the value of r and of $\sin \phi$. Thus, the solution for each case is

(*a*) $\phi = 0$, $\sin \phi = 0$ and therefore $|B| = 0$.

(*b*) $\phi = 90°$, $\sin \phi = 1$ and $r = 2.0 \times 10^{-3}$. Therefore $|B| = 6.0 \times 10^{-7}(1)/(2.0 \times 10^{-3})^2 = 0.15$ T.

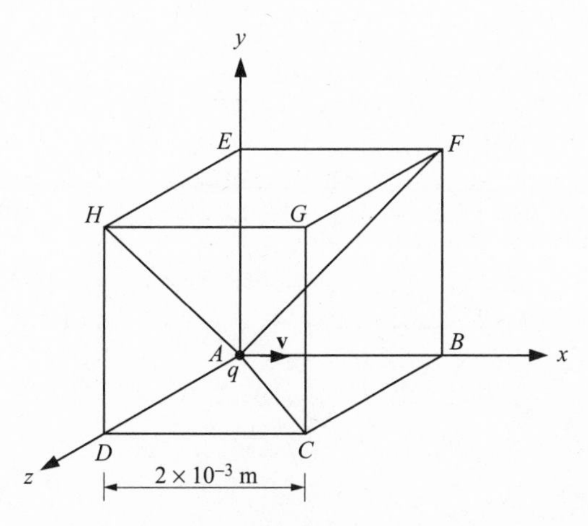

Fig. 7-2

(*c*) $\phi = 90°$, $\sin\phi = 1$ and $r = 2.0 \times 10^{-3}\,(\sqrt{2})$. Therefore $|B| = 0.075$ T.

(*d*) $\phi = 45°$, $\sin\phi = 0.707$ and $r = 2.0 \times 10^{-3}\,(\sqrt{2})$. Therefore $|B| = 0.053$ T.

(*e*) $\phi = 45°$, $\sin\phi = 0.707$ and $r = 2.0 \times 10^{-3}\,(\sqrt{2})$. Therefore $|B| = 0.053$ T.

Direction of the Field

The direction of the field is perpendicular to both **v** and **r**, and it is therefore perpendicular to the plane containing both **v** and **r**. This is illustrated in Fig. 7-3. Here, we call θ the angle between **v** and **r**. Again, there are two possible directions which are perpendicular to this plane, and we need a rule to select the correct direction. Once again this is the right-hand rule. In this case we apply the rule by placing our fingers in the direction to rotate **v** into **r**, and our thumb will then point in the direction of **B**. In Fig. 7-3, we draw the plane of **v** and **r**, and the perpendicular to that plane. The two possible perpendicular direction are up and down. Using the right-hand rule selects the downward direction as the correct one. This is the correct answer if the charge q is positive. For a negative charge, the direction of **B** is reversed.

There is a nice way to visualize this geometry. In Fig. 7-4 we draw the same vectors **v** and **r**. At the tip of **r** (point b) we draw the plane through b that is perpendicular to the line of the vector **v**, cutting that line at point O. Thus the line from b to O is perpendicular to line aO, and equals $r\,\sin\theta$. The magnetic field at b lies in this plane and is perpendicular to bO at b. In fact, if one draws a circle in this plane, whose center is at O and whose circumference passes through point b, the direction of the magnetic field at point b is tangent to the circle at that point. The circle through b is shown in the figure. It

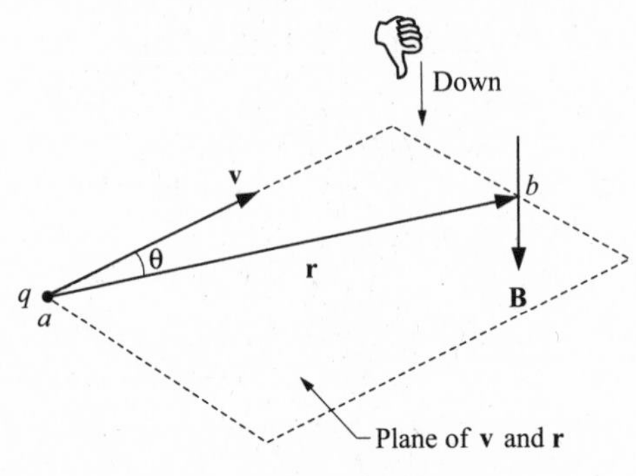

Fig. 7-3

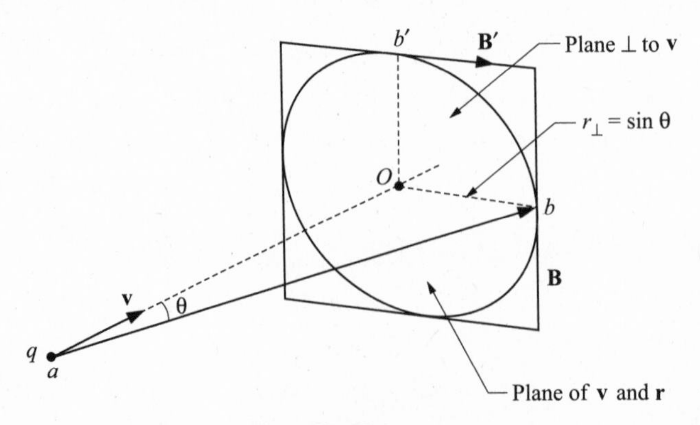

Fig. 7-4

is not hard to see that the magnetic field at any point on this circle, has a common magnitude and a direction tangent to the circle at that point. To determine which way the magnetic field points along the tangent (for instance, at b the direction could be either up or down), we use an equivalent right-hand rule to that defined above. Put the thumb of your right hand along the direction of $\mathbf{v}$, and your fingers will circle around that line in the direction of $\mathbf{B}$. In the case depicted, the direction of $\mathbf{B}$ at point b is down. At point b' the tangent to the circle is horizontal, and the magnetic field points to the right as can be seen by applying the right-hand rule. This picture is often very useful, since it shows that if one traces the magnetic field lines, (in a manner similar to tracing electric field lines), they form concentric circles around the direction of $\mathbf{v}$. The direction of this circling is determined by the right-hand rule. Again, if the charge q is negative, we reverse the direction of $\mathbf{B}$, which means that that the magnetic field lines will now be circling in the opposite direction. It is often useful to view the field lines by looking toward the charge along the direction of the velocity. This is shown in Fig. 7-5, where we view the charge coming out of the paper. The magnetic field lines are circles in the plane of the paper, with centers on the line of $\mathbf{v}$. The direction of the field is tangent to the circle at any point, with the sense determined by the right-hand rule. In the case shown, the field lines circle in the counter-clockwise direction for a positive charge, and clockwise for a negative charge.

Problem 7.3. Determine the direction of the field in Problem 7.2.

Solution

(*a*) Since the field is zero at B, there is obviously no direction to determine.

(*b*) Here r is the vector from A to E. The plane containing $\mathbf{v}$ and $\mathbf{r}$ is the x–y plane, and the perpendicular to that plane is the $\pm z$ direction. If we place our fingers in the direction rotating $\mathbf{v}$ into $\mathbf{r}$, our thumb points in the $+z$ direction, which is therefore the direction of $\mathbf{B}$.

Alternatively, we could draw the view as seen by looking toward $\mathbf{v}$, i.e. by looking down the x axis. This view is shown in Fig. 7-6(*a*). We draw a circle through E, with its center at A. The tangent to this circle at any point is the direction of $\mathbf{B}$ at that point. By the right-hand rule we know that the field lines circle in the counter-clockwise direction (for the positive charge). At E, the tangent to the circle, and therefore the direction of $\mathbf{B}$, points toward H, or in the positive z direction.

(*c*) Here it is simplest to use the alternate approach discussed in part (*b*). The same Fig. 7-6(*a*) also contains the point H. The circle through H, with center at A, is also shown in the figure. The tangent to this circle at H is perpendicular to AH and pointing in the direction ED, i.e. 45° below the positive z axis, as shown.

(*d*) Again, we draw the plane that we see as we look toward $\mathbf{v}$, but this time we draw the plane through point C [see Fig. 7-6(*b*)]. The point B is where the extension of the vector $\mathbf{v}$ pierces this plane. The circle centered on B and passing through C is shown in the figure, again with the field lines circling counter-clockwise. At C, the tangent to the circle is in the $-y$ direction, which is the direction of $\mathbf{B}$.

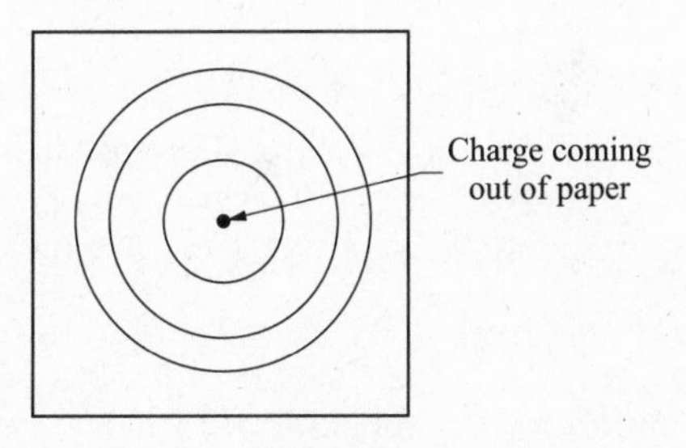

Fig. 7-5

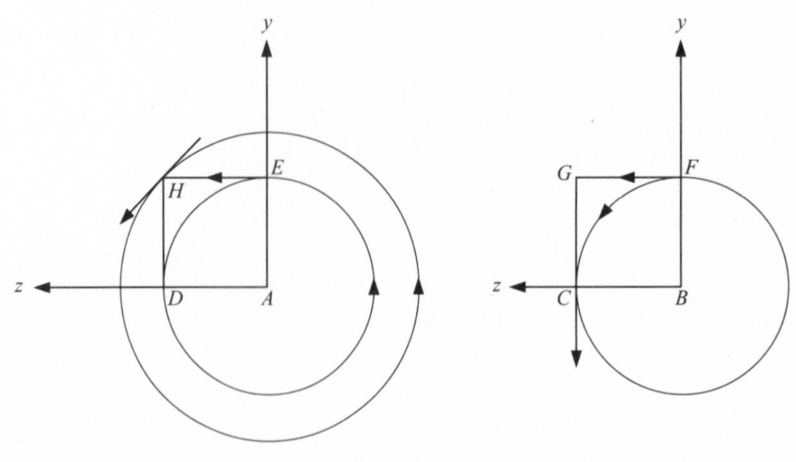

Fig. 7-6

(*e*) The same drawing that we used for part (*d*) can be used for this part. The circle through F has a tangent at F which points in the positive z direction which is therefore the direction of **B**.

Problem 7.4. A charge q_1 is moving with velocity $\mathbf{v}_1$ along the positive x axis. Another charge q_2 is moving with velocity $\mathbf{v}_2$ parallel to $\mathbf{v}_1$ at a distance d from q_1 as shown in Fig. 7-7(*a*). Find an expression for the magnitude of the magnetic force that q_1 exerts on q_2, in terms of q_1, q_2, $\mathbf{v}_1$, $\mathbf{v}_2$, and d, and find the direction of this force.

Solution

Charge q_1 produces a magnetic field at the position of q_2 as we just learned. This magnetic field exerts a force on q_2 as we learned in Chap. 6. We therefore solve this problem in two steps. First we calculate the field $\mathbf{B}_{1 \text{ at } 2}$ produced by q_1 at the position of q_2, and then we calculate the force that this field exerts on q_2.

To calculate the field produced by q_1 at the position of q_2, we first calculate its magnitude. This is given by $|\mathbf{B}_{1 \text{ at } 2}| = \mu_0 q_1|\mathbf{v}_1|\sin\theta/4\pi r^2$, where $r = d$ and θ is the angle between $\mathbf{v}_1$ and **r**, which is 90° (**r** is the vector from point 1 to point 2). Thus,

$$|\mathbf{B}_{1 \text{ at } 2}| = (\mu_0/4\pi)(q_1|\mathbf{v}_1|/d^2) \tag{7.2}$$

To determine the direction of $\mathbf{B}_{1 \text{ at } 2}$ we draw the view with $\mathbf{v}_1$ coming out of the paper, as shown in Fig. 7-7(*b*). The circle centered on the line of $\mathbf{v}_1$ and going through q_2 is shown on the figure. The tangent to this circle at q_2 is in the $-y$ direction, which is the direction of $\mathbf{B}_{1 \text{ at } 2}$.

We now have to calculate the force exerted by $\mathbf{B}_{1 \text{ at } 2}$ on q_2. This is also done by calculating separately the magnitude and the direction of the force. We recall from Chap. 6 that the magnitude of the force is, in general, given by Eq. (*6.1*):

$$|F| = |qvB \sin\phi|$$

In our case, $q = q_2$, $v = v_2$, $B = B_{1 \text{ at } 2}$, and $\phi = 90°$ since the angle between **B** and $\mathbf{v}_2$ is 90°. Substituting for $B_{1 \text{ at } 2}$ in the equation, we get that

$$|F| = |q_2 v_2(\mu_0/4\pi)(q_1 v_1/d^2)| = |(\mu_0/4\pi)q_1 q_2 v_1 v_2/d^2| \tag{7.3}$$

To get the direction of the force, we recall that **F** is perpendicular to both $\mathbf{v}_2$ and to $\mathbf{B}_{1 \text{ at } 2}$. Since $\mathbf{B}_{1 \text{ at } 2}$ is in the $-y$ direction and $\mathbf{v}_2$ is in the $+x$ direction, the plane containing both these vectors is the x–y plane. Using the right-hand rule (rotate $\mathbf{v}_2$ into **B**), we find that the direction of **F** is $-z$. Thus q_2 is being attracted to q_1 via the magnetic force.

The above calculation was performed for the case of two positive charges moving parallel to each other in the same direction. In this case, we got an attractive force on q_2 due to q_1. We can deduce

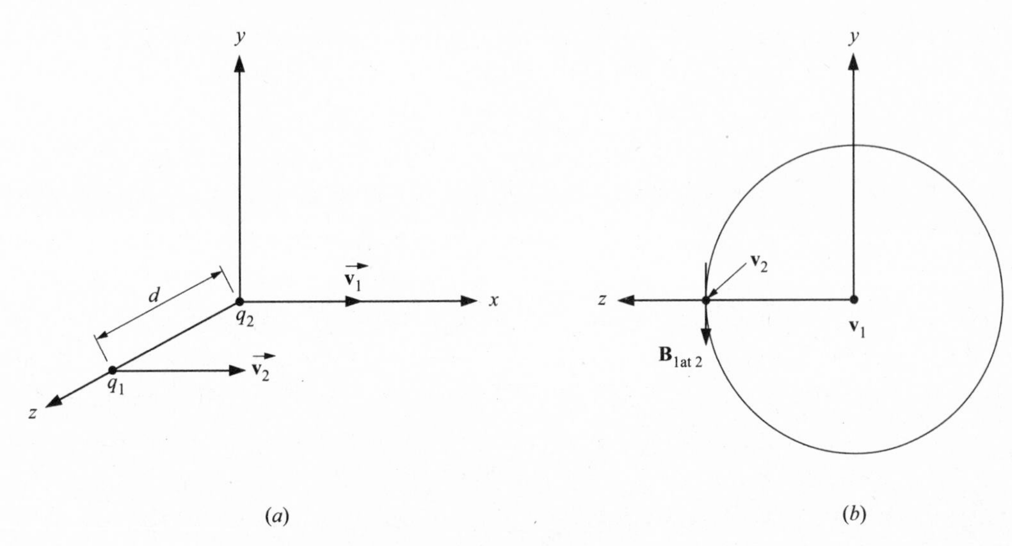

(a) (b)

Fig. 7-7

several generalizations from this example. Firstly, if we had calculated the force that q_2 exerts on q_1, we would also have found an attractive force of the same magnitude. Secondly if *one* of the charges had been negative, then the force would have become repulsive, since either $\mathbf{B}_{1 \text{ at } 2}$ (if q_1 were negative) or the right-hand rule for F (if q_2 were negative) would have been reversed. However, if both charges were negative, the force would still have been attractive. Thirdly, if the velocities of the charges had been in parallel, but *opposite* directions, then, for charges of the same sign, the force would be repulsive, since *one* of the velocity directions would be reversed.

It is also important to note that, in addition to the magnetic force between the charges, there is also an electric force between the charges. In fact, the electric force will generally be much greater than the magnetic force, unless the velocities of the charges are comparable to the velocity of light. This is more fully explored in one of the supplementary problems.

7.3 FIELD PRODUCED BY CURRENTS

As we learned in Chap. 6, current flowing in a wire is equivalent to moving charge. There we discussed the force exerted on a current flowing in a small length of wire. We showed there that we could use the same formula that we used for a charge q moving with a velocity v, if we replaced $q\mathbf{v}$ with $I\mathbf{L}$, where $\mathbf{L}$ is the small, directed length of wire carrying current I. The same is fundamentally true when one wishes to calculate the field produced by a current element $\mathbf{L}$ carrying current I. However, any current element must be part of a continuous circuit, and each part of that circuit produces its own magnetic field at every point. The actual field at any point is the vector sum of the contributions from all elements of the circuit. This will clearly pose calculation problems, which we will have to discuss.

To calculate the magnetic field produced by a current I flowing in a wire of length ΔL (since we are talking about a small portion of a longer wire we use the designation ΔL), we use the same formula that we developed in Sec. 7.2, replacing $q\mathbf{v}$ with $I\Delta\mathbf{L}$. Thus, the magnitude of the part of the field produced by this current element at a point located at a distance r from the element is (see Fig. 7-8)

$$|\Delta B| = (\mu_0/4\pi)|I\Delta L \sin\phi/r^2| \tag{7.4}$$

where the terms in the formula have the same meaning as previously, and as labeled in the figure. The direction of the field is calculated in the same manner as for the moving charge, and would be into the paper for the case in Fig. 7-8.

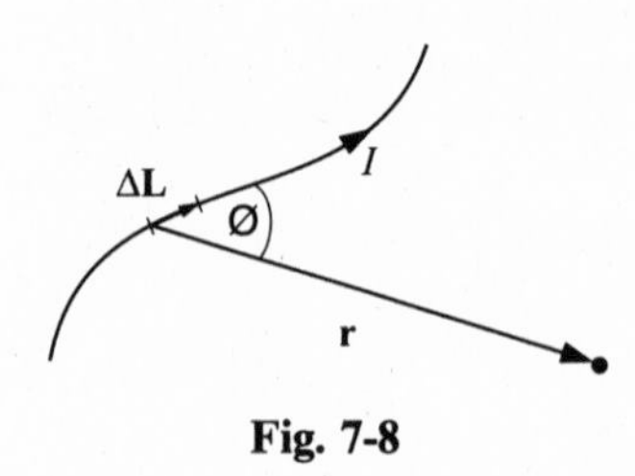

Fig. 7-8

In order to calculate the field at this point, we would have to add together, vectorially, the contributions from the entire length of the wire. This can be done for certain special cases without the use of advanced mathematics, but for the more general case, the use of calculus is needed. Let us discuss some special cases which are often used in practice.

Field at the Center of a Current Carrying Ring

Consider a ring, of radius R, which carries a current I, flowing in the clockwise direction in Fig. 7-9. We want to calculate the field at the center of the ring due to this current. We proceed by calculating the field produced by the segment ΔL at the center (point O). The magnitude of the field is $\Delta B = (\mu_0/4\pi)\,|I\Delta L/R^2|$, since the angle between $\Delta\mathbf{L}$ and $\mathbf{R}$ is 90°. This magnitude is the same for any segment of the ring, independent of the location of ΔL along the ring. The direction of the field is perpendicular to the plane of $\Delta\mathbf{L}$ and $\mathbf{R}$, which means that it is in or out of the plane of the ring. Using the right-hand rule, we determine that it is into the paper in the figure. Again, this direction is independent of the location of ΔL along the ring, so that every ΔL along the wire contributes the same ΔB, and in the same direction. To get the total magnetic field from all the ΔL, we have to add vectors which are all in the same direction, so we have only to add the magnitudes. The total field will therefore be $B = (\mu_0/4\pi)\,|IL/R^2| = (\mu_0/4\pi)I(2\pi R)/R^2 = \mu_0 I/2R$. Thus, in general, the field at the center of a ring is given by:

$$B = \mu_0 I/2R \tag{7.5}$$

The direction (into the paper) can be deduced from the following right-hand rule. Wrap the fingers of your right hand around the circle in the direction of the current, and your thumb points in the direction of the field.

Problem 7.5.

(*a*) Calculate the magnitude and direction of the magnetic field at the center of the circle in Fig. 7-10, whose radius is 1.5 m and which carries a current of 2 A in the direction shown.

(*b*) Suppose that instead of a single circle we had a tightly wound coil of $N = 50$ turns (same current, same radius); find B.

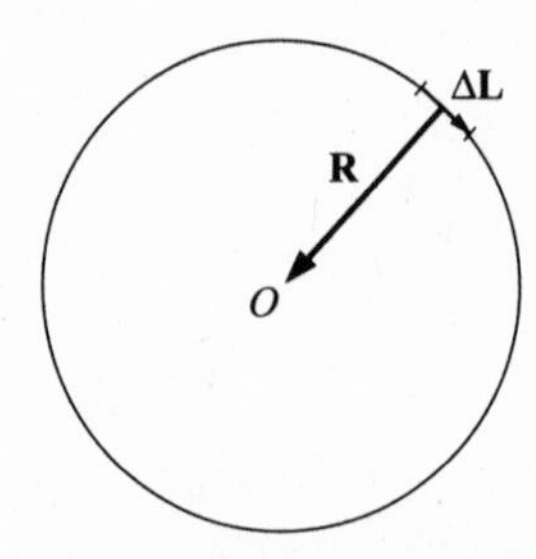

Fig. 7-9

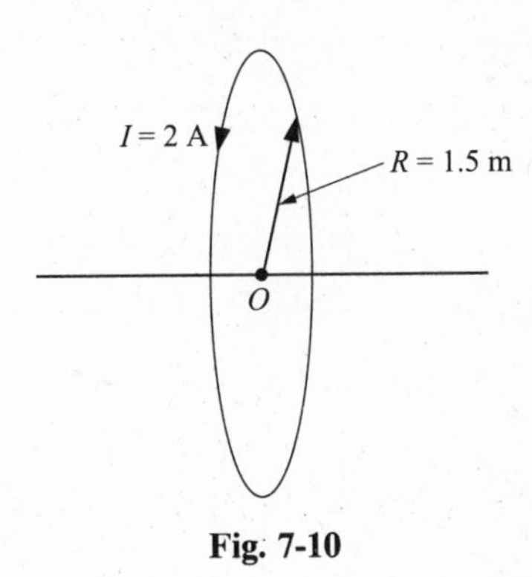

Fig. 7-10

Solution

(*a*) Using Eq. (7.5), $B = \mu_0 I/2R = \mu_0(2)/2(1.5) = 4\pi \times 10^{-7}/1.5 = 8.38 \times 10^{-7}$ T. To get the direction, we wrap our fingers around the ring in the direction of I, and determine that the direction is to the right.

(*b*) If the coil consists of N turns, all of the same radius and carrying the same current, we just add the contributions of each loop of the coil:

$$B = N\mu_0 I/2R = 50(8.38 \times 10^{-7}) = 4.19 \times 10^{-5}\text{T} \tag{7.6}$$

Field Along the Axis of a Ring

In Fig. 7-11, we draw a ring, of radius R, carrying a current I in the direction shown. The current is out of the paper at the top, and into the paper at the bottom. Along the axis of the ring (the line OP), we seek the magnetic field produced by this current at an arbitrary distance, x, from the center of the ring, O. We choose a current element ΔL at the top of the ring, where the current is coming out of the paper. The field at P due to this element is

$$|\Delta B| = (\mu_0/4\pi)|I\Delta L \sin \phi/r^2| \tag{7.7}$$

Here ϕ is the angle between $\mathbf{\Delta L}$ (out of the paper) and $\mathbf{r}$, which is 90°, and $r = \sqrt{R^2 + x^2}$. Thus

$$|\Delta B| = (\mu_0/4\pi)I\Delta L/(R^2 + x^2) \tag{7.8}$$

The direction of the field is more difficult to visualize. It is shown if Fig. 7-11. This direction is clearly perpendicular to $\mathbf{r}$ and to $\mathbf{\Delta L}$, and conforms to the right-hand rule. Note that θ, the angle between $\mathbf{R}$ and $\mathbf{r}$, is also the angle between $\mathbf{B}$ and the x axis. If we now consider an equivalent current element at the bottom of the ring, we get a field of the same magnitude and with a direction below the axis at the

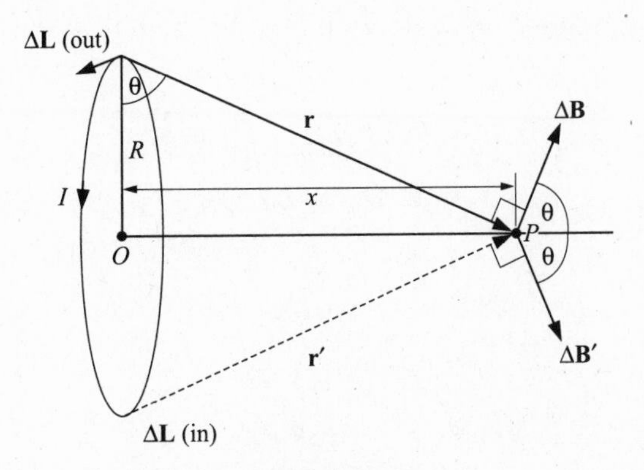

Fig. 7-11

same angle as this ΔB (see Fig. 7-11). If we add those two fields together, the vertical components will cancel and we will be left with only the horizontal components. This horizontal component will equal $\Delta B \cos\theta$

$$|\Delta B_x| = (\mu_0/4\pi)I\Delta L \cos\theta/(R^2 + x^2),$$

and with $\cos\theta = R/\sqrt{R^2 + x^2}$, we get

$$|\Delta B_x| = (\mu_0/4\pi)I\Delta LR/(R^2 + x^2)^{3/2} \tag{7.9}$$

Problem 7.6. Find an expression for the field due to a current carrying coil at any point along its axis of symmetry.

Solution

Using Eq. (*7.9*), and noting that for every segment in the upper part of the ring, there will be an opposite segment in the lower part which will cancel other components than the horizontal component, we have only to add together the horizontal components due to all the ΔLs around the loop. Since the magnitude of B_x is the same for all the segments, we can add them together very easily. Adding all the ΔL together gives the circumference of the circle, $2\pi R$, and therefore

$$B = (\mu_0/4\pi)I(2\pi R)R/(R^2 + x^2)^{3/2} = \mu_0 IR^2/2(R^2 + x^2)^{3/2} \tag{7.10}$$

Again, for N turns in the coil, the field is multiplied by N. The direction is along the axis to the right for the direction of current chosen. In general, the direction along the axis can be determined by the same right-hand rule as for the center of the circle.

Note. Obtaining the magnetic field of the loop off the x axis is much harder, since we don't have the symmetry that allowed us to solve the on axis problem.

Problem 7.7. Calculate the magnitude and direction of the magnetic field at point P in Fig. 7-11, if $R = 1.5$ m, $I = 2$ A and $x = 2$ m.

Solution

Using Eq. (*7.10*), $B = 4\pi \times 10^{-7}(2)(1.5)^2/2(1.5^2 + 2^2)^{3/2} = 1.8 \times 10^{-7}$ T. The direction is to the right.

Problem 7.8. Two identical coils, each having 1000 turns, are separated by a distance of 1.6 m along a common axis, as in Fig. 7-12. For each coil $R = 1.5$ m and $I = 2$ A. Calculate the field at a point on the axis midway between the coils.

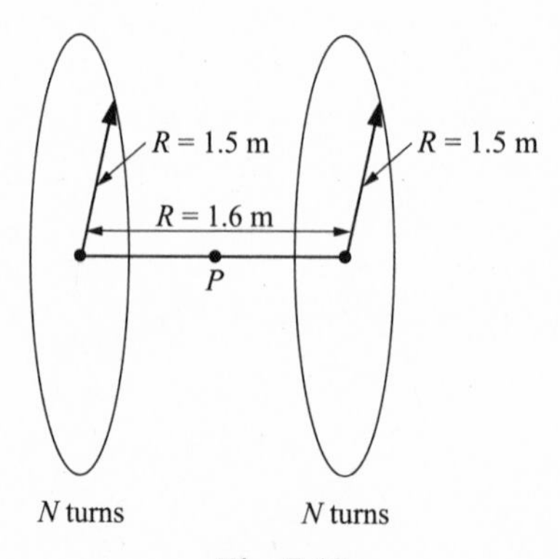

Fig. 7-12

Solution

For each coil, we can use Eq. (*7.10*) multiplying by N for the number of turns. Therefore, for each coil, $B = 10^3\,(4\pi \times 10^{-7})(2)(1.5)^2/2(1.5^2 + 0.8^2)^{3/2} = 5.8 \times 10^{-4}$ T. The direction is to the right. For both coils together, the field is 1.16×10^{-3} T.

Field of a Long Straight Wire

Suppose that a long straight wire carries a current I to the right, as in Fig. 7-13(*a*). We want to know the magnitude and direction of the field produced by the current in this wire at a point located at a distance R from the wire. The method required is to take segments of length ΔL along the wire, calculate the field ΔB produced by that segment at the point and add the contributions from each segment together vectorially. We will carry out part of this process, then indicate how to complete it and then give the final answer. Choose a segment ΔL as shown. At point P, we calculate the magnitude of ΔB to be

$$\Delta B = (\mu_0/4\pi)I\Delta L \sin\theta/r^2$$

The direction of the field is perpendicular to $\mathbf{\Delta L}$ and to $\mathbf{r}$, i.e. perpendicular to the paper, and by the right-hand rule the direction is into the paper. This direction is the same for all segments of the wire, so that we can deduce that the direction of the field of the entire wire will be in this same direction. However, the magnitude of the field from each segment will vary since r and $\sin\theta$ is different for each segment. To add up all the contributions from the segments requires the use of calculus. When we do this, we come up with the result that

$$B = (\mu_0/4\pi)2I/R \tag{7.11}$$

In this equation, I is the current in the wire and R is the perpendicular distance of the point P from the wire. The magnitude of the field depends only on these two variables. To get the direction of the magnetic field, we make use of the same picture that we developed for the direction of the magnetic field produced by a moving charge. The field lines are circles about the axis of the long wire. This is easiest to visualize if we draw the straight wire as coming out of or going into the paper, as in Fig. 7-13(*b*). The magnetic field lines are then circles in the plane of the paper, with their center at the wire. The magnetic field at any point is tangent to the circle through that point and the direction of circling is obtained from the right-hand rule, i.e. put your thumb in the direction of the current and the fingers circle in the direction of the magnetic field.

Problem 7.9. A current of 4 A flows in a long, straight wire along the x axis in Fig. 7-14. Calculate the magnitude and direction of the magnetic field at the following corners of the cube, whose side is 0.8 m: (*a*) corner d; (*b*) corner f; and (*c*) corner h.

Solution

(*a*) It is useful to draw a picture of the situation looking at the current coming out of the paper. We therefore draw Fig. 7-14(*b*), with the current coming out at the origin, and the sides *bcgh* and *adhe* in

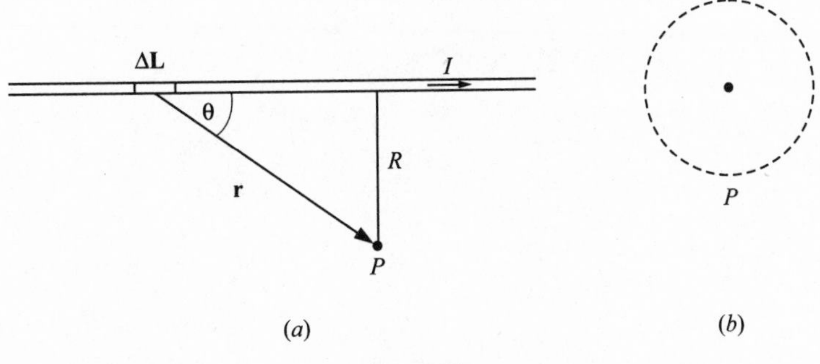

Fig. 7-13

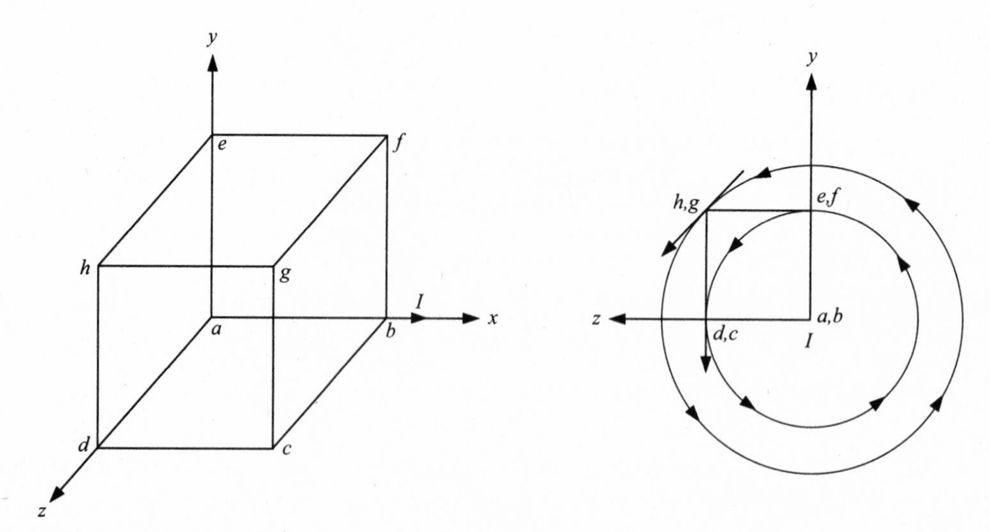

Fig. 7-14

the plane of the paper. We know that the magnitude of the field at any point is given by $B = (\mu_0/4\pi)2I/R$, where R is the perpendicular distance from the point to the line of current, which in our case means to the x axis.

For corner d, $|B| = 10^{-7}\,(2)(4)/0.8 = 10^{-6}$ T, and the direction is tangent to the circle drawn with center at a and going through point d. This circle is shown on the figure. It is seen from the figure that this tangent is in the $\pm y$ direction at point d. To choose between these two possibilities, we use the right-hand rule which tells us that the field lines circle the axis in a counter-clockwise direction. Therefore the magnetic field at c is in the $-y$ direction.

(*b*) For corner f the magnitude of B is also $B = 10^{-6}$ T, since R is the same as for part (*a*), and the same circle drawn for part (*a*) also goes through point e. Note that the fact that point e is further out along the x axis is totally irrelevant to this calculation. The tangent to the field line at this point is in the $+z$ direction.

(*c*) For corner h the perpendicular distance to the x axis is $0.8\sqrt{2}$, and the magnetic field is $B = 7.07 \times 10^{-5}$ T. The circle through h is also shown on the figure, and the tangent to that circle at h is at an angle of 45° below the $+z$ axis (the direction from e to d). This direction is also shown on the figure.

Problem 7.10. A current of 4 A flows in a long, straight wire out of the paper, as in Fig. 7-15. A charge of 2×10^{-4} C is located at point P, a distance of 3 mm from the wire. This charge is moving with a velocity of magnitude 4×10^{6} m/s. Calculate the force exerted on the charge if the direction of its velocity is (*a*) into the paper; (*b*) to the right; and (*c*) upward.

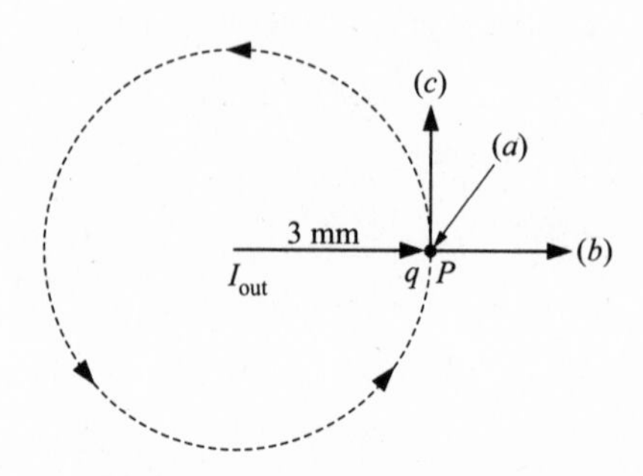

Fig. 7-15

Solution

(*a*) The force exerted on the charge comes from the magnetic field produced by the current in the long straight wire. We will therefore first calculate the magnetic field produced by the wire at the position of the wire, i.e. at point P. The magnitude of the field is $B = 10^{-7}\ (2)(4)/3 \times 10^{-3} = 2.67 \times 10^{-4}$ T. Drawing the circle through P around the current, we find the direction of B to be upward. Note that this magnetic field exists at P whether or not there is a charge at this point. Since a moving charge does exist at P, the field exerts a force on this moving charge. This force depends on the magnitude and direction of **v**.

The magnitude of the force is given by $F = |qvB \sin \phi|$. If **v** is into the paper, then $\phi = 90°$, and $F = 2 \times 10^{-4}\ (4 \times 10^6)(2.67 \times 10^{-4}) = 0.21$ N. The direction of the force is perpendicular to the plane of **v** and **B**, so that it is either toward or away from the current line. Using the right-hand rule (rotating **v** into **B**) we find that the force is away from the current.

(*b*) Using the same value for **B** that we calculated in (*a*), and noting that $\phi = 90°$, we get that $F = 0.21$ N again. The direction is now either into or out of the paper (perpendicular to **v** and **B**), and the right-hand rule chooses out of the paper.

(*c*) Here v is in the same direction as the magnetic field, so that $\phi = 0$. The force is therefore zero.

Field in a Long Solenoid

We now consider the field produced within a long, tightly wound **solenoid**. This case is depicted in Fig. 7-16. A wire is continuously wound around a long pipe with adjacent windings close to each other. This is similar to the case of the field produced by a ring along its axis, except that we have to add together the fields of many parallel rings. This calculation can be performed using the calculus. Furthermore, we want to know the field everywhere within the solenoid, not just on its axis. This is an extremely difficult calculation, but can be derived using Ampere's law—see the next section. When this calculation is performed, we find that the field within a long solenoid is the same at any point within the solenoid, and is zero (or very small) outside the solenoid. The magnitude of the field inside the solenoid is given by $|B| = \mu_0 nI$, where n is the number of turns per meter. The direction of the field is parallel to the axis with the direction given by the same right-hand rule used for the ring (the fingers circle in the direction of the current and the thumb points in the direction of the field). The result is the reason that solenoids are so very useful for producing magnetic fields. The field produced is *uniform*, with the same magnitude and direction everywhere within the solenoid. Furthermore, this **uniform field** does not depend on the radius of the solenoid, only on the number of windings per unit length. One can, for instance, wind several layers of turns, one on top of the other, to increase n, and each layer will contribute the same field, independent of the radius (as long as the solenoid is truly long).

Problem 7.11. Calculate the magnetic field produced by the solenoid in Fig. 7-16, if the current is 25 A, the radius of the winding is 3 cm, and if there are 700 turns per meter.

Solution

The magnitude of the field is $B = \mu_0 nI = (4\pi \times 10^{-7})(700)(25) = 0.022$ T. The radius did not enter into the calculation. The direction, using the right-hand rule, is to the right.

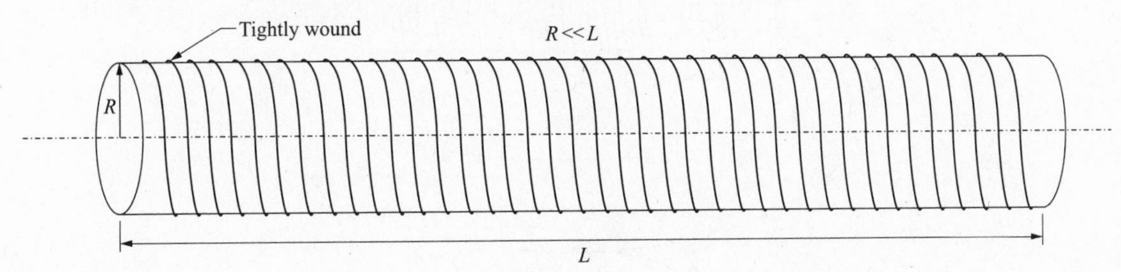

Fig. 7-16

If the solenoid is not infinitely long, but the length is much greater than the radius, then the above result is still nearly true as long as one is not too near to the end of the windings. The field lines inside the solenoid are straight lines, parallel to the axis, until one approaches the ends. Outside the solenoid, the field is no longer zero, and the field lines are as shown in Fig. 7-17. This happens to be the same field line configuration as for a permanent bar magnet, which we have already described at the end of Sec. 6.5 as consisting of many circulating currents. It is therefore not surprising that a bar magnet produces the same field as that of a solenoid.

Note again that **field lines** form closed loops, unlike electric field lines that begin or end at a point charge. The fact that magnetic field lines don't converge to or diverge from a point is a fundamental property of the magnetic field and can be stated as a general law: "**Magnetic field lines** never converge to a point or diverge from a point".

Composite Fields

If several wires each produce magnetic fields, then the actual **magnetic field** at any point is the vector sum of the fields produced by each wire. This is illustrated by the following examples.

Problem 7.12. Calculate the magnetic field produced by the two long parallel wires, each carrying a current of 25 A, which are separated by 0.6 m, as in Fig. 7-18.

(*a*) At point *P*, between the two wires and at a distance of 0.2 m from the first wire.

(*b*) At point *Q*, to the right of the wires, and at a distance of 0.4 m from the second wire.

Solution

(*a*) The magnitude of the field from each wire is $B = (\mu_0/4\pi)(2I/R)$. Therefore,

$$B_1 = 10^{-7}(2)(25)/(0.2) = 2.5 \times 10^{-5}\text{T},$$

and

$$B_2 = 10^{-7}(2)(25)/(0.4) = 1.25 \times 10^{-5}\text{T}.$$

The direction of B_1 is into the paper, whereas the direction of B_2 is out of the paper. Adding the two vectorially, we get that $B = (2.5 - 1.25) \times 10^{-5}$ T into the paper.

(*b*) Here $B_1 = 10^{-7}(2)(25)/(1) = 5 \times 10^{-6}$ T, and $B_2 = 10^{-7}(2)(25)/(0.4) = 1.25 \times 10^{-5}$ T. The direction of both B_1 and B_2 is into the paper. Thus $B = (1.25 + 0.5) \times 10^{-5}$ T $= 1.75 \times 10^{-5}$ T, into the paper.

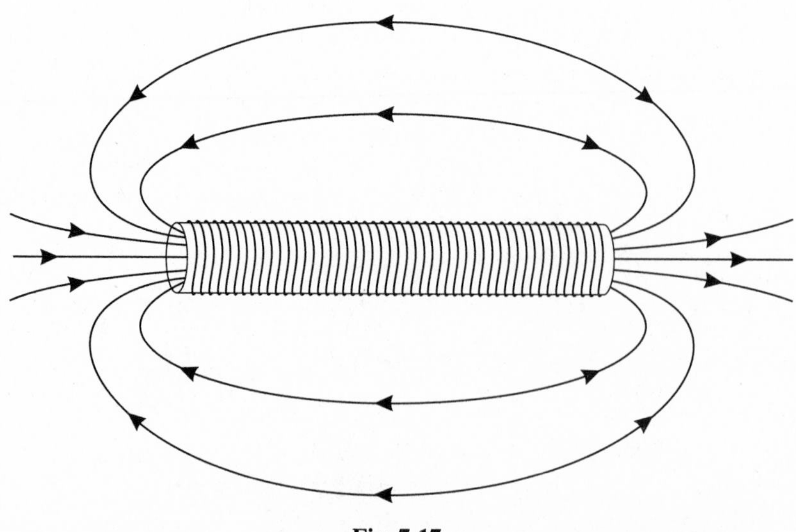

Fig. 7-17

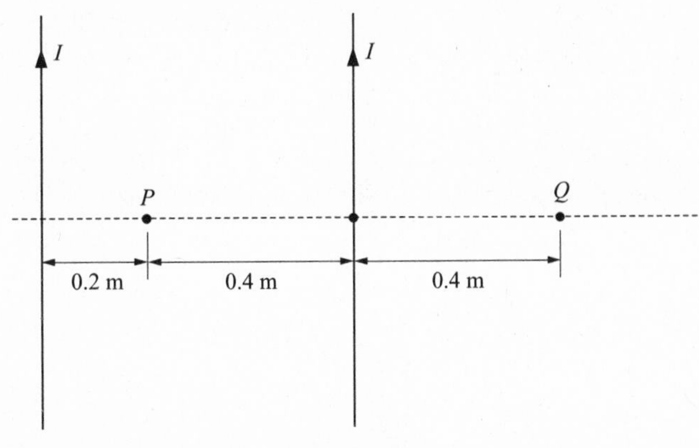

Fig. 7-18

Problem 7.13. Calculate the magnetic field produced by the long parallel wire carrying a current of 25 A, and the ring carrying a current of 2.5 A, at the center of the ring. The center of the ring is 0.5 m from the wire, and the ring has a radius of 0.2 m. The currents are in the direction shown in Fig. 7-19.

Solution

The magnitude of the field from the wire is $B = (\mu_0/4\pi)(2I/R)$. Therefore,

$$B_w = 10^{-7}(2)(25)/(0.5) = 10^{-5}\text{T}.$$

The magnitude of the field from the ring at its center is $B = \mu_0 I/2R$, and therefore

$$B_R = 4\pi \times 10^{-7}(2.5)/(2)(0.2) = 7.9 \times 10^{-6}\text{T}.$$

The direction of B_w is into the paper, whereas the direction of B_R is out of the paper. Adding these together vectorially gives $B = (7.9 - 1) \times 10^{-6} = 6.9 \times 10^{-6}$ T out of the paper.

7.4 AMPERE'S LAW

In Sec. 7.3, we learned how to calculate the field produced by a current. In that formulation, we added together the contributions of the various segments of the wire to get the total field. As we saw, except for the simplest situations, obtaining the field is very difficult. There is a powerful general law relating the magnetic field and the current, which often gives insight into the behavior of the magnetic field, and, in certain circumstances, allows for the compete determination of the field without lengthy calculation. This relationship is given by **Ampere's law**. This mathematical relationship between the current and the magnetic field is similar in spirit to Gauss's law in electrostatics, but quite different mathematically. We will first show the basis for arriving at this law in a very special case, then state the law in general and apply it to calculating the magnetic fields of special current configurations.

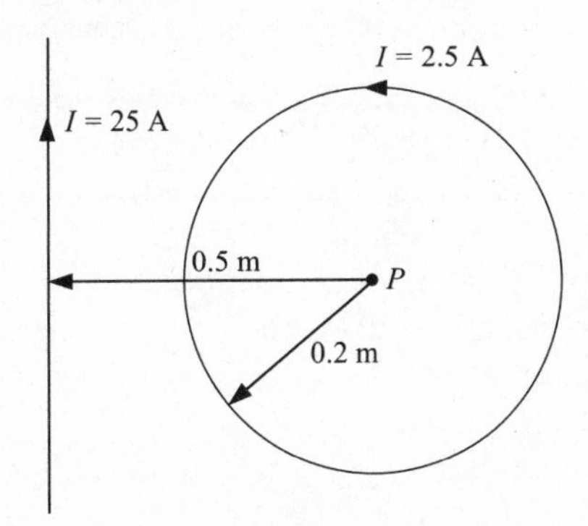

Fig. 7-19

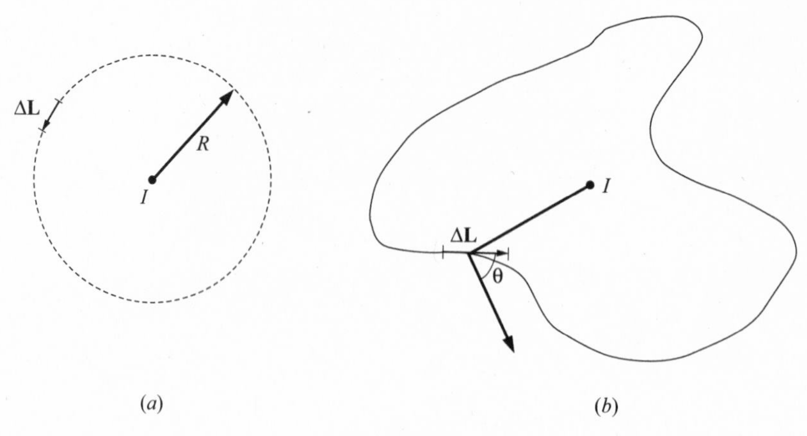

Fig. 7-20

Consider the case of a long straight wire, as drawn in Fig. 7-20(*a*). We have already shown that the magnetic field circles around the wire. If we draw a circular path around the wire as shown, at a radius R, the magnetic field at any point on this wire is the same as at any other point and has the value $B = (\mu_0/4\pi)2I/R$. Furthermore, the direction of $\mathbf{B}$ is tangent to the circle, and parallel to the $\mathbf{\Delta L}$ segments making up the circumference. If we add up all the $B\Delta L$ products around the circle, we get: $[(\mu_0/4\pi)2I/R](2\pi R) = \mu_0 I$. What makes this interesting is the fact that it is generalizable to any shape closed path contour surrounding the wire, such as that of Fig. 7-20(*b*). As shown in this figure, $\mathbf{B}$ is no longer tangent to the contour and indeed varies both in magnitude and in angle to the curve from point to point. Nonetheless, if we multiply the tangential component of $\mathbf{B}$ by $\mathbf{\Delta L}$ for each infinitesimal segment, and add them up around the contour, the result is still $\mu_0 I$. This is analogous to calculating the work done by a variable force as a particle moves along its path (Chapter 6, Section 6.2). We thus calculate the quantity ($B \cos \theta \Delta L$) for each segment of the path, where θ is the angle between $\mathbf{B}$ and $\mathbf{\Delta L}$. As you recall, this is the same as taking the component of $\mathbf{B}$ along $\mathbf{\Delta L}$ at any point, and multiplying by ΔL. The contribution from each segment can be positive or negative, depending on whether θ is less than or greater than 90°. To get Ampere's law, we assume that the $\mathbf{\Delta L}$ directions obey the right-hand rule with thumb pointing in the direction of the current and fingers circling in the direction of the $\mathbf{\Delta L}$'s. If we now add the contributions to ($B \cos \theta \Delta L$) from all the segments of the path, then the sum will still equal $\mu_0 I$. For any path around the wire, as long as it is a closed path, and it is directed by the right-hand rule, we get that the sum of ($B \cos \theta \Delta L$) for the entire closed path will equal $\mu_0 I$. If several currents flow in different wires going through the area enclosed by the path, each will contribute its own $\mu_0 I$, and the sum of ($B \cos \theta \Delta L = B_t \Delta L$) for the entire closed path will equal $\mu_0 I_{\text{total}}$, where I_{total} is the total current flowing through the area enclosed by the path. Using the terminology of the calculus, we say that the *line integral* of $B \cos \theta \Delta L$ around a closed path equals the total current flowing through the area enclosed by that path. This very important result is **Ampere's law**!

$$\lim \Delta L \to 0, \quad \sum_{\text{closed loop}} B_t \Delta L = \mu_0 I_{\text{total}} \tag{7.12}$$

Ampere's law is a very general result, valid in all circumstances of magnetostatics. It depends on the fact that, in contrast to the case of electric field lines in electrostatics, magnetic field lines do not start or end on charges. There are no point sources of magnetic field lines. Instead, magnetic field lines close upon themselves. The amount of magnetic field along these closed paths depends on the enclosed current, in the manner given by Ampere's law.

Note. If we were calculating the work due to a conservative force around a closed loop, we would get zero, in contrast to the non-zero result of Ampere's law. In this regard, B behaves like a non-conservative "force".

In order to be able to use Ampere's law to evaluate magnetic fields, one has to be able to evaluate the sum of ($B \cos \theta \Delta L$) along some closed path. This is usually possible only for cases of special symmetry, where one knows that the field has the same value at every point along the path. In that case, the sum is just equal to the value of B times the length of the path. This is similar to the case of Gauss's law which can be used to evaluate the electric field in cases of special symmetry. We will discuss several such symmetry cases, where Ampere's law is often used to evaluate the magnetic field.

Long Straight Wire

This case was actually used by us in deriving the simplest special case of Ampere's law. We now use Ampere's law to derive the field in this case just to introduce the technique that is generally used in applying Ampere's law.

In Fig. 7-20(*a*), the current in the wire is coming out of the paper. We draw the circle around this wire at a radius R, and will use this circle as the path for adding up all the contributions to ($B \cos \theta \Delta L$). The symmetry of the problem immediately tells us that B has the same value at all points on the path, since each point at the same R looks identical to the wire. Furthermore, there cannot be a radial component to the magnetic field, or a component in or out of the paper, because the magnetic field has to be perpendicular to both the current direction and the displacement from a current element to a point of interest on the circle. Thus B must be everywhere tangent to the circle and $B \cos \theta$ will equal the constant B at every point on the path. The sum of ($B \cos \theta \Delta L$) along the whole circle will therefore be $B(2\pi R)$, and, by Ampere's law, this equals $\mu_0 I$, or

$$B(2\pi R) = \mu_0 I, \qquad \text{or} \qquad B = \mu_0 I/2\pi R,$$

which is the correct result.

Coaxial Cable

A **coaxial cable** consists of an inner solid conductor of radius R_1, carrying current I out of the paper, and an outer, concentric hollow cylinder, of radius R_2, carrying the same current I into the paper, the current being distributed uniformly around the cylinder, as in Fig. 7-21(*a*). We will calculate the magnetic field produced by these currents in the region between the conductors ($R_1 < r < R_2$), and in the region outside both conductors ($r > R_2$). Again, because of the circular symmetry, the field will be the same at all points at the same distance from the wires. Therefore, if we choose as our path a circle of radius r, centered on the axis of the wires, the field will be the same at all points on the path. Also, for the same reason discussed in the previous section the field is confined to the plane of the circle. In

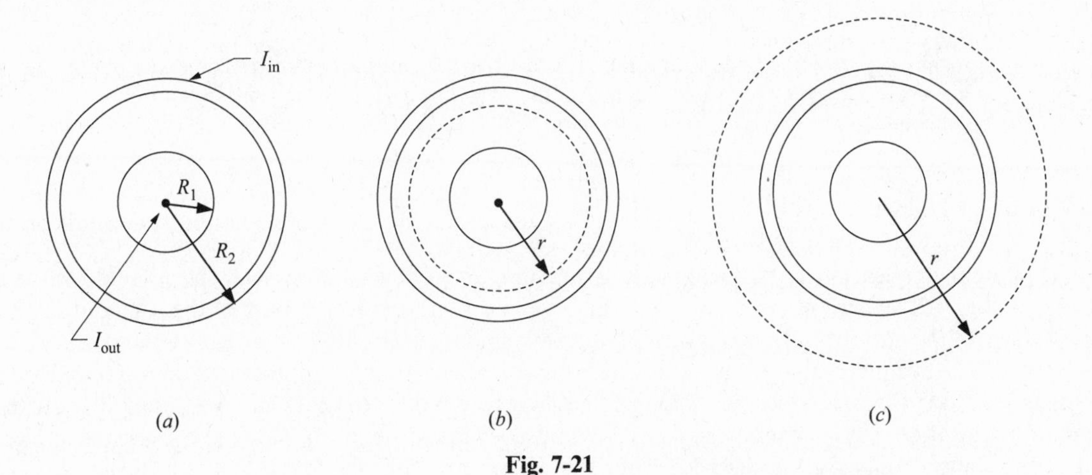

Fig. 7-21

addition the field will not have a radial component since, by symmetry, this would imply field lines meeting at a point at the center of the circle which, as noted earlier, violates a key law of magnetic fields. Thus $\cos\theta = 1$ everywhere along the circle.

To evaluate the magnetic field between the wires, we draw a circular path of radius r between the wires ($R_1 < r < R_2$), as in Fig. 7-21(*b*). Evaluating ($B\cos\theta\Delta L$) along this path gives $B(2\pi r)$, which must equal $\mu_0 I_{total}$, where I_{total} is the current flowing through the area enclosed by the path. Since $r < R_2$, the only current flowing through this area is that in the inner conductor, which is I. This, once again, gives that $B = \mu_0 I/2\pi r$.

To evaluate the field outside of both cylinders, we draw a circular path with radius greater the $R_2(r > R_2)$, as in Fig. 7-21(*c*). Again, the sum of ($B\cos\theta\Delta L$) along this path gives $B(2\pi r)$, which must equal $\mu_0 I_{total}$, where I_{total} is the current flowing through the area enclosed by the path. Now, however, the circular path encloses both cylinders, and both currents flow through the enclosed area. Since the currents flow in opposite directions, the total current will be zero. Therefore, we find that $B = 0$ in this region.

Problem 7.14. A coaxial cable consists of an inner conductor of radius 0.02 m and a thin outer conductor of radius 0.06 m. What current is needed in this cable to produce a magnetic field of 10^{-5} T at a point located at a distance of 0.03 m from the axis of the cable?

Solution

Since this point is located between the cylinders ($0.02 < 0.03 < 0.06$), we use the formula derived for that case $B = \mu_0 I/2\pi r$. Thus $B = (4\pi \times 10^{-7})(I)/2\pi(0.03)$, or $10^{-6} = 2 \times 10^{-7}(I)/0.03$, which gives $I = 0.15$ A.

Problem 7.15. A special coaxial cable consists of an inner conductor of radius 0.02 m and an outer conductor of radius 0.06 m. The inner conductor carries a current of 2 A out of the paper, while the outer conductor carries a current of only 1.5 A into the paper. What is the magnetic field at (*a*) a point between the cylinders, at a distance r from the axis; and (*b*) a point outside both cylinders at a distance r from the axis?

Solution

Since the two currents are not equal in magnitude, we cannot just use the previous result. Instead, we have to start with Ampere's law and apply it to this case. We still have the circular symmetry, and can use a circular path to evaluate the sum of ($B\cos\theta\Delta L$).

(*a*) For this case we use a circle with radius between 0.02 m and 0.06 m, with radius r. The evaluation of the sum is identical to our previous case. The I_{total} in this case is just the current in the inner conductor. This results in $B = \mu_0 I/2\pi r = 4 \times 10^{-7}/r$.

(*b*) For this case we use a circle with radius greater than 0.06 m, with radius r. The evaluation of the sum is identical to our previous case. The I_{total} in this case, however, is the current in both the inner and the outer conductor. The total current is therefore (2.0 − 1.5)A. This results in $B = \mu_0 I_{total}/2\pi r = 10^{-7}/r$.

Long Solenoid

A different application of Ampere's law is in the case of a very long **solenoid**, pictured in cross-section in Fig. 7-22, where we assume $R \ll L$, and we are interested in the part of the solenoid far from the ends. Here the current in the coils circling the solenoid, I, is coming out at the top and going in at the bottom. The symmetry that we have for this long solenoid requires that the field does not depend on the distance along the axis, since, for a long solenoid, every point along its length is identical. Furthermore, if one draws a circle around the axis, the field does not depend on where one is on the circle. In fact it can depend only on the distance from the axis, r (we will see that it actually turns out to be

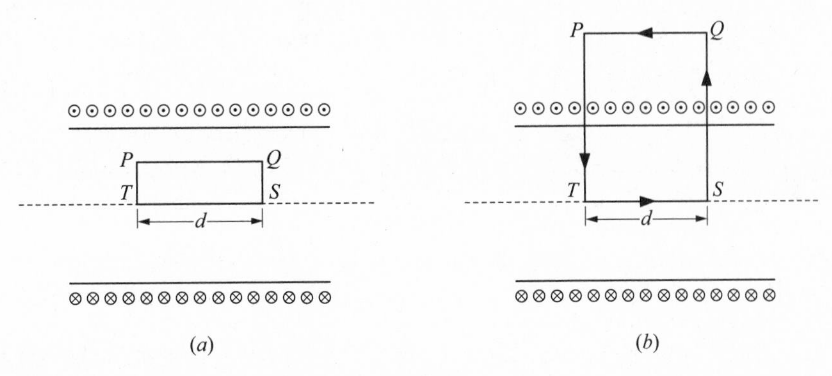

Fig. 7-22

independent of r as well). At any point P, at a distance r from the axis, the field cannot have a radial component for the same reason mentioned when we discussed a long straight wire. Similarly it cannot have a component tangent to the circle since, by symmetry, that would be the same everywhere along the circle and Ampere's law would give a non-zero result even though no current flows through the circle. It has only a component along the direction of the axis. Thus the field lines within the solenoid are parallel to the axis of the solenoid.

Problem 7.16. Use Ampere's law to show that the magnetic field anywhere within the solenoid is the same as that along the axis of the solenoid, and therefore is uniform.

Solution

To show that the field is uniform, we choose a rectangular path $PQST$ shown in Fig. 7-22(*a*), where TS is along the axis. Since there is no radial component of the field, then, along the segments TP and QS, there is no component of B along the path, and the contribution from these segments will be zero. Along the segments PQ and ST, the field is parallel to the path, let us assume to the right. Then for segment PQ the contribution to ($B \cos \theta \Delta L$) will be $B_r d$, where B_r is the field at a distance r from the axis, and d is the length of the path PQ. For the segment ST, the contribution will $-B_0 d$, where B_0 is the field at the axis, and the minus sign comes from the fact that we are moving along that path in a direction opposite to the field. The sum over the entire path is therefore $(B_r - B_0)d$, which, by Ampere's law must equal $\mu_0 I_{\text{total}}$, where I_{total} is the current going through the area of $PQST$. There is no current going through this area, since the only current in the problem exists in the wires circling the solenoid. Thus,

$$(B_r - B_0)d = 0, \qquad \text{and} \qquad B_r = B_0.$$

This shows that at all r within the solenoid the field is the same and equal to the value on the axis.

Problem 7.17. Use Ampere's law to calculate the value of the uniform field within the solenoid.

Solution

To calculate the magnitude of the field, we note that the field *outside* the solenoid will be very small as we go far away from the solenoid. We choose a path $PQST$, shown in Fig. 7-22(*b*), which has current flowing through it in the direction out of the paper due to the coil lines at the top of the solenoid. We choose the segment PQ to be very far from the solenoid, so the contribution from that segment will be zero. The direction around the loop is chosen counter-clockwise by the right-hand rule. Again, the contribution from TP and QS will be zero, and the only segment contributing a non-zero value is ST. From this segment we get $B_0 d$ as in the previous paragraphs. This must equal $\mu_0 I_{\text{total}}$, where I_{total} is the current flowing through the area of $TPRS$. The total current equals the number of wires between U and V times I, the current in each wire. This is ndI, where n is the number of turns per meter. We therefore conclude that

$$B_0 d = \mu_0 ndI, \qquad \text{or} \qquad B_0 = \mu_0 nI,$$

which is the result we quoted in Sec. 7.3.

Problem 7.18. A solenoid is made from 2000 windings on a length of 2 m. The radius of the solenoid is 0.3 m. If the windings carry a current of 3 A, what is the magnetic field near the middle of the solenoid?

Solution

Since the length of the solenoid is much greater than the radius, it can be considered to be a long solenoid when one calculates the field far from the ends of the solenoid. Then the field is $B = \mu_0 nI = 4\pi \times 10^{-7}(2000/2)(3) = 3.77 \times 10^{-3}$T.

Toroidal Solenoid

A **toroidal solenoid** consists of wires wound around a toroid, which is a doughnut shape usually with a circular cross-section, as in Fig. 7-23. The mean radius of the toroid is r. We assume that the current flows into the paper on the outside of the toroid, and out of the paper on the inside of the toroid. In order to use Ampere's law, we draw a circular path through the toroid at its mean radius, r. We will go around this circle in the counter clockwise direction, since the positive direction for current going through the area of this circle is out of the paper. Every point on this path is identical to any other point on the path, so the magnetic field along this direction will not vary as we move along the path. Furthermore, by arguments similar to those for the long solenoid, there is no component of B in the plane of any cross-section perpendicular to the solenoid. When we add all the contributions to the sum, we get $B(2\pi r)$, where B is the component of the field along the path at the radius r. This must equal μ_0 times the current through the area. The only current in the problem is the current in the wires wound around the toroid. Only the wires on the inside of the toroid go through the area of our path, so the total current through the path is NI, where N is the total number of windings. Furthermore, this current is positive, since it is coming out of the paper, which is our positive direction. Equating the sum to the total current gives

$$B(2\pi r) = \mu_0 NI, \qquad \text{or} \qquad B = \mu_0 NI/(2\pi r)$$

This is the field within the toroid at a point located at a distance r from the axis of the toroid. Note that $N/2\pi r$ is the number of turns per unit length, if the length is measured at the center of the ring. For a case where the radius of the cross-sectional area of the toroid is much less than the mean radius of the toroid, the toroid is nearly like a long solenoid, and the formula for the magnetic field is identical with the one for the solenoid.

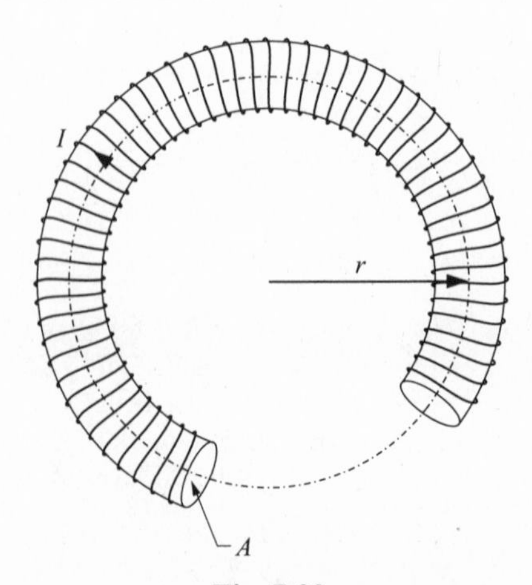

Fig. 7-23

Problem 7.19. A toroid has a mean radius of 1.5 m, and a circular cross section with a radius of 5 cm. There are 3000 windings on the toroid, carrying a current of 2 A. What is the magnetic field at (*a*) the center of the cross-section; and (*b*) within the toroid, just at the inner edge?

Solution

(*a*) The field is given by $B = \mu_0 NI/(2\pi r) = 4\pi \times 10^{-7}(3000)(2)/(2\pi)(1.5) = 9 \times 10^4$T.

(*b*) Here, the radius of the path is $(1.5 - 0.05)$ m $= 1.45$ m, so that $B = 9.3 \times 10^{-4}$T. This does not differ too much from the field at the center since the radius of the cross-section is small compared to the mean radius of the toroid.

This case of the toroid illustrates the usefulness of Ampere's law for calculating the magnetic field, since, for this case, any other kind of calculation would be very difficult.

Problems for Review and Mind Stretching

Problem 7.20. In Fig. 7-24, a charge of -6×10^{-4} C is moving up (in the $+y$ direction) with a velocity of 3×10^6 m/s. The charge is instantaneously at the point on the y axis at a distance of 1.5 m from the origin. What magnetic field does this charge produce (*a*) on the positive x axis, at a distance of 2 m from the origin; and (*b*) on the positive z axis, at a distance of 3.6 m from the origin?

Solution

(*a*) The formula for the magnitude of the field is

$$|B| = (\mu_0/4\pi)|qv \sin \phi/r^2| \tag{7.1}$$

Here $r = (1.5^2 + 2^2)^{1/2} = 2.5$, and $\sin \phi = \sin(180° - \phi) = 2/2.5 = 0.8$. Thus

$$B = 10^{-7}(6 \times 10^{-4})(3 \times 10^6)(0.8)/(2.5)^2 = 2.3 \times 10^{-5}\text{T}$$

The direction is obtained by drawing a circle about the y axis (the line of **v**), going through the point, as in the figure. The tangent to the point on the x axis is in the $\pm z$ direction. Using the right-hand rule (thumb along **v** and the fingers curl around the circle), the direction is in $-z$, *if* q were positive. Since q is negative, the correct direction is in $+z$.

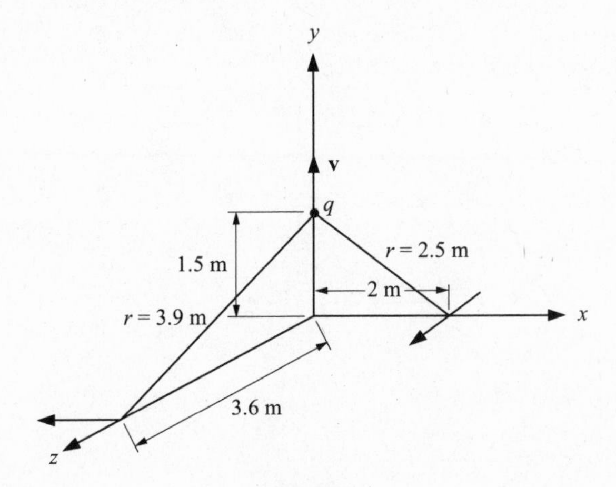

Fig. 7-24

(*b*) Here $r = (1.5^2 + 3.6^2)^{1/2} = 3.9$ m, and $\sin\phi = \sin(180° - \phi) = 3.6/3.9 = 0.923$. Thus

$$B = 10^{-7}(6 \times 10^{-4})(3 \times 10^6)(0.923)/(3.9)^2 = 1.1 \times 10^{-5}\text{T}$$

The direction is obtained by drawing a circle about the y axis (the line of **v**), going through the point, as in the figure. The tangent to the point on the z axis is in the $\pm x$ direction. Using the right-hand rule (thumb along **v** and the fingers curl around the circle), the direction is in $+x$, *if* q were positive. Since q is negative, the correct direction is in $-x$.

Problem 7.21. In Problem 7.4, calculate the electric force between the two charges, and compare it with the magnetic force between the same charges.

Solution

In Problem 7.4, we calculated the force between two charges, q_1 and q_2, moving parallel to each other with velocities v_1 and v_2, and separated by a distance d. The result was an attractive force of

$$|F_{\text{mag}}| = (\mu_0/4\pi)q_1 q_2 v_1 v_2/d^2 = 10^{-7} q_1 q_2 v_1 v_2/d^2 \qquad (i)$$

To calculate the electric force, we use Eq. (*3.1*),

$$|F_{\text{elec}}| = [1/(4\pi\varepsilon_0)]q_1 q_2/d^2 = 9 \times 10^9 q_1 q_2/d^2 \qquad (ii)$$

The ratio of these forces $F_{\text{mag}}/F_{\text{elec}} = 10^{-7}\, v_1 v_2/9 \times 10^9 = v_1 v_2/(3 \times 10^8)^2$. This is a small number, unless v_1 and v_2 are comparable to 3×10^8 m/s, which is the speed of light.

Problem 7.22. Two identical coils, each having 1000 turns, are separated by a distance of 1.6 m along a common axis, as in Fig. 7-25. For each coil $R = 1.5$ m and $I = 2$ A. The currents flow in *opposite* directions in the two coils, as shown. Calculate the field (*a*) at a point P_1 on the axis midway between the coils; and (*b*) at the center P_2 of the first coil.

Solution

(*a*) For each coil, we can use Eq. (*7.10*) multiplying by N for the number of turns. Therefore, for each coil, $B = 10^3\,(4\pi \times 10^{-7})(2)(1.5)^2/2(1.5^2 + 0.8^2)^{3/2} = 5.8 \times 10^{-4}$ T. The direction is to the right for the field from the first coil and to the left for the field from the second coil. For both coils together, the field is therefore zero.

(*b*) For the first coil, $B = 10^3\,(4\pi \times 10^{-7})(2)/2(1.5) = 8.38 \times 10^{-4}$ T, and points to the right. For the second coil, $B = 10^3(4\pi \times 10^{-7})(2)(1.5)^2/2(1.5^2 + 1.6^2)^{3/2} = 2.68 \times 10^{-4}$ T. The direction of this field is to the left. Therefore, the total field is $(8.38 - 2.68) \times 10^{-4} = 5.7 \times 10^{-4}$ T, to the right.

Problem 7.23. Two long, straight, parallel wires carry the same current, I, in the same direction. Calculate the force on a one meter length of the second wire due to the magnetic field produced by the first wire, if the wires are separated by a distance of 1 m.

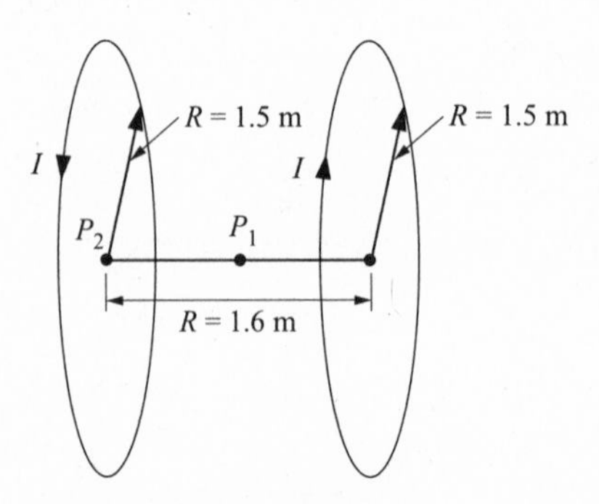

Fig. 7-25

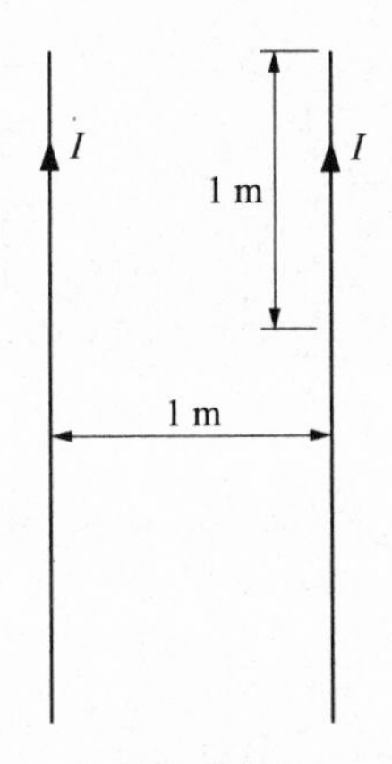

Fig. 7-26

Solution

We draw the situation in Fig. 7-26. The first wire produces a magnetic field of

$$B = (\mu_0/4\pi)2I/R = 2 \times 10^{-7}I, \text{ and the direction is into the paper} \qquad (7.11)$$

The force on the length ΔL $(=1)$ of the second wire is $I\Delta LB = 2 \times 10^{-7}I^2$, and the direction is toward the first wire. This is actually the way we define the unit of current (ampere), and from the ampere we define the unit of charge (coulomb). The ampere is defined as the current needed in this setup so that a force of 2×10^{-7} N is exerted on a 1 m length of the second wire.

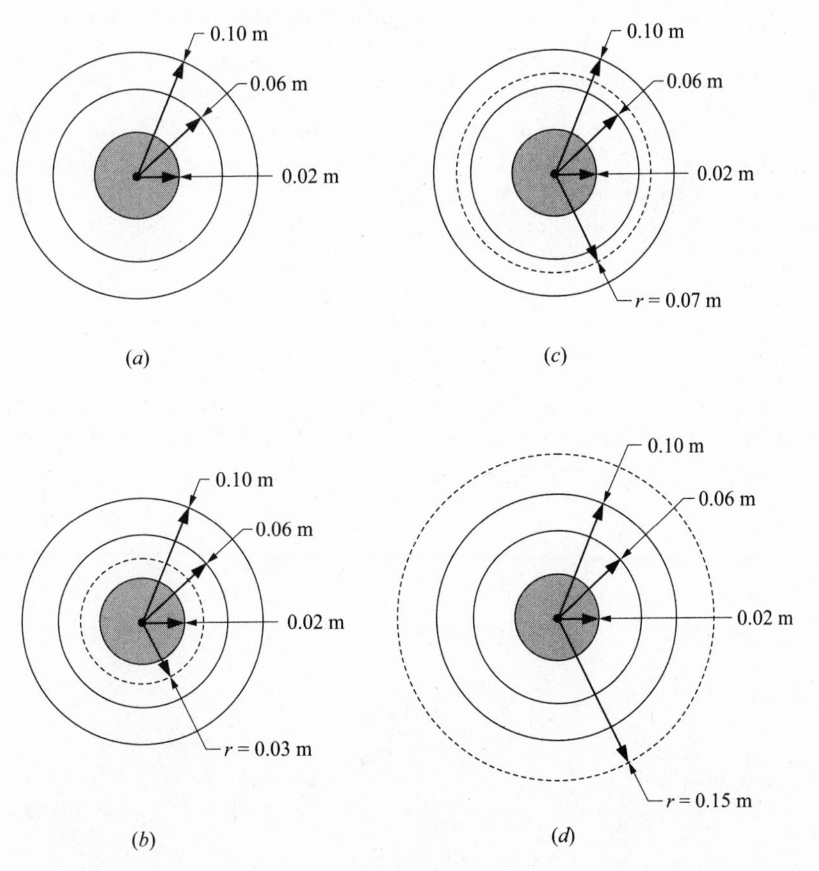

Fig. 7-27

Problem 7.24. A special coaxial cable consists of three concentric cylinders, as in Fig. 7-27(*a*). The inner cylinder is a solid conductor, of radius 0.02 m. The outer two cylinders are thin conducting hollow cylinders, with radii of 0.06 and 0.10 m, respectively. The inner cylinder carries a current of 2 A out of the paper, the second carries a uniformly distributed current of 1.5 A into the paper, and the third carries a uniformly distributed current of 0.5 A into the paper. Calculate the magnetic field produced at (*a*) $r = 0.03$ m; (*b*) $r = 0.07$ m; and (*c*) $r = 0.15$ m.

Solution

(*a*) We draw a circular path at $r = 0.03$ m, as in Fig. 7-27(*b*). This circle lies between the two inner conductors. The sum along the path gives $2\pi(0.03)B$, which we equate to $\mu_0 I_{\text{total}}$. Only the inner conductor carries current through the area of the path, so that $I_{\text{total}} = 2$ A. Thus $B = 4\pi \times 10^{-7}(2)/(2\pi)(0.03) = 1.33 \times 10^{-5}$ T, and points counter-clockwise about the symmetry axis.

(*b*) We draw a circular path at $r = 0.07$ m, as in Fig. 7-27(*c*). This circle lies between the two outer conductors. Again, the sum along the path, $2\pi(0.07)B$, equals $\mu_0 I_{\text{total}}$, where I_{total} is now the current in the two innermost conductors. This current equals $(2 - 1.5)$ A out of the paper. Therefore, $B = 4\pi \times 10^{-7}(0.5)/(2\pi)(0.07) = 1.41 \times 10^{-6}$ T.

(*c*) We draw a circular path at $r = 0.15$ m, as in Fig. 7-27(*d*). This circle lies outside of all the conductors. Again, the sum along the path, $2\pi(0.15)B$, equals $\mu_0 I_{\text{total}}$, where I_{total} is now the current in all three conductors. This current equals $(2 - 1.5 - 0.5) = 0$. Therefore, $B = 0$.

Supplementary Problems

Problem 7.25. A charge of 1.7×10^{-3} C is moving north with a velocity of 3×10^5 m/s. What magnetic field (magnitude and direction) does it produce at a point due east which is 1.2×10^{-2} m away?

Ans. 0.35 T, vertically down

Problem 7.26. A charge of -1.7×10^{-3} C is moving south with a velocity of 3×10^5 m/s. What magnetic field (magnitude and direction) does it produce at a point due west which is 1.2×10^{-2} m away?

Ans. 0.35 T, vertically up

Problem 7.27. A charge of -1.7×10^{-3} C is moving west with a velocity of 3×10^5 m/s. What magnetic field (magnitude and direction) does it produce at a point, *P*, which is reached by going north 1.2×10^{-2} m and then west by 0.9×10^{-2} m? (See Fig. 7-28.)

Ans. 0.18 T, vertically up

Problem 7.28. An elevator, carrying a charge of 0.2 C, is moving down with a velocity of 4×10^3 m/s. The elevator is 10 m from the bottom and 3 m horizontally from point *P* in Fig. 7-29. What magnetic field does it produce at point *P*?

Ans. 2.1×10^{-5} T, out

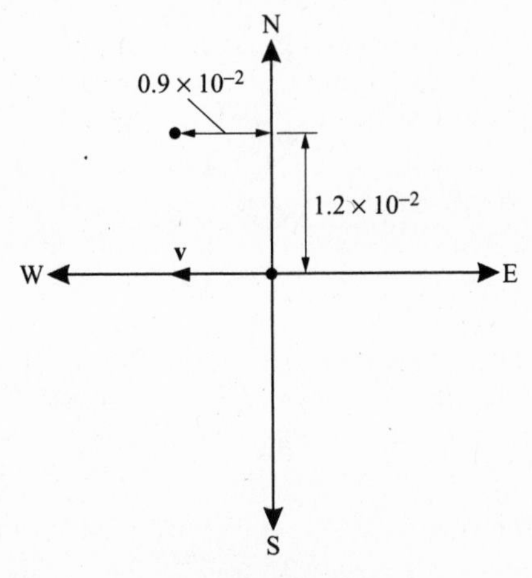

Fig. 7-28

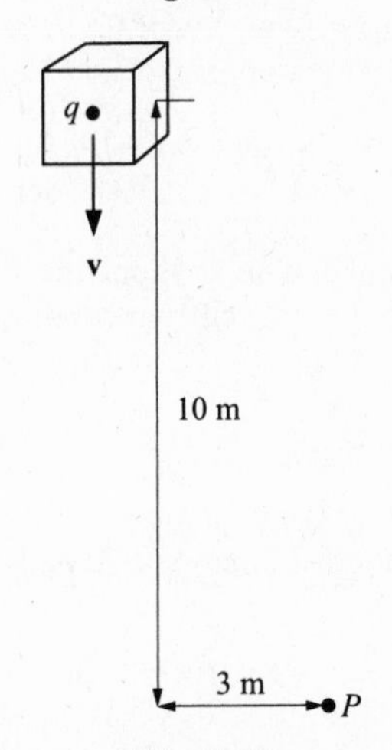

Fig. 7-29

Problem 7.29. One charged particle of -1.3×10^{-6} C is moving north with a velocity of 5×10^6 m/s. Another charged particle, of -2×10^{-6} C is moving south, on a parallel path, with a velocity of 3×10^6 m/s, at a distance of 0.11 m (see Fig. 7-30). What force is exerted between the particles?

Ans. 3.22×10^{-4} N, repulsion

Problem 7.30. A magnetic field at the center of a ring of radius 0.6 m, due to the current in the ring, is 1.2×10^{-4} T. If there are 175 turns in the ring, what current is flowing in the ring?

Ans. 0.65 A

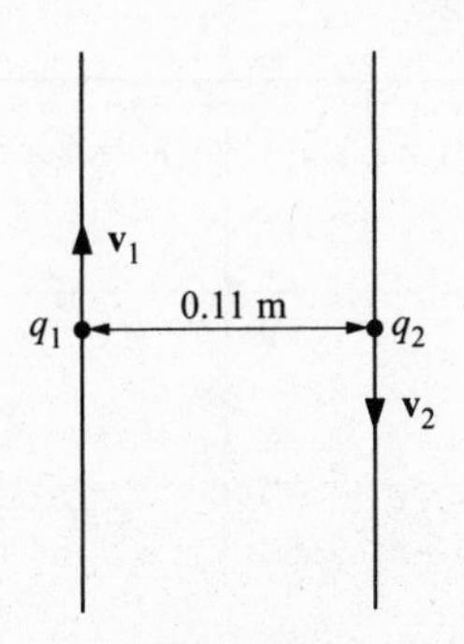

Fig. 7-30

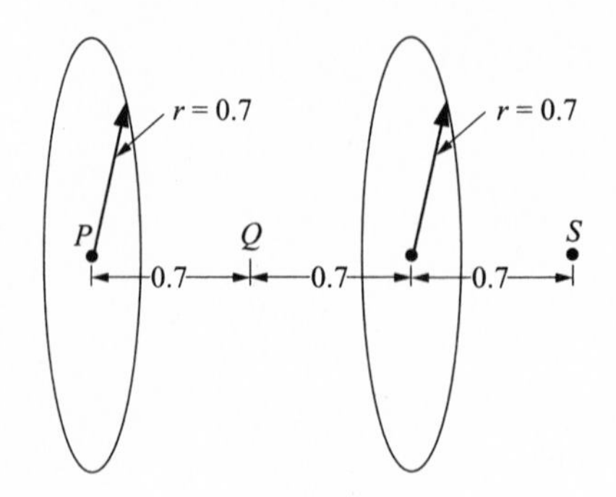

Fig. 7-31

Problem 7.31. Two identical coils, of radius 0.7 m, and having 1200 turns are parallel to each other on the same common axis, as in Fig. 7-31. They each carry a current of 0.8 A in the direction shown. Calculate the magnetic field produced by the two coils (*a*) at point P; (*b*) at point Q; and (*c*) at point S.

Ans. (*a*) 9.38×10^{-4} T; (*b*) 6.09×10^{-4} T; (*c*) 3.32×10^{-4} T

Problem 7.32. An overhead electric transmission line, supplying current to the houses on the street, carries a current of 2000 A. What is the magnetic field that this current produces on the street, 4 m below the line?

Ans. 10^{-4} T

Problem 7.33. A long wire carries current into the paper at the center of a rectangle of sides 6 m × 8 m (Fig. 7-32). The current in the wire is 6 A. What magnetic field (magnitude and direction) is produced at (*a*) point P; (*b*) point Q; and (*c*) point S?

Ans. (*a*) 4×10^{-7} T, in $-x$ direction; (*b*) 3×10^{-7} T, in $-y$ direction; (*c*) 2.4×10^{-7} T, at an angle of 53° below the $-x$ direction

Problem 7.34. Two long wires carry currents of 1.2 A into the paper. The wires are 0.2 m apart, as in Fig. 7-33. Calculate the magnetic field (magnitude and direction) that the wires produce at (*a*) point P; (*b*) point Q; and (*c*) point S.

Ans. (*a*) 0; (*b*) 3.2×10^{-6} T, in $-y$ direction; (*c*) 2.4×10^{-6} T, in $+x$ direction

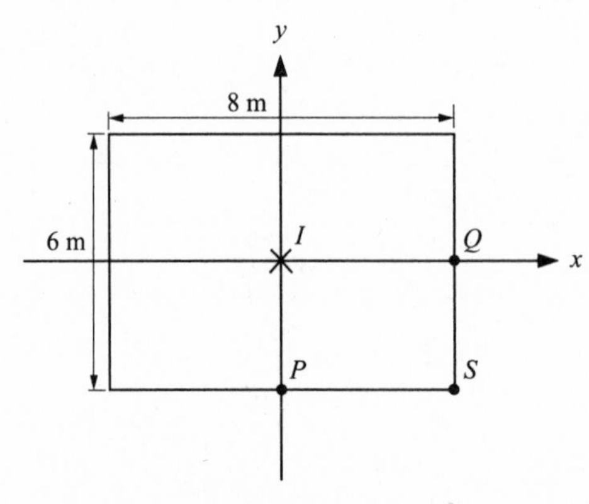

Fig. 7-32

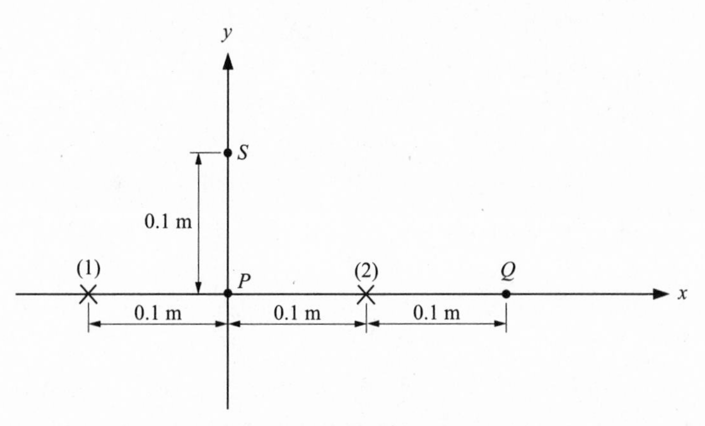

Fig. 7-33

Problem 7.35. Two long wires carry currents of 1.2 A, the first into the paper, and the second out of the paper. The wires are 0.2 m apart, as in Fig. 7-34. Calculate the magnetic field (magnitude and direction) that the wires produce at (*a*) point P; (*b*) point Q; and (*c*) point S.

Ans. (*a*) 4.8×10^{-6} T, in $-y$ direction; (*b*) 1.6×10^{-6} T, in $+y$ direction; (*c*) 2.4×10^{-6} T, in $-y$ direction

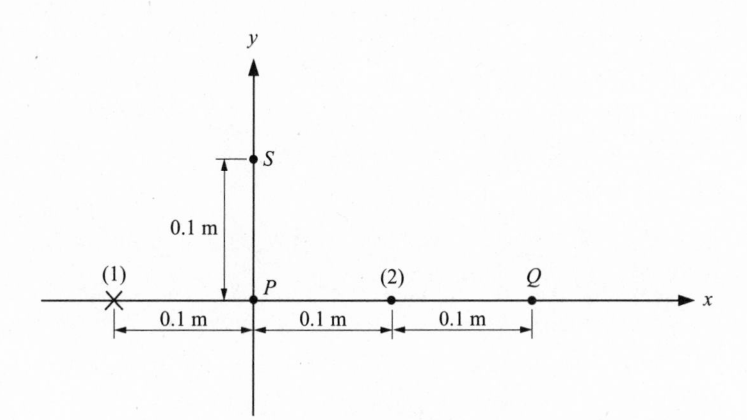

Fig. 7-34

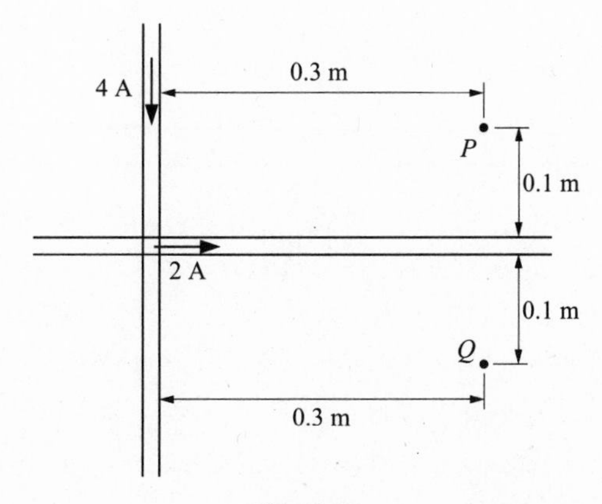

Fig. 7-35

Problem 7.36. Two perpendicular wires carry currents of 2 A and 4 A, respectively, as in Fig. 7-35. Calculate the magnetic field (magnitude and direction) at points P and Q from the two wires.

Ans. 6.67×10^{-6} T out for P; 1.33×10^{-6} T in for Q

Problem 7.37. A long solenoid is made by winding 1500 turns per meter on a radius of 0.3 m, and a second winding of 3500 windings per meter on a radius of 0.5 m, as in Fig. 7-36. Each winding carries a current of 2 A, and the direction of the current in the inner winding is shown on the figure.

(*a*) Calculate the field inside the inner coil if (*i*) the currents flow in the same direction in both windings; and (*ii*) the currents flow in opposite directions in both windings.

(*b*) Calculate the field in the region between the windings if (*i*) the currents flow in the same direction in both windings; and (*ii*) the currents flow in opposite directions in both windings.

Ans. (*a*) (*i*) 12.57×10^{-3} T to the left; (*ii*) 5.03×10^{-3} T to the right; (*b*) (*i*) 8.80×10^{-3} T to the left; (*ii*) 8.80×10^{-3} T to the right

Problem 7.38. A long solenoid has a length of 7 m and has 8400 windings on it. The field inside is 2×10^{-3} T. What current is flowing in the windings?

Ans. 1.33 A

Problem 7.39. A coaxial cable consists of a long, solid inner cylinder of radius 0.02 m and a long, concentric, hollow, conducting cylinder of inner radius 0.08 m. The current is 5 A in the opposite direction in the two conductors. What is the magnetic field at a radius of 0.03 m?

Ans. 3.33×10^{-5} T

Problem 7.40. A coaxial cable consists of a long, solid inner cylinder of radius 0.02 m and a long, concentric, hollow, conducting cylinder of inner radius 0.08 m. The current is 5 A in the *same* direction in the two conductors.

(*a*) What is the magnetic field at a radius of 0.03 m?

(*b*) What is the magnetic field outside of both conductors, at a radius of 0.10 m?

Ans. (*a*) 3.33×10^{-5} T; (*b*) 2×10^{-5} T

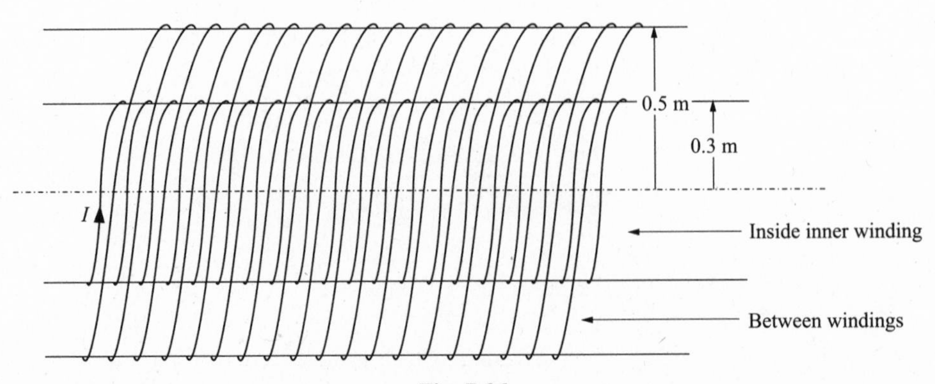

Fig. 7-36

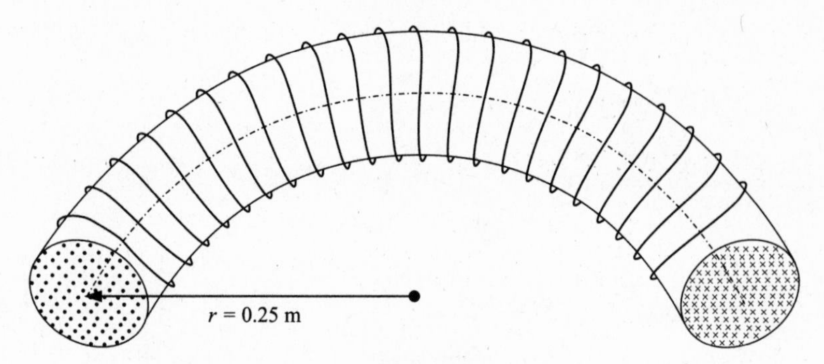

Fig. 7-37

Problem 7.41. A toroidal solenoid has a mean radius of 0.25 m. There are 800 windings around the solenoid, producing a magnetic field of 6×10^{-6} T at that radius. Figure 7-37 shows a cross-sectional view taken at the center of the toroid. The field is coming out at the left side, and going in at the right side. What current flows in the wire, and is the flow clockwise or counter-clockwise at the left?

Ans. 9.4×10^{-3} A, counter clockwise

Problem 7.42. A toroidal solenoid has a rectangular cross-section, of 0.01 m × 0.02 m, as in Fig. 7-38. The mean diameter of the toroid is 2.3 m. The current in the 7500 windings is 3 A, and flows clockwise around the left cross-section, as shown. Calculate the magnetic field (including the direction) at (*a*) the center of the rectangle (point P); and (*b*) the outer edge of the rectangle (point Q).

Ans. (*a*) 3.91×10^{-3} T, in at the left; (*b*) 3.88×10^{-3} T, in at the left

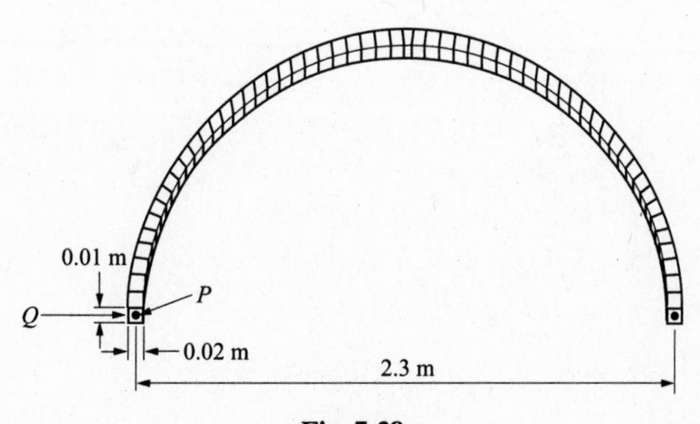

Fig. 7-38

Problem 7.43. What is the *direction* of the magnetic field at point P for the current configurations shown in Fig. 7-39?

Ans. to the right in all cases

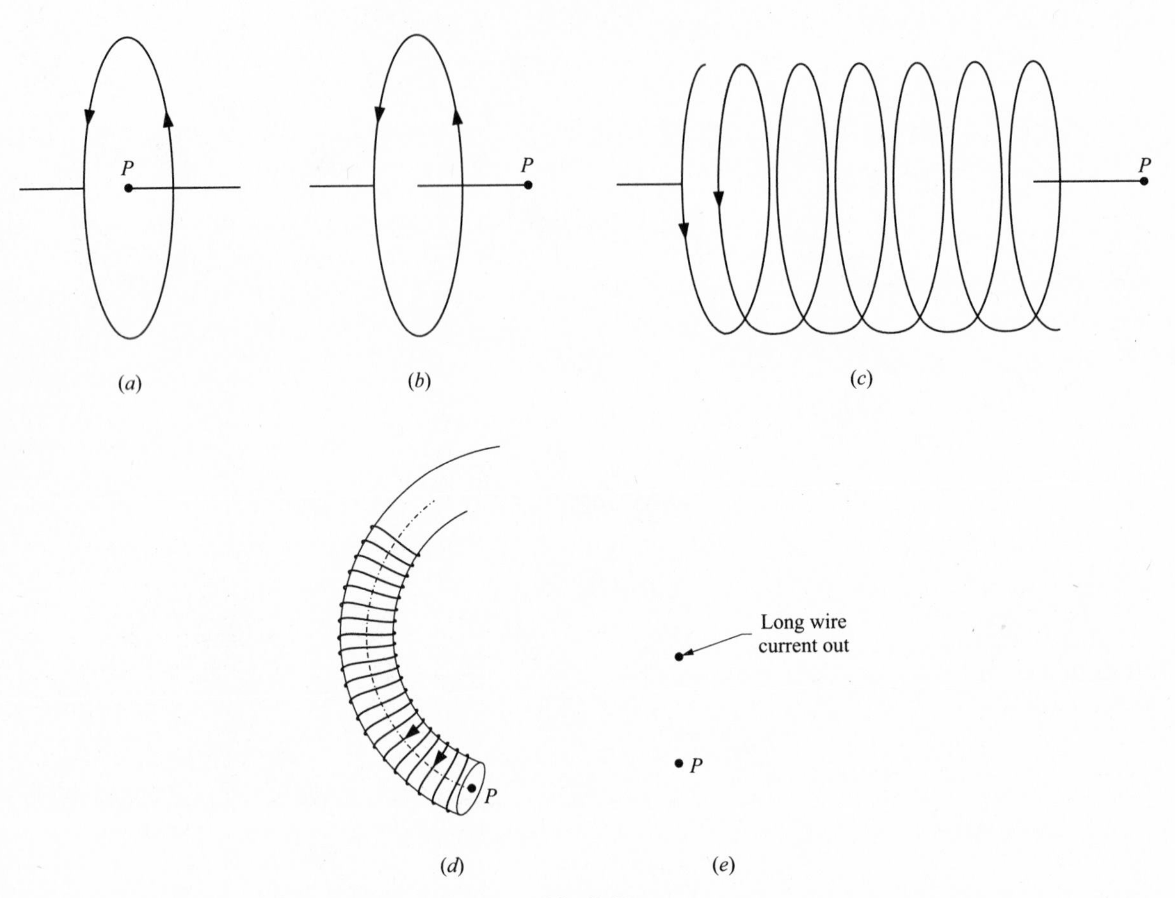

Fig. 7-39

Chapter 8

Magnetic Properties of Matter

8.1 INTRODUCTION

In the previous chapter, we learned about the production of magnetic fields in space, due to moving charges and due to currents. We did not include the possibility that this **magnetic field** could change the properties of the material in which it was created, and result in having the material produce its own field, which has to be added to the original field. We had a similar situation in the case of electric fields, where the electric field induced a polarization of the material, which, in turn, produced its own electric field. Let us try to understand how a magnetic field can modify the properties of a material and induce a "**magnetization**" of the material to produce its own field.

All material consist of a collection of atoms and molecules. These atoms, in turn, consist of positively and negatively charged particles, held together mainly by electric forces. In order for a magnetic field to have any effect on these particles, the particles must be moving, since a magnetic force is exerted only on *moving* particles. There are two types of motion that exist for these particles, that we classify as orbital motion and spin. The **orbital motion** can be thought of as the motion that occurs when an electron circulates about a nucleus, similar to the way a planet circulates about the sun. Actually, with our present knowledge of quantum mechanics, we no longer consider the picture of electrons moving in orbits about the nucleus as being accurate, but such a picture still provides an intuitive guide that is useful and we will explore it. The circulating or orbiting electron has angular momentum and, indeed, even in quantum mechanics a certain amount of angular momentum is carried by the moving electrons, creating effective current loops. In many atoms, the angular momentum of the electrons averages to zero as does the current, while in others there is a net angular momentum and a net current. In addition to orbiting the nucleus, the electrons **spin** on their axes and, as a consequence, have an additional "spin" angular momentum and "spin" electric current loops. As with orbital motion, in many atoms the spin angular momentum of the various electrons averages to zero, while in others there is a net angular momentum and a net current. In either case, orbital or spin, magnetic fields can be set up by the atomic current loops, and an external magnetic field can exert forces on the electrons, and thereby modify their motion. It can be shown that, because of the concept of magnetic induction and Lenz's law, which will be discussed in the next chapter, one effect of an external magnetic field is to induce currents and associated magnetic moments in the atoms of a material. These magnetic moments, in turn, produce their own magnetic fields, which, by Lenz's law, are in a direction opposite to the original field. The materials in which this is the dominant effect are called ***di*amagnetic** materials, in the same manner as materials that produce electric fields opposed to the original electric field are *di*electric materials. In general, such induced magnetic fields in an atom are very small and the external field is reduced by a tiny amount as a consequence of diamagnetism. While diamagnetism occurs in all atoms, it is often overshadowed by another effect of the external magnetic field on the electrons of an atom. This other effect only occurs in atoms in which there is a net orbital and/or spin angular momentum and a net effective current loop for the atom. Such an "effective" current loop gives the atom an definite overall magnetic dipole moment. As we learned in the previous chapter an external magnetic field exerts a torque on such a magnetic moment and the torque tries to line up the moment parallel to the magnetic field. The lined up moments will then produce their own magnetic field in the same direction as the original field, thus increasing the magnetic field. Materials in which this is the dominant effect are called **paramagnetic** materials. They are more common than diamagnetic materials, especially since they dominate in materials where both effects are present. Thus, while diamagnetism is present in all atoms, it dominates only in those atoms in which the orbital and spin angular moments average out to zero. In other atoms there is a permanent magnetic dipole moment set up by the net current loops and paramagnetism is dominant. Even so, while paramagnetism is stronger than diamagnetism, it is still a relatively small effect. This is because only some of the atoms with permanent moments line up with the external magnetic field. Thermal effects tend to fight against the alignment and assure that the net increase in magnetic field is not too strong.

There is an important exception to this limitation. In some materials the field produced by the moment on one atom is strong enough to rotate the moment on another, neighboring atom and cause it to align itself with the first atom. This second atom, in turn, causes its neighbor to align its moment parallel with its own moment, and this can continue from atom to atom. In this manner, even in the absence of an external magnetic field, the magnetic moments will not be randomly aligned, but rather large groups of atoms will all align themselves together in a certain direction. These aligned regions are called magnetic domains and are huge on the atomic scale but still microscopic on the human dimension. Usually there are many microscopic domains in a material and the direction of alignment of the various domains is random. Then the average magnetization of the entire material will still be zero in the absence of an external field, but will consist of many regions that are locally magnetized. An external magnetic field can then cause the domains to align themselves parallel to the magnetic field, resulting in a large magnetization, and the production of a field from this magnetization that adds greatly to the initial field. These materials are ferromagnetic materials. If one removes the magnetic field there will again be a tendency for the magnetic domains to randomize their directions. However, there will usually be some remaining net magnetization, which we call the **remanent magnetization**. If the remanent magnetization is large, there will remain a large "permanent" magnetization, and the material has become a **permanent magnet**. Common magnets are made from iron which can be made to have a large remanence.

The intrinsic magnets we discussed in the previous paragraph were assumed to tend to line up parallel to each other. While this is the usual situation there are cases where the neighboring moments tend to line up antiparallel to each other. The domains will then have moments are opposite directions. This requires materials in which the neighboring atoms are from different elements, with different moments. There can then be a net magnetic moment for the material which depends on the difference in the magnetic moments of these elements. We will not be discussing this rather exotic type of material any further.

8.2 FERROMAGNETISM

Certain materials, notably iron, nickel and cobalt, exhibit **ferromagnetism** at room temperature. This means that the magnetic interactions between the magnetic moments of neighboring atoms is strong enough, even at room temperature, to align the moments in the same direction. Since this cooperative behavior results in "macroscopic" domains (on the atomic scale) in which the moments are aligned, the magnetic field produced by these domain moments can be quite large. If an external field is applied to the ferromagnetic material it has the effect of causing the domains of aligned moments to rotate and point in the same direction, the direction of the magnetic field. The maximum field that they can produce occurs if one is successful in aligning all the domains. When this condition is reached, the material is said to be saturated, and the field being produced is the saturation field. For iron, this saturation field is of the order of 1T. Once they have been aligned, they tend to remain aligned even if the external field that originally caused them to align is removed. The material has now become a permanent magnet.

When the material is magnetized, there are many electrons that are all rotating in the same direction, as in Fig. 8-1. This is equivalent to a large amount of charge going around a circle in the same direction. Indeed, we will show in Sec. 8.3, that this is equivalent to a current flowing around the surface of the material, as in the case of a solenoid. The field produced by such a solenoid in the region outside the solenoid is shown in Fig. 8-2. Indeed the field on the outside is nearly the same in shape as the electric field produced by oppositely charged particles located at the ends of the bar. We therefore often

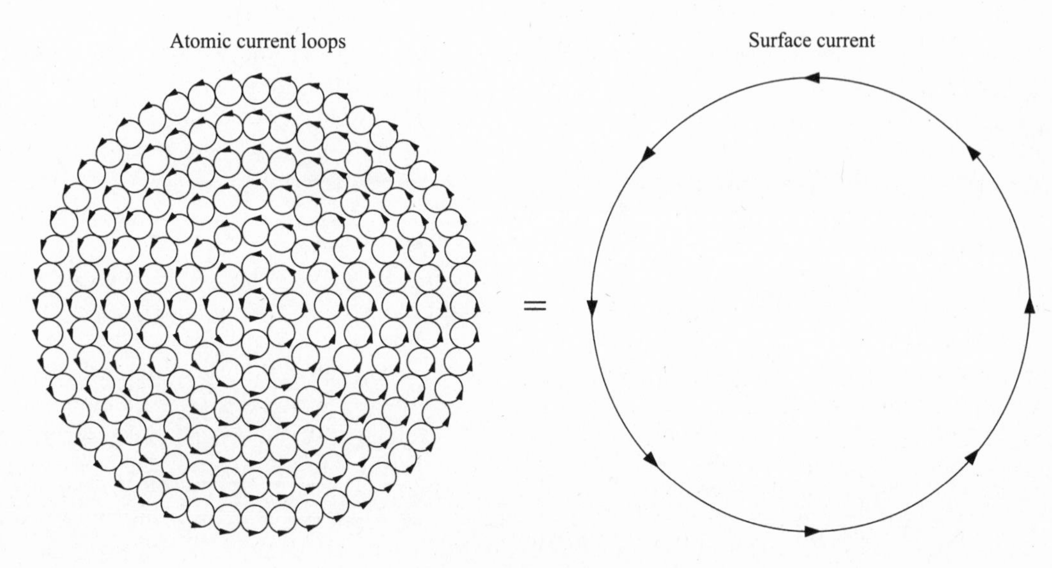

Cross-section of magnetized cyliner

Fig. 8-1

talk of the bar as being composed of two opposite magnetic poles (the substitute for electric charges), one called a north pole and the other called a south pole. The **north pole** is the apparent source of magnetic field lines (as is a positive charge for electric field lines), and the **south pole** is a sink for the lines. In actuality the lines do not terminate at the poles, but continue in straight lines within the material, forming closed loops. The designation of north or south pole arises from the fact that the bar tends to line up in the magnetic field of the earth with the north pole facing in the northerly direction. The poles are designated as "north" and "south" poles, the north pole being positive, and the south pole being negative. There is, however, a big difference between the electrical case and the magnetic case. In the case of opposite electric charges at the ends of a bar, if one were to cut the bar in half, then one piece would be charged positively, and the other piece would be charged negatively. For the magnetic case, if one divides the bar, then each piece would still be a (smaller) bar magnet, with its own smaller north and south pole. One never finds a piece of magnetic material with only a north pole, or only a south pole. Thus, these poles are just a convenient artifact for describing the magnetic field outside a magnet, unlike electric charges which are real and can be isolated. (While these magnetic "monopoles" do not seem to exist in nature, there are certain speculative theories that are currently under consideration which assume that magnetic monopoles might indeed exist. To date, no such monopoles have been observed.) Indeed, the absence of magnetic monopoles is often written as a principle in the same form as Gauss' law for electrostatics. This law states that if we take the magnetic flux (the total number of magnetic field lines, which we will define more carefully later) through a closed surface, then the result will always equal zero, because there are no magnetic monopoles. For electrostatics, the result of taking the electric flux through a closed surface gives us a result proportional to the total charge enclosed by that surface. For magnetism, that total "charge" is zero. Nevertheless the use of the concept of magnetic poles is convenient in discussing the interaction of magnets with a magnetic field, and with other magnets.

As we learned in Chap. 6, magnetic poles will tend to align themselves parallel to the magnetic field, with their magnetic moments in the same direction as the field. Since a magnet will line up in the magnetic field of the earth with the north pole of a magnet pointing north, this means that the earth's magnetic field points in the northerly direction, with the magnetic field lines entering the earth at the magnetic North Pole and the lines emanating from the earth at the magnetic South Pole. If we think of the earth as the equivalent of a bar magnet, then the geometric North Pole of the earth is actually a magnetic south pole, where field lines enter. Indeed, the magnetic field of the earth bears a striking resemblance to the magnetic field of a bar magnet.

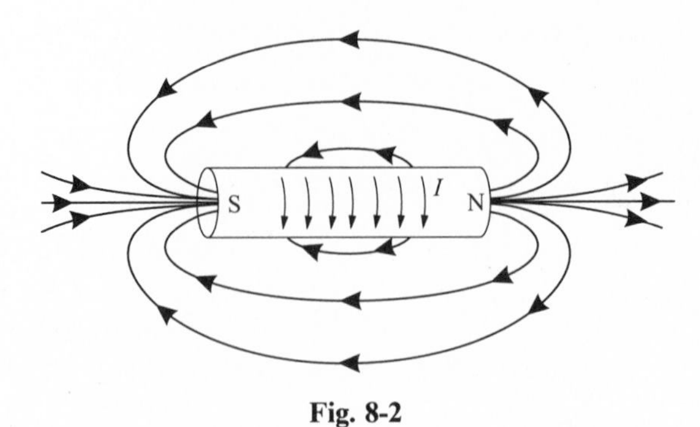

Fig. 8-2

If one places two bar magnets near each other, we can consider one magnet as producing a magnetic field, which then affects the other magnet. The magnetic field of a bar magnet is shown in Fig. 8-2, with the field emanating from the north pole and entering at the south pole. If one places another magnet in the region of this field, this second magnet will tend to line up parallel to the field, with its south to north direction in the same direction as the magnetic field. Thus, near the north pole of the first magnet, the second magnet will line up with the south pole nearer the first north pole. The north pole attracts a different south pole, and repels a north pole. Again, as in the case of electric charges, opposite poles attract, and similar poles repel each other. Near the sides of the first magnet, the magnetic field is parallel to the magnet, but pointing in the direction from north to south. The second pole will therefore line up with its own poles pointing opposite to the poles of the first magnet. Again, the opposite poles are attracted to each other. Magnets tend to attract each other and to cling together with opposite poles near each other.

If one puts a magnet near a piece of magnetic material, such as a needle, the magnet will cause the magnetic material to become **magnetized** with its moment parallel to that of the magnet itself. Thus the needle will have a south pole near the north pole of the magnet, and be attracted to the magnet. If the magnetic force of attraction is sufficiently great to overcome the force of gravity, the magnet will be able to lift up the needle. This explains the tendency of magnets to be able to locate and pick up pieces of magnetic material.

8.3 MAGNETIZATION

When we described the properties of a magnetic material, we talked about the microscopic interactions which gave rise to a "**magnetization**" of the material, which is the counterpart of polarization for electric fields. This magnetization arises from many small circulating currents, all circulating in the same sense about the direction of the external magnetic field. In some cases these circulating currents are due to the orbital motion of the electrons, but in most cases the source of the currents are the spins of the electrons. If we look at these currents as they fill space, the picture looks somewhat like what is shown in Fig. 8-1. At any point within the interior of a uniform material in a constant external magnetic field, there are always adjoining current loops, one above and one below the point. Since the currents are circulating in the same direction, their directions are opposite at the top and at the bottom. Thus, at any point in the interior, there will be no net current flowing in any direction. Only at the edge of the material, where there is no cancellation from currents on the other side of the surface, will we have a net current. (We have met an analogous situation for dielectric materials, where the polarization of the charge does not produce net charge anywhere except at the surface.) The net current in the case of magnetic materials will be flowing around the surface of the material in the same circulatory direction as the individual currents. For a cylindrically shaped bar, with the external field along its axis, the current flowing around the surface is the same as that for a solenoid. Thus, the field produced by the magnetization of the uniform cylindrical bar is identical to that produced by a solenoid, uniform on the inside, and looking like Fig. 8-2 on the outside.

We now define the magnetization vector, $\boldsymbol{M}$, of our material as the total magnetic moment per unit volume (We previously used $\mathbf{M}$ as the magnetic moment vector, so we shall use $\boldsymbol{M}$ as the magnetization vector). Thus, a material in a magnetic field can become magnetized, with a magnetization $\boldsymbol{M} = \sum \mathbf{M}/V$. The magnetic field, B_{M}, produced by the magnetic dipoles in the material is related to the magnetization in a straightforward way. We consider the case of the uniform cylindrical magnetic material, where, as we have seen, the effect of a uniform external field on the material is effectively the establishment of a surface current around the cylinder. This current creates its own field in the cylinder which mimics that of a solenoid. Recalling the equations for the field inside a solenoid, we have for our equivalent solenoid: $B_{\mathrm{M}} = B_{\mathrm{equiv}} = \mu_0 nI$ where n is the number of turns/unit length and I is the equivalent current in each turn. For a solenoid of length d and cross-sectional area A, we have: $B_{\mathrm{M}} = B_{\mathrm{equiv}} = \mu_0 nI$ $\mu_0(N/d)I = \mu_0 NIA/dA$ where N is the total number of turns in the solenoid. Since IA is the magnetic moment of each turn, NIA = total magnetic moment. Since $dA = V$, the volume of the solenoid, we have

$$B_{\mathrm{M}} = \mu_0 \mathbf{M}/V = \mu_0 \boldsymbol{M} \tag{8.1}$$

Note also that

$$\boldsymbol{M} = NI = \text{surface current/unit length.} \tag{8.2}$$

If our solenoid has a current equivalent to our magnetic cylinder we conclude that the extra field produced by the material is just $\mu_0\boldsymbol{M}$ with direction along the cylinder. We would therefore expect that the actual field that is produced in a material is the sum of the field that would be produced in the absence of the material, $\mathbf{B}_0$, plus the field produced by the magnetization, $\mu_0 \boldsymbol{M}$. This can be written as $\mathbf{B} = \mathbf{B}_0 + \mu_0 \boldsymbol{M}$, or $\mathbf{B}_0/\mu_0 = \mathbf{B}/\mu_0 - \boldsymbol{M}$. We define a new vector $\mathbf{H}$, the magnetic intensity vector, as $\mathbf{B}_0/\mu_0 = \mathbf{H}$. This magnetic intensity vector is a quantity that, in the absence of permanent magnets, is dependent on only the external currents that produce the magnetic field, and not on the magnetic moment currents of the atoms of the material that produce the magnetization field. The unit for H is A/m, as is the unit for $\boldsymbol{M}$. While the above was derived for the special case of a uniform cylinder, one can indeed show that it is true in general that

$$\mathbf{H} = \mathbf{B} + \mu_0 \boldsymbol{M}. \tag{8.3}$$

where $\mathbf{H}$ can be calculated from the external currents alone. In fact, the best way to write Ampere's law is in terms of the magnetic intensity, as $\sum (H \cos \theta \, \Delta L) = I_{\mathrm{ext}}$, where the sum is taken around a closed circuit, and the I_{ext} is the external current flowing through the circuit. The currents induced as magnetization in the material are not included in this equation, only the external currents. This means that for all the cases that we were able to calculate the magnetic field using Ampere's law, we can just as simply calculate the magnetic intensity, $\mathbf{H}$, even in the presence of material that can be magnetized. If we find a method of converting from $\mathbf{H}$ to $\mathbf{B}$, then we will also be able to calculate $\mathbf{B}$ in those cases.

In general, except for material that becomes a permanent magnet, the magnetization is proportional to the magnetic field, and therefore to the magnetic intensity as well. We can therefore write that

$$\boldsymbol{M} = \chi \mathrm{H}, \tag{8.4}$$

where χ is called the **magnetic susceptibility** of the material. Then

$$B = \mu_0(H + \boldsymbol{M}) = \mu_0 H(1 + \chi) = \mu_0 \kappa_{\mathrm{m}} H = \mu H, \tag{8.5}$$

where μ is the **permeability** of the material, κ_{m} is the relative permeability of the material, and $\mu = \mu_0$ κ_{m}., with $\kappa_{\mathrm{m}}. = 1 + \chi$. This means that for these materials, we can calculate $\mathbf{B}$ if we know $\mathbf{H}$, merely by multiplying $\mathbf{H}$ by μ.

For a vacuum, χ is zero, κ_m is one, and μ equals μ_0. The quantity μ_0 is sometimes called the permeability of free space. For a paramagnetic material, the magnetization is in the same direction as H (and as B), and therefore χ is positive, κ_m is greater than one, and μ is greater than μ_0. For a diamagnetic material, $\boldsymbol{M}$ is opposite to H, so χ is negative, κ_m is less than one and μ is less than μ_0. (Since χ is typically very small, κ_m and μ are always positive and B is parallel to H.) Table 8.1 lists some typical values of the magnetic susceptibility of diamagnetic and paramagnetic materials.

Problem 8.1.

(*a*) Calculate the magnetic field in the inside of a long air filled solenoid, with 150 turns per meter, and carrying a current of 2 A.

(*b*) How does this change if the solenoid is filled with material that has a magnetic susceptibility of 60×10^{-5}?

Solution

(*a*) Using Ampere's law we can calculate B within the solenoid. In Chap. 3 we used Ampere's law for B in a vacuum in the form $\sum (B \cos \theta\, \Delta L) = \mu_0 I_{\text{total}}$, to get that the magnetic field inside the solenoid was $B = \mu_0 nI$. (Or we can use Ampere's law for H in the form $\sum (H \cos \theta\, \Delta L) = I_{\text{total}}$, to get that $H = nI$.) Thus $H = 150(2) = 300$ A/m $\Rightarrow B = 4\pi \times 10^{-7}(300) = 3.7699 \times 10^{-4}$ T.

(*b*) With the material inside, H doesn't change. However, now there is an additional magnetic field due to the induced surface current on the surface of the material. Given $\chi = 6 \times 10^{-4}$ and $\mu = \mu_0(1 + \chi)$ we have $B = \mu H = \mu nI$. Thus the field is altered by replacing μ_0 by μ. Now $\mu = \mu_0(1 + \chi) = 4\pi \times 10^{-7}$ $(1 + 6 \times 10^{-4}) = 4\pi \times 10^{-7}(1.0006)$. Then $B = 4\pi \times 10^{-7}(1.0006)(300) = 3.7722 \times 10^{-4}$ T, which is almost identical to the air filled solenoid. Only with ferromagnetic material, which has a very much higher equivalent magnetic susceptibility, will the magnetic field differ substantially from that of an air filled solenoid.

Table 8.1. Magnetic Susceptibilities of Paramagnetic and Diamagnetic Materials

Materials	$\chi = \kappa_m - 1$
Paramagnetic:	
Iron ammonium alum	66×10^{-5}
Oxygen, liquid	152
Uranium	40
Platinum	26
Aluminum	2.2
Sodium	0.72
Oxygen gas	0.19
Diamagnetic:	
Bismuth	-16.6×10^{-5}
Mercury	−2.9
Silver	−2.6
Carbon (diamond)	−2.1
Lead	−1.8
Rock salt	−1.4
Copper	−1.0

Problem 8.2. A long solenoid has 1800 turns per meter, carrying a current of 2 A. It is filled with a paramagnetic material with $\chi = 66 \times 10^{-5}$.

(*a*) What is the magnetic intensity, H, in the material?

(*b*) What is the magnetic field, B, in the material?

(*c*) What is the magnetization in the material?

(*d*) What surface current per unit length flows around the material?

Solution

(*a*) As we showed in Problem 8.1, the magnetic intensity inside a long solenoid is $H = nI$. Thus, $H = 1800(2) = 3600$ A/m.

(*b*) We know that $B = \mu H = \mu_0(1 + \chi)H = 4\pi \times 10^{-7}\ (1 + 66 \times 10^{-5})(3600) = 4.53 \times 10^{-3}$ T. Since χ was very small, this is not very different than the field produced in a vacuum.

(*c*) The magnetization, $\boldsymbol{M} = \chi_{\mathrm{H}} = 66 \times 10^{-5}(3600) = 2.38$ A/m.

(*d*) The surface current per unit length equals $\boldsymbol{M}$, and therefore equals 2.38 A/m.

Problem 8.3. A long solenoid has 150 turns/meter. If it is filled with air, it produces a magnetic field of 0.05 T inside the solenoid.

(*a*) How much current is needed to produce this field?

(*b*) If it is filled with a ferromagnetic material, which has a relative permeability of 1.5×10^4, how much current is needed to produce the same field?

Solution

(*a*) We know that the magnetic field inside a long solenoid is $B = \mu_0 nI$, and $H = nI$. Thus $I = B/\mu_0 n = 0.05/(4\pi \times 10^{-7})(150) = 265$ A.

(*b*) If we fill the solenoid with ferromagnetic material, we have $H = nI$ and $B = \mu_0 \kappa_m H = 0.05$. Thus $I = H/n = B/\mu_0 \kappa_m n = I_{\mathrm{air}}/\kappa_m = 265/1.5 \times 10^4 = 0.018$ A.

8.4 SUPERCONDUCTORS

There is a special type of material that illustrates the idea of a surface current very vividly. It has been discovered since the beginning of the 1900s, that some materials, at sufficiently low temperatures lose all their resistivity. These materials are called superconductors. Any current flowing in these materials will not result in the absorption of any energy, since there is no resistance and the power absorbed is $I^2R = 0$. These materials have another important property, known as the Meissner Effect, that they expel any magnetic field from their interior. They do this by setting up surface currents which themselves produce an exactly opposite field, and thereby cancel any field which tries to be established in its interior. Thus, a **superconductor** can be considered to be a perfect diamagnet. For instance, if one has a cylindrical bar of superconducting material, and puts this bar in a uniform magnetic field parallel to its axis, as in Fig. 8-3, a surface current will flow around the bar in the magnitude and direction needed to produce an opposite, canceling field. Then there will be no net field within the superconductor.

Problem 8.4. For a superconductor, calculate its magnetic susceptibility and its relative permeability.

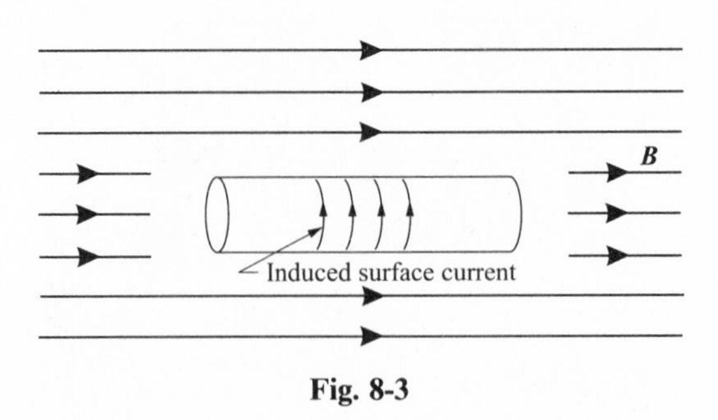

Fig. 8-3

Solution

Since a superconductor can have no magnetic field in its interior, $B = 0$. But $B = \mu_0(H + \mathbf{M}) = 0$, implies that $\mathbf{M} = -H = \chi H$, or $\chi = -1$. Thus, the susceptibility equals -1. Since $\kappa_m = 1 + \chi$, $\kappa_m = 0$. Thus the relative permeability is 0.

Problems for Review and Mind Stretching

Problem 8.5. A toroidal solenoid has 18,000 turns carrying a current of 2 A. It is filled with a paramagnetic material with $\chi = 66 \times 10^{-5}$. The mean radius of the toroid is 0.9 m.

(*a*) What is the magnetic intensity, H, in the material, at its mean radius?

(*b*) What is the magnetic field, B, in the material, at its mean radius?

(*c*) What is the magnetization in the material, at its mean radius?

(*d*) What total surface current flows around the toroid, assuming that the field is uniform inside?

Solution

(*a*) As we showed in Problem 8.1, the magnetic intensity can be calculated using Ampere's law for H. Applying this to a toroid, as we did in Chap. 7, we get that $H = NI/2\pi r = 18{,}000(2)/(2)(\pi)(0.9) = 6.37 \times 10^3$ A/m.

(*b*) We know that $B = \mu H = \mu_0(1 + \chi)H = 4\pi \times 10^{-7}\ (1 + 66 \times 10^{-5})(6.37 \times 10^3) = 8.01$ T. Since χ was very small, this is not very different than the field produced in a vacuum.

(*c*) The magnetization, $M = \chi_H = 66 \times 10^{-5}\ (6.37 \times 10^3) = 4.20$ A/m.

(*d*) The surface current per unit length equals M, and therefore equals 4.20 A/m. The total current equals $2\pi r(M) = 2\pi(0.9)(4.20) = 23.8$ A.

Problem 8.6. A bar magnet is brought near a superconducting cylinder, with its north pole facing the cylinder, as in Fig. 8-4.

(*a*) How will the superconducting cylinder be magnetized?

(*b*) Will the magnet attract or repel the cylinder?

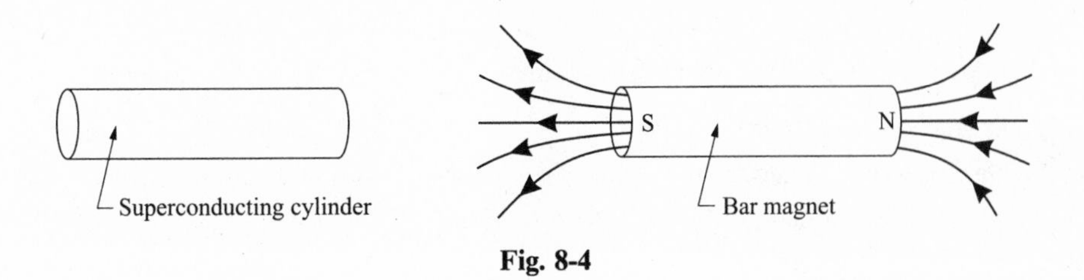

Fig. 8-4

Solution

(*a*) The magnetic field of the bar magnet attempts to establish a magnetic field in the cylinder pointing away from the north pole. To prevent a magnetic field from entering the superconducting material, a surface charge will flow around the cylinder setting up a magnetic field in the opposite direction. Thus the end of the cylinder near the bar magnet will become a north pole.

(*b*) Since like poles repel, the bar magnet and superconducting cylinders will repel each other.

Supplementary Problems

Problem 8.7. A long solenoid, with a radius of 0.6 m, has 2500 turns per meter, carrying a current of 3 A. It is filled with material having a relative permeability of 150.

(*a*) What is the magnetic intensity, H, in the material?

(*b*) What is the magnetic field, B, in the material?

(*c*) What is the magnetization in the material?

(*d*) What surface current per meter flows around the solenoid?

Ans. (*a*) 7500 A/m; (*b*) 1.42 T; (*c*) 1.13×10^6 A/m; (*d*) 1.13×10^6 A/m

Problem 8.8. For the solenoid in Problem 8.7, calculate the flux through a cross-section of the solenoid.

Ans. 1.61 Wb

Problem 8.9. A long solenoid, with a radius of 0.6 m, has 2500 turns per meter, carrying a current of 3 A. It is filled with ferromagnetic material having a relative permeability of 1.5×10^4. If one removes the ferromagnetic material and it is replaced with air, how many turns per meter would the solenoid require to produce the same magnetic field?

Ans. 3.75×10^7 turns/m

Problem 8.10. A certain material has a magnetization $\boldsymbol{M}$ of 2.5 A/m. The molecular weight of the material is 195, and the mass density of the material is 8×10^3 kg/m^3.

(*a*) What is the average magnetic moment per molecule in the material?

(*b*) The standard unit for atomic magnetic moments is the Bohr magneton, $\mu_B = 9.274 \times 10^{-24}$ A · m^2, which is the smallest magnetic moment an atom can have. How many Bohr magnetons are in this average moment?

(c) How can the average magnetic moment per molecule be less than μ_B?

Ans. (a) 1.11×10^{-28} A · m^2; (*b*) 1.09×10^{-5}; (*c*) Only a small fraction of the molecular magnetic moments line up and contribute to the magnetization. The vast majority remain random. Thus, the "average" over all molecules is very small.

Problem 8.11. A solenoid has 1800 turns/m carrying a current of 2 A. The solenoid is filled with a magnetic material. It is found that the magnetic field, B, in the solenoid is 5×10^{-3} T.

(*a*) What is the magnetic intensity, H, in the material?

(*b*) What is the magnetization in the material?

(c) What is the relative permeability of the magnetic material?

Ans. (*a*) 3600 A/m; (*b*) 379 A/m; (*c*) 1.11

Chapter 9

Induced EMF

9.1 INTRODUCTION

In the previous chapters, we learned about the effect of a magnetic field on a moving charge (or on a current-carrying wire), and about sources of magnetic fields. In this chapter we will explore further the effects that magnetic fields can have on the charges, and in particular on the charges in conductors. In Chap. 6 we saw that magnetic fields can exert forces on moving charges and on current-carrying wires. Here we will explore the effect the magnetic field can have on the charges in a wire, whether or not there is a current. We will see that when a wire moves through a magnetic field an **EMF** is generated in the wire, which has the ability to move charges through the wire. This means that it is possible to build an apparatus that makes use of magnetic effects to produce EMFs that drive electrical circuits connected to the apparatus. The apparatus is called a generator and, like a battery (described in Chap. 5), it pumps positive charges within the apparatus toward the high-potential end of the apparatus, so that, in an external circuit, the charges produce a current flowing from the high- to the low-voltage terminals. As in a battery, the voltage produced by the generator on open circuit is its EMF.

We start our discussion of how magnetic fields produce an EMF by examining wires moving through a magnetic field. The EMF produced in such moving wires is called **motional EMF** and can be understood by using concepts that we have already developed in the previous chapters. Next we introduce the situation where the wires are not moving, but instead the magnetic fields are changing. In such cases EMFs, called **induced EMFs**, can be produced in the wires. The generation of such EMFs, is the subject of Faraday's law. As we shall see, Faraday's law can be viewed as a generalization required by ideas about relative motion. When stated in the form of Faraday's law, both motional EMF and induced EMF can be described by the same mathematical statement.

9.2 MOTIONAL EMF

The basic idea of motional EMF can be understood from the experiment illustrated in Fig. 9-1. Here, a nonmagnetic conducting bar is moving, with a velocity **v**, in a perpendicular magnetic field, **B**. Since a conducting bar contains charges that are free to move in the presence of forces, the force exerted on these charges by the magnetic field will cause them to move. For a velocity to the right and a

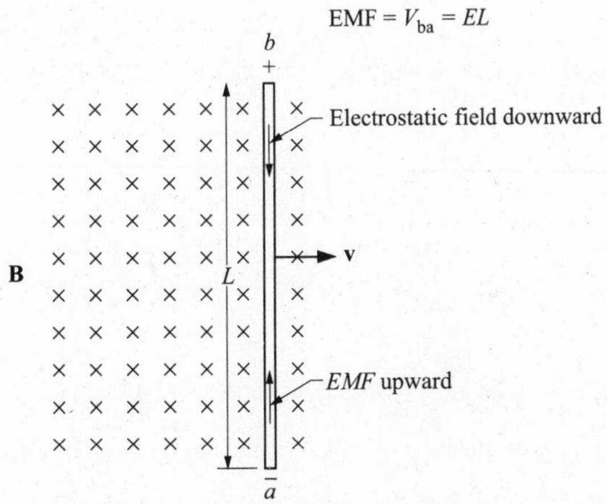

Fig. 9-1

magnetic field into the plane of the paper, the force on positive charges will be $F = qvB$, and will be directed up on the bar (toward b). Negative charges would experience a downward force (toward a). Thus, the result of the magnetic field is to drive positive charges toward b and negative charges toward a. As the charges accumulate at the ends of the bar, the positive charges at b will exert repulsive forces on any additional charges that the magnetic field tries to force in that direction, and the same will be true for the negative charges at a. An equilibrium will be established when the electrostatic force exerted by these charges balances the magnetic force exerted by the magnetic field. This occurs when the accumulated charges produce a uniform electrostatic field, E, inside the bar such that $qE = qvB$, or $E = vB$. The uniform electric field will produce a potential difference, V_{ba}, between the ends of the bar, with the positively charged end (point b in Fig. 9-1) at the higher potential. This potential difference tries to push positive charge away from the positive terminal and toward the negative terminal. The magnetic force counterbalances the electric force as long as the bar moves in the magnetic field, and the charges will not move within the bar. If one were to connect the terminals a and b with a resistor to form a circuit, this potential difference will cause a current to flow from b to a through the "external" part of the circuit, the resistor. This is the same effect as if the moving wire were replaced by a battery with an equivalent EMF in the circuit. Recall that the EMF of the battery represents the positive work done per unit charge by the chemical forces in the battery in moving positive charge from the negative to the positive terminal. The electrostatic charges oppose this motion in the battery and thus the charges gain electrostatic potential energy as they move to the positive terminal. This potential energy is lost as the charges move from the positive terminal back to the negative terminal through the external circuit. In a similar fashion, the moving wire through the magnetic field causes non-electrostatic forces (a combination of the wire and the magnetic field) to do work in pushing charges through the wire, opposed to the electrostatic field. Again the charges gain electrostatic potential energy, which they then lose as they move through the external circuit. In both the battery and the moving wire the EMF is positive from the lower to the higher potential, and equals the open-circuit potential difference between the terminals, V_{ba} (Fig. 9-1). Recalling that for a wire of length L, with uniform electric field $\mathbf{E}$, the potential difference is $V_{ba} = EL$, and using our result $E = vB$, we have for our moving wire:

$$\text{EMF}_{\text{motional}} = vBL \tag{9.1}$$

If we complete the circuit with our resistor, with resistance R, then a current of magnitude $I = V_{ba}/R = vBL/R$ will flow through the resistor from b to a.

> ***Note.*** $V_{ba} = \text{EMF}$ is strictly true only on open circuit. If a current flows, then the equation remains true only if there are no thermal losses (resistance) in the moving wire. If there is loss, due to a resistance r, then $\text{EMF} = V_{ba} + Ir$, as in the case of a battery with effective internal resistance r.

Problem 9.1. A bar of length 0.8 m is moving to the right with a velocity of 500 m/s in a magnetic field of 0.3 T going into the paper (see Fig. 9-1).

(*a*) What is the motional EMF produced?

(*b*) Which end of the bar is at the high potential?

(*c*) If the magnetic field pointed out of the paper, would the answer to (*a*) or (*b*) change, and if so, how?

Solution

(*a*) Substituting into Eq. (*9.1*), we get EMF = (500 m/s)(0.3 T)(0.8 m) = 120 V.

(*b*) Since the magnetic force on positive charges is up (from a to b), the high-potential end is b.

(*c*) The magnitude remains the same. The force on positive charges is now down (from b to a); therefore, the high-potential end is now a.

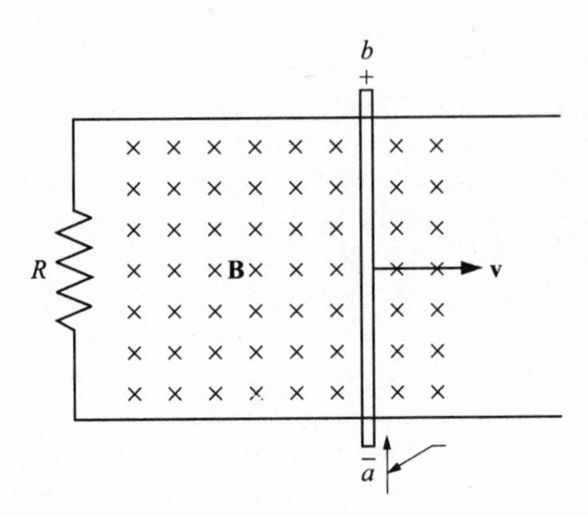

Fig. 9-2

Problem 9.2. The bar in the previous problem is sliding with the same constant velocity along two frictionless conducting railings, connected to a resistor R of 60 Ω, as shown in Fig. 9-2.

(*a*) What current flows in the resistor?

(*b*) In what direction does the current flow through the bar?

(*c*) How much power is absorbed in the resistor?

Solution

(*a*) The EMF is 120 V, as calculated in the previous problem. The current is (120 V)/(60 Ω) = 2 A.

(*b*) The current flows through the external circuit (resistor) from b to a, so it flows from a to b in the bar.

(*c*) The power is $P = I^2R$ (or IV) = (2 A)^2(60 Ω), [or P = (2 A)(120 V)] = 240 W.

Problem 9.3. Referring to Problem 9.2:

(*a*) While the current I is flowing, calculate the magnitude and direction of the magnetic force on the bar.

(*b*) What force must be exerted on the bar by some outside source to keep the bar moving at constant speed?

(*c*) How much power must this outside force deliver?

Solution

(*a*) The induced current is 2 A. Current flowing in a length L in a magnetic field B experiences a force of $F = ILB \sin\phi$ = (2 A)(0.8 m)(0.3 T) = 0.48 N. Here ϕ is the usual angle between the direction of the current through $\mathbf{L}$ and the direction of $\mathbf{B}$, which in this case is 90°, so $\sin\phi = 1$. The direction of the force is perpendicular to $\mathbf{L}$ and to $\mathbf{B}$ and therefore along the horizontal direction. By using the right-hand rule, we determine that the direction is to the left.

(*b*) The outside source must exert a counterbalancing force of 0.48 N in the direction to the right.

(*c*) The power is given by Volume I formula 7.2, $P = Fv$ = (0.48 N)(500 m/s) = 240 W. This is positive, since the force is in the same direction as v.

Problem 9.4.

(*a*) Compare the answers to part (*c*) of Problems 9.2 and 9.3. Why are these the same?

(*b*) What happens if the force pushing the bar [Problem 9.3(*b*)] were removed?

Solution

(*a*) From Problem 9.2(*c*) we note that energy is being dissipated in the resistor. The energy must be supplied by some source. The magnetic field does not deliver energy to the system since the force of the magnetic field on a moving charge is perpendicular to the velocity and hence does no work. The bar itself is moving at constant speed so its kinetic energy is constant. The only source of energy is therefore the work done by the external source; this energy is completely dissipated as heat in the resistor. From an energy point of view, the magnetic field is the necessary agent for converting the energy of the external source into electrical energy that ultimately turns into thermal energy.

(*b*) If the external force were removed, the magnetic force on the bar [Problem 9.3(*a*)] would be unbalanced and the bar would slow down; this in turn would decrease the EMF in the bar and the current would also drop, until ultimately the bar would stop moving and the current would cease. From an energy point of view this corresponds to the kinetic energy of the bar being the only source of energy for dissipation in the resistor; when the kinetic energy drops to zero there is no more energy available and the current becomes zero. Again, the magnetic field acts as the agent—this time for converting the kinetic energy into electrical energy that ultimately turns into thermal energy.

Another example of a motional EMF is the case of a single loop coil turning in a uniform magnetic field. Consider the rectangular coil *abcd* in Fig. 9-3, which is rotating in the uniform magnetic field **B**. At the time when **B** is parallel to the plane of the coil, as in the figure, side *ad* is moving out of the plane of the paper, and side *bc* is moving into the paper. Both sides are moving perpendicular to the field, and will develop EMFs as a result of this motion. For side *ad*, the force on the free charges in this side will be up (from *d* to *a*) and therefore the induced EMF will drive current around the coil in a clockwise direction. On side *bc*, the force on positive charges is down (from *b* to *c*), and the induced EMF will also drive current around the coil in a clockwise direction. From Eq. (*9.1*), the EMF developed in each side is vBh, and $v = \omega(w/2)$, giving EMF $= \omega(w/2)Bh$. The total EMF is therefore $\omega Bwh = \omega BA$, where A is the area of the coil wh. The other two sides will not contribute any EMF, since the magnetic force on charges in these wires are perpendicular to the wires and therefore do not contribute an EMF along the wire (i.e. no charges are pushed along the wires). The result of this calculation is that this rotating coil, in the position shown, produces a total motional EMF around the coil equal to ωBA. Although we derived this result for a rectangular coil, it is actually true for any other shape planar coil as well. As the angle between the plane of the coil and the magnetic field increases above zero, the component of the velocity of each side perpendicular to the magnetic field decreases and the EMF decreases correspondingly. We will discuss this further in a later section. However, we want to mention at this time that the above example is the basic configuration that is used to generate voltages, and is therefore the prototype of a **generator**.

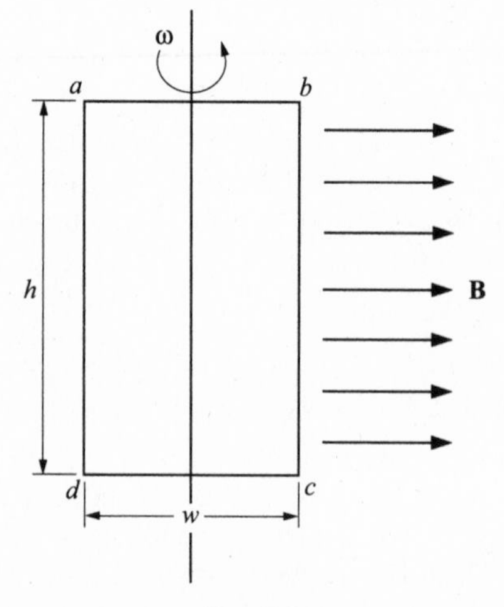

Fig. 9-3

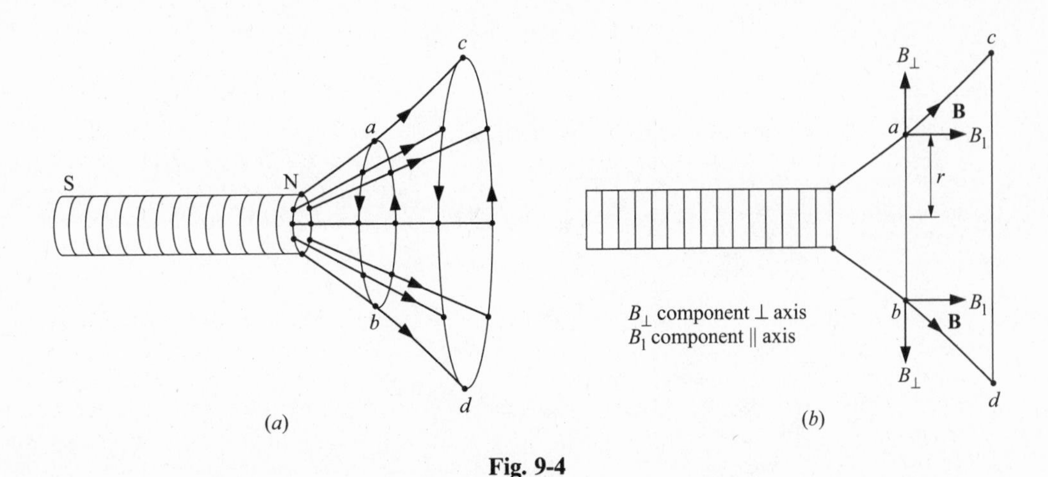

Fig. 9-4

A third example of a motional EMF is the case of a coil moving in the magnetic field produced by a bar magnet. Consider the bar magnet in Fig. 9-4(*a*), and the field it produces near its north pole. The field lines seem to come out of the pole, and spread out equally in all directions. A few lines coming from the edge of the pole are drawn in the figure. Consider the circular loop of wire going through points a and b. At every point on this loop the magnetic field has a longitudinal component along the direction of the axis of the magnet, and a component perpendicular to the axis of the magnet. The fields at a and b are shown in Fig. 9-4(*b*), with these two components clearly delineated. If the loop moved without tilting parallel to the axis of the magnet, then its velocity would be parallel to B_l and this component of the field would not exert a force on the free charges moving with the coil. However, the $B_\perp$ component does exert a force on the charges in the coil. At point a this force is out of the paper and at point b the force is into the paper. This force is in the direction driving charge around the coil in the direction of the arrows in Fig. 9-4(*a*). At any other point on the coil the force acts in the same manner, causing an EMF which induces current to flow around the coil. The magnitude of the force would be $qvB_\perp$, and the balancing electric field would be $E = vB_\perp$. The EMF would therefore be $vB_\perp(2\pi r)$, where r is the radius of the coil. If one reverses the direction of the motion of the coil, the direction of the EMF will reverse. Also, if one reverses the direction of the magnetic field, by using a south pole, the direction of the EMF would also reverse.

We now raise an interesting question. Suppose that instead of the bar magnet being fixed and the loop moving to the right, the loop is kept fixed and the magnet is moving to the left. Will there still be an induced EMF in the coil? If such an EMF exists, it clearly cannot be motional EMF, since the coil is not moving, and magnetic forces require a *moving* charge. If an EMF exists, it will therefore be due to a new phenomenon that we have not yet discussed. There is an interesting argument in favor of the existence of such an EMF. In Volume I Chap. 3, we learned that velocity has to be measured relative to some system that is considered to be at rest. It can be shown that the laws of mechanics, as stated by Newton, are applicable without modification in any inertial system that is not accelerating. The laws in an elevator that is moving at constant speed are identical to the laws in an elevator at rest. There is no mechanical experiment that can be done within the elevator that would allow a person to determine whether the elevator is moving and the building is standing still, or whether the elevator is standing still and the building is moving in the opposite direction. If we assume that it is also true that the person in the elevator cannot use an experiment in electricity and magnetism to distinguish between these choices, then we can show that there *must* be an induced EMF even if one moves the bar magnet rather than the loop. To show this let us attach a loop to the elevator (see Fig. 9-5), and a bar magnet to the elevator shaft. If the elevator moves and the building is at rest, there will be a motional EMF induced in the coil. On the other hand, if the building moves and the elevator is at rest, there will be no *motional* EMF induced in the coil. If no other EMF is induced in the coil, we will have determined through this experiment that the elevator is the moving system. Therefore, if we believe that the laws of physics do

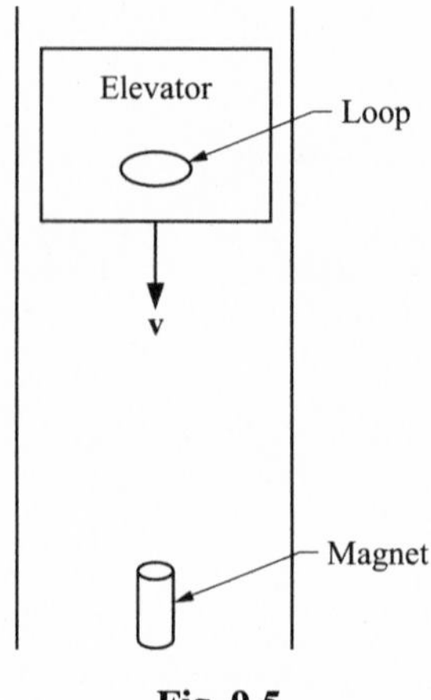

Fig. 9-5

not allow us to make such a determination, then there must be an EMF even if the magnet, rather than the coil, moves.

While this is a very nice argument, it does not prove that such an EMF does indeed exist. The only way to find out if there is such an EMF is to do so experimentally, and we can carry out an easy experiment using our bar magnet and loop. We find that when we move the magnet instead of the loop in Fig. 9-4(*b*) we get the identical EMF in the loop that we get when we move the loop. As long as the relative velocity of the loop and magnet is the same, we get an identical EMF. This new EMF is induced in the loop by the moving magnet (and associated magnetic field) and is therefore called the **induced EMF**; its characteristics are given by Faraday's law and are developed in the next section.

9.3 INDUCED EMF

Magnetic Flux

Before we can state Faraday's law, we must develop the concept of magnetic flux. This is a concept which is similar to the concept of electric flux, developed in Chap. 3. Consider a small planar area A as shown in Fig. 9-6(*a*). This area can be represented by a vector $\mathbf{A}$ that has a magnitude equal to the area, and a direction perpendicular to the plane of the area. Since there are two possible directions for $\mathbf{A}$ we have to make a choice. This was discussed in Sec. 6.5 for the case of a magnetic moment, and in Sec. 3.6 for the case of electric flux. In those cases we were able to fix the positive direction of the area vector, either by noting the direction of the circulating current and using the right-hand rule (for the magnetic moment) or by choosing the vector to point from inside a closed surface to the outside (for electric flux in Gauss' law). Here, the positive direction is arbitrary, so we can choose it in either of the two possible directions. With this choice made, we define the **magnetic flux** that passes through the area in the positive direction as

$$\Phi_m = BA \cos \theta, \tag{9.2}$$

where $\mathbf{B}$ is the magnetic field in the region and θ is the angle between $\mathbf{B}$ and $\mathbf{A}$. The flux therefore depends on three variables, B, A and θ. If Φ_m is positive, then the flux is passing through the area in the positive direction, and the reverse is true for negative Φ_m. If the field is perpendicular to the plane of the area, then the angle θ is 0°, and Φ_m has its maximum value of BA. If B is parallel to the plane of the area, then no flux passes through the area ($\theta = 90°$). The unit for magnetic flux is $T \cdot m^2$, which is given the name Weber (Wb). As in the case of electric flux one can visualize the magnetic flux by drawing magnetic field lines, with the number of field lines passing through a unit area perpendicular to the lines proportional to $\mathbf{B}$ at that location. By tracing field lines to other locations the number of field lines

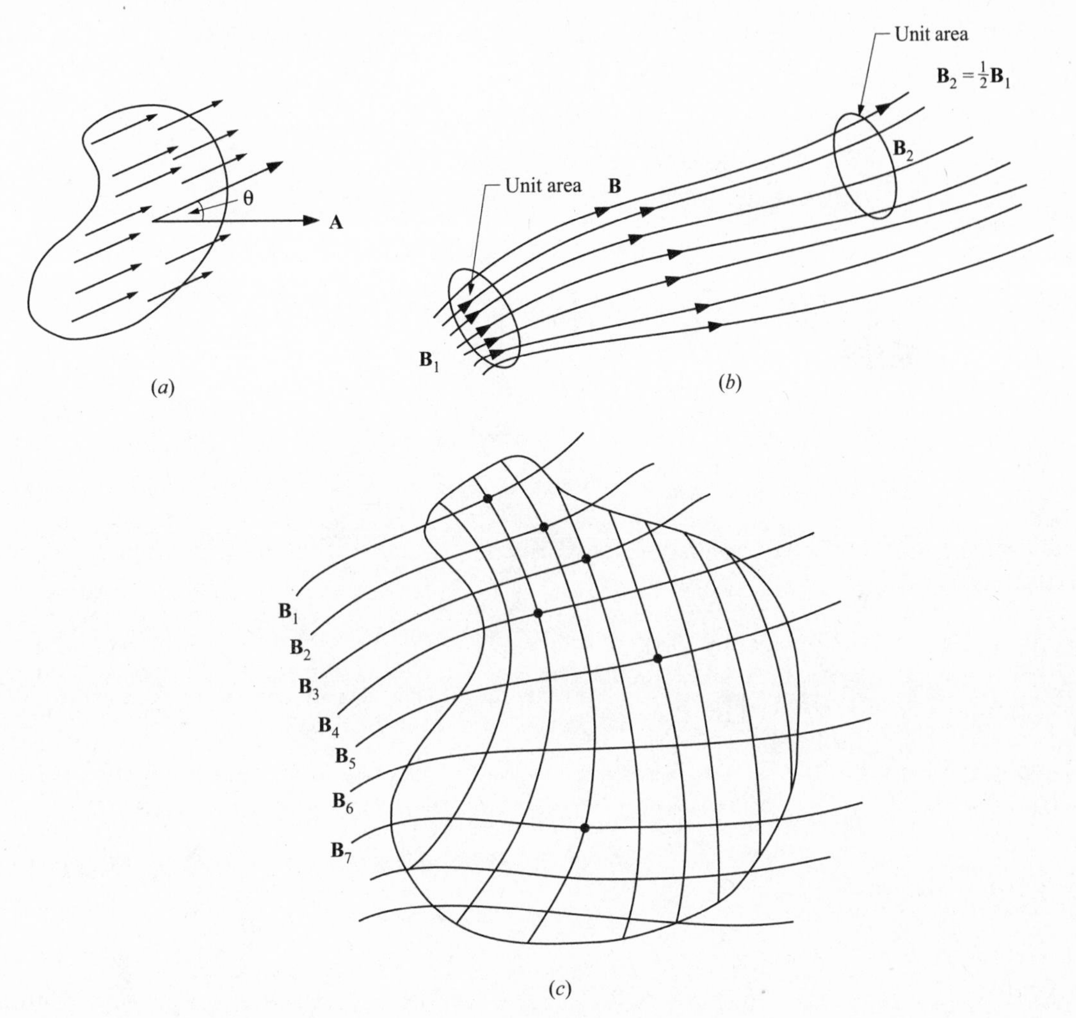

Fig. 9-6

passing through unit perpendicular area at these locations will be proportional to **B** at that location with the same constant of proportionality. This follows because magnetic field lines never stop or start at a point but rather form closed loops. The total magnetic flux through any area is then just proportional to the total number of field lines through that area, again with the same proportionality constant [see Fig. 9-6(*b*) and (*c*)]. Note that for a large curved surface, as in Fig. 9-6(*c*), one breaks the surface up into many small sections, each of which is almost planar. Then one applies Eq. (*9.2*) to each section and adds them up to get the total flux:

$$\Phi_M = \sum B_i A_i \cos\theta_i \qquad \text{i (all subsections of } S) \tag{9.3}$$

Problem 9.5. A rectangular coil, with sides $h = 0.5$ m and $w = 0.3$ m, is in the y–z plane, as in Fig. 9-7. There is a uniform magnetic field **B** in the region of 0.25 T. Calculate the magnetic flux going through the coil in the $+x$ direction, if the magnetic field points (see Fig. 9-7) in: (*a*) the $+x$ direction; (*b*) the $-y$ direction; (*c*) at an angle of 30° with the $+x$ axis, in the x–y plane; and (*d*) at an angle of 45° with the $-x$ axis, in the x–y plane.

Solution

(*a*) The direction of **A** is the positive x direction, since this is perpendicular to the plane of the area, and we want the flux in $+x$. Substituting into Eq. (*9.2*), we get $\Phi_m = (0.25\ \text{T})(0.3\ \text{m})(0.5\ \text{m}) = 0.0375$ Wb.

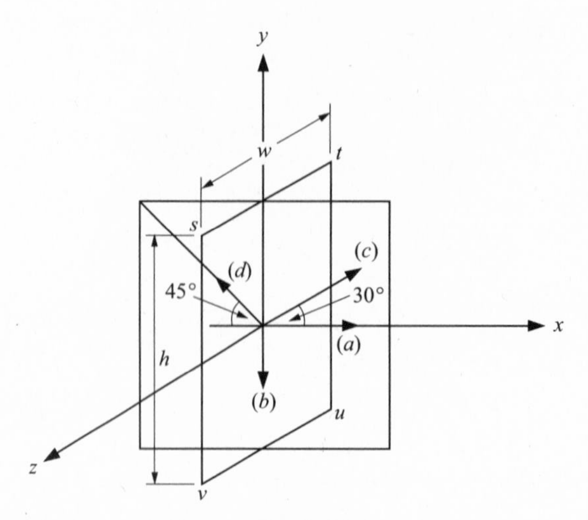

Fig. 9-7

(*b*) Since the magnetic field is perpendicular to **A**, the flux is zero (i.e. no field lines pass through the area).

(*c*) The angle between **B** and **A** is 30°, so $\Phi_m = (0.25\text{T})(0.3\text{ m})(0.5\text{ m})\cos 30° = 0.0325$ Wb.

> ***Note.*** This is less than in part (*a*) because for the same density of field lines fewer pass through the area; indeed the cosine accounts for this reduction in flux.

(*d*) The angle between B and **A** is 135°, so $\Phi_m = (0.25\text{ T})(0.3\text{ m})(0.5\text{ m})\cos 135° = -0.0265$ Wb.

> ***Note.*** Not only the number of field lines passing through the area is reduced, but they pass through in the opposite sense to our choice of positive A, hence the minus sign.

Change in Flux

Now that we have defined the magnetic flux, we are nearly ready to define Faraday's law. This law makes use of the *change in flux*, rather than the flux itself. The flux can be changed by altering any one (or several) of the three variables that enter into the definition of flux, B, A and θ. As usual the change means the final minus the initial, so that

$$\Delta\Phi = \Phi_f - \Phi_i, \text{ where we have dropped the subscript m on the flux.} \tag{9.4}$$

The following problem illustrates the calculation of changes in flux as one varies either the field, the area or the angle involved in the flux.

Problem 9.6. A rectangular coil, with sides $h = 0.50$ m and $w = 0.30$ m, is in the y–z plane, as in Fig. 9-7. There is a uniform magnetic field **B** in the region of 0.25 T, pointing in the $+x$ direction.

(*a*) Calculate the change in flux if the field increases to 0.30 T.

(*b*) Calculate the change in flux if the field decreases to 0.20 T.

(*c*) Calculate the change in flux if the magnitude remains at 0.25 T, but its direction changes to the $-y$ direction.

(*d*) Calculate the change in flux if the magnitude remains at 0.25 T, but its direction changes to the $-x$ direction.

(*e*) Calculate the change in flux if the magnetic field does not change, but the width of the coil decreases to 0.1 m.

Solution

(*a*) The initial and the final fields are both in $+x$. Substituting into Eq. (*9.2*), we get $\Phi_i = (0.25\ \text{T})(0.3\ \text{m})(0.5\ \text{m}) = 0.0375$ Wb, and $\Phi_f = (0.30\ \text{T})(0.30\ \text{m})(0.50\ \text{m}) = 0.045$ Wb. Thus, $\Delta\Phi = \Phi_f - \Phi_i = 0.045 - 0.0375 = 0.0075$ Wb. This change is an *increase* in the flux in the $+x$ direction.

(*b*) Using the same technique as in (*a*), we find that $\Phi_f = (0.20\ \text{T})(0.30\ \text{m})(0.50\ \text{m}) = 0.030$ Wb. Thus, $\Delta\Phi = \Phi_f - \Phi_i = 0.030 - 0.0375 = -0.0075$ Wb. This change is a *decrease* in the flux in the $+x$ direction.

(*c*) Here we must calculate the final flux using the new value of θ, which results in a final flux of zero. Thus, $\Delta\Phi = \Phi_f - \Phi_i = 0 - 0.0375 = -0.0375$ Wb. This change is a *decrease* in the flux in the $+x$ direction.

(*d*) The final flux in this case is -0.375 Wb, since the field has now changed to the $-x$ direction which is opposite to **B** (which we chose as the positive direction). Thus, $\Delta\Phi = \Phi_f - \Phi_i = -0.0375 - 0.0375 = -0.075$ Wb. This change is a *decrease* in the flux in the $+x$ direction.

(*e*) The change in flux is now due to a change in area. The final flux $\Phi_f = (0.30\ \text{T})(0.10\ \text{m})(0.50\text{m}) = 0.015$ Wb. Thus, $\Delta\Phi = \Phi_f - \Phi_i = 0.015 - 0.0375 = -0.0225$ Wb. This change is a *decrease* in the flux in the $+x$ direction.

We will be calculating many more examples of changes in flux in future problems.

Faraday's Law

Now that we have defined the basic concepts needed for **Faraday's law**, we are ready to state that law. This law says that whenever there is a change in flux within a circuit there will be an EMF induced in the circuit. This EMF depends on the time rate of change of the flux through the circuit

$$\text{EMF} = -\Delta\Phi/\Delta t \tag{9.5}$$

where $\Delta\Phi$ is the change in magnetic flux through the circuit in a short time interval, Δt. Eq. (*9.5*) is true no matter what the cause of the change in flux: the circuit moves, the magnetic field changes while the circuit stays fixed, or any combination of these. Thus, as we will see later (Problem 9.8), motional EMF is included in Faraday's law. The minus sign in Eq. (*9.5*) is necessary to assure that the correct direction is given for the EMF. To understand the sign convention in Eq. (*9.5*), consider Problem 9.6(*a*) (Fig. 9-7). Since the direction of **A** is to the right (positive x direction) $\Delta\Phi$ is positive when the flux increases to the right. The usual convention for positive circulation in a loop of area **A** is given by the right-hand rule: when the thumb of the right hand points in the direction of **A** then the fingers curl in the positive direction of flow in the loop. Without the minus sign in Eq. (*9.5*) this would imply that for $\Delta\Phi$ positive the induced EMF would be in the direction $t \to s \to v \to u$ in the loop, causing a current to circulate in the same direction. In fact, the induced EMF for this case points in the *opposite* sense $u \to v \to s \to t$. At first glance it may seem that nature arbitrarily decided which way the EMF will act in the loop—that it could have just as easily been in the other direction [i.e. no minus sign in Eq. (*9.5*)], but that is not the case. Without the minus sign, crazy things happen that violate the foundations of physical law, including conservation of energy.

The requirement of the minus sign is called Lenz's law.

Problem 9.7.

(*a*) Using the situation in Fig. 9-7 and Problem 9.6(*a*), show that without the minus sign in Eq. (*9.5*) we would get non-physical results.

(*b*) Show that the same would be true in the case of Problem 9.5(*d*).

(*c*) What is the connection between Lenz's law and the conservation of energy?

Solution

(*a*) Here $\Delta\Phi$ is positive to the right, so $\Delta\Phi/\Delta t$ is to the right. As noted in the text, without the minus sign in Eq. (*9.5*), the right-hand rule gives the induced EMF around the loop in the direction $t \to s \to v \to u$. This EMF in turn creates a current in the same direction. Since a current in a loop creates its own magnetic field, the current itself will create additional flux through the loop. Again using the right hand rule to find the direction of this additional flux, we see that it points in the direction of the original $\Delta\Phi$, i.e. to the right. According to Eq. (*9.5*) (without the minus sign) this will cause a further EMF which will cause a further increase in current, which will cause further increase in flux, etc. In this way the current will continue increasing and so will the magnetic field without any source of energy beyond the source of the original $\Delta\Phi$, which could have been quite small. This kind of "perpetual motion" machine is clearly non physical. With the minus sign in Eq. (*9.5*), however, the EMF is in the direction $u \to v \to s \to t$ and the current produced is in the same direction. Now the right-hand rule shows that the extra flux through the loop induced by this current is to the left, and tends to reduce the original $\Delta\Phi$ to the right. This in turn tends to decrease the EMF and the current rather than letting it increase in run-away fashion.

(*b*) In Problem 9.5(*d*) $\Delta\Phi$ is negative, so without the minus sign in Eq. (*9.5*) the right-hand side of the equation would be negative and the induced EMF and hence the induced current would be in direction $u \to v \to s \to t$. By the right-hand rule this current produces a magnetic field (and flux) which is also negative through the loop. Again, the additional change in flux would increase the current in a perpetual motion way. On the other hand, if we had the minus sign in Eq. (*9.5*) the induced current would create a flux that was opposite to the original $\Delta\Phi$ tending to decrease the change in flux and hence damp down the current.

(*c*) Without the minus sign (Lenz's law) we have a run-away increase in current in the loop. We will see later that a loop has magnetic energy that is proportional to I^2. If the current keeps increasing, so will this magnetic energy even though there is no source of energy, thus violating the law of conservation of energy.

Lenz's Law

From Problem 9.7 we can now see the explicit meaning of the minus sign in Faraday's law Eq. (*9.5*). A change in flux causes an induced EMF in a circuit, which, in turn causes an induced current in the circuit. If the change in flux is positive, then the induced current will flow in that direction that sets up an induced flux that is negative. If the change in flux is negative then the induced current will set up an induced flux that is positive. This means that the current will always flow in such a direction that opposes the change that created it in the first place. **Lenz's law** thus states that the EMF produced by a changing flux is always in a direction to produce a current whose own flux is in the opposite direction to the initial change in flux. If one is careful about minus signs and the interpretation of positive and negative directions, one can use the minus sign in Faraday's law to determine the direction of the current immediately. In practice, you will find it easier to determine just the magnitude of the induced EMF from Faraday's law, and use Lenz's law to find the direction of the induced EMF and current. As noted above if there actually is induced current flowing, then this induced current will itself induce an additional change in flux, Φ_{induced}, which can further modify the EMF and the current. These effects will be discussed in the next chapter. *In doing the problems in this chapter* we will neglect this self induced change in flux (except for determining current directions by Lenz's law) and include only the current induced by the externally caused changes in flux. We will illustrate Faraday's law and the above procedure for Lenz's law in some more detail in the following problems.

Applications

Problem 9.8. A metal bar of length L slides along two railings with a velocity of v to the right, as in Fig. 9-8. A magnetic field of B is into the paper throughout the area. Show that by applying Faraday's

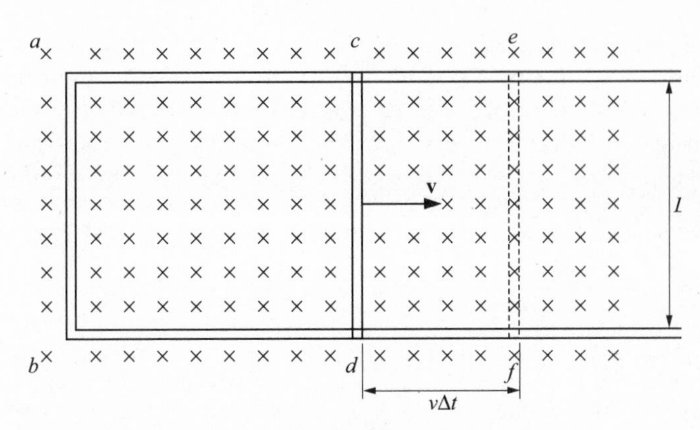

Fig. 9-8

law to the closed circuit shown in Fig. 9-9 one gets the same EMF as the motional EMF formula given by Eq. (*9.1*).

Solution

To use Faraday's law, we must calculate the change in flux per unit time. Consider the circuit of the bar and the stationary railings (circuit *cabd*). When the bar is in position *cd*, the flux equals B times the area A enclosed by circuit *cabd*, and is into the paper. After a time Δt, the bar has moved a distance $v\Delta t$, and there is an additional flux of $B\Delta A = B(v\Delta tL)$. The change in flux is therefore $\Delta\Phi = BvL\Delta t$. Since the magnitude of the EMF is equal to $\Delta\Phi/\Delta t$, we see that EMF $= BvL\Delta t/\Delta t = BvL$, as we calculated previously using the concept of motional EMF. To get the direction of the EMF let us choose $\mathbf{A}$ into the paper. Then Φ_i and $\Delta\Phi$ are both positive (into the paper), and so is $\Delta\Phi/\Delta t$. $-\Delta\Phi/\Delta t$ is therefore out of the paper, and by the right-hand rule the EMF is counter clockwise which is the same direction calculated using motional EMF (from $d \to c$).

We could also get the direction from Lenz's law using the following table:

Φ_{initial}	into paper
$\Delta\Phi$	into paper (since the flux increased)
Φ_{induced}	out of paper (by Lenz's law)
Induced current	around *cabd* (since must produce a field out of paper)

Again, this direction is the same as the direction calculated using motional EMF.

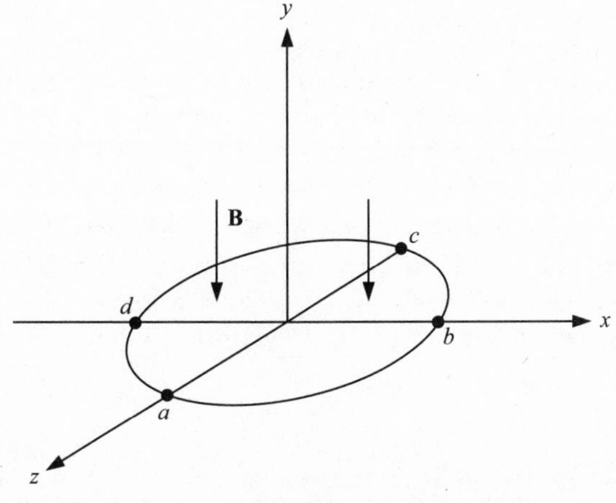

Fig. 9-9

Problem 9.9. A circular loop, with area $A = 0.24\ \text{m}^2$, is in the x–z plane, as in Fig. 9-9. There is a uniform magnetic field B in the region of 0.25 T, pointing in the $-y$ direction. In all of the following parts, a change takes place in a time of 3.0×10^{-3} s. Calculate the induced EMF in each case and determine the direction of the induced current around the loop. (Remember that only this externally induced current is considered here, and we will neglect the additional current induced by the externally induced current.)

(*a*) The field increases to 0.30 T.

(*b*) The field reverses to the $+y$ direction.

(*c*) The field stays fixed but the loop rotates about the x-axis so that point a moves up by 30°.

Solution

(*a*) The initial and the final fields are both in $-y$. We choose **A** to point downward as well. Substituting into Eq. (*9.2*), and noting $\theta = 0$, we get $\Phi_i = (0.25\ \text{T})(0.24\ \text{m}^2) = 0.060$ Wb, and $\Phi_f = (0.30\ \text{T})(0.24\ \text{m}^2) = 0.072$ Wb. Thus, $\Delta\Phi = \Phi_f - \Phi_i = 0.072 - 0.06 = 0.012$ Wb. The magnitude of the induced EMF is therefore EMF = (0.012 Wb)/(0.0030 s) = 4.0 V.

To get the direction, we make the following table

Φ_{initial}	down
$\Delta\Phi$	down (since the flux increased in the initial direction)
Φ_{induced}	up (by Lenz's law)
Induced EMF (and current)	around *abcd* (since must produce a field up)

(*b*) The initial and final fluxes are both 0.060 Wb, but the final flux is up, while the initial flux is down. This time choosing **A** as positive upward, $\Delta\Phi = \Phi_f - \Phi_i = 0.060 - (-0.060) = 0.12$ Wb. The magnitude of the induced EMF is therefore EMF = (0.12 Wb)/(0.0030 s) = 40 V.

To get the direction, we make the following table

Φ_{initial}	down
$\Delta\Phi$	up (since the flux changed to the opposite direction)
Φ_{induced}	down (by Lenz's law)
Induced EMF (and current)	around *adcb* (since must produce a field down)

(*c*) Choosing **A** downward, we have $\Phi_f = (0.25\ \text{T})(0.24\ \text{m}^2)\cos 30° = 0.052$ Wb, while $\Phi_i = 0.060$. Both go through the coil in the same downward direction. The change in flux is $\Delta\Phi = \Phi_f - \Phi_f = 0.052 - 0.060 = -0.0080$ Wb. The magnitude of the induced EMF is therefore EMF = (0.0080 Wb)/(0.0030 s) = 2.67 V.

To get the direction, we make the following table

Φ_{initial}	down
$\Delta\Phi$	up (since the downward flux decreased)
Φ_{induced}	down (by Lenz's law)
Induced EMF and current	around *adcb* (since must produce a field down)

Problem 9.10. The circular coil in Fig. 9-9 is rotating about the x-axis with an angular velocity of 130 radians/s, with point a moving upward from the position shown. The coil has an area of $0.24\ \text{m}^2$ and is initially in the x–y plane. There is a uniform magnetic field B in the region of 0.25 T, pointing in the $-y$ direction. Calculate the average EMF induced in the coil during a time of 5×10^{-3} s.

Solution

During the time of $\Delta t = 5 \times 10^{-3}$ s, the coil rotates through an angle of $\omega\Delta t = 130\Delta t = 0.65$ radians. The final flux is therefore the flux after point a has moved up through this angle. In Problem 9.5(*c*), we did this calculation for an angle of 30°. Repeating that calculation for an angle of 0.65 rad, we get that $\Phi_f =$

$(0.25\ \text{T})(0.24\ \text{m}^2)\cos 0.65 = 0.048$ Wb. Also, $\Phi_i = (0.25\ \text{T})(0.24\ \text{m}^2) = 0.060$ Wb. Thus $\Delta\Phi = \Phi_f - \Phi_i = 0.048 - 0.060 = -0.012$ Wb. Therefore, the magnitude of the EMF is $(0.012\ \text{Wb})/(0.0050\ \text{s}) = 2.4$ V.

To get the direction, we make the following table

Φ_{initial}	down
$\Delta\Phi$	up (since the downward flux decreased)
Φ_{induced}	down (by Lenz's law)
Induced EMF and current	around *adcb* (since must produce a field down)

Problem 9.11. An elastic circular conducting loop, is at the equator of an air filled balloon, a hemispherical cross section of which is shown in Fig. 9-10. The sphere has a radius of 0.60 m. There is a uniform magnetic field **B** in the region of 0.25 T, pointing in the $+y$ direction. During a time of 5.0×10^{-2} s, the balloon is deflated to a radius of 0.30 m. What is the average EMF induced in the coil during this time?

Solution

The initial flux is $\Phi_i = (0.25\ \text{T})(\pi)(0.60\ \text{m})^2 = 0.28$ Wb. The final flux is $\Phi_f = (0.25\ \text{T})(\pi)(0.30\ \text{m})^2 = 0.070$ Wb. Thus, $\Delta\Phi = \Phi_f - \Phi_i = 0.070 - 0.28 = -0.21$ Wb, and the magnitude of the EMF $= (0.21\ \text{Wb})/(0.050\ \text{s}) = 4.2$ V.

To get the direction, we make the following table

Φ_{initial}	up
$\Delta\Phi$	down (since the downward flux decreased)
Φ_{induced}	up (by Lenz's law)
Induced EMF and current	around *abcd* (since must produce a field up)

Problem 9.12. A long solenoid, with 2500 turns/m, and a cross-sectional area of 0.70 m², carries a current of 2.0 A, in the direction shown in Fig. 9-11. Another wire is wound around the solenoid in the same direction, near its center, with 15 turns to form a coil. This coil is connected in a circuit containing a resistance of 3.0 Ω. The current in the solenoid is turned off in a time of 0.0030 s.

(*a*) What average EMF is induced in the coil?

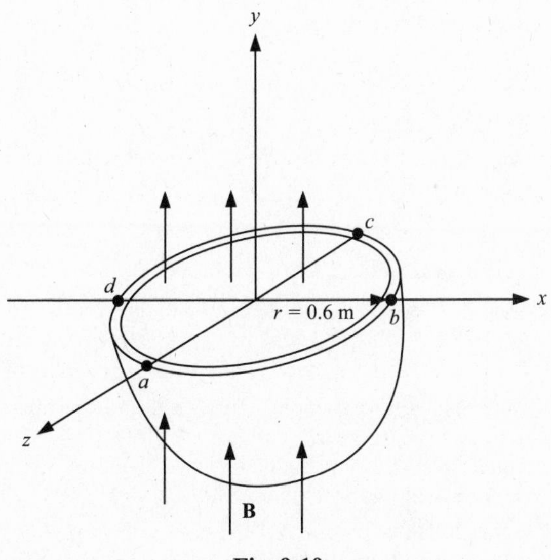

Fig. 9-10

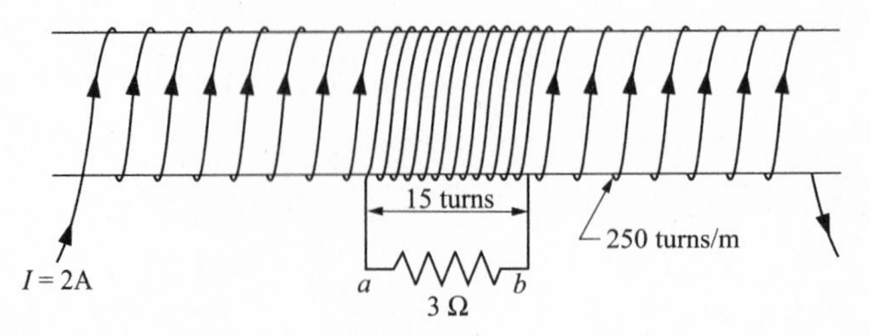

Fig. 9-11

(*b*) What average current flows through the resistor?

(*c*) Does the induced current flow through the resistor from a to b or from b to a?

(*d*) How would the answer change if the coil were wound in the opposite direction to that of the solenoid?

Solution

(*a*) Since for a long solenoid, the field is uniform inside, the flux through *one turn* of the second wire is $\Phi = BA$, where $B = \mu_0 nI = 4\pi \times 10^{-7}(2500)(2.0\ \text{A}) = 6.28 \times 10^{-3}$ T, and A is the cross-sectional of the solenoid and equals 0.70 m^2.

> ***Note.*** A is also the area of a tightly wound second coil; however even if the area of the coil were larger than the area of the solenoid, the flux through a turn of the coil would still be the same, since the field of the solenoid is zero in the area between it and the coil.
>
> The flux through the entire coil is 15 times this amount, since there are 15 turns in the wire. Thus $\Phi_i = 9.42 \times 10^{-2}$ Wb. The final flux is 0. Thus, $\Delta\Phi = \Phi_f - \Phi_i = 0\ 9.42 \times 10^{-2}$ Wb, and the magnitude of the average EMF $= (9.42 \times 10^{-2}\ \text{Wb})/(0.0030\ \text{s}) = 31.4$ V.

(*b*) Assuming that the coil itself has no resistance, the EMF = terminal voltage of the coil. From Ohm's law the average current is $V/R = (31.4\ \text{V})/(3.0\ \Omega) = 10.5$ A.

(*c*) To get the direction, we make the following table

Φ_{initial}	to left (using right-hand rule) for current in the first wire on the solenoid
$\Delta\Phi$	to right (since the flux to the left decreased)
Φ_{induced}	to left (by Lenz's law)
Induced EMF	from a to b through coil; then b is at a higher voltage than a and the current in the resistor is from b to a, and from a to b in the coil, assuring Φ_{induced} is toward the left.

(*d*) In this case the current would have to flow from a to b through the resistor (and b to a through the coil) to insure Φ_{induced} is to the left.

Problem 9.13. Repeat Problem 9.12 if the solenoid is filled with a magnetic material that has a relative permeability of 1500.

Solution

(*a*) As in Problem 9.12 the field in the solenoid is uniform, and the flux through one turn of the second wire is $\Phi = BA$. Now, however, $B = \mu nI = 4\pi \times 10^{-7}\ (1500)(2500)(2.0\ \text{A}) = 9.42$ T; the area A is still the area of the now filled solenoid and equals 0.70 m^2. The total flux in the second coil is then 15(0.70 m^2)(9.42 T) = 99 Wb. The final flux is 0. Thus, $\Delta\Phi = \Phi_f - \Phi_i = 0 - 99$ Wb, and the magnitude of the average EMF $= (99\ \text{Wb})/(0.0030\ \text{s}) = 3.30 \times 10^4$ V.

(*b*) The average current is again EMF/R $= (3.3 \times 10^4\ \text{V})/(3.0\ \Omega) = 1.1 \times 10^7$ A.

(*c*) To get the direction, we use the same table as in Problem 9.12(*c*) with the same result: the induced current is from b to a through the resistor to ensure that the induced flux is to the left (for the case of the coil wound in the same direction as the solenoid).

(*d*) Same as in 9.13(*d*)

Problem 9.14. A hollow, long solenoid has 1500 turns/m, and carries a current of 3 A. At the center of the solenoid there is a small rectangle, of sides $h = 0.020$ m and $w = 0.040$ m, whose area is parallel to the axis of the solenoid, as in Fig. 9-12(*a*). In a time of 4.0×10^{-4} s, the rectangle rotates so that the plane is now perpendicular to the axis, as in Fig. 9-12(*b*). Calculate the average EMF induced in the rectangle during this time.

Solution

Since for a solenoid, the field inside is uniform, the flux through the rectangle is $\Phi = BA \cos \theta$, where $B = \mu_0 nI = 4\pi \times 10^{-7}(1500)(3.0 \text{ A}) = 5.65 \times 10^{-3}$ T, and the field points along the axis to the left. The area of the rectangle is $A = (0.020)(0.040)\text{m}^2 = 8.0 \times 10^{-4}\ \text{m}^2$. The flux through the rectangle is initially zero, since the area vector is perpendicular to the field ($\theta = 90°$). After the rotation, $\theta = 0°$, and the flux is $\Phi_f = (5.65 \times 10^{-3}\ \text{T})(8.0 \times 10^{-4}\ \text{m}^2) = 4.52 \times 10^{-6}$ Wb. Then $|\text{EMF}| = \Delta\Phi/\Delta t = (4.52 \times 10^{-6}\ \text{Wb})/(4.0 \times 10^{-4}\ \text{s}) = 1.13 \times 10^{-3}$ V.

Problem 9.15. A rectangle *abcd* is located at the origin, as in Fig. 9-13. A uniform magnetic field exists only in the first quadrant, and points into the paper. Calculate the direction of flow of the induced current, if (*a*) the field is increased; (*b*) the rectangle moves to the right; (*c*) the rectangle moves to the left; (*d*) side *bc* of the rectangle is pulled to the right while side *ad* is held fixed, and the rectangle's sides *dc* and *ab* are stretched without snapping; (*e*) the rectangle rotates in the plane about point a in the counter-clockwise direction; in the clockwise direction; and (*f*) the rectangle rotates about the y-axis into the paper.

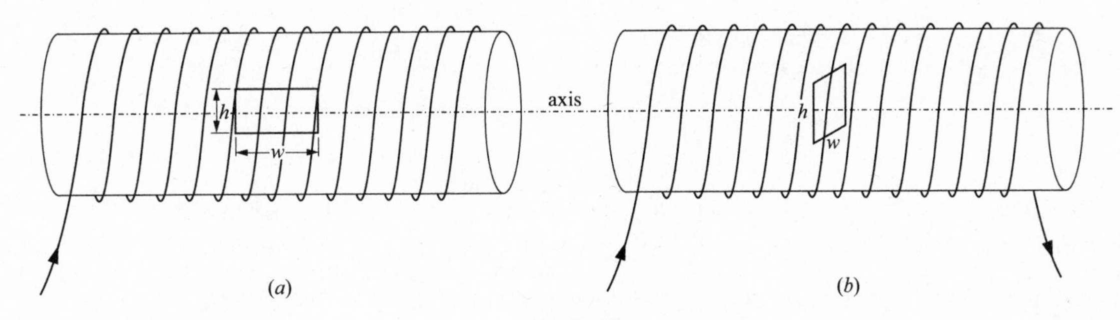

Fig. 9-12

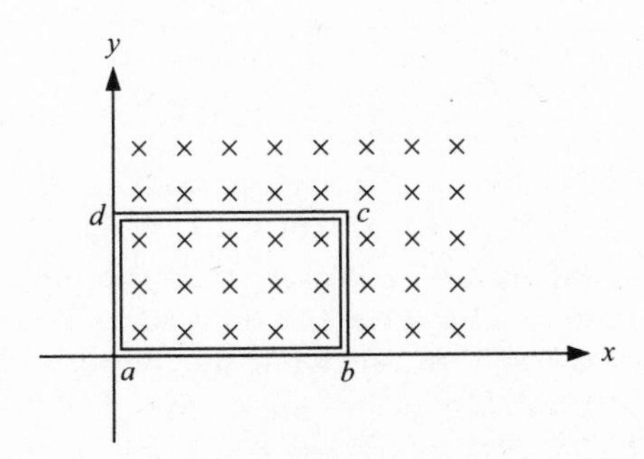

Fig. 9-13

Solution

To get the direction in each case we make the usual table:

(*a*)	Φ_{initial}	in
	$\Delta\Phi$	in (since the inward flux increased)
	Φ_{induced}	out (by Lenz's law)
	Induced current	counter-clockwise (ccw) Ω (since must produce field out)
(*b*)	Φ_{initial}	in
	$\Delta\Phi$	0 (since the flux doesn't change)
	Φ_{induced}	0
	Induced current	0
(*c*)	Φ_{initial}	in
	$\Delta\Phi$	out (since the inward flux decreased)
	Φ_{induced}	in (by Lenz's law)
	Induced current	clockwise (cw) (since must produce field in)
(*d*)	Φ_{initial}	in
	$\Delta\Phi$	in (since the inward flux increased—area inside first quadrant increased
	Φ_{induced}	out (by Lenz's law)
	Induced current	ccw (since must produce field out)

(*e*) For either clockwise or counter clockwise rotation

Φ_{initial}	in
$\Delta\Phi$	out (since the inward flux decreased—part of the rectangle moves outside the first quadrant where there are no field lines)
Φ_{induced}	in (by Lenz's law)
Induced current	cw (since must produce field in)

(*f*)	Φ_{initial}	in
	$\Delta\Phi$	out (since the inward flux decreased—the number of field lines through the rectangle decreases as the rectangle rotates ($\cos\theta$ effect))
	Φ_{induced}	in (by Lenz's law)
	Induced current	cw (since must produce field in)

9.4 GENERATORS

We have previously mentioned that one can generate an EMF by rotating a coil in a magnetic field. In Fig. 9-14(*a*), we have a tightly wound rectangular coil of area A rotating about the z-axis with an angular velocity ω, in a magnetic field B which points in the $+x$ direction. In Fig. 9-14(*b*), we show this coil in projection. The flux through a single turn of the coil is given by $BA\cos\theta$, where θ is the angle between the normal to the rectangle and the field. Assuming the position shown in (*a*) is at $t = 0$, θ is zero, and then θ changes as ωt, i.e. $\theta = \omega t$. Therefore we can write the flux as $\Phi = BA\cos\omega t$. In order to calculate the EMF we would have to be able to calculate $\text{EMF} = -\Delta\Phi/\Delta t$ at each instant of time. This can be easily done using calculus, with the result that

$$\text{EMF} = \omega BA \sin\omega t \tag{9.6}$$

Let us examine this result carefully. The EMF varies as $\sin\omega t$. This is sometimes positive and sometimes negative. When the EMF is positive the current flows around the coil in the "positive" direction. This direction is defined by our choice of the positive **A** vector [Fig. 9-14(*b*)]. The direction chosen implies that the positive direction for circulation around the coil is from a to b to c to d to a. This is because if you wind your fingers in that direction, your thumb points in the positive **A** direction. Note in the figure that EMF is positive when **A** makes an angle less than 180° with **B**, but turns negative when that angle is between 180° and 360°. A negative EMF will cause an induced current to flow in the direction *adcb*. The EMF produced in this manner will change its direction and then change back again

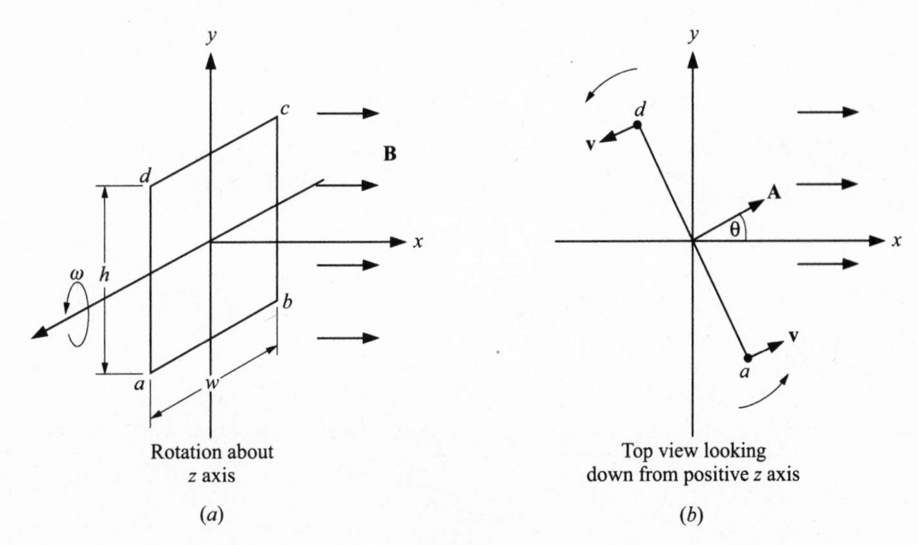

Fig. 9-14

at an angular frequency ω, or at a frequency $f = \omega/2\pi$ and period $T = 2\pi/\omega$. This is what we call an alternating voltage which produces an **alternating current** (AC). Thus, by rotating a coil in a magnetic field, we can easily generate an AC voltage. The magnitude of the voltage can be increased by constructing the coil out of many turns, N, of wire, in which case the voltage becomes

$$\text{EMF} = \omega NBA \sin \omega t \tag{9.7}$$

Generally the frequency used in the USA for AC is 60 Hz (in Europe they use 50 Hz). The time variation of the EMF [representing Eq. (*9.6*)] is shown in Fig. 9-15(*a*). The rotating coil is the basic principle behind the AC generator.

From Eq. (*9.7*) we see that the maximum terminal voltage that is generated is ωNBA. This maximum occurs when $\sin \omega t = 1$, and the voltage varies between $\pm\omega NBA$. Instead of defining an AC voltage by this maximum value, we will see, in a later chapter, that one uses the "rms" or "root mean square" value, which, for a sinusoidal voltage, is just $V_{max}/\sqrt{2} = 0.707\ V_{max}$. This V_{rms} equals 120 volts in the USA, and 220 volts in much of Europe.

Problem 9.16. A generator is built to provide AC at $f = 60$ Hz, with a voltage of 120 V. If a magnet is used that has a magnetic field of 0.20 T, what area coil is required, if the coil contains 200 turns?

Solution

We know that EMF $= \omega NBA \sin \omega t$, and therefore $V_{max} = \omega NBA$, and $V_{rms} = 0.707\ \omega NBA$. Substituting in the equation gives us that (recalling $\omega = 2\pi f$):

$$120\ \text{V} = 0.707(2\pi)(60\ \text{Hz})(200)0.20\ \text{T})A, \qquad A = 2.25\ \text{m}^2$$

In practice, the utility companies produce much higher voltages, which they send on transmission lines to substations, where the voltages are reduced before being transmitted to individual homes.

It is also possible to construct generators that do not give rise to alternating current. To produce a DC voltage (DC means **direct current**, which was discussed previously in Chap. 5), we add two modifications. First, we reverse the connection to the outside wires every time the direction of the EMF in the coil reverses direction. Then the current will always move in the same direction in the outside circuit. The resultant EMF in the outside circuit will then take the form shown in Fig. 9-15(*b*). Secondly, we use several coils, which are wound around frames fixed on a common rotating shaft with the frames at fixed

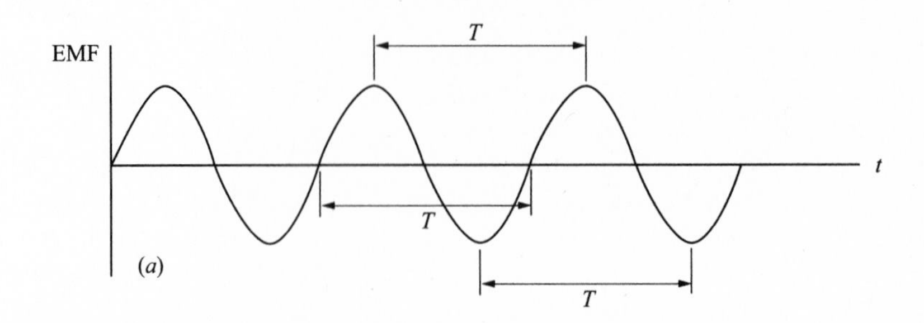

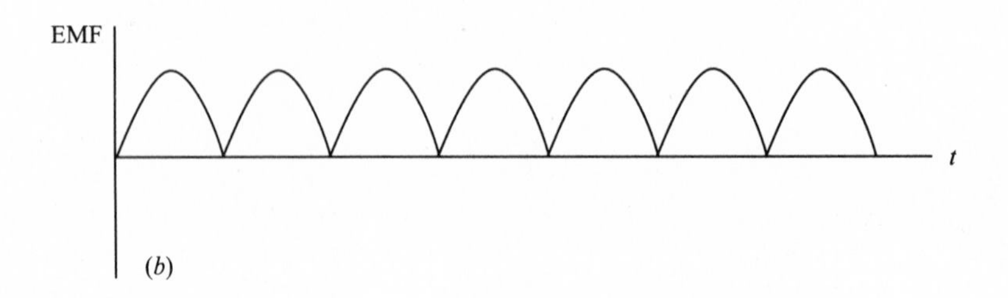

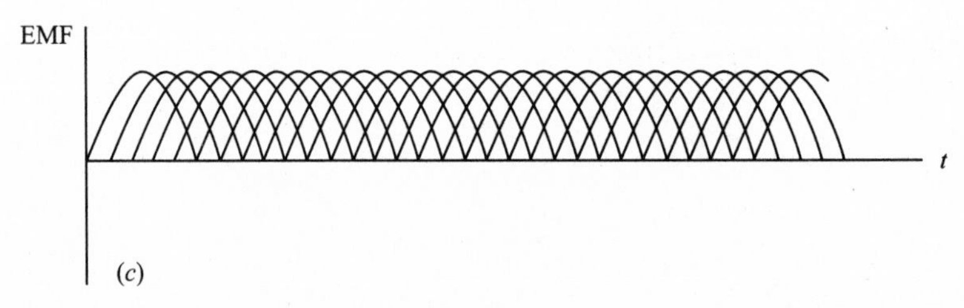

Fig. 9-15

angles to each other. Such a structure is called a turning "armature" (Fig. 9-16). Each coil will produce a voltage which reaches its maximum at a different time, and the total voltage will vary very little with time. Fig. 9-15(c) shows the voltage if there are four coils spaced at 45° from each other around the armature.

9.5 INDUCED ELECTRIC FIELDS

When we produce an EMF by induction as a result of a changing magnetic field with the circuit itself fixed, there are no magnetic forces on the charges in the circuit since the circuit is not moving.

y
B
x

Cross section of four coil armature
seen from above (z axis)

Fig. 9-16

What then exerts the force on the charges in the circuit that produces the EMF? The answer is that a new type of electric field is produced by the time varying magnetic field in the vicinity of the circuit that pushes the charges and creates the EMF. This electric field will be equal to what is required to create the EMF predicted by Faraday's law as we go around the circuit. This new electric field is fundamentally different from the "electrostatic" field produced by point charges, because the field of point charges will do zero work as they push a charge around a closed circuit. This latter property is responsible for the fact that the "electrostatic" electric force of point charges is conservative. This new electric field will not be conservative, because an EMF in a stationary circuit means that net work will be done on charges that are moved around a closed circuit. The amount of work done per unit charge will, in fact, just equal the EMF.

We can show that any changing magnetic field is a source of a non conservative electric field. Thus, Faraday's law has profound implications for our concept of the electric field. We will see later on that a changing electric field will also be a source of a magnetic field. These two new sources of fields, when added to the previous sources, form the basis of the complete laws describing electro-magnetic fields, and the resulting equations are known as Maxwell's equations.

Problems for Review and Mind Stretching

As stated previously, in all the problems we include only currents induced by external changes in flux, and ignore any additional changes in current due to Φ_{induced}, the flux induced by currents themselves.

Problem 9.17. A square coil of side 0.10 m is moving to the right at a velocity of 5.0 m/s. It is 0.10 m from a region that contains a uniform magnetic field of 0.90 T into the paper, and extends for a distance of 0.30 m, as in Fig. 9-17.

(*a*) What motional EMF is induced in the coil in the original position? Does the induced current flow clockwise or counter-clockwise?

(*b*) What motional EMF is induced in the coil when the right side of the coil is in the field, while the left side is still outside? Does the induced current flow clockwise or counter-clockwise?

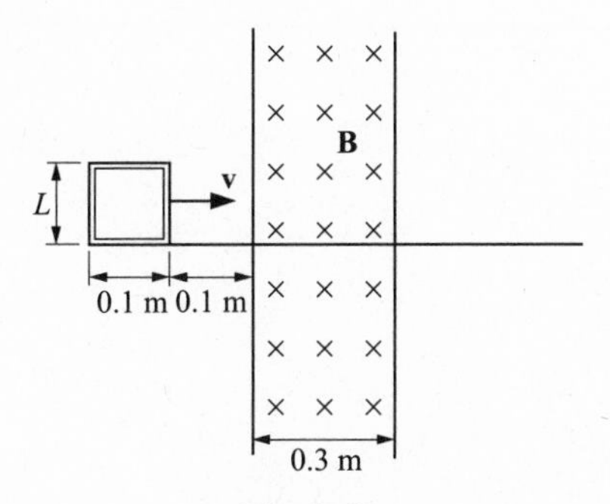

Fig. 9-17

(*c*) What motional EMF is induced in the coil when the entire coil is in the magnetic field? Does the induced current flow clockwise or counter-clockwise?

(*d*) What motional EMF is induced in the coil when the right side is outside of the field while the left side is still in the field? Does the induced current flow clockwise or counter-clockwise?

Solution

(*a*) Since there is no motion within a magnetic field there is no induced EMF.

(*b*) The EMF on the right side is EMF $= vLB = (5.0\ \text{m/s})(0.10\ \text{m})(0.90\ \text{T}) = 0.45$ V. The induced current flows counter-clockwise.

(*c*) The EMF on the right side is still 0.45 V, but the induced EMF on the left side is also 0.45 V. Both EMFs point upward and tend to induce currents in opposite directions. Thus, they cancel each other and the net EMF in the coil is zero.

(*d*) There is no EMF on the right side, and a clockwise EMF of 0.45 V on the left side.

Problem 9.18. A long straight wire carries a current of 2.0 A as in Fig. 9-18. A small circular coil of diameter 0.20 m is located with its center at a distance of 5.0 m from the wire. Assume that the field within the coil is uniform, and has the value of the field at its center.

(*a*) Calculate the flux in the coil.

(*b*) If the current in the long wire decreases to 1.5 A in 0.0030 s, what average EMF is induced in the coil?

(*c*) Will the induced current flow clockwise or counter-clockwise?

Solution

(*a*) The magnetic field is $B = (\mu_0/2\pi)(I/r) = 2 \times 10^{-7}(2.0\ \text{A})/(5.0\ \text{m}) = 8.0 \times 10^{-8}$ T. The flux is $BA = (8.0 \times 10^{-8}\ \text{T})(\pi)(0.10\ \text{m})^2 = 2.51 \times 10^{-9}$ Wb, into the paper.

(*b*) The new magnetic field is $B = (2 \times 10^{-7})(1.5\ \text{A})/(5.0\ \text{m}) = 6.0 \times 10^{-8}$ T, and the new flux is 1.88×10^{-9} Wb. The change in flux is -6.3×10^{-10} Wb, and the average EMF is $(6.3 \times 10^{-10}\ \text{Wb})/(0.0030\ \text{s}) = 2.1 \times 10^{-7}$ V.

(*c*) To get the direction, we make the following table

Φ_{initial}	in (since field is in)
$\Delta\Phi$	out (since the flux decreased)
Φ_{induced}	in (by Lenz's law)
Induced current	clockwise (since must produce Φ_{induced})

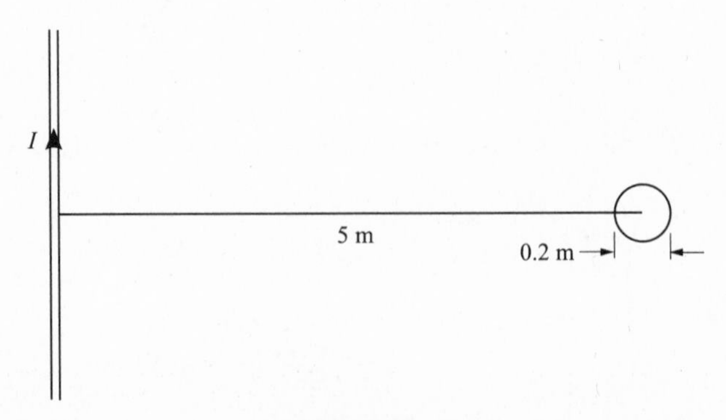

Fig. 9-18

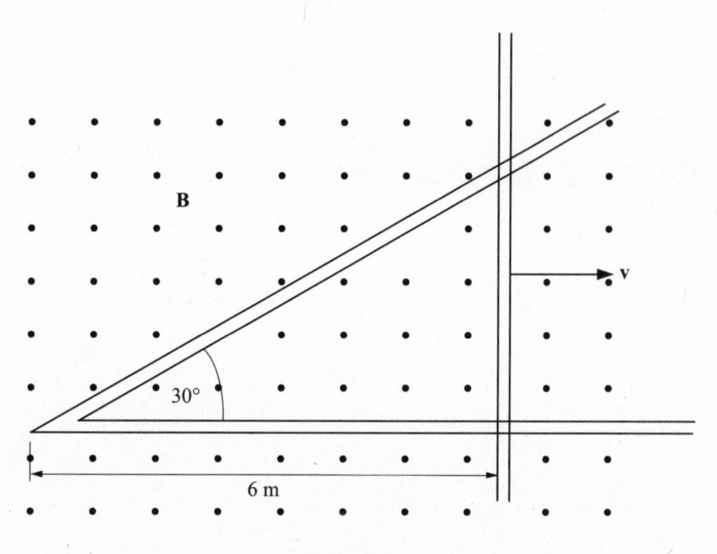

Fig. 9-19

Problem 9.19. A bar is moving along a triangular track as seen in Fig. 9-19, with a velocity of 8.0 m/s. The bar is 6.0 m from the start of the track at time $t = t_0$. There is a uniform magnetic field of 0.80 T out of the paper in the entire region. Use the concept of induced EMF to solve this problem.

(*a*) What is the average EMF induced in the circuit between $t = t_0$ and $t = t_0 + 2$ s?

(*b*) What is the average EMF induced in the circuit between $t = t_0$ and $t = t_0 + 1$ s?

(*c*) What is the average EMF induced in the circuit between $t = t_0$ and $t = t_0 + 0.25$ s?

(*d*) What is the average EMF induced in the circuit between $t = t_0$ and $t = t_0 + 0.1$ s?

(*e*) Plotting EMF_{AV} vs. time, determine the instantaneous EMF in the initial position.

(*f*) In which direction does the induced current flow?

Solution

(*a*) Starting from the initial position after 2.0 s the bar has moved 16 m. The sides of the triangle are then 22 m and 11 m. Initially, the sides of the triangle were 6.0 m and 3.0 m. The initial area is (6.0 m)(3.0 m)/2 = 9.0 m^2 and the final area is 22(11)/2 = 121 m^2. The initial flux is (9.0 m^2)(0.80 T) = 7.2 Wb and the final flux is (121 m^2)(0.80 T) = 96.8 Wb. The change in flux is 96.8 − 7.2 = 89.6 Wb and the average EMF is (89.6 Wb)/(2.0 s) = 44.8 V.

(*b*) After 1s the bar has moved only 8.0 m. The sides of the triangle are then 14 m and 7.0 m, giving an area of 49 m^2, and a flux of 39.2 Wb. The change in flux is 39.2 − 7.2 = 32 Wb and the average EMF is (32 Wb)/(1.0 s) = 32 V.

(*c*) After 0.25 s the bar moved 2.0 m. The sides of the triangle are then 8.0 and 4.0 m, giving an area of 16 m^2, and a flux of 12.8 Wb. The change in flux is 12.8 − 7.2 = 5.6 Wb and the average EMF is (5.6 Wb)/(0.25 s) = 22.4 V.

(*d*) After 0.10 s, the bar has moved 0.80 m. The sides of the triangle are 6.8 and 3.4 m, giving an area of 11.56 m^2 and a flux of 9.248 Wb. The change is flux is 9.248 − 7.2 = 2.048 Wb and an average EMF of (2.048 Wb)/(0.1 s) = 20.5 V.

(e)

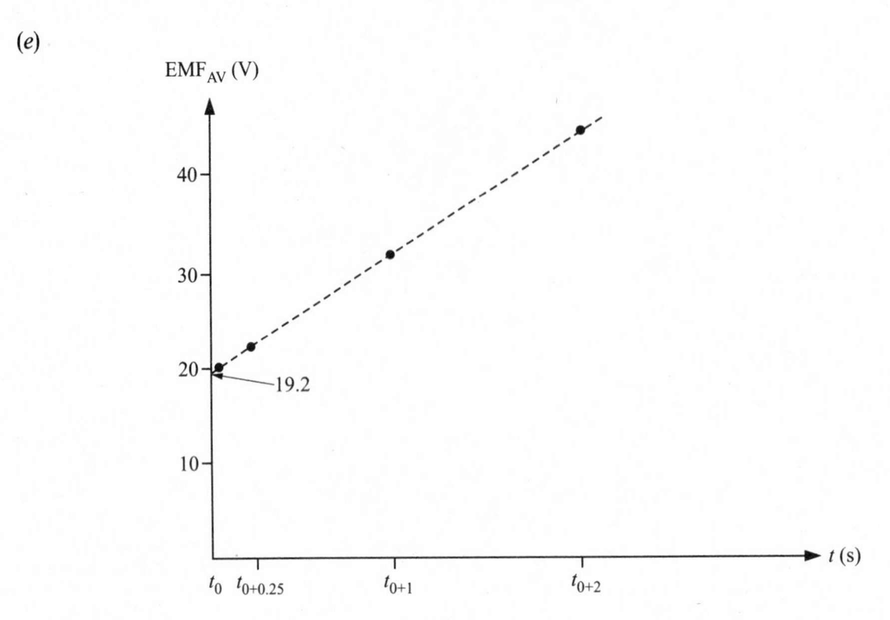

The instantaneous voltage is 19.2 V.

(f) The current flows clockwise.

Problem 9.20. A circular coil of radius 0.20 m is in a uniform magnetic field of $B = 0.60$ T coming out of the paper as in Fig. 9-20(a). In a time of 0.0020 s, the circle is deformed into a square with the same perimeter, Fig. 9-20(b). What EMF is induced in the coil, and in which direction will the induced current flow?

Solution

The initial flux is (0.60 T)(π) (0.20 m)^2 = 0.075 Wb. The new square will have sides of length h, such that $4\ h = 2\pi r$, or $h = 2\pi(0.20)/(4) = 0.314$ m and area $(0.314)^2 = 0.099$ m^2 giving a final flux $\Phi_\phi = 0.059$ Wb. The induced EMF is therefore (0.059 Wb − 0.075 Wb)/(0.0020 s) = −8.0 V.

To get the direction, we make the following table

Φ_{initial}	out
$\Delta\Phi$	in (since the flux decreased)
Φ_{induced}	out (by Lenz's law)
Induced current	ccw (since must produce field out)

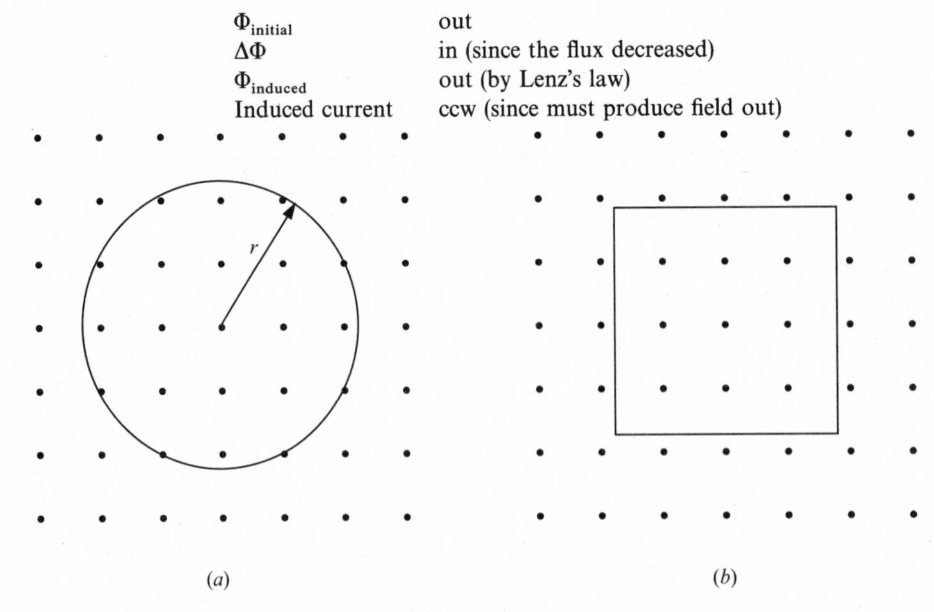

Fig. 9-20

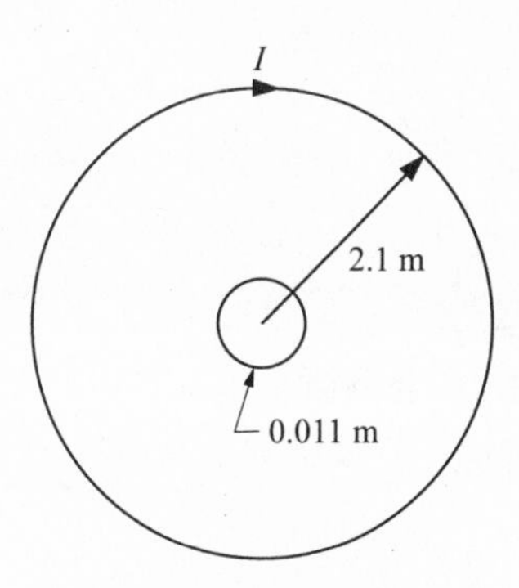

Fig. 9-21

Problem 9.21. A circular coil, of radius 2.1 m carries a clockwise current of 2.0 A, as in Fig. 9-21. A small circular coil at the center of the outside coil has a radius of 0.011 m. Assume that the magnetic field produced by the outside coil is uniform within the area of the small coil. The current in the outer coil increases at the steady rate $\Delta I/\Delta t = 3.0 \times 10^{-2}$ A/s. What EMF is induced in the small coil?

Solution

The field produced by the outer coil at its center is $B = \mu_0 I/2r$, and the flux in the small coil is $\Phi = \mu_0 IA/2r$, where A is the area of the small coil. The EMF then has magnitude $\Delta\Phi/\Delta t = (\mu_0 A/2r)(\Delta I/\Delta t)$, since only the current changes in the problem. Thus,

$$\text{EMF} = [(4\pi \times 10^{-7})(\pi)(0.011\ \text{m})^2/2(2.1\ \text{m})][3 \times 10^{-2}\ \text{A/s}] = 3.41 \times 10^{-12}\ \text{V}.$$

To get the direction, we make the following table

Φ_{initial}	out
$\Delta\Phi$	out (since the flux increased)
Φ_{induced}	in (by Lenz's law)
Induced current	cw (since must produce field in)

Problem 9.22. A long straight wire carries a current in the direction shown in Fig. 9-22. A small coil is located to the right of the wire. In what direction would current be induced in this coil if it were moving (*a*) toward the wire?; (*b*) away from the wire; and (*c*) parallel to the wire?

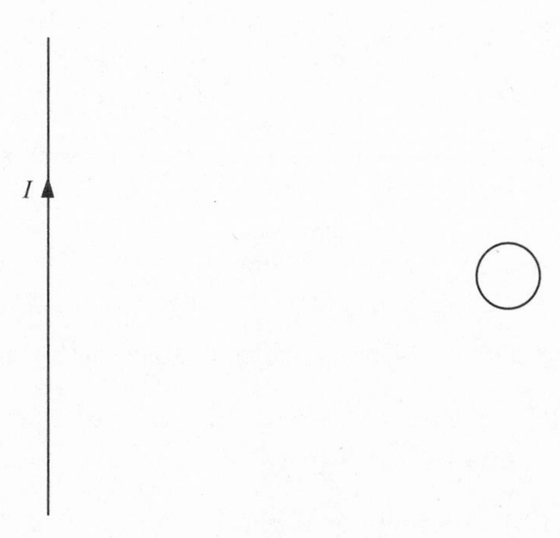

Fig. 9-22

Solution

(*a*) To get the direction, we make the following table

Φ_{initial}	in (since field is into the paper on the right)
$\Delta\Phi$	in (since the field increases as move toward wire)
Φ_{induced}	out (by Lenz's law)
Induced current	ccw (since must produce field out)

(*b*) To get the direction, we make the following table

Φ_{initial}	in (since field is into the paper on the right)
$\Delta\Phi$	out (since the field decreases as move away from wire)
Φ_{induced}	in (by Lenz's law)
Induced current	cw (since must produce field in)

(*c*) To get the direction, we make the following table

Φ_{initial}	in (since field is into the paper on the right)
$\Delta\Phi$	0 (since the field does not change)
Induced current	0

Problem 9.23. The coil in Fig. 9-14 has dimensions 0.60 m × 0.80 m, and is turning at frequency 60 Hz in a magnetic field of 0.50 T. The coil has an electrical resistance of 0.20 Ω.

(*a*) What is the current in the coil as a function of time?

(*b*) What power is consumed in the coil as a function of time?

(*c*) What is the magnetic moment of the coil as a function of time?

(*d*) What torque is exerted on the coil by the magnetic field as a function of time?

(*e*) What power must be exerted from the outside to keep the coil turning at a constant angular velocity?

Solution

(*a*) We learned in Sec. 9.4 that EMF $= \omega BA \sin \omega t = 2\pi(60 \text{ Hz})(0.50 \text{ T})(0.60 \text{ m})(0.80 \text{ m}) \sin [2\pi(60)t] = 90.5 \sin 377t$. Thus, the current is $I = \text{EMF}/R = (452 \text{ A}) \sin 377t$. The direction of the induced current is, as shown in Sec. 9.4, around the coil in the direction *abcd*. Of course, when I is negative, the meaning of this negative current is that it flows in the opposite direction.

(*b*) The power consumed is $I^2R = (453 \sin 377t)^2 \, (0.20 \, \Omega) = (4.1 \times 10^4 \text{ W})\sin^2 377t$.

(*c*) The magnetic moment, M is $IA = (217 \text{ A} \cdot \text{m}^2) \sin 377t$. Since the positive current flows around *abcd*, the direction of the moment is the vector direction of A pictured in the figure.

(*d*) The torque is given by $\Gamma = MB \sin\theta = (217 \text{ A} \cdot \text{m}^2)(\sin^2 377t)(0.50 \text{ T}) = (108 \text{ N} \cdot \text{m})\sin^2 377t$. The direction of the torque is in $+z$, which means that it opposes the rotation.

(*e*) The outside torque required to keep the coil rotating at constant speed is $108 \sin^2 377t$ in the $-z$ direction. The power needed is $\omega\Gamma = 4.1 \times 10^4 \sin^2 377t$. This is just equal to the power consumed in the circuit. Therefore, the source of the power dissipated in the circuit is the source of the external torque that turns the coil. This could be a person turning a hand crank, a diesel engine turning a shaft connected to the coil, or a water paddle turned by a waterfall (hydroelectric power).

Supplementary Problems

Problem 9.24. A conducting rod is moving to the left with a velocity of 8.0 m/s, in a magnetic field of 0.09 T, going into the paper, as in Fig. 9-23. The rod has a length of 1.2 m.

(*a*) What EMF is induced in the bar?

(*b*) Which end of the bar is at the high potential?

Ans. (*a*) 0.86 V; (*b*) *a*

Problem 9.25. A coil *abcd* is in the *y–z* plane as in Fig. 9-24. A bar magnet is parallel to the *x*-axis, to the right of the coil. What is the direction of the induced current in the coil if the magnet is moving (*a*) toward the coil; and (*b*) away from the coil?

Ans. (*a*) *abcd*; (*b*) *adcb*

Problem 9.26. Two coils are parallel to each other as in Fig. 9-25. Coil 1 has a current I_1 in the direction shown. What is the direction of the induced current in coil 2, if (*a*) it moves toward coil 1; (*b*) it moves away from coil 1; and (*c*) if the current in coil 1 is turned off?

Ans. (*a*) Opposite to I_1; (*b*) same as I_1; (*c*) same as I_1

Problem 9.27. In Fig. 9-26, a magnetic field of 0.40 T is in the *x* direction. A five sided object, *abcdef*, with dimensions shown in the figure, is placed in the magnetic field. Consider the positive direction of the area vector for each side to be from inside the object to the outside. What magnetic flux goes through sides (*a*) *abcd*; (*b*) *bcf*; (*c*) *abfe*; and (*d*) *cdef*?

Ans. (*a*) -1.6 Wb; (*b*) 0; (*c*) 1.6 Wb; (*d*) 0

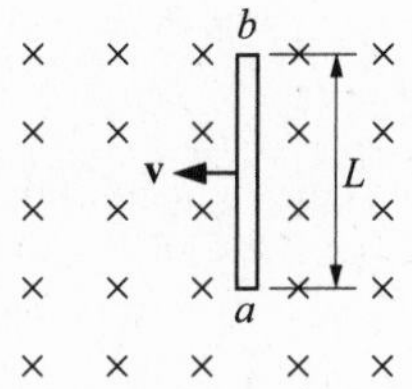

Fig. 9-23

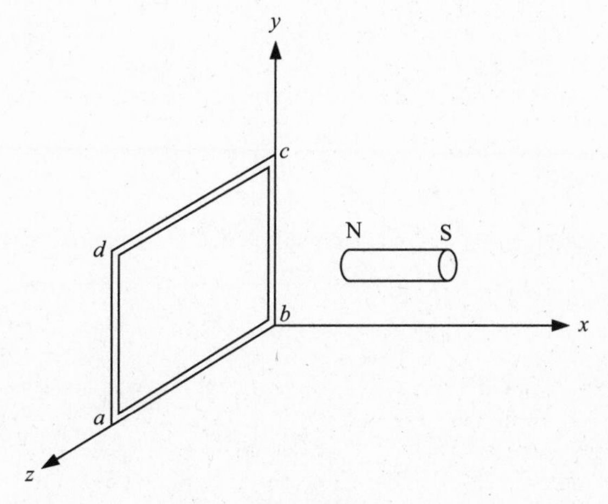

Fig. 9-24

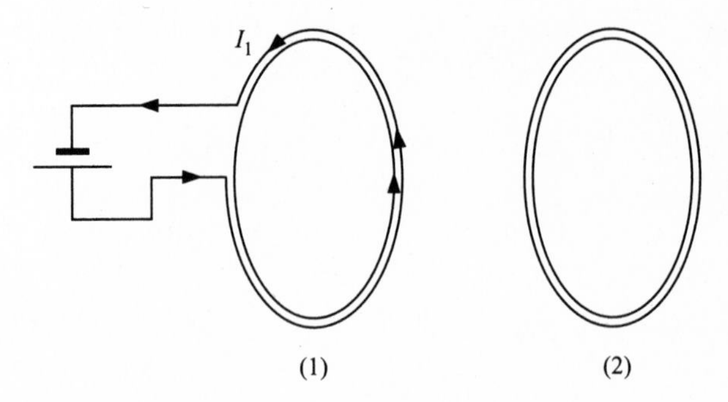

Fig. 9-25

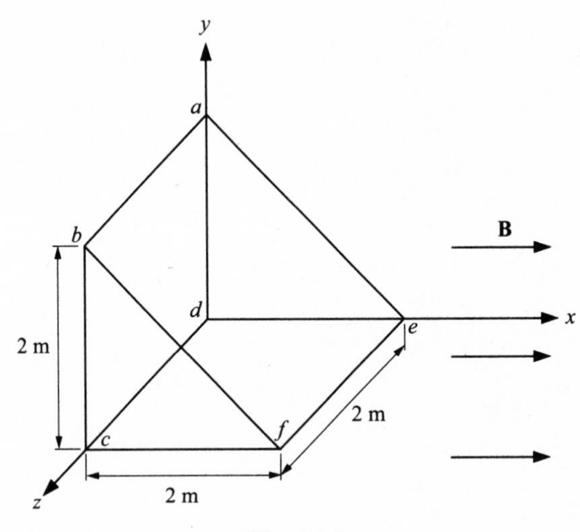

Fig. 9-26

Problem 9.28. For the object in Fig. 9-26, the magnetic field of 0.40 T is changed to the $+z$ direction. What is the magnetic flux through sides (*a*) *abcd*; (*b*) *bcf*; (*c*) *abfe*; and (*d*) *cdef*?

Ans. (*a*) 0; (*b*) 0.8 Wb; (*c*) 0

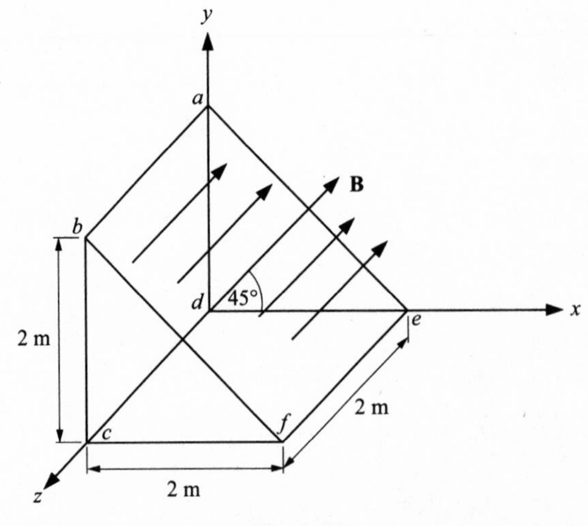

Fig. 9-27

Problem 9.29. For the object in Fig. 9-26, the magnetic field of 0.40 T is now acting at an angle of 45° to the x-axis and parallel to the x–y plane, as in Fig. 9-27. What is the magnetic flux through sides (*a*) *abcd*; (*b*) *bcf*; and (*c*) *abfe*?

Ans. (*a*) – 1.13 Wb; (*b*) 0; (*c*) 2.26 Wb

Problem 9.30. For the object in Fig. 9-26, the magnetic field, of 0.40 T, is now at an angle of 30° to the x-axis and parallel to the x–y plane, as shown in Fig. 9-28. What is the magnetic flux through sides (*a*) *abcd*; (*b*) *bcf*; and (*c*) *abfe*?

Ans. (*a*) –1.38 Wb; (*b*) 0; (*c*) 2.19 Wb

Problem 9.31. In Fig. 9-29, a magnetic field, of 0.40 T, is in the x-direction. The field is increased to 0.45 T. What is the change in flux through circuit (*a*) *abcd*; (*b*) *bcf*; and (*c*) *abfe*?

Ans. (*a*) – 0.45 Wb; (*b*) 0; (*c*) 0.45 Wb

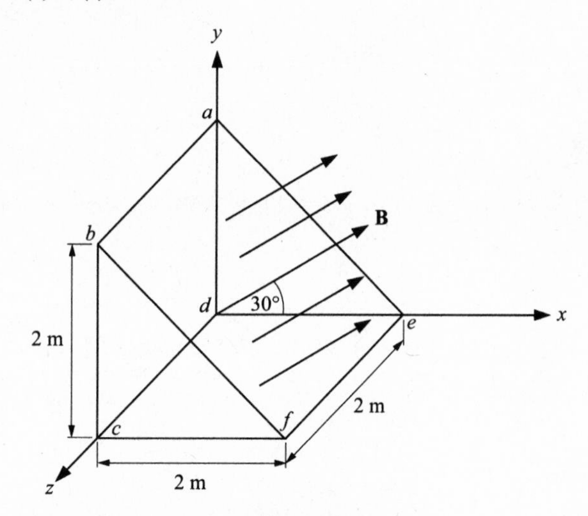

Fig. 9-28

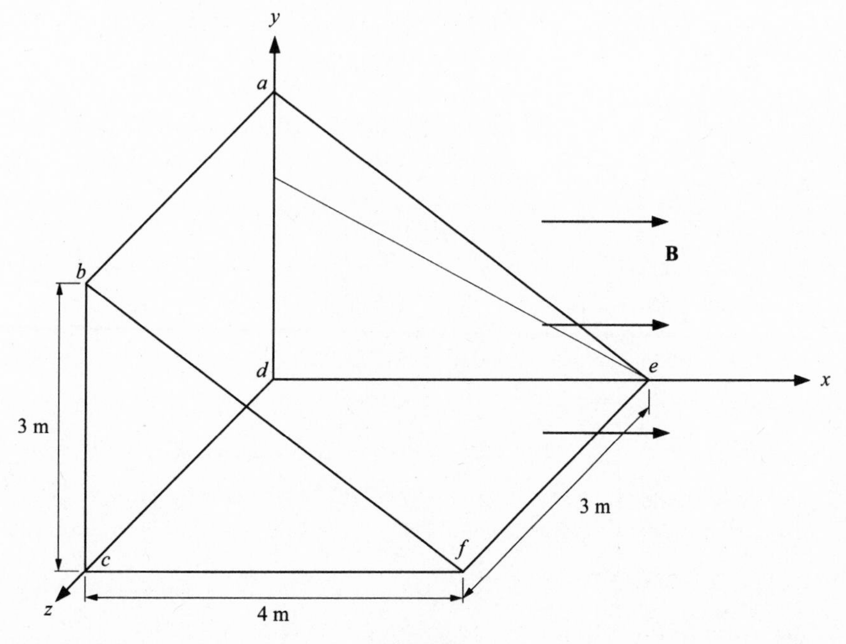

Fig. 9-29

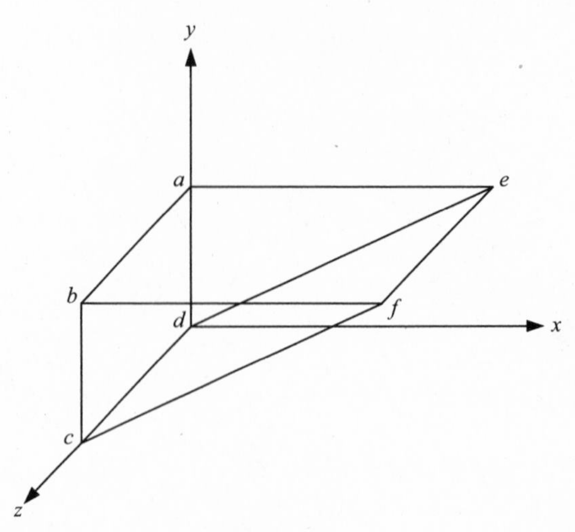

Fig. 9-30

Problem 9.32. In Fig. 9-29, a magnetic field, of 0.40 T, is in the x-direction. Side *abfe* is lifted up until it is parallel to the x–z plane, as in Fig. 9-30. What is the change in flux through the circuit *abfe*?

Ans. −3.6 Wb

Problem 9.33. In Problem 9.28, the change in field occurred during a time of 4.0×10^{-3} s. What average EMF is induced, and in which direction does it tend to induce a current flow if *abcd* were a circuit?

Ans. 112 V, direction *adcb*

Problem 9.34. In Problem 9.28, the change in field occurred during a time of 4.0×10^{-3} s. What average EMF is induced, and in which direction does it tend to induce a current flow if now *abfe* were a circuit?

Ans. 112 V, in direction *aefb*

Problem 9.35. In Fig. 9-29, a magnetic field, of 0.40 T, is in the x-direction. It is reduced to zero in a time of 4.0×10^{-4} s. What average EMF is induced, and in which direction will the induced current flow in (*a*) a circuit along *abcd*; and (*b*) a circuit along *abfe*?

Ans. (*a*) 9000 V, direction *abcd*; (*b*) 9000 V, direction *abfe*

Problem 9.36. In Problem 9.28, side *abfe* is lifted up in a time of 6.0×10^{-2} s. What average EMF is induced in circuit *abfe*, and in which direction will the induced current flow?

Ans. 60 V, direction *abfe*

Problem 9.37. In Fig. 9-26, the field of 0.40 T is changed in a time of 4.0×10^{-3} s. An average EMF of 150 V is induced in a circuit around *abcd*, with induced current flowing around the circuit in the direction *abcd*.

(*a*) What was the change in flux that occurred?

(*b*) What was the final magnetic field?

Ans. (*a*) 0.6 Wb; (*b*) 0.25 T

Problem 9.38. A triangular coil, shown in Fig. 9-31, has a magnetic field of 0.20 T, into the paper. The triangle collapses to zero area in a time of 4.0×10^{-2} s. What average EMF is induced in the coil, and in which direction does the induced current flow?

Ans. 21.9 V, direction *abc*

Problem 9.39. A toroidal solenoid (Fig. 9-32) has a mean radius of 2.1 m, and there are 1900 turns wound on its circular cross-section, of radius 0.020 m. Assume that the magnetic field is the same everywhere within the toroid, and equal to its value at the mean radius. The windings carry a current of 3.0 A. A small section of the toroid has a secondary winding of 10 turns, attached to a resistor, R, of 10 Ω.

(*a*) What is the flux through one turn of the secondary coil?

(*b*) If the current in the primary winding is increased to 3.5 A in 0.002 s, what current is induced in the secondary winding?

(*c*) If the toroid is filled with a material of relative permeability 150, what is the answer to part (*b*)?

Ans. (*a*) 6.82×10^{-7} Wb; (*b*) 5.68×10^{-5} A; (*c*) 8.53×10^{-3} A

Problem 9.40. A circular coil, of radius 1.6 m, carries a current of 4.0 A. At a point on its axis, at a distance of 0.90 m, there is a small coil, of radius 0.15 m, with 150 turns of wire, as in Fig. 9-33.

(*a*) What is the flux through the small coil, assuming the magnetic field is uniform throughout its area?

(*b*) If the coil is turned by 90° in 1.3×10^{-5} s, what EMF is induced in it?

(*c*) If the coil has a resistance of 15 Ω, what is the induced current in it?

Ans. (*a*) 1.10×10^{-5} Wb; (*b*) 0.85 V; (*c*) 0.057 A

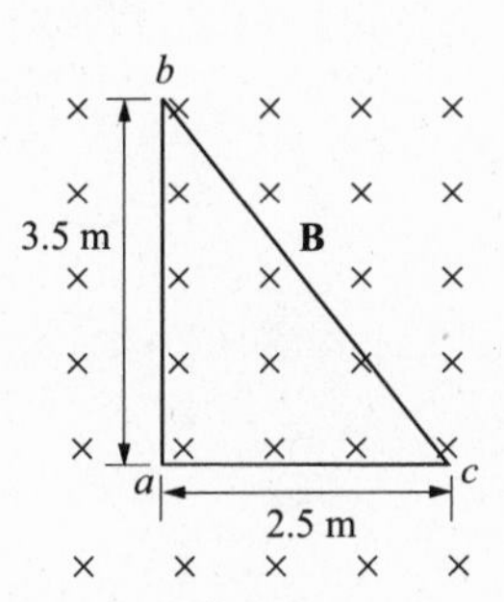

Fig. 9-31

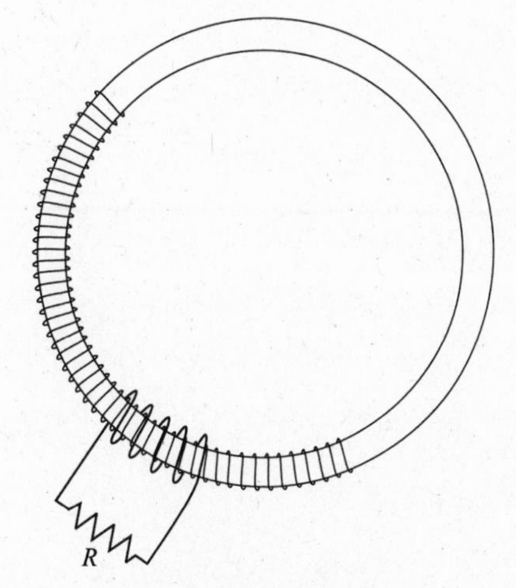

Fig. 9-32

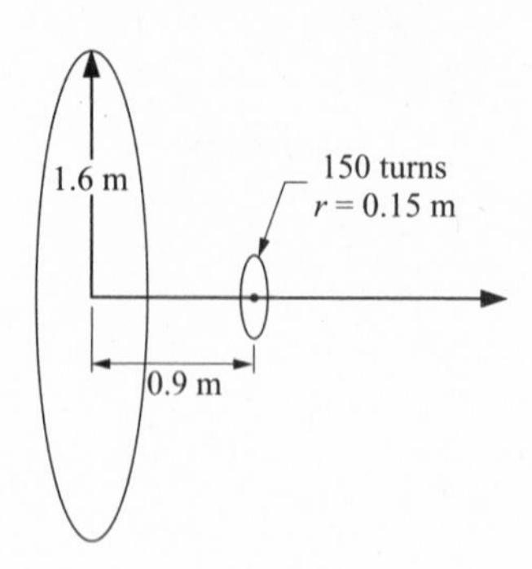

Fig. 9-33

Problem 9.41. A rectangular coil, with 300 turns each of area 6.0 m^2, rotates in a magnetic field of 0.90 T. It is connected to a circuit with a resistance of 15,000 Ω.

(*a*) What rms current flows if the coil rotates in the magnetic field at a frequency of 60 Hz?

(*b*) What rms current flows if the coil rotates in the magnetic field at a frequency of 50 Hz?

Ans. (*a*) 29 A; (*b*) 24 A

Problem 9.42. Two coils are near each other, with the first coil connected to a source of current, and the second coil connected to a voltmeter. When the first coil has a current of 2 A, the flux through the second coil is 1.3×10^{-4} Wb. If the current in the first coil is changed to 2.5 A in a time of 1.1×10^{-3} s, what voltage will be measured on the voltmeter?

Ans. 0.030 V

Chapter 10

Inductance

10.1 INTRODUCTION

In the previous chapters, we learned about the creation of a magnetic field by a current in a wire, about magnetic flux, and about the EMF produced if the magnetic flux changes. It is clear that whenever a circuit carries a current, I, a magnetic field is produced in space, and specifically in the area surrounded by the circuit. Thus there will be a certain amount of magnetic flux through the circuit, due to the current in the circuit itself. This flux depends on the magnetic field produced by the current, as well as on the geometry of the circuit. This magnetic field is always proportional to the current in the circuit, and therefore the flux through the circuit is proportional to the current as well. The flux will therefore be given by some factor times the current, where that factor will depend on the detailed geometry of the circuit. That factor is given the name **self inductance**, L.

Similarly, if we have two circuits in close proximity, as in Fig. 10-1, the current in each will produce a magnetic field in the area of the other, and therefore a flux through the other circuit. The flux through each circuit is proportional to the current in the other circuit and the proportionality constants are called mutual inductances. We will first discuss self inductance and then address the issue of mutual inductance.

10.2 SELF INDUCTANCE

As was stated in the introduction, self inductance arises from the flux that a current circuit produces within its own area. The self inductance, which depends only on the geometry of the circuit, connects this flux with the current, and is defined as

$$L = \Phi/I, \text{ or } \Phi = LI \tag{10.1}$$

The terminology "self" inductance arises from the fact that it involves only the flux through a circuit caused by the current in that circuit itself. In general there may be additional flux through a circuit which originates from currents in other circuits or from permanent magnets. The self inductance is usually just called inductance, unless one wishes to distinguish it from the mutual inductance. The

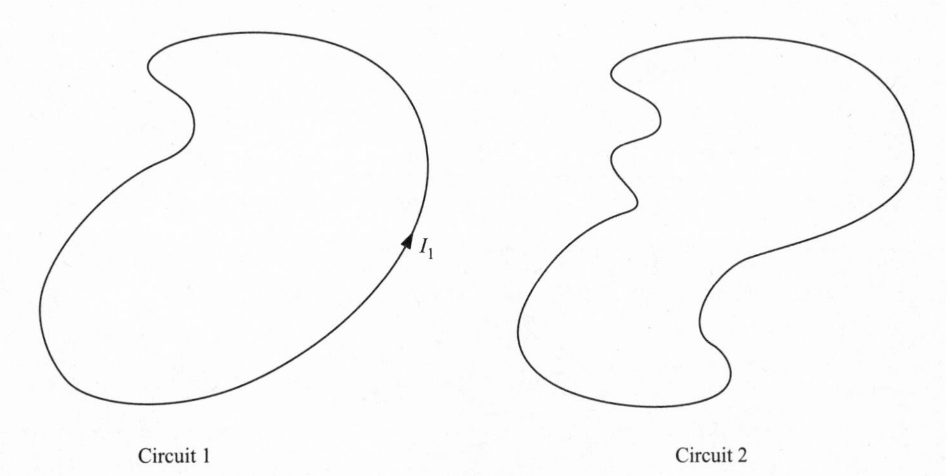

Fig. 10-1

unit for inductance is Wb/A, which is given the name **henry**. Practical circuits have inductance much smaller than one henry, more in the range of millihenries. The main use of the concept of inductance will be in circuits where the current changes, thus causing a proportional change in flux. This changing flux induces an EMF:

$$\text{EMF} = -\Delta\Phi/\Delta t = -L(\Delta I/\Delta t) \tag{10.2}$$

The fact that this EMF is "induced" by the changing flux is the source of the name inductance. The minus sign is a reminder of Lenz's law, that the induced current tries to oppose the change in current.

The procedure for calculating the self inductance is simple in principle, but in practice it may involve complicated calculations. The procedure is as follows. First, we calculate the magnetic field produced by a current, I, at every point within the area of the circuit. Then, using this field, we calculate the flux through the area of the circuit, taking account of the fact that the field is likely to vary from point to point in the area. Once we have calculated the flux through the area, we divide this flux by the current, resulting in the self inductance. The application of this procedure is best illustrated by some examples.

Solenoid

Suppose we have a long solenoid of length d which has n turns per meter, as in Fig. 10-2. We want to know the self inductance of the solenoid. Following the procedure outlined above, we first calculate the magnetic field inside the solenoid. This has previously been calculated to be $B = \mu_0 nI$. This field is uniform everywhere within the solenoid, and therefore the flux passing a single turn of the solenoid is just $\Phi_1 = BA\cos\theta = \mu_0 nIA$. In the length, d, of the solenoid, there are nd turns, so the total flux is $\Phi_T = \mu_0 n^2 IAd$ through the circuit. The self inductance, L, of the solenoid is therefore

$$L = \mu_0 n^2 Ad \tag{10.3a}$$

and the inductance per unit length is

$$L/d = \mu_0 n^2 A \tag{10.3b}$$

Problem 10.1.

(*a*) For the solenoid shown in Fig. 10-2, calculate the inductance per unit length if there are 180 turns/m, and the radius of the solenoid is 0.60 m.

(*b*) How does this answer change if the number of turns/m is 360?

Solution

(*a*) Using the formula that we just developed, we get $L/d = 4\pi \times 10^{-7}(180)^2(\pi)(0.60)^2 = 4.6 \times 10^{-2}$H.

(*b*) The inductance varies as the square of the number of turns/length, n; therefore $L/d = 4 \times$ (result of part *a*) $= 18.4 \times 10^{-2}$ H.

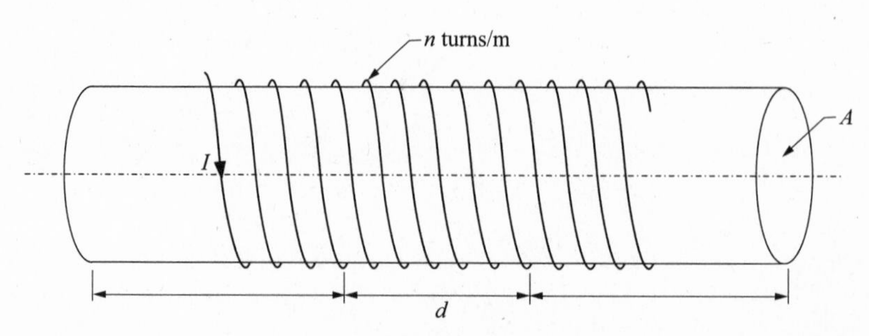

Fig. 10-2

Problem 10.2. For the solenoid shown in Fig. 10-2, calculate the inductance per unit length if there are 180 turns/m, the radius of the solenoid is 0.60 m and the solenoid is filled with a material of permeability of 5.2.

Solution

The field through the solenoid is now larger, as is the flux, so we expect an increase in the inductance. The formula that we just developed must be modified to account for the different material inside the solenoid. The magnetic field is modified by substituting μ for μ_0, and with this one change, we get $L/d = \mu n^2 A$. Thus $L/d = 4\pi \times 10^{-7}(5.2)(180)^2(\pi)(0.60)^2 = 2.4 \times 10^{-1}$ H.

Often an electric circuit will contain a solenoid, with the only significant magnetic field being produced by the solenoid itself. In that case a knowledge of the self inductance of the solenoid is necessary in order to fully understand the response of the circuit to changes in current.

Problem 10.3. A circuit contains a solenoid with an inductance of 3.0 mH, and carries a current of 2.0 A.

(*a*) How much flux passes through the solenoid?

(*b*) If the current is increased to 4.0 A in a 2 s interval, what is the average EMF induced in the circuit?

Solution

(*a*) We know that $\Phi = LI$, so therefore $\Phi = 3.0 \times 10^{-3}(2.0) = 6.0 \times 10^{-3}$ Wb.

(*b*) $\mathrm{EMF_{AVE}} = -\Delta\Phi/\Delta t = -(6.0 \times 10^{-3}\ \mathrm{Wb} - 3.0 \times 10^{-3}\ \mathrm{Wb})/2.0\ \mathrm{s} = -1.5 \times 10^{-3}$ V.

Problem 10.4. The circuit in the previous problem is changing its current at a steady rate of $\Delta I/\Delta t = 0.15$ A/s. How much EMF is induced in this circuit by this changing current?

Solution

We know that $\Phi = LI$, so therefore $\Delta\Phi = L\Delta I$, and $\Delta\Phi/\Delta t = L\Delta I/\Delta t$. Thus, $\mathrm{EMF} = -\Delta\Phi/\Delta t = -L\Delta I/\Delta t = -3.0 \times 10^{-3}(0.15) = -4.4 \times 10^{-4}$ V. The minus sign reminds us that, in accordance with Lenz's law, the voltage is a "**back EMF**", opposing the change in current.

Problem 10.5. A circuit is changing its current at the rate of $\Delta I/\Delta t = 0.45$ A/m, and produces a back EMF of 3.0×10^{-5} V. What is the inductance of the circuit?

Solution

As in the previous problem, $\mathrm{EMF} = -\Delta\Phi/\Delta t = -L\Delta I/\Delta t$. Therefore, $3.0 \times 10^{-3} = L(0.45)$ or $L = 6.7 \times 10^{-3}$H.

Toroid

To get the inductance of a toroid (Fig. 10-3), we follow the procedure developed previously. The field of a toroid at its mean radius, r, is $B = \mu_0 NI/2\pi r$ (see Chap. 7), where N is the total number of turns on the toroid. If the radius of the cross-sectional area is much less than the mean radius, r, then the field is practically uniform over the area of each turn. The flux through each turn is BA, and the total flux through all the turns of the toroid is $NBA = NA(\mu_0 NI/2\pi r) = \mu_0 N^2 IA/2\pi r$. Dividing by I gives

$$L = \mu_0 N^2 A/2\pi r \tag{10.4a}$$

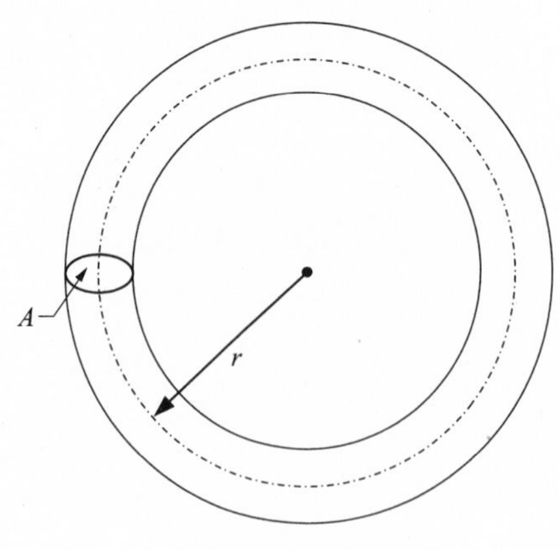

Fig. 10-3

as the inductance of the toroid. If the toroid is filled with material of permeability μ, then

$$L = \mu N^2 A/2\pi r. \tag{10.4b}$$

Problem 10.6. A toroid, of mean radius $r = 1.1$ m, has a cross-sectional area of 3.0×10^{-3} m^2.

(*a*) How many turns are needed if one wants to have an inductance of 8.0 mH?

(*b*) If the toroid were now stretched out to form a straight solenoid, what would the inductance be?

Solution

(*a*) We showed that $L = \mu_0 N^2 A/2\pi r$. Therefore $8.0 \times 10^{-3} = 4\pi \times 10^{-7}(\mathrm{N}^2)(3.0 \times 10^{-3})/2\pi(1.1)$, or $N^2 = 1.47 \times 10^7$, or $N = 3.8 \times 10^3$ turns.

(*b*) The solenoid would have length $d = 2\pi r$, and the same small cross-section A. From Eq. (*10.4a*), $L_{\mathrm{SOL}} = \mu_0 n^2 A/d$ with $n = N/d \Rightarrow L = \mu_0 N^2 A/d = \mu_0 N^2 A/2\pi r$ which is the same as the inductance of the toroid so the inductance is the same.

10.3 MUTUAL INDUCTANCE

Whenever one has two circuits near each other, it will be possible for a current which exists in one circuit to produce flux through the second circuit. We define a **mutual inductance** between the two circuits in the same way that we define the self inductance for a single circuit. If Φ_{12} is the flux in circuit 2 caused by a current I_1 in circuit 1, Then M_{12} is the factor that connect these two quantities, i.e.

$$\Phi_{12} = M_{12} I_1 \tag{10.5}$$

The exact value of M_{12} is determined by the geometrical relationship between the two circuits, just like the self inductance is determined by the geometry of the single circuit. If we manage to deduce the value of the mutual inductance, then we can always calculate the flux in circuit 2 produced by the current in circuit 1. Furthermore, if the current in circuit 1 changes, then the flux in circuit 2 changes proportionally, which means that $\Delta\Phi = M_{12}\Delta I$. Using Faraday's law, we know that this change in flux produces an EMF in circuit 2, given by

$$\mathrm{EMF} = -\Delta\Phi/\Delta t = -M_{12}\Delta I_1/\Delta t \tag{10.6}$$

Therefore, the mutual inductance is also the link between the induced EMF and the changing current in the other circuit.

Note that if we were to consider the effect of a current in circuit 2 on circuit 1, we would obtain the analogous equations to Eqs. (*10.5*) and (*10.6*) by interchanging all subscripts 1 and 2. Then M_{21} would be the mutual inductance and Φ_{21} the flux in circuit 1 due to the current in circuit 2. It can be shown that

$$M_{12} = M_{21} \tag{10.7}$$

so that there is only one mutual inductance for the two circuits. We can measure the **mutual inductance** by measuring the induced EMF produced in one circuit by a known rate of change in current in the other circuit.

Problem 10.7. Two circuits are near each other, so that a current change in one circuit produces an induced EMF in the other circuit. The current in circuit 1 is 3.0 A, and changes to 3.5 A in a time of 2.0×10^{-2} s. The average EMF induced in the second circuit during this time is 3.4 V. What is the mutual inductance of the two circuits?

Solution

We know that EMF $= -M_{12}\,\Delta I_1/\Delta t$. Thus, $3.4 = M_{12}(0.5)/0.0020$, or $M_{12} = 0.014$ H.

To calculate the mutual inductance for a particular combination of circuits, the procedure is the same as for calculating the self inductance. First we calculate the magnetic field produced by the current I_1 in circuit 1 at the position of circuit 2. Using this field, we calculate the flux, Φ_{12}, enclosed by circuit 2. The mutual inductance is then Φ_{12}/I_1. We will follow this procedure in the examples below to calculate the mutual inductance for several special cases.

Coil on Solenoid

Problem 10.8. Consider a long solenoid, with n_1 turns/m wound on a radius r_1. This solenoid is part of circuit 1. Another coil is wound around the outside of the solenoid, with a total of N_2 turns. This is part of circuit 2. Find a formula for the mutual inductance between these circuits. The setup is shown in Fig. 10-4.

Solution

Following the definition of mutual inductance we first calculate the field produced in the region of circuit 2 by a current I_1 in circuit 1. This field is uniform and equal to $B_{12} = \mu_0 n_1 I_1$. Since the coil is wrapped tightly on the solenoid it has the same radius r_1. Each turn of circuit 2 thus has an area $\pi r_1{}^2$, and

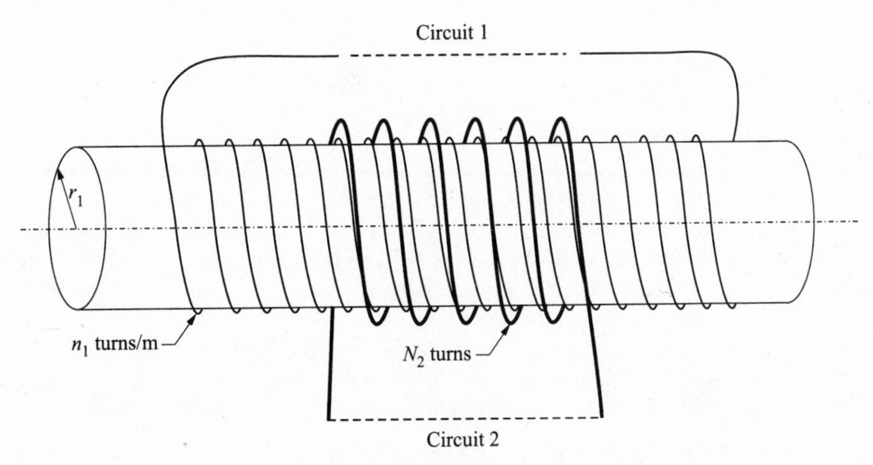

Fig. 10-4

encloses a flux of $B_{12}\pi r_1{}^2$. Therefore, the total flux through circuit 2 is $\Phi_{12} = N_2 B_{12}\pi r_1{}^2 = \mu_0 n_1 N_2 \pi r_1{}^2 I_1$. Finally, $M_{12} = \Phi_{12}/I_1 = \mu_0 n_1 N_2 \pi r_1{}^2$.

Note. The rest of circuit 1 would typically involve a single loop of wire and other circuit elements (not shown in the figure). The same is true of circuit 2. Some additional flux may pass through the single loop as well as other circuit elements further away. In general the flux through these is much smaller than that through the multiple wound elements that are close to each other, and we ignore these small additional contributions to flux in our considerations.

Problem 10.9. A long solenoid has 1800 turns/m, wound on a radius of 0.90 m. A second coil of 25 turns is wound on top of this solenoid. The first winding carries a current of 2.0 A.

(*a*) What is the flux through one turn of the second winding?

(*b*) What is the mutual inductance of the two circuits?

Solution

(*a*) The field produced by the current in circuit 1 is $B_{12} = \mu_0 n_1 I_1 = 4\pi \times 10^{-7}$ (1800)(2.0) = 4.52×10^{-3} T. The flux through one turn of the second winding is $B_{12}\pi r^2 = 4.52 \times 10^{-3}(\pi)(0.90)^2 = 0.0115$ Wb.

(*b*) The total flux through the second circuit is 25(0.115) = 0.288 Wb. The mutual inductance is 0.288/2 = 0.144 H. Alternatively, we could have directly used the formula for the mutual inductance of this geometry from Problem 10.8: $M_{12} = \mu_0 n_1 N_2 \pi r_1^2 = 4\pi \times 10^{-7}$ (1800)(25)(π)(0.90)^2 = 0.144 H.

Problem 10.10. Suppose the coil in Problem 10.9 has a current of 3.0 A. How much flux is produced in the solenoid due to this current?

Solution

We could proceed as in Problem 10.9 and find the magnetic field everywhere in the solenoid due to the current in the coil, but this would be difficult since, unlike the field due to the current in the solenoid, the field produced by the coil varies in magnitude and direction at different locations. Instead, we take advantage of Eq. (*10.7*): $M_{21} = M_{12}$. Then the flux through the solenoid (circuit 2) is $\Phi_{21} = M_{21}I_2 = M_{12}I_1 =$ 0.144(3.0) = 0.432 Wb.

Problem 10.11. A long solenoid has 1800 turns/m, wound on a radius of $r_1 = 0.90$ m. A second coil of 25 turns is wound on top of this solenoid, but at a larger radius of $r_2 = 1.6$ m, as in Fig. 10-5. The first winding carries a current of 2.0 A.

(*a*) What is the flux through one turn of the second winding?

(*b*) What is the mutual inductance of the two circuits?

Solution

(*a*) The field produced by the current in circuit 1 is $B_{12} = \mu_0 n_1 I_1 = 4\pi \times 10^{-7}$ (1800)(2.0) = 4.52×10^{-3} T. This field exists only within the first winding. Outside the radius of the first winding, the field is zero (see Sec. 7.4.3). The flux through one turn of the second winding is the sum of the fluxes within the radius r_1 and in the area between r_1 and r_2. Since the field is zero in that part of the area outside the first winding, this part of the area will not contribute anything to the sum. Therefore, the flux is just equal to the field inside the first winding, multiplied by the area of the first winding. This means that the flux equals $B_{12}\pi r_1{}^2 = 4.52 \times 10^{-3}(\pi)(0.90)^2 = 0.0115$ Wb, which is the same for whatever radius the second coil may have, provided it is greater than the radius of the first winding.

(*b*) The mutual inductance is the ratio of the total flux divided by the primary current. This equals $M_{12} =$ 25(0.0115)/2.0 = 0.144 H.

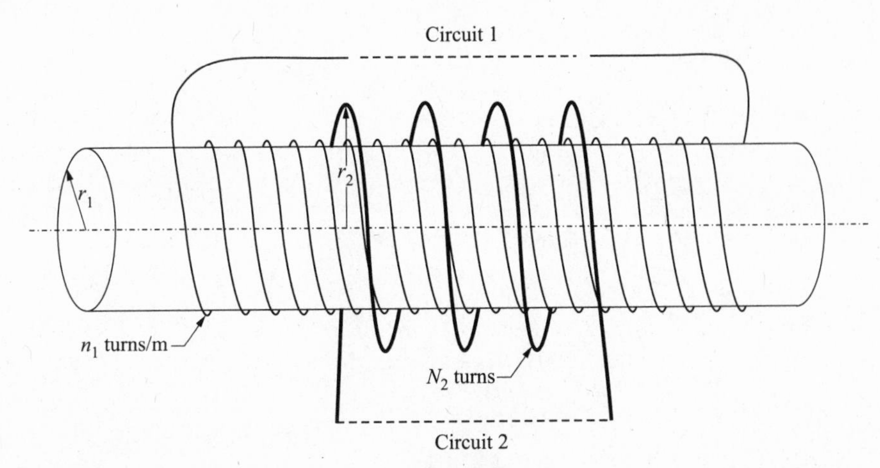

Fig. 10-5

Problem 10.12. A long solenoid has 1800 turns/m, wound around a material of relative permeability $\kappa_M = 150$, at a radius of 0.90 m. A second coil of 25 turns is wound on top of this solenoid. The first winding carries a current of 2.0 A.

(*a*) What is the flux through one turn of the second winding?

(*b*) What is the mutual inductance of the two circuits?

Solution

(*a*) The field produced by the current in circuit 1 is (recalling that $\mu = \chi_M \mu_0$) $B_{12} = \mu n_1 I_1 = 4\pi \times 10^{-7}(150)(1800)(2.0) = 0.68$ T. The flux through one turn of the second winding is $B_{12}\pi r^2 = 0.68(\pi)(0.90)^2 = 1.73$ Wb.

(*b*) $M_{12} = N_2/I_1 = 25(1.73)/2.0 = 21.6$ H.

Note. The results of (*a*) and (*b*) are just those of Problem 10.9 multiplied by the relative permeability, $\kappa_M = 150$.

Coil on Toroid

Consider the case of a toroid, which has a primary winding of N_1 turns, as in Fig. 10-6. The mean radius of the toroid is r, and the cross-sectional area of the toroid is A. A secondary winding of N_2 turns is wound on top of the primary, as shown in the figure. We wish to calculate the mutual inductance of these two circuits.

First we calculate the field produced by a current I_1 in the primary coil, in the region of the secondary coil. The field of a toroid at its mean radius is given by (see Sec. 10.2.2), $B = \mu_0 N_1 I/2\pi r$. As noted earlier, if the radius of the cross-sectional area, A, is small compared to the mean radius, r, then the field will be nearly uniform within the toroid. Then the flux through one turn of the secondary winding will be $\Phi = \mu_0 N_1 I_1 A/2\pi r$, and the total flux through the N_2 turns of the secondary winding will be $\Phi_{total} = \mu_0 N_1 N_2 I_1 A/2\pi r$. Therefore, the mutual inductance will be $M_{12} = \mu_0 N_1 N_2 A/2\pi r$. If the toroid is filled with material of permeability μ, then the mutual inductance will be

$$M_{12} = \mu N_1 N_2 A/2\pi r \tag{10.8}$$

Problem 10.13. A toroid has 550 turns, wound on a material of permeability 15, and has a mean radius of 2.5 m. A second coil of 25 turns is wound on top of this toroid. The cross-sectional area of the toroid is 0.56 m^2, and the first winding carries a current of 2.0 A.

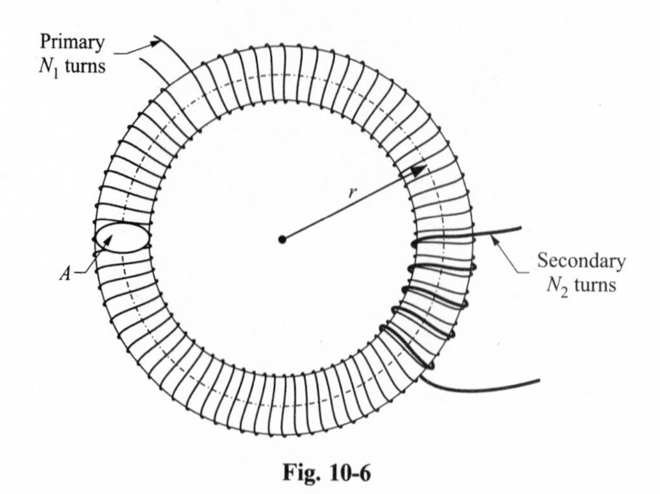

Fig. 10-6

(*a*) What is the flux through one turn of the second winding?

(*b*) What is the mutual inductance, M_{12}, of the two circuits?

(*c*) Find the flux in the toroid when a current of 7.0 A flows in the secondary winding.

Solution

(*a*) The field produced by the current in circuit 1 is $B_{12} = \mu N_1 I_1 / 2\pi r = 4\pi \times 10^{-7}$ (15)(550)(2.0)/(2π)(2.5) $= 1.32 \times 10^{-3}$ T. The flux through one turn of the second winding is $B_{12} A = 1.32 \times 10^{-3}(0.56) = 7.39 \times 10^{-4}$Wb.

(*b*) $M_{12} = N_2 \Phi / I_1 = 25(2.32 \times 10^{-3})/2.0 = 9.24 \times 10^{-3}$ H.

(*c*) In the case we just discussed in this problem, we calculated M_{12}, which is the ratio of the flux in circuit 2 to the current in circuit 1. As in Problem 10.10, to calculate M_{21}, which is the ratio of the flux in circuit 1 to the current in circuit 2, would be much more difficult. Instead, we use the fact that $M_{21} = M_{12}$ to get $\Phi_{21} = M_{21} I_2 = 9.24 \times 10^{-3}(7.0) = 6.47$ Wb.

Coil Near Long Wire

Suppose the primary circuit involves a long straight wire, and the secondary circuit includes a small coil of area, A, with N_2 turns, located at a distance, r, from the wire (see Fig. 10-7). The long wire produces a magnetic field at the position of the small coil, and therefore, a flux through each turn of the coil. To calculate the mutual inductance of these two circuits, we again follow the prescribed procedure.

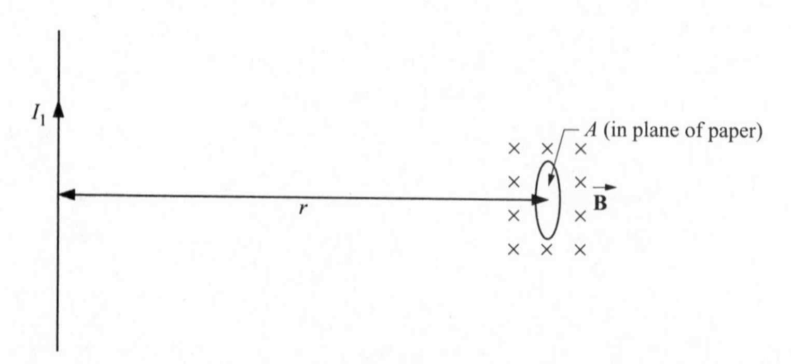

Fig. 10-7

First we calculate the field produced by the primary circuit (the long wire) at the position of the secondary circuit (the small coil). From Eq. (7.11) we know that this field is $B_{12} = (\mu_0/4\pi)\, 2I_1/r$. If the area of the coil is small enough so that all parts of the coil are at nearly the same distance, r, from the wire, then the field is uniform over this area, and the flux is just $\Phi_{12} = B_{12}A$. Then the mutual inductance will be $M_{12} = M_{21} = N_2\Phi_{12}/I_1 = (\mu_0/4\pi)2I_1N_2A/rI_1 = (\mu_0/2\pi)N_2A/r$.

Problem 10.14. A long straight wire carries a current of 3.0 A. A small rectangular coil, of sides 4.0 cm × 3.0 cm, is located at a distance of 1.3 m from the wire, and contains 246 turns.

(*a*) What is the flux through one turn of the rectangle?

(*b*) What is the mutual inductance of the two circuits?

Solution

(*a*) The field produced by the current in the long wire is $B_{12} = (\mu_0/4\pi)2I_1/r = 10^{-7}(2)(3.0)/(1.3) = 4.62 \times 10^{-7}$ T. The flux through one turn of the coil is $B_{12}A = 4.62 \times 10^{-7}(0.040)(0.030) = 5.54 \times 10^{-10}$ Wb.

(*b*) $M_{12} = N_2\Phi/I_1 = 246(5.54 \times 10^{-10})/3.0 = 4.54 \times 10^{-8}$ H.

Coil at Center of Loop

Problem 10.15. Suppose we have two concentric single loop circular coils, as in Fig. 10-8. If the small, inner coil (1) carries a current I_1, find an expression for the flux through the large, outer coil (2).

Solution

We realize that this flux can be calculated from $\Phi_{12} = M_{12}I_1$, if we know M_{12}. It would be very difficult to calculate M_{12} directly by the methods we used in the previous examples, since the magnetic field of the inner circuit varies considerably within the area of the outer circuit. However, since we know that $M_{12} = M_{21}$, we can calculate M_{21} instead, using the techniques we have used in the previous examples. To calculate M_{21} we first note that coil 1 is very small and right at the center of large coil 2. Therefore, to find the flux through coil 1 due to a current in coil 2, we need only find the magnetic field due to a current I_2 in coil 2 at the center of the coil. We have already done this in Chap. 7, and the result for a single loop is given by Eq. (7.5), $B_{21} = \mu_0 I_2/2r_2$. Since the radius of the small coil, r_1, is very small compared with r_2, the radius of the outer coil, the field is nearly uniform over the area of the small coil. The flux through the single loop small coil is then $\Phi_{21} = B_{21}\pi r_1^2 = \mu_0 I_2\pi r_1^2/2r_2$, and $M_{21} = \mu_0\pi r_1^2/2r_2 = M_{12}$. Finally we get $\Phi_{12} = M_{12}I_1 = \mu_0\pi r_1^2 I_1/2r_2$.

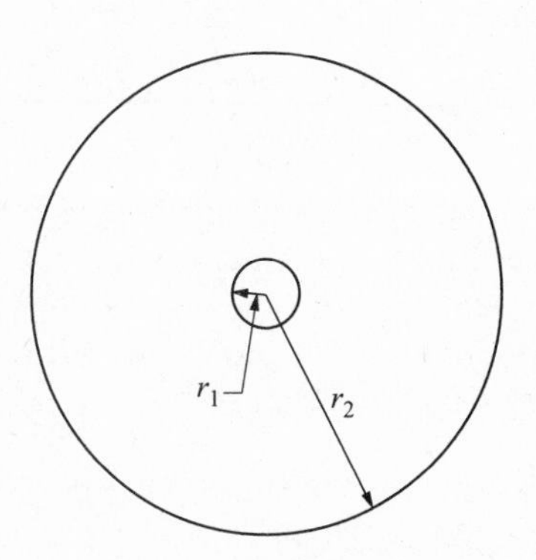

Fig. 10-8

Problem 10.16. Two concentric coils, of radii 0.80 m and 0.015 m, each have 250 turns. If the current in the inner coil changes at the average rate of 0.30 A/s, what is the average EMF induced in the outer coil?

Solution

The EMF in the outer coil is given by $N_2 \Delta\Phi_{12}/\Delta t = N_2 M_{12} \Delta I_1/\Delta t$. We have just shown (Problem 10.15) that for one turn of the inner coil, $M_{12} = \mu_0 \pi r_1{}^2/2r_2$; and therefore for N_1 turns, $M_{12} = N_1 \mu_0 \pi r_1{}^2/2r_2$. Then our EMF in the outer coil is: $\text{EMF} = N_2(N_1 \mu_0 \pi r_1{}^2/2r_2)\Delta I_1/\Delta t$. Using the values given, we get that $\text{EMF} = [250^2(4\pi \times 10^{-7})\pi(0.015)^2/2(0.80)](0.30) = 1.04 \times 10^{-5}$ V.

10.4 ENERGY IN AN INDUCTOR

Any circuit element that generates an inductance when current flows through it (e.g. a coil, a solenoid, a toroid) is called an inductor. Whenever one has an inductor which initially has no current, it takes energy to make current flow in the inductor. This can be seen from the fact that if one wants to increase the current, a **back EMF** is produced which attempts to stop the increase. In order to increase the current, an external driving voltage must be imposed on the circuit to overcome the back EMF, and this voltage will do work against the resisting EMF. The voltage will continue to do work until the current reaches its final value, at which time the current is no longer changing and no back EMF is being produced. During the time that the current is building up from zero to its final value, however, work must be done on the inductor. The work can be calculated if we remember that the **power** delivered to a system is the current at that time, I, times the voltage, V, at that same time. The power, $P = IV$, is the rate at which energy is being delivered to the system, $P = \Delta W/\Delta t$, where ΔW is the energy added to the system during the time interval, Δt.

The voltage imposed on the inductor is the negative of the back EMF. Since the back EMF equals $-L\Delta I/\Delta t$, the driving voltage must equal $L\Delta I/\Delta t$, and then $P = \Delta W/\Delta t = LI\Delta I/\Delta t$. We must solve this equation to get the total work needed, i.e. the total energy added to the system. By using the methods of calculus, one can show that the result is the same no matter how one changes the current in the inductor. It is always true that the energy stored in an inductor by virtue of the current that we have induced to flow in the inductor is

$$\text{Energy} = (\tfrac{1}{2})LI^2 \tag{10.9}$$

This result is similar to the case of storing energy in a capacitor by virtue of the charge that we have placed on the plates of the capacitor. There the energy was $\text{Energy} = (\frac{1}{2})Q^2/C$. We will make more use of these relationships in the future.

Problem 10.17. A solenoid, with an inductance of 55 mH stores an energy of 3.0 J. How much current is flowing in the solenoid?

Solution

The energy equals $(\frac{1}{2})LI^2 = 3.0 = (\frac{1}{2})(55 \times 10^{-3})I^2$, and therefore $I^2 = 109$, and $I = 10.4$ A.

Problem 10.18. A superconducting magnet carries a current of 500 A, and has a self inductance of 5.0 H. While the wires in the magnet are superconducting, the current does not decrease, since there is zero resistance and no energy is being dissipated. If the wires are heated and lose their superconductivity, the current rapidly becomes reduced to zero. How much energy would be released in the process of reducing the current to zero? Where does this energy go?

Solution

The energy released equals $(\frac{1}{2})LI^2 = (\frac{1}{2})(5.0)(500)^2 = 6.25 \times 10^5$ J. The energy would be dissipated as heat (RI^2) since the wires now have resistance.

We have shown that when current flows in an inductor, energy is stored, and we have interpreted this energy as being due to the current flow that we have induced in the inductor. There is another way to interpret this stored energy. Whenever current flows in an inductor, magnetic fields are set up in space. These magnetic fields are directly related to the currents, and the energy needed to set up the currents could equally well have been interpreted as the energy needed to set up the magnetic fields. This is analogous to the case of a capacitor, where the energy needed to charge the plates could equally well be interpreted as the energy needed to set up the electric fields due to these charges. Let us calculate the energy stored in the inductor in terms of the magnetic fields rather than in terms of the current.

To carry out this calculation we will take the case of a long solenoid, as in Fig. 10-2. In the case of a solenoid we know that the magnetic field, B, equals $\mu_0 nI$ inside the solenoid, and is zero outside. We also calculated previously that the inductance per unit length of the solenoid is $L/d = \mu_0 n^2 A$. The energy stored in length d is therefore $E = (\frac{1}{2})LI^2 = (\frac{1}{2})\mu_0 n^2 AdI^2$. But $I = B/\mu_0 n$, and therefore the energy equals $(\frac{1}{2})\mu_0 n^2 Ad(B/\mu_0 n)^2 = (\frac{1}{2})B^2(Ad)/\mu_0$, where Ad is the volume of the length d of the solenoid. Thus the energy density, or energy per unit volume, is

$$\text{Energy density} = (\tfrac{1}{2})B^2/\mu_0 \qquad (10.10)$$

In this form, the energy stored in the solenoid is considered as being due to the magnetic fields that have been set up in space. At any point in space, where there is a magnetic field, a certain amount of energy is stored. This energy equals the energy density times the volume of space being considered. Although this calculation was for the special case of a solenoid, the result is true for any other configuration as well. We have previously shown that the same general consideration holds for electric fields as well and indeed the electric field energy density is given by $(\frac{1}{2})\varepsilon_0 E^2$. In other words, wherever electric or magnetic fields exist in space, energy is being stored in the form of these fields. The total energy density at any point in space is the sum of the electric and the magnetic field energy densities. Since the units for energy density are the same irrespective of their source, this offers a means of comparing the relative magnitudes of electric and magnetic fields. Electric and magnetic fields with the same energy density can be considered to be comparable to each other. In fact, we will see that in electromagnetic waves, which we will discuss in a later chapter, the electric and the magnetic fields associated with the wave have equal energy densities. These considerations lend credence to the idea that these fields are real physical quantities that actually exist in space, and are not merely mathematical contrivances that make it easier to calculate the forces exerted by the electric and magnetic interactions.

Problem 10.19. An electromagnetic wave in free space has an electric field of 100 V/m. If there is also a magnetic field associated with this wave, and the energy density of the magnetic field is the same as the energy density of the electric field, what is the magnitude of the magnetic field?

Solution

The energy density for the electric field is $(\frac{1}{2})\varepsilon_0 E^2$, and the energy density for the magnetic field is $(\frac{1}{2})B^2/\mu_0$. Equating these two expressions gives $(\frac{1}{2})B^2/\mu_0 = (\frac{1}{2})\varepsilon_0 E^2$, or $B^2 = \mu_0 \varepsilon_0 E^2 = (4\pi \times 10^{-7})(8.85 \times 10^{-12})(100)^2 = 1.12 \times 10^{-13}$. Thus, $B = 3.33 \times 10^{-7}$ T.

10.5 TRANSFORMERS

From what we have learned in the previous sections, it is clear that we can induce EMFs in one circuit by changing the current in another circuit. This forms the basis of the **transformer**, which is used to transform voltage in one circuit into a different voltage in a second circuit. We have already developed the ideas for this in our discussion of the mutual inductance of two windings on a solenoid in Sec. 10.3. In that case, all the magnetic flux established by the first winding, called the **primary coil**, passes through the turns of the other winding, called the **secondary coil**. In order to get large fluxes, it is useful to place ferromagnetic material within the solenoid that has a large permeability, such as iron. Using such a ferromagnetic material has another advantage. When one magnetizes the iron inside the solenoid by applying a current to the primary winding, that iron, being ferromagnetic, will cause the rest of the iron atoms to align their magnetic moments in the same direction, and become magnetized as well. Furthermore, most of the flux will be confined within the core, so that a wire wound around one part of the core will experience the same flux as that wound around another part. It is then possible to wind the secondary coil on a different part of the iron core, not necessarily on top of the primary winding. It is even possible to bend the iron into a different shape, such as the often used shape shown in Fig. 10-9. Here, the primary winding, with N_1 turns, is wound on one side of the rectangular ring, and the secondary winding, with N_2 turns, is wound on the other side of the ring. This is a typical transformer. If one changes the voltage in the primary circuit, the current in the primary circuit will change, and therefore the flux. For a perfect transformer, the flux through one turn of the secondary is the same as the flux through one turn of the primary. Therefore, the total EMF developed in each winding will depend on the number of turns in that circuit.

If one has DC in the primary, the current does not change, and there is no *change* in the flux. Then, there will be no EMF induced in the secondary. A transformer is useful only with currents that are changing, as with AC. In that case, it is possible to use a transformer to convert a voltage applied to the primary circuit into a larger or smaller voltage in the secondary circuit. This ability to easily convert (transform) voltages in AC, which is much more difficult for DC, is the main reason why AC is the primary source of power throughout the world.

By analyzing the transformer shown in Fig. 10-9 in more detail, one can relate the EMF induced in the secondary circuit and the applied voltage in the primary, V_p to the relative number of turns in these circuits. The result is that:

$$V_s/V_p = N_s/N_p \tag{10.11}$$

If $N_s > N_p$, then $V_s > V_p$, and we will have a step-up transformer. This is useful, for instance if one wants to use an appliance built for 220 volts in an area where only 110 volts are available. If $N_s < N_p$, then $V_s < V_p$, and we have a step-down transformer. This is used by power generating companies, who transmit power along transmission lines at very high voltages, and transform them down to safe levels before they enter one's home.

Problem 10.20. A power company generates electricity at a voltage of 12,000 V, and steps up this voltage to 240,000 V, using a transformer (transformer 1). The electricity is transmitted at this voltage to a substation, where it is stepped down to 8000 V (transformer 2) before being transmitted further. Before entering a house, the voltage is stepped down further to 240 V (transformer 3). What are the turns ratio of each of these five transformers?

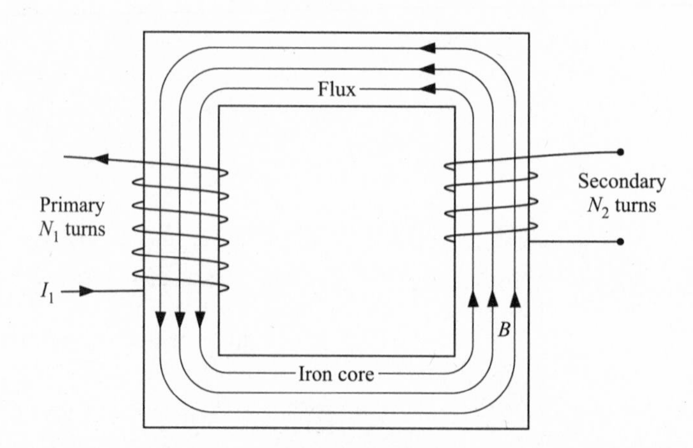

Fig. 10-9

Solution

The turn ratio $N_s/N_p = V_s/V_p$. Therefore, (*a*) for transformer 1, $N_s/N_p = 240{,}000/12{,}000 = 20$; (*b*) for transformer 2, $N_s/N_p = 8000/240{,}000 = 1/30$; and (*c*) for transformer 3, $N_s/N_p = 240/8000 = 3/100 = 0.030$.

Problems for Review and Mind Stretching

Problem 10.21. A coaxial cable consists of an inner conductor of radius, $R = 0.50$ m, separated from a hollow outer conductor by a distance, $\Delta R = 0.0020$ m, as in Fig. 10-10. The inner conductor carries a current, $I = 5.0$ A to the right, and the outer conductor carries the same current to the left. Consider the shaded region between the conductors, of length *d*. At any point in this region, *r* is between 0.500 m and 0.502 m.

(*a*) What is the magnetic field at a point in the shaded region, at a distance *r* from the axis?

(*b*) What is the flux through the shaded region of length *d*?

(*c*) What is the self inductance of the length *d* of the coaxial cable, and the inductance per unit length of the cable?

Solution

(*a*) In Sec. 7.4.2, we calculated the field produced by a coaxial cable in the region between the conductors. We found that $B = \mu_0 I/2\pi r$. Substituting in this equation gives, $B = (4\pi \times 10^{-7})(5.0)/2\pi r = 10^{-6}/r$ T. Since *r* is between 0.500 and 0.502 m, the field hardly varies in this region and we can substitute either value to get $B = 2.0 \times 10^{-6}$ T. The direction of the field is out of the paper in the shaded region, since the field lines circle about the center conductor in this direction (the right-hand rule).

(*b*) In general the flux through the area is $BA \cos\theta$, where θ is the angle between **B** and the normal to the area. For our case $\theta = 0°$, and $A = d\Delta R$. Thus, the flux, Φ, is $\Phi = Bd\Delta R = (\mu_0 I/2\pi R)\Delta R d = (\mu_0 I/2\pi)(\Delta R/R)d = 4.0 \times 10^{-9}$ d.

(*c*) The self inductance, *L*, is $\Phi/I = (\mu_0 I/2\pi R)\Delta R d/I = (\mu_0/2\pi R)\Delta R d = 8.0 \times 10^{-10}$ d. The inductance per unit length is $L/d = (\mu_0/2\pi R)\Delta R = (\mu_0/2\pi)(\Delta R/R) = 8.0 \times 10^{-10}$ H/m.

Note. A more accurate calculation, taking account of the variation of the field within the region between the inner and outer conductors of a coaxial cable yields the inductance per unit length to be $L/d = (\mu_0/2\pi) \ln (R_2/R_1)$; this is valid even if the difference between R_1 and R_2 is large.

Problem 10.22. An inductor with inductance *L* is connected in series with a resistor *R*. A battery completes the circuit with terminal voltage V_0 as shown in Fig. 10-11. At a certain instant, labeled $t = 0$,

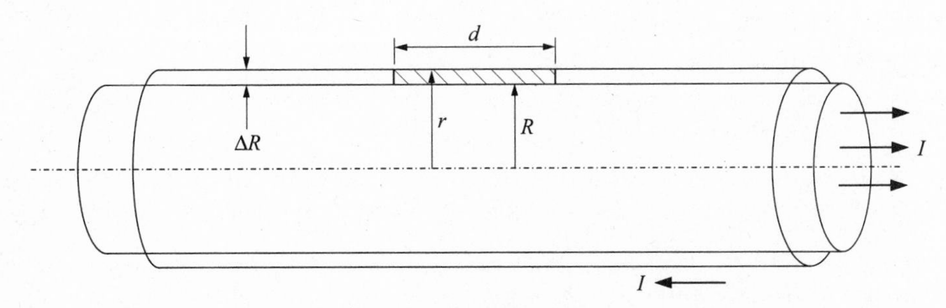

Fig. 10-10

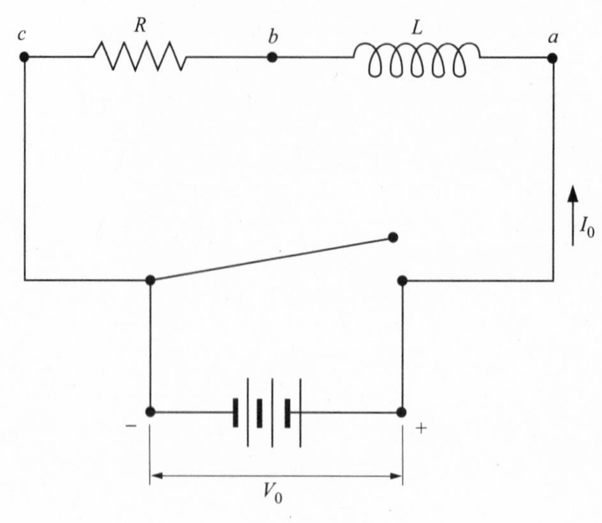

Fig. 10-11

the short circuit switch is closed eliminating the EMF of the battery. Assume the inductor coils have negligible resistance.

(*a*) What is the voltage drop from *b* to *c*, V_{bc}, across the resistance before the switch is closed, and what is the voltage drop from *a* to *b*, V_{ab}? Explain.

(*b*) At the instant the switch is closed explain qualitatively what happens to the current and voltages across the two elements.

Solution

(*a*) Since there is a steady current I_0, there is no change in flux and no EMF in the inductor. Therefore, since the inductor has zero resistance, V_{ab} is zero. Thus the entire terminal voltage of the battery appears across the resistance and we have $V_{bc} = V_0 = IR$ from Ohm's law.

(*b*) When the switch is closed there is no longer a voltage between points *a* and *c*. The current through the resistance starts to collapse. If not for the inductor this collapse, by Ohm's law, would be essentially instantaneous. Because of the inductor the collapse in the current is immediately opposed by an induced EMF in the inductor. This EMF opposes the change in the current and thus tries to maintain the status quo. The collapse in current is thus slowed down and occurs over a finite time interval.

Problem 10.23. Referring to Problem 10.22(*b*), and assuming ΔI is the change in current in an infinitesimal time interval Δt after the switch is closed and I is the current during that time interval:

(*a*) Following the reasoning behind Problem 10.22(*b*) what must the induced EMF in the inductor be in any infinitesimal time interval after the switch is closed? What is the direction of the EMF and what is the voltage V_{ab}?

(*b*) Find a relationship between L, R, ΔI, I and Δt.

Solution

(*a*) Since immediately after the switch closing the voltage drops around the circuit are zero, and the voltage across the resistor is still $V_{bc} = IR$, where I is the current at any time t after the switch closed, the induced EMF in the inductor must take the place of the battery to support the current. The direction of the EMF is from *a* to *b*, since it opposes the decrease in current in that direction (Lenz's law), and has the magnitude $\text{EMF}_{ab} = -L\Delta I/\Delta t$ for any infinitesimal time interval Δt.

Note. ΔI is negative from a to b so the EMF indeed points from a to b. The voltage from a to b is opposite to the EMF since the EMF points from higher to lower electrostatic voltage, as for a battery. In other words, the potential at b is higher than the potential at a, so $V_{ab} = -\text{EMF}_{ab} = L\Delta I/\Delta t$ (which is negative).

(*b*) We must have $V_{ab} + V_{bc} = 0$ at every instant. So:

$$L\Delta I/\Delta t + RI = 0 \qquad (i)$$

(In the infinitesimal time interval Δt immediately after the switch is closed, from $t = 0$ to $t = \Delta t$, this is $L\Delta I/\Delta t + RI_0 = 0$.) Turning the equation around we have, in general,

$$-(L/R)\Delta I/I = \Delta t \qquad (ii)$$

This equation can be solved using the calculus to give an expression for how the current falls to zero over time. While this will be discussed further in the next chapter, we note here that if L/R is large for a given Δt, $\Delta I/I$ will be small and the current will fall slowly; if L/R is small then for the same time interval Δt, $\Delta I/I$ will be larger and the current will fall more quickly. L/R is therefore called the time constant of the circuit.

Problem 10.24. An inductor, with inductance L, is connected in series with a resistor, R. A battery completes the circuit with terminal voltage V_0, and causes a current I_0 to flow in the circuit, as in Fig. 10-11. The voltage V_0 is increased by an amount ΔV_0. The current does not increase instantaneously since the inductor produces an induced EMF which tries to prevent any change in current. After a time Δt, however, the current has increased by ΔI, and the voltage across the resistor has increased by ΔV, where $\Delta V = R\Delta I$. This voltage, plus the voltage across the inductor, must equal the voltage ΔV_0.

(*a*) What is the magnitude of the average EMF induced in the circuit by the inductor during this time?

(*b*) What is the ratio of this average induced EMF in the inductor to the voltage increase that appears across the resistor after this time?

(*c*) If $L = 10$ mH and $R = 100\ \Omega$, at what value of Δt would the average induced EMF across the inductor equal the change in voltage ΔV?

Solution

(*a*) The average EMF induced by the inductor equals, in magnitude, $L\Delta I/\Delta t$.

(*b*) The change in voltage across the resistor is $\Delta V = R\Delta I$. Thus $\text{EMF}/\Delta V = L(\Delta I/\Delta t)/(R\Delta I) = (L/R)/\Delta t$.

(*c*) If $\text{EMF} = \Delta V$, then the ratio in part (*b*) is 1, and $L/R = \Delta t = 1.0 \times 10^{-4}$ s. Initially, all the increase in voltage of the battery appeared across the inductor, since the current has not yet changed, and there is therefore as yet no increase in voltage across the resistor. After a long time has elapsed, the current reaches its final constant value and there is no longer any change in current and therefore no voltage across the inductor. After a time equal to the time constant, the increase in voltage across the resistor approximately equals the voltage across the inductor.

The time constant, L/R, is thus a measure of how quickly the circuit approaches its final value. The time constant will be discussed further in the next chapter.

Problem 10.25. A long solenoid has 300 turns/m, and is wound on a radius of 0.90 m. A smaller coil, of radius 0.60 m, is inside the solenoid, with its plane perpendicular to the axis of the solenoid, as in Fig. 10-12.

(*a*) Determine the mutual inductance between the solenoid and the coil.

(*b*) If the coil is rotated by 90° in a time interval of 2.0×10^{-3} s while a current of 4.0 A is flowing in the solenoid, what is the average induced EMF in the coil during that time interval?

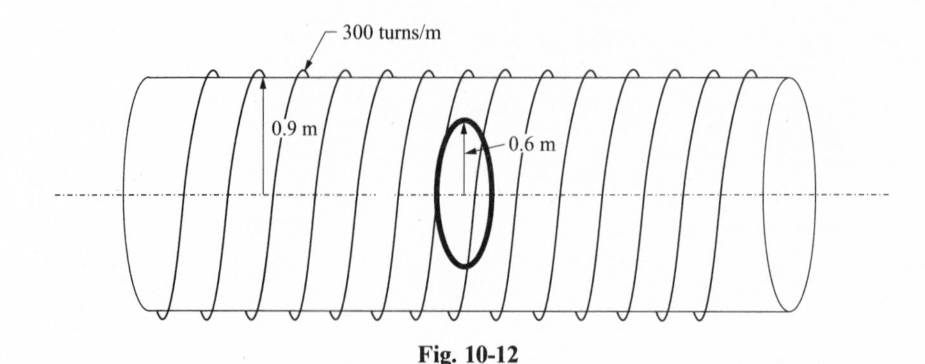

Fig. 10-12

Solution

(*a*) The field established by the long solenoid is $B = \mu_0 nI$, and is uniform within the solenoid. The flux through the small coil is $\Phi = BA = \mu_0 nI(\pi r^2)$, where r is the radius of the small coil. The mutual inductance is $M_{12} = M_{21} = M = \Phi/I = \mu_0 nI(\pi r^2)/I = \mu_0 n(\pi r^2) = (4\pi \times 10^{-7})(300)\pi(0.60)^2 = 4.26 \times 10^{-4}$ H.

(*b*) Here the flux drops to zero since the coil becomes parallel to the flux lines in the solenoid. EMF $= \Delta\Phi/\Delta t = \Phi/\Delta t = MI/\Delta t = 4.26 \times 10^{-4}(4.0)/2.0 \times 10^{-3} = 0.852$ V.

Problem 10.26. A toroid with a mean radius of 1.2 m and a cross-sectional area of 0.050 m^2, carries a current of 5.0 A in 750 turns.

(*a*) What is the self inductance of this toroid?

(*b*) How much energy is stored in the toroid?

(*c*) What is the energy density within the toroid?

Solution

(*a*) The inductance of a toroid was calculated in Sec. 10.2.2, and given by [Eq. (*10.4a*)] as $L = \mu_0 N^2 A/2\pi r$. Therefore, $L = (4\pi \times 10^{-7})(750)^2(0.050)/2\pi(1.2) = 4.7 \times 10^{-3}$ H.

(*b*) The energy stored is [Eq. (*10.9*)] $E = (1/2)LI^2 = 0.5(4.7 \times 10^{-3})(5.0)^2 = 0.059$ J.

(*c*) The energy density is [Eq. (*10.10*)] $(1/2)B^2/\mu_0 = (1/2)(\mu_0 NI/2\pi r)^2/\mu_0 = 0.155$ J/m^3. This result could also have been obtained by taking the total energy (0.059 J) and dividing by the volume of the toroid ($2\pi rA = 0.38$ m^3), or $0.059/0.38 = 0.155$.

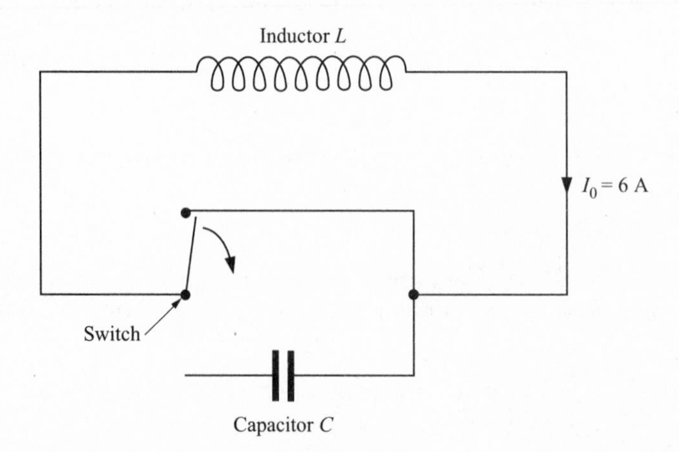

Fig. 10-13

Problem 10.27. A current of 6.0 A is initially flowing through an inductor which has an inductance of 0.15 H in a resistanceless circuit, as shown in Fig. 10-13. The switch is thrown introducing a capacitor into the circuit. The capacitor has a capacitance of 1.22×10^{-3} f, and is initially uncharged.

(*a*) What is the energy stored in the inductor initially?

(*b*) As the current flows into the capacitor, the capacitor becomes charged and the current decrease. When the current becomes zero, the capacitor is charged to its maximum. Assuming that no energy has been lost in this time, what is the voltage across the capacitor at this instant in time, and how much charge is stored on the capacitor?

Solution

(*a*) The energy stored in the inductor is $(1/2)LI^2$. Therefore Energy $= 0.5(0.15)6.0^2 = 2.7$ J.

(*b*) The inductor has no energy at this time, since the current is zero. All the energy is in the capacitor, and that energy is $(1/2)CV^2 = 2.7$ J. Thus, $V^2 = 2.7(2)/1.22 \times 10^{-3} = 4426$, $V = 66.5$ V. The charge stored is $Q = CV = 1.22 \times 10^{-3}\,(66.5) = 0.081$ C. Alternatively, one could have used Energy $= (\frac{1}{2})Q^2/C = 2.7$ J, which also yields $Q = 0.081$ C.

Supplementary Problems

Problem 10.28. A coil has an inductance of 25 mH.

(*a*) If the current in the coil is 2.0 A, what is the flux in the coil?

(*b*) If there is a back EMF of 0.30 V, what is the average rate of change of the current?

Ans. (*a*) 0.050 Wb; (*b*) 12 A/s

Problem 10.29. A coil starts with a current of 5.0 A, which is changing at the rate of 0.90 A/s. This change produces a back EMF of 0.030 V.

(*a*) What is the inductance of the coil?

(*b*) What is the flux in the coil at the start?

Ans. (*a*) 33 mH; (*b*) 0.167 Wb

Problem 10.30. A length of 0.50 m of a long solenoid, with a cross-sectional area of 0.030 m^2, has an inductance of 0.080 H. How many turns/length are on the solenoid?

Ans. 2060 turns/m

Problem 10.31. A toroid has 1500 turns, a mean radius of 1.1 m, a cross-sectional area of 0.95 m^2 and carries a current of 7.0 A.

(*a*) What is the magnetic field at the mean radius?

(*b*) What is the inductance of the toroid?

Ans. (*a*) 1.91×10^{-3} T; (*b*) 0.39 H

Problem 10.32. A coil, with an inductance of 0.50 H, has a uniform magnetic field in its area of 0.40 m^2. It carries a current of 0.50 A. What is the magnetic field in its area?

Ans. 0.625 T

Problem 10.33. A circuit, consisting of one coil, has an inductance of 5.0 mH and carries a current of 1.5 A.

(*a*) What is the flux through the coil?

(*b*) If one increases the circuit to 15 coils, with the same current, what would be the flux through *each* coil? What is the flux through all 15 coils?

(*c*) What is the inductance of the 15 coils?

Ans. (*a*) 7.5×10^{-3} Wb; (*b*) 0.113 Wb, 1.69 Wb; (*c*) 1.13 H

Problem 10.34. Two coils are near each other, so that when one changes the current in the first at the rate of 5.0 A/s, there is an EMF of 2.0 mV induced in the second.

(*a*) What is the mutual inductance between the two coils?

(*b*) What flux goes through the first coil, if there is a current of 8.0 A in the second coil?

Ans. (*a*) 0.40 mH; (*b*) 3.2×10^{-3} Wb

Problem 10.35. A long wire carries a current of 5.0 A. Nearby, there is a loop of area 0.80 m^2, as in Fig. 10-7. The mutual inductance between the wire and loop is 6.0×10^{-5} H.

(*a*) What is the flux through the loop?

(*b*) What is the average field within the loop?

(*c*) What is the average distance between the wire and loop?

Ans. (*a*) 3.0×10^{-4} Wb; (*b*) 3.75×10^{-4} T; (*c*) 2.7×10^{-3} m

Problem 10.36. A long solenoid has 500 turns/m, and produces a magnetic field of 4.0×10^{-3} T inside the solenoid. The solenoid has a cross-sectional area of 0.020 m^2.

(*a*) What current is flowing in the wires?

(*b*) What is the flux through the area of one turn of the solenoid?

(*c*) If one winds 25 turns on the outside of the solenoid, what is the mutual inductance between the solenoid and the turns?

Ans. (*a*) 6.4 A; (*b*) 8.0×10^{-5} Wb; (*c*) 3.1×10^{-4} H

Problem 10.37. A toroid has 750 turns, and a secondary winding on the toroid has 25 turns. The cross-sectional area of the toroid is 0.0090 m^2, and the mean radius is 0.72 m.

(*a*) What is the mutual inductance between the toroid and the secondary winding?

(*b*) If one fills the toroid with material of magnetic permeability 75, what is the mutual inductance?

(*c*) If, in part (*b*), a current of 5.0 A in the toroid is removed in 1.0×10^{-3} s, what average voltage is induced in the secondary?

Ans. (*a*) 4.69×10^{-5} H; (*b*) 3.5×10^{-3} H; (*c*) 17.6 V

Problem 10.38. A small coil, of area 5.0×10^{-4} m^2, is on the axis of a large coil at a distance of 2.1 m from the center of the large coil, as in Fig. 10-14. The large coil has a radius of 0.60 m, and contains 2000 turns. The large coil has a current of 10 A.

(*a*) What is the magnetic field at the center of the small coil?

(*b*) If the field is uniform within the small coil, what is the flux in the small coil?

(*c*) What is the mutual inductance between the coils?

Ans. (*a*) 4.34×10^{-4} T; (*b*) 2.17×10^{-7} Wb; (*c*) 2.17×10^{-8} H

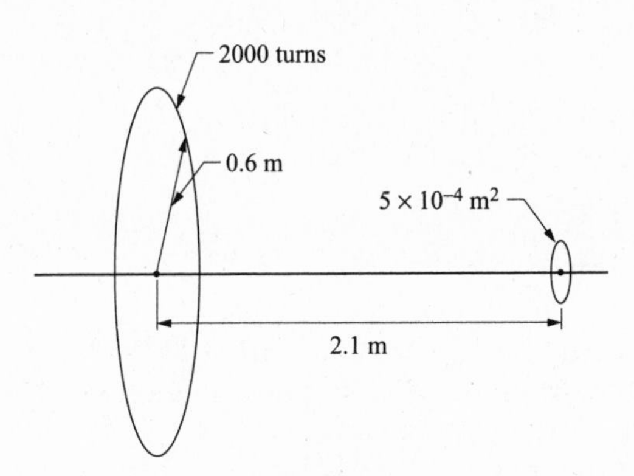

Fig. 10-14

Problem 10.39. A coil with a single turn has a self inductance of 1.5 mH. A tightly wound coil of the same radius, but with 16 turns, is placed flush against the first coil. A current of 0.60 A flows through the first coil.

(*a*) What is the flux through the area of the first coil?

(*b*) What flux goes through the area of each turn of the second coil?

(*c*) What is the mutual inductance between the coils?

Ans. (*a*) 9.0×10^{-4} Wb; (*b*) 9.0×10^{-4} Wb; (*c*) 0.024 H

Problem 10.40. A coil has a self inductance of 250 mH. How much work has to be done to cause a current of 1.1 A to flow in the coil?

Ans. 0.15 J

Problem 10.41. A coil has a self inductance of 250 mH. How much work has to be done to increase the current from 1.1 A to 2.2 A?

Ans. 0.45 J

Problem 10.42. A coil, with an inductance of 150 mH, is carrying a current of 2.1 A. The current is reduced to zero in a time of 2.0×10^{-2} s.

(*a*) How much energy was released to some external circuit?

(*b*) What average power was applied to that external circuit?

Ans. (*a*) 0.33 J; (*b*) 16.5 W

Problem 10.43. A capacitor, with capacitance 2.0×10^{-3} f, is charged to 110 V. The capacitor is then connected to an inductor, of inductance 0.20 H, and current begins to flow. The energy stored in the capacitor is transferred to the inductor.

(*a*) How much energy is initially stored in the capacitor?

(*b*) When the voltage is reduced to 50 V, what current is flowing in the inductor?

(*c*) What is the maximum current in the inductor?

Ans. (*a*) 12.1 J; (*b*) 9.8 A; (*c*) 11 A

Problem 10.44. A toroid has an inductance of 0.15 H, and a volume of 0.80 m^3. It carries a current of 0.50 A.

(*a*) How much energy is stored in the toroid?

(*b*) What is the average energy density in the toroid?

(*c*) What is the average magnetic field in the toroid?

Ans. (*a*) 1.88×10^{-2} J; (*b*) 2.34×10^{-2} J/m^3; (*c*) 2.43×10^{-4} T

Problem 10.45. An air conditioner is built to work at a voltage of 208 V. If a 110 V circuit is to be used to bring power to the air conditioner, does one need a step-up or a step-down transformer, and what ratio of turns is required?

Ans. step-up with turns ratio of 1.89 : 1

Problem 10.46. A step-down transformer is used in Switzerland, where the voltage is 220 V, to run an American appliance rated at 110 V, 250 W. What is the turns ratio in the transformer?

Ans. 1 : 2

Problem 10.47. A transformer is built from a rectangular ring, as in Fig. 10-9. The iron core is not perfect (some flux lines leave the core), and the flux in the iron at the secondary is only 75% of the flux at the primary. If the turns ratio is $N_2/N_1 = 20$, what is the voltage ratio V_2/V_1?

Ans. 15

Chapter 11

Time Varying Electric Circuits

11.1 INTRODUCTION

In Chap. 5, we learned about the behavior of DC circuits, in which the only elements were batteries and resistors. We found that, at all times, the voltage across a resistor was $V = RI$, where R was the resistance of the resistor and I the current through the resistor. Since then we have learned about two other circuit elements, a capacitor and an inductor. An inductor has the property that it produces a back EMF if the current is changing, but does nothing if the current is steady. A capacitor has the property that there is no current flow through it, so that, in the steady state it acts like an open circuit. However, the capacitor can become charged as a result of current flowing towards its positive plate, and discharged as a result of charge flowing away from its positive plate. Therefore, current can flow in a DC circuit containing a capacitor, during the time that the capacitor is charging or discharging. While an inductor will not affect a DC circuit once a current has been established it will be of great importance during the time that the current is being turned on or off. Thus, both capacitors and inductors play an important role when the current is adjusting or transitioning to a steady state. Such phenomena are called the transient response of a circuit, and will be discussed in Sec. 11.2.

Instead of having a battery supply a steady voltage and setting up a steady current (after the transient phenomena have stopped) we can have a generator supplying a constantly varying sinusoidal current (AC). Then we have to consider the response of all three circuit elements (resistor, capacitor and inductor) to this voltage. In this case, the inductor will have an important effect at all times, since the current is continually changing. Similarly, the capacitor will be continually charging and discharging, and it will not act as an open circuit for AC. In Sec. 11.3, we will discuss how to deal with these circuit elements when the current is sinusoidally varying.

11.2 TRANSIENT RESPONSE IN DC CIRCUITS

A DC circuit generally consists of a battery acting as the energy source, through its EMF, and causing a steady voltage to act across one or more resistors, capacitors or inductors. In the steady state, after all transient phenomena have stopped, there is no voltage across an inductor, since the current is no longer changing. There is a steady voltage across a capacitor, equal to Q/C, but there is no current flowing to or from the capacitor. The voltage across a resistor will equal $V = RI$. Steady state phenomena are fairly easy to treat, using these principles. Since transient phenomena are more complicated, we will treat only the simplest cases, consisting of a battery providing an EMF to series combinations of either R and C, or R and L, or L and C. We will ignore internal resistance in the battery so the EMF will equal the voltage across the terminals of the battery.

> ***Note.*** EMF of the battery relates to chemical energy and voltage to electrical energy. It is the voltage that is connected with the behavior of resistance, capacitance and inductance. Since the EMF and the terminal voltages are the same they are sometimes used interchangeably in text as a short hand.

Resistor and Capacitor

Consider the circuit shown in Fig. 11-1. A battery produces an EMF of V volts, and, after closing the switch S, the associated terminal voltage is applied to the series combination of R and C. We seek to determine what happens during the time that the capacitor C is being charged through the resistor R. We know that, at the time that the switch is first closed (which we call time $t = 0$) there is no charge as

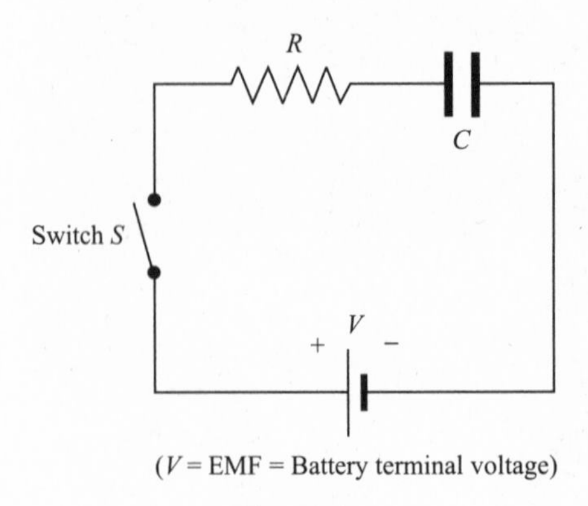

(V = EMF = Battery terminal voltage)

Fig. 11-1

yet on the capacitor, so that the voltage across the capacitor is zero. Therefore, all the terminal voltage of the battery will appear across the resistor, and there will be an initial current of $I = V/R$ flowing in the circuit. This current will bring charge to the capacitor, which will begin to accumulate charge, and the voltage across the capacitor will increase as the charge increases. As voltage appears across the capacitor, the voltage across the resistor will decrease, since the sum of the voltage across the resistor and the capacitor will always equal V.

$$V = V_R + V_C = IR + Q/C \tag{11.1}$$

As the voltage across R decreases, less and less current will flow in the circuit, and the capacitor will be charging more slowly. The charging of the capacitor continues until the current decreases to zero, and all the voltage V appears across the capacitor. In this final state, the charge on the capacitor, Q_f, will equal CV, and the current (and voltage across the resistor) will be zero. The situation at $t = 0$ and at $t = t_f$ is given by

$$V = V_{Ri} = I_i R, \qquad Q_i = 0, \qquad V_{Ci} = 0, \quad I_i = V/R \tag{11.2}$$

$$V = V_{Cf} = Q_f/C, \qquad I = 0, \qquad V_{Rf} = 0 \tag{11.3}$$

At other times, the current will be decreasing from its initial value to its final value of zero, while the charge on the capacitor will be increasing from zero to its final value of CV.

Problem 11.1. A battery with an EMF of 12 V is connected to a series circuit consisting of a resistor of 100 Ω and a capacitor of 0.25 mF.

(*a*) What initial current flows in the circuit?

(*b*) What final charge is on the capacitor?

(*c*) When the current is 0.080 A, what is the charge on the capacitor?

Solution

(*a*) Initially, all the voltage is across the resistor, so $V = I_i R = 12\text{ V} = I_i\,(100\ \Omega)$, or $I_i = 0.12$ A.

(*b*) Finally, all the voltage is across the capacitor, so $V = Q_f/C \rightarrow 12\text{ V} = Q_f/(2.5 \times 10^{-4}\text{ F})$, or $Q_f = 3 \times 10^{-3}$ C.

(*c*) At all times $V = V_R + V_C = 12$ V. Here, $V_R = IR = (0.080\text{ A})(100\ \Omega) = 8.0$ V, so $V_C = 12 - 8 = 4$ V and $Q = CV = (2.5 \times 10^{-4}\text{ F})(4\ V) = 10^{-3}$ C.

During the time that the capacitor is charging, the current is decreasing to zero, while the charge is increasing to Q_f. To calculate how the current and charge depend on time it is best to use the methods

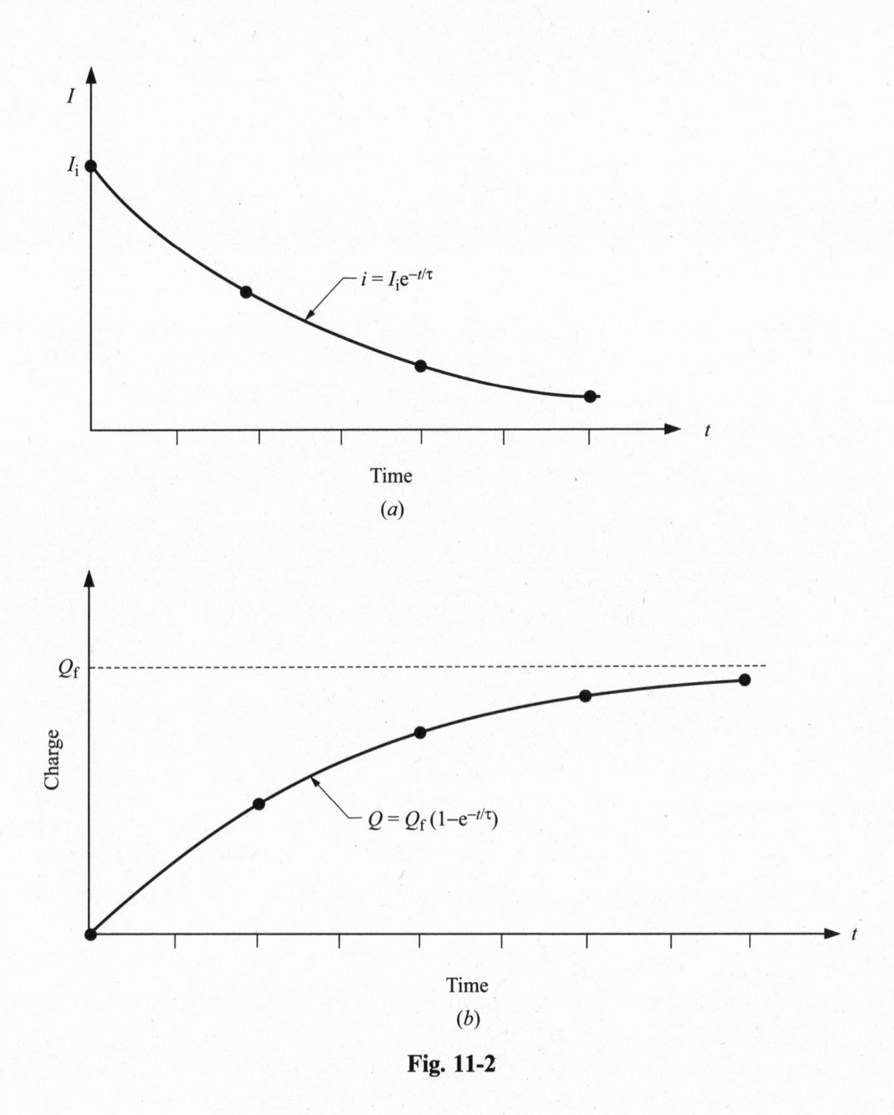

Fig. 11-2

of calculus. We can, however, get a qualitative understanding of the dependence by use of the graphs in Fig. 11-2. In Fig. 11-2(*b*), the charge on the capacitor is plotted as a function of the time. The charge starts at zero at the beginning, and approaches Q_f as time increases. The rate at which the charge increases is the current ($I = \Delta Q/\Delta t$), and this gets smaller and smaller as the current decreases with time. Therefore the slope of the curve gets smaller and smaller with time as is indeed the case for the curve drawn in Fig. 11-2(*b*). Similarly, in Fig. 11-2(*a*), we sketch the variation of the current with time. Here, the current starts at a value of $I_i = V/R$, and decreases to zero. Again, the curve in the figure depicts a decrease that gets slower with time, approaching zero at large times. Both of these curves are examples of what are called ***exponential decays***. In an exponential decay the dependent variable [such as the current I, plotted on the vertical axis in Fig. 11-2(*a*)] decreases by a fixed multiplicative factor for each constant-step increase in the independent variable [e.g. for our example the time, t, plotted on the horizontal axis in Fig. 11-2(*a*)]. As an example consider the case where, for each one second increase in t the current drops to 1/3 of its previous value. Then at $t = 0$ s: $I = I_i$, at $t = 1$, $I = I_i/3$, at $t = 2$ s, $I = I_i/9$, at $t = 3$ s, $I = I_i/27$, etc. One way of expressing this mathematically is by the formula $I = I_i(3)^{-t/t'}$ where t' is a constant = 1 s. Recalling that any number to the zeroth power = 1, we have at $t = 0, I = I_i 3^{-0}$. At $t = 1$ s, $I = I_i 3^{-1} = I_i/3$; at $t = 2$ s, $I = I_i 3^{-2} = I_i/9$; at $t = 3$ s, $I = I_i 3^{-3} = I_i/27$ and so on. The behavior of exponential functions is discussed in some detail in Chap. 2. As implied there, any function A^x (with A constant) can be re-expressed in terms of the natural exponential function. Thus

$(3)^{-t/t'} = e^{-t/\tau}$ for some constant τ. Thus, in general, any exponentially decaying current can be expressed as:

$$I = I_i e^{-t/\tau} \equiv I_i \exp(-t/\tau) \tag{11.4}$$

where τ is the appropriate constant in seconds. Different values of τ will control how fast I drops with each 1 s increase in t. For small τ, I will drop rapidly because (t/τ) builds up fast. For large τ, (t/τ) builds up more slowly with increasing t. Indeed, the time $t = \tau$ always corresponds to the time it takes for I to drop to 1/e of its original value, from I_i to $I_i/e = 0.368\, I_i$ (i.e., $I = I_i \exp(-t/\tau) = I_i \exp(-1) = I_i/e$). For these reasons τ is called the time constant of the decay. Eq. (*11.4*) can be considered the prototype for all exponential decay situations we encounter in this chapter and elsewhere in the book. While the curve for the charge on the capacitor, Fig. 11-2(*b*), involves Q increasing, it really involves the same kind of exponential decay we have been discussing. Here we have:

$$Q = Q_f - Q_f \exp(-t/\tau) = Q_f[1 - \exp(-t/\tau)] \tag{11.5}$$

where we have substituted the decay from the constant final value, Q_f, so that at $t = 0$, $Q = Q_f - Q_f = 0$. However, as time goes on, the term we subtract off decays exponentially leaving the value Q_f as the final value for very large times. For the case of our simple circuit with the capacitor C and the resistor R (Fig. 11-1), the time constant τ, for both I and Q is the same, and can be shown to be given by

$$\tau = RC \tag{11.6}$$

Problem 11.2.

(*a*) Show that $\tau = RC$ indeed has units of time.

(*b*) If $I_i = 3$ A, and $\tau = 0.80$ s, find the value of I in Eq. (*11.4*) at $t = 0$, $t = 1$ s, and $t = 2$ s.

(*c*) At what value of t will the current of part (*b*) equal 1 A?

Solution

(*a*) RC = (volts/amps)(coulombs/volts) = (coulombs/amps) = coulombs/(coulombs/seconds) = seconds.

(*b*) $I = (3.0\text{ A}) \exp(-t/0.80\text{ s})$; at $t = 0$, $I = (3.0\text{ A}) \exp(0) = 3.0$ A; at $t = 1$ s, $I = (3.0\text{ A}) \exp(-1/0.80) = (3.0\text{ A}) \exp(-1.25) = (3.0\text{ A})(0.287) = 0.860$ A; at $t = 2$ s, $I = (3.0\text{ A}) \exp(-2/0.80) = (3.0\text{ A}) \exp(-2.5) = (3.0\text{ A})(0.0821) = 0.246$ A; where, in the last two cases we use a calculator. Note that $\exp(-2.5) = [\exp(-1.25)]^2$ as expected, so that each 1 s increase corresponds to a multiplicative decrease in I by the same factor of (0.287).

(*c*) $1.0\text{ A} = (3.0\text{ A}) \exp(-t/0.80\text{ s})$, or: $\exp(-t/0.80\text{ s}) = 1/3 \rightarrow \exp(t/0.80\text{ s}) = 3 \rightarrow t/0.80\text{ s} = \ln(3.0) \rightarrow t = 0.80\text{ s} \ln(3.0) = (0.80\text{ s})(1.09) = 0.879$ s, where we have used the natural logarithm function on a calculator, to obtain ln (3).

Problem 11.3. A battery with an EMF of 12 V is connected to a series circuit consisting of a resistor of 100 Ω and a capacitor of 0.25 mF.

(*a*) What is the time constant of the circuit?

(*b*) How long does it take for the current to reach 1/2 of its original value?

(*c*) What is the charge on the capacitor after a time of 0.01 s?

Solution

(*a*) Using Eq. (*11.6*), $\tau = RC = (100\ \Omega)(0.25 \times 10^{-3}\text{ F}) = 0.025$ s.

(*b*) Using Eq. (*11.4*), $I = I_i \exp[-(t/\tau)]$, $0.5\, I_i = I_i \exp[-(t/\tau)]$, $0.5 = \exp[-(t/\tau)]$, $-(t/0.025) = \ln(0.5) = -0.693$, $t = 0.0173$ s.

As can be seen from the calculation, this time is $t = 0.693\tau$, and is called the "half-life" when applied to problems of decay. The **half-life** $\tau_{1/2}$ is the time it takes to reach 1/2 of the final value, so in general $\tau_{1/2} = 0.693\tau$.

(*c*) Using Eq. (*11.5*), $Q = Q_f(1 - \exp[-(t/\tau)]) = CV(1 - \exp[-(t/\tau)]) = 0.25 \times 10^{-3}$ (12) $(1 - \exp[-(0.01/0.025)]) = 9.9 \times 10^{-4}$ C.

The above discussion, related to Fig. 11-2 and Eqs. (*11.4* and *11.5*), concerned the case in which the capacitor was being charged (starting at time $t = 0$). The situation is very similar in the case of the discharge of a capacitor. Suppose a capacitor, with capacitance, C, has been charged to a voltage, V. If this capacitor is now allowed to discharge through a resistor, R, as in Fig. 11-3 when one closes the switch, the charge decays exponentially to zero. This is illustrated in the next problem.

Problem 11.4. A capacitor, with capacitance $C = 25\ \mu$F, is initially charged to a voltage $V_0 = 12$ V. The capacitor is then connected across a resistor $R = 1500\ \Omega$, and the capacitor is discharged through the resistor.

(*a*) What is the initial charge, Q_0, on the capacitor?

(*b*) What is the initial current, I_0, through the resistor, when the switch is closed?

(*c*) What is the current in the resistor when only $\frac{1}{4}$ of the initial charge remains on the capacitor?

(*d*) What is the formula for the current as a function of time? Plot the current as a function of time.

(*e*) At what time does the capacitor have only $\frac{1}{4}$ of the initial charge?

Solution

(*a*) For a capacitor, $Q = CV$. Therefore, initially, $Q_0 = (25 \times 10^{-6}\ \text{F})(12\ \text{V}) = 3 \times 10^{-4}$ C.

(*b*) When the circuit is closed the capacitor and resistor have the same voltage across them, $V = (V_a - V_b)$. Therefore, the initial voltage across the resistor is also 12 V, and the initial current in the resistor is $I_0 = V/R = (12\ \text{V})/(1500\ \Omega) = 8.0 \times 10^{-3}$ A.

(*c*) If $\frac{1}{4}$ of the initial charge remains on the capacitor, then the voltage across the capacitor is $\frac{1}{4}$ of the initial voltage, or $(12\ \text{V})/4 = 3.0$ V. The same voltage is across the resistor, so the current in the resistor is $I = V/R = (3.0\ \text{V})/(1500\ \Omega) = 2.0 \times 10^{-3}$ A. This is, of course just $\frac{1}{4}$ of the initial current. Thus, unlike the case of charging, current decreases with time in the same direct proportion to the charge on the capacitor: $I = V/R = Q/CR$.

(*d*) Assuming that discharging obeys an exponential decay with the same time constant as charging, the formula for the current would be $I = I_0 \exp[-(t/\tau)]$, where $\tau = RC$. In our case, this means that $\tau = (1500\ \Omega)(25 \times 10^{-6}\ \text{F}) = 0.0375$ s. Thus, $I = (8 \times 10^{-3}\ \text{A}) \exp[-(t/0.0375)]$. This is plotted in Fig. 11-4.

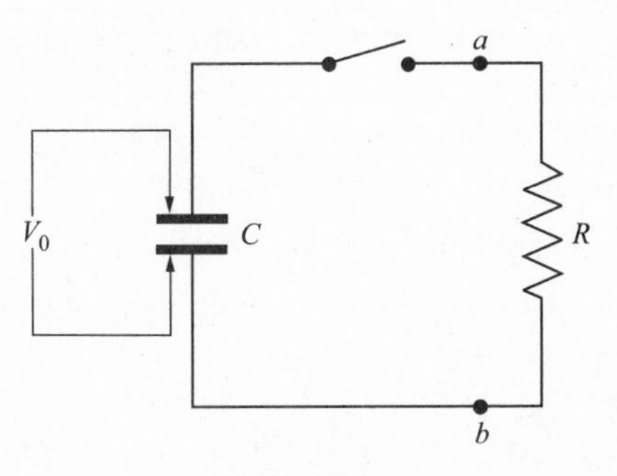

Fig. 11-3

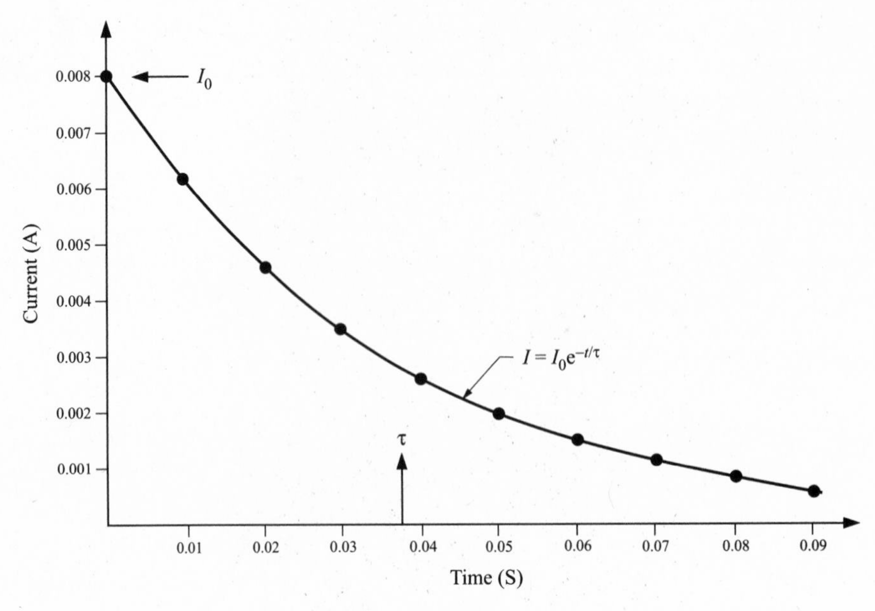

Fig. 11-4

(*e*) If $I = (\frac{1}{4})I_0$, then $(\frac{1}{4})I_0 = I_0 \exp[-(t/0.0375)]$. Thus $\ln(0.25) = -t/0.0375$, or $t = 0.052$ s. [Equivalently, we would have $Q = Q_0 \exp(-t/\tau)$ and $Q = (1/4)Q_0 \rightarrow \exp(-t/\tau) = 0.25$, as before.]

In summary, we have learned that for an R–C circuit, the currents, voltages and charges all move from initial to final values via an exponential $\exp[-(t/\tau)]$, where $\tau = RC$.

Resistor and Inductor

Consider the circuit consisting of a resistor, R, in series with an inductor, L, connected to a battery producing a voltage, V, through the switch, S, as in Fig. 11-5. Before the switch is closed, there is obviously no current flowing in the circuit. When we close the switch the battery will attempt to send current through the circuit. Without the inductance the current would instantly rise to $I = V_{bat}/R$. Because of the inductance, however, we have a back EMF opposing the increasing current: $\text{EMF} = -L\Delta I/\Delta t$; the faster the current tries to rise the greater the back EMF. Thus, when the switch is shut and the current tries to surge, it is stopped cold by a back EMF across L that instantly rises to

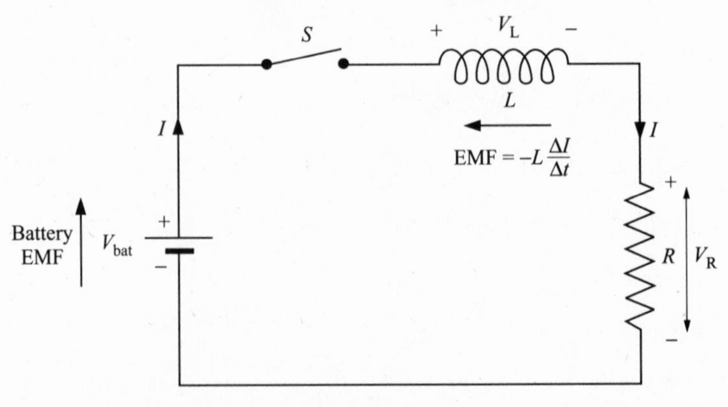

Fig. 11-5

be equal and opposite to the battery EMF. Therefore, the initial current will be zero as will the initial voltage across the resistor. Note that the electrostatic potential across the inductor, V_L, tends to push charge through the inductance in the opposite sense to the EMF, as shown in Fig. 11-8. This is completely analogous to the same situation in the battery. At $t = 0$ we thus have:

$$V_{bat} = V_L;\ V_R = 0, \qquad \text{or} \qquad t = 0:\ V_{bat} = +L\Delta I/\Delta t \qquad \text{and} \qquad I = V_R/R = 0 \tag{11.7}$$

From the laws of electrostatics, the sum of the electrostatic potential drops around the circuit must algebraically sum to zero, as a glance at Fig. 11-8 shows, the magnitudes of the voltages obey at all times:

$$V_{bat} = V_L + V_R \tag{11.8}$$

where

$$V_L = L\Delta I/\Delta t, \qquad V_R = IR \tag{11.9}$$

Substituting into Eq. (*11.8*) we get for all times t:

$$V_{bat} = L\Delta I/\Delta t + IR \tag{11.10}$$

Note. In magnitude, V_{bat} = EMF as long as we can ignore the internal resistance of the battery.

From Eqs. (*11.9*) and (*11.10*) we see that $\Delta I/\Delta t$ is positive, so current is building up in time, and will continue to rise until it reaches its final value. This final value is easy to calculate. When the current stops changing, there will be no back EMF and the voltage across the inductor will be zero. All the battery voltage is then across the resistor, and the current will equal V/R. At times between the initial and the final state, there will be voltage across both the inductor and the resistor, and both terms on the right of Eq. (*11.10*) will contribute. As can be seen, as the current increases, V_R gets larger and V_L gets smaller so the change in current, $\Delta I/\Delta t$, gets smaller. This means that it takes longer and longer for the current to increase toward its final value. This is analogous to the rise of charge on a capacitor in an R, C circuit. as shown in Fig. 11-2(*b*). It should be no surprise that an exponential decay law again holds for our L, R circuit, this time with a time constant,

$$\tau = L/R \tag{11.11}$$

Thus, (letting $V_{bat} = V$) we can show that V_L exponentially decays according to:

$$V_L = V \exp\left[-(t/\tau)\right] \tag{11.12}$$

Then, from Eq. (*11.8*):

$$V_R = V - V_L = 0 = V[1 - \exp(-t/\tau)] \tag{11.13}$$

and

$$I = V_R/R = (V/R)[1 - \exp(-t/\tau)] = (I_f)[1 - \exp(-t/\tau)] \tag{11.14}$$

at

$$t = \infty, \qquad I = V/R, \qquad V_R = V, \qquad V_L = 0, \qquad \Delta I/\Delta t = 0 \tag{11.15}$$

Note that the time constant does indeed have the dimensions of time. This is most easily deduced from the fact that $L\Delta I/\Delta t$ has the same dimensions as RI (i.e. volts), so L/R has the dimension of $I/(\Delta I/\Delta t)$ = seconds. Furthermore, the formula is physically reasonable. Thus a larger L means an increased time constant, because larger L means a greater ability to stop the current from rising quickly. Similarly, larger R implies a smaller time constant, since the final current is smaller and easier to reach.

Eqs. (*11.12*) and (*11.14*) are plotted in Figs. 11-6(*a*) and (*b*).

Problem 11.5. Consider the L–R circuit in Fig. 11-5. Let $L = 30$ mH, $R = 10\ \Omega$ and $V = 12$ V.

(*a*) What is the final current?

(*b*) What is the initial rate of change of current?

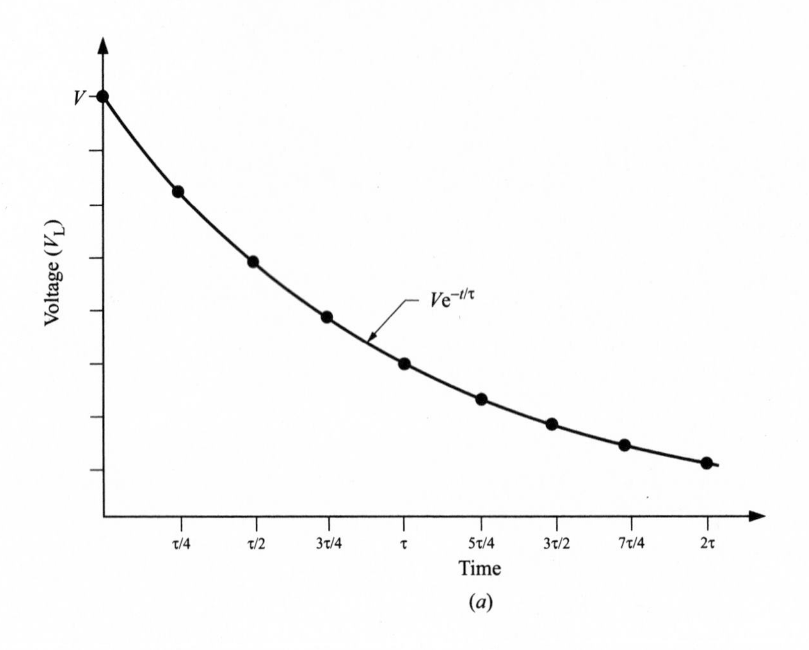

(*a*)

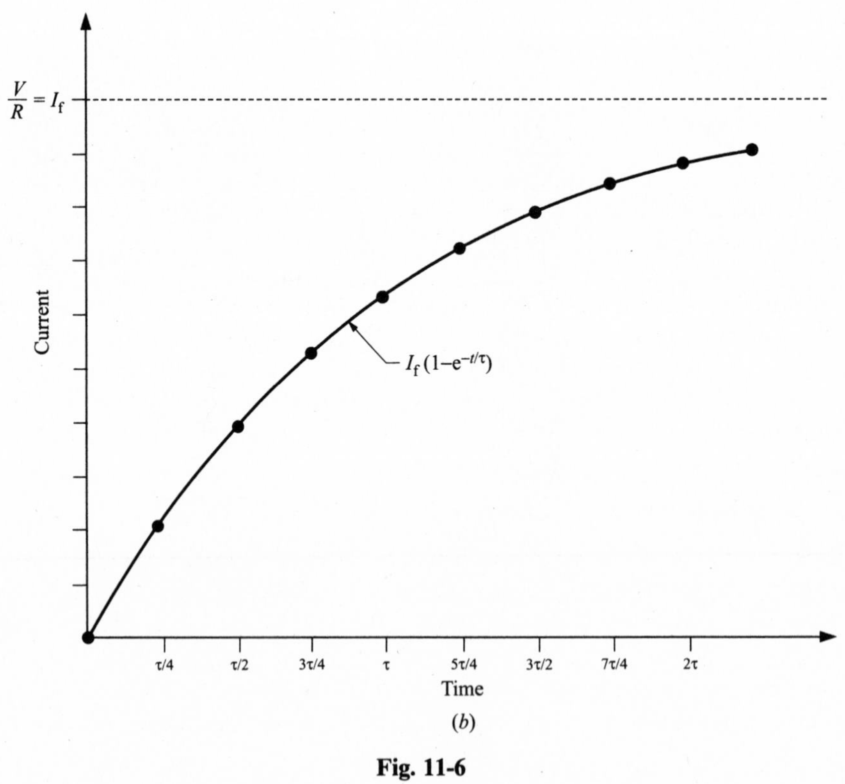

(*b*)

Fig. 11-6

(*c*) When the current is 0.70 A, what is the voltage across the inductor?

(*d*) What is the time constant of the circuit?

(*e*) What is the current at $t = 0.001$ s?

(*f*) At what time will the current equal 1.0 A?

Solution

(*a*) From Eq. (*11.13*), $I_f = V/R = (12\ \text{V})/(10\ \Omega) = 1.2\ \text{A}$.

(*b*) From Eq. (*11.7*), $(\Delta I/\Delta t)_i = V/L = (12\ \text{V})/(0.03\ \text{H}) = 400\ \text{A/s}$.

(*c*) If $I = 0.70$ A, $V_R = IR = (0.70\ \text{A})(10\ \Omega) = 7.0$ V, and $V_L = 12\ \text{V} - 7\ \text{V} = 5$ V.

(*d*) Using Eq. (*11.14*), $\tau = L/R = (0.03\ \text{H})/(10\ \Omega) = 0.0030$ s.

(*e*) We use Eq. (*11.12*) and substitute $t = 0.001$ s:

$$I = I_f(1 - \exp[-(0.001/0.003)] = (1.2\ \text{A})(1 - 0.716) = 0.34\ \text{A}.$$

(*f*) Again using Eq. (*11.12*), $I = I_f(1 - \exp[-(t/0.003\ \text{s})] \to (1.0\ \text{A}) = (1.2\ \text{A})(1 - \exp[-(t/0.003\ \text{s})]$. Therefore, $1.0/1.2 = 0.833 = (1 - \exp[-(t/0.003)])$, or $0.167 = \exp[-(t/0.003)]$. Then $\ln(0.167) = -t/0.003$, or $t = 0.0054$ s.

In a similar manner, if current is decreasing in an L–R circuit, it will decay exponentially with a time constant L/R.

Capacitor and Inductor

When we consider a circuit consisting of a capacitor and inductor alone, without a resistor, we are considering a fundamentally different problem than the two previous cases. The reason is that a resistor is an element that absorbs energy, converting it to heat, whereas the power supplied to a capacitor or inductor is used to store energy in these elements. In the case of a capacitor, the energy is stored in the form of potential energy of separated charges (or alternatively, in the form of energy stored in the associated electric fields), while for an inductor, the energy is stored in the form of moving charges (currents) (or alternatively in the form of energy stored in the associated magnetic fields). The energy is not dissipated, and remains in the system. Therefore, in the case of a pure L–C circuit, as depicted in Fig. 11-7, the energy due to the charges originally stored on the capacitor, and due to the currents do not decay away, rather they are interchanged, with the energy of the separated charges being converted to the energy of currents (moving charges), and vice versa. Unless there is some resistance (as there always is in real situations except in the case of superconductors), there will be an oscillatory situation, with a repetitive interchange of energy between the capacitor and the inductor *ad infinitum.* The total energy U_{total} (we use the symbol U for the energy as we did in thermodynamics to distinguish it from the symbol E that we use for the electric field) equals the sum of the capacitive and inductive energies, $U_C + U_L$. Thus, recalling the results of earlier chapters,

$$U_{\text{total}} = (\tfrac{1}{2})Q^2/C + (\tfrac{1}{2})LI^2 \tag{11.16}$$

Consider the circuit in Fig. 11-7. We first close switch S_1 while S_2 is open, and charge the capacitor to a voltage V, and charge Q_{max}. We then open S_1, with the capacitor charged, and close S_2 to permit the

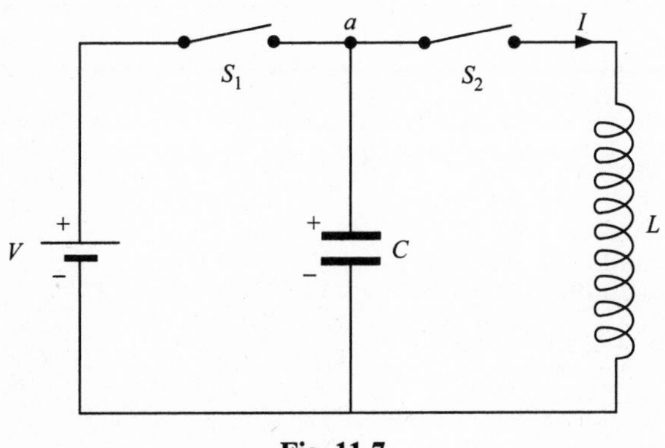

Fig. 11-7

capacitor to discharge through the inductor. Whenever current is flowing in the circuit, the capacitor is either charging or discharging. For positive I (clockwise in the figure), positive charge is leaving the positive plate and entering the negative plate, and for negative I, the positive charge is entering the positive plate and leaving the negative plate. When the current I is zero, the capacitor is instantaneously neither charging nor discharging, which means that it has reached its maximum (positive or negative) charge. This is similar to the case of simple harmonic motion (SHM) where the velocity is zero when the mass at the end of a spring is at its maximum positive or negative extension from equilibrium. Similarly, when the charge on the capacitor is zero, the voltage across the capacitor is zero ($V = Q/C$), and so is the voltage across the inductor. If there is no voltage across the inductor, then there is no back EMF, and $\Delta I/\Delta t$ is instantaneously zero, so the current is not changing and has reached a maximum positive or negative value. This shows clearly that when Q is zero, all the energy is in the inductor and equals $(\frac{1}{2})LI_{max}^2$. Also, when $I = 0$, all the energy is in the capacitor, and equals $(\frac{1}{2})Q_{max}^2/C$. We can then rewrite Eq. (*11.16*) as

$$U_{total} = (\tfrac{1}{2})Q^2/C + (\tfrac{1}{2})LI^2 = (\tfrac{1}{2})LI_{max}^2 = (\tfrac{1}{2})Q_{max}^2/C \tag{11.17}$$

To proceed further, we will make use of an analogy to the case of simple harmonic motion. There we had that (Ibid. Eq. (*12.5*))

$$U_{tot} = (\tfrac{1}{2})kx^2 + (\tfrac{1}{2})mv^2 = (\tfrac{1}{2})v_{max}^2 = (\tfrac{1}{2})kA^2 \text{ (with } x_{max} = A) \tag{12.5}$$

Here, $v = \Delta x/\Delta t$, just like $I = \Delta Q/\Delta t$ in our circuit equation. Equations (*11.17*) and (*12.5*) are identical mathematical relationships, as can be seen if we associate Q with x, $I = \Delta Q/\Delta t$ with $v = \Delta x/\Delta t$, k with $1/C$ and m with L. We can then use the results of simple harmonic motion to write equivalent equations for our circuit. This means that the charge and the current will undergo periodic sinusoidal behavior. Recalling that (for $x = A$ at $t = 0$):

$x = A \cos \omega t$, $v = -\omega A \sin \omega t = V_{max} \sin \omega t$, we have from our analogy:

$$Q = Q_{max} \cos \omega t, \qquad I = -I_{max} \sin \omega t = -\omega Q_{max} \sin \omega t \tag{11.18}$$

Similarly from $v = \pm\omega(A^2 - x^2)^{1/2}$ we have

$$I = \pm\omega(Q_{max}^2 - Q^2)^{1/2} \tag{11.19}$$

where $\omega = \sqrt{k/m}$ becomes

$$\omega = 1/\sqrt{LC}, \qquad f = 1/(2\pi\sqrt{LC}), \qquad T = 2\pi\sqrt{LC} \tag{11.20}$$

This frequency, f, is called the resonance frequency of the circuit.

Problem 11.6. Use Eq. (*11.17*) to: (*a*) show that $I_{max} = \omega Q_{max}$ [as used in Eq. (*11.18*)]; and (*b*) derive Eq. (*11.19*).

Solution

(*a*) From Eq. (*11.17*): $(\frac{1}{2})LI_{max}^2 = (\frac{1}{2})Q_{max}^2 \rightarrow I_{max} = (1/LC)Q_{max} = \omega Q_{max}$.

(*b*) From Eq. (*11.17*): $(\frac{1}{2})Q_{max}^2/C = (\frac{1}{2})Q^2/C + (\frac{1}{2})LI^2 \rightarrow LI^2 = (1/C)(Q_{max}^2 - Q^2) \rightarrow I = \pm(1/LC)^{1/2}(Q_{max}^2 - Q^2)^{1/2} = \pm\omega(Q_{max}^2 - Q^2)^{1/2}$.

Problem 11.7. What capacitance is needed in an L–C circuit to provide an oscillation of 1.5 MHz if the inductor has an inductance of 0.3 mH?

Solution

Using Eq. (*11.20*), $f = 1.5 \times 10^6$ Hz $= (1/2\pi)/\sqrt{LC}$, $LC = (2\pi)^2/f^{-2} = 1.75 \times 10^{-11}$ s^2, $C = 5.85 \times 10^{-8}$ F

Problem 11.8. Consider the L–C circuit in Fig. 11-7. Let $L = 30$ mH, $C = 10$ μF and $V = 12$ V. The capacitor is charged by closing switch S_1, and when it is fully charged, S_1 is opened and S_2 is closed. The initial time is when S_2 is closed.

(*a*) What is the initial charge on the capacitor, and the initial current through the inductor?

(*b*) What is the initial total energy of the circuit?

(*c*) What is the maximum current in the circuit?

(*d*) What is the frequency of oscillation of the circuit? What is the period of the oscillation?

(*e*) What is the current in the circuit when the capacitor has a charge of 8×10^{-5} C?

(*f*) Plot the voltage across the capacitor as a function of time.

Solution

(*a*) The initial conditions are that $I = 0$, and the capacitor is fully charged. Thus, $I_i = 0$, $Q_i = Q_{max} = CV = (10 \times 10^{-6}\text{ F})(12\text{ V}) = 1.2 \times 10^{-4}$ C.

(*b*) Using Eq. (*11.12*), $U = \text{constant} = (\frac{1}{2})Q_{max}^2/C = 0.5(1.2 \times 10^{-4}\text{ C})^2/(10^{-5}\text{ F}) = 7.2 \times 10^{-4}$ J.

(*c*) The maximum current occurs when the charge on the capacitor is zero. Using the energy equation, $(\frac{1}{2})LI_{max}^2 = U = 7.2 \times 10^{-4}$ J, $I_{max} = [(7.2 \times 10^{-4}\text{ J})(2)/(30 \times 10^{-3}\text{ H})]^{1/2} = 0.22$ A.

(*d*) Using Eq. (*11.20*), $\omega = 1/\sqrt{LC} = 1/[(30 \times 10^{-3})(10^{-5})]^{1/2} = 1.83 \times 10^3\ \text{s}^{-1}$, $f = \omega/2\pi = 291$ Hz, $T = 1/f = 3.44 \times 10^{-3}$ s.

(*e*) The total energy is constant, and equals $7.2 \times 10^{-4}\text{ J} = (1/2)Q^2/C + (1/2)LI^2 = 0.5(8 \times 10^{-5}\text{ C})^2/(10^{-5}\text{ F}) + 0.5(30 \times 10^{-3}\text{ H})I^2$, or $I^2 = 0.0267\text{ A}^2$, $I = 0.16$ A.

(*f*) The voltage across the capacitor is given by $Q/C = (Q_{max}/C)\cos\omega t = (12\text{ A})\cos\omega t$. This is plotted in Fig. 11-8.

Let us briefly summarize what we have learned regarding transient phenomena in DC circuits containing various combinations of resistors, capacitors and inductors. Whenever resistors are present, the energy is dissipated by virtue of the absorption of energy by those resistors. This results in an exponential decay toward the final value of all the relevant quantities (charges, currents, voltages, etc.), with a time constant whose magnitude is determined by the values of R, L and C. Whenever inductors and capacitors are together in a circuit, there will be an interchange of energy between these circuit elements,

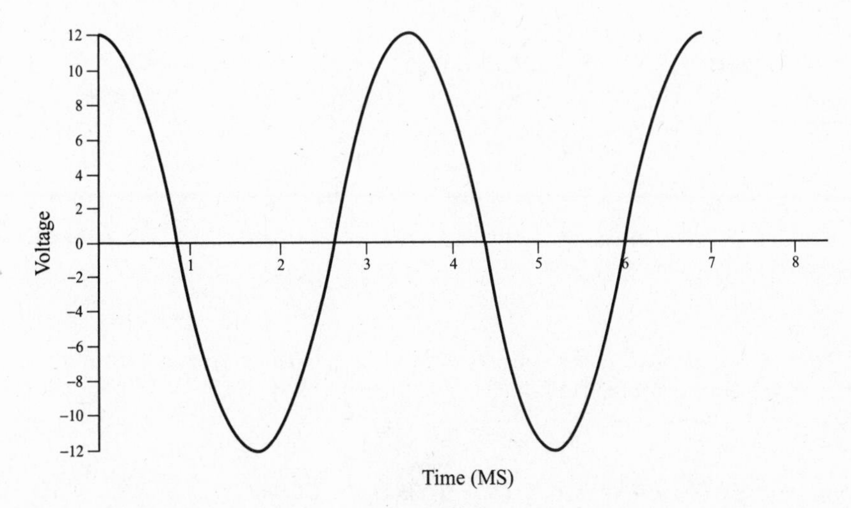

Fig. 11-8

resulting in an oscillation at a frequency determined by the values of the circuit elements. In the absence of a resistor, there will be no loss in energy, and there will be no decay in the charges, currents or voltages, merely an interchange at this frequency. In practice, there is always some resistance, which will then result in a decay of the oscillation.

11.3 STEADY STATE PHENOMENA IN AC CIRCUITS

So far in this chapter, we have discussed the transient phenomena that take place during the time that a circuit approaches a constant final state. In those cases the charges, currents and voltages no longer change after reaching this state. Those cases occur when the source of voltage is a battery or some other source of steady voltage. In this section we will discuss the phenomena that occur when the source of voltage is varying sinusoidally, as in the circuit of Fig. 11-9. In that case we expect that the variables of the circuit will also vary sinusoidally, after there has been sufficient time for the circuit to adjust to this state. In other words, although we expect transient phenomena to occur, these phenomena are expected to decay exponentially with time, and lead to a "steady state" in which all quantities that can vary are varying sinusoidally. This is indeed what we find happens. We will not discuss any details of the transient phenomena, and will discuss only the situation that results after sufficient time has elapsed that we reach the steady state. This time is usually short enough that the effect of the transients can be neglected.

Whenever we have sinusoidal variation, we can express the variables as sine or cosine functions of time. The frequency, f, of the sinusoidal variation can be expressed in terms of an angular frequency, ω, which simplifies the equations. Of course, the frequency can also be related to the period, T, by the relationship $f = 1/T$. As a matter of convenience we will take the current to vary as $i = I_0 \cos \omega t$. In this equation, as well as in future equations, we shall use the lower case (i) to represent something that varies with time, and the upper case (I) to represent a constant. As we have seen previously (for example in simple harmonic motion), the factor in front of the cosine function, I_0, represents the maximum value of the variable, i. This is the amplitude of the variation, and it represents the maximum value the variable can have. We shall see later that there is another value of the variable, related to the magnitude, known as the "RMS" value of the variable, which is often useful to use. Both the amplitude and the RMS values are constants, with $I_{RMS} = I_0/\sqrt{2}$, and both are represented by upper case letters. In what follows we will examine a series R–L–C circuit driven by a sinusoidal EMF (e.g. Fig. 11-9). Since the current will be the same everywhere in the circuit at each instant of time, it is easiest to start with a simple sinusoidal description of the current and then to examine the related voltages across each element and combinations of elements. We will start with such a current through a simple resistor.

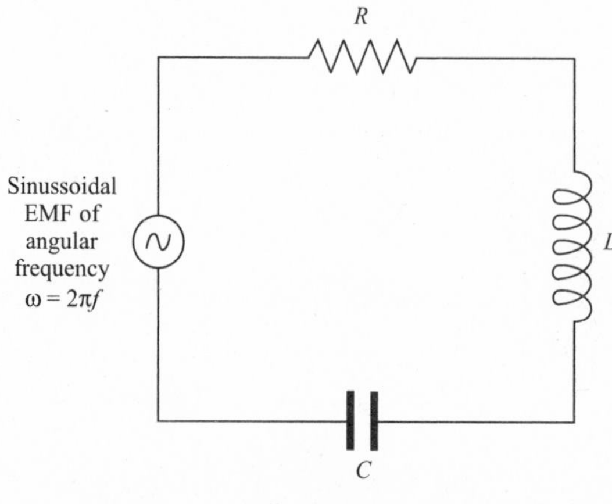

Fig. 11-9

Resistor

Consider a circuit such as that in Fig. 11-9 in which an AC voltage source produces a steady state current, $i = I_0 \cos \omega t$ through a resistor R. We know, by the definition of the resistance, that the voltage across the resistor at any time is just the current times the resistance, or $v = iR$. Here v is the time varying voltage across the resistor, which will equal $v = I_0 R \cos \omega t = V_0 \cos \omega t$, where the maximum voltage across the resistor, V_0, equals $I_0 R$. Because both i and v vary identically with time as $\cos \omega t$, we say that the two are "in phase". This means that they both attain their maximum value at the same time, and both go through zero at the same time. We will see that this is true only for a resistor, while for capacitors and inductors, the voltage will not be in phase with the current.

If we plot the current and voltage as functions of time, we get the familiar variation such as that seen in Fig. 11-10. The fact that they vary in phase with each other is easily seen on this graph, since the voltage at any time is just a fixed multiple of the current.

There is another convenient way to represent the time variation of the current and voltage. This is especially useful for cases in which the variables are not in phase, and we will introduce this method here. The method is based on the same idea that was used in simple harmonic motion to describe the SHM. There we showed that for a particle moving around a circle at constant speed, the projection on any axis, and specifically on the x axis is one of SHM. In a similar manner if we picture a vector of magnitude I_0 rotating, at constant angular velocity ω, about the origin, as in Fig. 11-11, the x-component of the vector varies sinusoidally. If the vector starts on the x axis at time zero, then the angle it makes with the x axis will be $\theta = \omega t$, and $i = I_0 \cos \omega t$. Thus, the x component of the rotating vector will give us the time varying current, i. Similarly, the vector representing the voltage will be a vector always at the same angle as the current, with a magnitude of RI_0. Being in phase means that the two vectors are at the same angle, and rotate about the origin together. When used in this way for electric circuits these "vectors" are called phasors. The general idea is that the projection on the x axis will give us the proper time variation, and that we can represent these currents and voltages as phasors rotating about the origin at the same ω.

The result for a pure resistance is:

$$i = I_0 \cos \omega t \qquad v = V_0 \cos \omega t \qquad V_0 = I_0 R \tag{11.21}$$

Our next project is to calculate the energy absorbed by the resistor when the current is sinusoidal. To do this, we recall that the power used at any time is the product of current and voltage. Therefore

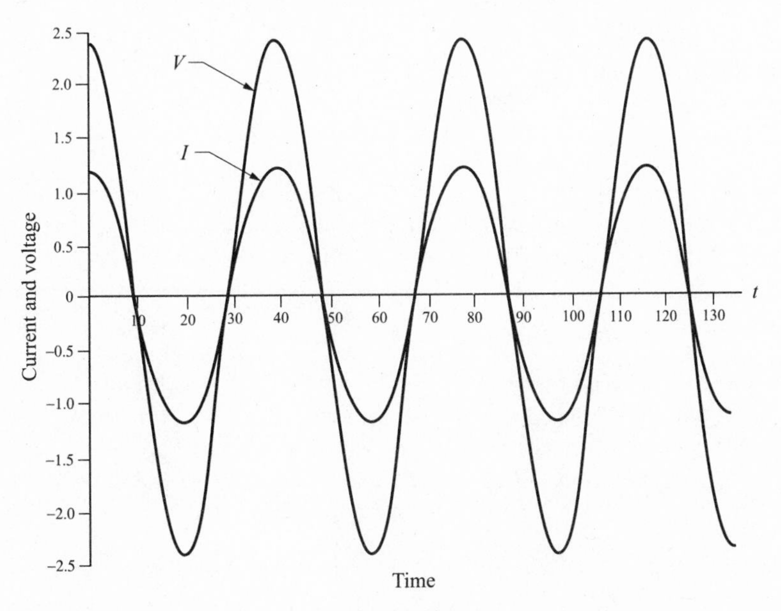

Fig. 11-10

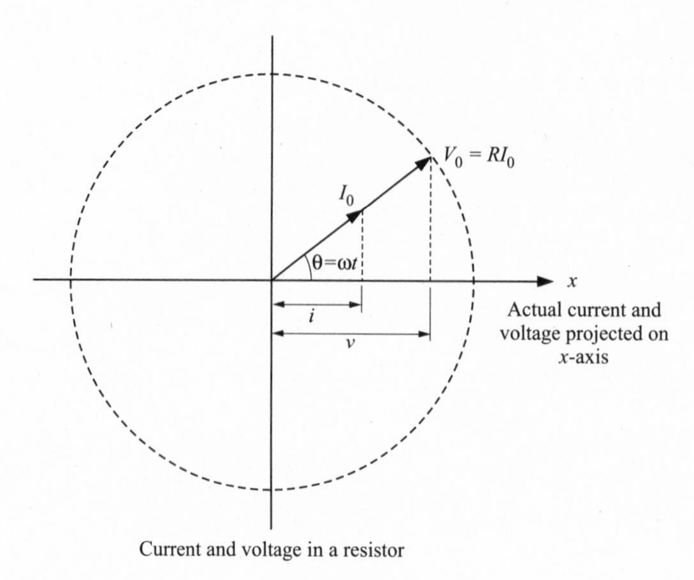

Current and voltage in a resistor

Fig. 11-11

$p = iv = I_0 V_0 \cos^2 \omega t$. This power is always positive, since $\cos^2 \omega t$ is always positive, even when $\cos \omega t$ is negative. Thus power is always being provided to the circuit, and the resistor is, at all times, absorbing energy.

Problem 11.9. Calculate the time-average power absorbed by the resistor.

Solution

The average power will be the average of $P_{av} = (I_0 V_0 \cos^2 \omega t)_{av} = I_0 V_0 (\cos^2 \omega t)_{av}$ over time. Since the cosine function repeats every 2π radians, it is adequate to get the average of $\cos^2 \omega t$ over one complete cycle.

To do this we recall that $(\cos^2 \omega t + \sin^2 \omega t) = 1$, and recognize that the averages of $\cos^2 \omega t$ and $\sin^2 \omega t$ over a *complete* cycle must be equal to each other. Then, noting that the average of a constant over time (such as "1") is just the constant itself, we have $(\cos^2 \omega t + \sin^2 \omega t)_{av} = 1, \rightarrow (\cos^2 \omega t)_{av} + (\sin^2 \omega t)_{av} = 1, \rightarrow 2(\cos^2 \omega t)_{av} = 1, \rightarrow (\cos^2 \omega t)_{av} = 1/2$. The average power is therefore $P_{av} = I_0 V_0(\frac{1}{2})$. This can be rewritten as $P_{av} = (I_0/\sqrt{2})(V_0/\sqrt{2}) = I_{RMS} V_{RMS}$, where, as noted earlier, $I_{RMS} = I_0/\sqrt{2} = 0.707 I_0$, and $V_{RMS} = V_0/\sqrt{2} = 0.707 V_0$.

We have derived a very important result. We have shown that the average power in AC can be written in the same form as in DC, provided that we use RMS values for the current and voltage. Note that, since $V_0 = RI_0$, therefore $V_{RMS} = RI_{RMS}$, and we can therefore write

$$P_{av} = I_{RMS} V_{RMS} = I_{RMS}^2 R = V_{RMS}^2/R \tag{11.2}$$

In most formulas used in AC circuits, the quantity we use for the "magnitude" of currents and voltages will be the RMS value, and therefore when we write just *I or V we will refer to the RMS values.* The term **RMS** actually stands for "**root-mean-square**", which refers to the method used to determine its value. To get the RMS value of a variable, we have to take the square root (root) of the average (mean) of the square (square) of the quantity. Thus to get the RMS value of the current ($I_0 \cos \omega t$), we first square the current ($I_0^2 \cos^2 \omega t$), then take the average $[(I_0^2 \cos^2 \omega t)_{av} = I_0^2/2)$, and then take the square root ($= I_0/\sqrt{2}$), which gives I_{RMS}.

Problem 11.10. An AC source of voltage produces a current of 2.0 A in a resistor of 100 Ω.

(*a*) What voltage is across the resistor?

(*b*) What is the average power supplied to the resistor?

(*c*) What is the peak voltage across the resistor?

(*d*) What is the peak power supplied to the resistor?

Solution

(*a*) Since the current was given as 2.0 A, we can assume that this refers to the RMS current. Then the voltage (RMS) will equal $V = IR = (2.0\ \text{A})(100\ \Omega) = 200\ \text{V}$.

(*b*) The average power is given by Eq. (*11.38*). $P_{av} = I^2R = (2.0\ \text{A})^2(100\ \Omega) = 400\ \text{W}$. Alternatively we could have used $P_{av} = IV = (2.0\ \text{A})(200\ \text{V}) = 400\ \text{W}$.

(*c*) The peak voltage is V_0, which is related to the RMS voltage by $V_{RMS} = V_0/\sqrt{2}$. Thus $V_0 = V_{RMS}(\sqrt{2}) = 200(1.414) = 283\ \text{V}$.

(*d*) The power is given by iv (or i^2R). This is $p = I_0V_0 \cos^2 \omega t = (\sqrt{2}I)(\sqrt{2}V) \cos^2 \omega t$, whose peak value occurs when $\cos^2 \omega t = 1$. Thus the peak value of the power is $2IV = 2(2.0\ \text{A})(200\ \text{V}) = 800\ \text{W}$.

The case of a resistor is relatively simple because the voltage and the current are in phase. The other cases we will discuss are ones in which the current and voltage are not in phase, and we will now learn how to handle those cases.

Capacitor

Again we will assume (Fig. 11-9) that a generator causes a sinusoidal current to flow through the capacitor given by $i = I_0 \cos \omega t$. When we refer to current flowing "through" the capacitor, we mean that charge is flowing toward one plate of the capacitor and away from the other plate.

Note. Current flowing away from an initially positive plate includes discharging positive charge from that plate and then effectively charging that plate with negative charge; similarly current flowing toward an initially negative plate first neutralizes that plate and then charges it positively.

Since the current is alternating between positive and negative values at a frequency of $f = \omega/2\pi$, the capacitor is alternately charging each plate positively and negatively. For purposes of sign convention for the voltage across the capacitor we define the "**positive**" plate as the one into which positive charges flow when the current is positive. Thus, if clockwise current is chosen as positive, the right side of the capacitor in Fig. 11-9 is the "positive" plate. Q is then defined as the charge on the "positive" plate. The voltage across the capacitor is defined to be from positive to negative plate and is always given as $v = q/C$, where v and q are the variable voltage and charge on the capacitor, and C is the constant capacitance of the capacitor. Both v and q will vary sinusoidally at the same frequency as the current. By our convention, when the positive charge is increasing on the positive plate, the current is positive. When the positive charge on that plate is decreasing, the current is negative. Just before the charge reaches its positive maximum value, the current is positive, and just after the maximum the current is negative. At the maximum, the current is temporarily zero. Since the capacitor voltage equals q/C the voltage and charge reach maximum at the same time. Thus, when the voltage reaches its peak, the current is zero. This is in contrast to the case of the resistor where the current and voltage reach their peak at the same time. In the case of the capacitor, the current and voltage are said to be 90° or $\pi/2$ out of phase, because the sine (or cosine) changes from maximum to zero in 90°. Before the voltage reaches its positive peak, the current is positive but decreasing to zero. Therefore, the current reached its positive peak before the voltage, and we say that the current "**leads**" the voltage, and the voltage "**lags**" behind the current. It is easy to verify that a voltage $v = V_0 \sin \omega t$ has just the properties required. It has the same frequency as the current, and it reaches a positive maximum when $\omega t = \pi/2$ while the current already reached its maximum at $\omega t = 0$. We thus know that, for a capacitor, we can represent the voltage by $v = V_0 \sin \omega t$ if the current is given by $i = I_0 \cos \omega t$. We must still develop the relationship between the magnitudes of the current and voltage. We expect that if the maximum current is increased then the maximum charge on the capacitor will increase proportionally, and therefore also the maximum voltage. Consequently, we can write that $V_0 = X_C I_0$, where the constant of proportionality X_C is called the **capacitive reactance** of the capacitor. Similarly, $V_{RMS} = X_C I_{RMS}$ or $V = X_C I$. This capacitive reactance depends on the capacitance and on the frequency. If X_C is large, it means that it is easy to get a large voltage from a small current. This occurs for small capacitance C (since $v = q/C$) or for small f (since there is a long time to charge up before the current reverses). We are therefore not surprised to find that

$$X_C = 1/\omega C = 1/(2\pi f C), \tag{11.23}$$

and

$$V_C = IX_C = I/(2\pi f C) \tag{11.24}$$

$$v_C = V_{0C} \sin \omega t = I_0 X_C \sin \omega t \tag{11.25}$$

Eq. (*11.23*) can be rigorously demonstrated using the calculus.

In Fig. 11-12, we draw the current and voltage for the case of a capacitor, showing that they reach their peaks 90° apart, with the current leading the voltage, i.e. reaching its positive peak earlier. While the *magnitudes* of the current and voltage are proportional to each other, it cannot be said that the current is proportional to the voltage since their peaks occur at different times.

If we again represent the current by a rotating phasor so that its projection along the x axis will be $I_0 \cos \omega t$, then the voltage can be represented by a phasor 90° behind it. The projection of the voltage phasor on the x axis will then be $V_0 \sin \omega t$. This is depicted in Fig. 11-13.

Note that unless otherwise indicated, *currents and voltages* in the problems will refer to RMS values (e.g. "find the current" means find I_{RMS}, etc.)

Problem 11.11. An AC source of voltage at a frequency of 1500 Hz produces a current of 2.0 A in a capacitor of 100 μF.

(*a*) What is the reactance of the capacitor?

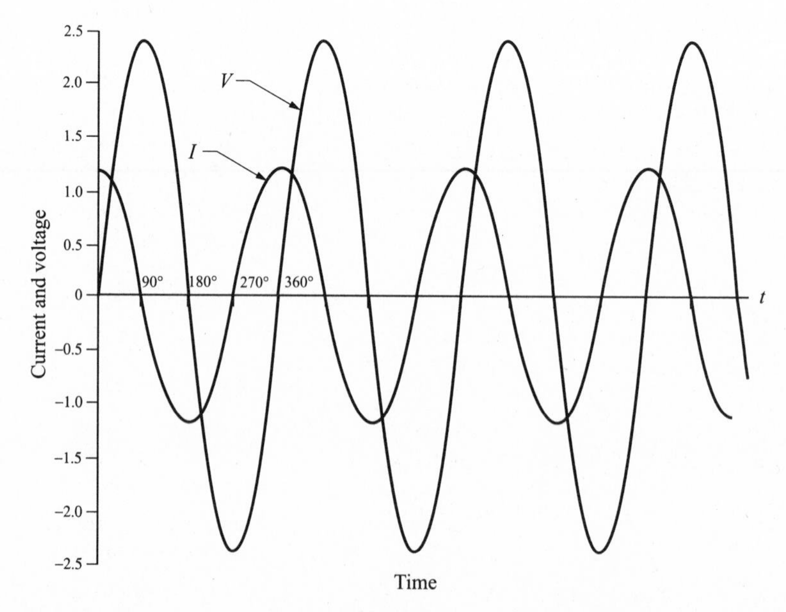

Fig. 11-12

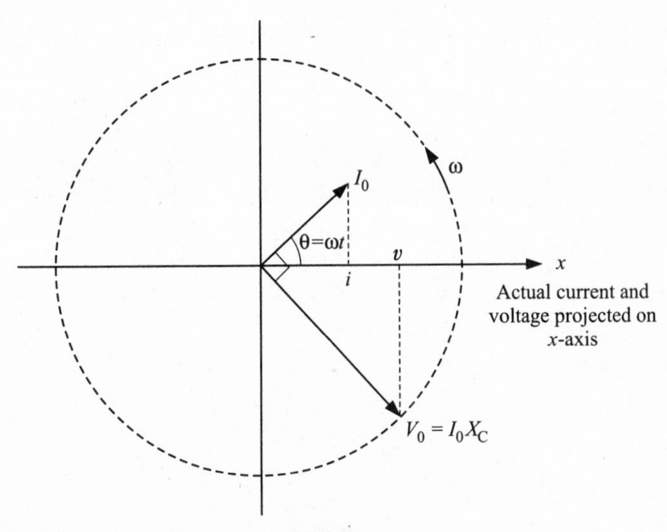

Fig. 11-13

(*b*) What is the voltage across the capacitor?

(*c*) What is the peak voltage across the capacitor?

Solution

(*a*) The reactance of a capacitor is $1/(2\pi fC) = 1/(2\pi)(1500\text{ Hz})(100 \times 10^{-6}\text{ C}) = 1.06\ \Omega$.

(*b*) The voltage across the capacitor is $V = IX_C = (2.0\text{ A})(1.06\ \Omega) = 2.12$ V.

(*c*) The peak voltage is $V\sqrt{2} = 3.00$ V.

We now calculate the power consumption of the capacitor. Again, we know that at any instant the power equals the current times the voltage, $p = iv = (I_0 \cos \omega t)(V_0 \sin \omega t)$. In contrast to the case of the resistor, this power is not always positive. Indeed (see, e.g. Fig. 11-12), in the first and third quadrants ($0° < \omega t < 90°$ and $180° < \omega t < 270°$) the power is positive since the sine and cosine have the same sign. However, in the second and fourth quadrants ($90° < \omega t < 180°$ and $270° < \omega t < 360°$), the power is negative, since the sine and cosine have opposite signs. The *average* power is therefore zero. We can understand this result if we remember that it takes energy to charge a capacitor, but that energy is regained when the capacitor discharges. Therefore, as long as there is no resistor, the energy is alternately stored and released, but none of it is dissipated as heat or otherwise transferred outside the circuit. A pure capacitor therefore does not consume any energy, and the average power supplied to a capacitor is zero. This result actually occurs whenever the current and voltage are 90° apart.

Problem 11.12. An AC source of voltage at a frequency of 150 Hz produces a current of 2 A in a capacitor of 15 μF.

(*a*) Write an expression for the current in the capacitor.

(*b*) Write an expression for the voltage across the capacitor.

(*c*) What is the average power consumed by the capacitor?

(*d*) What is the peak power supplied to the capacitor?

Solution

(*a*) The current is given by $i = I_0 \cos \omega t = \sqrt{2}(I \cos \omega t) = (2.83\text{ A}) \cos(942t)$, assuming, as usual, that we arbitrarily set $t = 0$ when the current has its positive peak value.

(*b*) The voltage is given by $V_0 \sin \omega t = \sqrt{2}(V \sin \omega t) = \sqrt{2}(IX_C) \sin \omega t = [\sqrt{2}(2.0\ \text{A})/\omega C] \sin \omega t = (200\ \text{V}) \sin (942t)$.

(c) The average power is zero.

(*d*) The power is given by $p = iv = (I_0 \cos \omega t)(V_0 \sin \omega t) = 2IV \cos \omega t \sin \omega t$). To get the peak power, we must determine the maximum value of the expression $[\cos \omega t \sin \omega t]$. We cannot say that the maximum of $\cos \omega t \sin \omega t$ is the maximum of the sine times the maximum of the cosine, since they don't reach their maxima at the same time. To get the correct answer, we make use of the trigonometric relationship $\sin (2x) = 2 \sin (x) \cos (x)$, or $\sin (x) \cos (x) = 0.5 \sin (2x)$. With $x = \omega t$, we therefore have $[\cos \omega t \sin \omega t] = 0.5 \sin (2\omega t)$, and the maximum value of this is 0.5. Therefore, the maximum power is $P_{MAX} = 2IV(0.5) = 283$ W. This shows that, although the average power is indeed zero, there can be large power inputs during each cycle.

Capacitor and Resistor in Series (R–C Circuit)

Let us now consider a case in which an AC generator produces a sinusoidal current through a series combination of a resistor and capacitor (see Fig. 11-14). The current is given by $i = I_0 \cos \omega t$, and is the same in both circuit elements since they are in series. From the previous sections we know how to calculate the voltage across each of the circuit elements, and our task in this section is to learn how to combine these voltages and calculate the voltage across the whole circuit. The voltage across the resistor (v at point a minus v at point b in the figure) will be $v_R = RI_0 \cos \omega t$, as shown in Eq. (*11.36*). (Recall that v_R and i are in phase.) Similarly the voltage across the capacitor (v at point b minus v at point c in the figure) is given by $v_C = X_C I_0 \sin \omega t$ [Eq. (*11.42*), voltage lags current by 90°]. At any instant of time, the voltage across the entire circuit (v at point a minus v at point c in the figure) is just the sum of $v_R + v_C = RI_0 \cos \omega t + X_C I_0 \sin \omega t = v_T$, the total voltage. Adding these voltages by using trigonometric identities is possible, but complicated, and there is a simpler method making use of the phasor diagrams (see Fig. 11-15). If we draw the phasor for the current at the moment shown if Fig. 11-15(*a*), then the phasor for the voltage across the resistor will be along the same direction, while the phasor for the voltage across the capacitor will lag by 90°. The true total voltage, v_T, is the sum of these

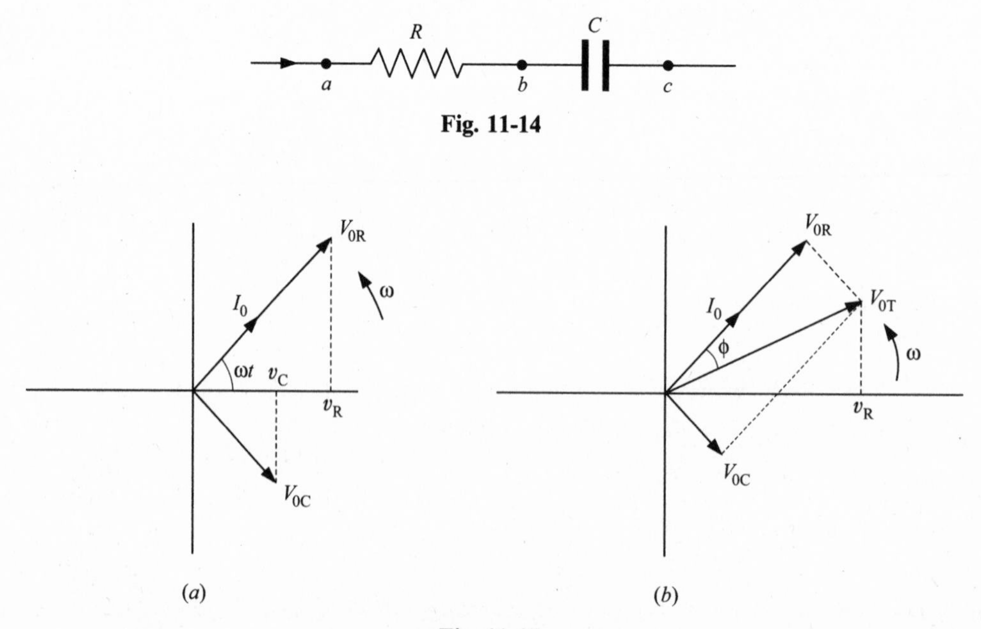

Fig. 11-14

Fig. 11-15

components. The actual voltages at time t are the projections, or components, of the phasors on the x axis. Since the x component of the sum of vectors is just the sum of the x components of the vectors, v_T is just the x component of a new phasor which is the vector sum of the v_{OR} and v_{OC} phasors. This is shown in Fig. 11-15(*b*). From the figure we can see that (using the Pythagorean theorem)

$$V_{OT} = (V_{OR}{}^2 + V_{OC}{}^2)^{1/2} = I_0(R^2 + X_C{}^2)^{1/2} = I_0 Z \tag{11.26a}$$

or in terms of RMS values

$$V_T = (V_R{}^2 + V_C{}^2)^{1/2} = I(R^2 + X_C{}^2)^{1/2} = IZ \tag{11.26b}$$

where Z is called the "**impedance**" of the circuit and

$$Z = (R^2 + X_C{}^2)^{1/2} \tag{11.27}$$

The angle ϕ in the figure is known as the "**phase angle**", and represents the angle between the current and the voltage. In the case we are discussing, the voltage is seen to lag behind the current so we treat this angle as negative, and the angle is given by $\tan\phi = -V_{OC}/V_{OR} = -(I_0 X_C)/(I_0 X_R)$, or

$$\tan\phi = -X_C/R. \tag{11.28}$$

Problem 11.13. Show that the formula for $\tan\phi$ gives the correct result in the limit of a pure resistor or a pure capacitor.

Solution

For a pure resistor, $\tan\phi = -X_C/R = 0$, giving $\phi = 0$. For a pure capacitor, $\tan\phi = -X_C/R = -X_C/0 = -\infty$, giving $\phi = -90°$. This is exactly what we know is the case for those situations.

Problem 11.14. A circuit consists of a resistor, $R = 300\ \Omega$, in series with a capacitor with a reactance of $400\ \Omega$. The current is 2.0 A at a frequency of 500 Hz.

(*a*) What is the capacitance of the capacitor?

(*b*) What is the impedance of the circuit?

(*c*) What is the voltage across the resistor? Across the capacitor? Across the series circuit?

(*d*) What is the phase angle between the current and the total voltage?

Solution

(*a*) We know that $X_C = 1/(2\pi fC)$. Therefore $C = 1/(2\pi fX_C) = 1/[(2\pi)(500)(400)] = 8.0 \times 10^{-7}$ F.

(*b*) The impedance is given by Eq. (*11.44*), $Z = (R^2 + X_C^2)^{1/2} = (300^2 + 400^2)^{1/2} = 500\ \Omega$.

(*c*) $V_R = IR = (2.0\ \text{A})(300\ \Omega) = 600$ V.

$V_C = IX_C = (2.0\ \text{A})(400\ \Omega) = 800$ V.

$V_T = IZ = (2.0\ \text{A})(500\ \Omega) = 1000$ V.

Note. V_T does *not* equal the sum of $V_R + V_C$, since those voltages are not in phase. This is equivalent to the statement about vectors that the magnitude of the resultant vector is *not* equal to the sum of the magnitudes of the vectors that are being added.

(*d*) We know that $\tan\phi = -X_C/R = -400/300 = -1.333$. Taking the inverse tangent results in $\phi = -53.1°$.

Problem 11.15. A circuit consists of a resistor, $R = 30\ \Omega$, in series with a capacitor with a capacitance of $40\ \mu$F. A generator produces a voltage of 120 V at a frequency of 60 Hz across the series circuit.

(*a*) What is the reactance of the capacitor?

(*b*) What is the impedance of the circuit?

(*c*) What is the current in the circuit?

(*d*) What is the phase angle between the current and the total voltage?

Solution

(*a*) We know that $X_C = 1/(2\pi fC) = 1/[(2\pi)(60)(40 \times 10^{-6}) = 66.3\ \Omega$.

(*b*) The impedance is given by Eq. (*11.44*), $Z = (R^2 + X_C{}^2)^{1/2} = (30^2 + 66.3^2)^{1/2} = 72.8\ \Omega$.

(*c*) $V_T = IZ = 120\ \text{V} = \text{I}(72.8\ \Omega)$. Therefore, $I = 1.65$ A.

(*d*) We know that $\tan\phi = -X_C/R = -66.3/30 = -2.21$. Taking the inverse tangent results in $\phi = -65.7°$.

We now turn our attention to calculating the energy absorbed in the circuit. The simplest way to analyze the situation is to note that we have already determined that a capacitor does not absorb any energy. Therefore, the total energy absorbed by the circuit is the energy absorbed by the resistor. The average power absorbed in the resistor is equal to $P = I^2R$, which is therefore the total average power supplied to the circuit. This can be rewritten in terms of the current and total voltage by substituting $I = V_T/Z$ for one of the currents, to give $P = IV_T(R/Z)$. Recalling that $I_0 R = V_{0R}$ and $I_0 Z = V_{0T}$, and using Fig. 11-15(*b*) to obtain $R/Z = \cos\phi$, we finally get:

$$P = IV_T \cos\phi \qquad (11.29)$$

The formula for power is the same as in DC except for the factor $\cos\phi$, which is appropriately called the power factor. Of course, since only the resistor absorbs energy we can write the average power directly as:

$$P = I^2R = IV_R = V_R{}^2/R \qquad (11.30)$$

We could have used our basic method for calculating average power, namely calculating the average of iv_T, and we would have gotten the same results as Eqs. (*11.29*) and (*11.30*).

Problem 11.16. Calculate the average power in a series *R–C* circuit by averaging the time varying power.

Solution

We know that $i = I_0 \cos\omega t$. We must write down a similar equation for v_T. We know that the magnitude of the total voltage is $V_0 = I_0 Z$, and that it lags the current by ϕ. Thus, the correct formula for v_T is $v_T = V_0 \cos(\omega t + \phi)$, where the lag of the voltage is included by virtue of the fact that ϕ is negative. Since the power is iv_T, we get $p = iv_T = (I_0 \cos\omega t)[V_0 \cos(\omega t + \phi)] = I_0 V_0 (\cos\omega t)\cos(\omega t + \phi)$. We must now average $(\cos\omega t)\cos(\omega t + \phi)$. Using the trigonometric relationship that $\cos(a + b) = \cos(a)\cos(b) - \sin(a)\sin(b)$, we write $\cos(\omega t + \phi) = (\cos\omega t)(\cos\phi) - (\sin\omega t)(\sin\phi)$. If we multiply this by $\cos\omega t$, the first term is $\cos^2\omega t \cos\phi$, and we have already shown that $\cos^2\omega t$ averages to $\frac{1}{2}$, and the second term has $[\cos\omega t \sin\omega t \sin\phi]$, which we have shown averages to zero. Therefore, $[(\cos\omega t)\cos(\omega t + \phi)]_{av} = (\frac{1}{2})\cos\phi$. Thus, $P = [iv_T]_{av} = I_0 V_0(\frac{1}{2})\cos\phi = IV\cos\phi$, as we derived previously in a much simpler manner.

Problem 11.17. In Problem 11.15, calculate (*a*) the power in the resistor; (*b*) the power factor of the circuit; and (*c*) the power in the entire circuit.

Solution

(*a*) The power in the resistor is $I^2R = (1.65\ \text{A})^2(30\ \Omega) = 81.7$ W.

(*b*) The power factor is $\cos\phi = R/Z = 30/72.8 = 0.412$.

(*c*) The power in the circuit is the power in the resistor = 81.7 W. Alternatively, the total power is $IV \cos\phi = (1.65\text{ A})(120\text{ V})(0.412) = 81.6$ W, which is the result of part (*a*) to within rounding errors.

Problem 11.18. In a series *R–C* circuit, a generating voltage of 120 V at a frequency of 75 Hz produces a current of 1.5 A and uses energy at the rate of 50 W. Calculate (*a*) the impedance of the circuit; (*b*) the power factor of the circuit; (*c*) the resistance of the resistor in the circuit; and (*d*) the capacitance of the capacitor in the circuit.

Solution

(*a*) The impedance is given by $Z = V_T/I = (120\text{ V})/(1.5\text{ A}) = 80\ \Omega$.

(*b*) The power factor is given by $P = IV_T \cos\phi = (1.5\text{ A})(120\text{ V})\cos\phi = 50$ W. Thus $\cos\phi = 0.278$.

(*c*) In deriving Eq. (*11.29*) in the text, we showed that the power factor is also given by $\cos\phi = R/Z$. Thus $0.278 = R/80$, or $R = 22.2\ \Omega$. (Or, $P = I^2R \rightarrow 50\text{ W} = (1.5)^2R \rightarrow R = 22.2\ \Omega$.)

(*d*) To get *C* we first must calculate X_C. We know that $Z = (R^2 + X_C{}^2)^{1/2} = 80\ \Omega$. Substituting for *R* and solving, yields $X_C = 76.9\ \Omega = 1/(2\pi fC)$. Then $C = 27.6\ \mu$F.

Pure Inductance

We now turn our attention to the last of the three circuit elements we will discuss, the inductor (see Fig. 11-16). Again we assume that an AC generator produces a current, $i = I_0 \cos\omega t$ in the inductor. As we know, the inductor opposes changes in the current that flows through it, by producing a back EMF equal to $(-L\Delta I/\Delta t)$. Since this current is changing sinusoidally, there will be a corresponding sinusoidal back EMF. This back EMF is balanced by the electrostatic voltage across the inductor, v_L, as shown in the figure. This voltage is large when the current is changing quickly, and zero when the current is instantaneously not changing. Whenever the current is at a positive maximum, as at time $t = 0$, the current is instantaneously not changing. This can be seen from the following argument. Before reaching the maximum the current is increasing, and after reaching the maximum the current is decreasing. At the maximum it is therefore neither increasing nor decreasing, but rather it is instantaneously not changing. Therefore, when the current is at a maximum, the back EMF and voltage are zero. This is also true when the current reaches its maximum negative value. It is therefore clear that the voltage across the inductor is 90° out of phase with the current. In order to determine whether the current leads or lags the voltage we will have to examine the situation in more detail. In Fig. 11-16 the current is positive when flowing from point *a* to point *b* through the inductor, and v_L is positive for $v_a > v_b$. Suppose we are at a time when the current has reached its positive maximum (e.g. at $t = 0$), and as noted above, $v_L = 0$. A short time later the current is smaller, and therefore ΔI is negative. The back EMF opposes this change, and, indeed, for negative ΔI points from *a* to *b*, trying to stop the change from occurring. The opposing electrostatic voltage, v_L, will at this instant be negative (see Fig. 11-16). Since the voltage across the inductor at time $t = 0$ is zero, and a short time later becomes negative, its positive peak occurred before $t = 0$, and it decreased to zero at $t = 0$ when the current just reached its

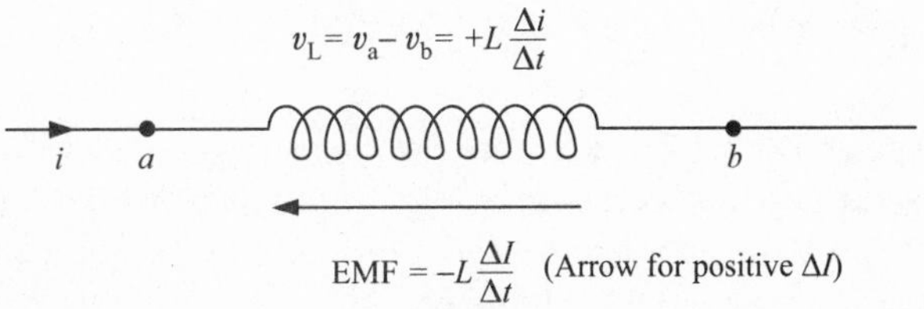

$v_L = v_a - v_b = +L\dfrac{\Delta i}{\Delta t}$

i *a* *b*

$\text{EMF} = -L\dfrac{\Delta I}{\Delta t}$ (Arrow for positive ΔI)

Fig. 11-16

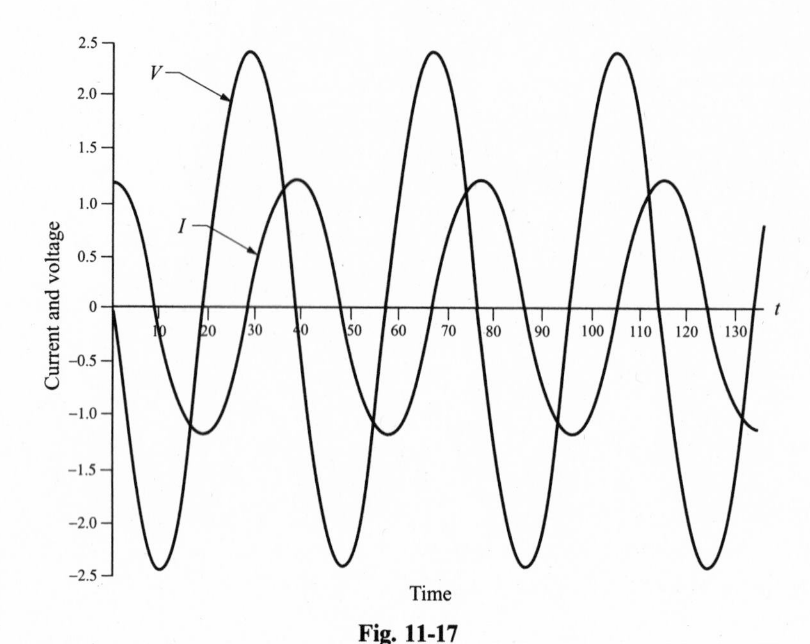

Fig. 11-17

positive peak. The voltage therefore leads the current by 90°. We can therefore write the voltage as $v = V_0 \cos(\omega t + 90°) = -V_0 \sin \omega t$. The current and voltage are plotted in Fig. 11-17.

We must still obtain a relationship between the magnitudes of the current and the voltage. This is given by

$$V_L = IX_L \tag{11.31}$$

where X_L is called the inductive reactance. As always, this relationship is true for the amplitudes as well as for the RMS values. The **inductive reactance** is given by:

$$X_L = \omega L = 2\pi f L \tag{11.32}$$

We can understand this relationship in the following manner. A large reactance means that, even for a small current, the back EMF is large. This is because $L\Delta I/\Delta t$ is large. If L is large, one would therefore expect a large X_L. Furthermore, if f is large then the current changes is rapidly changing making $\Delta I/\Delta t$ large. This would also increase the EMF and therefore X_L. This is exactly what the formula gives as the dependence of X_L on L and on f. Eq. (*11.32*) can be derived rigorously using the calculus.

Problem 11.19. In a circuit, a voltage of 120 V at a frequency of 75 Hz produces a current of 1.5 A in an inductor. Calculate (*a*) the reactance of the circuit; and (*b*) the inductance of the inductor.

Solution

(*a*) Since we aren't given the inductance, we can't use Eq. (*11.32*). Instead we use Eq. (*11.31*), and the reactance is given by $X_L = V_L/I = (120\text{ V})/(1.5\text{ A}) = 80\ \Omega$.

(*b*) The reactance is given by $X_L = 2\pi f L = 80\ \Omega = 2\pi(75\text{ Hz})L$. Thus, L = 0.17 H.

To represent the voltage across the inductor we can, once again, make use of the phasor diagrams. If the phasor for the current at time t is as shown in Fig. 11-18, then the phasor representing the voltage will be, as shown, 90° ahead of the current. As the voltage phasor rotates, the x component of this phasor will give the voltage, v_L, across the inductor. The magnitude of the voltage phasor will be $V_0 = I_0 X_L$, and the RMS voltage will be $V = IX_L$.

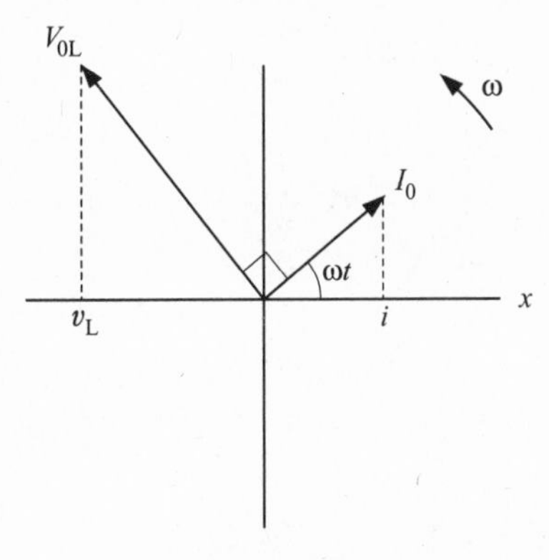

Fig. 11-18

We now turn our attention to calculating the power consumed by the inductor. To calculate this average power, we again have to write the instantaneous power, and calculate its average. The instantaneous power is $iv = (I_0 \cos \omega t)(-V_0 \sin \omega t) = -V_0 I_0 \cos \omega t \sin \omega t$. To get the average we must take the average of $\cos \omega t \sin \omega t$, which we have previously shown to be zero. Thus, the average power consumed by an inductor is zero, just as the average power consumed by the capacitor was zero. The reason is that energy is stored in an inductor by the currents that are established, and this energy is released when the currents are decreased. No energy is dissipated in the process if the inductor is a pure inductor without any resistance.

Inductor and Resistor in Series (R–L Circuit).

Our next case is one in which a resistor and inductor are in series, again with a sinusoidal current of $i = I_0 \cos \omega t$ (see Fig. 11-19). Any real inductor has some amount of resistance, unless it is made from superconducting material. Such a real inductor can be considered as a pure inductance in series with a resistance, the case we are presently discussing. Applying our previously acquired knowledge, we know that the magnitude of the voltage across the resistor (measured from a to b) is $V_{0R} = I_0 R$ and this voltage is in phase with the current, while the magnitude of the voltage across the inductor (measured from b to c) is $V_{0L} = I_0 X_L$ and this voltage leads the current by 90°. The phasors for the various voltages and for the current are shown in Fig. 11-20. As was the case for the R–C circuit, we can obtain the magnitude of total voltage V_{0T}, (V at point a, minus V at point c) by adding the phasors as if they were vectors, giving the total voltage as shown in the figure. From that figure we deduce that

$$V_{0T} = (V_{0R}{}^2 + V_{0L}{}^2)^{1/2} = I_0(R^2 + X_L{}^2)^{1/2} \tag{11.33}$$

$$V_{0T} = I_0 Z, \qquad \text{where} \qquad Z = (R^2 + X_L{}^2)^{1/2} \tag{11.34}$$

$$\tan \phi = V_{0L}/V_{0R} = X_L/R \tag{11.35}$$

This is identical to the case of the R–C circuit, with the exception that X_L replaces X_C and that the formula for the phase angle is chosen positive rather than negative, to insure that ϕ is positive, indicating the fact that the voltage across the inductor leads the current. The same relationship expressed in Eqs. (*11-33*) and (*11-34*) hold for RMS values, i.e. by replacing I_0, V_{0T}, V_{0R} and V_{0L} by I, V_T, V_R and V_L, respectively.

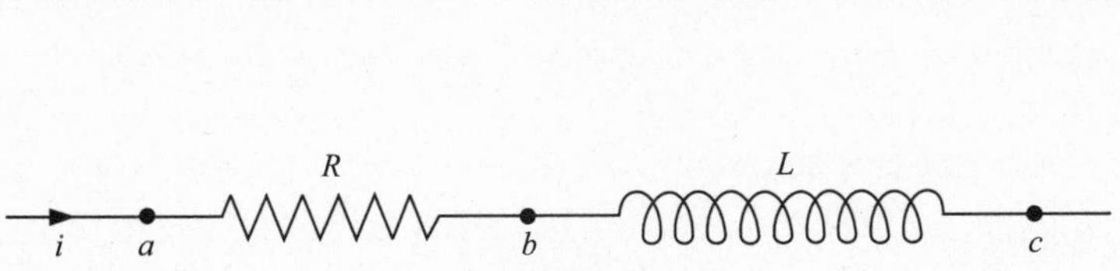

R L
i a b c

Fig. 11-19

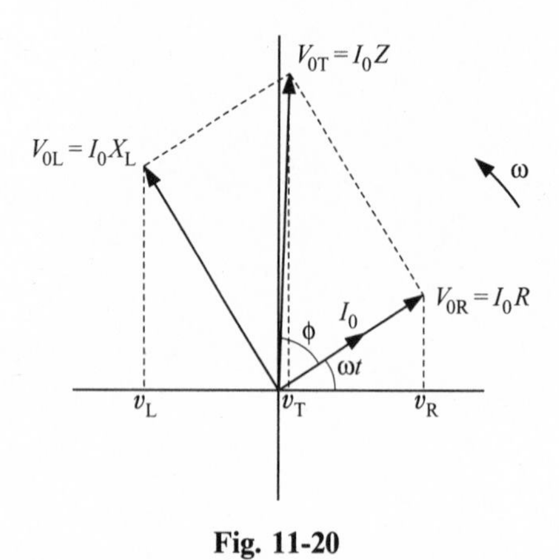

Fig. 11-20

Problem 11.20. Show that the average power consumed by an R–L circuit is given by $P = IV_T \cos \phi$, the same relationship as for an R–C circuit [Eq. (*11.29*)] except with the R–L definition of ϕ.

Solution

We already proved that an inductor does not absorb any energy on the average. Thus, all the power consumption is due to the resistor, and that equals $P = I^2R$. Now $V_T = IZ$, so we can substitute $I = V_T/Z$ for one of the currents, giving $P = IV_T R/Z$. From Fig. 11-20 we see that $R/Z = \cos \phi$.

From problem 11.20 we see that $\cos \phi$ is again the power factor for an R–L circuit as it was for an R–C circuit. We can generally write that $\cos \phi = X/R$, where X is the reactance of the circuit, and equals X_L for a circuit with inductance and $-X_C$ for a circuit with capacitance. Similarly, the impedance can then be written as $Z = (R^2 + X^2)^{1/2}$, which will be valid for both R–C and R–L circuits. Additionally, we can write that the total voltage will vary with time as $v_T = V_{0T} \cos (\omega t + \phi)$, both for the case of the R–C and the R–L circuits. For the R–C circuit, ϕ is negative, and in the R–L circuit, ϕ is positive. We will find that we can extend these ideas to the last case, the R–L–C circuit also.

Problem 11.21. A series R–L circuit has an inductance of 20 mH, and a resistance of 90 Ω. There is an AC voltage of 120 V across the inductor at a frequency of 1500 Hz. Calculate (*a*) the reactance of the inductor; (*b*) the impedance of the circuit; (*c*) the current in the circuit; (*d*) the voltage across the resistor; (*e*) the voltage across the entire circuit; (*f*) the power factor of the circuit; (*g*) the average power consumed by the circuit; and (*h*) the maximum voltage across the inductor.

Solution

(*a*) The inductive reactance is $X_L = 2\pi fL = 2\pi(1500\ \text{Hz})(20 \times 10^{-3}\ \text{H}) = 188\ \Omega$.

(*b*) The impedance is $Z = (R^2 + X_L{}^2)^{1/2} = (90^2 + 188^2)^{1/2} = 209\ \Omega$.

(*c*) The current can be deduced from the voltage across the inductor, $V_L = IX_L$. Thus, 120 V = I(188 Ω), or I = 0.64 A.

(*d*) Now that we know the current, we can get the voltages everywhere. Specifically, $V_R = IR = (0.64\ \text{A})(90\ \Omega) = 57\ \text{V}$.

(*e*) $V_T = IZ = (0.64\ \text{A})(209\ \Omega) = 134\ \text{V}$.

(*f*) The power factor $\cos \phi = R/Z = 90/209 = 0.43$. (The phase angle would be $\phi = +64.5°$.)

(*g*) The average power is $IV_T \cos \phi = (0.64\ \text{A})(134\ \text{V})(0.43) = 36.9\ \text{W}$.

(h) The maximum voltage across the inductor is $V_{0L} = \sqrt{2}(V_L) = 170$ V.

Resistor, Inductor and Capacitor in Series (R–L–C Circuit)

The last case that we will discuss is the circuit in which all three elements are in series (see Fig. 11-21). Again, the current is $i = I_0 \cos \omega t$. We can calculate the RMS voltages across each individual circuit element by using $V = IR$ for the resistor, and $V = IX$ for the inductor and for the capacitor. We also know that v_R is in phase with the current, that v_L leads the current by 90°, and that v_C lags the current by 90°. We can therefore write the equations giving these respective voltages as functions of time as:

$$\begin{aligned} v_R &= I_0 R \cos \omega t \\ v_L &= I_0 X_L \cos(\omega t + 90°) = -I_0 X_L \sin \omega t \\ v_C &= I_0 X_C \cos(\omega t - 90°) = I_0 X_C \sin \omega t \end{aligned} \tag{11.36}$$

To get the total voltage across the entire circuit, v_T, we have to add these three voltages together at each instant of time. Again this can be done using trigonometric identities, but we will make use of the much simpler technique employing phasors that we have used for the other case. We will come to the conclusion that the total voltage can be written as $v_T = I_0 Z \cos(\omega t + \phi)$, with Z and ϕ determined from the phasor diagram (see Fig. 11-22). In this figure, we have again drawn the current phasor at some time t. The phasor for V_R is then in phase with the current, the phasor for V_L is 90° ahead of the current, and the phasor for V_C lags the current by 90°. To get the total voltage we must add all three phasors. We do this in two stages. First we add the two anti-parallel reactive phasors (V_{0L} and V_{0C}), giving $V_{0L} - V_{0C} = I_0(X_L - X_C) = I_0 X$, where $X = X_L - X_C$. This voltage either leads or lags V_{0R} by 90°, depending on whether the inductive reactance is greater than or smaller than the capacitive reactance. In the figure we have drawn the case where $X_L > X_C$. To get the total voltage phasor, V_{0T}, we now add the phasor for

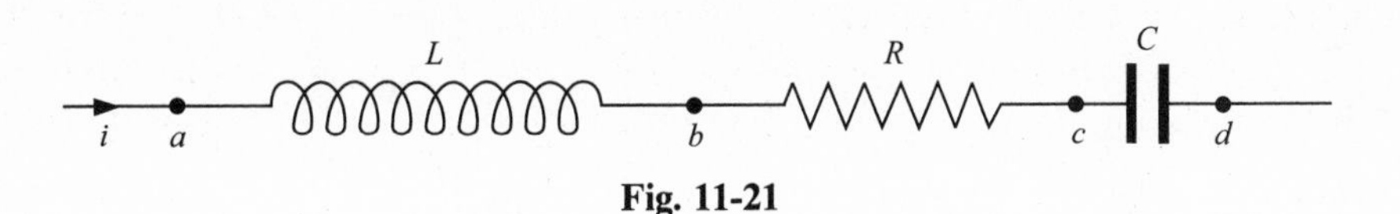

Fig. 11-21

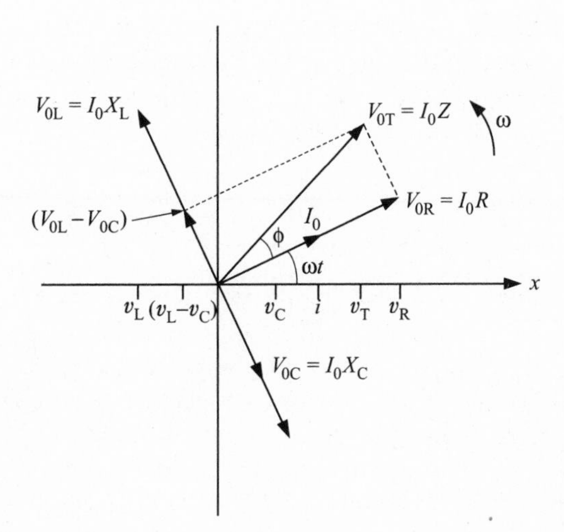

Fig. 11-22

the voltage across the resistor, V_{OR} to the $(V_{OL} - V_{OC})$ phasor. From the diagram we can deduce that:

$$V_{OT} = [V_{OR}^2 + (V_{OL} - V_{OC})^2]^{1/2} = I_0[R^2 + (X_L - X_C)^2]^{1/2} = I_0[R^2 + X^2]^{1/2} \quad (11.37)$$

or

$$V_{OT} = I_0 Z, \quad (11.38)$$

where

$$Z = [R^2 + X^2]^{1/2}, \quad \text{and} \quad X = X_L - X_C \quad (11.39)$$

$$\tan\phi = (X_L - X_C)/R \quad (11.40)$$

Again, we can replace the magnitudes by RMS values in Eqs. (*11.37*) $(V_{OT}, V_{OR}, V_{OL}, V_{OC}, I_0) \rightarrow (V_T, V_R, V_L, V_C, I)$

Problem 11.22. A series *R–L–C* circuit has an inductance of 20 mH, a capacitance of 30 μF and a resistance of 40 Ω. Calculate the impedance and phase angle for this circuit at a frequency of (*a*) 120 Hz; and (*b*) 500 Hz.

Solution

(*a*) The inductive reactance is $X_L = 2\pi fL = 2\pi(120\text{ Hz})(20 \times 10^{-3}\text{ H}) = 15.1\ \Omega$. The capacitive reactance is $X_C = 1/(2\pi fC) = 1/[(2\pi)(120\text{ Hz})(30 \times 10^{-6}\text{ F}) = 44.2\ \Omega$. The reactance is $X_L - X_C = -29.1\ \Omega$. The impedance is $(40^2 + 29.1^2)^{1/2} = 49.5\ \Omega$. To get the phase angle, we use $\tan\phi = X/R = -29.1/40 = -0.728$, and $\phi = -36°$. This means that the total voltage lags the current by 36°.

(*b*) The inductive reactance is $X_L = 2\pi fL = 2\pi(500)(20 \times 10^{-3}) = 62.8\ \Omega$. The capacitive reactance is $X_C = 1/(2\pi fC) = 1/[(2\pi)(500)(30 \times 10^{-6}) = 10.6\ \Omega$. The reactance is $X_L - X_C = 52.2\ \Omega$. The impedance is $(40^2 + 52.2^2)^{1/2} = 65.8\ \Omega$. To get the phase angle, we use $\tan\phi = X/R = 52.2/40 = 1.31$, and $\phi = 52.5°$. This means that the total voltage leads the current by 52.5°.

Problem 11.23. A series *R–L–C* circuit has an inductance of 20 mH, a capacitance of 30 μF and a resistance of 40 Ω. A generator, at a frequency of 120 Hz, provides a voltage of 220 V to the circuit, as in Fig. 11-21.

(*a*) What current flows in the circuit?

(*b*) What is the voltage across: the resistor; the inductor; and the capacitor?

(*c*) What power is consumed in the circuit?

Solution

(*a*) The situation is the same as in Problem 11.22 part (*a*), so $X_L = 15.1\ \Omega$, $X_C = 44.2\ \Omega$, $Z = 49.5\ \Omega$. The current is therefore $I = V/Z = (220\text{ V})/(49.5\ \Omega) = 4.44$ A.

(*b*) The voltage across the resistor is $V_R = IR = 4.44(40) = 178$ V.
The voltage across the inductor is $V_L = IX_L = 4.44(15.1) = 67.1$ V.
The voltage across the capacitor is $V_C = IX_C = 4.44(44.2) = 196$ V.

(*c*) The power consumed can be calculated from $P = I^2R$, or from $P = IV_T\cos\phi$. Using the first formula we get $P = (4.44\text{ A})^2(40\ \Omega) = 789$ W. Using the second formula, and the fact that $\cos\phi = R/Z = 40/49.5$, we get $P = (4.44\text{ A})(220\text{ V})(40/49.5) = 789$ W.

In Problem 11.22 we saw that the reactance $(X_L - X_C)$ changes from negative to positive as the frequency increases. This means that there is some frequency at which the reactance is zero. This frequency is known as the "**resonance frequency**" of the circuit, and corresponds to the lowest possible impedance, *Z*, for a given *R–L–C* circuit, and therefore the largest current for a given generator voltage *V*. This frequency can be shown to be given by:

$$2\pi f_r = 1/\sqrt{LC} \quad (11.41)$$

Problem 11.24. A series R–L–C circuit has inductance L, capacitance C and resistance R, and is connected to a generator producing a voltage V at a variable frequency f.

(*a*) What is the resonance frequency of the circuit, in terms of R, L and C?

(*b*) Show that, as one varies f, the current reaches its maximum value at the resonance frequency.

(*c*) Calculate the maximum current in the circuit as one varies f, in terms of R, L, C, V and f.

(*d*) What is the phase angle of the circuit at the resonance frequency?

Solution

(*a*) The resonance frequency has been defined as the frequency at which $X_L = X_C$. Therefore, $2\pi f_r L = 1/(2\pi f_r C)$, or $(2\pi f_r)^2 = 1/LC$, or $2\pi f_r = \omega_r = 1/\sqrt{LC}$, which verifies Eq. (11.41).

(*b*) The current is given by $I = V_T/Z$. Since $V_T = V$ (the RMS magnitude) is fixed (it doesn't vary with f), the maximum current occurs when Z is a minimum as one varies f. Now, $Z = (R^2 + X^2)^{1/2}$, and R is fixed, so the minimum Z occurs when X^2 is a minimum. Since X^2 is always positive, even though X can be negative, the minimum X is zero, which occurs when the circuit is at its resonance frequency. Thus the current is maximum at this frequency.

(*c*) At the resonance frequency, X is zero. The impedance is then $Z = R$, and the current is V/R. Note that the value of the resonance current does not depend on the value of the resonance frequency.

(*d*) The phase angle is given by $\tan\phi = X/R$. At the resonance frequency, $X = 0$, and therefore $\phi = 0$. This means that the total voltage is in phase with the current at this frequency. This is shown in Fig. 11-23, where we draw the phasor diagram for the case of resonance.

Problem 11.25. A series R–L–C circuit has an inductance L, a capacitance C and a resistance R, and is connected to a generator producing a voltage V at a variable frequency f, as in Fig. 11-21. In terms of V, R, L and C, at the resonance frequency, find (*a*) the voltage across the resistor; (*b*) the voltage across the inductor; (*c*) the voltage across the capacitor; and (*d*) the power consumed by the circuit. Show that this power is a maximum as one varies the frequency.

Solution

(*a*) At the resonance frequency $X_L = X_C$, $X = 0$, $Z = R$, and $I = V_T/R$. The voltage across the resistor is therefore $V_R = IR = V_T = V$.

(*b*) The voltage across the inductor is $V_L = IX_L = (V/R)(2\pi fL)$. But $f = (\frac{1}{2}\pi\sqrt{LC})$ [Eq. (*11.41*)], and therefore $V_L = (V/R)(\sqrt{L/C})$.

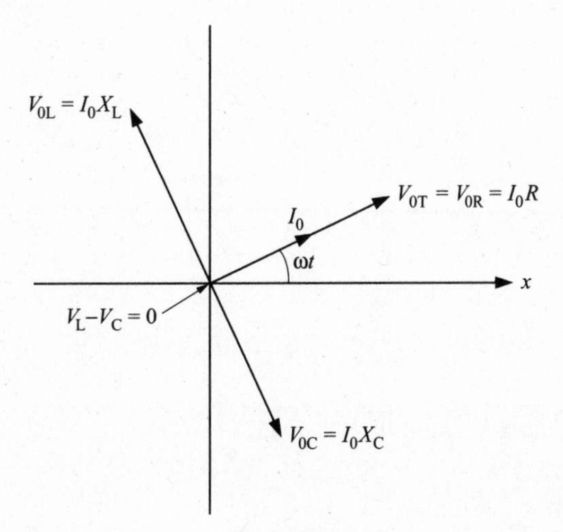

Fig. 11-23

(c) The voltage across the capacitor is IX_C, which must equal the voltage across the inductor, since $X_C = X_L$. Calculating V_C directly, $V_C = (V/R)[1/(2\pi fC)]$. Using the fact that $f = (\frac{1}{2}\pi\sqrt{LC})$, we get that $V_C = (V/R)(\sqrt{L/C})$.

(d) The power consumed is I^2R, or $IV \cos \phi$. At the resonance frequency, I is at a maximum, as we showed earlier. Also since ϕ is zero, $\cos \phi = 1$, which is the maximum possible value for the power factor. Since V and R do not vary with f, both I and $\cos \phi$ are at a maximum at the resonance frequency, and therefore the power is maximum at that frequency.

Problem 11.26. A series R–L–C circuit has an inductance $L = 20$ mH, a capacitance $C = 30\ \mu$F and a resistance $R = 40\ \Omega$, and is connected to a generator producing a voltage $V = 120$ V at a variable frequency f.

(a) What is the resonance frequency of the circuit?

(b) At the resonance frequency, what is the voltage across the inductor?

(c) At the resonance frequency, what is the voltage across the capacitor?

(d) At the resonance frequency of the circuit, what is the power consumed by the circuit?

(e) At what frequency above the resonance frequency is the current equal to half its value at the resonance frequency?

Solution

(a) From Eq. (*11.41*), the resonance frequency is given by $2\pi f = 1/\sqrt{LC} = 1/[(20 \times 10^{-3}$ H)$(30 \times 10^{-6}$ C)$]^{1/2} = 1.29 \times 10^3$, and $f = 205$ Hz.

(b) The voltage across the inductor is $V_L = IX_L = (V_T/R)(2\pi fL)$. From Problem 11.25(*b*) this leads to $V_L = (V/R)(\sqrt{L/C}) = (120\text{ V})/(40\ \Omega)(20 \times 10^{-3}\text{ H}/30 \times 10^{-6}\text{ C})^{1/2} = 77.5$ V.

(c) The voltage across the capacitor is IX_C, which must equal the voltage across the inductor, since $X_C = X_L$. Therefore, $V_C = 77.5$ V.

(d) The power consumed is $P = I^2R$, or, since $\phi = 0$, $P = IV \cos \phi = IV$. Recalling that at resonance $V = IR$ [Problem 11.25(*a*)], we get $P = V^2/R = (120\text{ V})^2/(40\ \Omega) = 360$ W.

(e) The current equals V/Z. At resonance, the current is $I_r = V/R$. If $I = (\frac{1}{2})I_r$, then $V/Z = (\frac{1}{2})V/R$, or $Z = 2R$. Thus, $(R^2 + X^2)^{1/2} = 2R \rightarrow R^2 + X^2 = 4R^2$ or $X^2 = 3R^2$, and $X = X_L - X_C = \pm R\sqrt{3} = 2\pi fL - \frac{1}{2}\pi fC$. Substituting numbers, we get: $40\sqrt{3} = 2\pi f(20 \times 10^{-3}\text{ H}) - 1/[(2\pi f)(30 \times 10^{-6}\text{ F})]$ $= 69.3 = 0.126f - 5.31 \times 10^3/f$. This is a quadratic equation in f, after multiplying by f and rearranging,

$$0.126f^2 - 69.3f - 5.31 \times 10^3 = 0.$$

We use the formula for solving a quadratic equation $x = [-b \pm \sqrt{(b^2 - 4ac)}]/2a$, to get, $f = \{+69.3 \pm \sqrt{[69.3^2 + 4(0.126)(5.31 \times 10^3)]}\}/2(0.126) = (69.3 \pm 86.5)/0.252 = 618$ Hz, since only the positive frequency makes sense.

Problems for Review and Mind Stretching

Problem 11.27. A battery has a voltage of 15 V, and is connected across a series R–C circuit with an initially uncharged capacitor. The initial current in the circuit is observed to be 7.5×10^{-2} A, and the time constant of the circuit is 2×10^{-3} s.

(a) What is the resistance of the resistor?

(*b*) What is the capacitance of the capacitor?

(*c*) What is the final charge that accumulates on the capacitor?

(*d*) After a time of 10^{-3} s, what is the current flowing, and what is the charge on the capacitor?

Solution

(*a*) Initially, all the voltage is across the resistor, and $V_R = I_i R = 15\ V = (7.5 \times 10^{-2}\ A)R$. Thus $R = 200\ \Omega$.

(*b*) The time constant equals $RC = 2 \times 10^{-3}\ s = (200\ \Omega)C$. Therefore, $C = 10^{-5}$ F.

(*c*) Finally, all the voltage is across the capacitor, and equals $Q_f/C = 15$ V. Thus $Q_f = 1.5 \times 10^{-4}$ C.

(*d*) At $t = 10^{-3}$ s, $I = I_i\ e^{-(t/\tau)} = 7.5 \times 10^{-2}\ e^{-(1/2)} = 4.55 \times 10^{-2}$ A. The charge on the capacitor is $Q = Q_f(1 - e^{-(t/\tau)}) = 1.5 \times 10^{-4}\ (1 - e^{-(1/2)}) = 5.9 \times 10^{-5}$ C.

Problem 11.28. A battery with a voltage of 15 V is connected across a series *L–R* circuit. The final current flowing in the circuit is 2.5 A, and the time constant of the circuit is 2.0×10^{-3} s.

(*a*) What is the resistance of the resistor?

(*b*) What is the inductance of the inductor?

(*c*) What is the voltage across the inductor when the current is 1.5 A?

(*d*) At what time is the current equal to 1.5 A?

Solution

(*a*) Finally, all the voltage is across the resistor, and $V_R = I_f R = 15\ V = 2.5R$. Thus $R = 6.0\ \Omega$.

(*b*) The time constant equals $L/R = 2.0 \times 10^{-3}\ s = L/6$. Therefore, $L = 1.2 \times 10^{-2}$ H.

(*c*) When the current is 1.5 A, the voltage across the resistor is (1.5 A)(6.0 Ω) = 9.0 V. Thus, the voltage across the inductor is 15 − 9.0 = 6.0 V.

(*d*) When the current equals 1.5 A, $I = I_f(1 - \exp[-(t/\tau)]) = 2.5(1 - \exp[-(t/\tau)]) = 1.5$. Then $1.5/2.5 = 0.60 = (1 - \exp[-(t/\tau)])$, and $0.40 = \exp[-(t/\tau)]$, $\ln 0.40 = -t/\tau$, $t = -\tau \ln 0.40 = 1.83 \times 10^{-3}$ s.

Problem 11.29. An *L–C* circuit has an inductance of 25 mH and a capacitance of 2.0 μF. The capacitance is initially charged, and stores energy of 1.25×10^{-4} J.

(*a*) What is resonance frequency and the period of this circuit?

(*b*) What is the initial charge on the capacitor?

(*c*) What is the maximum current in the circuit?

Solution

(*a*) We know that the frequency is $\omega/2\pi$, where $\omega = 1/\sqrt{LC} = [1/(25 \times 10^{-3}\ H)(2.0 \times 10^{-6}\ F)]^{1/2} = 4.47 \times 10^3$. Therefore, $f = 4.47 \times 10^3/2\pi = 712$ Hz, and $T = 1/f = 1.40 \times 10^{-3}$ s.

(*b*) The energy stored in the capacitor is $(\frac{1}{2})Q_i^2/C = 1.25 \times 10^{-4}$ J. Thus $Q_i = [2(1.25 \times 10^{-4}\ J)(2.0 \times 10^{-6}\ F)]^{1/2} = 2.24 \times 10^{-5}$ C.

(*c*) The maximum current occurs when all the energy is in the inductor. Then, $U = (\frac{1}{2})LI_{max}^2 = 1.25 \times 10^{-4}\ J = 0.5(25 \times 10^{-3}\ H)I_{max}^2$, or $I_{max} = 0.10$ A.

Problem 11.30. An *R–C* circuit has a capacitance of 20 μF and a resistance of 100 Ω. An AC generator supplies a voltage of 180 V to the circuit at a frequency, *f*. The phase angle between the current and the voltage is measured to be 45°.

(*a*) What is the frequency of the voltage being supplied to the circuit?

(*b*) What is the current in the circuit?

(*c*) What is the voltage across the resistor in the circuit? Across the capacitor?

(*d*) What power is being supplied to the circuit?

Solution

(*a*) The phase angle is given by $\tan\phi = X_C/R = \tan 45° = 1$. Thus $X_C = R = 100\ \Omega = 1/\omega C$. Thus, $\omega = 1/[(100)(20 \times 10^{-6}\ \text{F})] = 500$, or $f = 79.6$ Hz.

(*b*) The current is V/Z, and $Z = (X_C{}^2 + R^2)^{1/2} = 100\sqrt{2}\ \Omega$ (since $X_C = R = 100\ \Omega$). Thus, $I = (180\ \text{V})/(141\ \Omega) = 1.27$ A.

(*c*) We know that $V_R = IR = (1.27\ \text{A})(100\ \Omega) = 127$ V, and $V_C = IX_C = 1.27(100) = 127$ V.

(*d*) The power is $I^2R = (1.27\ \text{A})^2(100\ \Omega) = 162$ J. Alternatively, $P = IV_T \cos\phi = (1.27\ \text{A})(180\ \text{V}) \cos 45° = 162$ J.

Problem 11.31. An *R–L–C* circuit is supplied by an AC generator with a voltage of 85 V at a frequency of 1500 Hz. The resistor has a resistance of 120 Ω, the capacitor has a reactance of 90 Ω at this frequency, and the inductor has a reactance of 140 Ω at this frequency.

(*a*) What is the impedance of the circuit at this frequency?

(*b*) What is the current in the circuit?

(*c*) At what other frequency does one have the same current?

Solution

(*a*) The impedance is $Z = (R^2 + X^2)^{1/2}$, and $X = X_L - X_C = 140 - 90 = 50\ \Omega$. Thus, $Z = (120^2 + 50^2)^{1/2} = 130\ \Omega$.

(*b*) The current is $V/Z = (85\ \text{V})/(130\ \Omega) = 0.65$ A.

(*c*) To get the same current at a different frequency, we require that the impedance have the same value as at the original frequency. Since the only quantity in Z that varies with f is X^2, we can get the same Z if, at the new frequency, $X \to -X$. This means that instead of $X = X_L - X_C = 50\ \Omega$, we have $X_L - X_C = -50\ \Omega$ so now X_C will exceed X_L by 50 Ω. Denoting the new angular frequency as ω', we have that $[1/(\omega' C) - \omega' L] = 50$. To solve for ω' we need L and C. Now $L = X_L/\omega = 140/2\pi(1500) = 14.9$ mH, and $C = 1/\omega X_C = 1.179\ \mu$F. Therefore, $1/[(\omega')(1.179 \times 10^{-6})] - \omega'(14.9 \times 10^{-3}) = 50$. This is a quadratic equation in ω', which we can solve, to get $\omega' = 6050$, or $f = 963$ Hz. We can check this result by calculating X_C and X_L at this frequency. This yields $X_C = 140\ \Omega$ and $X_L = 90\ \Omega$, giving $X = 90 - 140 = -50$, as desired.

Problem 11.32. An *R–L–C* circuit is supplied by an AC generator and is in resonance at the frequency of radio station WINS (1010 on the AM dial). The inductance in the circuit is 2.0×10^{-6} H, and the resistance is 0.80 Ω. The variable capacitor is tuned to resonance at this frequency.

(*a*) What is the capacitance at this frequency?

(*b*) With this same capacitance, at what frequency does the impedance become twice as great as at resonance?

Solution

(*a*) The AM station numbers represent the frequency in kHz so $f = 1010 \times 10^3$ Hz. The resonance frequency is given by $\omega^2 = 1/LC$. Thus, $C = 1/\omega^2 L = 1/[(2\pi)(1010 \times 10^3\ \text{Hz})]^2[2 \times 10^{-6}\ \text{H}] = 1.24 \times 10^{-8}$ F.

(*b*) At resonance, the impedance is equal to the resistance, since $X = X_L - X_C = 0$. Thus $Z = 0.80\ \Omega$. At the new frequency f', $Z = 2(0.80) = 1.6\ \Omega$. Since $Z^2 = R^2 + X^2$, we have that $X^2 = 1.6^2 - 0.8^2 = 1.92$, and $X = \pm 1.39$. There are two solutions, one for $X = +1.39$ and one for $X = -1.39$. Let us calculate the solution for $X = +1.39$. Here, $\omega' L - 1/\omega' C = 1.39$. When we substitute for L and C, we get a quadratic equation for ω', which we solve to get $\omega' = 6.71 \times 10^6$, or $f' = 1067$ kHz. Repeating for $X = -1.39$ we get $\omega' = 6.01 \times 10^6$, or $f' = 957$ kHz.

Supplementary Problems

Problem 11.33. A battery is connected to a series combination of a resistor with $R = 100\ \Omega$ and an uncharged capacitor with $C = 25\ \mu F$ After a long time the capacitor attains a charge of 2.25×10^{-4} C.

(*a*) What is the time constant of the circuit?

(*b*) What is the initial current in the circuit?

(*c*) What is the current in the circuit after a time of 0.003 s?

Ans. (*a*) 0.0025 s; (*b*) 0.09 A; (*c*) 0.0027 A

Problem 11.34. A battery with a voltage of 12 V is connected to a series combination of a resistor and uncharged capacitor. The time constant of the circuit is 1.3×10^{-4} s, and the initial current is 0.15 A.

(*a*) What is the resistance in the circuit?

(*b*) What is the capacitance in the circuit?

(*c*) What is the final charge on the capacitor?

Ans. (*a*) 80 Ω; (*b*) 1.625 μF; (*c*) 1.95×10^{-5} C

Problem 11.35. A battery of 15 V is connected to a series combination of a resistor and uncharged capacitor. After a long time the capacitor attains a charge of 5.5×10^{-5} C. After a time of 0.0020 s the charge on the capacitor is 2.0×10^{-5} C.

(*a*) What is the capacitance in the circuit?

(*b*) What is the time constant of the circuit?

(*c*) What is the initial current in the circuit?

(*d*) What is the resistance in the circuit?

Ans. (*a*) 3.67 μF; (*b*) 4.42 ms; (*c*) 0.0125 A; (*d*) 1.20 kΩ

Problem 11.36. A battery, of 12 V, is connected across a series circuit consisting of a resistance of 100 Ω and an inductance of 0.30 mH.

(*a*) What is the time constant of the circuit?

(*b*) What is the final current in the circuit?

(*c*) What is the current in the circuit after a time of 1.5×10^{-6} s?

Ans. (*a*) 3×10^{-6} s; (*b*) 0.12 A; (*c*) 0.0472 A

Problem 11.37. A battery, of 12 V, is connected across a series circuit consisting of a resistor and an inductor. The final current in the circuit is 0.60 A, and the circuit has a time constant of 4.0×10^{-5} s.

(*a*) What is the resistance in the circuit?

(*b*) What is the inductance in the circuit?

(*c*) What is the voltage across the inductor after a time of 4.0×10^{-5} s?

Ans. (*a*) 20 Ω; (*b*) 0.80 mH; (*c*) 4.41 V

Problem 11.38. A battery, of 50 V, is connected across a series circuit consisting of a resistor and an inductor. After 2.0×10^{-4} s, the current is 2.0 A and the voltage across the inductor is 20 V.

(*a*) What is the time constant of the circuit?

(*b*) What is the resistance in the circuit?

(*c*) What is the inductance in the circuit?

Ans. (*a*) 2.18×10^{-4} s; (*b*) 15 Ω; (*c*) 3.27 mH

Problem 11.39. A capacitor with a capacitance of 0.20 μF is charged until it stores electrical energy of 3.0×10^{-3} J. It is then disconnected from the battery and connected to an inductor of inductance 0.50 mH.

(*a*) What is resonance frequency of the circuit?

(*b*) What is the energy stored in the inductor and the energy stored in the capacitor after a time equal to one quarter of a period (90°)?

(*c*) What is the energy stored in the inductor and the energy stored in the capacitor after a time equal to one half of a period (180°)?

Ans. (*a*) 1.59×10^4 Hz; (*b*) $U_C = 0$, $U_L = 3 \times 10^{-3}$ J; (*c*) $U_C = 3 \times 10^{-3}$ J, $U_L = 0$

Problem 11.40. Consider the same situation as in Problem 11.39. At a time *t* after the capacitor is connected to the inductor the capacitor stores an energy of 1.3×10^{-3} J and the inductor stores an energy of 1.7×10^{-3} J.

(*a*) What is maximum current in the circuit?

(*b*) What is the maximum charge stored on the capacitor?

(*c*) What is the smallest value of *t*?

Ans. (*a*) 3.46 A; (*b*) 3.46×10^{-5} C; (*c*) 8.83×10^{-3} s

Problem 11.41. A capacitor is charged until it stores electrical energy of 3×10^{-4} J holding a charge of 9×10^{-6} C. It is then disconnected from the battery and connected to an inductor, and the resultant circuit has a resonance frequency of 2×10^6 Hz.

(*a*) What is the capacitance in the circuit?

(*b*) What is the inductance in the circuit?

(*c*) What is the charge on the capacitor after a time of one eighth the period?

Ans. (*a*) 1.35×10^{-7} F; (*b*) 4.69×10^{-8} H; (*c*) 6.4×10^{-6} C

Problem 11.42. A capacitor is first charged, then disconnected from the battery and then connected to an inductor. At a later time the current in the circuit is 0.05 A, and the capacitor has a charge of 8×10^{-5} C. The resonance frequency of the circuit is 400 Hz.

(*a*) What is the initial charge on the capacitor?

(*b*) What is maximum current in the circuit?

(*c*) At what time did the circuit have the given charge and current?

Ans. (*a*) 8.24×10^{-5} C; (*b*) 0.207 A; (*c*) 9.63×10^{-5} s

Problem 11.43. An AC generator, at a frequency of 60 Hz, supplies a current of 0.50 A to a series circuit containing a 110 Ω resistor, and a 25 μF capacitor. (As usual, current and voltage are understood to be RMS values unless otherwise specified.)

(*a*) What is the reactance (magnitude and sign) in the circuit?
(*b*) What is the impedance of the circuit?
(*c*) What is the voltage across the capacitor?
(*d*) What is the voltage across the resistor?
(*e*) What is the voltage of the generator?
(*f*) What power is consumed by the circuit?
(*g*) What is the phase angle between the current and voltage in the circuit?

Ans. (*a*) −106 Ω; (*b*) 153 Ω; (*c*) 53 V; (*d*) 55 V; (*e*) 76.5 V; (*f*) 27.5 W; (*g*) −44°

Problem 11.44. An AC generator, at a frequency of 60 Hz, supplies a current of 0.50 A to a series circuit containing a 110 Ω resistor, and a 0.30 H inductor.

(*a*) What is the reactance in the circuit?
(*b*) What is the impedance of the circuit?
(*c*) What is the voltage across the inductor?
(*d*) What is the voltage across the resistor?
(*e*) What is the voltage of the generator?
(*f*) What power is consumed by the circuit?
(*g*) What is the phase angle between the current and voltage in the circuit?

Ans. (*a*) 113 Ω; (*b*) 158 Ω; (*c*) 56.5 V; (*d*) 55 V; (*e*) 79 V; (*f*) 27.5 W; (*g*) 46°

Problem 11.45. An AC generator supplies power of 110 W to a series circuit containing a 120 Ω resistor, and a capacitor with a reactance of 50 Ω.

(*a*) What is the impedance of the circuit?
(*b*) What is the current in the circuit?
(*c*) What is the phase angle between the current and voltage in the circuit?

Ans. (*a*) 130 Ω; (*b*) 0.957 A; (*c*) −22.6°

Problem 11.46. A coil can be considered to be a resistance in series with an inductance. A particular coil has an resistance of 40 Ω and an inductance of $(1/4\pi)$ H. One can apply either a DC voltage or an AC voltage at 60 Hz, to the coil.

(*a*) If one applies a DC voltage of 120 V to the coil, what final current will flow?
(*b*) If one applies an AC voltage of 120 V to the coil, what current (RMS) will flow?
(*c*) What is the *maximum* voltage across the coil [for case (*b*)]?
(*d*) Again, for the voltage of (*b*), what power is consumed in the circuit?

Ans. (*a*) 3 A; (*b*) 2.4 A; (*c*) 170 V; (d) 230 W

Problem 11.47. An AC circuit consists of two parallel resistors, R_1 and R_2, which are then in series with a capacitor C, as shown in Fig. 11-24. The voltage across R_1 is 105 V, and the frequency is 60 Hz. The circuit elements have the following values: $R_1 = 75\ \Omega$, $R_2 = 90\ \Omega$, $C = 2.0 \times 10^{-5}$ F.

(*a*) What is the power generated in resistor R_1?
(*b*) What is the current in resistor R_1?

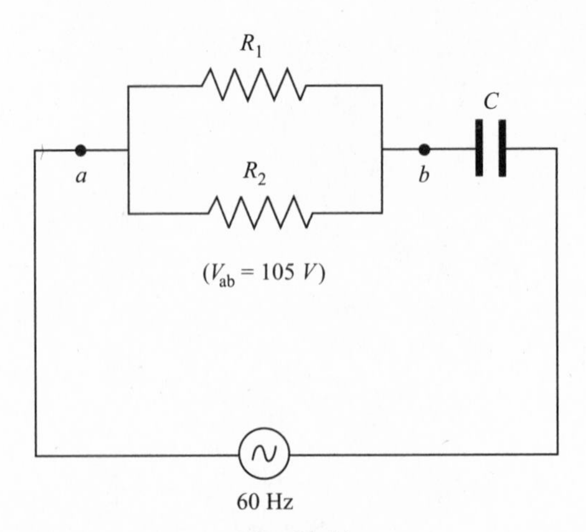

Fig. 11-24

(*c*) What is the current in resistor R_2 ?

(*d*) What is the voltage across the capacitor?

(*e*) What is the voltage across the entire circuit?

Ans. (*a*) 147 W; (*b*) 1.40 A; (*c*) 1.17 A; (*d*) 341 V; (*e*) 358 V

Problem 11.48. An AC generator, at a frequency of 8.0 kHz, supplies a voltage of 20 V to a series circuit containing a 75 Ω resistor, a 0.30 mH inductor and a 0.25 μF capacitor.

(*a*) What is the impedance of the circuit?

(*b*) What is the current in the circuit?

(*c*) What power is consumed by the circuit?

(*d*) What is the phase angle between the current and the voltage?

Ans. (*a*) 98.9 Ω; (*b*) 0.202 A; (*c*) 3.07 W; (*d*) $-40.7°$

Problem 11.49. A purely inductive coil with a reactance of 75 Ω is connected in series with a capacitor having a reactance of 25 Ω and a resistor with a resistance of 120 Ω. A current of 2.0 A flows in the circuit.

(*a*) What is the voltage across the capacitor; the coil; and the resistor?

(*b*) What is the impedance of the circuit?

(*c*) What power is consumed by the circuit?

(*d*) What is the voltage across the whole circuit?

Ans. (*a*) $V_C = 50$ V, $V_L = 150$ V, $V_R = 240$ V; (*b*) 130 Ω; (*c*) 480 W; (*d*) 260 V

Problem 11.50. A series circuit, consisting of a resistor R, an inductor L and a capacitor C, is connected to an AC generator producing a voltage of 200 V across the entire circuit. This results in a current of 4.0 A and a power consumption of 640 W. The reactance of the capacitor is 100 Ω, and is larger than the reactance of the inductor.

(*a*) What is the voltage across the resistor?

(*b*) What is the voltage across the capacitor?

(*c*) What is the impedance of the circuit?

(*d*) What is the voltage across the inductor?

Ans. (*a*) 160 V; (*b*) 400 V; (*c*) 50 Ω; (*d*) 280 V

Problem 11.51. An AC generator supplies a voltage of 86 V to a series circuit containing a resistor, an inductor and a capacitor. The current in the circuit is 2.0 A, and the phase angle between the current and voltage is $+18°$.

(*a*) What is the impedance of the circuit?

(*b*) What is the resistance in the circuit?

(*c*) What is the reactance in the circuit?

(*d*) What power is consumed by the circuit?

Ans. (*a*) 43 Ω; (*b*) 40.9 Ω; (*c*) 13.3 Ω; (*d*) 164 W

Problem 11.52. An AC generator supplies a voltage to a series circuit containing a 110 Ω resistor, a 0.30 mH inductor and a 0.25 μF capacitor.

(*a*) What is the resonance frequency of the circuit?

(*b*) If the frequency being supplied is twice this resonance frequency, what is the impedance and phase angle of the circuit?

Ans. (*a*) 1.84×10^4 Hz; (*b*) $Z = 122\ \Omega$, $\phi = +25.6°$

Problem 11.53. An AC generator supplies a voltage of 8.0 V to a series circuit containing a resistor, an inductor and a capacitor. At the resonance frequency of 1.53×10^3 Hz, the current is 0.020 A, and the reactance of the capacitor is 1200 Ω.

(*a*) What is the resistance in the circuit?

(*b*) What is the inductance in the circuit?

(*c*) What is the capacitance in the circuit?

(*d*) What power is consumed by the circuit?

Ans. (*a*) 400 Ω; (*b*) 0.125 H; (*c*) 8.67×10^{-8} F; (*d*) 0.16 W

Chapter 12

Electromagnetic Waves

12.1 INTRODUCTION

In the previous chapters, we learned about the production of electric and magnetic fields in space, due to charges and due to currents, respectively. We then learned that an electric field would also be produced by a changing magnetic field. In this chapter, we will show that these laws are not yet complete, and require the addition of one further new concept, the concept of displacement current. With the addition of this concept, we will then summarize the laws for the production of electric and magnetic fields in the form of the four Maxwell equations. In turn this will lead to the use of these equations to predict the existence and the properties of electromagnetic waves, which we will then discuss.

12.2 DISPLACEMENT CURRENT

In Chap. 7, we learned about the production of a magnetic field in space, due to currents in a wire. The law that we developed that must be followed is **Ampere's law**, which states that when we go around a closed loop (such as a circle) and add together the component of the magnetic field along the loop times Δl (the infinitesimal length along the loop), i.e. $\sum B \cos \theta \Delta l$, this sum equals μ_0 times the current through the closed loop. Let us analyze this current in more detail. Consider the circular loop a in Fig. 12-1 and a wire, w, with current I perpendicular to the loop. What do we mean by the current flowing *through* the loop a? A possible answer is that the current is that which flows through the area A_1 (shown in the figure) which is bounded by the loop. Since in our case the only current is in the wire, which passes through A_1, the current through the loop is I. A_1, however, is not the only area bounded by the loop a. For instance, consider the cylindrical shaped surface formed by disk surface A_2 and cylinder surface A_3. The combination of these two surfaces is bounded by loop a, just as area A_1 is. Indeed it is easy to see that there are an infinite number of surfaces bound by the loop a. This is generally true for any closed loop whether circular or not, whether in a plane or not. For the case of our loop a and our $A_2 + A_3$ combined surface bound by a we can see that there is no current flowing through area A_3, and the current flowing through A_2 is just the current in the wire, I. We therefore get the same current whether we use A_1 for the area, or the $(A_2 + A_3)$ surface for the area. It would therefore seem that there is no ambiguity about which surface bounded by the loop one should use. This is generally true as long as the wires don't end, and, if they do end, then we know that for DC currents, there can be no current in such a wire and therefore no problem either. However, we do run into a problem if the wires terminate, and we have AC currents. Consider, for instance, a circuit with a

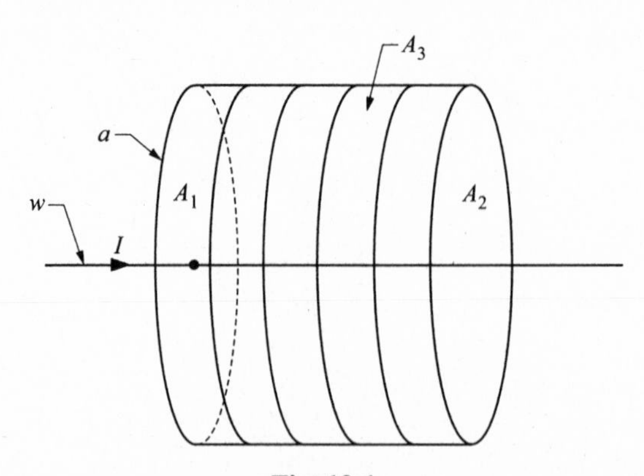

Fig. 12-1

wire entering one plate of a capacitor. During the time that the capacitor is being charged, a current flows in the wire but no current flows from one capacitor plate to the other. This creates a situation in which a closed loop can bound one surface with current flowing through it and bound another surface with no current through it. This can be seen for the case depicted in Fig. 12-2. Here a wire w is carrying a current I to the capacitor plate during the time that the capacitor is being charged. We use the circle a as the closed loop and consider $\sum (B \cos \theta \, \Delta l)$ around the loop. By Ampere's law, this should equal the current through the area bounded by the curve. But the answer we get is quite different if we use the area A_1, or the area $A_2 + A_3$. For area A_1, the current is I, the current in the wire, which flows through A_1. For the area $A_2 + A_3$, we would get zero as the answer, since no current is flowing between the capacitor plates. We are thus faced with a dilemma, since we get contradictory answers for different surfaces bounded by the same loop. Which answer should we use in Ampere's law? It would be nice if the current through any surface bounded by the loop were the same. The capacitor plate situation affords the opportunity to accomplish this by broadening the definition of current for Ampere's law. We use the term conduction current, which is current conducted in a wire or some conducting medium, to distinguish it from the displacement current which we shall shortly define. Even though there is no conduction current between the capacitor plates, there is something there that is not present at area A_1. That something is a changing electric field. Perhaps this changing electric field can be associated with a new "displacement" current that contributes to Ampere's law in just such a way that the "current" through A_2 equals the current through A_1 in Fig. 12-2. Let us calculate how this could happen.

We know that the field within a parallel plate capacitor is uniform and is equal to $E = q/\varepsilon_0 A$, where q is the charge on the capacitor and A is the area of the capacitor. Strictly speaking, this is only true for the field within large plates, and the field varies as one approaches the edge of the plates. However, a more exact calculation shows that this won't change our conclusions. If the capacitor is being charged, then both q and E are changing, and we can write that $\Delta E/\Delta t = (\Delta q/\Delta t)/\varepsilon_0 A = I/\varepsilon_0 A$. Thus,

$$I = \varepsilon_0 A(\Delta E/\Delta t) \tag{12.1a}$$

If we define a "**displacement current density**" as

$$J_D = \varepsilon_0(\Delta E/\Delta t) \tag{12.1b}$$

then we find that the current through surface A_2 is $I_D = J_D A = I$, the same result as for area A_1. The **displacement current** I_D can also be written as $I_D = \varepsilon_0(\Delta EA/\Delta t) = \varepsilon_0 \Delta\Psi/\Delta t$, where Ψ is the electric flux through the area. By modifying Ampere's law to include displacement current, we eliminate the contradiction that we had previously discussed. Ampere's law would then state that

$$\sum B \cos \theta \, \Delta l = \mu_0(I + I_D) \tag{12.2}$$

In the case of our wire and capacitor, $I_D = 0$ through A_1, while $I = 0$ through A_2. While Eq. (*12.2*) eliminates our ambiguity, we must still demonstrate that it is true. We must find if this concept of displacement current predicts something new, and then test this prediction experimentally.

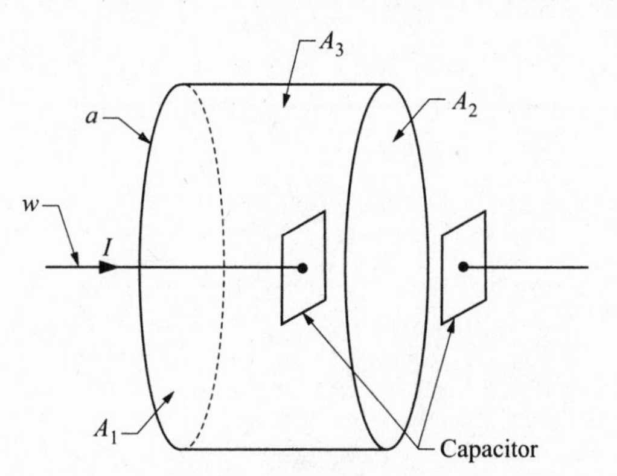

Fig. 12-2

Problem 12.1. Using Ampere's law, calculate the magnetic field between circular parallel plate capacitors, at a distance r from the center, when the capacitor is being charged at the rate of I coulombs/second.

Solution

In Fig. 12-3, we draw the circular capacitor plates which are being charged. At a distance, r, from the center, we draw a circular path that we will use for Ampere's law. The electric field points to the right if the left plate is positively charged, and that charge is being increased at the rate of $I = \Delta q/\Delta t$ C/s. The electric field is uniform, and equal to $q/\varepsilon_0 A$, and increasing at the rate of $\Delta E/\Delta t = (\Delta q/\Delta t)/\varepsilon_0 A = I/\varepsilon_0 A$, where A is the area of the plates. The displacement current density is equal to $\varepsilon_0 \Delta E/\Delta t$, and is uniform within the region. The magnetic field is the same at every point on the circular path, by symmetry. We go around the path in the direction shown, using the right-hand rule with our thumb in the direction of **E**, so our fingers curl about the path in this direction. Adding the magnetic field along the curve we get $\sum B \cos\theta \, \Delta l = B(2\pi r)$. We must now calculate the total current going through this loop. There is no conduction current, so the only contribution comes from the displacement current. The total displacement current through this curve is $J_D A_r = (\varepsilon_0 \Delta E/\Delta t)(\pi r^2) = \varepsilon_0(I/\varepsilon_0 A)\pi r^2 = I(\pi r^2/A)$. By Ampere's law, we therefore have that $B(2\pi r) = \mu_0 I(\pi r^2/A)$, or $B = \mu_0 \pi I r/A$. This shows that the field increases linearly with r as one moves from the center to the edge of the plate. This can, of course, be tested experimentally and the result agrees with prediction.

Outside the capacitor plates, we can also calculate the field using Ampere's law. In that case, the displacement current density would be $(\varepsilon_0 \Delta E/\Delta t)$ within the plates, and zero outside the plates. The total current through the area would then be $J_D A = (\varepsilon_0 \Delta E/\Delta t)A = I$, giving $B(2\pi r) = \mu_0 I$, or $B = \mu_0 I/2\pi r$, as for the field of a long straight wire.

Problem 12.2. A rectangle *abcd*, with sides of 60 cm × 80 cm, is in an electric field of 10^3 V/m directed into the paper, as in Fig. 12-4. The field is increasing at the rate of 300 V/m · s.

(*a*) What is the displacement current through this area?

(*b*) What is the direction of the component of the magnetic field along *ab*?

Solution

(*a*) The displacement current density is $J_D = \varepsilon_0 \Delta E/\Delta t$, and is constant within the area. Since the field is increasing, $\Delta E/\Delta t$ is also into the paper, as is the direction of J_D. The displacement current is therefore $J_D A = \varepsilon_0 A \, \Delta E/\Delta t = 8.85 \times 10^{-12}(0.48)(3 \times 10^2) = 1.27 \times 10^{-9}$ A, into the paper.

(*b*) Since I_D is into the paper, the positive direction for going around *abcd* is clockwise ($a \to b \to c \to d$). Ampere's law states that $\sum B \cos\theta \, \Delta l = +\mu_0(I + I_D) = \mu_0 I_D$, in this case. Therefore, the components of B are in the positive direction of circling around the rectangle, and along *ab* the field is directed from *a* to *b*.

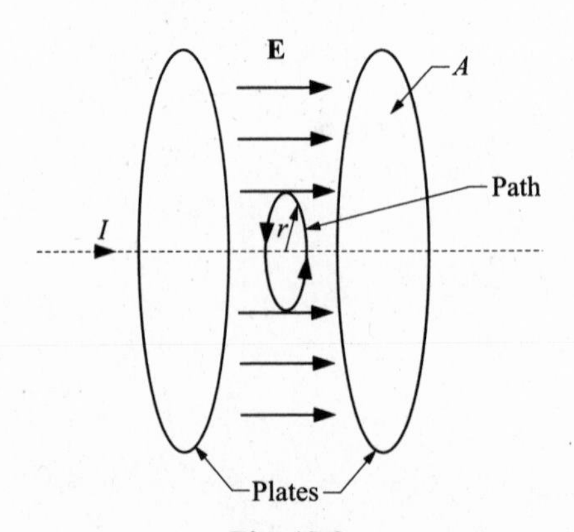

Fig. 12-3

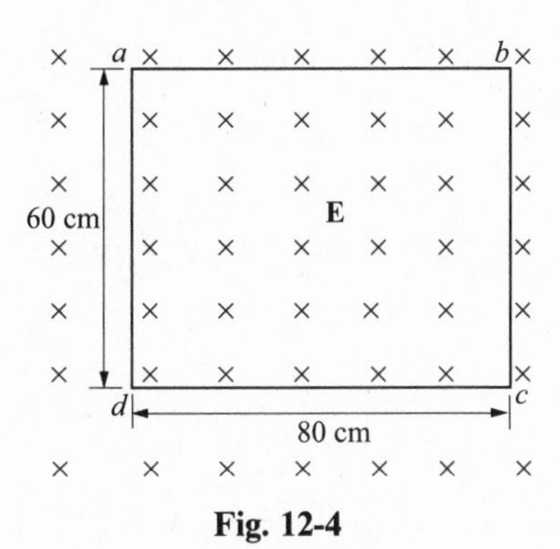

Fig. 12-4

Problem 12.3. For the same case as in Problem 12.2, what would the answers be if the rectangle were located in a region of dielectric constant 3.5?

Solution

Using the same reasoning that we used in defining the displacement current originally, but for a capacitor filled with a dielectric material of dielectric constant κ, we would define the displacement current in this case as $J_D = \varepsilon \Delta E/\Delta t$, where $\varepsilon = \kappa\varepsilon_0$. Thus we need only replace ε_0 by ε to get the answer for our problem. The current is therefore $1.27 \times 10^{-9}(3.5) = 4.45 \times 10^{-9}$ A, and the direction is the same as for Problem 12.2.

In order to determine whether the concept of displacement current is generally valid, we seek some prediction which is different with or without this concept. This is what Maxwell did after summarizing the laws of electricity and magnetism in the famous **Maxwell equations.**

12.3 MAXWELL'S EQUATIONS

We have already stated that, after including the displacement current, it is possible to summarize the laws governing the creation of electrical and magnetic fields in four fundamental equations.

The first equation is **Gauss' law**, which states that electric fields can be established by free charges. This law is written in terms of the electric flux that passes through a closed surface, and depends on the understanding that all electric field lines start at positive charges and end on negative charges (lines can also go to infinity, such as those of an isolated point charge, where they are presumed to land on opposite charges at that distance). By convention the number of electric field lines per unit area, the electric flux density, at a given point is chosen equal to the magnitude of the electric field at that point. As discussed in Chap. 3 it then equals the electric field at every other point as well. Gauss' law then relates the total charge within a closed surface to the net number of electric field lines that pass through the surface. Gauss' law can be written as

$$\sum \Psi_E = \sum Q/\varepsilon_0 \tag{12.3}$$

where Ψ_E = flux through an infinitesimal surface area $A = E(\cos\theta)A$ with θ the angle below **E** and the outward normal to the surface element A, and the sum goes over all elements A making up the closed surface. The sum over charges includes all charges within the closed surface.

The second (Maxwell) equation is based on how this same concept applies to magnetic fields. We have learned that there are no magnetic monopoles that act as sources for a magnetic field. Therefore magnetic fields do not have poles where they begin or end. All magnetic field lines must therefore close on themselves. This means that any magnetic field line that passes through a closed surface must necessarily pass through the surface again in the opposite direction, in order to close on itself. This means that the net total magnetic flux which passes through a surface is zero. This is written as

$$\sum \Phi_B = 0 \tag{12.4}$$

where Φ_B = (magnetic flux through an infinitesimal surface area A) = $B(\cos\theta)A$ with θ the angle between **B** and the outward normal to the surface element A, and the sum goes over all elements A making up the closed surface.

The third equation states that an electric field can also be produced by a changing magnetic flux. This is **Faraday's law**, which can be written as

$$\sum E(\cos\theta)\Delta l = -\Delta\Phi_B/\Delta t \tag{12.5}$$

where Δl is an infinitesimal length along a closed curve, θ is the angle between **E** and the tangent to the curve at Δl and the sum is taken over all the elements Δl of the closed curve, and the flux Φ_B is the total magnetic flux through the area bounded by the curve.

The fourth, and last equation, states that magnetic fields are created by currents, either conduction current or displacement current. This can be written in the form of **Ampere's law**, including displacement current, as

$$\sum B(\cos\theta)\Delta l = \mu_0(I + I_D) = \mu_0(I + \varepsilon_0\,\Delta\Psi_E/\Delta t) \tag{12.6}$$

where Δl is again an infinitesimal length along a closed curve, θ is the angle between **B** and the tangent to the curve at Δl, and the sum is taken over all elements of the closed curve. The current I is the total current passing through a surface bounded by the curve, and Ψ_E is the total electric flux through the same surface.

These four equations are relationships between the electric and magnetic fields and their sources, charges and currents. The electric and magnetic fluxes are determined directly from the electric and magnetic fields and are not separate variables. Thus these equations tell us how to calculate the electric and magnetic fields that are produced by charges, both at rest and moving. The particular form that we have used for these equations is not the most useful for actual calculations, but is the easiest to understand conceptually. For purposes of calculations, these equations are expressed more formally in the language of the integral and differential calculus, which can then be solved for specific cases.

We have written these equations for the case of free space, and not for the situation in which there is dielectric or magnetic material present. For the case of materials, one must modify these equations using the concepts of **electric displacement** (D) and **magnetic intensity** (H). We will not write down these modified equations, but will point out where changes occur in the solutions that we will discuss.

These four equations constitute Maxwell's equations, which are the fundamental laws governing the existence of electric and magnetic fields, which are jointly called electromagnetic fields. We see clearly from these equations that electric and magnetic fields are not really independent quantities, but are rather quantities that are bound together by these relationships. Changes in one produce or modify the other. These equations are remarkable in that, unlike Newton's laws, they do not require fundamental modification as a result of the theory of relativity. In fact, the solutions to these equations gave rise to questions that required the theory of relativity for their resolution. Furthermore, the quantum theory also accepts these equations as the fundamental ones describing electromagnetic phenomena, requiring only a proper interpretation in light of the fundamentally new concepts of quantum mechanics.

In order to give a complete description, in theory, of electromagnetic phenomena, we must add the laws that tell us what effect these fields have on objects. This is given by the statement that electric fields exert forces on any electrical charges, while magnetic fields exert forces on moving charges. The magnitudes and directions of these forces were discussed previously in Chaps. 3 and 6.

12.4 ELECTROMAGNETIC WAVES

Maxwell was able to show that there were solutions to these equations that corresponded to waves propagating in free space, i.e. in regions where there are no charges or currents. These waves, which he called **electromagnetic waves** (EM) had special properties, which could be derived from these equations. In all the waves previously discussed the wave was a consequence of the vibration of molecules of a medium about their equilibrium positions—their displacement—and the propagation of this disturbance with a velocity characteristic of the medium. In the case of electromagnetic waves the time varying quantity is not the displacement but rather the electric and magnetic fields at a point in space. Indeed, for electromagnetic waves one does not even need a medium—they can travel through empty space. Like ordinary waves, however, they do have a characteristic velocity which depends on the material the wave travels through and has an especially significant value in empty space. Maxwell was able to show that these waves were transverse, and that their speed, in free space was equal to $1/\sqrt{\varepsilon_0 \mu_0} = c$.

For a wave traveling in the x direction, this means that the electric and magnetic fields associated with this wave are in the y–z plane. Indeed, one can show that these fields are also perpendicular to each other, and that their magnitudes are given by:

$$E = cB \tag{12.7}$$

It is important to note that these results would not be true if one left out the term for displacement current. That EM waves exist is the strongest evidence that the displacement current should be included in Ampere's law. Furthermore, Maxwell's prediction that these electromagnetic waves travel with speed $1/\sqrt{\mu_0 \varepsilon_0}$, which numerically equals 3×10^8 m/s, matches the measured value for the **speed of light**. This quickly led to the realization that light consists of electromagnetic waves in a certain frequency range to which the eye is sensitive and can "see". Approximately 22 years after Maxwell predicted the existence of these electromagnetic waves, and delineated their properties, Henry produced and detected these waves in the radio range of frequencies.

If the medium in which this wave propagates is not free space, but rather a material with a **dielectric constant** κ and **magnetic permeability** κ_M, then the velocity of the waves will become

$$v = 1/\sqrt{\kappa\kappa_\mu \mu_0 \varepsilon_0} = 1/\sqrt{\mu\varepsilon}. \tag{12.8}$$

Problem 12.4. For an electromagnetic wave, traveling in a medium with dielectric constant 3.5, and magnetic permeability 1.2, what is the speed of this wave?

Solution

Using Eq. (*12.8*), and knowing that $1/\sqrt{\mu_0 \varepsilon_0} = 3 \times 10^8$ m/s, we get $v = 3 \times 10^8/\sqrt{(3.5)(1.2)} = 1.46 \times 10^8$ m/s.

Problem 12.5. For an electromagnetic wave, traveling in the x direction in free space, the electric field has a magnitude of 1.5 V/m, and is in the y direction. What is the magnitude and direction of the magnetic field?

Solution

Using Eq. (*12.7*), the magnitude of B is $B = E/c = (1.5\ \text{V/m})/(3 \times 10^8\ \text{m/s}) = 0.5 \times 10^{-8}$ T. The direction of the magnetic field is in the $+z$ direction, perpendicular to both **B** and the direction of propagation.

Problem 12.6. For an electromagnetic wave, traveling in the $-x$ direction in free space, the electric field has a magnitude of 1.5 V/m, and is in the $+y$ direction. What is the magnitude and direction of the magnetic field?

Solution

Using Eq. (*12.7*), the magnitude of B is $B = E/c = (1.5\ \text{V/m})/(3 \times 10^8\ \text{m/s}) = 0.5 \times 10^{-8}$ T. To get the direction we note that it can be shown that, in general, the three perpendicular directions, **E**, **B** and **c**, are related like **v**, **B** and **F** in the magnetic force on a charge. This means that if your fingers are in the direction that rotates **E** into **B**, the thumb will point in the direction of the velocity, **c**. Therefore, the direction of **B** in our problem is in $-z$.

Problem 12.7. For an electromagnetic wave, traveling in the $+x$ direction in free space, the electric field has a magnitude of 1.5 V/m, and is in the $+z$ direction. What is the magnitude and direction of the magnetic field?

Solution

Using Eq. (*12.7*), the magnitude of B is $B = E/c = 1.5/3 \times 10^8 = 0.5 \times 10^{-8}$ T. We know that the magnetic field must be perpendicular to both **E** and the velocity, **c**, and therefore is in the $\pm y$ direction. We showed in Problem 12.6, that the three perpendicular directions, **E**, **B** and **c**, are related so that if your fingers are in the direction that rotates **E** into **B**, the thumb will point in the direction of the velocity, **c**. Applying this to our case, we see that the magnetic field is in the $-y$ direction.

As noted, it can be demonstrated that light is one form of electromagnetic wave. The fact that a light wave travels with the speed predicted for an electromagnetic wave is one of the reasons that it was quickly suspected that this was true; given Maxwell's theoretical result. How can one measure the speed of light? One method is illustrated in the next problem.

Problem 12.8. Light passes through an opening in a rim of a notched wheel, as in Fig. 12-5. Light travels to a mirror at a distance of 5×10^4 m and is reflected back to the wheel. There are 50 notches in the wheel. At what angular speed must the wheel turn so that the reflected light passes through the adjacent opening?

Solution

The light that passes through one notch travels to the mirror and then back to the wheel in a time equal to $2L/c$, where L is the distance to the mirror. During this same time, the wheel has to turn just far enough that the next notch is now in the position of the first notch. Since there are 50 notches on the wheel, the wheel has to rotate through 1/50 of a full rotation, or through an angle of $2\pi/50$. The time that this takes is $(2\pi/50)/\omega$. Therefore $(2\pi/50)/\omega = 2L/c$, or $\omega = (2\pi/50)c/2L = 2\pi(3 \times 10^8)/(50)(2)(5 \times 10^4) = 377$ rad/s.

Just as in the case of sound waves, it is useful to consider electromagnetic waves that are sinusoidal. This means that if we take a picture of the wave at any time, the disturbance will vary sinusoidally in space along the direction of propagation. Furthermore, at any position is space, the disturbance will vary sinusoidally in time. As in the case of sound, one can have different shaped electromagnetic waves, such as spherical waves emitting from a local region, but far away they appear nearly planar over a region small compared to the distance from the disturbance. In such a region the light wave has the same value of electric and magnetic field at all points in the plane perpendicular to the direction of propagation, and varying in lock step. It is important to keep in mind that the disturbance associated with an electromagnetic wave is the electric and magnetic field along the wave. In Fig. 12-6, we show a

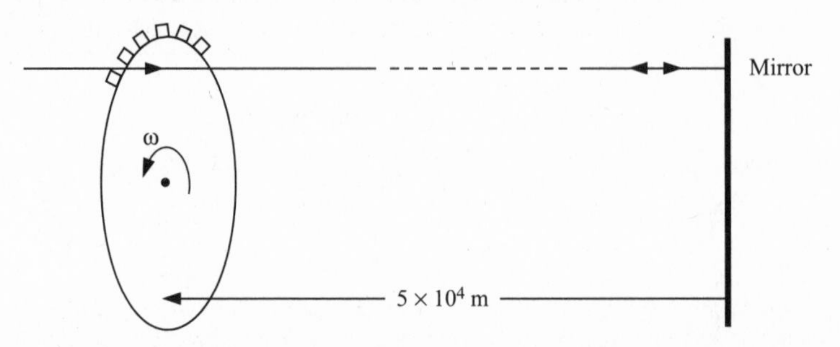

Fig. 12-5

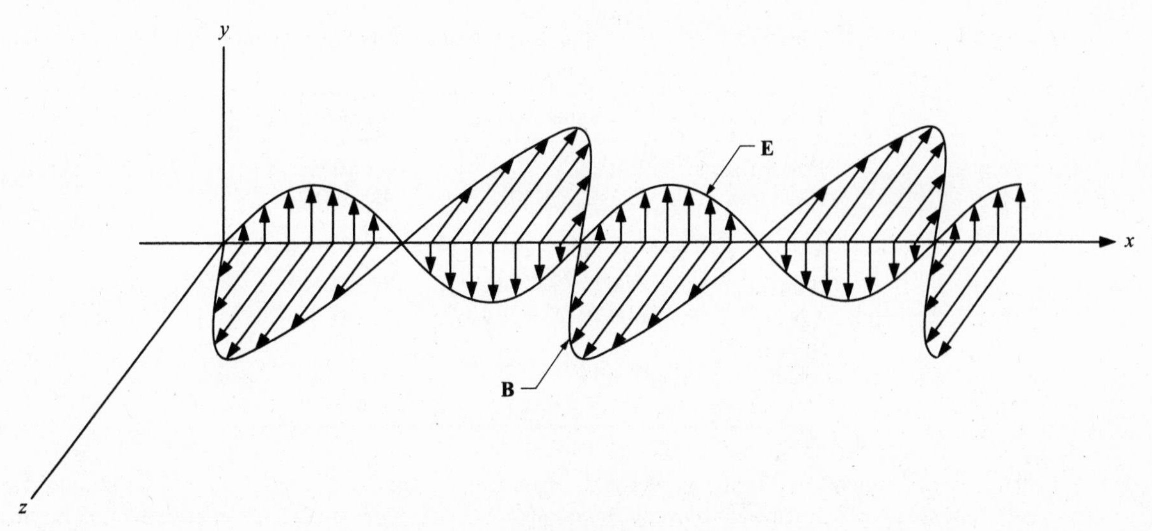

Fig. 12-6

plane wave as it varies in space at a particular instant of time. The wave is traveling in the positive x direction, and both the electric and magnetic fields are transverse to this direction. We draw the electric field in the y direction, and then the magnetic field must be in the z direction as we showed in the previous problems. This wave is said to be linearly polarized in the y direction, which is the direction of the electric field. The electric and magnetic fields have magnitudes that are related by $E = cB$ at every point, and the wave travels with a velocity c. The distance between successive crests or between successive troughs is the wavelength, λ, of the wave. The time that it takes for the wave to travel a distance of one wavelength is the period, T, of the wave, which is also the time for the wave to go from crest to crest at any given point in space. As with all waves, this leads to the relationship that $c = \lambda/T = \lambda f$, where f is the frequency.

Electromagnetic waves exist with wavelengths ranging from very small to very large (and corresponding frequencies from very large to very small). The various possible wavelength (and frequency) ranges constitute the electromagnetic spectrum. For small frequencies the wave is usually denoted by its frequency, and for short wavelength it is denoted by its wavelength. In Table 12.1, we list the frequency range for some common types of electromagnetic waves with small frequencies, and others with small wavelengths. In the next problem, we will complete the table.

Problem 12.9. For the electromagnetic waves in the table, calculate the missing wavelengths and frequencies to complete the table.

Table 12.1 Frequency Range of Common Types of Electromagnetic Waves

Type of wave	Frequency	Wavelength (m)
Power line	60	
AM radio	$(0.5\text{–}1.5) \times 10^6$	
FM radio	10^8	
Microwaves	$10^9\text{–}10^{11}$	
Infrared		$10^{-3}\text{–}10^{-6}$
Visible		$(8\text{–}4) \times 10^{-7}$
Ultraviolet		$4 \times 10^{-7}\text{–}10^{-9}$
X-rays		$10^{-8}\text{–}10^{-11}$
Gamma rays		$<10^{-11}$

Table 12.2 Frequency Range of Common Types of Electromagnetic Waves

Type of wave	Frequency (Hz)	Wavelength (m)
Power line	60	5×10^6
AM radio	$(0.5–1.5) \times 10^6$	600–200
FM radio	10^8	3
Microwaves	$10^9–10^{11}$	$0.3–3 \times 10^{-3}$
Infrared	$3 \times 10^{11}–3 \times 10^{14}$	$10^{-3}–10^{-6}$
Visible	$4 \times 10^{-14}–8 \times 10^{14}$	$(8–4) \times 10^{-7}$
Ultraviolet	$8 \times 10^{14}–3 \times 10^{17}$	$4 \times 10^{-7}–10^{-9}$
X-rays	$3 \times 10^{16}–3 \times 10^{19}$	$10^{-8}–10^{-11}$
Gamma rays	$>3 \times 10^{19}$	$<10^{-11}$

Solution

The relationship between frequency and wavelength is $c = \lambda f$. For power line frequencies of 60 Hz, the wavelength will be $\lambda = 3 \times 10^8/60 = 5 \times 10^6$ m. For am radio, the wavelength will vary between $\lambda = 3 \times 10^8/0.5 \times 10^6 = 600$ m and $\lambda = 3 \times 10^8/1.5 \times 10^6 = 200$ m. Repeating this calculation for all the given frequencies allows us to complete the table where wavelengths are missing. Where wavelengths are given, we calculate the frequencies using $f = 3 \times 10^8/\lambda$. For instance, in the case of infrared radiation, the frequency ranges from $3 \times 10^8/10^{-3} = 3 \times 10^{11}$ Hz to $3 \times 10^8/10^{-6} = 3 \times 10^{14}$ Hz. Table 12.1 then becomes as shown in Table 12.2.

The limits given in the Table 12.2 are only approximate, and the various types actually overlap considerably. For instance, microwaves and infrared radiation include the wavelengths around 10^{-3} m, and are identical waves irrespective of whether they are called microwaves or infrared. One usually distinguishes between them on the basis of how they were produced. If they were produced electronically, they are called microwaves. If they are produced from heat, they are called infrared. Similar distinctions are made at the boundaries of the different types of radiation. All of these waves travel with a speed of c, all are transverse, and all carry perpendicular electric and magnetic fields with them.

12.5 MATHEMATICAL DESCRIPTION OF ELECTROMAGNETIC WAVES

As in the case of a sound wave (and any other type of wave), the disturbance that is carried by the wave varies with both time and space. A plane wave travels in one dimension with its disturbance depending only on time and the position along the direction of travel. Suppose the wave is traveling in the x direction. At any point x and instant t every point in the plane parallel to the y–z plane at that x, has the same disturbance. As time changes the disturbance at all points in this plane change in lock step, i.e in phase.

The equation for the disturbance of a electromagnetic sinusoidal plane wave, traveling in the $+x$ direction, is given in terms of its disturbance (an electric field in the y direction) by

$$E = E_0 \cos 2\pi(ft - x/\lambda) = E_0 \cos(\omega t - kx), \tag{12.9}$$

where

$$\omega = 2\pi f \quad \text{and} \quad k = 2\pi/\lambda \tag{12.10}$$

Here ω is the angular frequency of the wave, and k is the "wavenumber" of the wave, and has units of m^{-1}. E_0 is the maximum value of the electric field, and is thus the amplitude of the wave. For a wave traveling in the $-x$ direction, the equation for the electric field is

$$E = E_0 \cos 2\pi(ft + x/\lambda) = E_0 \cos(\omega t + kx) \tag{12.11}$$

This equation will suffice for any single plane wave, since we can choose the direction of travel to be the x direction.

Problem 12.10. An electromagnetic plane wave is traveling in the x direction. The electric field is given by $E = 1.5 \cos(6 \times 10^4 t - 2 \times 10^{-4} x)$. Assume standard units.

(*a*) What is the amplitude of the wave?

(*b*) What is the frequency of the wave?

(*c*) What is the wavelength of the wave?

Solution

(*a*) The amplitude is the maximum electric field in the wave, which is given by the factor before the cosine function. Thus, the amplitude is $E_0 = 1.5$ V/m.

(*b*) We can see by comparing the general formula [Eq. (*12.9*)] to the specific equation given in this problem, that $\omega t = 6 \times 10^4 t$, or $\omega = 6 \times 10^4$ rad/s. Then $f = \omega/2\pi = 9.5 \times 10^3$ Hz.

(*c*) Again, comparing the general equation to the specific numbers in our equation, $kx = 2 \times 10^{-4}x$, or $k = 2 \times 10^{-4}\ \text{m}^{-1}$. Then $\lambda = 2\pi/k = 2\pi/2 \times 10^{-4}$, or $\lambda = 3.14 \times 10^4$ m. Note that this is consistent with the requirement that $f\lambda = c = 3 \times 10^8$ m/s.

Problem 12.11. An electromagnetic plane wave is traveling in the $-x$ direction. The electric field has an amplitude of 2 V/m, and a frequency of 1000 Hz. Write down an equation for the wave as a function of time and distance.

Solution

The general equation for an electromagnetic wave traveling in the $-x$ direction, is given by Eq. (*12.9*), $E = E_0 \cos 2\pi(ft + x/\lambda) = E_0 \cos(\omega t + kx)$. In our case, $E_0 = 2, f = 10^3$ Hz, and $\lambda = c/f = 3 \times 10^5$ m. Thus this wave is given by $E = 2 \cos [2\pi(1000t + x/3 \times 10^5)] = 2 \cos (6.28 \times 10^3 t + 2.09 \times 10^{-5}x)$.

Problem 12.12. An electromagnetic plane wave is traveling in the x direction. The electric field has an amplitude of 0.5 V/m, and a frequency of 2500 Hz. Write down an equation for the magnetic field component of the wave as a function of time and distance.

Solution

The general equation for the electric field of an electromagnetic wave traveling in the x direction, is given by Eq. (*12.9*), $E = E_0 \cos 2\pi(ft - x/\lambda) = E_0 \cos(\omega t - kx)$. For the magnetic field part of the wave, we will have the same general equation, except that the amplitude will be different and the direction of the magnetic field given by the formula, will be in the z direction rather than the y direction. Then, $B = B_0 \cos 2\pi(ft - x/\lambda) = B_0 \cos(\omega t - kx)$. In our case, $E_0 = 0.5$ V/m, $f = 2500$ Hz, $\lambda = c/f = 1.2 \times 10^5$ m and $B_0 = E_0/c = 1.67 \times 10^{-9}$ T. Thus the magnetic field of this wave is given by $B = 1.67 \times 10^{-9} \cos 2\pi(2500t - x/1.2 \times 10^5) = 1.67 \times 10^{-9} \cos(1.57 \times 10^4 t - 5.23 \times 10^{-5}x)$.

While plane waves are particularly simple to describe (they are essentially one-dimensional) other relatively simple sinusoidal waves are also worth noting. One such wave is a cylindrical wave. In this case, the wave travels radially away from a straight line with the same speed in all radial directions. Now the surfaces of constant phase in the electric (or magnetic) fields are concentric cylinders about the line. Thus if one point on a given cylindrical surface corresponds to maximum amplitude (a crest) of electric field, all other points on the surface are also crests of the electric field. If the line, which is the symmetry axis for the concentric cylinders is along the z axis, then the electric and magnetic fields will depend only on x and y. Actually the magnitude of the fields depends only on the radial distance, r, from the line. The direction of the fields does depend on where in the x–y plane one is, and the direction of propagation is along the radial direction, which is different for different x, y positions. The overall effect, however, is much like the expanding ripple in a pond, except that it is now a cylindrical surface expanding at speed c rather than a circle. Another simple sinusoidal wave, discussed briefly in the context of sound waves, is the spherical wave. In this case the source of the electromagnetic disturbance is a small region approximated by a point, and the disturbance expands out in a spherical shell. These concentric spherical surfaces correspond to constant phase in the electric and magnetic fields. The directions of the fields are always tangent to the spherical surfaces and the direction of propagation is

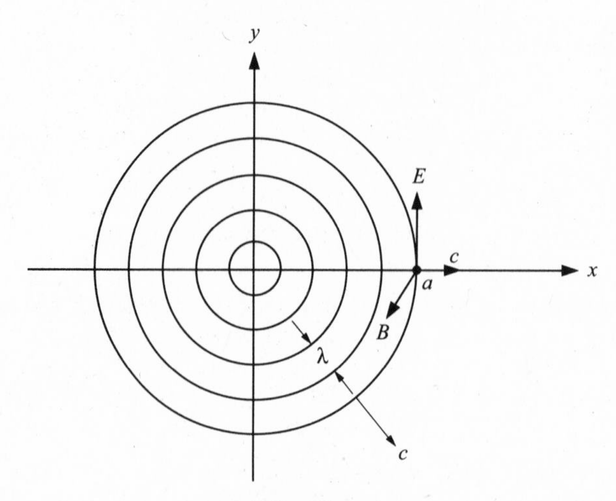

Fig. 12-7

always radially outward. While the direction of the electric field thus depends on the point in space (x, y, z), the magnitude depends only on the radial distance, r, to the point. The spherical shells of constant phase will be expanding in all directions with the speed c, and the radial distance between shells of adjacent crests will be the wavelength of the wave. We will still have the relationship between frequency, wavelength and speed given by $c = \lambda f$. As a given spherical shell expands out at speed c, its surface area will continually increase, and the energy of the wave (we will discuss this energy in the next section) which is uniformly spread over the surface will pass through larger and larger surfaces. The intensity of the wave, I, which is the energy per unit area perpendicular to the direction of propagation, therefore falls off. Since the area of a spherical surface is $4\pi r^2$, the intensity falls off as $1/r^2$. The amplitude of the wave, A, is related to the intensity by $I \propto A^2$, and therefore A falls off as $1/r$. In Fig. 12-7, we draw the intersection of constant phase spherical shells (corresponding to successive crests) centered on the origin with the x–y plane. The intersections correspond to concentric circles, separated by the wavelength λ. At every point of the spheres, the wave is moving radially outward with speed c. In the x–y intersection plane, the circles are moving outward with this same speed at each point. In particular, at point a, the wave is moving in the x direction with speed c. The direction of the electric field (the disturbance) is perpendicular to this direction, and in the y–z plane. Let us take it to be in the y direction. Then the magnetic field of the wave will be in the z direction. As one gets further and further away from the central point, the spheres have less and less curvature, and the wave begin to look more and more like plane waves (see e.g., Chap. 2 for an equivalent discussion for sound waves).

The spherical wave has a mathematical form similar to the plane wave. A major difference is that the electric field is now a function of distance, r, from a point, rather than of x, and that the amplitude gets smaller as r increases. The magnitude of E is constant over the surface of the spherical shell, just as it is constant over the surface of the plane for plane waves. However, the direction varies over the spherical surface so that it is always perpendicular to the radial direction. The formula for the magnitude of E is given by

$$E = (A/r) \cos 2\pi(ft - r/\lambda) = (A/r) \cos (\omega t - kr) \tag{12.12}$$

We will use this relationship later on when we discuss the energy and momentum carried by electromagnetic waves.

12.6 ENERGY AND MOMENTUM FLUX OF ELECTROMAGNETIC WAVES

We already showed previously that electric and magnetic fields contain energy. The energy density was shown to be $u_E = \varepsilon_0 E^2/2$ for electric fields and $u_B = B^2/2\mu_0$ for magnetic fields. The electromagnetic energy of an electromagnetic wave is just the sum of the energies of its electric and magnetic fields. The maximum energy is located at those points where the fields are at their maxima, which occurs at the crests of these waves. But these crests move with time at a speed of c, and therefore the energy is transported in the direction that the wave travels at this speed. An electromagnetic wave, therefore carries energy with it, just as do sound waves or other waves traveling through a medium, and in fact it also carries momentum.

To calculate the energy carried by an plane electromagnetic wave we would proceed in two steps. First we would calculate the total energy contained between successive crests of an electromagnetic wave. Then we would calculate the average energy that this wave transports per unit area and time as it travels with speed c in the x direction, which is defined as the intensity, I, of this wave.

The result is that:

$$I = c^2\varepsilon_0 E_0 B_0/2 = E_0 B_0/2\mu_0 = c\varepsilon_0 E_0{}^2/2 \tag{12.13}$$

where we have used the fact that $E_0 = cB_0$. We see that the intensity of an electromagnetic wave is proportional to the amplitude squared, as was true for sound waves.

This quantity is often assigned a direction, **the direction of propagation**, and called the average **Poynting vector**. This vector is the average of the instantaneous Poynting vector, **S**, whose magnitude is EB/μ_0, and whose direction is perpendicular to **E** and **B**, and obeying the right-hand rule with finger curling from **E** to **B**. This is the direction of **c**, the propagation velocity. The Poynting vector, **S**, represents the instantaneous energy transported through unit area per unit time by the wave. It is actually more general than that, representing the energy transport even if the fields do not represent waves. The unit for intensity is $\text{J/m}^2 \cdot \text{s} = \text{W/m}^2$, since energy/s is power, or $\text{J/s} = \text{W}$.

Problem 12.13. An electromagnetic plane wave is traveling in the x direction. The formula for the wave is given by $E = (1.5\ \text{V/m}) \cos [(6 \times 10^4\ \text{s}^{-1})t - (2 \times 10^{-4}\ \text{m}^{-1})x)]$. Calculate the intensity of this wave.

Solution

The intensity of a wave is $I = c\varepsilon_0 E_0{}^2/2$. For the above wave, $E_0 = 1.5$ V/m, and therefore the intensity is $I = 3 \times 10^8(8.85 \times 10^{-12})(1.5^2)/2 = 3.0 \times 10^{-3}\ \text{W/m}^2$.

Problem 12.14. An electromagnetic plane wave is traveling in the x direction. The intensity of this wave is $5 \times 10^{-3}\ \text{W/m}^2$. Calculate the maximum electric and magnetic fields of this wave.

Solution

The intensity of a wave is $I = c\varepsilon_0 E_0{}^2/2 = 5 \times 10^{-3}$. Thus $E_0 = [2(5 \times 10^{-3})/(3 \times 10^8)(8.85 \times 10^{-12})]^{1/2} = 1.94$ V/m. The maximum magnetic field, B_0 is $E_0/c = 6.5 \times 10^{-9}$ T.

Problem 12.15. An electromagnetic spherical wave is traveling outward from a point source. The intensity of this wave is $5 \times 10^{-3}\ \text{W/m}^2$, when the distance from the source is 2 m. Calculate the intensity of this wave when the distance is 5 m.

Solution

The intensity of a plane wave is $I = c\varepsilon_0 E_0{}^2/2$. For a spherical wave, Eq. (*12.12*), at any point in space the amplitude is $E_{max} = A/r$, instead of a constant. Therefore, the intensity of a spherical wave is given by $I = c\varepsilon_0(A/r)^2/2 = c\varepsilon_0 A^2/2r^2$. The intensity is therefore seen to decrease with distance from the source as $1/r^2$.

The intensity at $r = 2$ is given as 5×10^{-3}, so that $I_2 = 5 \times 10^{-3} = (c\varepsilon_0 A^2/2)(1/2)^2$. The intensity at $r = 5$ is $I_5 = (c\varepsilon_0 A^2/2)(1/5)^2$. Thus $I_5/I_2 = (2/5)^2 = 0.16$, or $I_5 = 0.16(5 \times 10^{-3}) = 8 \times 10^{-4}\ \text{W/m}^2$.

The $1/r^2$ dependence of the intensity could have been derived in a different manner. The energy per unit time, or power, from the source travels away uniformly in all directions. All the power emitted by the source flows through the surface of a sphere enclosing the source. For concentric spheres at radii r_1 and r_2, the power is evenly distributed over a surface area of πr_1^2 and πr_2^2, respectively. The intensity, which is the power per unit area, is therefore dependent on the distance as $1/r^2$, since we divide the power by the area to get the intensity.

Problem 12.16. An electromagnetic spherical wave is traveling outward from a point source. The intensity of this wave is 5×10^{-3} W/m^2, when the distance from the source is 2 m. Calculate the maximum electric field of this wave when the distance is 2 m and when the distance is 5 m.

Solution

For a spherical wave at any point in space the amplitude is $E_{max} = A/r$, the maximum electric field at that distance. Therefore, the intensity of a spherical wave is given by $I = c\varepsilon_0(E_{max})^2/2$. Since intensity decreases with distance from the source as $1/r^2$, (Problem 12.20), E_{max} decreases as $1/r$.

At $r = 2$, the intensity is given as $5 \times 10^{-3} = 3 \times 10^8(8.85 \times 10^{-12})E_{max}^2/2$, so $E_{max} = 1.94$ V/m, as obtained in Problem 12.19. To get E_{max} at $r = 5$, we use the fact that E_{max} depends on r as $1/r$, giving $E_5 = (\frac{2}{5})E_2 = 0.4(1.94) = 0.78$ V/m.

Whenever energy moves in a certain direction, there is also a certain amount of momentum in that direction. For instance, a particle with kinetic energy $(\frac{1}{2})mv^2$ has a momentum given by mv. One can show, using the equations of electromagnetic theory, that an electromagnetic wave also has momentum, but the calculation is not simple. We therefore present only the result of the calculation. We get that the average momentum density, which is just the average momentum per unit volume in the region where the plane wave exists is:

$$\text{momentum density} = S_{av}/c^2 = E_0 B_0/2c^2\mu_0 \tag{12.14}$$

with the direction of **S** in the direction of the momentum of the wave.

From Eq. (*12.13*) we see that the intensity I can be expressed as:

$$\text{Energy flux density} = I_{av} = S_{av} = E_0 B_0/2\mu_0 \tag{12.15}$$

and the energy density (from earlier work) is just

$$\text{energy density} = u = S_{av}/c = E_0 B_0/2\mu_0 c \tag{12.16}$$

Finally, the momentum that passes through an area A perpendicular to the direction of propagation in time Δt is given by:

$$\text{momentum flux density} = S_{av}/c = E_0 B_0/2c\mu_0 \tag{12.7}$$

These formulas give the average values for these quantities for a sinusoidal wave, where we have S is $E_0 B_0/2\mu_0$. If we replace $E_0 B_0/2$ by EB and S_{av} by S in Eqs. (*12.14*)–(*12.17*) we get the instantaneous values for the given quantities.

Problem 12.17. Sun light above the earth's atmosphere has an average intensity of approximately 1.4 kW/m^2. If this sunlight is absorbed by a solar panel with an area of 5 m^2, oriented perpendicular to the radiation direction, calculate (*a*) the energy absorbed per second; (*b*) the momentum absorbed per second; and (*c*) the force exerted on the solar panel by the sunlight.

Solution

(*a*) The intensity of the electromagnetic wave is $I = S_{av} = 1.4 \times 10^3$ W/m^2. The total energy absorbed, per second, by the panel is $IA = 1.4 \times 10^3(5) = 7 \times 10^3$ W.

(*b*) To get the momentum absorbed by this area per second, we multiply the momentum flux density by the area. Thus the momentum absorbed is $(S_{av}/c)A = 1.4 \times 10^3(5)/3 \times 10^8 = 2.33 \times 10^{-5}$ N.

(*c*) Whenever there is a change in momentum, there is a force causing this change in momentum. We know that for an object of mass m the force equals $ma = m\Delta v/\Delta t = \Delta p/\Delta t$. Thus, the force equals the rate of change of the momentum. If the radiation has a momentum flux of S/c, then the solar panel is absorbing SA/c units of momentum per second. This is the rate at which the momentum of the wave is decreased. The panel exerts the force that causes this change in momentum which equals $F = \Delta p/\Delta t = SA/c = 2.33 \times 10^{-5}$ N in the direction opposite to **S**. By **Newton's third law**, there is a reaction force of the wave on the panel of equal magnitude. In effect, the sunlight exerts a force on the panel.

This example shows how one can use sunlight in space to not only supply power but to exert a force on a spacecraft. The force that is exerted on the surface can best be characterized by the force exerted per unit area, or pressure, $P = F/A$. This "**radiation pressure**" equals, for a totally absorbing surface, $(SA/c)/A = S/c =$ momentum flux density. We will see in the supplementary problems that a totally reflecting surface is subject to twice this force (but it absorbs no energy).

For electromagnetic waves in ordinary matter, all the above equations still apply, provided that we use ε for ε_0, and μ for μ_0.

Problems for Review and Mind Stretching

Problem 12.18. A parallel plate capacitor is being charged at a rate of 5×10^{-3} C/s. Calculate the displacement current between the plates.

Solution

We know that the answer has to be 5×10^{-3} A, since the current in Ampere's law must be the same inside the plates as it is outside the plates. We will nevertheless perform the calculation to show that this is true. The electric field between the plates is V/d, where V is the potential difference across the plates, and d is the distance between the plates. But $V = Q/C$, and therefore $E = Q/Cd$. Now $C = \varepsilon_0 A/d$, giving $E = Qd/A\varepsilon_0 d = Q/A\varepsilon_0$. The displacement current is $\varepsilon_0 A\Delta E/\Delta t = \varepsilon_0 A(\Delta Q/\Delta t)/\varepsilon_0 A = \Delta Q/\Delta t = I = 5 \times 10^{-3}$ A.

Problem 12.19. A lightning bolt produces a flash of light and associated peal of thunder. The light is an electromagnetic radiation traveling at the speed of electromagnetic waves, while the thunder is a sound wave traveling with a speed of 345 m/s. If an observer hears the thunder 7.5 s after he sees the lightning, how far away did the lightning strike?

Solution

If the distance is called D, then the time it takes the lightning to reach the observer is $D/3.0 \times 10^8$, and the time for the thunder is $D/345$. The difference in time is 7.5 s, and equals $D[1/345 - 1/3.0 \times 10^8] = D/345$, since $3.0 \times 10^8 \gg 345$. Thus $7.5 = D/345$, $D = 2.6 \times 10^3$ m.

Note. The time for the light to travel this distance is $2.6 \times 10^3/3.0 \times 10^8 = 8.7 \times 10^{-6}$ s, confirming our assumption that the full 7.5 s represented the time for the sound wave to reach the observer.

Problem 12.20. How far does light travel in one year?

Solution

The distance is $ct = 3.0 \times 10^8 \times [365 \times 24 \times 60 \times 60] = (3.0 \times 10^8 \text{ m/s})(31.536 \times 10^6 \text{ s}) = 8.95 \times 10^{31}$ m $= 9.46 \times 10^{15}$ km. This distance is called a "**light-year**".

Problem 12.21. A powerful laser produces an electromagnetic plane wave. The power given to the wave is 10 MW, and the light beam is confined to an area of 2 mm^2. What is the intensity of the laser beam?

Solution

The intensity of any electromagnetic wave is the power passing unit area. Thus, the intensity of this beam is $(1.0 \times 10^7 \text{ W})/(2 \times 10^{-6} \text{ m}^2) = 5 \times 10^{12} \text{ W/m}^2$.

Problem 12.22. Sun light above the earth's atmosphere has an average intensity of approximately 1.4 kW/m^2.

(*a*) What is the maximum electric and magnetic field in this wave?

(*b*) What is the maximum force exerted by these fields on an electron moving with a velocity of 10^6 m/s in this sunlight?

Solution

(*a*) The intensity is given by $I = c\varepsilon_0 E_0{}^2/2 = 1.4 \times 10^3 \text{ W/m}^2$. Thus $E_0 = [2(1.4 \times 10^3)/(3 \times 10^8)(8.85 \times 10^{-12})]^{1/2} = 1.03 \times 10^3$ V/m. The magnetic field is $B_0 = E_0/c = 3.4 \times 10^{-6}$ T.

(*b*) The maximum force (magnitude) exerted by the electric field on the electron is $eE_0 = (1.6 \times 10^{-19} \text{ C})(1.03 \times 10^3 \text{ V/m}) = 1.6 \times 10^{-16}$ N. The maximum magnetic force is $evB_0 = 1.6 \times 10^{-19} (10^6)(3.4 \times 10^{-6}) = 5.4 \times 10^{-19} \text{ N} \ll$ electric force.

Note. For the magnetic force on the electron to be the same as the electric force the electron would have to be traveling at exactly the speed of light—which is not possible according to the theory of relativity.

Supplementary Problems

Problem 12.23. The displacement current through an area of 5×10^{-4} m^2 is 3 mA. What is the rate at which the electric field is changing in this region?

Ans. 6.78×10^{11} V/m · s

Problem 12.24. How long does it take for light to travel to (*a*) the moon; (*b*) the sun; and (*c*) a star at a distance of 3 light years?

Ans. (*a*) 1.27 s; (*b*) 497 s = 8.3 min; (*c*) 3 years

Problem 12.25. In air, light travels with a velocity that is 0.03% smaller than in vacuum. What is the difference in time that it takes for light to travel 10^3 m in air and in vacuum?

Ans. 1.00×10^{-9} s

Problem 12.26. Water has a dielectric constant of 1.77 at the frequencies of visible light, and essentially no magnetic properties. What is the velocity of light in water?

Ans. 2.25×10^8 m/s

Problem 12.27. An electromagnetic wave is traveling in the $+y$ direction with its electric field in the $+x$ direction. What is the direction of the magnetic field?

Ans. $-z$ direction

Problem 12.28. What is the frequency of an electromagnetic wave whose wavelength equals the diameter of the earth?

Ans. 23.4 Hz

Problem 12.29. An electromagnetic wave is given by $E = 20 \cos 2\pi(ft - 3.3 \times 10^{-7}x)$ in standard units.

(*a*) What is the amplitude of the wave?

(*b*) What is the wavelength of the wave?

(*c*) What is the frequency of the wave?

Ans. (*a*) 20 V/m; (*b*) 3.03×10^6 m; (*c*) 99 Hz

Problem 12.30. Sunlight near the surface of the earth has an intensity of 1.1×10^3 W/m^2. A lens, of diameter 6 cm, concentrates the sunlight it collects onto a circle of diameter of 1.5 mm, as in Fig. 12-8. What is the intensity of the light at this small circle?

Ans. 1.76×10^6 W/m^2

Problem 12.31. A 100 watt light bulb is 25% efficient in converting electrical energy into light. What is the intensity of the light from the bulb at a distance of 2 m?

Ans. 0.50 W/m^2

Problem 12.32. Sunlight has an intensity of 1.4×10^3 W/m^2. The light is totally reflected from a surface of area 6.5 m^2.

(*a*) What is the force on the surface?

(*b*) What is the radiation pressure on the surface?

Ans. (*a*) 6.07×10^{-5} N; (*b*) 9.33×10^{-6}

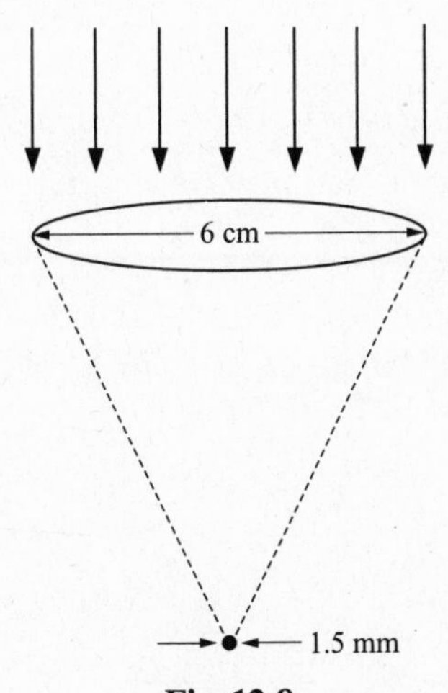

Fig. 12-8

Problem 12.33. An electromagnetic wave has a frequency of 3×10^6 Hz, and an intensity of 1.8 mW/m^2.

(*a*) What is the amplitude of the wave?

(*b*) What is the wavelength of the wave?

(*c*) What is the radiation pressure of the wave on a totally absorbing surface?

Ans. (*a*) 1.16 V/m; (*b*) 100 m; (*c*) 6×10^{-12} kg · m/s

Chapter 13

Light and Optical Phenomena

13.1 INTRODUCTION

In the previous chapter we learned that light was one form of electromagnetic wave. As such it has a wavelength and frequency whose product equals the speed of propagation for all electromagnetic waves. This speed was predicted and verified to be $c = 2.9979 \times 10^8$ m/s in vacuum (it is known to many more significant figures, but in our calculations we will generally use the value of 3.00×10^8 m/s). In a dielectric, the speed of light differs from its value in vacuum, giving rise to some of the phenomena we discuss in this and the next chapter. We will assume that the speed of light in vacuum equals c in our frame of reference, and defer to Chap. 15 the question of what speed light would have in a frame of reference moving relative to us. The wavelength of visible light ranges roughly from 400 nm (violet) to 800 nm (red).

Discussions of optical phenomena are usually divided into two areas, "geometrical optics" and "physical optics". In **physical optics** we treat the phenomena that arise due to the wave nature of the light, insofar as they produce the type of interference already discussed briefly in Chaps. 1 and 2, as well as other similar interference phenomena. We will discuss those effects in Chap. 14. In Chaps. 12 and 13 we will discuss **geometrical optics**, the phenomena that arise when light can be considered to be adequately described by rays traveling in straight lines (perpendicular to the wave fronts) that change speed in moving from one medium to another. This is the case as long as the objects through which the light travels have dimensions that are much larger than the wavelength of the wave.

In Sec. 12.5 of the previous chapter we showed that in three dimensions waves can be spherical (near a point source) or cylindrical (near a line source) or planar (far away from the source). These descriptions refer to the geometrical shape of the wave front surfaces in space at any instant, i.e. the surface over which the electric and magnetic field displacements are everywhere in phase. For plane waves, these wave front surfaces are planes perpendicular to the direction of travel of the wave. The geometry is illustrated in Fig. 13-1. The distance between adjacent wave fronts is the wavelength, and has been exaggerated in size in the figure for clarity (typically $\lambda \ll$ other human scale dimensions). As long as this wave continues to travel in the same medium, there will be no change in this pattern. However, if the wave encounters another medium, in which the speed is different, there will obviously be changes in the pattern. These changes are the subject of the remainder of this chapter.

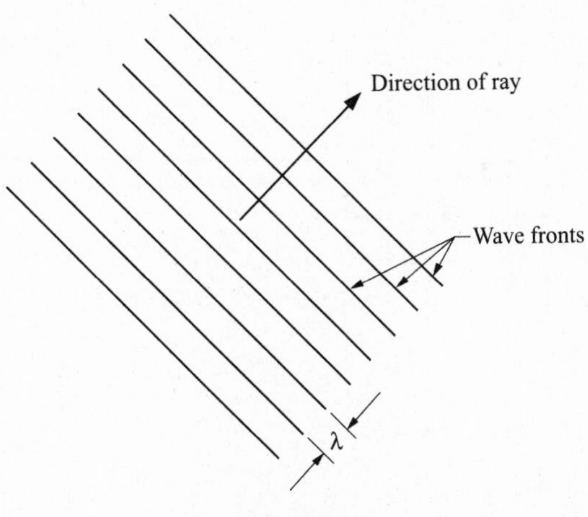

Fig. 13-1

13.2 REFLECTION AND REFRACTION

In Chap. 1 we discussed what happens when a wave on a string encounters the end of the string. We found that the wave is reflected back along the string, either with a change in phase or without such a change. In that case, the wave had no opportunity to be transmitted past the boundary, since the string ended at that point. Furthermore, the string was one-dimensional, allowing for travel only along the axis of the string. We also considered the possibility of a wave on a string encountering a junction where the speed of the wave changes. In that case, we stated that the wave will be partially transmitted and partially reflected. The same general ideas are applicable for the case of a wave of light, with the additional generalization that the light waves can travel in three dimensions and are not restricted to one-dimensional travel. This was already discussed in Chap. 2 in the context of sound waves and we will repeat it here in greater detail in the specific context of light waves. We will first discuss the case which is most nearly identical to the one-dimensional case of a wave on a string, namely the case of "normal incidence".

Suppose we have a planar boundary between two materials in which light has a different speed, as in Fig. 13-2. We define a quantity called the "**index of refraction**", n, in terms of the velocity of light in the material, v, relative to its velocity, c, in a vacuum:

$$n = c/v \tag{13.1}$$

In Table 13.1 we list the index of refraction of some typical materials.

Table 13.1 Indices of Refraction

Material	Index of refraction, n
Vacuum	1.000,000,0
Air	1.000,29
Water	1.33
Alcohol	1.36
Glass	1.4–1.6
Diamond	2.42

In the figure, the incident wave is in vacuum, with a wavelength λ, and traveling in the direction of the normal to the surface. The reflected wave also travels in the direction normal to the surface, but

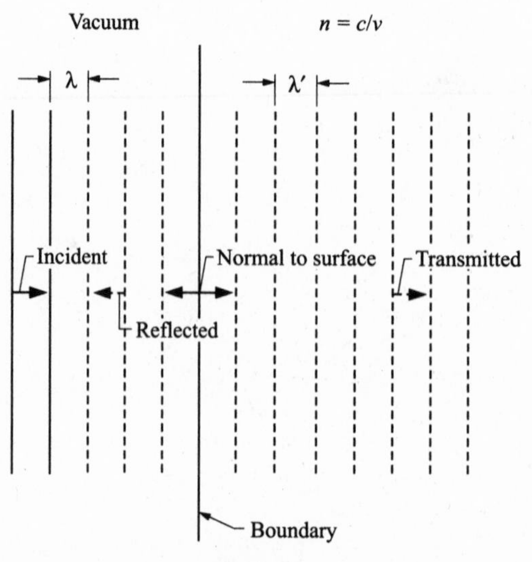

Fig. 13-2

away from the surface, and has the same wavelength as the incident wave. Similarly the transmitted wave also travels in the direction normal to the surface, but with a different speed. Since in general $v = \lambda f = \lambda/T$, it follows that if v changes λ must change and our transmitted light wave will have a different wavelength, λ'. The frequencies of the transmitted and reflected waves are the same as that of the incident waves since the rate of oscillation in the disturbance is precisely what is propagated from one location to the next. Since $T = 1/f$ is constant we can calculate the new wavelength as the distance traveled during one period, or

$$\lambda' = vT = v\lambda/c = \lambda/n \tag{13.2a}$$

We will show later that the speed of light in a material (or the speed of anything carrying mass or energy) is less than c, and therefore $n > 1$. Then $\lambda' < \lambda$, as we have drawn in the figure. The relative intensity of the reflected and transmitted light can also be calculated in terms of the indices of refraction, by using Maxwell's equations, but we will not concern ourselves with that calculation at this time. Although Fig. 13-2 is for a case in which the incident light was in a vacuum, this is not necessary. The same phenomena would occur if the incident material had index of refraction n_1 and the second material had index of refraction n_2. In that case we would have $v_2 = c/n_2$ and $v_1 = c/n_1$, so $\lambda_2 = v_2 T = v_2\lambda_1/v_1$ and

$$\lambda_2 = \lambda_1(n_1/n_2) \tag{13.2b}$$

The relative intensities would also change, but not the basic ideas.

We now turn our attention to the case of a wave which is incident on the planar surface bounding the two regions, but not at normal incidence. Since the direction of propagation is always perpendicular to the wave front the orientation of the wave is completely determined by this direction. It is therefore convenient to draw a directed line segment along the direction of propagation called the "light ray" or just ray, and determine what happens to the ray at reflection and refraction. In Fig. 13-3, we show an incident wave represented by a ray whose direction of travel is at an angle θ_1 to the normal to the surface. The plane defined by the incident ray and the normal to the surface at the point of intersection with the ray is called the plane of incidence, which we have drawn as the plane of the figure. The incident ray is in a material with an index of refraction n_1, so that its wavelength is $\lambda_1 = \lambda/n_1$, where λ is the wavelength in vacuum. The angle θ_1 between the ray direction and the direction of the normal is called the **angle of incidence**. When the incident wave impinges on the surface it is partially reflected and partially transmitted. We have drawn the reflected ray as traveling in a direction making an angle θ_r to the direction of the normal, and the transmitted ray (called the refracted ray) traveling in a direction making an angle θ_2 to the normal. The incident ray, reflected ray, transmitted ray and normal to the surface as drawn in the figure all lie in the same plane, so the plane of reflection and plane of refraction are the same as the plane of incidence. The wavelength of the reflected ray is clearly λ_1 since it is traveling in the same material as the incident ray. The refracted ray, however, has a different wavelength, $\lambda_2 = \lambda/n_2$, since it is in a different material with an index of refraction n_2. We have chosen $n_2 > n_1$ in this figure, and therefore $\lambda_2 < \lambda_1$, as drawn in the figure. We will use this figure to show that the **angle of reflection** equals the angle of incidence, i.e.:

$$\theta_r = \theta_1 \tag{13.3}$$

and that the **angle of refraction** is given by "**Snell's law**":

$$n_1 \sin\theta_1 = n_2 \sin\theta_2 \tag{13.4}$$

It is important to note, as can be seen from the geometry in the figure, that the angle of incidence, reflection and refraction also represent the angles that the wave fronts of the incident, reflected and refracted waves respectively make with the surface.

Problem 13.1. Use Fig. 13-3 to show that the angle of reflection equals the angle of incidence.

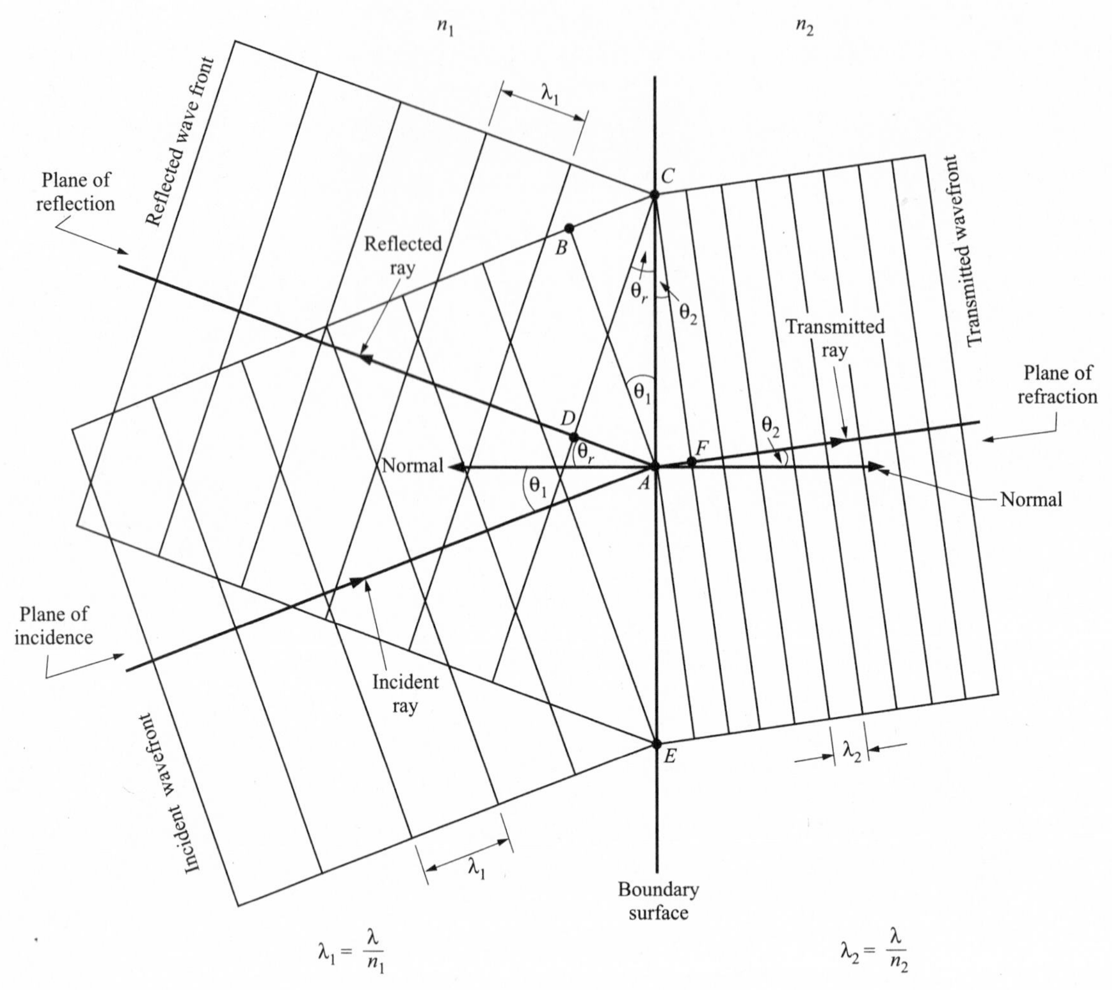

Fig. 13-3

Solution

The distance between adjacent wave fronts is the wavelength and is the distance traveled during one period. The width of the incident ray has been chosen so that when point A of the wavefront AB of the incident wave reaches the surface, point B is still one wavelength away from the surface. During the time of one period that it takes for this wavefront to reach the surface at C, the part of the wavefront that was reflected at A has traveled one wavelength to D. The wavefront of the reflected ray is now at CD. Both AD and BC are equal to one wavelength, and therefore equal to each other. Since right triangles ABC and CDA have a common hypotenuse, AC, as well as one equal side $AD = BC$, the other side is also equal by the Pythagorean theorem ($AB = CD$) and the two triangles are congruent. Therefore, $\theta_1 = \theta_r$.

Problem 13.2. Use Fig. 13-3 to show that $n_1 \sin\theta_1 = n_2 \sin\theta_2$ (Snell's law).

Solution

As in the previous problem, we note that when wavefront AB of the incident wave reaches the surface at A, point B is still one wavelength away from the surface at C, so $BC = \lambda_1$. During the time that it takes for this part of the wavefront to reach C, the refracted wavefront has traveled one wavelength from A to F in material 2. Thus, the distance AF will equal $\lambda_2 = \lambda_1(n_1/n_2)$ from Eq. (*13.2b*). In right triangle AFC the angle ACF equals θ_2. Then $\sin\theta_2 = AF/AC = \lambda_1(n_1/n_2)/(AC)$. Similarly, in right triangle ABC, the angle BAC equals θ_1, and $\sin\theta_1 = BC/AC = \lambda_1/(AC)$. Dividing $\sin\theta_1$ by $\sin\theta_2$ we get: $\sin\theta_1/\sin\theta_2 = \lambda_1/[\lambda_1(n_1/n_2)] = n_2/n_1$ or $n_1 \sin\theta_1 = n_2 \sin\theta_2$.

In the figure we have assumed that $n_1 < n_2$, and therefore $\lambda_1 > \lambda_2$ and $\theta_1 > \theta_2$. The derivation of Snell's law could just as easily have been deduced from a figure in which $n_1 < n_2$, and therefore $\lambda_1 < \lambda_2$. For the case of $n_1 < n_2$ which we used, the refracted ray was bent toward the normal ($\theta_1 > \theta_2$). If $n_1 > n_2$, then the refracted ray would be bent away from the normal and $\theta_1 < \theta_2$. The reflected ray is not dependent on the relative sizes of the indices of refraction, as is evident from Eq. (*13.3*) as derived in Problem 13.1.

In deriving the relationship for the angles of reflection and refraction we used a figure in which the wave fronts were clearly depicted. In general, we do not need to draw the figures in such detail, and we usually draw only the direction of the incident, reflected and refracted rays, as shown in Fig. 13-4(*a*) for $n_1 < n_2$ and in Fig. 13-4(*b*) for $n_1 > n_2$.

Problem 13.3. Light is incident from air to glass ($n = 1.51$). The light has a wavelength of 541 nm (1 nm $= 1.0 \times 10^{-9}$ m) in air.

(*a*) What is the wavelength of the light in the glass?

(*b*) If the angle of incidence is 37°, what are the angles of reflection and of refraction?

(*c*) If the angle of incidence is 85°, what are the angles of reflection and of refraction?

Solution

(*a*) The wavelength in the glass will equal $\lambda_2 = \lambda/n_2 = (541 \text{ nm})/1.51 = 358$ nm, since the wavelength in air is essentially the same as the wavelength in vacuum ($n \approx 1$).

(*b*) The angle of reflection equals the angle of incidence, so $\theta_r = 37°$. The angle of refraction can be calculated using Snell's law, $n_1 \sin \theta_1 = n_2 \sin \theta_2 \rightarrow (1) \sin 37° = (1.51) \sin \theta_2 \rightarrow \sin \theta_2 = 0.399 \rightarrow \theta_2 = 23.5°$.

(*c*) The angle of reflection equals the angle of incidence, so $\theta_r = 85°$. The angle of refraction can be calculated using Snell's law, $n_1 \sin \theta_1 = n_2 \sin \theta_2 \rightarrow (1) \sin 85° = (1.51) \sin \theta_2 \rightarrow \sin \theta_2 = 0.0.660 \rightarrow \theta_2 = 41.3°$.

We see in the last problem that, since $n_1 < n_2$, the angle of refraction is less than the angle of incidence. This can be seen in general from Snell's law [Eq. (*13.4*)] which implies that $n_1 < n_2 \rightarrow \sin \theta_2 < \sin \theta_1$. Thus, for any angle of incidence ($0 \leq \theta_1 \leq 90°$), $0 \leq \theta_2 \leq 90°$ as well. However, if $n_2 < n_1$, then Snell's law implies $\sin \theta_2 > \sin \theta_1$, and for certain angles of θ_1 this will require $\sin \theta_2 > 1$ and there will be no solution for θ_2. This is investigated in the next problem.

Problem 13.4. Light is incident from glass ($n = 1.51$) to air ($n = 1$).

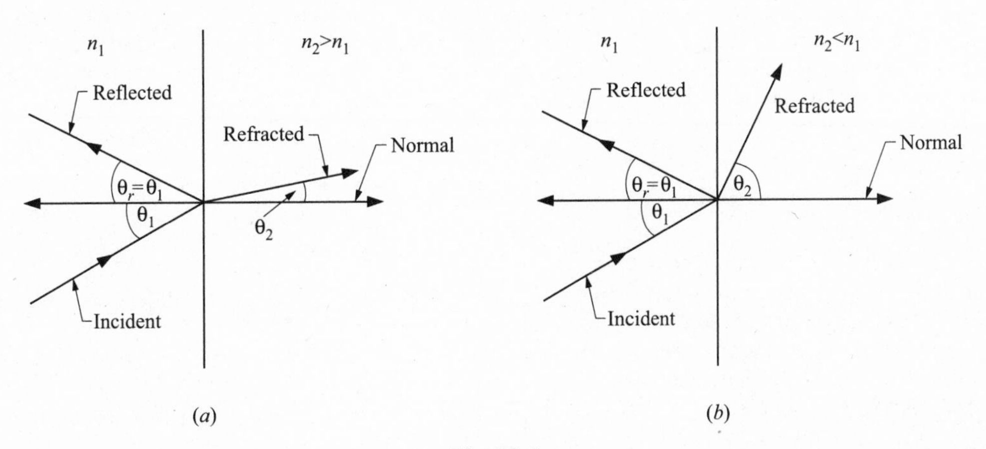

Fig. 13-4

(*a*) If the angle of incidence is 37°, what are the angles of reflection and of refraction?

(*b*) If the angle of incidence is 85°, is there a solution for the angle of refraction?

(*c*) What is the largest angle of incidence for which a solution exists for the angle of refraction?

(*d*) Derive a formula for the largest angle of incidence that produces a solution for θ_2, in terms of the indices of refraction.

Solution

(*a*) The angle of reflection is always equal to the angle of incidence, and therefore equals 37°. The angle of refraction is given by $n_1 \sin \theta_1 = n_2 \sin \theta_2 \rightarrow (1.51) \sin 37° = (1) \sin \theta_2 = 0.909 \rightarrow \theta_2 = 65.3°$.

(*b*) The angle of reflection is the same as the angle of incidence, and therefore equals 85°. When we try to calculate the angle of refraction we find that $(1.51) \sin 85° = \sin \theta_2 = 1.50$, which exceeds 1, and therefore there is no solution for θ_2.

(*c*) The largest value that we can have for $\sin \theta_2$ is 1. This occurs when $(1.51) \sin \theta_1 = 1$, or $\sin \theta_1 = 1/1.51 = 0.662$, $\theta_1 = 41.5°$. At this angle of incidence, θ_2 will equal 90°.

(*d*) The largest value that we can have for $\sin \theta_2$ is 1, which occurs when $\theta_2 = 90°$. At this angle of incidence, which is called the critical angle, θ_c, we have $n_1 \sin \theta_c = n_2(1)$, or

$$\sin \theta_c = n_2/n_1 \tag{13.5}$$

This formula for the critical angle shows that, as noted earlier, such an angle exists only for the case of $n_2 < n_1$, where $\sin \theta_c < 1$. When light is incident at an angle greater than the critical angle, no light is refracted so all the light must be reflected. We call this case one of "**total reflection**". Total reflection is very useful for bending light at a surface without losing any of the energy to transmission through the surface.

Problem 13.5. A transparent cylindrical bowl closed at both ends and sitting on its side is half filled with water ($n = 1.33$) with the other half containing air ($n = 1$), as in Fig. 13-5. A ray of light is directed to the center of the bowl from a source that can be rotated to point to the center from any angle

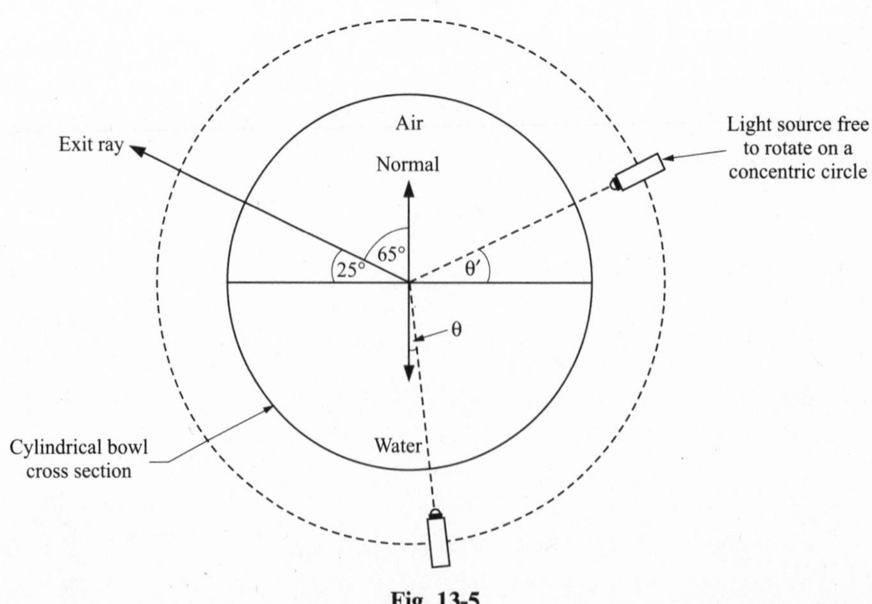

Fig. 13-5

desired, as shown in the figure. We want light to exit the bowl at an angle of 25° above the air–water surface, as shown.

(*a*) At what angle must one place the light source so that the light exits at this angle after reflection from the surface?

(*b*) At what angle must one place the light source so that the light exits at this angle after refraction from the surface?

(*c*) At what minimum angle with the vertical must one place the light source so that there is no refracted ray?

Solution

(*a*) The light must enter from the air side in order to exit as required after reflection. Since the angle of incidence equals the angle of reflection, the incident ray must make the same angle with the normal as the reflected ray. The normal direction is the vertical direction, and the angle of the reflected ray with the normal is therefore 65°. The incident ray must make the same angle with the normal, namely 65°, and the angle θ' is therefore 25°.

Note. For reflection, we could have equated the angles between the rays and the surface instead of the angles between the rays and the normal, and gotten the correct answer. However, for refraction we *must* use the angles between the rays and the normal in Snell's law to get the correct answer, and it is advisable to always use these angles, even for reflection.

(*b*) In this case the light is incident from the water onto the surface, and refracted into the air. The angle of incidence is labeled θ in the figure, and the angle of refraction is 65°. To obtain θ we use Snell's law, $(1.33) \sin \theta = (1) \sin 65° \rightarrow \sin \theta = 0.681 \rightarrow \theta = 43.0°$.

(*c*) The angle of incidence must equal the critical angle. There is a critical angle only for incidence from the water to the air, since we require that $n_1 > n_2$. For that case, we need $\sin \theta_c = n_2/n_1 = 1/1.33 = 0.752 \rightarrow \theta_c = 48.8°$.

Problem 13.6. Light is incident on a prism of glass ($n = 1.46$) at an angle ϕ with the surface, as shown in Fig. 13-6. The prism has base angles of 65°, and the ray is refracted so that it moves parallel to the base within the prism.

(*a*) Calculate the angle ϕ.

(*b*) When the light leaves the prism its direction of motion is at an angle δ from the incident direction. Calculate this "**angle of deviation**".

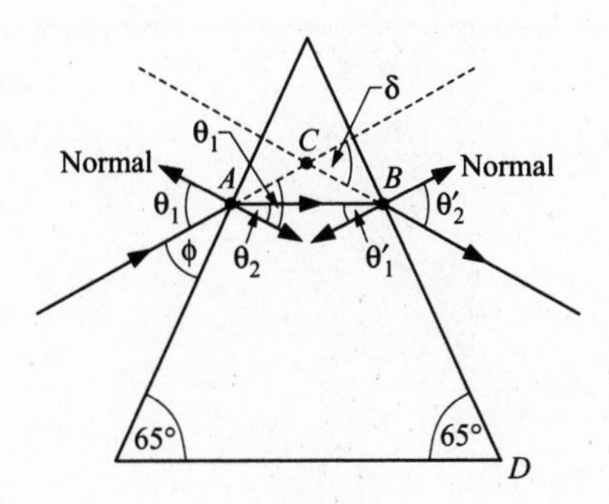

Fig. 13-6

(*c*) If the angle at D were equal to 60° (instead of 65°) all else being the same, what would the angle of deviation be?

Solution

(*a*) At point A we draw the normal to the surface. The angle of incidence, θ_1, will equal $90° - \phi$. Since the path of the light in the prism is parallel to the base, the angle of refraction, θ_2, will equal $90 - 65 = 25°$. Using Snell's law, we have: $(1) \sin \theta_1 = (1.46) \sin 25° = 0.617 \rightarrow \theta_1 = 38.1° \rightarrow \phi = 51.9°$.

(*b*) We now draw the normal at point B where the light impinges on the second surface of the prism. Since the path of the light is parallel to the base, the angle of incidence is $\theta'_1 = 25°$. Using Snell's law again, we have $(1.46) \sin 25 = (1) \sin \theta'_2$. Thus $\sin \theta'_2 = 0.617 \rightarrow \theta'_2 = 38.1°$. To get the angle of deviation we note that the angle $CAB = \theta_1 - \theta_2 = 13.1°$. Similarly, the angle CBA will equal $\theta'_2 - \theta'_1 = 13.1°$. The angle of deviation, δ, is just the sum of these two angles, so $\delta = 26.2°$.

(*c*) Changing the angle at D does not alter the refraction at A. Thus the angle ϕ is still equal to 51.9°, and the angle CAB is still 13.1°. But the angle of incidence at B, θ'_1, is now 30° (instead of 25°), and therefore $(1.46) \sin 30 = (1) \sin \theta'_2 = 0.73 \rightarrow \theta'_2 = 46.9°$. Now, angle $CBA = \theta'_2 - \theta'_1 = 16.9°$. Then $\delta = 13.1 + 16.9 = 30°$.

Problem 13.7. A source of light is located at a distance of 0.72 m below the surface of water ($n = 1.33$), as in Fig. 13-7. The light emerges at the surface after undergoing refraction at the surface. Show that all the emerging light is all within a circle at the surface, and calculate the radius of the circle.

Solution

Consider the ray SA from the source which impinges on the surface at A. It is refracted at this surface with an angle of refraction that is greater than the angle of incidence since the index of refraction of water is greater than that of air. If we choose rays with larger angles of incidence that impinge on the surface further from the point O which is vertically above the source, the emerging ray will be refracted further, until, at the critical angle of incidence, there will no longer be any emerging light. At that distance from O in all directions along the surface the light ceases to emerge. Thus the light is restricted to a region within a circle, whose radius is the distance OB from O at which the incident ray is at the critical angle. We note that $OB = (OS) \tan \theta_c$, since angle OSB equals θ_c. Thus the circle has a radius $r = OB = (0.72 \text{ m}) \tan \theta_c$, where $\sin \theta_c = n_2/n_1 = 1/1.33 \rightarrow \theta_c = 48.8°$, $\tan \theta_c = 1.14$, $r = 0.82$ m.

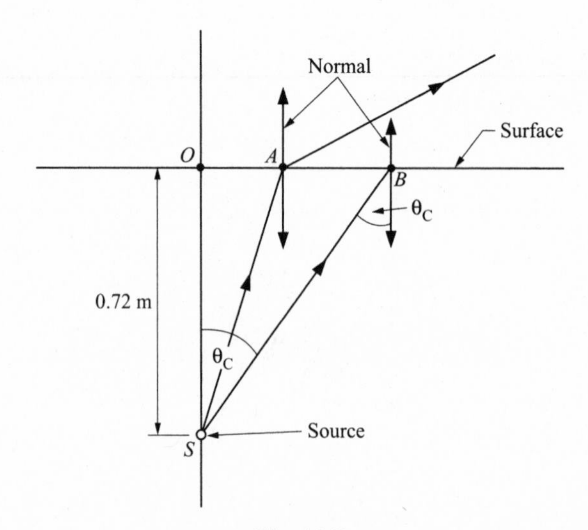

Fig. 13-7

Problem 13.8. A flat slab of glass, of thickness D, has an index of refraction of 1.41. Light is incident on the slab at an angle of θ, as shown in Fig. 13-8(a). Prove that the emerging ray is parallel to the incident ray, i.e. that $\theta'' = \theta$ and show how one can calculate the displacement, d, of the emerging ray from the path of the incident ray. Then carry out the calculation for $\theta = 18°$, and $D = 0.0192$ m.

Solution

The angle of refraction at the bottom surface is given by (1) sin $\theta = (1.41)$ sin θ'. Since the two sides of the slab are parallel, the angle of incidence at the top surface will equal the angle of refraction, θ', at the bottom surface. Therefore the angle of refraction at the top, θ'', is given by (1.41) sin $\theta' =$ (1) sin θ'' and from above this equals (1) sin θ. Thus $\theta'' = \theta$.

To calculate the value of the displacement, we note from Fig. 13-8(b), that $d = DE$. In turn $DE = DC \cos\theta$, and $DC = BC - BD$. Further, $BC = D \tan\theta$ and $BD = D \tan\theta'$. Thus $d = D(\tan\theta - \tan\theta')\cos\theta$. If we are given θ, we can calculate θ', and given D we can obtain the displacement d. For the numbers given, we get $\theta' = 12.66°$, $\tan\theta = 0.325$, $\tan\theta' = 0.225$, $\cos\theta = 0.951$ and finally $d = 1.83 \times 10^{-3}$ m.

13.3 DISPERSION AND COLOR

Thus far we have assumed that the index of refraction of a material is a constant which does not depend on the wavelength of the light that is incident. This is true only for vacuum, where the speed c is the same for all wavelengths and therefore n equals 1 for all wavelengths. However, for materials there is a small dependence of the velocity of light, and therefore n, on the wavelength. This property is called "**dispersion**" since it can be used to disperse the various wavelengths that are included in a beam of light into different refractive paths, creating a "**spectrum**".

In Table 13.2 we list some typical values showing the variation of n with wavelength in a particular glass.

Table 13.2 Variation of n with Wavelength

Color	Wavelength in vacuum, nm	Index of refraction, n
Red	660	1.520
Orange	610	1.522
Yellow	580	1.523
Green	550	1.526
Blue	470	1.531
Violet	410	1.538

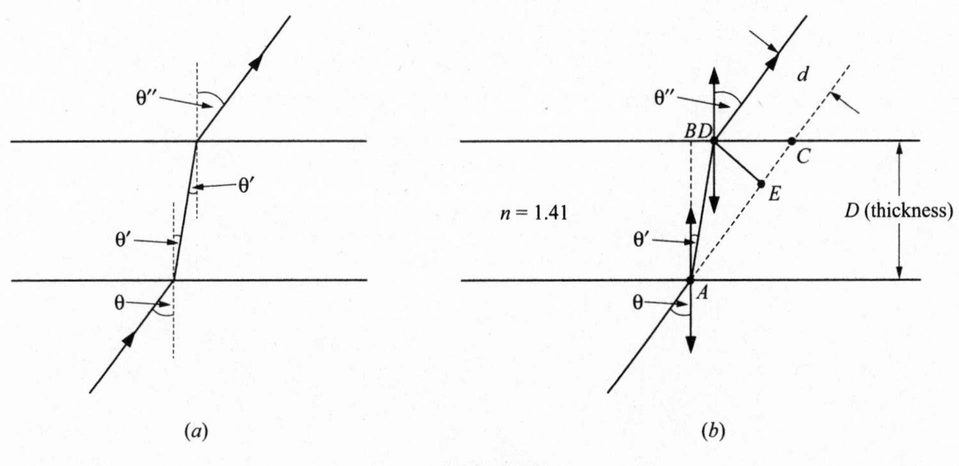

Fig. 13-8

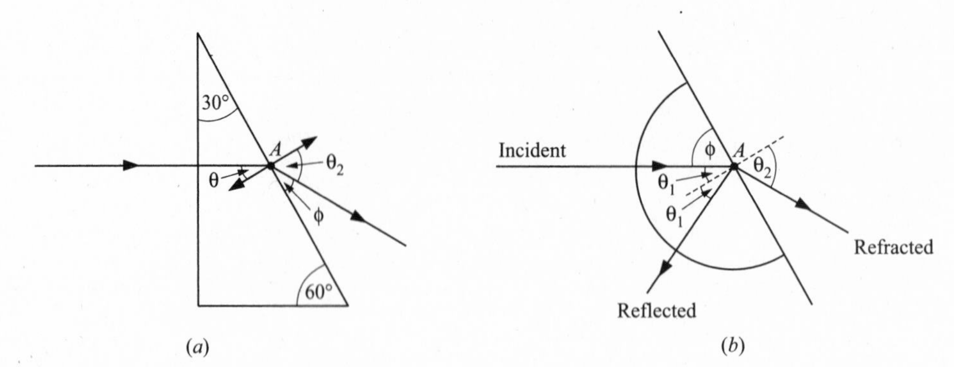

Fig. 13-9

As a result of the dispersion, white light refracted in a prism will be separated into its constituent wavelengths after passing through the prism. An example is given in the next problem.

Problem 13.9. A prism is made out of glass whose index of refraction varies with wavelength as given in Table 13.2. The light enters the prism normally, as shown in Fig. 13-9, and exits the prism at A at an angle ϕ. Calculate the value for ϕ at the wavelengths given in the table.

Solution

For the geometry of this prism the light is incident on the glass–air surface at A at an angle of incidence θ of 30°. The angle of refraction is given by: $n \sin 30 = \sin \theta_2$, where $\phi = 90 - \theta_2$. We use the values for n given in the table for the various wavelengths. At $\lambda = 660$ nm (red), we get $\sin \theta_2 = (1.520) \sin 30 = 0.760 \rightarrow \theta_2 = 49.46° \rightarrow \phi = 40.53°$. Doing the same at the other wavelengths, we find the values given in Table 13.3.

We note that the spectrum ranges over angles from 40.53 to 39.74, a spread of 0.79°, with the longest wavelengths deviated the least. It is generally the case that a prism deviates the wavelengths in this order, i.e. n increases as λ decreases.

One way to measure the variation of the index of refraction with wavelength accurately is to use the phenomenon of total reflection. This is illustrated in the next problem.

Problem 13.10. A prism in the form of a half circle is made of glass. The prism can be rotated about the symmetry axis of the full cylinder, as shown in cross-section in Fig. 13-10. Here point A is on the symmetry axis and is at the center of the circular cross-section. Light of a particular wavelength is incident at A after entering the prism shown. Because of the circular surface presented to the incident ray, light enters the prism at right angles for any orientation about point A. It then hits the planar surface at incident angle θ_1 and refracts at larger angle θ_2. Some light reflects at angle θ_1. The prism is rotated about A until $\theta_2 = 90°$, at which point the refracted ray disappears and the reflected ray has

Table 13.3 Variation of ϕ with angle

Color	Wavelength, nm	Index	ϕ
Red	660	1.520	40.53
Orange	610	1.522	40.45
Yellow	580	1.523	40.40
Green	550	1.526	40.27
Blue	470	1.531	40.05
Violet	410	1.538	39.74

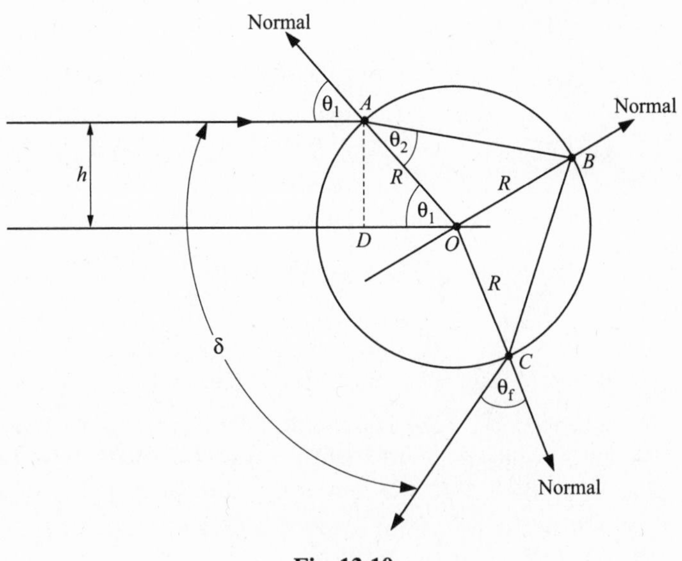

Fig. 13-10

maximum intensity, giving total reflection. Determine the index of refraction from the angle ϕ shown in the figure.

Solution

The angle of incidence at A is equal to $90 - \phi$. Thus $n \sin(90 - \phi) = (1) \sin \theta_2$, and $\theta_2 = 90°$ at total reflection. Thus $n \sin(90 - \phi) = n \cos \phi = 1$, giving $n = 1/\cos \phi$.

A **rainbow** is another case in which the variation of index of refraction with wavelength leads to a spectrum. The detailed analysis is quite complicated, and we will merely give one example which can show how this dispersion can lead to the effect. The rainbow is seen when light encounters a region of water drops and the light emerging from the drops is viewed from the general direction of the original light. In the next problem we investigate the angle at which the light returns in one possible such case.

Problem 13.11. A ray of light is traveling horizontally, and is incident on a spherical droplet of water at a height of h above the center, as shown in Fig. 13-10. The light is first refracted into the water at A, then reflected at B, and finally refracted into the air at C. We ultimately wish to calculate the angle δ, through which the light has been deviated.

(*a*) Calculate the angle θ_2 through which the light has been refracted at A, in terms of n, the radius of the droplet R, and the parallel displacement, h, of the incident ray from the symmetry axis of the droplet.

(*b*) Calculate the angle of incidence at B (angle ABO), and the angle of reflection (angle CBO), in terms of θ_2.

(*c*) Calculate the angle of incidence at C (angle BCO) and the angle of refraction θ_f, in terms of θ_1 and θ_2.

(*d*) Calculate the angle of deviation, δ, in terms of θ_1 and θ_2.

(*e*) For $n = 1.33$, and $h/R = 0.66$, calculate the angle δ.

(*f*) Does this angle of deviation increase or decrease if one increases the index of refraction slightly?

Solution

(*a*) Using Snell's law, (1) $\sin\theta_1 = n\sin\theta_2 = (1)h/R$ (from triangle *ADO*). Thus, $\sin\theta_2 = h/nR$.

(*b*) Triangle *ABO* is an isosceles triangle, since two sides are equal to radii of the circle. Thus, the base angles are equal, and the angle of incidence at *B* (angle *ABO*) equals θ_2. The angle of reflection (angle *CBO*) equals the angle of incidence, or θ_2.

(*c*) Triangle *OBC* is also an isosceles triangle, with two equal sides as radii of the circle. Therefore the base angles are equal. Thus the angle of incidence at *C* (angle *OCB*) equals θ_2. The angle of refraction, θ_f, is given by $n\sin\theta_2 = (1)\sin\theta_f$. But we know from part (*a*) that $n\sin\theta_2 = (1)\sin\theta_1$. Thus $\theta_f = \theta_1$.

(*d*) The angle *AOC* is composed of two equal angles each of which is the exterior angle to one of the isosceles triangles *ABO* or *CBO*. Those exterior angles are therefore each equal to $2\theta_2$ and the angle *AOC* equals $4\theta_2$. The angle δ equals this angle minus $2\theta_1$ (since $\theta_f = \theta_1$), so $\delta = 4\theta_2 - 2\theta_1$.

Note. Equivalently, angle $ABC = 2\theta_2$, δ is less than this because the incident and final refracted rays are each tilted around each other relative to lines *BA* and *BC*, respectively, by $(\theta_1 - \theta_2)$. Thus $\delta = 2\theta_2 - 2(\theta_1 - \theta_2) = 4\theta_2 - 2\theta_1$.

(*e*) For $n = 1.33$ and $h/R = 0.66$, we have $\sin\theta_1 = 0.66$ and $\sin\theta_2 = 0.66/1.33 = 0.496 \rightarrow \theta_1 = 41.3°$ and $\theta_2 = 29.8°$. Then $\delta = 4(29.8) - 2(41.3) = 36.6°$.

(*f*) If *n* increases slightly, θ_2 will decrease slightly. In that case, δ will also decrease slightly, and the exiting ray will be bent somewhat more upwards in the figure.

We see from this problem that if *n* varies with λ, then the different colors will exit in slightly different directions. The violet, with a somewhat larger *n*, will be bent upwards more than the red. If a person on the ground looking at the droplets of water is at the correct angle to see the red light, the violet light will be bent too much for him to see (see Fig. 13-11). However, droplets that are lower in the sky will be at the correct angle for the violet light, and at that height, he will see violet and not red. Thus, there will be a spectrum, with the red on top and the violet on the bottom producing a rainbow. The arc of the rainbow is part of a circle. To understand this "arc" effect we note that the sun's rays are essentially parallel over the whole region of sky that a person can see, since the sun is so far away. Thus all the droplets that can contribute red light to the eye of an observer do so along lines that make the same angle δ with the direction of the sun's rays, hence sweeping out a cone of half angle δ about the

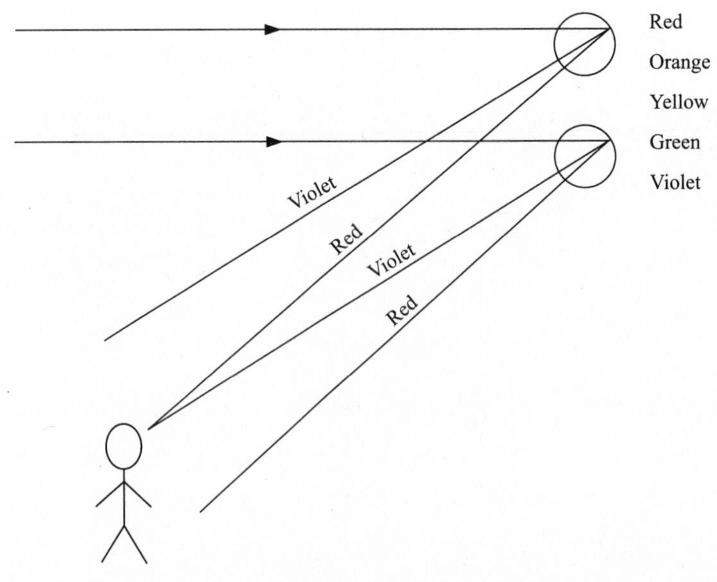

Fig. 13-11

sun's ray direction. The situation actually is more complicated, since we did not account for variations in h as light hits a single droplet, but the general idea is as has been discussed.

Problems for Review and Mind Stretching

Problem 13.12. Light is incident from air onto a flat boundary with material with index of refraction $n = 1.68$. The light is deviated through an angle of $\delta = 28.3°$, i.e. the angle between the incident ray direction and the refracted ray direction is 28.3°. What is the angle of incidence?

Solution

We know that $(1.00) \sin\theta_1 = (1.68) \sin\theta_2$. We are also given that $\theta_1 - \theta_2 = 28.3°$. Then $\theta_2 = \theta_1 - 28.3$, and $\sin\theta_1 = (1.68) \sin(\theta_1 - 28.3) = (1.68)[(\sin\theta_1)(\cos 28.3) - (\sin 28.3)(\cos\theta_1)] = 1.479 \sin\theta_1 - 0.796 \cos\theta_1$, where we have used the trigonometric identity for $\sin(A - B)$. Then $0.479 \sin\theta_1 = 0.796 \cos\theta_1 \rightarrow \tan\theta_1 = 1.66 \rightarrow \theta_1 = 59.0°$. We can check on this result by calculating θ_2. Since $\sin\theta_2 = (\sin 59.0)/1.68 = 0.510 \rightarrow \theta_2 = 30.7$, and $\delta = 59.0 - 30.7 = 28.3$ as required.

Problem 13.13. Light is incident from water ($n = 1.33$) onto glass ($n = 1.52$) at an angle of incidence of 30°. After traveling through the glass to the opposite parallel side, the light emerges into air ($n = 1.00$), as in Fig. 13-12.

(*a*) What is the angle of refraction, θ_2, in the glass at the first surface?

(*b*) What is the angle of refraction, θ_t, with which the light is transmitted into the air at the second surface?

(*c*) Light is also reflected back into the water at the first surface (see Fig. 13-12). Show that additional light also re-enters the water from the glass, and that the angle with which the rays enter the water is the same for all rays.

(*d*) Show that additional light is transmitted to the air from the glass, and that the angle with which the rays enter the air is the same for all the rays.

Solution

(*a*) The angle of incidence is 30°, and therefore the angle of refraction is given by $(1.33) \sin 30° = (1.52) \sin\theta_2 \rightarrow \sin\theta_2 = 0.4375 \rightarrow \theta_2 = 25.9°$.

(*b*) Since the surfaces are parallel, the angle of incidence at the second surface equals the angle of refraction at the first surface. Thus $(1.52) \sin 25.9° = (1.00) \sin\theta_t \rightarrow \sin\theta_t = 0.4375 \rightarrow \theta_t = 41.7°$.

(*c*) The light that is reflected at the first surface is reflected at the angle of incidence of 30°. The refracted light, at an angle of θ_2, after traveling through the slab, is partially reflected at surface 2, at the same angle θ_2. When this light reaches surface 1, part of it is reflected back into the glass, and part is transmitted into the water. The reflected part travels back to surface 2, where some of it is reflected back to surface 1. This process continues until the intensity is too small to consider further reflections. All the rays that return to surface 1 return at an angle of incidence equal to θ_2, since the back and forth reflections at surface 1 and at surface 2 all are at the same angle θ_2. When light is transmitted back into the water as a result of refraction from the glass, the angle of refraction into the water is given by $(1.52) \sin\theta_2 = (1.33) \sin\theta_r$, where θ_r equals the angle of refraction into the water. But we know from the initial refraction that $(1.33) \sin 30 = (1.52) \sin\theta_2$, so that $\theta_r = 30°$. Thus all the rays from the glass enter the water at the same angle of 30°.

(*d*) The rays that reach surface 2 after reflection from surface 1, as discussed in part (*c*) are incident at an angle of θ_2. Since all the rays are incident at this same angle, they are all partially refracted into the air at the same angle, θ_t.

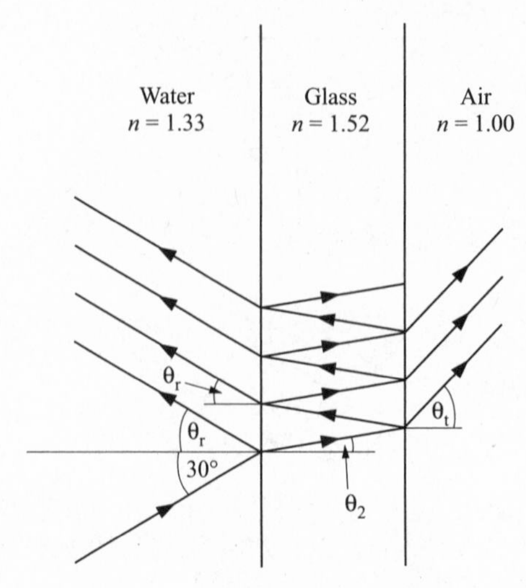

Fig. 13-12

Problem 13.14. Light is emitted from a source at a depth of D below the surface of water ($n = 1.33$). The light incident on the surface parallel to the normal emerges along the normal, while the light incident at a small angle to the normal is refracted at the surface. If one projects these two emerging rays backward, they meet at a point along the normal (see Fig. 13-13). The eye, when viewing the light from the source, projects these rays back automatically and thinks that the light came from the point of convergence. The distance of this point below the surface is called the "**apparent depth**" of the source.

(*a*) What is the angle of refraction for a ray incident on the surface at point B, at a small angle θ?

(*b*) What is the apparent depth that results from using a small enough angle that we can consider $\sin\theta \approx \tan\theta$?

(*c*) If a swimming pool is actually 5.4 m deep, how deep does it look to someone looking straight downward into the pool from the outside?

Solution

(*a*) The angle of refraction θ_2 is given by $\sin\theta_2 = (1.33)\sin\theta$.

(*b*) The distance, r, between the normal and point B, can be calculated from triangle SAB to equal $r = D\tan\theta$. The same distance can be calculated from triangle $S'AB$ to equal $r = d\tan\theta_2$. Then $D\tan\theta = d\tan\theta_2$. In our small angle approximation, $\tan\theta \approx \sin\theta$ and $\tan\theta_2 \approx \sin\theta_2$ so $D\sin\theta \approx d\sin\theta_2 = d(1.33)\sin\theta$. Thus $d = D/1.33$ (or, more generally, $d = D/n$).

(*c*) For $D = 5.4$ m, the apparent depth $d = (5.4\text{ m})/1.33 = 4.06$ m. This differs substantially from the actual depth.

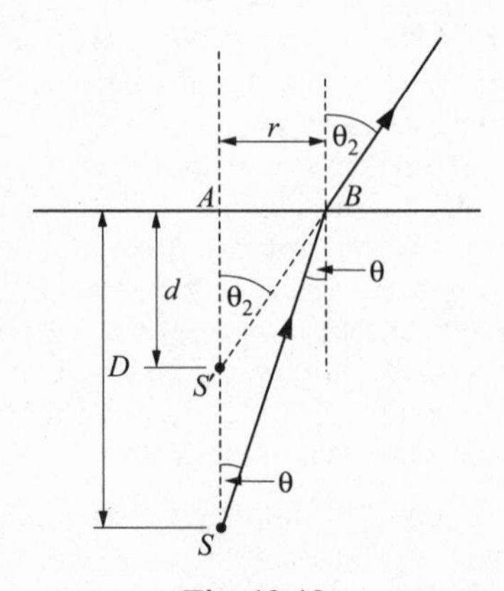

Fig. 13-13

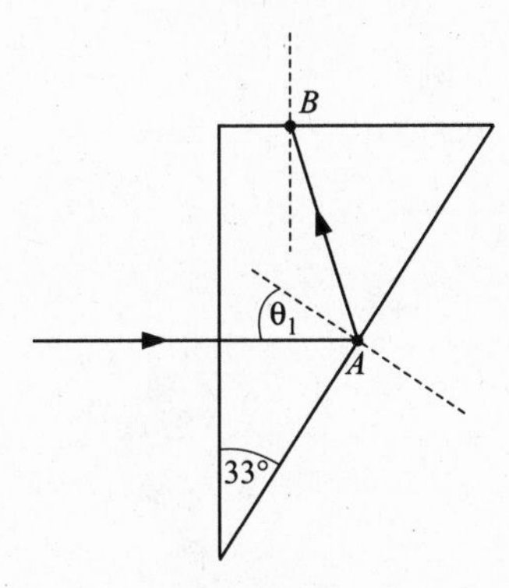

Fig. 13-14

Problem 13.15. A right angle prism is made of a material of index of refraction n. One of the base angles is 33° and light is incident on the prism normal to the side adjacent to this angle (see Fig. 13-14).

(*a*) What is the minimum value needed for n to get total reflection at A?

(*b*) For that n, will there also be total reflection at B?

(*c*) What is the minimum value of n to give total reflection at both A and B?

Solution

(*a*) The angle of incidence at A will equal the base angle, $\theta_1 = 33°$. To get total reflection this angle must be $\geq \theta_c$, where $\sin \theta_c = 1/n$. The minimum value needed for n is therefore that value for which $n_{\min} = 1/\sin 33 = 1.84$. Thus, the critical angle is 33°.

(*b*) The angle of reflection at A is also 33°, so from triangle BAD we see that the angle of incidence at B is $90 - 2(33) = 24°$. This is less than the critical angle for this value of n, and therefore there will not be total reflection at the point B.

(*c*) To get total reflection at B we must have n large enough so that the critical angle equals 24°. This requires $n = 1/\sin 24° = 2.46$. The angles of reflection, of, course, do not change as a consequence of changing n.

Problem 13.16. Light is incident on an isosceles prism with index of refraction n, and apex angle A. The angle of incidence is such that the light travels in the prism parallel to the base, as in Fig. 13-15. The light exits at B, and the angle of deviation, δ, is the angle between the direction of this ray and the direction of the incident ray.

(*a*) Show that the angle of incidence at D is the same as the angle of refraction at B, and that the angle of refraction at D is the same as the angle of incidence at B.

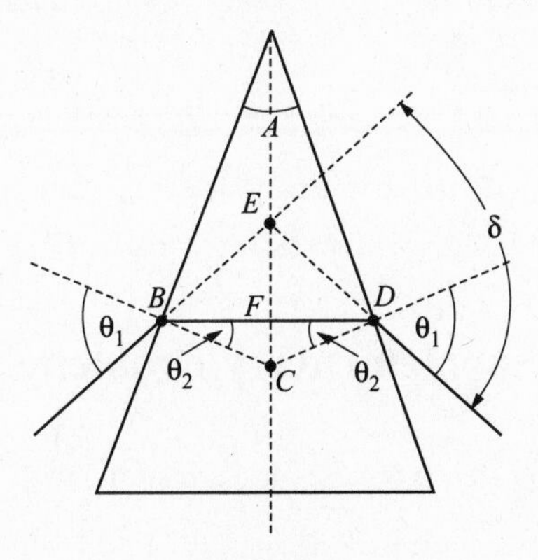

Fig. 13-15

(*b*) Show that, at B, the angle of refraction is, $\theta_2 = A/2$.

(*c*) Show that, at B, the angle of incidence is, $\theta_1 = (A + \delta)/2$.

(*d*) Show that the angle of deviation is given by $\sin[(A + \delta)/2] = n \sin[A/2]$.

Solution

(*a*) We already showed in Problem 13.6 that for the case of light in the prism parallel to the base of the prism, the angles BDC and DBC are equal. Thus both equal θ_2, the angle of refraction at B. The angle of incidence at B, θ_1, is given by $\sin \theta_1 = n \sin \theta_2$, and the angle of refraction at D, θ_D, is also given by the same equation $n \sin \theta_2 = \sin \theta_D$. Therefore $\theta_D = \theta_1$. (We have already labeled this angle as θ_1 in the figure.)

(*b*) The angle BAC in the right triangle BAC equals $A/2$. Angle ACB is therefore the complement of $A/2$. But θ_2 is also the complement of angle ACB (triangle FCB), and is therefore equal to $A/2$.

(*c*) The angle of deviation, δ, equals twice the base angles in triangle BED, so each of those angles equals $\delta/2$. But θ_1 equals angle EBC, and therefore $\theta_1 = \theta_2 + \delta/2 = (A + \delta)/2$.

(*d*) Using Snell's law at B we get $\sin[(A + \delta)/2] = n \sin[A/2]$, as expected.

Note. One can show that the angle of deviation is a minimum for this case, i.e. for the case where the light in the prism moves parallel to the base.

Problem 13.17. In Problem 13.16 the apex angle of the prism is 60°. Light of various wavelengths is incident on the prism, and, for each wavelength, the angle of incidence is adjusted so that the light in the prism moves parallel to the base. If the index of refraction is as given in Table 13.2 for those wavelengths, calculate the angle of deviation for each wavelength.

Solution

For each wavelength we have $\sin[(A + \delta)/2] = n \sin(A/2)$, as shown in the previous problem. Since $A = 60°$, this equation becomes $\sin(30 + \delta/2) = n \sin 30 = 0.50\, n$. For red light of wavelength 660 nm, this becomes $\sin(30 + \delta/2) = 0.50(1.520) = 0.760 \rightarrow (30 + \delta/2) = 49.46°$, $\delta = 38.9°$. We do the same calculation using the appropriate values for n for the other wavelengths. The results are given in Table 13.4.

Table 13.4 Calculating the Angle of Deviation

Color	Wavelength, nm	Index	δ
Red	660	1.520	38.9
Orange	610	1.522	39.1
Yellow	580	1.523	39.2
Green	550	1.526	39.5
Blue	470	1.531	39.9
Violet	410	1.538	40.5

Supplementary Problems

Problem 13.18. Light is incident from water ($n = 1.33$) onto a material with $n = 1.51$. The angle of incidence is 25°. What is the angle of refraction?

Ans. 21.9°

Problem 13.19. Light is incident from a material with $n = 1.51$ at an angle of 25° onto another material. The light is refracted at an angle of 38.6°. What is the index of refraction of the second material?

Ans. 1.02

Problem 13.20. Light in a material with an index of refraction of 1.57 is refracted into air, at an angle of refraction of 56°.

(*a*) What is the angle of incidence?

(*b*) By what angle is the incident ray deviated, i.e. what is the angle between the path of the incident and the refracted rays?

Ans. (*a*) 31.9°; (*b*) 24.1°

Problem 13.21. Light is incident from water ($n = 1.33$) onto air, and the ray is thereby deviated by 19.7°. What is the angle of incidence?

Ans. 40.9°

Problem 13.22. Light is incident on a right angle prism with one base angle equal to 53°. The prism has an index of refraction of 1.48. If the light is incident parallel to the base, as in Fig. 13–16 (*a*) what is the angle of refraction at the first surface; (*b*) what is the angle of refraction at the second surface; and (*c*) by what angle is the incident ray deviated, i.e. what is the angle between the path of the incident and the final refracted rays?

Ans. (*a*) 24.0°; (*b*) 19.4°; (*c*) 32.4°

Problem 13.23. For the same prism as in the previous problem, the light has an angle of incidence θ_1 at the first surface and then angle θ_1' at the second surface.

(*a*) What minimum angle of incidence is needed at the second surface in order to get total reflection?

(*b*) At that angle of incidence at the second surface, what is the angle of refraction at the first surface?

(*c*) At that angle of incidence at the second surface, what is the direction of incidence at the first surface?

Ans. (*a*) 42.5°; (*b*) 5.5°; (*c*) 8.16° above the normal

Problem 13.24. A fisherman sees a fish beneath him in a river at an apparent depth of 1.93 m. At what depth should he put his net?

Ans. 2.57 m

Problem 13.25. Light is incident on a glass surface at an angle of incidence of θ.

(*a*) If the index of refraction of the glass is 1.62, at what angle of incidence will the light have an angle of refraction of half that angle?

(*b*) If the angle of incidence is 56°, what index of refraction is needed to get an angle of refraction of half that size?

Ans. (*a*) 71.8°; (*b*) 1.77

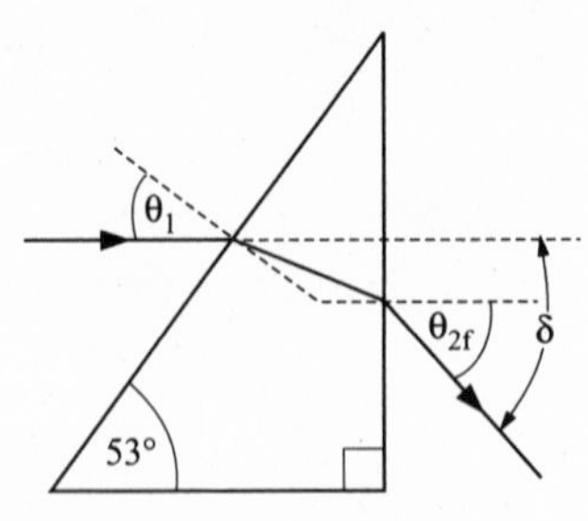

Fig. 13-16

Problem 13.26. Light is reflected from a plane surface, from an angle of incidence of 34.1°. The angle of deviation of the reflected light is the angle between the direction of the incident and the reflected rays.

(*a*) What is the angle of deviation?

(*b*) If the plane is rotated so that the angle of incidence increases by 11.4°, by what angle does the angle of deviation decrease?

Ans. (*a*) 111.8°; (*b*) 22.8°

Problem 13.27. Light is refracted into a 45–45–90 prism, as shown in Figs 13–17. The light is incident at an angle of 45° at a point where the width of the prism, h, is 6.2 cm. The light reaches the back surface at a distance y below the projection of the original ray on the back surface, as in the figure. From the figure one can see that the distance y is given by $y = h \tan(\theta_1 - \theta_2)$. The index of refraction depends on wavelength as given in Table 13.2. For the two extreme wavelengths in the table, red and violet, calculate (*a*) the angle of refraction; (*b*) the distance y; and (*c*) the vertical spread of the spectrum on the back surface.

Ans. (*a*) 27.72° and 27.37°; (*b*) 19.29 mm and 19.70 mm; (*c*) 0.41 mm

Problem 13.28. Light is incident on a glass surface at an angle of 60° to the normal. What must be the index of refraction so that the angle of deviation between the reflected and refracted rays is 90°?

Ans. 1.73

Problem 13.29. Light is incident from air onto a composite material made of a stack of three flat slabs of equal thickness d and indices of refraction of 1.3, 1.6 and 1.4. If the angle of incidence is 30°, find (*a*) the angle of refraction at the first interface surface; at the second interface surface; at the third interface surface; and (*b*) the angle of refraction into the air at the other side of the composite.

Ans. (*a*) 22.6°; 18.2°; 20.9°; (b) 30°

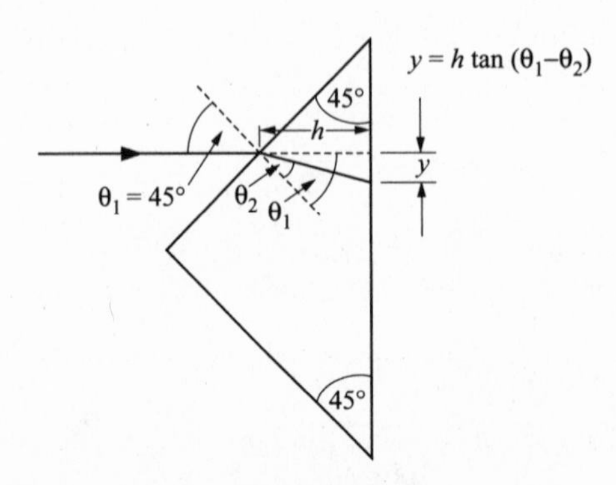

Fig. 13-17

Problem 13.30. Referring to Problem 13.29 what are the angles of reflection at the first surface; the second surface; the third surface; the fourth surface.

Ans. 30°; 22.6°; 18.2°; 20.9°

Problem 13.31. In Problem 13.30, consider the ray reflected off the back surface (fourth interface).

(*a*) What are the angles of refraction of this "back" ray from the third surface, second surface and first surface?

(*b*) What are the angles of deviation between the forward rays of Problem 13.29 and the backward rays of part (*a*), in each of the three slabs?

Ans. (*a*) 18.2°, 22.6° and 30°, respectively; (*b*) starting from the "back" slab: 138.2°, 143.6°, 134.8°

Problem 13.32. Referring to Problems 13.29 and 13.31, what is the lateral distance between the point of entry of the original ray through the first interface and the point of emergence of the back ray through the interface? Assume $d = 0.50$ cm.

Ans. 1.13 cm

Problem 13.33. Two plane mirrors are placed at right angles to each other so that light reflecting off one surface (over some range of angles) will have a second reflection off the second surface. For such a ray, having angle of incidence θ with one surface, find the angle of deviation between the incoming ray and the second reflected ray.

Ans. 180° (for all such rays)

Problem 13.34. For the prism of Problem 13.9 (Fig. 13-9) white light is incident normally and is dispersed by the prism into a spectrum of colors. Find the angle of deviation of the dispersed light across the spectrum—from red to violet (Table 13.2)—if the apex angle of the prism is 30° (the case shown in the figure); 35°; 40°.

Ans. 0.79; 1.23°; 3.65°

Chapter 14

Mirrors, Lenses and Optical Instruments

14.1 INTRODUCTION

Our discussion in the previous chapter showed how light is affected at the boundaries between materials of different index of refraction. In general, we saw that light is both reflected and refracted at such a surface. In this chapter we will show how these effects lead to instruments that can form images of objects, and how these images can be used in practice.

In order to understand the effects that we will describe, we have to first understand how the eye is able to take the light that reaches it from an object and to determine from that light where the object is located. Suppose that one has a point source of light located at a distance d from the eye, as in Fig. 14-1. The source sends out rays of light in all directions, some of which have been drawn in the figure. When these rays reach the eye, each of those that enter is moving in a slightly different direction. The eye has the physiological ability to trace these rays back to where they meet, and then assumes that the object is at that point of intersection. In the case shown in the figure, and in most other cases, this will give the true location of the source of light. However, if somehow the directions of the rays have been altered before they reach the eye, then the backward tracing that is done by the eye will result in an intersection of the rays at some different point, not at the actual source. This results in an "**image**" which is not at the position of the source, i.e. not at the "**object**". We have already seen in the previous chapter that light from a source under water seems to come from an "apparent depth" which is nearer the surface than the actual source. This is one example of the formation of an image at some position other than at the source. As we will find in our discussion, there are two possible types of image, labeled "**real**" and "**virtual**". In the case of a **real** image, rays from an object have been bent to actually converge on a point in space, and then to spread out again as they pass that point until they reach the eye. In that case, the rays really come from that point, although the original source (the object) is not at that point. For a **virtual** image, the eye traces the rays back to a point through which no rays actually pass. This was the case for the source under water. As we discuss specific examples this distinction will become clearer.

We will first discuss images that are formed using the reflection of light as a means of bending the paths of the rays, such as occurs with a mirror. We will then discuss the use of refraction for the purpose of forming images, which is the basis for the properties of lenses. Some applications to actual instruments will then be discussed.

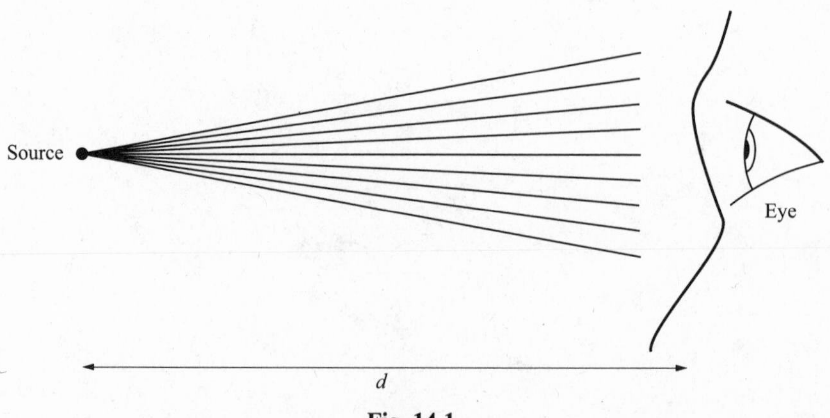

Fig. 14-1

14.2 MIRRORS

When light from a source reaches a mirror it is reflected back into the region from which it came. None of the light reaches the region behind the mirror. Consider a source whose light reaches a plane (flat) mirror, as in Fig. 14-2. The region to the left of the mirror is the "object" region since the light originates in that region. It is also the "image" region, since any real image will be formed there as a result of the reflection from the mirror into that region. If an image is formed in the region to the right of the mirror, that image would have to be a "virtual" image, since the light never actually was found in that region. We will, in fact, find that a plane mirror normally forms a virtual image.

In Fig. 14-2 we have drawn three examples of rays leaving the source (e.g. a point on an extended source) and being reflected by the mirror. For each we have used the fact that the angle of reflection equals the angle of incidence. (Recall that rays are perpendicular to the wave fronts, so the angle between the wavefronts and mirror equals the angle between the rays and the normal to the mirror.) Thus the central ray is reflected directly back, while the other two rays are reflected at angles that increase as their angles of incidence increase. The three reflected rays are moving in different directions and if they were to be viewed by an eye, the eye would trace the rays back till they meet. This is where the eye assumes the rays started and is the location of the image of the source. In the figure, we project these rays back and note that they meet at point I. The distance of the source from the mirror is called s and the distance of the image from the mirror is called s'. (By convention the distance s' is considered negative, since it is not in the region of the actual rays—sign conventions will be discussed below.) It is easy to see from the angles involved that $|s'| = s$.

Problem 14.1. Show that the point I is an image of the point source S and that the distance to the image in a plane mirror is equal to the distance from the source.

Solution

The ray that is incident on the mirror in the direction of the normal is reflecting directly backward, and the projection of that ray back into the region behind the mirror is along the normal. The next ray that has been drawn is incident at the angle θ, and reflected at that same angle θ. All the angles of triangle SBA equal the corresponding angles of triangle IBA, and they have a common equal side, AB. Therefore, all corresponding sides in the two triangles have equal length, and $SA = IA$. Since the same argument holds for any other rays (e.g. SC), all the reflected rays trace back to the same point I. We thus indeed have an image and $s = |s'|$.

We will introduce the following convention for the sign of lengths. Any object or image distance is positive if the object or image is in a region where light actually travels from the object or toward the image. If this is not the case, the distance is considered negative. For the case drawn in Fig. 14-2 this

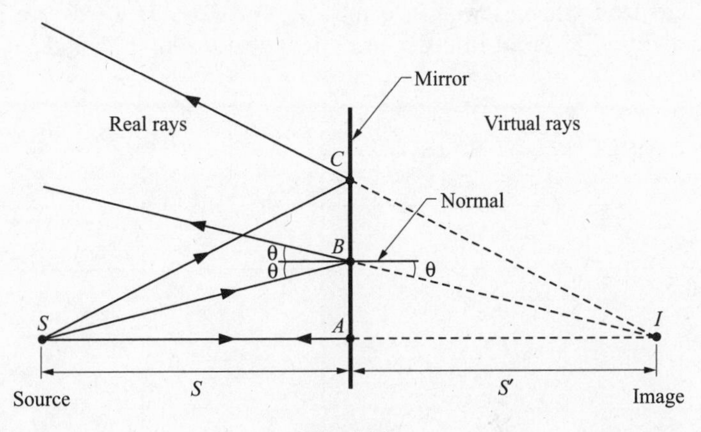

Fig. 14-2

means that any object or image distance is positive if the object or image is to the left of the mirror. Any object or image distance is negative if the object or image is on the other side of the mirror, where there is actually no light. (The question of how an object can be found in the region where there is no light—a "virtual" object—will be discussed in a later section. For the time being, we will always take the object distance to be positive, i.e. we will only consider "real" objects.) For the plane mirror we have therefore just proven that:

$$s' = -s \text{ (plane mirror)} \tag{14.1}$$

This means that a plane mirror forms a virtual image at a distance behind the mirror equal to the distance of the object in front of the mirror. We have so far considered a single point of an object. Let us see how to handle the extended object. Consider the vertical arrow in Fig. 14-3, at a distance s in front of the mirror. It is clear from the previous discussion that each point of the arrow is imaged at the same distance behind the mirror, so that the arrow will not be imaged as a bent or leaning arrow. It is therefore sufficient to image just the head and tail of the arrow, and draw the image at that location. In the case of the plane mirror, we see that the head of the arrow is above the tail (which is located on the base line in the figure) both for the object and for the image. We call this an upright or erect image. If the image were below the base line (for the case of an object above the base line) then the image would be "**inverted**". Furthermore, we see from the figure that the height of the image, y', equals the height of the object, y, and there is neither magnification nor demagnification. We define the ratio of the image height to the object height as the magnification, defined as:

$$M = y'/y \tag{14.2}$$

where the heights are positive if they are above the base line and negative if they are below the base line. Thus an upright image has a positive magnification and an inverted image has a negative magnification. For the plane mirror $y' = y$, and the magnification is one.

> ***Note.*** It is also true that $M = -s'/s$, since $s' = -s$. We will show that for spherical mirrors and lenses this latter formula still gives the magnification.

Thus:

$$M = 1 \text{ for a plane mirror} \tag{14.3}$$

We see therefore that the image in a plane mirror looks the same as the object as far as the height is concerned. Let us investigate what a mirror does to the "**left–right handedness**" of an object. Consider Fig. 14-4, where we show a child (as seen from above) facing the mirror with outstretched right and left hands. The image is also shown, and is clearly facing the child, with the front of the head nearer the mirror than the back. We note, however, that the right hand of the child is the left hand of the image. Thus the image has a reversed appearance concerning right and left handedness. Letters are therefore reversed and difficult to read when viewed in a mirror. Therefore, if words are generally read on a mirror they should be written reversed in order to be readable in the mirror. That is why the lettering

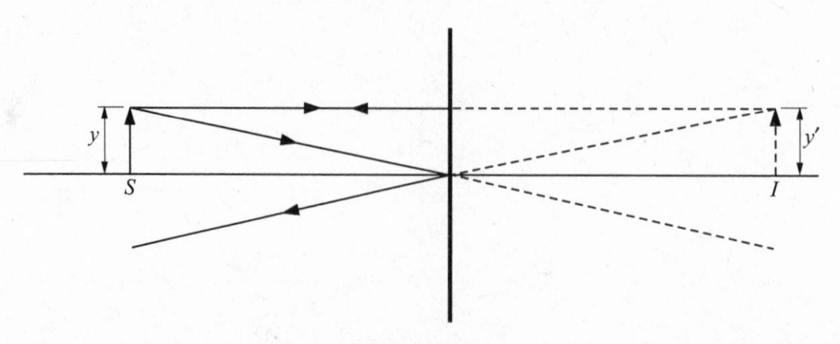

Fig. 14-3

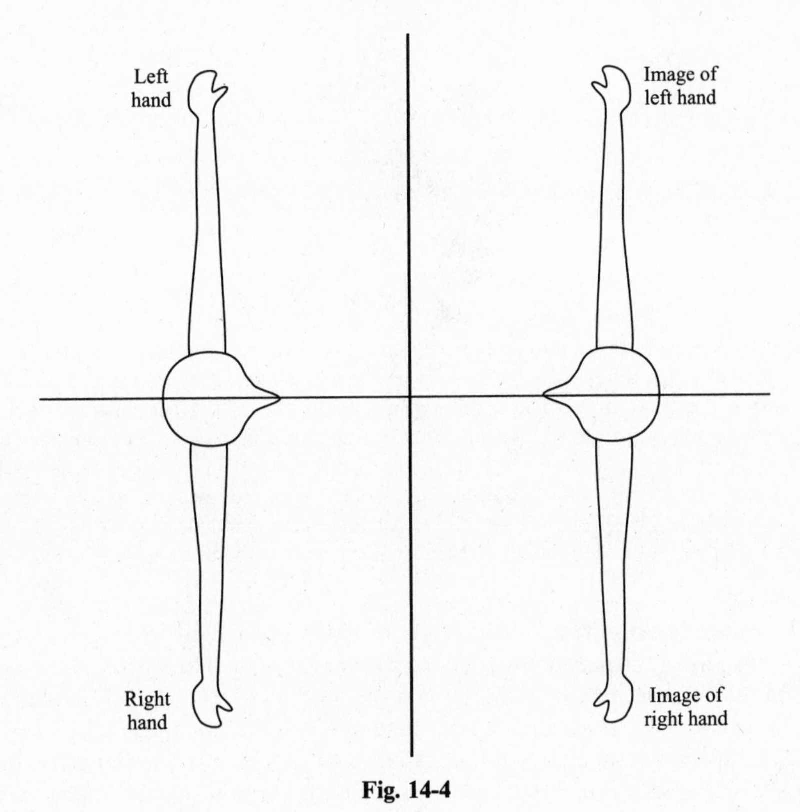

Fig. 14-4

on the front of an ambulance (which is generally read by cars that are ahead of the ambulance in their rear view mirrors) are often written reversed.

Plane mirrors are very simple in their properties, but many illusions can be created with the images of simple plane mirrors. Consider the mirror in Fig. 14-5, which is made of a piece of plane glass which is somewhat silvered so that it reflects a substantial amount of light as well as transmitting light (a half—silvered mirror). A candle at a distance s in front of the glass is hidden from the direct view of

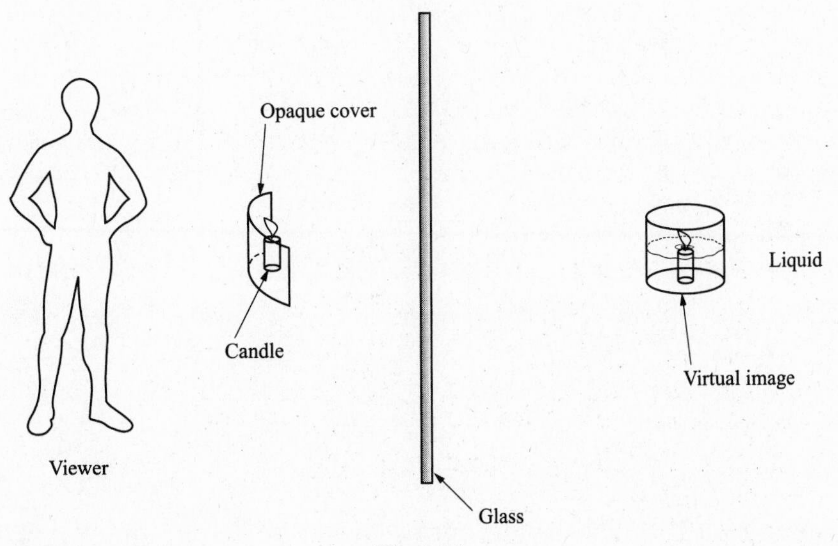

Fig. 14-5

persons in front of the glass, but its image can easily be seen by such viewers. If one places a beaker of liquid in back of the glass at the same distance s, the candle will be seen by these persons at the same position as the glass, i.e. one has created the illusion of a candle burning inside a liquid. Many "magic" tricks are performed in this manner.

Problem 14.2. Show that a half-length mirror is sufficient to produce a full length image of a person standing in front of a mirror.

Solution

In order for someone to see a full image of him/her self in a mirror, light from the bottom and from the top of the person must reflect off the mirror and reach his/her eyes (see Fig. 14-6). Since the angle of reflection equals the angle of incidence, the light from the top of the person that reaches the eyes of the person must have reflected from a point on the mirror halfway between the top and the eyes. Similarly, the light that reaches the eyes of the person from the bottom must have reflected off the mirror at a point halfway between the bottom and the eyes. The length of mirror needed to see from top to bottom is therefore half of the height of the person. Note that this is independent of the distance between the person and the mirror. (If you don't believe this, please try it out yourself with a mirror.)

When there are two plane mirrors at an angle to each other, light from a source can be reflected from just one mirror (either one) and produce a virtual image in that mirror. It is also possible that light, after reflecting from one mirror, reflects from the second mirror also. This can produce another image. If the angle between mirrors is arbitrary, there can, in fact, be many reflections, each producing another image. Tracing rays can be very complicated in that case, but an alternative exists. When rays are reflected from one mirror, all the rays appear to come from a point at the image of the object in that mirror. Therefore, this image acts as the (virtual) object for an image in the second mirror. One need merely determine where an image is produced in the second mirror for an object located at the original image location in the first mirror. This can be continued back and forth between mirrors. An example for a special case follows.

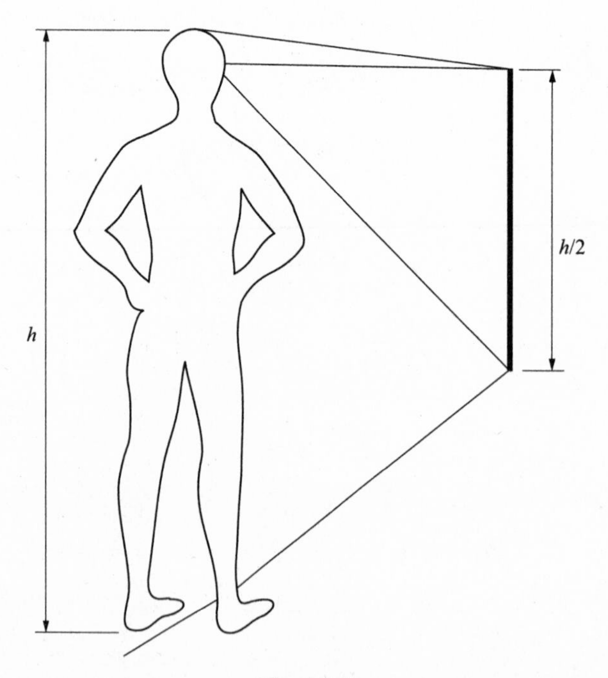

Fig. 14-6

Problem 14.3. Two mirrors are at an angle of 90° from each other, as shown in the top view in Fig. 14-7(*a*). An object is placed at a distance s_1 from mirror 1, and s_2 from mirror 2.

(*a*) Locate the image, I_1, of the object in mirror 1.

(*b*) Locate the image, I_2, of the object in mirror 2.

(c) Locate the image, I_{12}, of the object that has been first reflected from mirror 1 and then from mirror 2. Do the same for the image, I_{21}, that is formed by first reflecting from mirror 2 and then from mirror 1.

(*d*) Draw a ray diagram locating the image from one of the double reflections.

Solution

(*a*) The image from mirror 1 is located at a distance s_1 behind mirror 1, and shown in the figure as I_1.

(*b*) The image from mirror 2 is located at a distance s_2 behind mirror 2, and shown in the figure as I_2.

(*c*) The image I_{12} is determined by first finding the image, I_1, of the object in mirror 1, and then using this image as an object for mirror 2. I_1 is located at a distance s_2 above the plane of mirror 2, and s_1 behind mirror 1, as seen in the figure. The image of this in mirror 2 is at a distance s_2 below mirror 2, and labeled I_{12} in the figure. Similarly, to get I_{21}, one has to take the image, I_2, in mirror 2 and use this as the object for mirror 1. This results in I_{21} being located at the same position as I_{12}. Note that this is true only for the case of mirrors that are at an angle of 90° with each other. For other angles multiple images are generally formed.

(*d*) We can choose any two rays to trace out the double reflection, and then project these rays backward to locate the final image. This is done in Fig. 14-7(*b*). It requires care to draw this accurately and the procedure in part (*c*) is far preferable.

Our discussion so far has been for plane mirrors. We now enlarge our view to include spherical mirrors. By this we mean mirrors that are shaped to be part of the surface of a sphere of some radius *R*. These spherical mirrors can be "**concave**" if the center of the sphere is on the object side [see Fig. 14-8(*a*)] so that the inside surface of the sphere reflects light from the object, or "**convex**" if the center of the sphere is on the "negative" side [see Fig. 14-8(*b*)] so that the outside surface of the sphere reflects light from the object. In agreement with our previous convention, the concave surface is considered to have a positive radius, while the radius of the convex surface is considered to be negative. It is usual for the spherical mirror to be a symmetric portion of the surface of the sphere or cylinder, so that the rim of

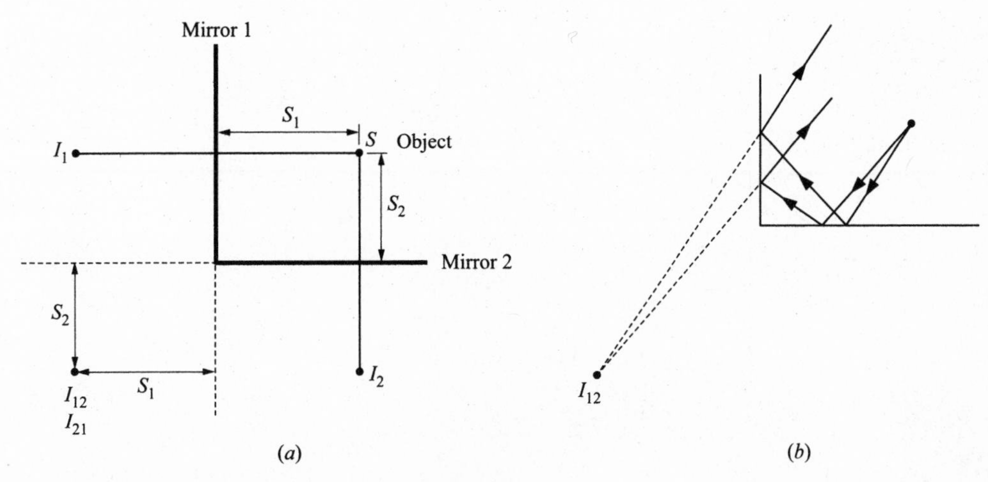

Fig. 14-7

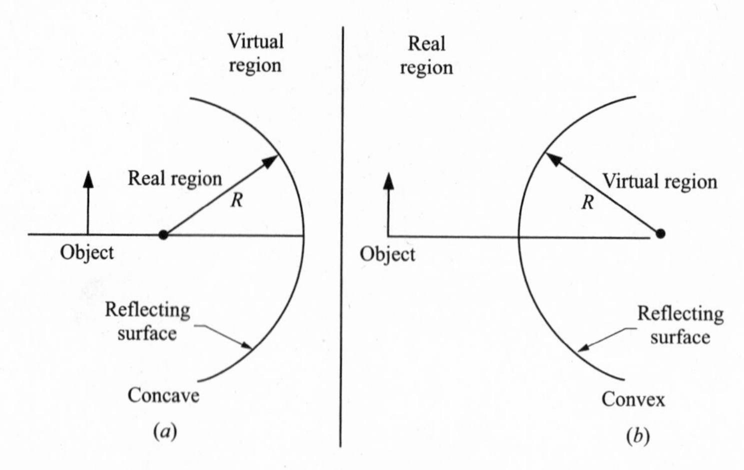

Fig. 14-8

the mirror is circular. The imaginary line through the center of the sphere and the center of the mirror is called the principal axis, as shown in Fig. 14-9(*a*) for the case of a concave mirror.

We must determine how these mirror surfaces reflect the light. The normal to the surface at any point is the radius drawn from the center C to that point. Consider a series of incident rays that are parallel to the principal axis and hence to each other. We take each incoming ray and equate the angle of reflection to the angle of incidence to get the reflected ray. The result for five equally spaced rays are shown in the figure. We note that the three parallel rays closest to the principal axis (including the one along the axis which is reflected straight back) have reflected rays that intersect at nearly the same point on the axis, but the outer two rays cross the axis nearer to the mirror. The common point where the rays near the principal axis meet is called the focal point, and labeled F. We will show mathematically that all parallel rays near the central ray (the ray through point c), called paraxial rays, are reflected through nearly the same point, the focal point, and we will in general consider only rays for which this is true (the paraxial approximation). If a mirror is large enough that the rays further away cannot be neglected, then the mirror will not give us the imaging properties that we want, and the mirror is said to have "**aberrations**". To avoid aberrations due to these non-paraxial rays, we would need a mirror formed in the shape of a paraboloid rather than a sphere. Such mirrors are used in large telescopes and

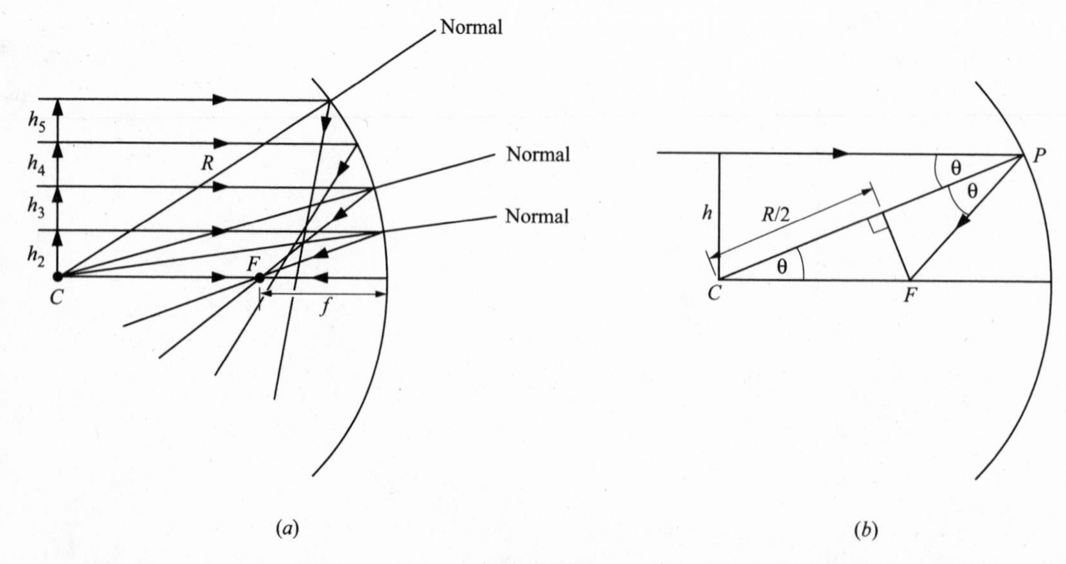

Fig. 14-9

other precision applications. In our discussion we will restrict our consideration to rays within the paraxial approximation.

Problem 14.4. For a concave mirror, of radius R, as in Fig. 14-9(*b*), consider the parallel ray at a height h above the central ray along the principal axis.

(*a*) Calculate the distance from the center, C, to the point, F, at which the reflected ray crosses the central line.

(*b*) Show that, for all rays with $h^2 \ll R^2$, the rays pass through a common point, F, and calculate the distance of F from the center C of the sphere.

Solution

(*a*) The ray at height h makes an angle θ with the normal at the surface of the mirror at P. That angle is given by $\sin\theta = h/R$. The angle of reflection is also equal to θ, and so is the angle PCF in triangle PCF. This triangle is therefore an isosceles triangle. We draw the perpendicular from F to the base of the isosceles triangle (the radius of the sphere), which will divide the base into two equal parts, each of length $R/2$. The distance we are seeking, CF, is therefore equal to $R/2(\cos\theta)$, where $\sin\theta = h/R$.

(*b*) When $h \ll R$, the angle θ is small. A quick look at triangle APC shows that for small θ, using the definition of radians, $\theta \approx h/R$. Then $\sin\theta \approx \theta$ in radians. Recalling that $\cos^2\theta = 1 - \sin^2\theta$ we have for small θ: $\cos^2\theta \approx 1 - \theta^2$. Since $(1 - \theta^2/2)^2 = 1 - \theta^2 + \theta^4/4$ and for small θ ($\ll 1$) $\theta^4 \ll \theta^2$, we have $\cos^2\theta \approx (1 - \theta^2/2)^2$ or $\cos\theta \approx 1 - \theta^2/2$. Since $\theta^2 \ll 1$, to a first approximation, $\cos\theta = 1$. Substituting h/R for $\sin\theta$, and using $\cos\theta \approx 1 - (h/R)^2/2 \approx 1$, the distance between C and F is $\approx R/2$. This is also the distance between the mirror and F, which is called the focal length, f, of the mirror, giving:

$$f = R/2 \tag{14.4}$$

We now know that any (paraxial) ray parallel to the principal axis will produce a reflected ray that goes through the focal point, which is located at a distance f from the mirror with $f = R/2$ and will make an small angle [2θ in Fig. 14-9(*b*)] with the principal axis. If we reverse the process and have a ray coming in toward the mirror along the same line as the above reflected ray (i.e., passing through F at an angle 2θ to the principal axis) its reflection will be back out along the line of the original parallel ray. It is follows that any ray passing through F at a small angle will reflect parallel to the principal axis. This fact is very useful in constructing the image formed by the mirror. Our discussion thus far has been for a concave mirror. What about a convex mirror? If we accept the statement made earlier that a convex mirror has a negative radius, and assume that our previous results are also correct for a convex mirror, then we expect to find a focal point at a negative distance of $R/2$ from the mirror, meaning at a point $|R/2|$ behind the mirror. Since no light actually goes through that point we would conclude that the light is reflected from the mirror in such a direction that a backward tracing of the reflected rays goes through that virtual point. This is the subject of the next problem.

Problem 14.5. For a convex mirror with a radius of $|R|$, show that the paraxial rays, when reflected from the mirror, project back to a point at a distance of $|R/2|$ in back of the mirror.

Solution

The mirror is shown in Fig. 14-10. For one ray parallel to and near to the principal axis the reflection is shown in the figure. Again CFP is an isosceles triangle because the base angles are equal, and the perpendicular to the base divides that base of length R into two equal parts of length $R/2$. The distance from C to F is therefore $R/2(\cos\theta)$, and for a paraxial ray $\cos\theta \approx 1$, giving the distance from C to R as $R/2$. Then the distance from the mirror to F is also $R/2 = f$. If we let R be negative, then this formula implies that f is also negative, as we expect for a point behind the mirror.

We are therefore led to the conclusion that Eq. (*14.4*) is valid for both concave and convex mirrors, where R and f are positive for a concave mirror and negative for a convex mirror. The negative values

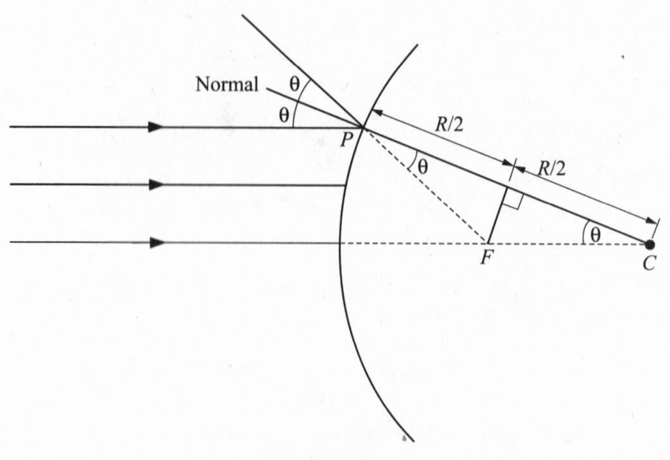

Fig. 14-10

mean that both the center of the sphere and the focal point are behind the mirror. As for our concave mirror, if we now consider a ray approaching the mirror at a small angle to the principal axis, and whose projection past the mirror passes through F, the associated reflected ray would be parallel to the principal axis.

We have shown that parallel paraxial rays of light that hit our spherical mirror have reflected rays that all pass through the focal point (concave mirror) or have virtual projections that pass through the virtual focal point (convex mirror). We have also shown that the reverse process is also true. If rays approach the mirror at small angles to the principal axis so as to pass through the focal point (concave mirror) or to have virtual projections through the virtual focal point (convex mirror) the corresponding reflected rays are all parallel to the principal axis. The validity of these results is dependent on the condition that $h/R \ll 1$ for incoming parallel rays and $\sin \theta \ll 1$ for rays at an angle passing through a focal point. The latter result explains why in a searchlight (or flashlight) we place the bulb at the focal point of a spherical reflector. The light hitting the reflector travels back out in approximate parallel lines forming a beam. If we wish to have a wide beam in the searchlight, and therefore use reflections further away from the principal axis, we must use a paraboloidal surface instead of a spherical one.

We are now in a position to determine the properties of the image of an extended object formed in a spherical mirror. Consider an object, such as an arrow, which we are imaging in the mirror. We examine the light coming from a given point on the arrow and approaching the mirror. When these rays are reflected from a concave mirror they converge on the axis and they intersect each other, the intersection is the position of the "real" image of the point on the arrow. If the rays from the point on the arrow reflect from a convex mirror they do not converge on the axis, but rather diverge, and we must project the rays back behind the mirror to their virtual meeting point which will be the virtual image of that point. In Fig. 14-11(*a*), we draw four rays from the tip of the arrow whose reflections can be easily drawn and used for finding the image formed by a concave mirror. We do the same for a convex mirror in Fig. 14-11(*b*). In both figures we draw the ray parallel to the principal axis and reflect it from the mirror through the focal point. Next we take a ray through the focal point (for the convex mirror this ray would reach the focal point if it were not reflected first) and it is reflected parallel to the principal axis. Next we take a ray through the center of the sphere, which, since it is along a radius is reflected back along the same path. Finally we take a ray to the point where the principal axis meets the mirror (central point). Since the principal axis is the normal at this point, the ray is reflected back at the same angle to the principal axis as it was incident. As long as the light reaching the mirror from the object makes small angles with the principal axis (paraxial approximation) these rays will all meet at the same point (or for the convex mirror will be projected back to the same point), which is where the image is produced. In our case, the concave mirror produced a real image, while the convex mirror produced a virtual image. While we have done this for the tip of the arrow we could do it for any other point on the arrow. The result would be the real inverted image of the arrow shown in Fig. 14-11(*a*) and

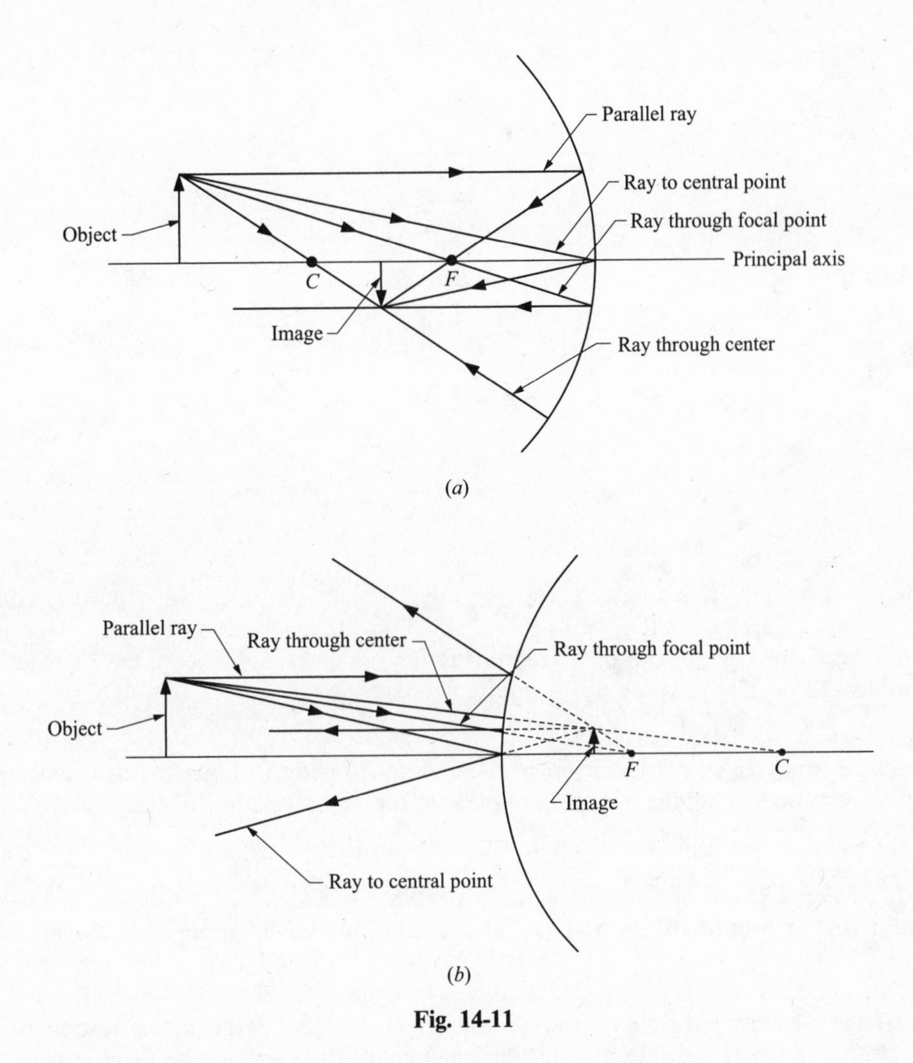

Fig. 14-11

the virtual upright image of the arrow shown in Fig. 14-11(*b*). As long as the paraxial approximation holds, the image may differ in size from the object but will not be distorted.

Problem 14.6. For the concave mirror shown in Fig. 14-11(*a*), draw the ray diagram for a case in which an object is at a distance of $f/2$ from the mirror, and describe the image that is formed.

Solution

In Fig. 14-12 we draw the same mirror as in Fig. 14-11(*a*). We then proceed to draw the four simple rays from the tip of the arrow discussed previously. The ray parallel to the principal axis is reflected through the focal point, and is drawn as ray (1). Next we draw the ray to the central point on the mirror which is reflected back at the same angle, and labeled (2) in the figure. These two rays are usually the simplest ones to draw in a ray diagram. The third ray that we draw is the one along a line through the focal point. This ray is along the straight line from the tip of the arrow to the mirror that traces back to F. It is reflected from the mirror into a path parallel to the principal axis, and labeled (3) on the figure. Lastly, we draw the ray that traces back through the center of the sphere. This ray is reflected back onto itself and then passes through C. It is labeled (4) on the figure. Unlike the case of Fig. 14-11(*a*) where the arrow was beyond the focal point and the rays from a point on the arrows reflected off the mirror so as to converge to a real image point, in the present case the reflected rays diverge, as in a convex mirror. All four reflected rays meet only when they are projected back into the virtual region behind the mirror, and form an upright, virtual image that, as can be seen, is larger than the object.

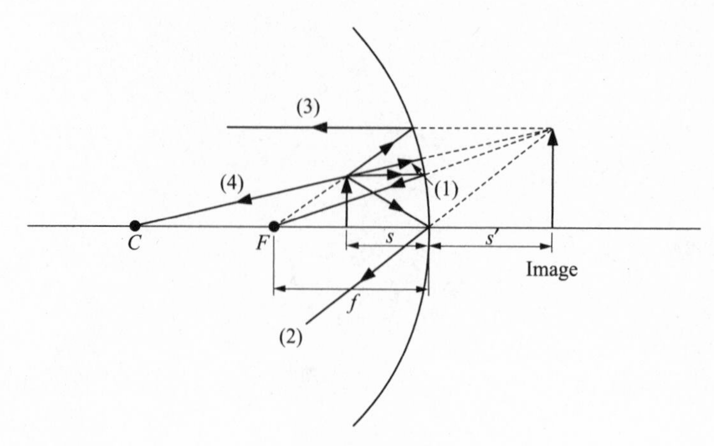

Fig. 14-12

This kind of construction can be used to solve any problem, but it is clearly not too convenient in practice. Therefore we seek an alternate analytical method for determining the position and magnification of an image. This method can be derived using the ray diagrams that we have already discussed. The next problem shows how this can be done for a concave mirror.

Problem 14.7. Consider a concave mirror, of focal length f, with an object at a distance s from the mirror. Assume that we can use the paraxial approximation, and take $s > f$ for simplicity.

(*a*) Draw the ray diagram using the ray parallel to the principal axis and the ray to the central point on the mirror to locate the image.

(*b*) Use the triangles formed by the ray to the central point to determine the magnification of the image.

Note. If the object has height y and the image has height y' the magnification is defined as $M = y'/y$. By convention if the image is inverted y' is considered to be negative.

(*c*) Use other similar triangles to derive a formula for s' in terms of s and f.

Solution

(*a*) The ray diagram is shown in Fig. 14-13. Here, $s = OA$, $s' = IA$, $f = FA$, and $y = O'O$. Since y' is negative, the distance $I'I = -y'$. We see that the image is real, inverted, and smaller than the object.

(*b*) Using right triangles $O'AO$ and $I'AI$, which have the same angles and are therefore similar, we equate the ratio of corresponding sides. This gives $-y'/y = IA/OA = s'/s$. Since the magnification is defined as $M = y'/y$, we have:

$$M = -(s'/s) \tag{14.5}$$

(*c*) If we can use the paraxial approximation, then AB can be considered a straight line perpendicular to the principal axis. Then triangle $I'IF$ and BAF are similar and we can equate the ratio of corresponding sides. This means that $I'I/BA = IF/AF$. Now $I'I = -y'$, $BA = y$, $AF = f$ and $IF = s' - f$. Substituting we get $-y'/y = (s' - f)/f$. In part (*b*) we already showed that $-y'/y = s'/s$, and therefore: $(s' - f)/f = s'/s$. Cross multiplying we get: $s's - sf = s'f \rightarrow s'f + sf = s's$. Finally, dividing by $(s'sf)$ yields:

$$1/s + 1/s' = 1/f \tag{14.6}$$

This equation was derived for the case of a concave mirror and for $s > f$. In fact it is valid for all cases, as long as we can use the paraxial approximation. We will consider another case in the review

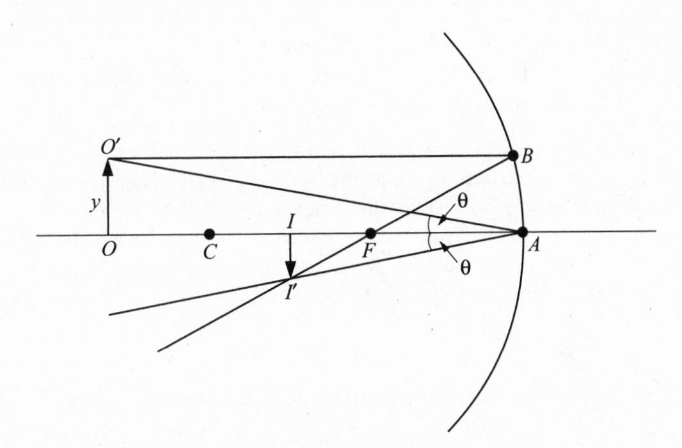

Fig. 14-13

problems. In this formula all distances are positive in the region where the light is actually found, and negative in the region behind the mirror. A real image has positive s' and is found in front of the mirror, while a virtual image has a negative s' and is found behind the mirror. A concave mirror has a positive radius and a positive focal length, while a convex mirror has a negative R and f. With these conventions and the Eqs. (*14.5*) and (*14.6*), we are equipped to handle problems associated with these spherical mirrors.

Note. We will see later that the identical formulas also are applicable to lenses with their appropriate conventions.

Problem 14.8.

(*a*) For the case of a concave mirror, describe qualitatively the size and location of the image of a vertical upright object as the object is moved from a point just beyond the focal point to infinity.

(*b*) Repeat part (*a*) as the object is moved from a point just within the focal point to the surface of the mirror.

(*c*) For the case of a convex mirror, describe qualitatively the size and location of the image as a vertical upright object is moved from the surface of the mirror out to infinity.

Solution

(*a*) We use Eqs. (*14.5*) and (*14.6*). In Eq. (*14.6*) f and s are positive. We start with s slightly greater than f so $1/s < \sim 1/f$. This implies that $1/s'$ is a very large positive number. The image is therefore real and far back from the mirror ($s' \to \infty$ as $s \to f$). Also, since $s' \gg s$, M is very large and the image is inverted [Eq. (*14.5*)]. As s gets larger, $1/s$ drops and $1/s'$ increases so s' decreases. When the object reaches the center of the sphere, $s = R$, and since $f = R/2$, Eq. (*14.6*) tells us that $s' = R$. Thus the image and object are at the same location (i.e., the image has been moving in as the object moves out and $s = R$ is the cross-over point). From Eq. (*14.5*) we see that the image is still inverted but the same size as the object. As s continues to increase, $1/s$ decreases further so $1/s'$ must increase and s' correspondingly decreases. Thus $s' < s$, and the image is real, inverted and smaller than the object. As $s \to \infty$, $s' \to f$, and since $M \to 0$, the image size shrinks to zero.

(*b*) Here s starts out slightly smaller than f so $1/s > 1/f$ and Eq. (*14.6*) implies that s' must be negative. The image is thus virtual and upright [Eq. (*14.5*)]. Furthermore, since s is only slightly smaller than f, $1/s'$ is a very small negative number and s' is very large. The upright, virtual image is thus very far behind the mirror and the magnitude is very large as well. As the object moves toward the mirror, $s \to 0$ and $1/s \to \infty$, so $1/s' \to -\infty$ and $s' \to 0$ (from the negative side). Since $1/f$ is a finite number, as $1/s \to \infty$ and

$1/s' \to -\infty$ we still must have $1/s - 1/|s'|$ is finite $\to s \approx |s'| \to M = 1$ for very small s. Thus at the mirror, $s \approx 0$ and we also have $|s'| \approx 0$, so the object and virtual image are the same size and at essentially the same location (the mirror surface).

(*c*) In this case, f is negative, so Eq. (14.6) tells us that for any positive s, s' is negative and $|s'| < s$. The image is therefore always virtual, upright and reduced in size [Eq. (*14.5*)]. For s close to zero, $|s'|$ is also close to zero and $M \approx 1$, just as for the concave mirror. For $s = |f|$, $s' = -|f|/2$ and $M = 1/2$. As $s \to \infty$ we have that $|s'| \to |f|$ from below (since $1/s'$ must always be more negative than $1/f$ for positive s), so the virtual image never gets further behind the mirror surface than the focal point, and the image size shrinks to zero since $|s'/s| \to 0$.

Problem 14.9. A concave mirror has a focal length of 25.3 cm. An object, of height 1.45 cm is placed 46 cm in front of the mirror.

(*a*) What is the radius of the mirror?

(*b*) Where is the image located? Is the image real or virtual?

(*c*) What is the size of the image? Is the image upright or inverted?

Solution

(*a*) For a mirror we know that the focal length is half the radius. Thus $R = 2f = 2(25.3 \text{ cm}) = 50.6$ cm.

(*b*) We use Eq. (*14.6*), $1/s + 1/s' = 1/f$, $\to 1/46 + 1/s' = 1/25.3 \to 1/s' = 0.178 \to s' = 56.2$ cm. Since this is positive, the image is real.

(*c*) We use Eq. (*14.5*) to get the magnification, $M = -(s'/s) = -(56.2/46) = -1.22$. Therefore the height of the image is $1.45(-1.22) = -1.77$ cm. The negative sign shows that the image is inverted.

Problem 14.10. Repeat the previous problem for a convex lens with a focal length of the same magnitude.

Solution

(*a*) For a convex lens the focal length is negative. Therefore the radius is -50.6 cm, indicating that the center of the sphere is behind the mirror.

(*b*) We again substitute in Eq. (*14.6*) using $f = -25.3$. Thus $1/46 + 1/s' = -1/25.3 \to 1/s' = -0.0613 \to s' = -16.3$ cm. Since this is negative, the image is virtual and is located 16.3 cm behind the mirror.

(*c*) The magnification is $M = -(-16.3/46) = 0.354$, and the size of the image is $0.354(1.45 \text{ cm}) = 0.514$ cm. Since the magnification is positive, the image is upright. The image is smaller than the object since the magnification has a magnitude less than one.

Problem 14.11. Repeat Problem 14.9 for the same concave mirror, but for the object at a distance of 20.1 cm from the mirror.

Solution

(*a*) The radius of the mirror is still 50.6 cm.

(*b*) We use the same equation, but with $s = 20.1$. Thus, $1/20.1 + 1/s' = 1/25.3 \to 1/s' = -0.0102 \to s' = -97.8$ cm. The image is virtual and at a distance of 97.8 cm behind the mirror.

(*c*) The magnification is $M = -(-97.8/20.1) = 4.87$, and the image has a height of 7.05 cm. It is upright and very much magnified.

14.3 THIN LENSES

We now turn our attention to a discussion of what occurs when light is incident on a "**thin lens**". By this we mean a thin piece of transparent material such as glass or plastic, with two spherical surfaces generally of different radii and a common principal axis. Light is incident on the lens from one side, called the object side, and is refracted at each of the two surfaces and emerges on the other side, called the image side. There are two kinds of lenses, **converging** and **diverging**. If rays of light are incident on the lens at some angle, the lens will bend the transmitted light, either toward the principal axis or away from the principal axis. We call the lens converging if it bends the light toward the principal axis and diverging if it bends the light away from the principal axis. For a converging lens it can be shown that (in the paraxial approximation) rays coming from a point on an object and passing through the lens converge (are focused) to a common point beyond the lens. If the object is very far away the rays are parallel as they enter the lens, so parallel rays also converge to a point. Rays that are parallel to the principal axis converge at a point along the principal axis called the **focal point**. This point is on the image side of the lens, and the distance of the focal point from the lens is called the **focal length**. Since light actually goes through this point, we choose to call this distance positive. This results in the convention that if we consider the position of light after it passes through the lens, the distance is positive on the image side of the lens. Thus our convention is that distances are positive for an object on the side where light travels from the object to the lens and for an image (and focal length) on the side where light travels from the lens. For a diverging lens rays from a point on the object diverge when they pass through the lens, but if we trace these rays back through the lens they appear to be coming from a common point on the object side, i.e. from a virtual image. For incoming parallel rays, the trace backs of the emergent rays meet at a point on the principal axis on the object side of the lens. This is called the focal point for this diverging lens, and the distance of the focal point from the lens is the focal length, which will consequently be negative. We will see in the next section how to determine the focal length of a lens from the radii of the two surfaces of the lens. It is clear that experimentally, these lengths can be determined by shining parallel light on the lens and seeing where the emerging rays focus, and whether the focus is of real rays converging or the trace back of virtual rays to a virtual convergence point.

In Figs 14-14(*a*) and (*b*), we draw the possible radii of lenses that produce each kind of lens. The determining factor for a converging lens is that the lens is wider at the center than at the ends, while for a converging lens it is wider at the ends than at the center. To get an idea of how the lens operates, let us consider the double convex lens in the figure. (A surface is called convex or concave depending on its shape when viewed from the outside of the lens.) As a first approximation, we can consider the lens to consist of two prisms, as in Fig. 14-15(*a*). We already learned in the previous chapter, that an incident ray is bent toward the normal when it enters the first surface. Thus the upper parallel ray in the figure is bent downwards. At the second surface the ray is bent away from the surface, which, in this case, means that it is bent further downwards. Thus the prism bends the upper ray toward the principal axis. The symmetrically placed lower parallel ray is similarly bent upwards toward the principal axis. Such a prism will not focus all parallel rays (i.e. for differing values of h) to one single point and cannot serve as a useful lens. However, if the surfaces are spherical rather than triangular, in the paraxial approximation all the parallel rays are focused to the same point. We will defer discussing a derivation of this statement until the next section. A diverging lens can be similarly viewed in a first approximation as two inverted prisms, as shown in Fig. 14-15(*b*), with similar conclusions for a virtual focal point.

If we can assume that a lens focuses all rays from an object point to a real or virtual image point, then we can determine the imaging properties of a thin lens, using the same procedure that we used for mirrors. We need to trace the path of at least two rays through the lens from a point on the object and the image of that point will be at the point where those two rays meet (or are projected back to meet). The first ray can be taken as the one parallel to the principal axis and the second ray can be taken to be the ray that passes through the center of the lens. At the center of the lens the two surfaces of the lens are both vertical and therefore parallel to each other. We learned in the previous chapter that a parallel plane piece of material will not change the direction of a ray that enters on one side. The ray will be

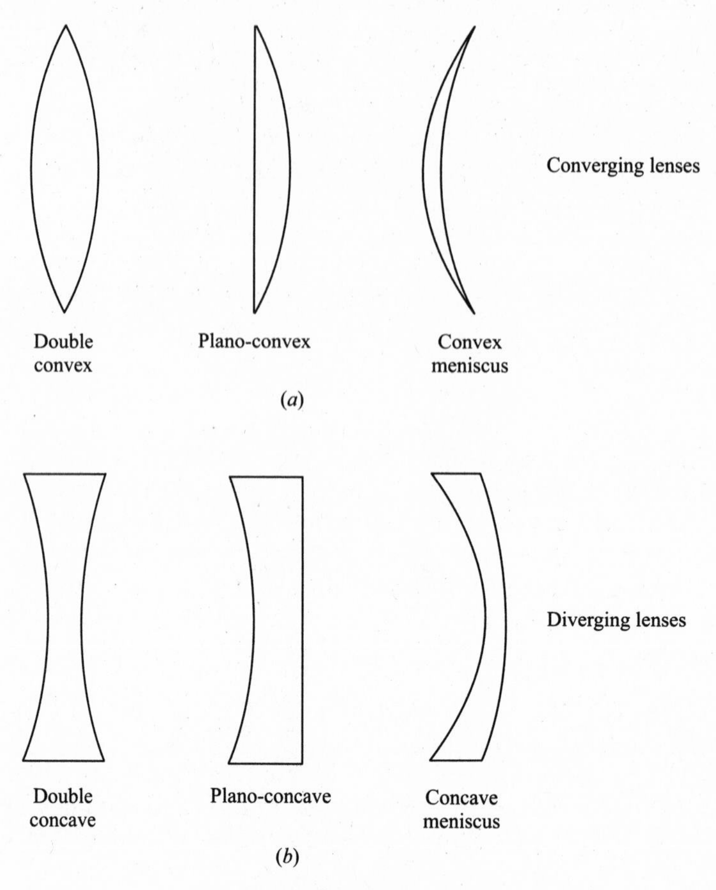

Fig. 14-14

displaced as a result of the passage through the plane, and this displacement will depend on the thickness of the plane. For a very thin plane, there will be hardly any displacement. When we say that we have a *thin* lens, we mean that the lens is thin enough that we can neglect this displacement. Therefore the ray through the center of the lens can be drawn as if it were neither displaced nor bent and is just a straight line through the center. This is true for both a converging and a diverging lens.

Note. Another property of a thin lens is that if we turn the lens around so the light from the object enters the "back" side the properties of the lens are unchanged. It thus makes no difference which side of the lens the light enters.

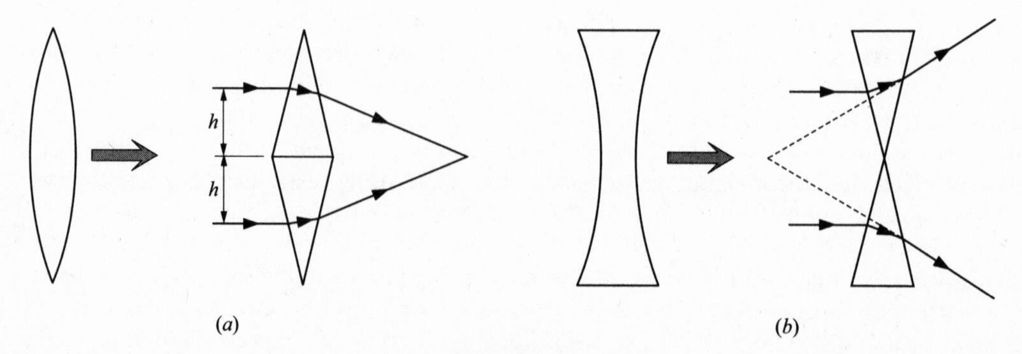

Fig. 14-15

Problem 14.12. Consider a converging lens of focal length f, and an object at a distance s from the lens. Assume $s > f$, but less than $2f$.

(a) Draw the ray diagram to locate the image. Describe the image (real or virtual, upright or inverted, larger or smaller than the object).

(b) Use the diagram to derive a formula for the magnification of the lens.

(c) Use the diagram to derive a formula for the position of the image.

(d) If the focal length is 4.75 cm, and the object is at a distance of 6.33 cm from the lens and of height 1.89 mm, determine the position and the size of the image.

Solution

(a) In Fig. 14-16 we draw the lens and the focal point at a distance f on the right side (the image side) of the lens. We then place the object (represented by an arrow) at a distance $s > f$ on the object side of the lens and examine the light coming from a point on the arrow, say the tip. All measurements are made from the center of the lens since we consider the lens to be thin enough that we can neglect its width. We then draw ray (1) from the arrow tip parallel to the principal axis. When it passes through the lens (point B) the ray is bent to intersect the focal point. Then we draw ray (2) through the center of the lens at O, and it passes through without any deviation. The two rays meet at I', which is the location of the image. (A similar analysis can be made for any other point on the arrow, including the base, which by symmetry has its image on the principal axis.) From the diagram we deduce that the image is real (on the image side), inverted (below the line) and bigger than the object (magnification has a magnitude greater than one). As with the case of mirrors, other than inversion and change in size, there is no distortion of the image in the paraxial approximation.

(b) Using triangles $S'OS$ and $I'OI$, which are similar since they have the same angles, we equate the ratios of corresponding sides. Thus $I'I/S'S = s'/s = -y'/y$ (As for mirrors, y' is negative for inverted images by convention). Thus:

$$M = y'/y = -(s'/s) \tag{14.7}$$

as we also found for mirrors.

(c) Using similar triangles BOF and $I'IF$, we equate the ratio of corresponding sides, $I'I/BO = (s' - f)/f = (-y')/y = -M = s'/s$. Cross multiplying we get $s's - fs = fs' \rightarrow$

$$1/s + 1/s' = 1/f \tag{14.8}$$

Again this is the same equation that we derived for the mirror. We must remember the convention for positive s (light from the object side), s' (light to the image side) and f (light to the image side).

(d) Substituting into Eq. (14.8), we get $1/6.33\text{ cm} + 1/s' = 1/4.74\text{ cm}$, and then $1/s' = 0.0525$, $s' = 19.0\text{ cm}$. Substituting into Eq. (14.7), $M = -(19.0/6.33) = (-3.01)$, and $y' = -3.01(1.89\text{ mm}) = -5.68\text{ mm}$.

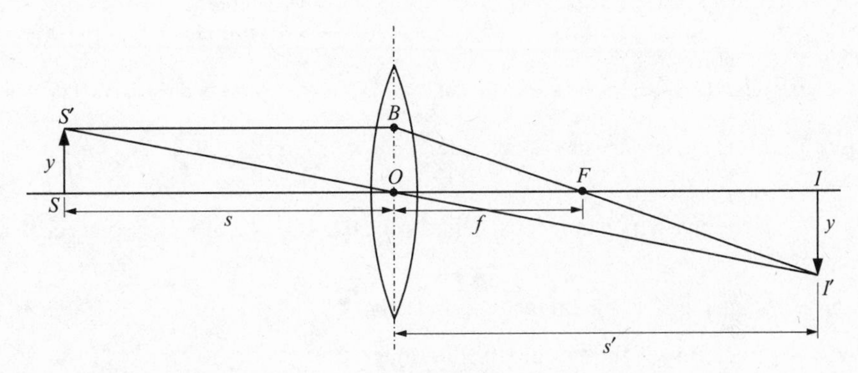

Fig. 14-16

Positive s' means that the image is real, negative M means that the image is inverted, and $|M| > 1$ means that the image is enlarged.

The above equations have been derived for the case of a converging lens with $s > f$. However the result is true for all cases, whether f is positive or negative, and for all values of s (even for "virtual" objects with $s < 0$). Thus, the equations that can be used for thin lenses are identical to those that can be used for mirrors, provided that we adhere to the sign conventions. Note that the converging lens behaves exactly like the concave mirror except that for the mirror real images are on the object side of the mirror and virtual images behind the mirror (where no real light goes), while for the lens the real images are beyond the lens while the virtual images are on the same side of the lens as the object (where no real light that has hit the lens goes). Similarly, a diverging lens behaves the same way as a concave mirror with suitable interpretation of the location of virtual images. A review of Problem 14.8 reinterpreted for lenses is recommended.

Problem 14.13. Show that the equations that we have derived for the spherical mirrors and for the thin lenses also give the correct result for a plane mirror.

Solution

A plane mirror can be considered as a spherical mirror with $R = \infty$. If $R = \infty$ then $f = R/2 = \infty$. Using Eq. (*14.8*) $1/s + 1/s' = 1/f = 0$, and $s' = -s$, as expected. Similarly, $M = -(s'/s) = 1$, again as expected. Therefore the plane mirror could be handled as a special case of a spherical mirror.

Problem 14.14. A diverging lens has a focal length of magnitude 47.9 cm. An object is placed 30.1 cm in front of the lens.

(*a*) Where is the image formed? Is the image real or virtual?

(*b*) What is the magnification of the image? Is the image upright or inverted? Is it larger or smaller than the object?

(*c*) Show that any diverging lens always produces a virtual, upright and smaller image.

Solution

(*a*) We substitute in Eq. (*14.8*), using $f = -47.9 \rightarrow 1/30.1 + 1/s' = -1/47.9 \rightarrow 1/s' = -0.0541 \rightarrow s' = -18.5$ cm. Since this is negative, the image is virtual, and forms on the object side of the lens.

(*b*) We substitute into Eq. (*14.7*), $M = -(s'/s) = -(-18.5/30.1) = 0.614$. Since this is positive, the image is upright. Since the magnification is less than one, the image is smaller than the object.

(*c*) We use Eq. (*14.8*), and rearrange $\rightarrow 1/s' = 1/f - 1/s$. Since f is negative for a diverging lens, both terms on the right-hand side of the equation are negative (since s is positive), and therefore s' is negative. Thus the image is virtual. Furthermore, with s' negative, the magnification is positive and the image is therefore upright. In the equation for $1/s'$, we note that the magnitude of $1/s'$ is greater than $1/s$, since adding two numbers of the same sign ($1/f$ and $-1/s$) produces a number with a larger magnitude than either term. Thus $|1/s'| > 1/s \rightarrow |s'| < s \rightarrow |M| < 1$. Thus the image is always smaller than the object.

Note. These are the same results obtained for a convex mirror [Problem 14.8(*c*)].

Problem 14.15. A converging lens has a focal length of 30.8 cm. An object is placed at a distance of 19.9 cm from the lens.

(*a*) Where is the image formed? Is the image real or virtual?

(*b*) What is the magnification of the image? Is the image upright or inverted? Is the image larger or smaller than the object?

(*c*) Show that for a converging lens, with the object distance less than the focal length, the image is always virtual, upright and larger than the object.

Solution

(*a*) We substitute in Eq. (*14.8*), using $f = 30.8 \rightarrow 1/19.9 + 1/s' = 1/30.8 \rightarrow 1/s' = -0.0178 \rightarrow s' = -56.2$ cm. Since this is negative, the image is virtual, and forms on the object side of the lens.

(*b*) We substitute into Eq. (*14.7*), $M = -(s'/s) = -(-56.2/19.9) = 2.83$. Since this is positive, the image is upright. Since the magnification is greater than one, the image is larger than the object.

(*c*) We use Eq. (*14.8*), and rearrange $\rightarrow 1/s' = 1/f - 1/s = (s - f)/fs$. Both f and s are positive, but $s < f$, and therefore $1/s'$ and s' are negative. Thus the image is virtual. Furthermore, with s' negative, the magnification is positive and the image is therefore upright. We rearrange the equation for $1/s'$ to give: $s'/s = f/(s - f)$. The right-hand side is negative, but its magnitude is greater than one, since $|s - f| < f$. Thus $|s'/s| > 1 \rightarrow |M| > 1$. Thus the image is always larger than the object.

Note. The same results were obtained for a concave mirror (Problem 14.8(*b*).

Problem 14.16. A real image is formed by a converging lens at a distance of 91.2 cm from the lens. The focal length of the lens is 76.9 cm.

(*a*) Where was the object placed?

(*b*) What is the magnification of the image?

Solution

(*a*) Again we use Eq. (*14.8*), and substitute for s' and f. We get $1/s = 1/f - 1/s' = 1/76.9 - 1/91.2 \rightarrow s =$ 490 cm.

(*b*) The magnification is $M = -(s'/s) = -(91.2/490) = -0.186$. The image is inverted and smaller than the object.

Problem 14.17. In Fig. 14-17(*a*), we show a lens with an object placed 1.5 cm from the lens. A partial ray diagram is drawn in the figure.

(*a*) Complete the ray diagram to obtain the image. State the properties of the image (real or virtual, upright or inverted, larger or smaller than the object).

(*b*) Determine the focal length of the lens, the position of the image and the magnification.

(*c*) Is the lens converging or diverging?

Solution

(*a*) In the figure we are given the paths of two rays that pass through the lens. These rays do not meet on the image side of the lens. In Fig. 14-17(*b*), we project the rays backwards until they meet on the object side of the lens. From the position of the image we deduce that the image is virtual, upright and larger than the object.

(*b*) The ray that started out parallel to the principal axis is bent to pass through the principal axis at a distance of 6.0 cm from the lens. Therefore $f = 6.0$ cm. Using Eq. (*14.8*) $1/s' = 1/6.0 - 1/1.5 = -0.50 \rightarrow s' = -2.0$ cm. The virtual image is 2.0 cm from the lens, and has a magnification of $M = -(-2.0/1.5) = 1.33$.

(*c*) Both converging and diverging lenses are capable of forming virtual, upright images, but, as shown in Problem 14.14(*c*), the image is always smaller than the object for a diverging lens. Since in our case $M > 1$ it must be a converging lens. A more direct approach is to note that the ray parallel to the principal axis bends toward that axis (and, indeed, crosses at the focal point) which is a key criterion for a converging as opposed to a diverging lens.

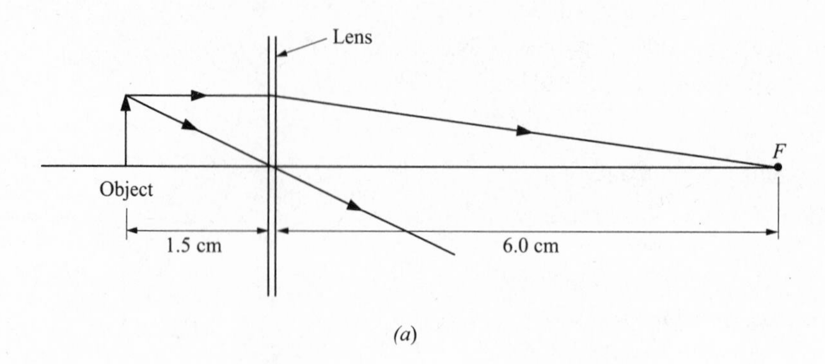

(a)

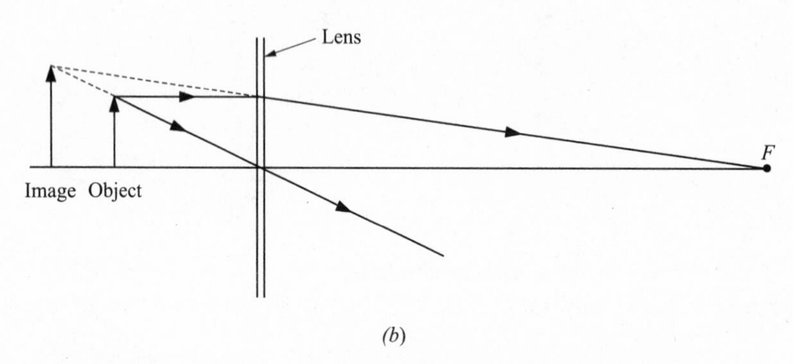

(b)

Fig. 14-17

Problem 14.18. A magnifying glass uses a lens whose focal length has a magnitude of 15.0 cm. The magnifying glass produces an upright image further from the magnifier than the object.

(a) Is the lens a converging or diverging lens?

(b) Is the image real or virtual?

(c) If $M = 3.0$, what are the object and image distances?

Solution

(a) We have shown in Problem 14.14 that a diverging lens always produces a smaller image. Since $|s'| > s$ we must have $|M| > 1$, so the lens must be a converging lens.

(b) Using Eq. (14.7), $M = -(s'/s)$, and the fact that M is positive (upright image) we must have that s' is negative. Therefore the image is virtual.

(c) From part (a) we know that f is positive (converging lens), from part (b) we know that $s' = -(3.0)s$, and from Eq. (14.8) we know that $1/s = 1/f - 1/s' = 1/15.0 + 1/3.0s \rightarrow (1/s)(1 - 1/3) = 1/15.0 \rightarrow s = 10.0$ cm. Then $s' = -30.0$ cm.

14.4 LENSMAKER'S EQUATION

When we discussed the image formation properties of lenses in the previous section we assumed that spherical lenses in the paraxial approximation focus light passing through the lens that comes from a given point on an object to a specific image point, and in case of light rays that are parallel to the principal axis, that point is the focal point of the lens. We wish to determine how the properties of the lens, its radii and its index of refraction influence the focal length. We can do this by calculating the position of the image formed by the first surface of a lens and using this image as the object for the second surface. For a "thin" lens we can neglect the distance between the surfaces, and the calculation is not too complicated. We can show that for a thin lens in air, made of material with an index of refraction n, and having radii R_1 and R_2:

$$1/f = (n - 1)[1/R_1 - 1/R_2] \tag{14.9}$$

where we assume light travels from left to right and radii are positive when light hits a convex surface and negative when light hits a concave surface. This is known as the **Lensmaker's equation.**

Problem 14.19. The spherical part of a thin plano-concave lens (see Fig. 14-14), made of glass of index of refraction 1.56, has a radius of magnitude 78.3 cm.

(*a*) Does it matter which of the two sides has the plane surface?

(*b*) What is the focal length of this lens?

(*c*) For any lens, does the focal length change if one reverses the lens?

Solution

(*a*) The plane side of the lens has a radius of ∞. Thus $1/R$ for the planar side equals zero. If the planar side is the first side, then the spherical side has a radius of $+78.3$ cm since its center is on the image side. If the spherical side is the first side, then its radius is negative, since its center is on the object side. When we use the Lensmaker's equation, Eq. (*14-9*), giving $1/f = (1.56 - 1)[1/R_1 - 1/R_2]$, we note that $[1/R_1 - 1/R_2]$ is negative in both cases. If the planar surface is first, then $1/R_1$ is zero, and $-1/R_2$ is negative. If the planar surface is last, then $1/R_1$ is negative and $1/R_2$ is zero. The focal length is therefore the same in both cases.

(*b*) Substituting in Eq. (*14-10*) leads to $1/f = (0.56)(-1/78.3 - 0) \to f = -140$ cm.

(*c*) When we reverse a lens, the radii both change sign, but we also interchange R_1 with R_2. This double change leaves $[1/R_1 - 1/R_2]$ unchanged, and therefore the focal length remains the same.

In Fig. 14-14 we draw the various possible shapes of converging and diverging lenses. It can be shown that if the lens is wider at the center than at the edge the lens is converging and if the lens is wider at the edge than at the center the lens is diverging.

Problem 14.20. We desire to make a converging lens from plastic, which has a refractive index of 1.31. The focal length desired is 29.3 cm.

(*a*) What radius is needed if we make it as a plano convex lens?

(*b*) What radius is needed if we make it out of a double convex lens of equal radii?

(*c*) What radius is needed for the smaller radius if we make it out of a meniscus with one radius double the other?

Solution

(*a*) We use Eq. (*14.9*) with $R_2 = \infty$. Then $1/f = 1/29.3 = (1.31 - 1)[1/R_1 - 0] \to R_1 = 9.08$ cm.

(*b*) If we have a double convex lens, then R_1 is positive and R_2 is negative. Since the radii are equal (in magnitude), $R_2 = -R_1$. Then $1/f = 1/29.3 = (1.31 - 1)[1/R_1 + 1/R_1] = (0.31)(2/R_1) \to R_1 = 18.2$ cm.

(*c*) Let us face the lens so that the first surface is convex. Then both radii are positive. (The focal length is the same no matter which side is first, so that we can choose our viewpoint to please.) Thus $1/f = 1/29.3 = (1.31 - 1)[1/R_1 - 1/2R_1]$, since we are given that one radius is twice the other. Solving for $R_1 \to R_1 = 4.54$ cm.

Although our discussion has been limited to a "thin" lens, one can handle cases of "thick" lenses as well. The first surface produces an image which is used as the object for the second surface. In the case of a **thick** lens, the object distance to the second surface must take account of the thickness of the lens. This may be simple to do in specific cases, but the general result can be quite complicated. A similar situation exists for multiple thin lens systems, which we discuss in the next section.

14.5 COMPOSITE LENS SYSTEMS

In many cases it is necessary to use more than one lens in an optical system. There are two categories that we will discuss, both using thin lenses. Category (1) is the case when the two lenses are not separated by any significant distance. This is equivalent to constructing a different thin lens from two lenses. The second category is where the lenses are separated by some distance which cannot be neglected. In both cases the method we will use is to consider the image produced by the first lens as the object for the second lens. In the case of the second category this will require two separate ray diagrams or two separate calculations, while for category (1) we will be able to consider the composite system as equivalent to a different single thin lens.

Problem 14.21. Consider two thin lenses, with focal length f_1 and f_2, that are separated by a negligible distance. Suppose an object is placed at a distance s from the first lens.

(*a*) What is the distance s'_1 from the first lens of the image caused by that lens alone?

(*b*) What is the distance to the second lens, s'_2 of the effective object it sees and what is the final image distance after passing through both lenses?

(*c*) What is the equivalent focal length of the combination of two lenses? Does it matter which lens is first?

Solution

(*a*) We use Eq. (*14.8*) for the first lens and get: $1/s + 1/s'_1 = 1/f_1 \rightarrow 1/s'_1 = 1/f_1 - 1/s$ where s is the object distance to the first lens. This equation will determine s'_1.

(*b*) The second lens sees the light refracted by the first lens, which is equivalent to it seeing light coming from the image formed by the first lens alone. Let the distance from that image to the second lens be s_2. Since the two lenses are at the same position (negligible separation), $s_2 = -s'_1$. Then, from part (*a*), $1/s_2 = [1/s - 1/f_1]$, and using Eq. (*14.8*) for lens (2) we get: $[1/s - 1/f_1] + 1/s' = 1/f_2 \rightarrow 1/s + 1/s' = 1/f_1 + 1/f_2$, where s' is the position of the final image after passing through the second lens. This last equation gives us the relationship between the original object distance s and the final image distance s' for our two lens system at a common location (pressed against each other).

(*c*) This last equation shows that the two lenses are equivalent in their effect to a single lens which has a focal length f given by:

$$1/f = 1/f_1 + 1/f_2 \tag{14.10}$$

It is clear from this formula that the focal length is the same no matter which lens was the first one. Note also that it is true for any combination of thin lenses, converging or diverging.

The composite lens has a focal length that can be calculated by adding together the reciprocals of the individual lenses. If there were more than two lenses in close proximity, we would just include the reciprocal of the focal length of that lens as well. The reciprocal of f has a special name. It is called the "**refractive power**" of the lens (or just the power of the lens), although it is not connected with the concept of power that we discussed in previous chapters.

$$\text{Refractive power of a lens} = 1/f \tag{14.11}$$

The unit for refractive power is clearly m^{-1}, which is called a diopter. Eq. (*14.10*) then states that the refractive power of a composite lens system, when the lenses are close together, is just equal to the sum of the power of the individual lenses. That is why an optometrist, when testing the effect of different lenses on the eyesight of a patient will add lenses in front of the eye to determine the best combination. When the best combination has been determined, he just adds together the power of all the lenses and prescribes a single lens with that total power. A lens with a negative focal length (a diverging lens) will have a negative power, while a converging lens has a positive power.

Problem 14.22. An optometrist finds that a patient sees best when he has a lens made from two lenses of power 2.3 diopter and −0.8 diopter.

(*a*) What is the focal length for each of the two lenses?

(*b*) What is the power and focal length of the lens that he should prescribe?

(*c*) If an object is placed 48.1 cm from the combination, where will the image be formed?

Solution

(*a*) The power equals $1/f$, as given by Eq. (*14.11*). Thus $f_1 = 1/(2.3\ \text{m}^{-1}) = 43.5$ cm and $f_2 = 1/(-0.8\ \text{m}^{-1}) = -125$ cm.

(*b*) The total power is the sum of the individual power, or $P = (2.3 - 0.8) = 1.5$ diopter. Thus $f = 1/(1.5) = 66.7$ cm.

(*c*) We use Eq. (*14.8*) for the combined lens of focal length 66.7 cm. Then $1/48.1 + 1/s' = 1/66.7 \rightarrow s' = -173$ cm. The image is virtual, upright and larger than the object.

If the lens combination is such that we cannot neglect the distance between lenses [category (2)], then we must handle each case separately to determine the effect of the combination. The principle to be applied is that the position of the image of the first lens is used as the position of the object for the second lens. If there are more than two lenses then the image of the immediately previous lens is always used as the object for the next lens, until we have accounted for all the lenses. Of course, we can use mirrors instead of lenses or in combination with lenses, using the same principle.

Problem 14.23. An object is placed 35.5 cm to the left of a converging lens of focal length 18.8 cm. A second lens, which is diverging and has a focal length of −88.8 cm is placed at a certain distance d to the right of the first lens.

(*a*) Where is the image of the object formed by the first lens and what is its magnification?

(*b*) If the distance, d, is 53.6 cm, where is the final image of the object? What are the properties of the image?

(*c*) If the distance, d, is 25.9 cm, where is the final image of the object? What are the properties of the image?

(*d*) If in part (*b*) the lenses are reversed, where is the final image of the object, and what are its properties?

Solution

(*a*) We use Eq. (*14.8*) for the first lens to get: $1/35.5 + 1/s'_1 = 1/18.8 \rightarrow s'_1 = 40.0$ cm. This is the position of the image through the first lens which is real and to the right of the lens. The magnification of the image is $M_1 = y'_1/y = -(40.0/35.5) = -1.13$.

(*b*) The distance of the image from the first lens is 40.0 cm to the right of the first lens. The second lens is 53.6 cm to the right of the first lens. Therefore the object distance to the second lens is $(53.6 - 40.0) = 13.6$ cm. Using Eq. (*14.8*) for the second lens we get: $1/13.6 + 1/s' = -1/88.8 \rightarrow s' = -11.8$ cm. Thus the final image is virtual and to the left of the second lens. The magnification produced by the second lens is $M_2 = y'_2/y'_1 = -(-11.8/13.6) = 0.868$. The total magnification

$M = y_2'/y = (y_2/y_1')(y_1'/y) = -0.98$. In general, we see that:

$$M = M_1 M_2 \text{ (for more lenses, } M_1 M_2 M_3 \ldots) \tag{14.12}$$

In this case, the total magnification is -0.98, indicating that the final image is inverted, and slightly smaller than the object.

(*c*) In this case the image through the first lens is 40.0 cm to the right of the first lens. The second lens, however, is only 25.9 cm to the right of the first lens. Thus the object for the second lens is at a distance of $(40.0 - 25.9) = 14.1$ cm to the right of the second lens. This is in the image region of the second lens, and thus the first image forms a virtual object for the second lens and the distance of this object from the second lens is therefore negative, $s_2 = -14.1$ cm. Using this value in Eq. (*14.8*) we get: $1/(-14.1) + 1/s' = -1/88.8 \rightarrow s' = 16.8$ cm. The image is therefore real at a distance of 16.8 cm to the right of the second lens. The reason that we obtained a real image from a diverging lens is that we had a virtual object. In other words the diverging lens intercepted the light before it had a chance to form the image of the first lens—which would have been indistinguishable from a real object to any subsequent lens if it were allowed to form. (For a real object we showed earlier that a diverging lens always produces a virtual image.) The magnification of the second lens is $M_2 = -(16.8/-14.1) = 1.19$, and the total magnification is $M = M_1 M_2 = (-1.13)(1.19) = -1.35$. The image is therefore inverted and somewhat larger than the object.

(*d*) If we reverse the order of the lenses, then $1/35.5 + 1/s_1' = -1/88.8 \rightarrow s_1' = -25.4$ cm, i.e. to the left of the diverging lens. Then $s_2 = 53.6 + 25.4 = 79.0$ cm, and $1/79.0 + 1/s' = 1/18.8 \rightarrow s' = 24.7$ cm. The image is real and 24.7 cm to the right of the converging lens. To get the magnification, we calculate M_1 and M_2. $M_1 = -(-25.4/35.5) = 0.715$, and $M_2 = -(24.7/79.0) = -0.312$, giving $M = -0.224$. Thus the image is inverted and much smaller than the object. Comparing the results of part (*b*) we see the resulting image is quite different and the order of the lenses is therefore important in this case.

In order to illustrate what happens in both of the cases (*b*) and (*c*) of this problem, we draw the ray diagrams for each case. In Fig. 14-18(*a*), we draw the diagram for case (*b*). The image from the first lens is at I_1, and serves as an object for the second lens. The focal point for the second lens is to the left (near the original object) and the parallel ray appears to be refracted from that point. The final image is to the left of the second lens, is virtual, inverted and slightly smaller than the object. In Fig. 14-18(*b*), we draw

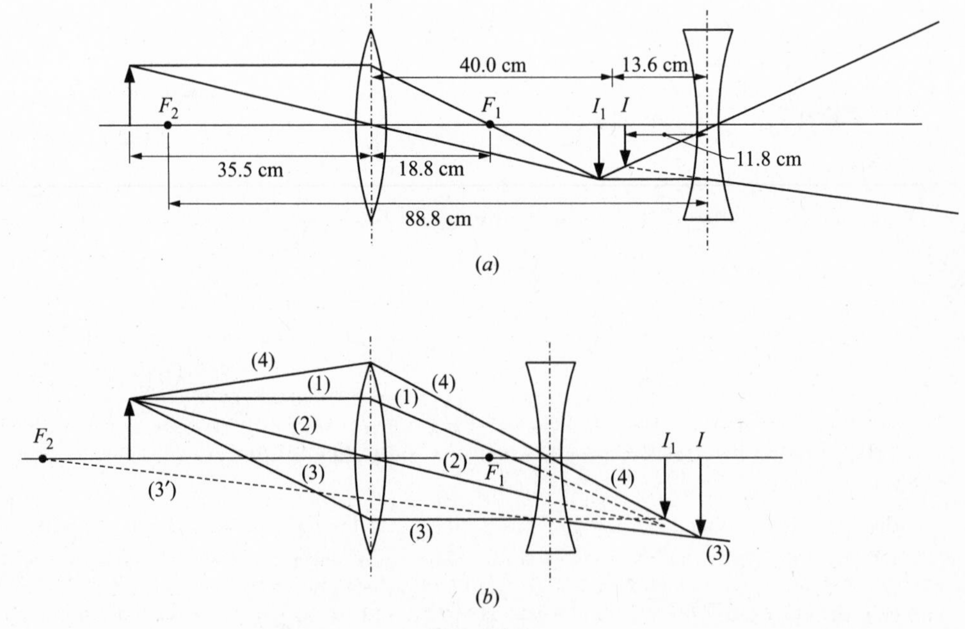

Fig. 14-18

the diagram for case (*c*). Again, the first lens refracts the light from the object to form an image at a distance of 40.0 cm from the first lens. Before reaching that point, the light encounters the second lens. The rays from the first lens would reach the position of the image I_1 were it not for the second lens. This image is therefore a virtual object for the second lens. Rays from the first lens that would have reached the image I_1 are instead refracted by the second lens. The light that emerges from lens 1 parallel to the principal axis (ray 3) is diverged by the second lens as if it had been traveling from F_2 (ray 3′), the focal point of the second lens. Light from the first lens that travels through the center of the second lens (ray 4) is undeflected. These two rays meet at the final image point I to form a real, inverted and larger image.

Problem 14.24. An object is placed 35.5 cm to the left of a converging lens of focal length 18.8 cm. A convex mirror that has a focal length of -88.8 cm is placed at a certain distance d to the right of the first lens.

(*a*) Where is the image of the object formed by the first lens and what is the magnification?

(*b*) If the distance, *d*, is 53.6 cm, where is the final image of the object? What are the properties of the image?

(*c*) If the distance, *d*, is 25.9 cm, where is the final image of the object? What are the properties of the image?

(*d*) If in part (*b*) the mirror was changed to one that is concave, with the same magnitude of focal length, where is the final image of the object, and what are its properties?

Solution

(*a*) We use Eq. (*14.8*) for the lens to get: $1/35.5 + 1/s'_1 = 1/18.8 \rightarrow s'_1 = 40.0$ cm. This is the position of the image to the right of the lens. The magnification of the image is $M_1 = y'_1/y = -(40.0/35.5) = -1.13$. [This is the same result as Problem 14.23(*a*).]

(*b*) The distance of the image from the lens is 40.0 cm to the right of the lens. The mirror is 53.6 cm to the right of the lens. Therefore the object distance to the mirror is $(53.6 - 40.0) = 13.6$ cm. Using Eq. (*14.6*) for the mirror we get: $1/13.6 + 1/s' = -1/88.8 \rightarrow s' = -11.8$ cm. Thus the final image is virtual, and therefore to the right of the mirror. The magnification produced by the mirror is $M_2 = y'_2/y'_1 = -(-11.8/13.6) = 0.868$. The total magnification is $M = M_1 M_2 = -0.98$, indicating that the final image is inverted, and slightly smaller than the object. These are the same results as Problem 14.23(*b*) except that the virtual image formed by the mirror is behind (to the right of) the mirror while the virtual image of the concave lens in Problem 14.23(*b*) is to the left of the lens (i.e. the side opposite to where the light emerges.)

(c) In this case the image through the lens is 40.0 cm to the right of the lens. The mirror, however, is only 25.9 cm from the lens. Thus the object for the mirror is at a distance of $(40.0 - 25.9) = 14.1$ cm to the right of the mirror. This is not in the object region of the mirror, and the object distance is therefore negative, $s_2 = -14.1$ cm. We call this a virtual object for the mirror. Using this value in Eq. (*14.6*) we get: $1/(-14.1) + 1/s' = -1/88.8 \rightarrow s' = 16.8$ cm. The image is therefore real, at a distance of 16.8 cm to the left of the mirror. (The reason that we obtained a real image from a convex mirror is that we had a virtual object. For a real object we know that a convex mirror always produces a virtual image.) The magnification of the mirror is $M_2 = -(16.8/-14.1) = 1.19$, and the total magnification is $M = M_1 M_2 = (-1.13)(1.19) = -1.35$. The image is therefore inverted and somewhat larger than the object. (Again, the same result as Problem 14.26(*c*) except that the real image of the mirror is to the left of the mirror while the real image of the concave lens is to the right of the lens.)

(d) The lens is still the same as in the previous parts and produces a real image at a distance of 40.0 cm to the right of the lens with an amplification of -1.13. The object distance to the mirror is $(53.6 - 40.0) = 13.6$ cm, as previously. The focal length of the mirror is now $+88.8$ cm, and therefore $1/13.6 + 1/s' = 1/88.8 \rightarrow s' = -16.1$ cm. The image is therefore virtual (to the right of the mirror). The magnification of the mirror is $M_2 = -(-16.1/13.6) = 1.18$, giving $M = (-1.13)(1.18) = -1.33$. The image is therefore inverted and larger than the object.

14.6 OPTICAL INSTRUMENTS

Many optical instruments, based on the principles that we have discussed, are part of our daily lives. They include simple single lens systems, such as magnifying glasses, the eyeglasses that we wear and the eye itself, and simple cameras. Most systems that are used in industry and in research are made of combinations of lenses. The calculation of their properties in actual cases is usually done by computers, using the fairly simple formulas that we have developed, and applying them to a host of applications. In this section we will discuss the general ideas involved in several sample instruments.

Magnifying Glass

A **magnifying glass** generally consists of a converging lens, with the object (a printed page, for instance) placed closer to the lens than the focal point. We showed in Problem 14.15 that a converging lens produces a larger, virtual and upright image whenever the object distance is less than the focal length. Therefore, a person looking at written material through a converging lens will see the writing behind the magnifying glass without inversion.

The magnification of magnifiers is usually expressed in terms of the "**angular magnification**". If the angle subtended at the eye by the magnified image of an object is θ', and the largest angle that the object can subtend at the naked eye and still be in focus is θ, then for small angles θ, θ' (less than about 1/4 rad or 15°) the angular magnification is expressed as θ/θ'. Angular magnification is a better measure of human perception of size increase than ordinary magnification because the size of an object, as seen by the eye, depends on the distance of the object from the eye. Thus all three arrows in Fig. 14-19(*a*) appear the same size to the eye even though they actually differ in size. If an object is nearer the eye, then the rays from the upper and lower parts of the object subtend a larger angle at the eye than the same object does when, as shown it is far away, as shown in Fig. 14-19(*b*). (The real image formed on the retina is roughly proportional in size to the angle subtended by these incoming rays). If one wishes to have the image appear larger, i.e. cover a larger area on the retina, one must increase the angle subtended by the incoming rays. This is what one does when one moves a page with writing as close to the eye as possible. There is, however, a closest point to which the eye can accommodate, called the near point of the eye. At smaller distances from the eye the eye cannot bring the object into focus. Thus the best that one can do without a magnifying glass is to have the object subtend the angle it would have if placed at the near point Fig. 14-19(*b*). For small angles this angle is (in radians) approximately equal to

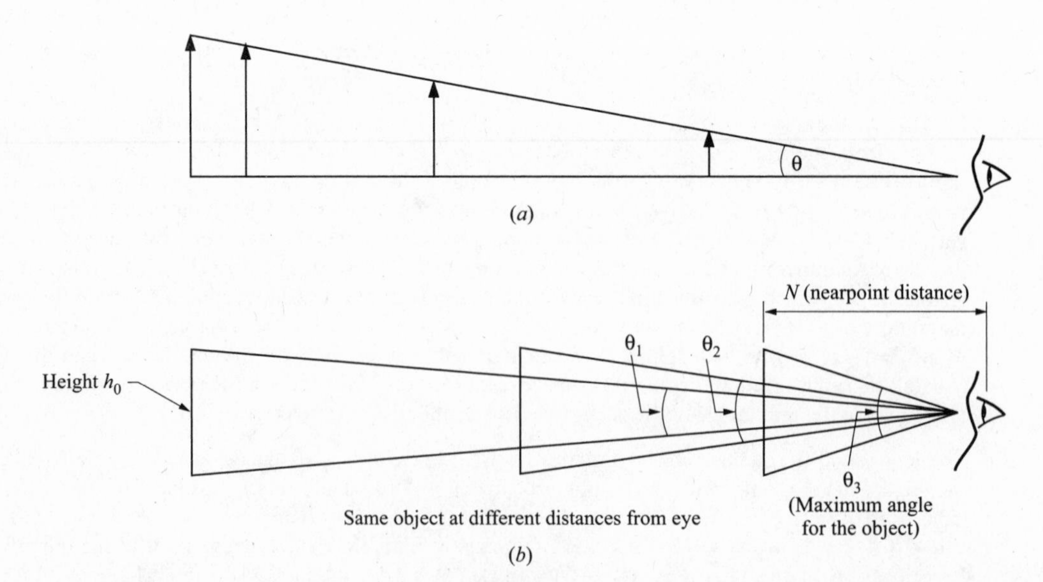

Fig. 14-19

h_0/N, where h_0 is the height of the object and N is the near point for the eye (≈ 25 cm for a young healthy eye). Note that the angle shown in Fig. 14-19(*b*) is not a small angle. When we place a magnifying glass in front of the eye, with the eye *very near* the magnifying glass, the rays entering the eye come from a virtual image of height $|y'|$ at a distance s' behind the lens (and the eye). The angle subtended by this image is $|y'/s'|$. Recalling that $|y'/y'| = M = |s'/s'|$ we have $|y'|/|s'| = |y|/|s| = h_0/s$, with h_0 the object height and s the object distance. The angular magnification is then $(h_0/s)/(h_0/N) = N/s$. Because of the magnifying glass, s can be made smaller than N since it is the image distance $|s'|$ that must be $\geq N$. We know that $1/s + 1/s' = 1/f$, or $1/s = 1/f - 1/s'$, with s' negative ($s < f$), or $1/s = 1/f + 1/|s'|$. To make s small we need to increase $1/s$, which corresponds to increasing $1/|s'|$ or decreasing $|s'|$. But the minimum $|s'|$ is N, the near point for the eye. Then $1/s = 1/f + 1/N$, and the maximum angular magnification is $N/s_{\min} = N(1/f + 1/N)$ or

$$\text{Angular magnification} = N/f + 1 \qquad (14.13a)$$

Often it is more relaxing to observe the image far away rather than at the near point. If we assume $|s'| = \infty$ then $s \approx f$ and $N/s \approx N/f$ so in this case:

$$\text{Angular magnification} = N/f \qquad (14.13b)$$

If $N \gg f$ the two formulas give results that are comparable so it is not worth the strain to maximize the angular magnification by having s' at the near point.

Problem 14.25. A jeweler examines a 2.0 cm diamond through a small magnifying glass (called a loupe), and wants to have an angular magnification of 3.50. The near point of his eye is 30.0 cm, and he wants the image of the diamond to be located at the near point.

(*a*) What is the angle subtended by the diamond without the loupe when the diamond is at the near point of his eye?

(*b*) What angle is subtended by the virtual image when he uses the magnifier, with the image at the near point?

(*c*) What focal length should the loupe have?

(*d*) For the same loupe, what would the angular magnification be if he adjusts the distances so that the image appears at a great distance (∞)?

Solution

(*a*) The angle subtended (in radians) is $\theta = h_0/N = (2.0 \text{ cm})/(30.0 \text{ cm}) = 0.0667$ rad. (The small angle approximation is thus valid for this case.)

(*b*) The angular magnification is 3.50, so the angle subtended by the image is $(3.50)(0.0667) = 0.233$ rad, still within the small angle approximation.

(*c*) We determined that the angular magnification when the virtual image is at the near point is $N/f + 1 = 3.50$. Thus, $N/f = 30.0/f = 2.50, f = 12.0$ cm.

(*d*) Now the angular magnification is given by $N/f = 30/12 = 2.5$. In this case there is a clear advantage to observing the image at the near point.

Camera

A **camera** basically consists of a converging lens that forms a real image on a film placed at a certain distance from the lens. Once the focal length of the lens has been chosen, the lens will form this image only for objects at the appropriate distance from the camera. By moving the lens back and forth slightly one can arrange to focus on objects at the desired distance in front of the lens. For most cameras the distance of the lens to the object, s, is much greater than the focal length, so $1/s \ll 1/f \to 1/$

$s' \approx 1/f$ or $s' \approx f$. Thus, the film is approximately at the focal point of the lens. Then $|y'/y| = |f/s|$ or $|y'| = |fy/s|$. For a given object and distance s, the linear dimensions of the image on the film is thus proportional to f. Most good cameras have multiple lenses in order to avoid some of the aberrations due to dispersion, the existence of non-paraxial rays, etc. Additionally, cameras generally have mechanisms to vary the time that a shutter is open to admit light (a fast exposure time allows capture of the image of an object in motion, but reduces the intensity of the light), and the size of the aperture that admits the light (called the diaphragm). A large diameter diaphragm allows more light to enter, but causes difficulty with aberrations. The camera is often characterized by an "***f*-number**", such as $f/8$ (f-number of 8), which is the ratio of the focal length to the diameter of the diaphragm. The f-number is a key factor in determining the proper exposure time needed to produce a good image on a given quality film. The exposure time needed depends on the intensity of the light hitting the film, I. For a given object, and set of lighting conditions, the intensity is the incoming energy/unit time divided by the area of film corresponding to the image. Since the linear dimensions of the image are proportional to the focal length, f, the intensity is inversely proportional to f^2. Similarly the incoming energy/unit time is proportional to the area of the aperture which goes as the diameter, D, squared. Thus, $I \sim D^2/f^2 = 1/(f\text{-number})^2$. Since I is inversely proportional to the time t needed for exposure we have: exposure time $\sim (f\text{-number})^2$. Thus a larger f-number means an increase in exposure time. TV cameras are basically similar to the photographic camera, except that the image is formed on an electronic medium rather than on a film.

The Eye

Although the eye is not a man made instrument, it clearly is an example of optics at work. The **eye** is very similar to a camera in the sense that a real image is formed by a lens on the retina, where nerve endings transform the light into impulses that are transmitted to the brain for analysis. The eye actually consists of several different parts that have different indices of refraction and are collectively responsible for forming an image on the retina. It is clear, however, that, in order to focus all the light onto the retina irrespective of the distance of the source from the eye, there must be some mechanism for changing the focal length of the lenses in the eye. To do this, there is one part of the eye, called the lens, that is attached to muscles that can stretch the lens and change the radii of its surface, thus altering the focal length. The eye automatically and rapidly changes this focal length, as needed, within its capacity to change. This capacity generally allows a healthy eye to focus on objects (this is called **accommodation**) from a far point (normally infinitely far away), to a near point (which increases with age). This near point is approximately 25 cm, and increases to about 50 cm at age 40, and several hundred cm at age 60, as the material of the lens stiffens with age. If the eye cannot focus on a point infinitely far away because the lens is too converging, then the eye is called **myopic** (nearsighted). In this case the far point has been decreased from infinity to some point nearer the eye. If the eye cannot focus on objects too near the eye because the lens is insufficiently converging, then the eye is called **hyperopic** (farsighted). In this case the near point has moved too far away, as generally happens with age. One can often easily correct for these defects with glasses (or contact lenses). For instance, the myopic eye can be helped by using a diverging lens to form a virtual image of the (infinitely far away) object at a point within the far point of the eye. Similarly, one can correct for the hyperopic eye by using a converging lens to form a virtual image further away than the nearby object (as a magnifying glass does), thereby effectively moving the object past the near point. The eye also adjusts for different light intensities by automatically decreasing or increasing the diameter of the exposed portion of the lens. This is caused by reflex action to protect the eye from sudden changes in intensity.

Problem 14.26. An elderly person has a hyperopic eye with a near point of 200 cm. He wishes to be able to read a book placed 25 cm from his eye. What is the power of the corrective lens that is needed to correct his vision for this purpose?

Solution

In order to correct his vision we need a lens that will take an object at a distance of 25 cm and form an image at 200 cm. Since the image is on the same side of the lens as the object it will be virtual and s' is negative. Thus we require Power $= 1/f = 1/s + 1/s' = 1/25 - 1/200 = 0.035\ \text{cm}^{-1} = 3.5$ diopter. (The focal length required is 28.6 cm, but the calculation is clearly most easily done in terms of power.)

Similar calculations can obviously be made for a myopic eye. There are many other possible defects of the eye, some of which can also be corrected with lenses.

Projector

In a camera, light from a large object is focused by a lens onto a photographic plate to form a real, inverted and smaller image. In the case of a **projector**, light from a slide is focused onto a screen to form a real, inverted (the slide is placed upside down in the projector) and larger image. (Usually a powerful light source is used to shine light through the slide.) The projector is thus, in some sense, the reverse of a camera. The focusing in the case of a projector is accomplished by moving the projection lens closer to or further away from the slide. In both cases, a converging lens is required, since a diverging lens always forms a virtual image of an (real) object. The difference is that a camera forms a smaller image and a projector forms a larger image.

Problem 14.27. A converging lens with a focal length of f is used to form a real image. We already showed (Problem 14.15) that if the object is nearer the lens than f, then the image is virtual.

(*a*) Show that the image will be real and smaller than the object if the object distance is greater than $2f$, but will be real and larger than the object if the object distance is between f and $2f$.

(*b*) What happens if the object distance is $2f$? What if it is f?

(*c*) A projector with a simple converging lens projects the image of a 2 cm × 2 cm slide so as to fill a 2 m × 2 m screen placed 10 m from the projector lens. Find the distance of the slide to the lens and the magnification.

(*d*) Find the focal length of the lens.

Solution

(*a*) The equation for s' can be written as $s' = fs/(s-f)$, as we did in Problem 14.15. If $s > 2f$, then $(s-f) > f$, and $f/(s-f) < 1$. Thus $s' < s$, and is positive. Therefore, the image is real, and the magnification $|M| = |s'/s| < 1$. The image is therefore smaller than the object. For $f < s < 2f$, we have $0 < (s-f) < f$. Then $fs/(s-f)$ is still positive giving a real image. However, $f/(s-f) > 1$, meaning that $s' > s$, and $|M| > 1$, giving a larger image.

(*b*) If $s = 2f$, then $s' = fs/(2f - f) = s$, and the image is real and the distance of the image past the lens is the same as the object distance in front of the lens. The magnification is $|M| = 1$, and the image has the same size as the object. If $s = f$ then our formula for s' has a denominator of zero. We therefore return to the original equation, $1/f = 1/s + 1/s' \rightarrow 1/f = 1/f + 1/s' \rightarrow 1/s' = 0$. Thus s' is infinite, meaning that the image has been projected to a distance infinitely far away. This is equivalent to saying that the rays never converge on a point, nor can they be projected back to a point, implying that the rays emerge from the lens as parallel rays.

(*c*) The magnification is the same along any linear dimension, so $|M| = 200\ \text{cm}/2\ \text{cm} = 100$. Also, $|M| = |s'/s| \rightarrow 100 = 10\ \text{m}/s \rightarrow s = 0.1\ \text{m} = 10\ \text{cm}$.

(*d*) $1/f = 1/s + 1/s' = 1/10 + 1/10{,}000 \rightarrow f = 9.99$ cm.

Note. As required for a real image, $s > f$, and for large magnification s is very close to f.

Microscope

The simple magnifying glass is limited in the angular magnification that it can usefully produce by the need to have a very small focal length [Eq. (*14.14*)] and therefore a very "fat" lens. Such a lens does not satisfy the paraxial approximation and gives rise to significant aberrations. To increase the magnification, one can use a "**compound microscope**", often called microscope for short, which consists of two or more lenses significantly separated from each other. In effect, we use a regular magnifying glass as an "**eyepiece**", and an earlier (closer to the object) converging lens of relatively small focal length as a premagnifier to form a larger real image that serves as the object for the eyepiece. This latter converging lens is called the "**objective**", and forms an image very near the focal point of the eyepiece. The eyepiece is then used to examine this real image and form a virtual image of it near infinity, which is then viewed by the eye placed against the eyepiece. The microscope is focused by moving both the objective and the eyepiece. Even in the compound microscope large magnification can only be obtained by using compound and corrected lenses to avoid distortion. To find the angular magnification of the microscope we consider Fig. 14-20(*a*), where the object of height h_0 is just beyond the focal length, f, of the objective and forms a real image of height h' just within the focal length, f_2, of the eyepiece. When the human eye looks through the eyepiece it sees a virtual image in the far distance. To obtain the angular magnification we define $\theta =$ angular size of the object as seen by the naked eye at its near point, so $\theta = h_0/N$, $\theta' =$ angular size of the real image as seen by the naked eye at its near point, so $\theta' = h'/N$, and $\theta'' =$ angular size of the final virtual image of the eyepiece as seen by the eye when looking through the eyepiece. We assume the virtual image is viewed at ∞. From our earlier discussion of the magnifying glass we know that $\theta''/\theta' = N/f_2$ [Eq. (*14.13b*)]. We also have that $\theta'/\theta = (h'/N)/(h_0/N) = h'/h_0 = |M_1| = |s'/s| \approx s'/f_1$ (since $s \approx f_1$). Then the angular magnification of the microscope is given by: angular magnification $= \theta''/\theta = (\theta''/\theta')(\theta'/\theta) = (N/f_2)(s'/f_1)$, or:

$$\text{Angular magnification} = Ns'/f_1 f_2 \qquad (14.14a)$$

If the distance between the lenses is L, then $L \approx s' + f_2$. If $f_2 \ll L$, then $s' \approx L$ and we have:

$$\text{Angular magnification} \approx NL/f_1 f_2 \qquad (14.14b)$$

Problem 14.28. The focal length of the objective and eyepiece of a compound microscope are 1.5 cm and 2.5 cm, respectively. The objective observes a plant cell 1.52 cm from the objective. Assume $N = 25$ cm.

(*a*) Find the real image distance s' due to the objective, and the lens separation L.

(*b*) Find the angular magnification using Eq. (*14.14a*); using approximate Eq. (*14.14b*).

Solution

(*a*) $1/s' = 1/f - 1/s = 1/1.5 - 1/1.52 \rightarrow s' = 114$ cm. $L \approx s' + f_2 = 116.5$ cm.

(*b*) Angular magnification = 760; Angular magnification ≈ 777, a fairly good approximation, since it is off by only 2%.

Telescope

A telescope is generally used for viewing objects that are very distant. Because of their distance, they appear very small to the naked eye, and subtend a small angle. In order to be able to distinguish features of the object one tries to magnify this angle so that the object will appear larger. A telescope is therefore rated by its angular magnification. Generally, the **telescope**, like the microscope, is composed of two optical elements. One is the eyepiece (also called the ocular) which is a magnifying glass. The element which first receives the light from the object is again called the objective, which forms a real image of this distant object. Since the object is distant, this image is formed at nearly the focal point of the objective. This image is then magnified by the ocular and viewed by the observer, either directly or

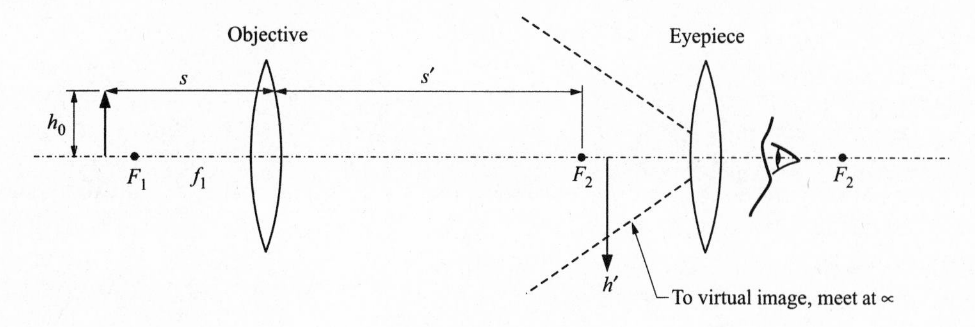

(*a*) Microscope schematic

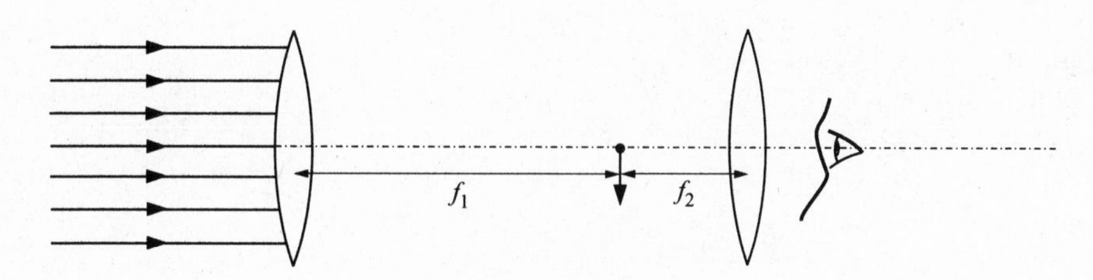

(*b*) Telescope schematic (refracting)

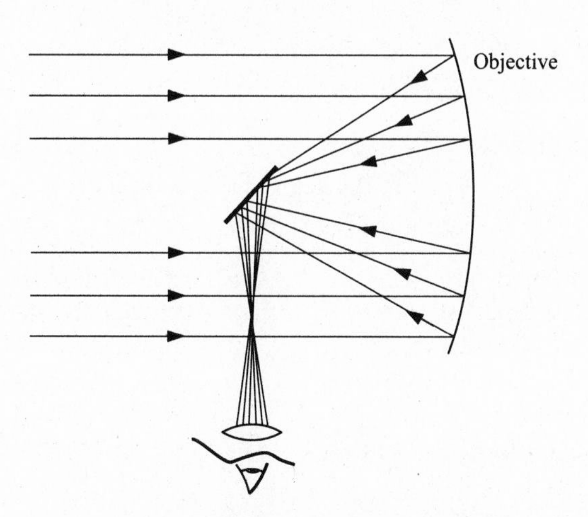

(*c*) Telescope schematic (reflecting)

Fig. 14-20

on some recording material (film or electronic detectors). The objective is often a concave mirror, since large mirrors are more easily constructed than large solid lenses. (We want a large diameter objective in order to gather lots of light to the image and to be able to see faint objects.) In order to avoid problems with the need for paraxial rays, large telescopes usually employ parabolic surfaces. Telescopes with converging lenses as objectives are called **refracting telescopes** while those with concave mirrors as objectives are called **reflecting telescopes**. In Fig. 14-20(*b*) we show a refracting telescope and in Fig. 14-20(*c*) a reflecting telescope. A small 45° flat mirror is used to allow the image to be observed by the ocular at 90° and out of the way of the incoming light.

For a telescope the angular magnification is defined, as before, as the ratio of the angular size seen with the instrument divided by that with the unaided eye. However, in this case the angular size of the object observed with the naked eye cannot be the largest possible value (that with the object at the near point) since the object being observed is far away. The angular size of the object seen with the unaided eye is thus the actual object size divided by the distance to the object, or the angle, θ, subtended at the eye by the rays coming from the top and the bottom of the object. Since the object is distant, this is the same angle the object subtends at the objective of the telescope. We assume that when the observer looks through the ocular of the telescope he will see the virtual image at ∞. Then the image formed by the objective is not only at the focal distance, f_1, from the objective, but also at the focal distance, f_2, of the ocular. (The distance between the two lenses is thus $L = f_1 + f_2$.) Suppose the telescope is aimed so that light from one end of the object travels parallel to the common principal axis of the two lenses. We follow one ray in particular, the one along the principal axis. This ray goes straight through both lenses to the eye, so the corresponding end of the final virtual image lies along that axis as well. Light from the other end of the object will enter the objective at the angle θ, and we consider the particular ray that travels through the center of the objective lens. That ray passes undeviated through the lens and continues to the point of the real image corresponding to that end of the real image formed by the objective. Assuming that image has height h, we then have $h/f_1 = \theta$ (for small θ). If we now consider a ray passing through the same point on the real image that happens to pass through the center of the ocular, that ray is also undeviated. Since that ray corresponds to the other end of the object, the angle it makes with the principal axis is the angular size of the final image, θ'. Then $\theta' = h/f_2$. The angular magnification is then $\theta'/\theta = (h/f_2)/(h/f_1) = f_1/f_2$. A minus sign can be inserted to indicate that the final image is inverted relative to the object and we have:

$$\text{Angular magnification} = -f_1/f_2 \tag{14.15}$$

Problem 14.29. A refracting telescope has a length of 1.20 m from objective to ocular. What is the angular magnification if the ocular has a focal length of 2.5 cm?

Solution

$f_1 = 120\text{ cm} - 2.5\text{ cm} = 117.5\text{ cm}$. Therefore, the angular magnification $= -117.5/2.5 = -47$. The image is inverted and looks 47 times larger in any linear direction.

Problems for Review and Mind Stretching

Problem 14.30. Two plane mirrors make an angle of 30°, as in Fig. 14-21. An object is placed at a distance L from the vertex, midway between the mirrors. Determine the position of all the images in the mirrors.

Solution

We proceed by locating the image in each of the mirrors. The image in mirror 1 is obtained by locating the point along the perpendicular to the mirror at the same distance on the other side of the mirror. This is point ① below the x axis. The image in mirror 2 is obtained by taking the point on the perpendicular to the mirror and at the same distance on the other side of the mirror. This is a point on a line at 30° behind the mirror and at a distance L from the vertex, and is labeled ②. We now take these images, and image them in both mirrors. Taking point ②, and imaging in mirror 1, gives point ③. Taking point ③ and imaging in mirror 2, gives a point on a line 30° behind the extension of mirror 2 and at a distance L from the vertex. This is point ④. Finally we get point ⑤ by imaging point ④ in mirror 1. We note that point ⑤ is also the image of point ③ in mirror 2, and there are therefore no other images. If the object had not been so symmetrically located, there would be many more images.

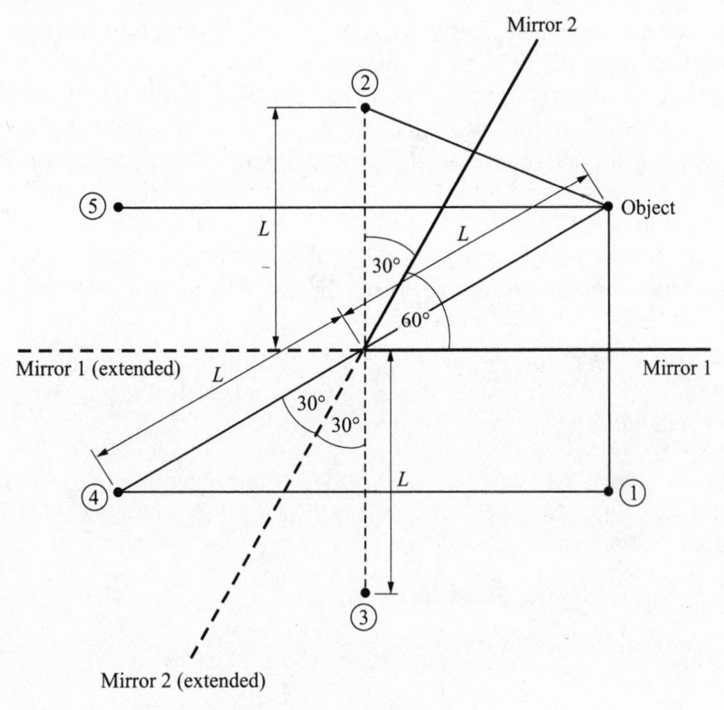

Fig. 14-21

Problem 14.31.

(*a*) For a thin diverging lens, with focal length *f*, draw a ray diagram for the image produced by an object at a distance s from the lens.

(*b*) Use this ray diagram to show that $1/s + 1/s' = 1/f$ for a diverging lens.

Solution

(*a*) The ray diagram is drawn in Fig. 14-22 for $s > f$. The image is virtual, upright and smaller than the object. In the diagram, the length of the distance $OF = -f$, since *f* is negative, and, similarly, the length of the distance OI is $-s'$, since s' is negative.

(*b*) From triangles $S'SO$ and $I'IO$ we can deduce that $S'S/I'I = s/(-s')$ (remember that s' is negative). Similarly, from triangles $O'OF$ and $I'IF$ we can deduce that $O'O/I'I = S'S/I'I = (-f)/(-f + s') \rightarrow s/(-s') = (-f)/(s' - f) \rightarrow ss' - sf = s'f \rightarrow ss' = sf + s'f \rightarrow 1/s + 1/s' = 1/f$.

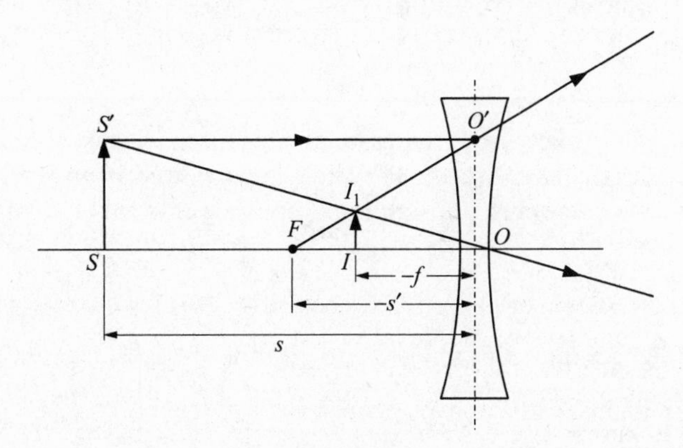

Fig. 14-22

Problem 14.32. We want to use a mirror to produce a real image 25.0 cm from the mirror, which is 1.90 times as large as the object.

(*a*) Where should we place the object?

(*b*) What focal length and what radius must the mirror have?

Solution

(*a*) Since we are using a real object, s must be positive, and since we desire a real image, s' must also be positive. Therefore the magnification, $M = -s'/s$ must be negative and the image will be inverted. Then $|M| = s'/s = 1.90 = 25.0/s \rightarrow s = 13.2$ cm.

(*b*) Using $1/s + 1/s' = 1/f$ for the mirror, we get: $1/13.2 + 1/25.0 = 1/f \rightarrow f = 8.6$ cm. We know that $f = r/2$ for a mirror, so $R = 17.2$ cm. Since R is positive, the mirror is concave. (We also know it is concave because it must converge light to a real image.)

Problem 14.33. A converging lens, with a focal length of 13.9 cm, is used to produce an image at a distance of 57 cm *from the object.*

(*a*) How far from the lens should we place the object?

(*b*) What is the magnification of the image?

Solution

(*a*) Since we are given the distance between the object and the image, we have $(s + s') = 57$. We use this together with Eq. (*14.8*) to determine s and s' separately. From the above, we have $s' = 57 - s$. We substitute this into $1/s + 1/s' = 1/f \rightarrow 1/s + 1/(57 - s) = 1/13.9$. Multiplying by $s(57 - s)(13.9)$ and simplifying, we get the quadratic equation: $s^2 - 57s + (57)(13.9) = 0$. We use the formula for solving a quadratic equation to get: $s = [57 \pm \sqrt{79.8}]/2 \rightarrow s = 33.0$ cm or $s = 24.0$ cm. The two solutions correspond to the fact that s and s' are symmetrical in the equations. If $s = 33.0$, then $s' = 24.0$, and if $s = 24.0$ then $s' = 33.0$.

(*b*) We know that $M = -s'/s$. Therefore, if $s = 33.0$ then $M = -(24.0/33.0) = -0.73$. If $s = 24.0$, then $M = -(33.0/24.0) = -1.38$.

Problem 14.34. A thin lens is made of glass ($n = 1.41$) and has a focal length of 76.7 cm. If the same lens had been made of sapphire ($n = 1.77$), what focal length would the lens have?

Solution

We know that for glass $1/f_g = (n_g - 1)(1/R_1 - 1/R_2) = 76.7$ with $n_g = 1.41$. For sapphire, $1/f_s = (n_s - 1)(1/R_1 - 1/R_2)$ with $n_s = 1.77$. Thus $f_s/f_g = (n_g - 1)/(n_s - 1) = (1.41 - 1)/(1.77 - 1) = 0.532 \rightarrow f_s = 0.532(76.7) = 40.8$ cm.

Problem 14.35. A compound microscope consists of an objective lens of focal length $f_O = 0.410$ cm, and an eyepiece of focal length $f_E = 3.00$ cm. The slide to be viewed is placed at a distance of 0.420 cm from the objective. The viewer, looking through the eyepiece wants to see a virtual image at his normal reading distance of 25 cm.

(*a*) How far from the objective is the image of the slide after passing through only the objective lens?

(*b*) How far from the eyepiece must this image be to give the virtual image that the viewer desires?

(*c*) What is the distance that one must have between the two lenses?

(*d*) What is the angular magnification for this case?

Solution

(*a*) We use Eq. (*14.8*) for the objective lens $\rightarrow 1/0.420 + 1/s'_O = 1/0.410 \rightarrow 1/s'_O = 1/0.410 - 1/0.420 \rightarrow 1/s'_O = (0.420 - 0.410)/(0.410)(0.420) = 0.581 \rightarrow s'_O = 17.22$ cm.

(*b*) We now use Eq. (*14.8*) for the eyepiece $\rightarrow 1/s_E + 1/(-25.0) = 1/3.00 \rightarrow s_E = 2.68$ cm.

(*c*) The real image distance from the objective, s'_O, plus the distance from that image to the eyepiece, s_E, equals L. Thus $L = 17.22 + 2.68 = 19.9$ cm.

(*d*) We use Eq. (*14.14a*) (as an approximation since the final image is at the near point rather than at ∞) to get: angular magnification $= Ns'/f_1 f_2 = (25 \text{ cm})(17.22 \text{ cm})/(3)(0.41) = 350$.

Supplementary Problems

Problem 14.36. An object is approaching a plane mirror along its normal with a speed of v.

(*a*) With what speed is the image moving and in which direction?

(*b*) With what speed is the image approaching the object?

Ans. (*a*) speed $= v$, toward mirror from the back; (*b*) $2v$

Problem 14.37. An object is moving in front of a plane mirror along a line parallel to its surface with a speed of v.

(*a*) With what speed is the image moving and in which direction?

(*b*) With what speed is the image approaching the object?

Ans. (*a*) v in same direction as object; (*b*) 0

Problem 14.38. An object is approaching the front of a plane mirror along a line at an angle of 53° to the normal with a speed of v.

(a) With what speed is the image moving and in which direction?

(*b*) With what speed is the image approaching the object?

Ans. (*a*) v, toward the mirror from the back at an angle of 53° to the normal; (*b*) $2v \cos 53°$

Problem 14.39. A plane mirror initially sits with its mirrored surface perpendicular to and facing the negative x axis, intersecting the axis at the origin. It is then rotated about the z axis so that its surface makes an angle of 53° with the negative x axis. An object is on the negative x axis at a distance of 3.7 cm from the origin.

(*a*) How far from the origin is the image?

(*b*) At what angle with the negative x axis is the line from the origin to the image?

(*c*) What are the x and y coordinates of the position of the image?

Ans. (*a*) 3.7 cm; (*b*) 106°; (*c*) $x = 1.02$ cm, $y = 3.56$ cm (or -3.56 cm depending on direction of rotation)

Problem 14.40. A convex mirror has a radius of 3.0 cm. An object is placed at a distance of 2.0 cm from the mirror.

(*a*) Where is the image formed?

(*b*) What is the magnification produced?

(*c*) Draw a ray diagram for this case.

Ans. (*a*) -0.86 cm; (*b*) 0.43; (*c*) see Fig. 14-23

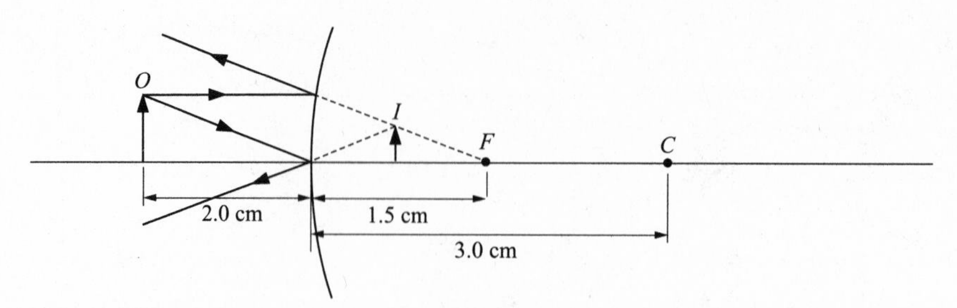

Fig. 14-23

Problem 14.41. A concave mirror has a radius of 10.0 cm. An object is placed 3.11 cm in front of the mirror.

(*a*) Where is the image formed?

(*b*) What is the magnification produced?

(*c*) Draw a ray diagram for this case.

Ans. (*a*) -8.22 cm; (*b*) 2.64; (*c*) see Fig. 14-24

Problem 14.42. A concave mirror has a radius of 17.8 cm. An object is placed in front of the mirror, and the mirror has a magnification of -2.5.

(*a*) Where was the object placed?

(*b*) Where was the image formed?

(*c*) Draw a ray diagram for this case.

Ans. (*a*) 12.5 cm; (*b*) 31.15 cm; (*c*) see Fig. 14-25

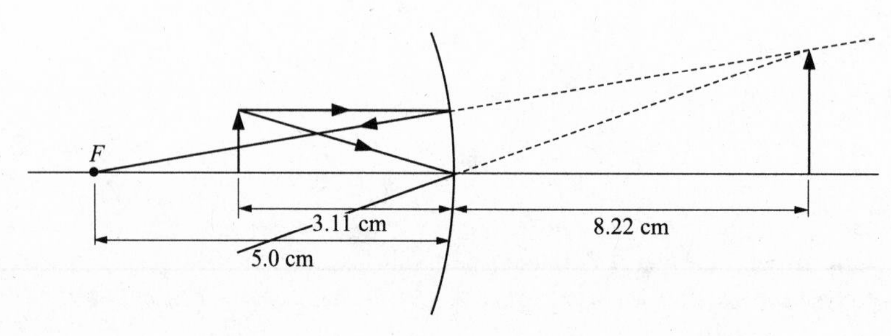

Fig. 14-24

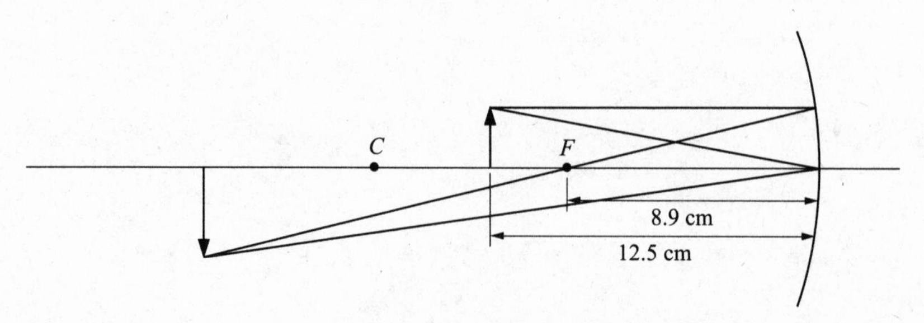

Fig. 14-25

Problem 14.43. A concave mirror has a focal length of 87 cm. A virtual image is formed 65 cm behind the mirror.

(*a*) What is the radius of the mirror?

(*b*) Where was the object placed?

(*c*) What is the magnification produced?

(*d*) Draw a ray diagram for this case.

Ans. (*a*) 174 cm; (*b*) 37.2; (*c*) 1.75; (*d*) see Fig. 14-26

Problem 14.44. A concave mirror has a focal length of f. Where is the image, and what is the magnification, if the object is placed at: (*a*) $s = 2f$; (*b*) $s = f/2$; (*c*) $s = 4f$; (*d*) s infinitesimally greater than f; and (*e*) s infinitesimally less than f?

Ans. (*a*) $s' = 2f$, $M = -1$; (*b*) $s' = -f$, $M = 2$; (*c*) $s' = 4f/3$, $M = -1/3$; (*d*) $s' \approx \infty$, $M \approx -\infty$; (*f*) $s' \approx -\infty$, $M \approx \infty$

Problem 14.45. A convex mirror has a focal length of f. Where is the image, and what is the magnification, if the object is placed at: (*a*) $s = 2|f|$; (*b*) $s = |f|/2$; and (*c*) $s = |f|$?

Ans. (*a*) $s' = -2|f|/3$, $M = 1/3$; (*b*) $s' = -|f|/3$, $M = 2/3$; (*c*) $s' = -|f|/2$, $M = 1/2$

Problem 14.46. A converging lens has a focal length of f.

(*a*) Where is the image of an object that is very far away ($s \to \infty$)?

(*b*) Where is an object whose image is very far away ($s' \to \infty$)?

(*c*) Where, in terms of f, should one place an object to get a magnification of 1/2?

Ans. (*a*) $s' = f$; (*b*) $s = f$; (*c*) $s = 3f$

Problem 14.47. A diverging lens has a focal length of -25.00 cm. An object is placed at $s = 36.00$ cm from the lens, and is then moved away slightly to 36.1 cm.

(*a*) Where is the original image of the object?

(*b*) Where is the final image of the object?

(*c*) What is Δs and what is $\Delta s'$?

(*d*) If the object is moving away at an instantaneous velocity of v, what is the instantaneous velocity of the image?

Ans. (*a*) -14.754 cm; (*b*) -14.771 cm; (*c*) $-\Delta s = 0.1$ cm, $\Delta s' = -0.017$ cm; (*d*) $0.17v$ away from the lens

Problem 14.48. A converging lens, with a focal length of 10 cm, produces a virtual image at a distance of 20 cm from the lens.

(*a*) What is the object distance?

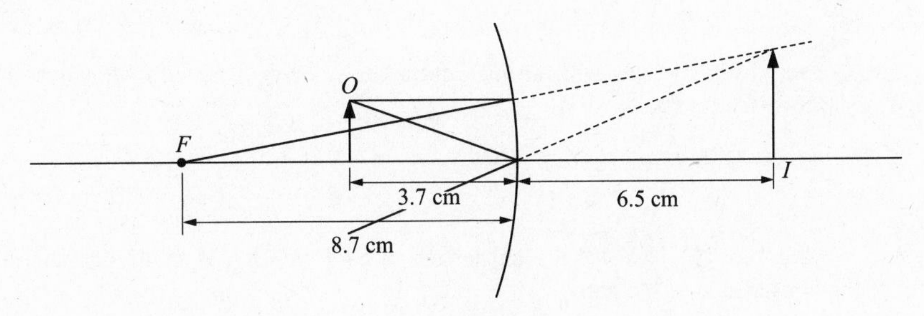

Fig. 14-26

(*b*) If the lens had been a diverging lens with a focal length of magnitude 40, producing an image at the same point, where is the object?

Ans. (*a*) 6.67 cm; (*b*) 40 cm

Problem 14.49. A converging lens, with a focal length of 15 cm is used to project the image of an object on a screen which is placed 80 cm from the *object*.

(*a*) Find one possible object distance.

(*b*) For this value, will the image be real or virtual, upright or inverted, larger or smaller?

Ans. (*a*) 60 cm (or 20 cm); (*b*) real, inverted, smaller (or real, inverted, larger)

Problem 14.50. A lens produces a magnification of 1.28 when an object is placed 2.5 cm from the lens.

(*a*) Where is the image located?

(*b*) Will the image be real or virtual, upright or inverted, larger or smaller?

(*c*) What is the focal length of the lens?

Ans. (*a*) -3.2 cm; (*b*) virtual, upright, larger; (*c*) 11.4 cm

Problem 14.51. A lens produces a magnification of -1.28 when an object is placed 2.5 cm from the lens.

(*a*) Where is the image located?

(*b*) Will the image be real or virtual, upright or inverted, larger or smaller?

(*c*) What is the focal length of the lens?

Ans. (*a*) 3.2 cm; (*b*) real, inverted, larger; (*c*) 1.40 cm

Problem 14.52. A lens, with a focal length of 22.3 cm produces a magnification of -0.68.

(*a*) Where is the object located?

(*b*) Where is the image located?

(*c*) Will the image be real or virtual, upright or inverted, larger or smaller?

Ans. (*a*) 55.1 cm; (*b*) 37.5 cm; (*c*) real, inverted, smaller

Problem 14.53. A converging lens has a focal length of 22.3 cm. An object is initially at $s = 44.10$ cm and then moves to $s = 44.20$ cm.

(*a*) Where is the image initially located?

(*b*) Where is the image located after the object moves?

(*c*) What is Δs and $\Delta s'$?

(*d*) If the object moves away from the lens with an instantaneous velocity of v, with what instantaneous velocity will the image move away from the lens?

Ans. (*a*) 45.111 cm; (*b*) 45.007 cm; (*c*) $\Delta s = 0.10$ cm, $\Delta s' = -0.104$ cm; (*d*) $-1.04v$

Problem 14.54. A thin plano convex lens has a focal length of 64.9 cm, and is made of material with $n = 1.52$. What is the radius of the convex side of the lens?

Ans. 33.7 cm

Problem 14.55. The radius of the convex side of a thin plano-convex lens is 64.9 cm. The lens is made of material with $n = 1.66$. What is the focal length of the lens?

Ans. 98.3 cm

Problem 14.56. A lens is made of material with $n = 1.55$. The focal length of the lens is 28.4 cm, and the first surface of the lens is concave with a radius of 66.8 cm. Is the second surface concave or convex, and what is its radius?

Ans. convex (from outside the lens) with $R = 12.7$ cm

Problem 14.57. A lens is made of material with $n = 1.81$. The two surfaces of the lens have radii of 55.1 cm and 41.7 cm, with the sign convention of the lensmaker's equation.

(*a*) What is the focal length of the lens?

(*b*) If one wants to reduce the focal length to half its value, to what radius should one change R_2?

Ans. (*a*) $-$ 212 cm; (*b*) 33.5 cm

Problem 14.58. A thin lens has a focal length of 61.7 cm. We need a focal length of 68.5 cm. If we attach another thin lens to the first lens: (*a*) what is the power needed for the added lens; and (*b*) What is the focal length of the added lens?

Ans. (*a*) -0.16 diopter; (*b*) -6.22 m

Problem 14.59. A person received glasses with a lens to correct for a hyperopic eye whose near point was 50 cm (see Problem 14.26). The near point has now advanced to 200 cm.

(*a*) How much added power is needed to correct his eyesight?

(*b*) What is the focal length of the added lens?

Ans. (*a*) 1.5 diopter; (*b*) 66.7 cm

Problem 14.60. A magnifying glass uses a converging lens whose focal length is 15 cm. The magnifying glass produces a virtual and upright image that is 3 times larger than the object.

(*a*) How far is the object from the lens?

(*b*) What is the image distance?

Ans. (*a*) 10 cm; (*b*) -30 cm

Problem 14.61. A myopic eye has a far point of only 550 cm. What power lens is needed so that an object that is very far away is imaged at the far point? What is the focal length of this lens?

Ans. -0.18 diopter, $f = -5.50$ m

Problem 14.62. Two lenses, the first with a power of 3.53 diopter and the second with a power of -4.81 diopter are separated by a distance of 75.6 cm. An object is placed 38.1 cm from the first lens.

(*a*) Where is the image of the object in the first lens?

(*b*) Where is the final image?

(*c*) What is the magnification of the combination?

Ans. (*a*) 110 cm from first lens; (*b*) -52.5 cm from second lens; (*c*) 4.41

Problem 14.63. Two lenses, the first with a power of 3.53 diopter and the second with a power of -4.81 diopter are separated by a distance of 75.6 cm. An object is placed 58.1 cm from the first lens.

(*a*) Where is the image of the object in the first lens?

(*b*) Where is the final image?

(*c*) What is the magnification of the combination?

Ans. (*a*) 55.3 cm from first lens; (*b*) -10.3 cm from second lens; (*c*) -0.48

Problem 14.64. Two lenses, the first with a power of 3.53 diopter and the second with a power of -4.81 diopter are separated by a distance of 75.6 cm. An object is placed 10.5 cm from the first lens.

(*a*) Where is the image of the object in the first lens?

(*b*) Where is the final image?

(*c*) What is the magnification of the combination?

Ans. (*a*) -16.7 cm from first lens; (*b*) -17.0 cm from second lens; (*c*) 0.29

Problem 14.65. A compound microscope has an objective of focal length 14.0 mm and an eyepiece of focal length 25.0 mm. Assume the observer has a near point of 25 cm.

(*a*) Find the angular magnification when the object to be observed is placed 1.5 cm from the objective. (Assume the final image is viewed at ∞.)

(*b*) What is the distance from objective to eyepiece for this case?

(*c*) If one wished to double the angular magnification by changing the eyepiece what would the new focal length have to be?

Ans. (*a*) 150; (*b*) 23.5 cm; (*c*) 12.5 mm

Problem 14.66. Suppose the microscope of Problem 14.65 had a fixed eyepiece of focal length 25 mm but a second eyepiece that can be rotated into position so that the distance between objective and eyepiece remains fixed as in Problem 14.65.

(*a*) What would the new objective focal length have to be to double the angular magnification of Problem 14.65(*b*)?

(*b*) What would be the new distance between the specimen and the objective?

Ans. (*a*) 7.0 mm; (*b*) 7.24 mm

Problem 14.67. A refracting telescope has an objective of focal length 2.40 m and an ocular of focal length 3.0 cm. The telescope is aimed at a far away object and the final image is viewed at ∞.

(*a*) What is the distance from ocular to objective?

(*b*) What is the angular magnification of the telescope?

(*c*) If another telescope had an objective with twice the width of the first one, but the focal lengths of objective and ocular were identical to the first telescope, how much brighter would the image be as seen through the second telescope?

Ans. (*a*) 2.43 m; (*b*) 80; (*c*) 4 $\times$ brighter

Chapter 15

Interference, Diffraction and Polarization

15.1 INTRODUCTION

Our discussion of light in the previous two chapters showed that for a variety of phenomena we can depict light as rays that propagate in straight lines, reflecting and refracting at interfaces, even if light is an electromagnetic wave. In this chapter we will discuss some of the phenomena that are uniquely associated with the wave nature of light, such as interference and diffraction of light.

In Chap. 1 we showed that a traveling sinusoidal wave can be described by a simple oscillating formula, Eq. (*1.7*). As noted in Chap. 12 [e.g. Eq. (*12.9*)] this can be rewritten for an electromagnetic plane wave, using the electric field E as the displacement, in the form of:

$$E(x, t) = E_0 \sin(\omega t - kx) \tag{15.1}$$

Here the electric field, E, which is the displacement, depends on the position in space, x, as well as the time, t. E_0 is the amplitude, or the maximum displacement of the wave; ω is the angular frequency $= 2\pi f$ of the wave; k is the propagation constant $= 2\pi/\lambda$ of the wave, while λ is the wavelength. At any point, x, in space, this wave is sinusoidal in time, repeating itself every period, $T = 1/f$. In this equation we have arbitrarily set the time such that at $t = 0$ and $x = 0$, the wave has zero displacement. The displacement that occurs at the position $x = 0$ also occurs at every other x that is distant from the origin by an integral multiple, m, of wavelengths, $m\lambda$. As t increases from zero, the wave moves to larger x at a speed of $c = 3.0 \times 10^8$ m/s. After a time $t = T$, the wave has moved one wavelength (recall that $c = f\lambda = \lambda/T$), and the wave again has the same displacement at all x that it had at $t = 0$. This is of course true for any multiple of T as well.

Suppose we have two waves, of the same frequency and amplitude, which look the same at time $t = 0$ (we say that they start with the same phase angle), and they both travel to another region of space. If they take the same time to reach a new point in space, then they will both have equal displacements at this second point at all time. We say that the two waves are "**in phase**" at this new position. This is illustrated in Fig. 15-1, where the two waves are shown as the solid and the dashed lines. The actual displacement is given at any time by the sum of these two displacements, using the principle of superposition discussed in Chap. 1. This sum is also shown on the figure. When the two waves are in phase,

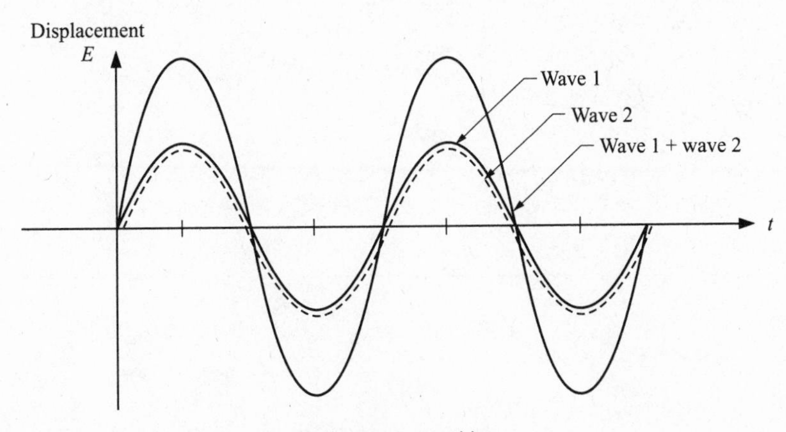

Fig. 15-1

their sum is just a doubling of the displacement at any time. The two waves are in phase not only if they take the same time to reach the new point, but also if one of them takes exactly one period longer than the other (or any integral multiple of the period longer). One can delay one of the waves relative to the other by making it travel in a medium where the speed of light is smaller, or by making it travel a longer path length. In either case the two waves are in phase if the time difference is an integral multiple of the period. Since the distance a wave travels in a period is the wavelength λ, a delay of one period corresponds to an increased path length of one wavelength. Therefore, for two waves to be in phase as they pass a common point (called "**constructive interference**") requires:

$$\Delta t = mT \quad \text{or} \quad \Delta l = m\lambda \rightarrow \text{constructive interference with } m \text{ integer} \tag{15.2}$$

where Δl is the difference in path length traveled by the two waves from the first to the second point in space.

In this extreme case, the sum of the two waves yields the greatest possible value. The other extreme case for the sum occurs if there is a time delay of half a period, $\Delta t = T/2$ (or any half integral multiple of T, $\Delta t = (m + 1/2)T$), between the two waves. Then the two waves at the second point will be as shown in Fig. 15-2. The sum of these two waves is zero, since the displacement of one is the negative of the displacement of the other. We call this "**destructive interference**" for the obvious reason that the two waves seem to destroy each other. The two waves are said to be "180° out of phase", or "out of phase by π (radians)" or simply "**out of phase**". A time delay of $T/2$ corresponds to a path difference of $\lambda/2$. Thus, for destructive interference we require:

$$\Delta t = (m + \tfrac{1}{2})T, \quad \text{or} \quad \Delta l = (m + \tfrac{1}{2})\lambda \rightarrow \text{destructive interference, with } m \text{ integer} \tag{15.3}$$

If the time delay (or added path length) is not one of these multiples, then the addition is more complicated. When we add the two waves we will get another wave that has an amplitude between the two extreme cases that we discussed, i.e. between zero and twice the amplitude of each wave. The two waves are said to be "θ degrees out of phase" if the time delay Δt divided by the period: $\Delta t/T = \Delta x/\lambda = \theta/360$. This is equivalent to saying that if the two waves described by Eq. (*15.1*) obey $E_1 = E_0 \sin \theta_1$, and $E_2 = E_0 \sin \theta_2$ with $\theta_1 = \theta_2$ at the initial time and location, at the new location at any given time they will obey $E_1 = E_0 \sin \theta_1'$, and $E_2 = E_0 \sin \theta_2'$ with $\theta_1' = \theta_2' \pm \theta$. Hence θ is called the phase difference. We often use radians instead of degrees, so $\theta(\text{radians})/2\pi = \Delta t/T$.

Problem 15.1. Suppose two waves of the same amplitude are "out of phase" by θ at a given point x in space.

(*a*) What is the sum of the two waves?

(*b*) Show that the result of (*a*) gives the correct answer when $\theta = 0$, and when $\theta = \pi$.

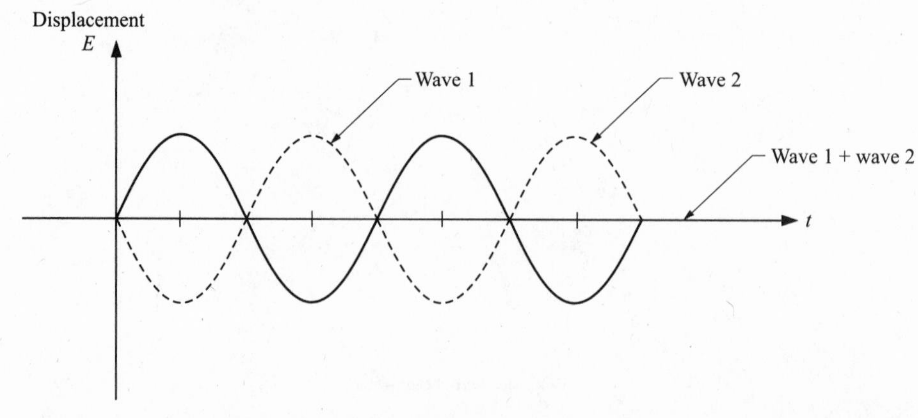

Fig. 15-2

Solution

(*a*) The straight forward solution involves some complicated trigonometric calculation. The first wave can be written as $E_1 = A \sin(\omega t - kx)$. The second wave will then be written as $E_2 = A \sin(\omega t - kx - \theta)$. [We choose a minus sign without loss of generality since we can let θ be positive or negative.] We must add these together. We expect to get a new wave with an amplitude A' and a phase angle θ', or $E' = A' \sin(\omega t - kx - \theta')$. Then $A \sin(\omega t - kx) + A \sin(\omega t - kx - \theta) = A' \sin(\omega t - kx - \theta')$. We let $\omega t - kx = z$, giving $A \sin(z) + A \sin(z - \theta) = A' \sin(z - \theta')$. Using the trigonometric identity $\sin(a \pm b) = (\sin a)(\cos b) \pm (\cos a)(\sin b)$, we get $A \sin z + A(\sin z)(\cos\theta) - A(\cos z)(\sin\theta) = A'(\sin z)(\cos\theta') - A'(\cos z)(\sin\theta')$. Since the equation is true for all values of $z = \omega t - kx$ as time varies, the coefficients of $\sin z$ on both sides of the equation must be equal and the coefficients of $\cos z$ on both sides of the equation must also be equal. Thus, $A(1 + \cos\theta) = A' \cos(\theta')$, and $A \sin(\theta) = A' \sin(\theta')$. Dividing the left and right sides of the last equations gives $(1 + \cos\theta)/\sin\theta = \cos\theta'/\sin\theta'$. Using the trigonometric identity $\cos(a \pm b) = \cos^2 a \mp \sin^2 b$, we have $\cos\theta = \cos^2\theta/2 - \sin^2\theta/2 = 2\cos^2\theta/2 - 1$, so $(1 + \cos\theta) = 2\cos^2(\theta/2)$ and from the $\sin(a \pm b)$ identity we get $\sin\theta = 2\cos(\theta/2)\sin(\theta/2)$, so that $(1 + \cos\theta)/\sin\theta = \cos\theta'/\sin\theta' \rightarrow \cos(\theta/2)/\sin(\theta/2) = \cos\theta'/\sin\theta'$, and $\underline{\theta' = \theta/2}$. Substituting in $A \sin(\theta) = A' \sin(\theta')$, and using the half angle formula for $\sin\theta$, we get $\underline{A' = 2A\cos(\theta/2)}$. This result was obtained by a complicated calculation. We could have arrived at the same result by using the phasor diagrams that we introduced in AC circuits. If we draw two phasors with the same amplitude but with a phase difference of θ, as in Fig. 15-3, we can immediately add these to get a new phasor with $\theta' = \theta/2$, and an amplitude $A' = 2A\cos(\theta/2)$. Thus $E' = [2A(\cos\theta/2)]\sin(\omega t - kx - \theta/2)$, as determined above.

(*b*) When the two waves are in phase, $\theta = 0$, and $E' = 2A\cos(0)\sin(\omega t - kx - 0)$, or $E' = 2A\sin(\omega t - kx)$, as expected. When the two waves are out of phase, $\theta = \pi$, and $E' = 2A\cos(\pi/2)\sin(\omega t - kx - \pi) = 0$, since $\cos(\pi/2) = 0$, again as expected.

Problem 15.2. A light wave has a wavelength of 550 nm.

(*a*) What is the period of the light?

(*b*) What is the smallest time delay required between two waves to obtain complete destructive interference?

(*c*) What is the smallest non-zero difference in path length required between two waves to obtain constructive interference?

Solution

(*a*) Using $c = \lambda/T$, we get $T = \lambda/c = (550 \times 10^{-9}\ \text{m})/(3.0 \times 10^8\ \text{m/s}) = 1.83 \times 10^{-15}$ s.

(*b*) For complete destructive interference we need [Eq. (*15.3*)] a time delay of $(m + 1/2)T$. The shortest time is $\Delta t = T/2 = 9.17 \times 10^{-16}$ s.

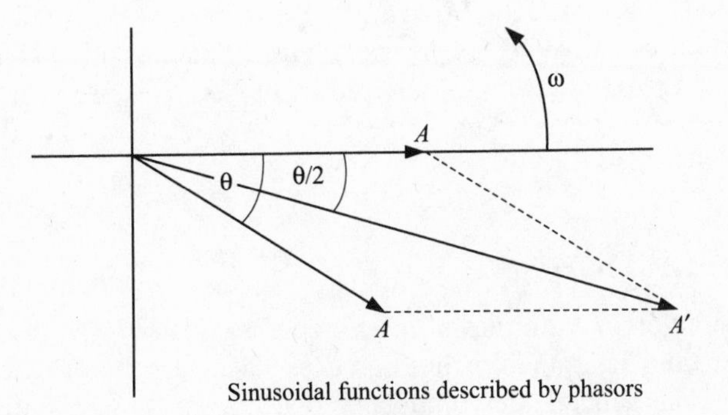

Sinusoidal functions described by phasors

Fig. 15-3

(c) For constructive interference we need [Eq. (*15.2*)] $\Delta x = m\lambda$, so the smallest difference in path length is $\Delta x = \lambda = 550 \times 10^{-9}$ m.

Problem 15.3. Two waves travel between two points along paths that have the same length of 4.3×10^{-7} m. One travels in a medium that has an index of refraction of 3.5, while the other travels in air or vacuum.

(*a*) How long does it take for each wave to travel between the points?

(*b*) What is the time delay between the waves?

(*c*) If the two waves start in phase at the first point and interfere destructively at the second point, what is the maximum wavelength of the light in air?

Solution

(*a*) The time of travel is $t = D/v$. For the wave in air, $v = c$, and for the wave in the other medium, $v = c/n$. Thus, the time of travel is $t_{air} = (4.3 \times 10^{-7}\ \text{m})/(3.0 \times 10^{8}\ \text{m/s}) = 1.43 \times 10^{-15}$ s, and $t_{med} = (4.3 \times 10^{-7}\ \text{m})(3.5)/(3.0 \times 10^{8}\ \text{m/s}) = 5.02 \times 10^{-15}$ s.

(*b*) The time delay is $t_{med} - t_{air} = 3.58 \times 10^{-15}$ s.

(*c*) For destructive interference we must have $\Delta t = (m + 1/2)T$. Thus $T_{max} = 2\Delta t = 7.17 \times 10^{-15}$ s. Since $\lambda = cT$, $\lambda_{max} = 2.15 \times 10^{-6}\ \text{m} = 2.15\ \mu$. This is in the infrared region.

We learned in Chap. 1 that the average energy transported by a traveling wave is proportional to the square of the amplitude [Eq. (*1.11*)]. In Chap. 12 we showed that for any electromagnetic wave the intensity (average energy transported per unit time per unit area) was also proportional to the square of the amplitude, i.e. to E_0^2. If we have two waves, each with an amplitude of E_0, then each wave will have an intensity proportional to E_0^2, and the total intensity would be expected to be proportional to $2E_0^2$. If we have destructive interference, then the intensity passing that point is equal to zero, and we get less energy flow than expected. On the other hand, when we have constructive interference, the resultant wave has an amplitude of $2E_0$, and the intensity at that point is proportional to $(2E_0)^2 = 4E_0^2$. This is more than the intensity of $2E_0^2$ that we expect from two waves. For this reason we call this **constructive** interference, since we get an increase in energy flow over that of two non-interfering waves. It can be shown that, if one considers all points in space, then the average intensity will be proportional to $2E_0^2$, and there will be conservation of energy overall. The excess energy flow at points of constructive interference will be offset by the depleted energy flow at points of destructive interference.

If we have more than two waves then we must add together all the waves to get the resultant wave. This will clearly result in a more complicated calculation in practice, although the underlying principle is still the same. Typically, when we consider waves traveling through openings in different barriers arriving at a common point we use the term **interference** to characterize the results of their addition. This often involves the addition of a relatively small number of waves. If we add together waves coming from different parts of a single aperture and arriving at a common point we usually refer to the effect as **diffraction**, and the addition can involve, in principle, large numbers of waves. Nonetheless, the basic principle for diffraction is the same as for interference. In Sec. 15.2 we will discuss various cases of interference with small numbers of waves, and in Sec. 15.3 we discuss cases of diffraction involving large numbers of waves.

15.2 INTERFERENCE OF LIGHT

We will discuss three different examples of interference in this section: (a) the "double slit", (b) the Michelson interferometer, and (c) thin film interference. There are many variations of each example which are based on the same general idea, and once one knows how to handle the example that we cover, it is usually easy to generalize to the other cases.

a Double Slit

Suppose we have plane wave light traveling in the x direction which is incident perpendicularly on a screen with two narrow-width, parallel slits, as shown in cross-section in Fig. 15-4. The slits, S_1 and S_2 (whose lengths are perpendicular to the paper) are separated by a small distance, d. The light travels to a screen which is at a distance of L from the slits. We assume that $L \gg d$. In the previous two chapters we assumed that light travels in a straight line and does not bend. In that case we would expect that the light which goes through each slit moves in the x direction directly to the screen and that two lines of light would be seen on the screen at points opposite to the slits. However we know from experience with sound waves and with water waves that the waves do bend upon going through an opening (such as a door for sound waves), and the waves travel in all directions from the opening. The reason for this will become clearer after we discuss diffraction of waves in Sec. 15.3. At this stage we will assume that the slits act as light sources of **Huygen's wavelets** as the plane wave hits them and thus the light waves spread in all directions after passing through each slit and investigate the consequences. In particular, consider the part of the wave from each slit that reaches an arbitrary point, P, on the screen. The point is at a distance of y on the screen measured from the point, O', opposite the midpoint, O, between the slits. The line from O to P makes an angle, θ, with the line OO', as seen in the figure. The light from each slit travels a slightly different distance to the point P, with $S_1P < OP < S_2P$. Since the light was initially in phase as the plane wave hit both slits, we can expect interference effects at point P resulting from this difference in path length. We recall that if the difference in path length equals $m\lambda$, then we will get constructive interference, while if that difference equals $(m + 1/2)\lambda$, then we will have destructive interference. We therefore calculate the difference in path length for our arbitrary point P at the angle θ, and then we will be able to determine for which angles we get destructive interference (when the difference equals $(m + 1/2)\lambda$) and constructive interference (when the difference equals $m\lambda$).

In our case we will calculate this difference using simple geometry. We draw the perpendicular lines S_1S_1' and S_2S_2' to the screen. We use the triangle $S_1S_1'P$ to calculate the distance S_1P. Noting that $S_1'P = (y - d/2)$, we get $(S_1P)^2 = L^2 + (y - d/2)^2 = L^2 + y^2 + (d/2)^2 - yd$. Similarly, using the triangle $S_2S_2'P$, we get $(S_2P)^2 = L^2 + y^2 + (d/2)^2 + yd$, since $S_2'P = (y + d/2)$. Subtracting we get $(S_2P)^2 - (S_1P)^2 = 2yd$. Using $a^2 - b^2 = (a - b)(a + b)$, we finally get $(S_2P - S_1P)(S_2P + S_1P) = 2yd$. So far there have been no approximations used. The first term in parenthesis is very sensitive to the values of S_2P and S_1P since it involves the difference of nearly equal lengths. The second term, being their sum, has no such sensitivity. We now note that both S_1P and S_2P are nearly equal to OP, with one of them slightly greater than OP and the other slightly less than OP. It is therefore a very good approximation to equate $(S_1P + S_2P)$ with $2OP$. Then the difference in path length that we are seeking, $(S_2P - S_1P) = (2yd)/2(OP)$. Noting that $y/(OP) = \sin\theta$, we get $S_2P - S_1P = d\sin\theta$. This result can be derived in many other ways as well, all of which prove that:

$$\Delta(\text{path length}) = d\sin\theta. \tag{15.4}$$

We can now determine the angles at which we have destructive and constructive interference by equating this difference in path length to the appropriate multiples of λ.

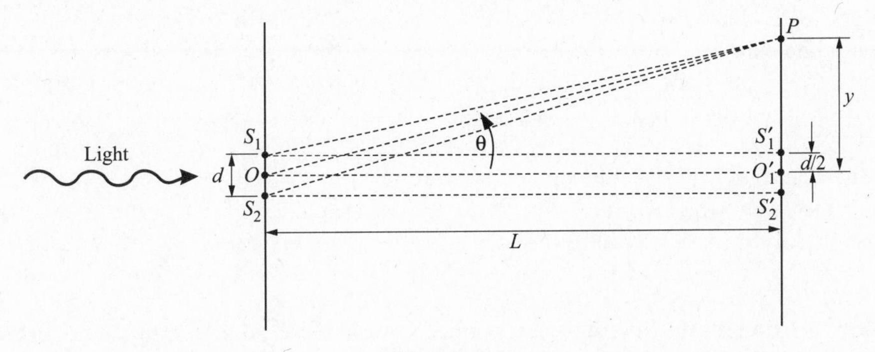

Fig. 15-4

Problem 15.4. For the case of two slits, separated by a distance d in the geometry of Fig. 15-4, determine: (*a*) the angles at which one gets constructive interference; (*b*) the angles at which one gets destructive interference; (*c*) the distance on a screen at which one gets constructive interference (the screen is distant from the slits by $L \gg d$); and (*d*) the distance on the screen between successive points where one has constructive interference.

Solution

(*a*) To get constructive interference we require that the difference in path length be equal to $m\lambda$. Since the difference in path length is given by $\Delta(\text{path length}) = d \sin\theta$, we see that constructive interference occurs when:

$$d \sin\theta = m\lambda \text{ (constructive interference), or } \sin\theta = m(\lambda/d), \qquad m = 0, 1, 2, 3, \ldots \qquad (15.5)$$

At the angles satisfying this equation the point on the screen will be bright, since at points of constructive interference the intensity is a maximum. As one moves away from this angle the intensity of the light will decrease, until we reach a point at which one has destructive interference. Actually, since the slits are usually relatively long (perpendicular to the paper) one will see bright lines parallel to the slits at the maximum intensity angles and dark lines where one has destructive interference.

(*b*) For destructive interference we require that the difference in path length equal $(m + 1/2)\lambda$. Thus, destructive interference occurs at angles such that:

$$d \sin\theta = (m + \tfrac{1}{2})\lambda \text{(destructive interference), or } \sin\theta = (m + \tfrac{1}{2})\lambda/d \qquad m = 0, 1, 2, 3, \ldots \qquad (15.6)$$

At the angles given by this equation, there will be no light on the screen because there is destructive interference. At the center, $(\theta = 0)$ there will be a bright line because there we have constructive interference. At a certain angle given by $m = 0$ in Eq. (*15.6*) there will be a dark line, followed by a bright line for $m = 1$ in Eq. (*15.5*). There will be alternate bright and dark lines as one increases the angle, leading to what are called "**fringes**" on the screen. Since θ could be measured below the center line as well as above it, the dark and light fringes will appear symmetrically above and below the central bright fringe on the screen.

(*c*) The distance y_1 to the first constructive interference occurs at an angle θ which satisfies Eq. (*15.5*) with $m = 1$. At that angle $y_1 = L \tan\theta$. One can calculate θ from Eq. (*15.5*) and substitute in the tangent to get y_1, but another approximation may be made if the angle θ is small, which is the case for $\lambda \ll d$. In this approximation, $\sin\theta \approx \tan\theta$. Consequently, $y_1 = L \tan\theta \approx L \sin\theta = L\lambda/d$. Similarly, the distance to the next constructive interference is $y_2 = 2L\lambda/d$, and to the mth constructive interference, $y_m = mL\lambda/d$, as long as $m\lambda/d = \sin\theta \ll 1$.

(*d*) The distance between successive maxima is $y_{m+1} - y_m = (m + 1)L\lambda/d - mL\lambda/d = L\lambda/d$. This is the distance δ between "fringes".

$$\delta = L\lambda/d \qquad (15.7)$$

As noted, the angle at which the first maximum occurs depends on λ/d. When λ/d is small, then $\sin\theta$ is small, as is θ. As long as the ratio is not too small, one can see these interference effects. However, if it is minuscule, then δ will be too small to observe. We will see in the next section that for wavelengths that are very short compared to the geometric dimensions of slits and apertures, there is little ability of the light to bend and there will be neither diffraction nor interference that is detectable. On the other hand, if λ/d gets bigger, the angle θ increases until, when $\lambda/d = 1$, the angle is 90°. To see fringes one needs this ratio to be much less than 1, but still not minuscule. In that region, one could use the double slit to measure the wavelength of some unknown radiation by measuring the angle at which interference occurs. In practice, one gets greater experimental accuracy if one uses a diffraction grating that contains thousands of slits rather than just two slits.

Problem 15.5. In a double slit the slits are separated by a distance of 0.36 mm, and the screen is 1.38 m from the slits. Light of wavelength 455 nm is incident on the slits.

(*a*) At what angle does the first (off central) maximum occur?

(*b*) At what angle does the third minimum occur?

(*c*) What is the spacing between adjacent fringes?

Solution

(*a*) The first maximum occurs when $\sin\theta = (1)\lambda/d = (455 \times 10^{-9}\text{ m})/(0.36 \times 10^{-3}\text{ m}) = 1.26 \times 10^{-3}$. Thus $\theta = 0.072° = 1.26 \times 10^{-3}$ rad.

(*b*) The third minimum occurs when $m = 2$ in Eq. (*15.6*), since the first minimum occurs when $m = 0$. Therefore, $\sin\theta = (2 + 1/2)\lambda/d = (5/2)(455 \times 10^{-9}\text{ m})/(0.36 \times 10^{-3}\text{ m}) = 3.16 \times 10^{-3}$. Then $\theta = 3.16 \times 10^{-3}\text{ rad} = 0.18$.

(*c*) For small angles we can use the approximation that $\tan\theta = \sin\theta$. Since the angles in this problem are very small this approximation is certainly valid. Then the fringe spacing is given by Eq. (*15.7*), $\delta = L\lambda/d = (1.38\text{ m})(455 \times 10^{-9}\text{ m})/(0.36 \times 10^{-3}\text{ m}) = 1.74 \times 10^{-3}\text{ m} = 1.74$ mm.

Problem 15.6. A double slit has a spacing of 0.99 mm between the slits. A screen is located at a distance of 1.08 m from the slits.

(*a*) What wavelength of light will have its first maximum at an angle of 0.11°?

(*b*) What wavelength of light will have its fourth minimum at an angle of 0.21°?

(*c*) What wavelength of light will have a fringe spacing of 2.35 mm on the screen at small angles?

Solution

(*a*) Using Eq. (*15.5*), $\sin\theta = (1)\lambda/d$, or $\lambda = d\sin\theta = (0.99 \times 10^{-3}\text{ m})(\sin 0.11°) = 1.90\ \mu$.

(*b*) Using Eq. (*15.6*), with $m = 3$ (see previous problem to see why we use 3 instead of 4), $d\sin\theta = (3 + 1/2)\lambda$, or $\lambda = (2/7)d\sin\theta = (2/7)(0.99 \times 10^{-3}\text{ m})\sin 0.21° = 1.04\ \mu$.

(*c*) We see from (*a*) and (*b*) that the angles are small and we can use Eq.(*15.7*), $\delta = L\lambda/d = 2.35 \times 10^{-3}$ m. Then $\lambda = (2.35 \times 10^{-3}\text{ m})(0.99 \times 10^{-3}\text{ m})/(1.08\text{ m}) = 2.15\ \mu$.

b The Michelson Interferometer

A schematic diagram of the **Michelson interferometer** is shown in Fig. 15-5. Here, light from a source, S, is incident on a "**half-silvered**" mirror, M_1. This mirror is coated with a thin film of some metal so that half the light is reflected and the other half is transmitted. With the mirror at an angle of 45° to the incident light, the reflected and transmitted light travel along different perpendicular paths, P_1 and P_2, until they each are reflected back from mirrors M_2 and M_3. When the reflected light on path P_1 is again incident on the half-silvered mirror after reflection from M_2, half of its light is transmitted down to the detector D. Similarly, half of the light traveling along path P_2, after reflection from M_3, is reflected from the half-silvered mirror and also reaches the detector D. The light reaching the detector D therefore consists of two rays that traveled along different paths. If the paths P_1 and P_2 are of exactly the same length and contain the same material (so that the velocity is the same), then the two rays will arrive at the detector in phase, resulting in constructive interference. However, if the lengths P_1 and P_2 are not equal, and differ by ΔD, then the rays will have traveled different distances before reaching the detector, with their path lengths differing by $2\Delta D$. Just as in the case of the double slit, this can give rise to constructive or destructive interference, or some intermediate amount of interference. The criterion for constructive interference remains that the difference in path length, $2\Delta D$, equals an integral multiple of the wavelength:

$$2\Delta D = m\lambda \text{ (constructive interference)} \tag{15.8}$$

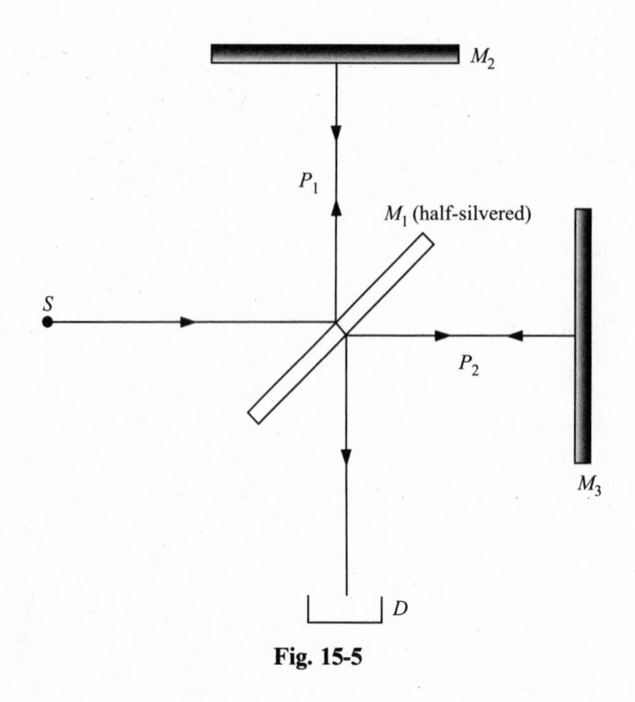

Fig. 15-5

For destructive interference we require that this difference in path length equal a half-integral multiple of λ:

$$2\Delta D = (m + \tfrac{1}{2})\lambda \text{ (destructive interference)} \tag{15.9}$$

As we change the distance to one of the mirrors, say to M_2, the detector will alternately record a large intensity and zero intensity as the interferometer changes from constructive to destructive interference and back again. This can be used to measure very small distances accurately.

Problem 15.7. Consider a Michelson interferometer, as shown in Fig. 15-5. The entire apparatus is in air, and the distances D_1 and D_2 are initially adjusted so that the detector records a maximum intensity of light at a wavelength λ.

(*a*) What is the minimum distance, *d*, that one has to move mirror M_2 so that the detector will again record a maximum intensity?

(*b*) If one moves the mirror M_2 a distance, *d*, and records a total of *m* maxima as one moves the mirror, what is the distance *d*?

(*c*) How far does one have to move the mirror M_2 so that the detector records a minimum intensity rather than a maximum intensity?

(*d*) If one can measure a shift of 1/10 of a fringe accurately, and one uses a wavelength of 408 nm, what is the minimum distance that one can measure accurately?

Solution

(*a*) Initially, one has constructive interference since the intensity at the detector is at a maximum. To get another maximum one must increase the difference in path length by $m\lambda$. The minimum distance one has to move the mirror will occur when $m = 1$. The added path length is $2d$, since the light travels back and forth along the path, and therefore $2d = \lambda$, or $d = \lambda/2$.

(*b*) Using the same logic as in (*a*), the distance traveled for each subsequent maximum is $d = \lambda/2$, and the total distance traveled for *m* maxima is $m\lambda/2$.

(c) To move from a maximum (constructive interference) to a minimum (destructive interference), the difference in path length must change from $m\lambda$ to $(m + \frac{1}{2})\lambda$. Thus the difference in path length, which equals $2d$, must increase by $\lambda/2$. Then $d = \lambda/4$.

(*d*) A shift of one full fringe occurs if one moves a distance of $\lambda/2$. A shift of $\frac{1}{10}$ of a fringe therefore occurs if one moves a distance of $d = (\frac{1}{10})\lambda/2 = 0.1(408 \text{ nm}/2) = 20.4$ nm. This is the minimum distance that can be measured. It corresponds to a time delay of $2d/c = 1.36 \times 10^{-16}$ s between the two waves reaching the detector.

The Michelson interferometer is clearly capable of measuring very small displacements, corresponding to fractions of a wavelength. Similarly, the interferometer can be used to measure very small time delays, as will be illustrated in the next problem. There are many practical applications of this ability, including measuring the number of wavelengths in the length of the standard meter (needed to calibrate the newer standard of length which was the length of a particular wavelength of light).

Problem 15.8. A Michelson interferometer is used to measure the index of refraction of air. To do this one inserts a tube, of length L, with transparent end faces, in one path of the interferometer (say between M_1 and M_2 in Fig. 15-5), and adjusts the interferometer to give constructive interference. Then, one gradually pumps out the air from the tube. As the air is removed the index of refraction of the material in the tube changes from that in air to that in a vacuum. Since light moves faster in vacuum than in air, the light in the branch with the tube will reach the detector earlier than the light in the other path. One counts the number of maxima that the detector records as the air is removed, and then calculates the index of refraction of air. The wavelength used is λ in vacuum.

(*a*) In terms of the index of refraction of air, what is the expression for the time that the light takes to pass through the tube (back and forth) when it is filled with air?

(*b*) What is the corresponding expression for the time that the light takes to pass through the tube when there is a vacuum in the tube?

(*c*) What is the expression for the time delay between the two paths caused by the removal of the air in the tube?

(*d*) If there are m maxima recorded by the detector as the air is removed, what is the formula for the index of refraction of air?

(*e*) When this experiment is performed using $\lambda_{vac} = 555$ nm and $L = 5.6$ cm one measures $m = 59.1$. What is the index of refraction of air?

(*f*) What error would be made if one used λ_{air} instead of λ_{vac} in the formula?

(*g*) What was the time delay in part (*e*)?

Solution

(*a*) The speed of light in air is c/n, where n is the index of refraction of air. The time taken to pass back and forth in the tube is $t_{air} = 2L/(c/n) = 2nL/c$.

(*b*) Similarly, the time taken to pass back and forth in the tube when there is a vacuum in the tube is $t_{vac} = 2L/c$.

(c) Before the air is removed from the tube the light from the two paths produced a constructive maximum. This occurred with light passing through air in the tube. Without the air in the tube the light traveled faster, and the light in this path now reaches the detector ahead of the other light by the difference in time taken in air and in vacuum. This time is $\Delta t = t_{air} - t_{vac} = 2nL/c - 2L/c = (2L/c)(n - 1)$.

(*d*) Each time a new maximum is recorded by the detector there is an added time delay of one period, $T = \lambda_{vac}/c$. If there are m maxima detected, then $\Delta t = m\lambda_{vac}/c = (n - 1)(2L/c)$. Thus $(n - 1) = m\lambda_{vac}/2L$.

(*e*) Using the formula developed in (*d*), we get $(n-1) = (59.1)(555 \text{ nm})/2(5.6 \text{ cm}) = 2.93 \times 10^{-4}$. Then $n = 1.000{,}293$.

(*f*) $\lambda_{air} = \lambda_{vac}/n \rightarrow \lambda_{air} = (555 \text{ nm})/(1.0002930 = 554.8 \text{ nm}$. From part (*e*) we see that $(n-1)$ would not even change in the third decimal place, so no significant loss of accuracy would occur.

(*g*) For each maximum there is a time delay of one period. Thus the total time delay was $59.1T = (59.1)(\lambda/c) = (59.1)(555 \text{ nm})/(3.0 \times 10^8 \text{ m/s}) = 1.09 \times 10^{-13}$ s.

This same experimental arrangement was used by Michelson and Morley to investigate the possibility that the velocity of light depended slightly on the direction of motion of the earth in its orbit around the sun, and the result obtained led to the theory of relativity.

c *Thin Film Interference*

Another possible method for producing two or more beams of light that travel different distances between two points is to use the reflection of light from the two surfaces of a thin transparent film. One possible example is shown in Fig. 15-6, where the light incident on the film is incident in a direction normal to the surface of the film. The principle involved would be the same for a different angle of incidence, but the calculation becomes more complicated, so we will restrict our discussion to the case of normal incidence. In the figure the incident ray is partially reflected at the top of the film as it travels from a material of index of refraction n_1 to the film that has an index of refraction of n_f. This reflected ray is drawn slightly displaced from the incident ray for purposes of clarity. The other part of the incident ray is transmitted through the film and reaches the bottom of the film. At that point it encounters another material with index of refraction n_2, and is partially reflected and partially transmitted to the new material. The reflected ray then travels back to the top surface. At that surface, it is partially reflected back down the film and partially transmitted to the material of original incidence. In actuality, the reflected part has multiple reflections in the film, being partially reflected and partially transmitted at each surface. It can be shown that, for the cases we will consider (constructive and destructive interference), we can add together all the rays that finally emerge into the incident material, and treat the sum as a single ray that has traveled only once through the film. This single ray is shown on the figure as the ray "from bottom of film". We then have two rays that travel together away from the film; the ray reflected from the top of the film and the ray reflected from the bottom of the film. These two rays will generally interfere with each other. There are two differences between the paths of these two rays before they meet again; (1) the ray from the bottom travels an additional distance of $2t$ within the film, where t is the thickness of the film, and (2) the two rays are reflected from different materials at the top and at the bottom. Each of these reflections can contribute to the phase difference of the two rays and hence to the interference that we detect. We first discuss the contribution of the additional distance

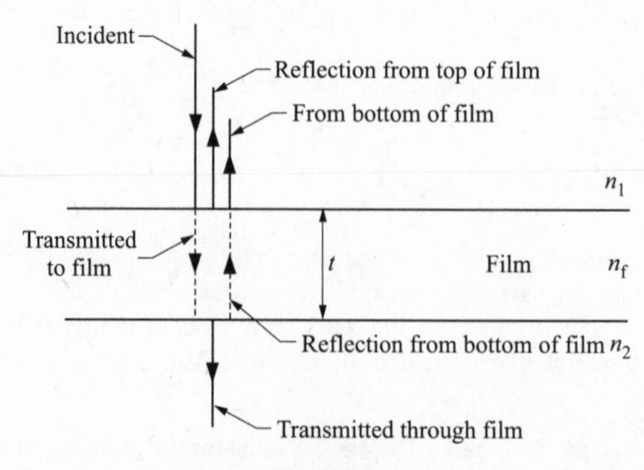

Fig. 15-6

traveled, which is just an extension of our previous discussion. We know that every additional path length of one wavelength contributes a "phase shift" of 2π radians (or 360°), which is the equivalent of no phase shift at all. However, every additional path length of a half wavelength contributes a phase shift of π radians (or 180°), and any other fraction of a wavelength of path length contributes that same fraction of 2π radians of phase shift. We will consider only cases where the additional path length equals an integral or half integral wavelength. The wavelength that must be used in these calculations is the wavelength within the material in which this additional path length occurred. In our case this is within the film, where the wavelength is $\lambda_f = \lambda_{vac}/n_f$. Therefore we know that, if the additional path length ($2t$) equals $m\lambda_{vac}/n_f$, then there will be effectively no phase shift introduced by the additional path length. On the other hand, if the additional path length ($2t$) equals $(m + \frac{1}{2})\lambda_{vac}n_f$, then there will be a phase shift of π radians between the two rays introduced by the thickness of the film. Thus we can write the phase shift due to the thickness as:

if $\quad t = m\lambda_{vac}/2n_f$, no phase shift, *(15.10a)*

and if $\quad t = (m + \frac{1}{2})\lambda_{vac}/2n_f$, phase shift of π radians (180°) *(15.10b)*

We now discuss the effect of the reflections at each surface. In general, whenever there is a reflection at a surface, the ray is traveling from a region of one index of refraction n_a to another region with a different index of refraction n_b. The reflected wave will have an amplitude that is determined by a formula involving the numerical values of these indices of refraction. However, the relative phase between the incident and the reflected ray is always either zero or π radians, and is determined only by whether the index of the incident material n_a is larger or smaller than the index of the reflecting material n_b. If $n_a < n_b$, then there is a phase difference of π radians, and if $n_a > n_b$, then there is no phase change. This result is both an experimental fact and can also be derived from Maxwell's equations for electromagnetic radiation. Thus the π shift occurs in going from a region of higher speed to one of lower speed. In our case the reflection at the top surface is from n_1 to n_f, and the reflection at the bottom is from n_f to n_2. If both reflections are from low index to high index, then there is a phase shift of π radians for both reflected rays and there is no difference in phase introduced between the two waves as a result of the different reflections. An example of this is when light is incident from air ($n_1 = 1$) to a thin film of water ($n_f = 1.33$) which is resting on glass ($n_2 = 1.5$). The same would be true if both reflections were from a high to a low index, where no phase shift is introduced at either reflection. However, if one reflection is from low to high index and the other is from high to low index, then one ray suffers a phase shift of π radians and the other ray has no phase shift at all. Then there will be a difference in phase of π radians introduced between the rays in addition to any phase shift due to the difference in path length. An example of this is when light is incident from air ($n = 1$) to a thin film of oil ($n = 1.5$) which is floating on water ($n = 1.33$). Here, the reflection at the top is from low to high index, and the reflection at the bottom is from high to low index.

The total difference in phase between the two rays is the sum of the shifts due to the different path lengths and that due to different reflections. Therefore, we can predict that, if $n_1 < n_f < n_2$, or $n_1 > n_f > n_2$ (where there is no difference in phase introduced between the rays by the reflections), then:

if $\quad t = m\lambda_{vac}/2n_f$, constructive interference, *(15.11a)*

and if $\quad t = (m + \frac{1}{2})\lambda_{vac}/2n_f$, destructive interference *(15.11b)*

For the other cases, if $n_1 < n_f > n_2$ or $n_1 > n_f < n_2$ (where there is a phase shift of π radians introduced between the two rays), then we get the exact opposite result, namely:

if $\quad t = m\lambda_{vac}/2n_f$, destructive interference, *(15.12a)*

and if $\quad t = (m + \frac{1}{2})\lambda_{vac}/2n_f$, constructive interference *(15.12b)*

Problem 15.9. A thin film of oil ($n = 1.50$) floats on a thick layer of water ($n = 1.33$). Light, whose wavelength is 487 nm in air (or vacuum) is incident normally on the film.

(*a*) What is the difference in phase between the ray reflected at the air–oil interface and the ray reflected at the oil–water interface due to the different reflection?

(*b*) What is the minimum thickness of film needed for constructive interference?

(*c*) What is the minimum thickness of film needed for destructive interference?

(*d*) If one has destructive interference at this wavelength, at what wavelength in the infrared would one have constructive interference?

Solution

(*a*) At the air–oil interface the reflection is from low to high index of reflection (from 1.00 to 1.50). For the ray reflected at this interface there is a phase shift of π radians. At the oil–water interface the reflection is from high to low index (from 1.50 to 1.33). The ray reflected at this interface is not shifted in phase at all by the reflection. Thus the two reflected rays will be out of phase by π radians due to the different reflections.

(*b*) For constructive interference we require a phase shift of zero or 2π radians. We already have a phase shift of π radians between the two rays, so the different path lengths must give us an additional phase shift of π radians. This is the case of Eq. (*15.12b*) with $m = 0$, $t = (m + \frac{1}{2})\lambda_{vac}/2n_f = (\frac{1}{2})(487 \text{ nm})/2(1.50) = 81.2 \text{ nm}$.

(c) For destructive interference we require a phase shift of π radians. We already have a phase shift of π radians between the two rays, so the different path lengths must give us either no additional phase shift (zero thickness) or an additional phase of a multiple of 2π radians. This is the case of Eq. (*15.12a*) with $m = 1$, $t = m\lambda_{vac}/2n_f = (487 \text{ nm})/2(1.50) = 162.3 \text{ nm}$.

(*d*) If one has destructive interference then the thickness of the film is 162.3 nm. To get constructive interference we need [Eq. (*15.12b*)]; $t = (m + 1/2)\lambda_{vac}/2n_f = 162.3$ nm. Then, for $m = 0$, $\lambda_{vac} = 4n_f t = 974$ nm, which is in the infrared. (Note that the wavelength giving constructive interference for $m = 1$ is $\lambda_{vac} = (4/3)n_f t = 325$ nm, which is not in the infrared.)

As can be seen from this problem, one can use these interference effects to measure the thickness of a thin film using a known wavelength, or one can measure an unknown wavelength if one has a known thickness. Furthermore, one can coat a surface with a specific thickness of film to produce destructive interference for a particular wavelength. In that case there will be no reflection from the surface at that wavelength. This is then called a **non-reflecting coating**. One can equally well produce constructive interference at a particular wavelength, resulting in a coating that provides nearly complete reflection. These are applications that are used in many actual situations.

Another similar case of thin film interference is illustrated by the case of a wedge. Consider two smooth glass plates that make a very small angle θ with each other, as in Fig. 15-7(*a*). In the figure the angle is exaggerated for purposes of clarity. The incident light is incident normally to both of the plates because this angle is so small. The air between the plates forms a "thin film" whose thickness increases as we move away from the corner. We refer to this arrangement as a wedge. For the wavelength of the incident light there will be a point along the bottom plate where the thickness of the film is just right to give constructive interference when one combines the light reflected at the top and at the bottom of the air film. At this point, labeled *a* in Fig. 15-7(*b*), there will be a bright line of light extending perpendicular to the plane of the figure. As one moves further away from this point, the thickness of the film increases, and there will not be constructive interference again until we reach point *b*, where the thickness is again just right for constructive interference. This must be due to the fact that the added path length, which is twice the added thickness Δt, is just equal to one wavelength of the light. Thus $2\Delta t = \lambda$. At this point *b*, there will again be a bright line. This pattern repeats itself as we move further from the corner. The result is a series of "fringes", or alternating bright and dark fringes along the glass plate. The pattern will be a series of regularly varying intensities illustrated in Fig. 15-7(*c*) by concentrations of vertical lines, where we show a view of the plates from the top.

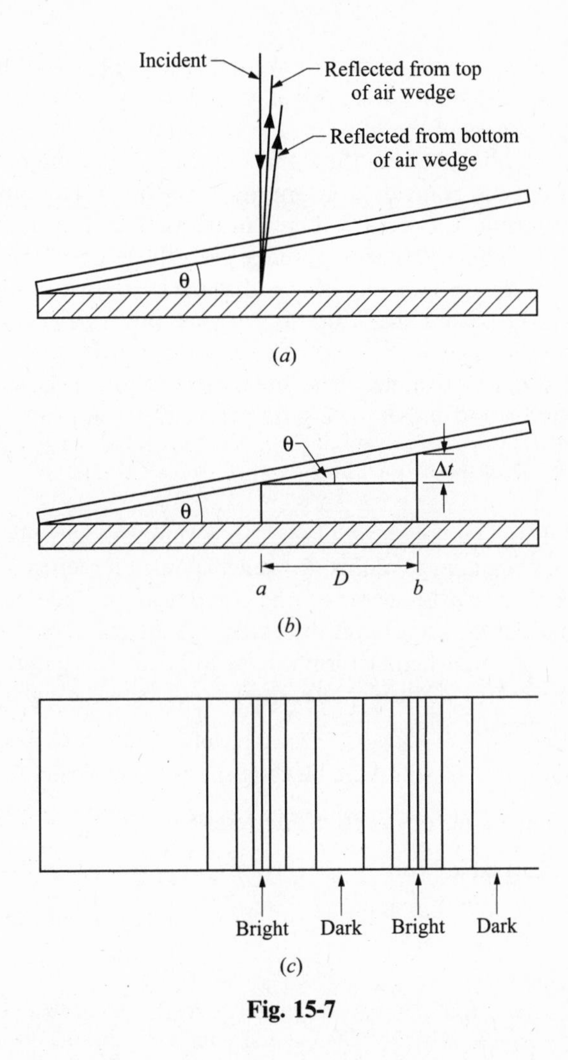

Fig. 15-7

Problem 15.10. For the wedge in Fig. 15-7, calculate: (*a*) the distance between adjacent fringes; and (*b*) the number of fringes per meter along the glass.

(*c*) For an angle θ of 1.23×10^{-4} rad, calculate the wavelength of light that will result in 444 fringes/m.

(*d*) At the corner of the wedge will there be a bright or a dark fringe?

Solution

(*a*) Examining Fig. 15-7(*b*), we note that the distance, D, between adjacent bright lines, which is the distance between the fringes, is given by $\tan\theta = \Delta t/D = \lambda/2D$ since $2\Delta t = \lambda$. Thus $D = \lambda/[2\tan\theta]$.

(*b*) The number of fringes per meter is just the reciprocal of this distance, or fringes/$m = 2(\tan\theta)/\lambda$.

(*c*) We use this formula to get $444 = 2(\tan\theta)/\lambda = 2(1.23 \times 10^{-4})/\lambda$, or $\lambda = 5.54 \times 10^{-7}$ m $= 554$ nm.

(*d*) At the corner the two rays will travel the same distance, so any difference in phase must arise from a phase difference in the two reflections. The reflection at the top of the air film is from glass to air, which means that the ray is going from a high to a low index. This wave will not suffer any phase shift. The ray reflected at the bottom of the air film is going from air to glass, i.e. from low to high index. Therefore this ray will suffer a phase shift of π radians. The two rays will therefore be out of phase by π radians as a result of the reflections. Since there is no other source of phase shift, the two rays will produce destructive interference, and there will be a dark line at the corner.

We see that everywhere along the line of a bright fringe the thickness of the air film is the same and corresponds to one of the values needed for constructive interference. If the glass is not exactly smooth, then the points where the thickness of the film has the correct value will not be a straight line, and the bright fringes will be somewhat distorted, as shown in Fig. 15-8. Since wherever the bright line wanders the thickness of the air film directly below is constant, the wavy line traces out points where the distance between the top and bottom plates is constant. This can be used to measure the smoothness of a plate. One takes a perfectly smooth plate, and forms a wedge with the plate whose smoothness one wants to determine. If one gets straight line fringes, one knows that the plate is smooth. If the lines are distorted, one knows where one has to remove a small amount of glass and can grind the glass there to achieve better smoothness. Since these fringes are sensitive to distances of fractions of a wavelength, this can assure smoothness to that order of distance. The same procedure can be employed for spherical surfaces, such as telescope mirrors and lenses. The next problem develops the basic concepts needed for this application.

Problem 15.11. A sector of a glass sphere is placed on a smooth glass plane, as in Fig. 15-9(*a*). This sector is cut from a sphere having a large radius, R, which is much larger than the height, h from the rim to the plate. In Fig. 15-7(*b*), we view the sector on the plane from the side, and note that there is an air film between the lower surface of the sector and the plane. The height of this film increases as one moves out from the center. Points of equal height form circles around the center. Light, of wavelength λ, is incident perpendicular to the plane. Consider the interference between light reflected at the top of the air film and at the bottom of the air film.

(*a*) At the center of the sector ($r = 0$) will there be constructive or destructive interference?

(*b*) At what values of r will there be constructive interference?

(*c*) What shape will the fringes have?

(*d*) If the sector of the sphere is not perfectly smooth, how will this affect the shape of the fringes?

Solution

(*a*) This is identical to part (*d*) of the previous problem. At the center there is no difference in path length between the two rays, so only the different reflection can cause a phase difference. There will be a difference in phase of π radians since the reflection at the top is from high to low index and that at the bottom is from low to high index. Therefore there will destructive interference at the center.

(*b*) Since there is already a phase difference of π radians due to the different reflections, we need a difference in path length of $(m + 1/2)(2\pi$ radians) so that the total phase difference will equal a multiple of 2π. This is the equivalent of Eq. (*15.12b*) where t is now the variable thickness of the air film. We must calculate this thickness in terms of r, the horizontal radial distance from the center. Referring to Fig. 15-9(*b*), we use the Pythagorean theorem on triangle *abc*. The hypotenuse is R, and the two legs are $(R - t)$ and r. Then, $R^2 = (R - t)^2 + r^2 = R^2 + t^2 - 2Rt + r^2$. This gives $r^2 = 2Rt - t^2$. We have

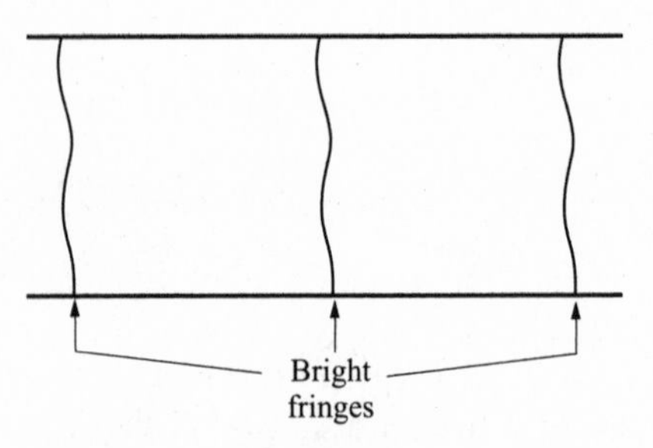

Fig. 15-8

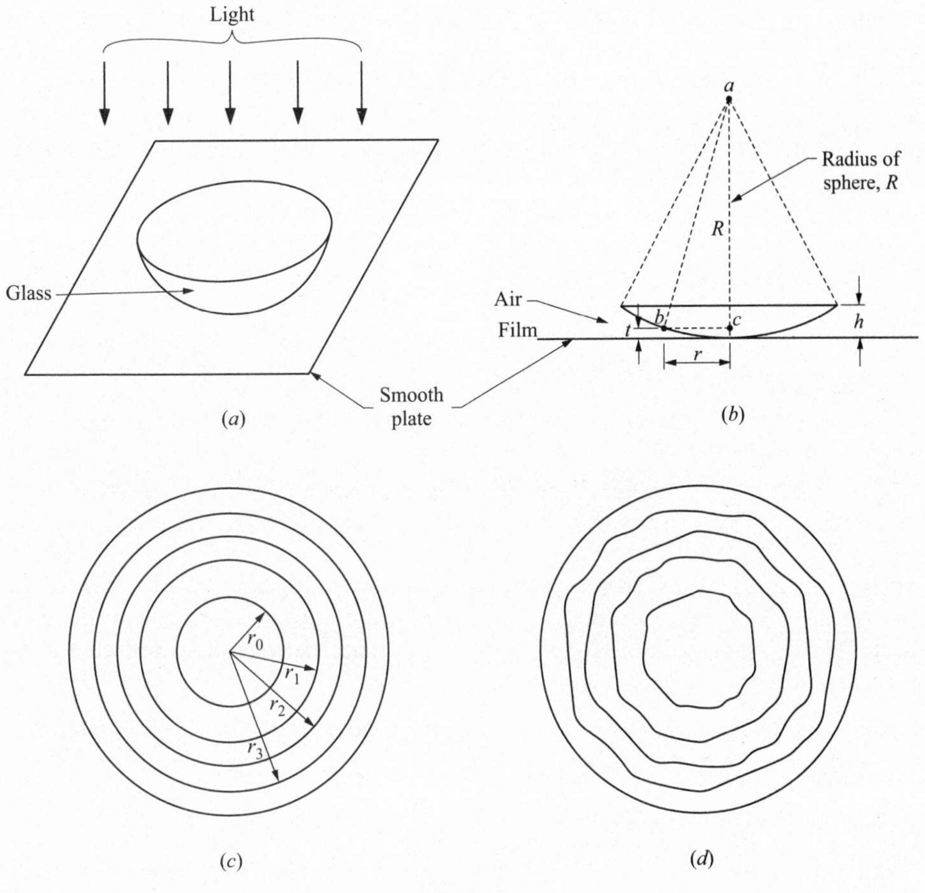

Fig. 15-9

assumed that the rim height $h \ll R$, and therefore $t \ll R$, and we can approximate $2Rt - t^2 \approx 2Rt$. Finally, this gives us that $t = r^2/2R$. Using Eq. (*15.12b*) there will be constructive interference when $r^2/2R = (m + 1/2)\lambda_{vac}/2n_f = (m + 1/2)\lambda/2$. Thus, $r = [(m + 1/2)\lambda R]^{1/2}$. The first maximum occurs when $r_0 = [\lambda R/2]^{1/2}$, the second at $r_1 = [3\lambda R/2]^{1/2}$, the third at $r_2 = [5\lambda R/2]^{1/2}$, etc.

(*c*) All points with the same r will have the same film thickness, and therefore the fringes will be circles, as shown in Fig. 15-9(*c*).

(*d*) If the sector of the sphere is not perfectly smooth then the points of equal thickness of the air film will no longer be circles. Then the fringes will be somewhat warped circles, as in Fig. 15-9(*d*).

The circles seen in the case described in this problem are called **Newton's rings**. This technique is used as standard procedure to test and adjust the smoothness and symmetry of spherical glass surfaces.

15.3 DIFFRACTION AND THE DIFFRACTION GRATING

We now return to a discussion of what occurs when light passes through a single small aperture, such as a slit. After discussing the case of a single slit, we will extend the discussion to a single circular aperture, and then to the result of these effects on our ability to distinguish small objects (resolving power). All of these are examples of diffraction. This will be followed by a discussion of the effect of having a large number of parallel identical slits, forming a "diffraction grating".

1 Single Slit

Suppose that plane-wave light is incident normally on a narrow slit of width w, as in Fig. 15-10. We can again think of the light as passing through the slit as Huygen's wavelets spreading in all directions. The light proceeds from the slit to a screen at a distance L from the slit, where $L \gg w$. We have drawn two rays that proceed to a point on the screen at a distance y from the center (at an angle θ), one from the top edge and one from the center. Actually, rays reach the screen from all parts of the slit, not just from these two points. In order to analyze this case properly one really has to add together the contributions from all the points on the slit, taking into account the difference in path length from each point. This requires the use of calculus. It is clear, however, that at the center (point O), all the rays are essentially in phase, since we have assumed that w is very small. Therefore the point O will be one of large intensity. As we move away from O to larger y, the rays from the bottom of the slit will be delayed relative to the rays from the top of the slit, and there will be some interference. However, since there are rays from all over the slit, it is difficult to determine just what will happen at any given y without doing an exact calculation using the calculus. There is, however, a simple method that can be used to find the particular angles, θ, at which we get complete destructive interference. This method is used in the following problem.

Problem 15.12. A single slit, of width w, is at a distance L from a screen, as in Fig. 15-10. Consider the point at a distance y along the screen from the center, O. Let us choose this point so that the difference in path length between light reaching this point from the upper edge of the slit and from the center of the slit is $\lambda/2$.

(*a*) Show that every point on the top half of the slit has a corresponding point on the bottom half of the slit whose path length to y is $\lambda/2$ greater.

(*b*) Show that this particular point on the screen will have complete destructive interference, and derive the relationship giving the angle θ at which this interference occurs.

(*c*) Use the same reasoning that led to this angle to find other angles at which there is destructive interference, and derive a formula for those angles.

Solution

(*a*) The difference in path length to the point y on the screen from two points separated by a distance d at the slit was already determined when we solved the two slit problem and is given by Eq. (*15.14*) as $\Delta(\text{path length}) = d \sin\theta$. In the case of two points, one at the top of the slit and the another at the center of the slit, the difference in path length to y is $(w/2)\sin\theta$. We are told that y is chosen so that this difference in path length is $\lambda/2$, so we have: $(w/2)\sin\theta = \lambda/2$. Let us now consider two other points each of which is, respectively, a distance Δd below each of the first two points. They, also, are separated by a distance $w/2$. The difference in path length to y from these two points is also $(w/2)\sin\theta$. Thus, any

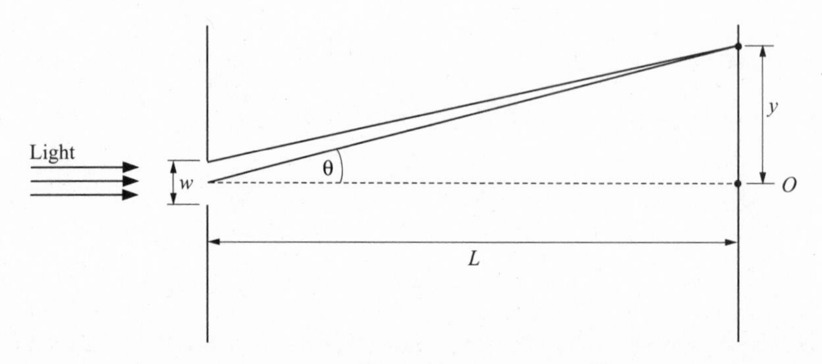

Fig. 15-10

point in the upper half of the slit has a corresponding point in the bottom half of the slit where the difference in path length is the same, namely $(w/2) \sin\theta = \lambda/2$.

(*b*) We assume the intensity of the entering plane wave is uniform across the slit. Since the first two points differ in path length to y by $\lambda/2$, they will destructively interfere with each other. For every point in the top half of the slit there will be another point in the bottom half of the slit that will destructively interfere with it. All points on the bottom will be included since the top and bottom parts have equal size. Thus there will be complete destructive interference when we add together the contributions from all the points in the slit. The angle θ is then given by $(w/2) \sin\theta = \lambda/2$, or:

$$w \sin\theta = \lambda, \text{ (destructive interference)} \tag{15.13}$$

(*c*) Suppose we divide the slit into four equal parts, rather than two parts. If the difference in path length between the point at the top of the slit and the point at a distance of $w/4$ from the top is Δ(path length) $= \lambda/2$, then every point in the upper quarter will have a point in the second quarter with a difference in path length of $\lambda/2$. Thus the first quarter will destructively interfere with the second quarter. By the same reasoning the third quarter will destructively interfere with the fourth quarter, and we will have complete destructive interference. Since the difference in path length is $(w/4) \sin\theta = \lambda/2$, we get destructive interference at $w \sin\theta = 2\lambda$. In general, if we divide the slit into any even number of equal parts, $2m$, and choose the angle θ so that the difference in path length between the top of the slit and the top of the next part is equal to $\lambda/2$, we will have an even number of adjacent parts interfere destructively with each other. The difference in path length equals $(w/2m) \sin\theta = \lambda/2$, so, in general we will have destructive interference whenever:

$$w \sin\theta = m\lambda, \qquad m = 1, 2, 3 \ldots \text{ (destructive interference)} \tag{15.14}$$

From this problem we determine the angles at which we get complete destructive interference. There will be maxima at angles approximately midway between these minima, but there is no simple way to determine their exact location without using calculus. Furthermore, the intensities at these maxima decrease as θ increases. The intensity at the center is far greater than the intensity at any of the secondary maxima. The intensity pattern on the screen is shown in Fig. 15-11. Additionally, the central maximum is wider than any of the other maxima, as we see from the next problem.

Problem 15.13. Consider a single slit of width w, located at a distance L from a screen, as in Fig. 15-10. Assume that $\lambda \ll w$, so that the angles for destructive interference at small m are small.

(*a*) What is the width of the central fringe?

(*b*) What is the width of the third fringe?

(*c*) What is the ratio of the width of the central fringe to the width of other fringes for small m?

Solution

(*a*) The angle θ at which we get the first minimum is given by $w \sin\theta = \lambda$. If the angle θ is small, then $\sin\theta \approx \tan\theta = y/L$, or $y = \lambda L/w$. At $\theta = 0$, there is a maximum, and by symmetry the minimum on the other side occurs at $y = -\lambda L/w$. The width of the central fringe, δ, is therefore given by $\delta = 2y = 2(L\lambda/w)$. Thus the central bright fringe has a width of

$$\delta_{\text{cent}} = 2L\lambda/w \tag{15.15}$$

(*b*) The width of the third fringe is the distance on the screen between the position of the third and fourth minima. Since the first minimum occurs for $m = 1$ in Eq. (*15.14*), the third minimum has $m = 3$, and the fourth has $m = 4$. The position on the screen for a particular m is $y_m = L \tan\theta \approx L \sin\theta = mL(\lambda/w)$. Then $y_3 - y_2 = L\lambda/w$, which is half the width of the central fringe.

(*c*) As long as θ is small, so that $\sin\theta \approx \tan\theta$, the width of a fringe is $L\lambda/w$. Thus the central bright fringe is twice as wide as the secondary fringes.

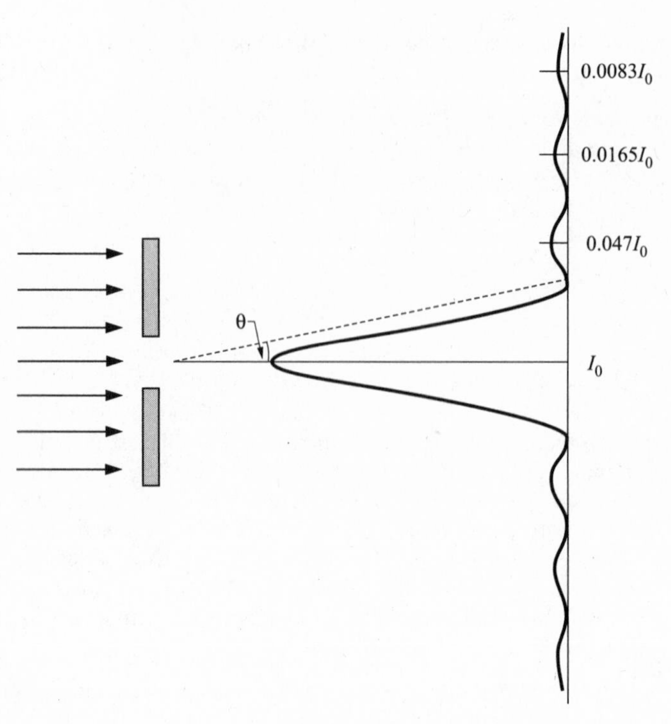

Fig. 15-11

Problem 15.14. A single slit is illuminated with light of 555 nm. At what angle will the first minimum occur if the slit width is: (*a*) 0.15 mm; (*b*) 1.5×10^{-5} m; and (*c*) 1.5 μ?

(*d*) Is the small angle approximation: $\sin\theta \approx \tan\theta \approx \theta$ (rad) valid for these cases?

Solution

(*a*) The angle is given by $w \sin\theta = \lambda$. Thus $\sin\theta = (\lambda/w) = (555 \times 10^{-9}\ \text{m})/(0.15 \times 10^{-3}\ \text{m}) = 3.7 \times 10^{-3}$, and $\theta = 0.21°$.

(*b*) The angle is now given by $\sin\theta = (555 \times 10^{-9}\ \text{m})/(1.5 \times 10^{-5}\ \text{m}) = 0.037$, $\theta = 2.12°$.

(*c*) The angle is now given by $\sin\theta = (555 \times 10^{-9}\ \text{m})/(1.5 \times 10^{-6}\ \text{m}) = 0.37$, $\theta = 21.7°$.

(*d*) The small angle approximation is valid to about 3% up to 15°. It is quite good for parts (*a*) and (*b*) and relatively poor for part (*c*).

Problem 15.15. A single slit, of width 2.3×10^{-6} m, is illuminated with light of wavelength 349 nm. A screen is located at a distance of 36 cm from the slit.

(*a*) What is the width of the central bright fringe?

(*b*) What is the largest value of *m* for which one can have a minimum?

(*c*) At what wavelength of light will there be no minimum for this slit?

Solution

(*a*) The angle at which the first minimum occurs is given by $\sin\theta = \lambda/w$. The distance to the first minimum is $y = L\tan\theta$. For this slit and wavelength, $\sin\theta = (349\ \text{nm})/(2.3 \times 10^{-6}\ \text{m}) = 0.152$, $\theta = 8.73°$, $\tan\theta = 0.154$, $y = 0.055\ \text{m} = 5.5$ cm. The central bright fringe is equal to twice this value because the minimum on the other side occurs at $y = -5.5$ cm. Thus $\delta = 11$ cm.

(*b*) As one increases *m*, the $\sin\theta$ at which the minimum occurs increases proportionally. However, $\sin\theta$ cannot exceed unity, since the maximum value of the sine is one. Thus, the maximum value of *m* corresponds to the angle at which $\sin\theta = 1$. Then $m_{max} = w(1)/\lambda$. In our case, $m_{max} = (2.3 \times 10^{-6}\ \text{m})/(349 \times 10^{-9}\ \text{m}) = 6.6$. Now, we know that m is always an integer. The result of $m = 6.6$ means that for m greater than 6.6 there is no possible value for θ. Thus $m_{max} = 6$.

Note. We do not round off m_{max} to the nearest integer. Even if the calculation gave $m_{max} = 6.97$, we would conclude that the maximum value of m that is possible is $m = 6$.

(*c*) If the value of $\sin\theta$ that we get for $m = 1$ is greater than one, then even the first minimum does not exist on the screen. For the first minimum $\sin\theta = \lambda/w$. If we set this equal to one, it will give the value of λ at which there is no minimum. Thus, $\lambda_{max} = w(1) = 2.3 \times 10^{-6}\ \text{m} = 2.3\ \mu$.

These problems illustrate the result that the angle of diffraction (the angle for the first minimum) increases with λ/w. The wave spreads out into a central bright fringe between $\pm$ (this angle). One can show that most of the light energy is contained in the central bright fringe, and one often thinks of the light as spreading out only into this area. As w gets smaller, approaching λ, the central bright fringe gets bigger, and the light spreads out more widely. We therefore see that one cannot consider the light as moving in a straight line if there are apertures comparable to the wavelength of the light. As long as $w \gg \lambda$, the light hardly spreads, and we *can* talk of the light as moving in a straight line. Since the wavelength of light is very small compared to macroscopic objects, we do not usually see the bending of light. In the case of water waves (or sound waves) which have macroscopic wavelengths, we nearly always see (or hear) bending. Thus we can hear someone speaking from the side of an open door even though we cannot see that person.

We now also understand why we could assume, in the case of a double slit, that the waves from each slit spread out onto the screen. In effect we assumed that the width of each slit was small compared to the distance, d, between the slits. For that case the central diffraction peak of both slits was so wide that the interference of the two slits took place within that broad diffraction peak. More accurately, while within the central bright fringe of the diffraction pattern of each slit there will be interference between the light from the two slits, there will also be interference within the secondary diffraction fringes, but with greatly reduced intensity. In general, the pattern that appears on the screen will be that of equally spaced interference fringes from the double slit, multiplied by the intensity pattern of the single slit. This is shown in Fig. 15-12 for the case of $w < d$. The width of the central bright fringe depends on the width w of the slits, while the spacing of the interference fringes depends on the distance d between the slits. As the width of the slit gets bigger, the central bright fringe gets smaller, and there will be less room for the interference fringes to be seen. If the fringes are closely spaced (λ/d small), there will be many interference fringes within the central bright diffraction fringe. As the spacing, d, approaches the width of the slits, there will be fewer and fewer fringes within the central fringe.

Problem 15.16. Two slits each have a width, w, and are separated by a distance d. How many interference fringes are there in the central bright diffraction fringe, if: (*a*) $d = 50w$; (*b*) $d = 10w$; and (*c*) $d = 5w$?

Solution

(*a*) The central bright diffraction fringe extends to the angle θ given by $\sin\theta = \lambda/w$. We want to know how many interference fringes there are between $\pm\theta$. We know that the mth interference fringe occurs at an angle $\sin\theta = m\lambda/d$. If we set this angle equal to the angle for the edge of the diffraction fringe, we find that $\lambda/w = m\lambda/d$, and $m = d/w$. On the other side, where θ is negative, we have an equal number of fringes. Therefore the total number of interference fringes within the central diffraction fringe equals (including the central interference fringe, $m = 0$) $2d/w + 1$. If $d = 50\ w$, then there are 101 interference fringes within the central bright diffraction fringe (including the outer two that are at the diffraction minimum — so one could argue that there are really 99).

(*b*) For $d = 10w$, there are 21 interference fringes within the central bright fringe.

(*c*) For $d = 5w$, there are 11 interference fringes within the central bright fringe.

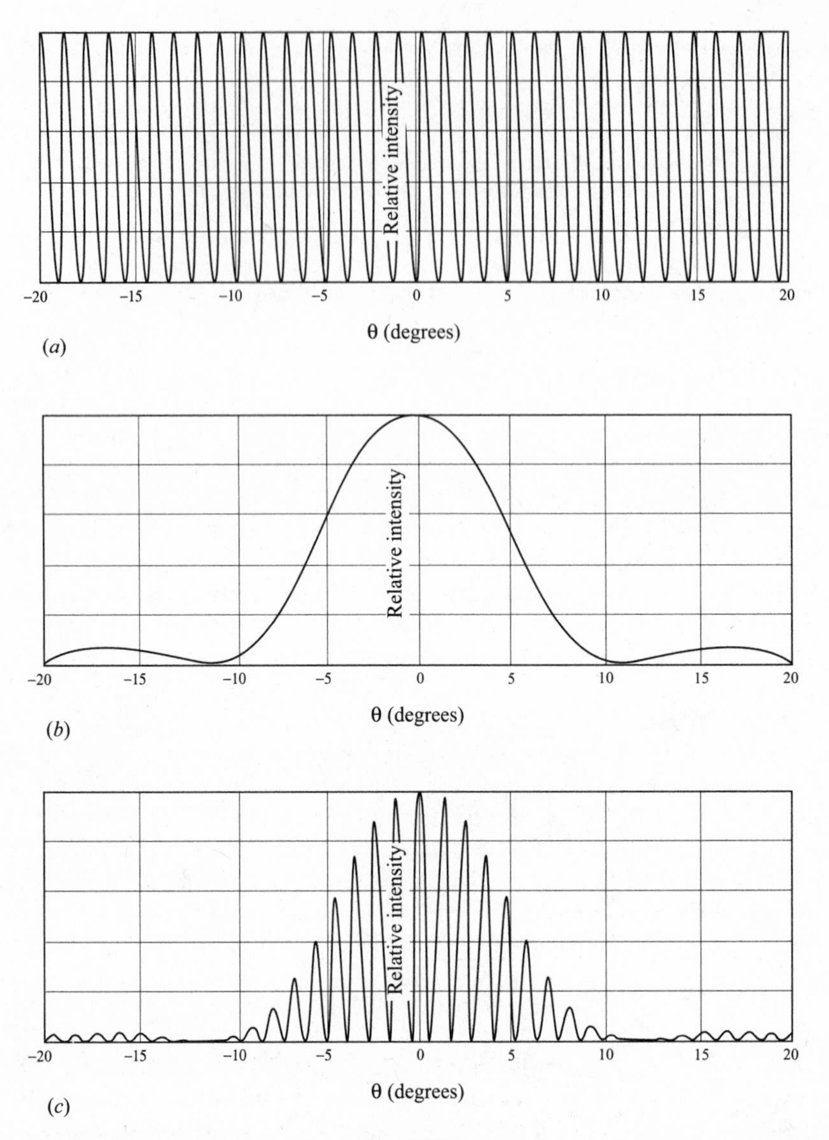

Fig. 15-12

Problem 15.17. A double slit consists of two slits, each of width 5.3×10^{-6} m, separated by a distance of 8.2×10^{-5} m. Light of wavelength 499 nm is incident on the slits.

(*a*) At what angle will the first diffraction minimum occur?

(*b*) How many interference fringes are contained within the central bright diffraction fringe?

(*c*) How many interference fringes are contained in the next diffraction fringe?

(*d*) How do the answers to these questions change, if the wavelength is doubled?

Solution

(*a*) The first diffraction minimum occurs at an angle given by $\sin\theta = \lambda/w = (499 \times 10^{-9}\ \text{m})/(5.3 \times 10^{-6}\ \text{m}) = 0.094$, $\theta = 5.40°$.

(*b*) In the previous problem we showed that the number of interference fringes within the central bright fringe is $2[d/w] + 1 = 2(8.2 \times 10^{-5}\ \text{m})/(5.3 \times 10^{-6}\ \text{m}) + 1 = 2 \times [15.5\} + 1$. Here, however, we must choose the next lowest integer for d/w, so we have $2 \times 15 + 1 = 31$.

(*c*) The second diffraction fringe extends from θ_1 to θ_2, where $\sin\theta_1 = \lambda/w$ and $\sin\theta_2 = 2\lambda/w$. The interference pattern results in fringe m_1 at θ_1 and m_2 at θ_2, where $\sin\theta_1 = m_1\lambda/d$ and $\sin\theta_2 = m_2\lambda/d$. The number of fringes between θ_1 and θ_2 is $(m_2 - m_1)$. Equating the angles in the formulas from diffraction and from interference we get $m_1 = d/w$ and $m_2 = 2d/w$. Then the number of fringes within the second diffraction fringe (on one side of the pattern only) is $(m_2 - m_1) = d/w = 15$.

(*d*) If the wavelength is doubled to 2λ the angle for the first diffraction minimum becomes $\sin\theta = 2\lambda/w = 0.188$, $\theta = 10.9°$ (the angle approximately doubles). However the number of interference fringes remains the same since the interference pattern expands proportionally. Thus the answers to (*b*) and (*c*) do not change.

2 *Single Hole*

In the case of a single slit we determined that the light widens into a larger area, with the area increasing as the width decreases. We were able to calculate the size of the central bright fringe, using simple mathematical techniques. If, instead of a slit, we have a small circular aperture, we would expect that the light passing through the hole would bend into a larger circle, with the angle of diffraction increasing as the diameter, D, of the hole decreases. This is indeed the case, although the mathematical analysis leading to an exact calculation of this spreading is quite complicated. The result of the calculation is that there is a bright central circular area (called the **Airy disk**), in which most of the energy of the light is concentrated, with secondary rings at larger angles containing smaller amounts of light energy. A small object viewed through a circular aperture expands into a larger circle, with the angle for the edge of the Airy disk given by:

$$\sin\theta = 1.22(\lambda/D) \qquad (15.16)$$

We note that this is the same as the formula for a single slit, except for the factor of 1.22. Therefore the image of a small object, viewed through a hole, expands into a larger circle on the screen. As (λ/D) becomes larger, either because D is small or because λ is large, the image size increases. If we view two neighboring tiny objects through the hole, then the circle produced by one object on the screen may overlap the circle produced by the other object on the screen. It will then become impossible to distinguish that there are two, rather than one object being observed. We can no longer "resolve" the objects from each other. Thus, diffraction imposes a limit on our ability to resolve nearby objects, and this limit is called the **diffraction limit of resolution**. There are several ways to define an exact limit, but it is clear that, for the best resolution, we must have large aperture, D, and/or small wavelength, λ. Given the above result the standard for concluding that objects cannot be resolved is that their angular separation is less than $1.22(\lambda/D)$. Thus, using visible light limits our ability to measure details less than the wavelength of light. In order to increase our resolving power we must use waves of shorter wavelength, such as ultraviolet light or X-rays, or "electron waves" that we will discuss in a later chapter. Furthermore, even for large D, there is a limit, although it is small. Thus, telescopes can resolve nearby stars only if they are further apart than the limit imposed by the diameter of the telescope. This is one of the reasons that telescopes are built with the largest practical diameter objective lens or mirror.

Problem 15.18. A telescope is used to view the moon which is at a distance of 3.8×10^8 m from the earth. The diameter of the spherical mirror used in the telescope is 1.35 m.

(*a*) If two points on the moon are separated by a small distance s, what is the angle between them, as seen from the earth?

(*b*) What is the minimum distance, s, between the points that can be resolved by this telescope, assuming that the limit is set by diffraction effects, and that the wavelength being used is 500 nm?

Solution

(*a*) We know that the arc length is given by $s = R\theta$, where θ is the angle measured in radians. The distance between two points at a large distance is the same as the arc length, provided that the angle is small.

Since the distance between the points in this problem is small, the angle is small, and $\theta = s/(3.8 \times 10^8 \text{ m})$ radians.

(*b*) The diffraction limit is given by $\theta = 1.22(\lambda/D) = s_{\min}/(3.8 \times 10^8 \text{ m}) \rightarrow$ $s = [1.22(500 \times 10^{-9} \text{ m})/(1.35 \text{ m})][3.8 \times 10^8 \text{ m}]$, or $s = 172$ m.

3 *Diffraction Grating*

If, instead of having two slits, we have many equally spaced slits, we will have to combine together the waves from all the slits when they reach the screen. It is clear that if the difference in path length between the first and the second slit is an integral number of wavelengths, then the difference between the first and the third slit will be twice as large, which is also equal to an integral number of wavelengths. The same is true for any other slit as well, so that the light from all the slits will be in phase. At those points on the screen we will have a very large maximum, since all the waves are in phase. These occur for $(d \sin \theta) = m\lambda$. However, when the difference in path length between the first and the second slit differ by even a small amount, the third, fourth, etc. slits will have path lengths that increasingly differ. If there are many slits the net effect of even a slight deviation will lead to many slits significantly canceling others that are far away. Therefore, at points between the large maxima, there will be a large amount of cancellation even if some of the waves are in phase, since other waves will be out of phase. The result is that the intensity of the light will become small very rapidly as we move away from the maxima, and the maxima will become very sharp and well defined. It will then become easy to obtain an accurate measurement of the angle θ at which the maximum occurs, and we can use this multiple slit arrangement to measure wavelengths very accurately, and to separate out different wavelengths in the source of light. When there are very many, closely spaced slits on a plate, we call the arrangement a "**diffraction grating**". Diffraction gratings were originally made by using very accurate machines to rule lines on plates at precisely spaced intervals, but nowadays they are made by casting replicas of these originals. Gratings can easily have 10,000 lines per cm, giving a spacing of $d = 10^{-6}$ m. [We assume that the width, w, of the individual lines is smaller yet, so that the grating pattern still appears within the central maximum of the single slit diffraction pattern.] The angles at which we get multiple slit maxima (the angle through which the light is "**diffracted**"), is given by:

$$d \sin \theta = m\lambda \tag{15.17}$$

The angle at which maxima occur clearly depends on the wavelength, and for multi-frequency light, for each value of m, there will be a spectrum of colors, from blue to red, as the angle θ increases. The number, m, is called the "**order**" of the spectrum.

Problem 15.19. A diffraction grating has 8985 lines/cm. Consider visible light to have wavelengths between 400 nm (blue) and 800 nm (red).

(*a*) At what angle will the blue line at 400 nm be diffracted in first order?

(*b*) Between what angles will the first order of visible light occur?

(*c*) At what angle will the blue line at 400 nm be diffracted in second order?

(*d*) What is the largest order that one can have with this grating for the blue light at 400 nm?

Solution

(*a*) The angle is given by $d \sin \theta = m\lambda$. For this grating, $d = (1/8985)$ cm $= 1.113 \times 10^{-6}$ m. Thus $\sin \theta = (1)(400 \times 10^{-9} \text{ m})/(1.113 \times 10^{-6} \text{ m}) = 0.359$, $\theta = 21.1°$.

(*b*) The angle for the first diffraction maximum for the red light at 800 nm, is given by $\sin \theta = (1)(800 \times 10^{-9} \text{ m})/(1.113 \times 10^{-6} \text{ m}) = 0.719$, $\theta = 46.0°$. Therefore, the first-order spectrum of visible light is between 21.1° and 46.0°.

(*c*) The angle is given by $\sin \theta = (2)(400 \times 10^{-9}\ \text{m})/(1.113 \times 10^{-6}\ \text{m}) = 0.719$, $\theta = 46.0°$.

(*d*) The maximum angle that can occur is 90°. This corresponds to $\sin \theta = 1$. Then the maximum order that can occur is $m_{max} = d/\lambda = (1.113 \times 10^{-6}\ \text{m})/(400 \times 10^{-9}\ \text{m}) = 2.8$. However, as we learned earlier, the correct answer is the integer less than this number, which is 2, since if $m = 3$, $\sin \theta$ would have to be greater than one.

Problem 15.20. A certain diffraction grating has N lines/cm. Using this grating, the second-order line for 695 nm falls at an angle of 17°.

(*a*) How many lines per cm does this grating have?

(*b*) At what angle does a wavelength at 453 nm have its maximum in third order?

(*c*) What is the range in angle of the third-order visible spectrum (between 400 nm and 800 nm)?

Solution

(*a*) At this angle we have $d \sin 17° = 2(695 \times 10^{-9}\ \text{m})$, $d = 4.754 \times 10^{-6}\ \text{m}$, $N = 2103$ lines/cm.

(*b*) Here $\sin \theta = 3(453 \times 10^{-9}\ \text{m})/(4.754 \times 10^{-6}\ \text{m}) = 0.286$, $\theta = 16.6°$.

(*c*) The blue wavelength at 400 nm has its third diffraction angle at $\sin \theta = 3(400\ \text{nm})/(4.754 \times 10^{-6}\ \text{m}) = 0.252$, $\theta = 14.6°$. The red wavelength at 800 nm has its third diffraction angle at $\sin \theta = 3(800\ \text{nm})/(4.754 \times 10^{-6}\ \text{m}) = 0.505$, $\theta = 30.3°$. Thus the third-order spectrum of the visible light extends from 14.6° to 30.3°.

15.4 POLARIZATION OF LIGHT

When we discussed electromagnetic radiation we noted that the displacement associated with those waves is the electric and magnetic field carried by the wave. These fields are vectors, and we showed that both vectors were perpendicular to the direction of motion of the wave, and they were also perpendicular to each other. The wave is therefore a transverse wave. The electric field (and the associated perpendicular magnetic field) can point in any direction in the plane perpendicular to the direction of motion. For definiteness we consider only the direction of the electric field since that of the magnetic field is then determined. We call the direction of the electric field in this plane the direction of "polarization". We assume that the direction of propagation is along the z axis of an x, y, z coordinate system. The electric field, or polarization, can be considered as having two components in the x, y plane, and the wave can, at any instant, be considered to be partially polarized in the x and y directions. The relative amounts in each direction can change with time, but the simplest, and most common, case of **polarization** is one in which the direction of polarization does not change with time, so that the electric field always points along one direction. Wherever possible, we will choose this direction as one of the two axes in the plane. This case is called linearly **polarized light**, or plane polarized light.

Most light that is produced by sources such as incandescent bulbs is **unpolarized**. To understand the difference between polarized and unpolarized light let us recall that, at any instant of time, the electric field at a point z will have an x and a y component. When added together, this results in an electric field pointing in some direction in the plane, which is then the instantaneous direction of polarization. If this remains the direction of polarization at point z for future times as well, or if the direction changes in a systematic manner to other directions, then we consider the wave as being polarized. However, if the direction changes randomly in time to other directions, so that, on average, no direction is more likely to be the direction of polarization at any instant, then we call the wave unpolarized. In the case of an incandescent bulb, or the sun, the light is emitted by the atoms and molecules of the source in a random fashion, and all polarizations are present in a random manner. Therefore the light is generally unpolarized. In the case of a laser (to be discussed in a later chapter), the light is usually "**coherent**", i.e. the light emitted at one time is correlated with the light emitted at a later time, and the polarization of all the light is coordinated. Therefore the light is generally polarized.

In order to obtain polarized light we can either produce it at the source, such as using a laser, or we can use some method of converting unpolarized light into polarized light. The latter is the more common method, and there are several methods which can be employed. All of them depend on using properties of a material which differentiates between light of different polarization. We will discuss materials where this property is absorption (dichroism), reflection, speed of propagation (birefringence) or scattering. In all cases the differentiation of behavior of the material for different polarizations is due to some geometrical asymmetry in the alignment of molecules in the material as often occurs with certain crystalline material.

Absorption (Dichroism)

We know that light has an electric field that can cause electric charges to accelerate, and thereby absorb energy. For instance, if we place a grid of wires in the path of the wave, with their direction being the same as that of the electric field, the electric field can cause the free electrons in the metal to oscillate along the wires and thereby absorb energy from the field of the wave. Then the wave will be absorbed by the grid and will not pass through the grid. If the electric field of the wave is perpendicular to the grid direction, and the wires are very thin (smaller than the wavelength of the wave), then the electrons cannot oscillate in the direction of the electric field and cannot absorb energy from the wave. If we can build materials with properties similar to this grid, so that electric fields in one direction are absorbed, while those in a perpendicular direction are minimally affected, then this will act as a polarizer, producing light that is polarized in the direction perpendicular to the "grid". An example of such a material is a sheet of "**polaroid**", which produces light polarized in one linear direction. Ideally, this material will absorb all the light polarized in one direction and transmit all the light polarized in the other direction. If we have unpolarized light incident on the polaroid, then half the light will be transmitted, since, for unpolarized light, there is as much light with one polarization as with another.

Reflection

When light is reflected from a surface, we have learned that the angle of reflection equals the angle of incidence. This result is correct no matter what polarization the incident light has. Consider the case of reflection shown in Fig. 15-13. The incident light is incident at an angle of θ_i, and the reflected light is reflected at the same angle θ_i. The two polarization directions for the incident light are chosen as in the plane of incidence (parallel to the paper) and perpendicular to the plane of incidence (in and out of the paper). Light of both polarizations have the same angle of reflection and the same angle of refraction.

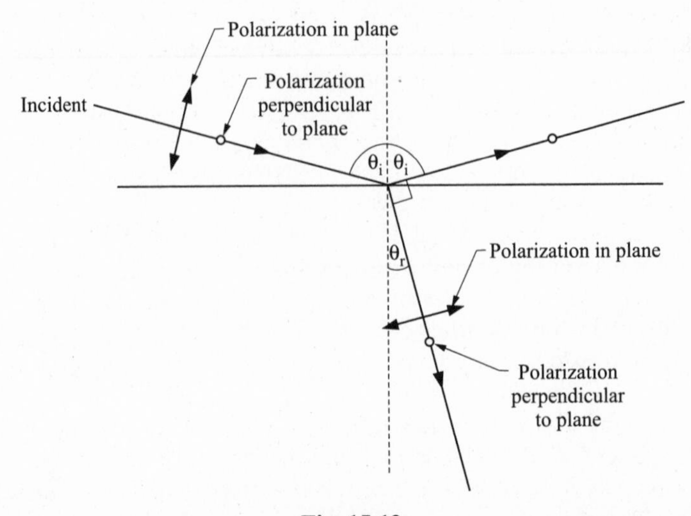

Fig. 15-13

However, the intensity of the reflected and the refracted light depends on the incident polarization. Indeed, there is a special angle, called the "**Brewster angle**", where the intensity of reflected light is zero for the incident polarization in the plane of incidence. At this angle only the part of the light polarized perpendicular to the plane is reflected, and the reflected light is therefore polarized in that direction. The Brewster angle is that angle of incidence, θ_B, for which the reflected and refracted light are at right angles, as shown in the figure. From the figure, one can see that this means that $(\theta_i + \theta_r) = 90°$. Using Snell's law for this case $(\theta_i = \theta_B)$, $n_i \sin \theta_i = n_r \sin \theta_r = n_r \sin (90 - \theta_i) = n_r \cos \theta_i$. Or, labeling $\theta_I = \theta_B$: so $n_i \sin \theta_B = n_r \cos \theta_B$, and

$$\tan \theta_B = (n_r/n_i) \tag{15.18}$$

We can understand why there is no reflected light at this angle for polarization in the plane of incidence. The reflected and refracted rays are the result of the oscillations of the electrons in the refracting material. These oscillations in turn are caused by the oscillating electric field in the incident light. At the Brewster angle the electrons, which oscillate perpendicular to the refraction direction, are also oscillating parallel to the reflection direction, as shown in the figure. In the reflection direction such oscillations can only cause an oscillating electric field along that direction. However, for the reflected light, this would correspond to a longitudinal wave, and we know that electromagnetic waves are transverse. Thus there is no reflection for that particular polarization at this angle. We see therefore that if we take unpolarized light and have it incident on a material at its Brewster angle, the reflected light will be linearly polarized perpendicular to the plane of incidence. This is how one can produce polarized light using reflection.

Speed of Propagation (Birefringence)

There are some materials that have an index of refraction that, for a certain direction of travel, depends on the polarization of the light. In that case, the critical angle, at which we get total internal reflection, will depend on the direction of polarization. If we arrange for the light to be incident at the critical angle for one polarization, and less than the critical angle for the other, then the refracted light will consist exclusively of one polarization. Clearly, one can develop instruments which use this phenomenon to produce polarized light.

Scattering

When light is incident on a collection of particles, such as those in a gas, the electric field of the light wave can cause the electrons in the particles to oscillate. These particles will oscillate in the same direction as the electric field, and then re-radiate with an electric field in the same direction as the incident electric field. This re-radiated light is called "**scattered**" light, and will be polarized in the same direction as the incident light. Consider the case of unpolarized light incident on a gas, as in Fig. 15-14. The incident light, being unpolarized, contains light polarized vertically, as well as light polarized horizontally (into/out of the paper). The light scattered downward, cannot be polarized vertically, since that would be along the direction of motion, as can be seen from the figure. Thus the vertically polarized

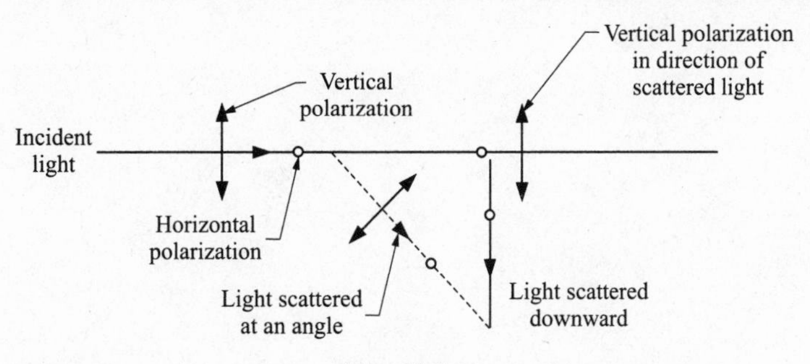

Fig. 15-14

light in the incident beam does not give rise to any light scattered downward. Only the horizontally polarized incident light can scatter downward, and that scattered light will be polarized in the direction of the horizontal polarization of the incident light (in and out of the paper in the figure). [This is very similar to the argument used in determining Brewster's angle.] Of course, the light scattered at an angle (see figure) can have some polarization with a component in the plane determined by the incident and scattered light, but the vast majority of the downward scattered light will be polarized in the horizontal direction. Therefore, scattering offers another means of producing polarized light. In fact, the light from the sun scattered by the atmosphere (the light we see during the day if we don't look at the sun), will be polarized, and we can selectively remove that light by using a polarizer that does not transmit that polarization (e.g. using polaroid sheets).

Problem 15.21.

(*a*) Light is incident on the reflecting surface of Fig. 15-13 at Brewster's angle. If there is no reflected light at all, what was the polarization of the incident light?

(*b*) Light is incident onto a scattering medium as in Fig. 15-14. If there is practically no light scattered in the direction perpendicular to the paper, what was the polarization of the incident light?

Solution

(*a*) The reflected light, if any, has polarization perpendicular to the plane of incidence. It acquired that polarization from that part of the incident light that was polarized in that direction. If the incident light consists of only light polarized in the plane of incidence, then there will be no reflected light at all. Thus, for the case of the problem, the incident light must have been polarized in the plane of incidence.

(*b*) The light scattered out of the paper in the figure will not have any electric field along its direction of motion, i.e. no polarization in or out of the paper. It will only have polarization in the vertical direction as defined in the figure. Therefore, if the incident light had no vertical polarization, then there would be no scattered light which moves in the direction perpendicular to the paper. In the case of the problem, the incident light must therefore have been only horizontally polarized, as defined in the figure.

We have learned that there are several methods that can be used to produce polarized light. These same methods can also be used to analyze the polarization of any incident light. This was already illustrated by Problem 15.21. The easiest way to polarize or to analyze light is to use the polaroid sheets mentioned previously. These sheets ideally have a preferred direction of polarization for which light normally incident on the sheet is completely transmitted (the polarizing axis), and the perpendicular direction for which normally incident light is completely absorbed. Unpolarized light directed at the sheet will be half transmitted and half absorbed and the light exiting the sheet will be plane polarized along the preferred direction. Suppose this light is now normally incident on another sheet of polaroid, which has a vertical polarizing axis, as in Fig. 15-15. Here the sheet is in the plane of the figure and the

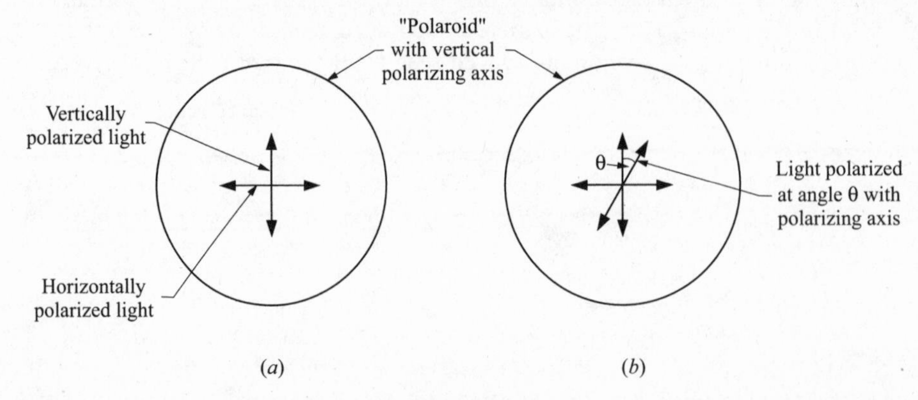

Fig. 15-15

light is incident normally from above. If the incident light is polarized vertically, then all the light will be transmitted, and the exiting light will be the same as the incident light. If the incident light is polarized horizontally, then none of it will be transmitted, and there will be no exiting light. This latter case is called "crossed polarizers" since the polarizer producing the incident light had a horizontal polarizing axis and the polarizer analyzing the light has a vertical polarizing axis. Suppose that the first polarizer has a preferred axis making an angle θ with that of the second polarizer whose preferred axis is along the vertical axis. Then the incident polarized light on the second vertical axis polarizer (often called the analyzer) also makes the angle θ with its polarizing axis. This means that the incident light has an electric field which is at an angle θ to the vertical polarizing axis of the analyzer. Using the vector nature of the electric field, we break it up into vertical and horizontal components. The component of the electric field in the vertical direction will be transmitted, while the horizontal component will be absorbed. The transmitted component is $E_i \cos\theta$, and therefore the transmitted light will have an electric field equal to $E_t = E_i \cos\theta$. The intensity, I, of the light is proportional to the square of the electric field, and therefore:

$$I_t = I_i \cos^2\theta, \qquad \text{or} \qquad I_t/I_i = \cos^2\theta \tag{15.19}$$

The fact that Eq. (*15.19*) is experimentally verified is further demonstration of the vector nature of the oscillating transverse displacement in electromagnetic waves. We see that this equation gives us the expected answer for vertically polarized light ($\theta = 0$) that $I_t = I_i$, and for horizontally polarized light ($\theta = 90°$) where $I_t = 0$.

Problem 15.22. Unpolarized light, of intensity I_i, is incident on a polarizer with a vertical polarizing axis. A second polarizer (the analyzer) is placed in the path of the light transmitted by the first polarizer. In terms of I_i, what is the intensity of the light transmitted by the analyzer if the polarizing axis of the analyzer is: (*a*) vertical; (*b*) horizontal; and (*c*) at an angle of 36° to the vertical?

Solution

(*a*) The intensity of the light transmitted by the first polarizer is $(\frac{1}{2})I_i$, since half of the incident unpolarized light is polarized along the vertical axis (e.g. over time the electric field of the incident light makes an angle with the vertical axis with equal probability and since $(\cos^2\theta)_{av} = \frac{1}{2}$, Eq. (*15.19*) yields the result). All of this light is transmitted by the analyzer, so that the transmitted light has intensity $I_t = (\frac{1}{2})I_i$.

(*b*) Again, the light incident on the analyzer has an intensity of $(\frac{1}{2})I_i$. However, since it is polarized vertically, none of this light is transmitted by the analyzer, whose polarizing axis is horizontal. Thus, $I_t = 0$.

(*c*) The light incident on the analyzer has intensity $I = (\frac{1}{2})I_i$, and is polarized vertically. The light transmitted by the analyzer is reduced by the factor $\cos^2\theta$, so the transmitted intensity is given by $I_t = (\frac{1}{2})I_i \cos^2\theta = 0.327\, I_i$.

Problem 15.23. Suppose unpolarized light is incident on three polarizers. The first and the third polarizers are "crossed". If the polarizing axis of the second polarizer is at an angle of θ with the axis of the first, what fraction of the original light will be transmitted through the last polarizer?

Solution

We know that the first polarizer transmits half of the incident unpolarized light. This light is polarized along the polarizing axis of the first polarizer, say along the vertical direction. The second polarizer has a polarizing axis at an angle θ to the vertical and therefore transmits only a fraction equal to $\cos^2\theta$ of the intensity of the incident light. The light incident on the third polarizer has an intensity of $(\frac{1}{2})\cos^2\theta$ of the original light and is polarized along the polarizing axis of the second polarizer. Since the third polarizer has its axis perpendicular to the first (they are crossed), the light is incident on the third polarizer at an angle of $(90 - \theta)$ to its polarizing axis. The fraction transmitted is $\cos^2(90 - \theta) = \sin^2\theta$. Thus the fraction of the original light that is transmitted will equal $(\frac{1}{2})(\cos^2\theta)(\sin^2\theta) = (\frac{1}{2})[(1/2)\sin 2\theta]^2 = (\frac{1}{8})\sin^2 2\theta$.

The effect of the second polarizer in this problem was to rotate the direction of polarization of the light (while also reducing its intensity), so that some of it can be transmitted by the third polarizer. Such an effect is again a demonstration of the vector nature of the electromagnetic disturbance (e.g. the electric field). Note that we have turned a vertical electric field into a horizontal one using the second polarizer! There are materials that have the ability to rotate the direction of polarization of the incident light, and their properties can be investigated using this apparatus.

Problems for Review and Mind Stretching

Problem 15.24. Light from point A can reach point B via two different paths. The path lengths differ by 250 nm. The light traveling on each path has an intensity of 3.4×10^{-5} W/m^2, and the light is in phase at point A.

(*a*) For what wavelength of visible light will there be complete destructive interference between the two rays at point B?

(*b*) For a wavelength of 250 nm what would be the intensity of the light at point B?

(*c*) If the paths are in air, how much extra time will the light spend on the longer path?

(*d*) If the paths are in water, with an index of refraction of 1.33, how much extra time would the light spend on the longer path?

Solution

(*a*) For destructive interference the difference in path length must equal $(m + \frac{1}{2})\lambda$. Therefore, $\lambda = (250 \text{ nm})/(m + \frac{1}{2})$. If $m = 0$, then $\lambda = 500$ nm, which is in the visible. If $m = 1$, then $\lambda = 167$ nm, which is in the ultraviolet. Therefore, the correct answer is $\lambda = 500$ nm.

(*b*) If the wavelength is equal to 250 nm, then the difference in path length is just one wavelength. This will give us the maximum constructive interference. Then the amplitude will be twice the amplitude of each individual wave, and the intensity will be four times the intensity of the individual wave. Therefore the intensity at B will equal $4(3.4 \times 10^{-5} \text{ W/m}^2) = 1.36 \times 10^{-4}$ W/m^2.

(*c*) The time delay will equal the difference in path length, divided by the velocity of the wave. Thus, $\Delta t = (250 \text{ nm})/(3.0 \times 10^8 \text{ m/s}) = 8.33 \times 10^{-16}$ s.

(*d*) Now, the speed of light is c/n, and the time delay is $\Delta t = (250 \text{ nm})/(3.0 \times 10^8/1.33 \text{ m/s}) = 1.11 \times 10^{-15}$ s.

Problem 15.25. Light passes through two slits that are 3.9×10^{-7} m apart, and strikes a screen at a distance of 0.62 m. The first minimum occurs at a distance of 2.5×10^{-2} m on the screen from the center.

(*a*) What is the width of each fringe?

(*b*) What wavelength of light was incident?

(*c*) At what distance on the screen (from the center) will the point be located where the difference in path length from the two slits equals $(3/4)\lambda$?

Solution

(*a*) The fringe width is the distance from a minimum to the next minimum (or a maximum to the next maximum). This distance is twice the distance from a maximum to a minimum. At the center we have a maximum and the distance to the minimum is given as 2.5×10^{-2} m. Thus the fringe width will equal twice this value, or $\delta = 5.0 \times 10^{-2}$ m.

(*b*) For a small angle we know that $\delta = \lambda L/d = 5.0 \times 10^{-2} \text{ m} = \lambda(0.62 \text{ m})/(3.9 \times 10^{-7} \text{ m})$, giving $\lambda = 3.15 \times 10^{-8}$ m.

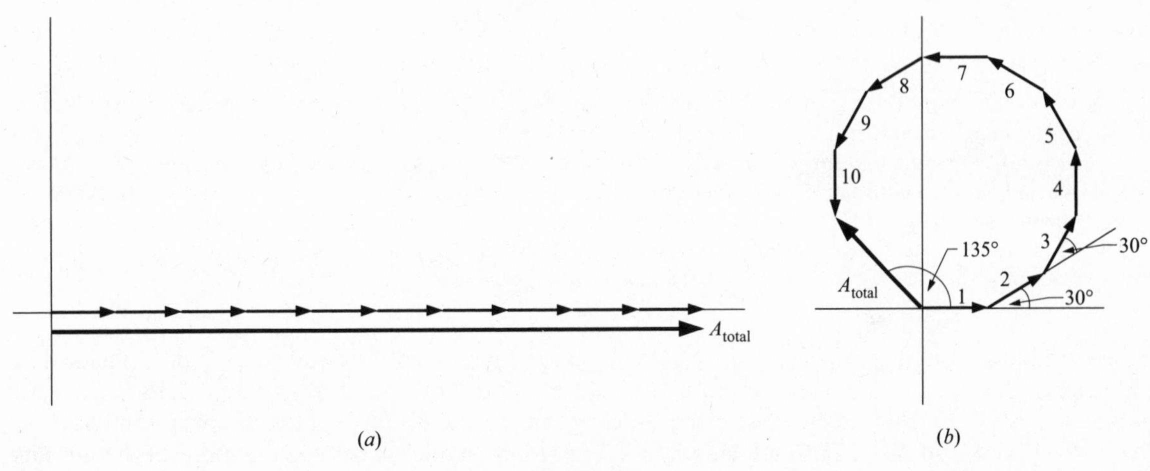

Fig. 15-16

(c) The difference in path length equals $d \sin \theta$ [Eq. (15.4)]. For small θ, the distance on the screen is $y = L \tan \theta \approx L \sin \theta = L\Delta(\text{path length})/d$. If this Δ(path length) is to equal $(3/4)\lambda$, then $y = (3/4)\lambda L/d = (0.75)(3.15 \times 10^{-8}\text{ m})(0.62\text{ m})/(3.9 \times 10^{-7}\text{ m}) = 3.75 \times 10^{-2}$ m. There is a simpler way to arrive at this same answer. For small θ, the distance on the screen is proportional to the difference in path length. For a difference in path length of one full wavelength, the distance on the screen is 5.0×10^{-2} m (which is the fringe width). Therefore, for a difference of path length of $\frac{3}{4}$ of a wavelength the distance on the screen will be $\frac{3}{4}$ of this distance or $(\frac{3}{4})(5.0 \times 10^{-2}\text{ m}) = 3.75 \times 10^{-2}$ m.

Problem 15.26. A double slit is illuminated with light of wavelength 649 nm. The slits are spaced apart a distance of 3.56 μ.

(a) If the light starts at the two slits in phase, at what two angles will we have the first minima?

(b) If the light starts at the two slits with a phase difference of 180°, at what non-zero angle will we have the first minimum?

(c) When the two rays start at the slits with no phase difference, the first minimum occurs at an angle θ. By how much must we delay one wave in order that there will be a maximum at this angle?

Solution

(a) When the waves start out at the slits in phase, which is the case that we have been discussing, the first minimum occurs at $\sin \theta = \lambda/2d = (649\text{ nm})/2(3.56\ \mu) = 0.0912$, $\theta = 5.23°$.

(b) If the waves start out with a phase difference of 180°, then the two waves will be able to interfere destructively without any further difference in path length. Therefore, we will get our first minimum at $\theta = 0$, where the central maximum would ordinarily be located. We will get additional minima when there is a difference in path length of an integral number of wavelengths, and a maximum when the difference in path length is a half integral multiple of wavelength. Thus the first non-zero angle at which we get a minimum is when the difference in path length is λ. This occurs when $d \sin \theta = \lambda$, or $\sin \theta = \lambda/d = 0.182$, $\theta = 10.5°$.

(c) We can convert a minimum to a maximum by adding a phase difference of 180°, or a time delay of $T/2$. Thus we need to delay one wave by $T/2 = \lambda/2c = 1.08 \times 10^{-15}$ s.

Problem 15.27. A film of oil ($n = 1.5$) is on water ($n = 1.33$). Light is incident at 90° to the oil. If we want to have the same order constructive interference for a wavelength (in oil) of 680 μ and destructive interference for a wavelength of 697 μ, what are the two thicknesses of oil that will accomplish this? What order, m, do they correspond to?

Solution

The reflection from air to oil is from low to high n, while the reflection from oil to water is from high to low n. Therefore there is a phase difference of 180° between the two reflected waves due to the different reflection. Then constructive interference occurs for $2t = (m + \frac{1}{2})\lambda_f$, while destructive interference occurs for $2t = m\lambda_f$. We note that for $m = 0$ we get destructive interference at any wavelength, including 697 μ. Then for our $m = 0$ constructive interference we set $2t = (\frac{1}{2})(680\ \mu) \to t = 170\ \mu$ for the smallest thickness to accomplish our goal. If we assume that the order of interference, m, is not zero for the two wavelengths, then we must have $(m + \frac{1}{2})(680\ \mu) = m(697\ \mu)$, or $(17m)\ \mu = 340\ \mu$, $m = 20$. Then $2t = 20(697\ \mu) \to t =$ 6.97 mm. Thus the two orders are $m = 0$ and $m = 20$, respectively.

Problem 15.28. A plate has ten equally spaced slits, each separated by a distance, d. Light is incident in phase on each slit, and travels to a screen at a distance of L from the slits. At an angle, θ, the light from adjacent slits travel distances to the screen differing by $\Delta(\text{path length}) = d \sin \theta$, corresponding to a difference in phase of $\Delta\phi = 2\pi\Delta(\text{path length})/\lambda$. Use phasors to add up the contributions of the ten slits if: (*a*) $d \sin \theta = \lambda$; (*b*) $d \sin \theta = \lambda/2$; and (*c*) $d \sin \theta = \lambda/12$.

Solution

(*a*) If $d \sin \theta = \lambda$, then all the phasors are in phase. Therefore, at any instant of time they are all drawn in the same direction (say horizontal), with the same amplitude A, and their sum is just $10A$ in the horizontal direction. It is easiest to draw them head to tail as we do for vectors that we want to add, as in Fig. 15-16(*a*). Therefore the total wave has just ten times the amplitude of the each individual wave, and is in phase with each individual wave.

(*b*) In this case, alternating phasors differ in phase by 180° (π radians) and hence are anti-parallel. If we draw the first phasor to the right, the next phasor will be to the left. Alternating phasors will be to the right and then the left, resulting in a sum that is equal to zero for an even number of slits. Thus the contributions of all the 10 slits adds up to zero.

(*c*) In this case, each phasor differs from the previous one by an angle of $2\pi/12 = 30°$. If we draw, at a given instant, the first horizontally to the right, the next is at an angle of $2\pi/12$ with the same amplitude. We draw all the head to tail phasors with a compass and ruler as in Fig. 15-16(*b*). After adding together ten phasors, each at an additional angle of 30°, we can see that the resultant phasor (at that instant) is at an angle of 135° to the positive horizontal and has an amplitude of $(2 \cos 15°)A = 1.932A$. This is much less than the amplitude of $10A$ that we calculated for part (*a*), and the difference is even greater when we calculate intensities which are proportional to the square of the amplitudes. We can use this technique to calculate the amplitude for any combination of slits and phase angles.

Supplementary Problems

Problem 15.29. Two waves, with parallel directions for their electric fields, and each with the same maximum intensity of $I_0 = 6.07 \times 10^{-4}\ \text{W/m}^2$, meet at some point. What is the resultant maximum intensity of the total wave if: (*a*) they are in phase; (*b*) they are 180° out of phase; and (*c*) they are 90° out of phase?

Ans. (*a*) $2.43 \times 10^{-3}\ \text{W/m}^2$; (*b*) 0; (*c*) $1.21 \times 10^{-3}\ \text{W/m}^2$

Problem 15.30. Two waves reach a point and add together. The amplitude of the first is twice as big as the amplitude of the second. $E_1 = 2E_2$ and both fields are parallel. In terms of the intensity of the first wave, I_1, calculate: (*a*) the intensity of the second wave; (*b*) the intensity of the sum if they are in phase; and (*c*) the intensity of the sum if they are 180° out of phase.

Ans. (*a*) $0.25\ I_1$; (*b*) $2.25\ I_1$; (*c*) $0.25\ I_1$

Problem 15.31. Two slits are separated by a distance of 3.6×10^{-5} m. Light at a wavelength of 541 nm illuminates the slits.

(*a*) At what angle does the third maximum occur?

(*b*) At what wavelength will the second maximum occur at this same angle?

Ans. (*a*) 2.58°; (*b*) 812 nm

Problem 15.32. Two slits produce fringes on a screen that is at a distance of 2.05 m from the slits. When the wavelength is 608 nm, the fringe spacing is 1.3 cm.

(*a*) What is the distance between the slits?

(*b*) For the same slits, what fringe spacing would occur for a wavelength of 497 nm?

Ans. (*a*) 9.59×10^{-5} m; (*b*) 1.06 cm

Problem 15.33. Two slits are illuminated with light of wavelength 637 nm. The slit spacing is 5.8 μ.

(*a*) At what angle will the first minimum occur, if the apparatus is in air?

(*b*) If the apparatus were placed in a material with an index of refraction of 1.41, at what angle would the first minimum occur?

Ans. (*a*) 3.15°; (*b*) 2.23°

Problem 15.34. A Michelson interferometer is used to calibrate the wavelength of light from Krypton. As one moves one mirror through a distance of 1/4 meter, one counts the number of fringe maxima that pass by. If one finds that there are 825,381.86 maxima, what is the wavelength of the Kr line?

Ans. 6.0578×10^{-7} m

Problem 15.35. The arms of a Michelson interferometer differ in length by 325 nm.

(*a*) If light of 650 nm is used in the interferometer, will there be destructive or constructive interference at the detector?

(*b*) At what *visible* wavelength will there be destructive interference at the detector?

Ans. (*a*) constructive; (*b*) 433 nm

Problem 15.36. An air wedge is formed by placing shims, of thickness 0.295 mm, between two smooth glass plates, on one end. The plates are 25 cm long, and the wedge is viewed with light of wavelength 493 nm impinging normally. How many fringes per meter are formed along the length of the plates?

Ans. 4790 fringes/m

Problem 15.37. An air wedge is formed between two plates. Using light with a wavelength of 576 nm, there are 886 fringes formed per meter.

(*a*) What is the angle of the wedge?

(*b*) If the air space is filled with material of index of refraction 1.43, how many fringes would be formed per meter?

Ans. (*a*) 0.0146°; (*b*) 1267 fringes/m

Problem 15.38. A thin film of water of thickness 275 nm forms a bubble, with air on the outside and on the inside. Light is reflected from the two sides of the water film ($n = 1.33$).

(*a*) What difference in phase results from the different interface reflections at the two surfaces?

(*b*) At what wavelength in air, in the visible region, will there be constructive interference?

(*c*) At what wavelength in air, in the visible region, will there be destructive interference?

Ans. (*a*) 180°; (*b*) 488 nm; (*c*) 732 nm

Problem 15.39. In an experiment one uses light of wavelength 500 nm in air.

(*a*) If the light is incident on a single slit of width 0.032 mm, at what angle will the second minimum occur?

(*b*) If the light is incident on a double slit of separation 0.051 mm, at what angle will the second minimum occur?

(*c*) If the light is incident on a grating with 1500 lines/cm, what is the maximum order that can be seen?

(*d*) If the light is incident normally on an oil film ($n = 1.2$) floating on water ($n = 1.33$), what is the minimum thickness of oil needed to get constructive interference in the reflected light?

Ans. (*a*) 1.79°; (*b*) 0.84°; (*c*) 13; (*d*) 2.08×10^{-7} m

Problem 15.40. Newton's rings are formed by a glass sector of radius 18 cm, placed on a smooth plane glass plate. When viewed normally with light of wavelength 597 nm, what are the radii of the first three bright circles?

Ans. 0.231 mm; 0.401 mm; 0.518 mm

Problem 15.41. In a double slit, the first diffraction minimum coincides with the 25th interference minimum.

(*a*) What is the ratio of the distance between slits to the slit width?

(*b*) How many interference fringes are within the central diffraction peak?

Ans. (*a*) 24.5; (*b*) 49

Problem 15.42. A small pinhole is illuminated with normally incident plane wave light, and forms a central maximum of radius 0.37 cm on a screen at a distance of 1.09 m. The wavelength of the light used was 440 nm. What is the diameter of the pinhole?

Ans. 1.58×10^{-4} m

Problem 15.43. With light of wavelength 636 nm:

(*a*) What diameter pinhole would produce only the central bright fringe on a screen?

(*b*) If one wanted to resolve two objects 11.5 m apart on the moon, which is at a distance of 3.80×10^8 m from the earth, what is the minimum diameter mirror that is needed?

Ans. (*a*) 7.76×10^{-7} m; (*b*) 25.6 m

Problem 15.44. A diffraction grating diffracts blue light (400 nm) at an angle of 8.3° in first order.

(*a*) At what angle is red light (800 nm) at the other end of the spectrum diffracted in first order?

(*b*) Between what angles is the second-order spectrum?

(*c*) What is the maximum order for this grating for the blue light?

Ans. (*a*) 16.8°; (*b*) 16.8° and 35.3°; (*c*) 6

Problem 15.45. A diffraction grating has 1500 lines/cm, and is diffracting light at a wavelength of 555 nm. A grating is capable of resolving a difference in wavelength, $\Delta\lambda = \lambda/Nm$, where m is the order of the spectrum and N is the total number of slits in the grating.

(*a*) At what angle is the light diffracted in second order?

(*b*) If the grating is 10 cm wide, what difference in wavelength can be resolved in this order?

(*c*) What fraction of the wavelength can be resolved in this order by this grating?

Ans. (*a*) 9.58°; (*b*) 1.85×10^{-2} nm; (*c*) 3.33×10^{-5}

Problem 15.46. Unpolarized light, of intensity 6.83×10^{-3} W/m^2, is incident normally on a linear polarizer in the x, y plane whose axis is in the y direction. The transmitted light is then incident on a linear polarizer whose axis makes an angle of 27° with the y axis. This is followed by another polarizer. What is the intensity of the light transmitted by this last polarizer, if the axis of this polarizer is: (*a*) along the x axis; (*b*) along the y axis; and (*c*) at an angle of 53° with the y axis (on the same side of the y axis as the second polarizer)?

Ans. (*a*) 5.59×10^{-4} W/m^2; (*b*) 2.15×10^{-3} W/m^2; (*c*) 2.19×10^{-3} W/m^2

Problem 15.47. What is the Brewster angle for light reflected from: (*a*) air to water, whose $n = 1.33$; (*b*) air to glass, whose $n = 1.51$; and (*c*) water to glass?

Ans. (*a*) 53.1° (*b*) 56.5° (*c*) 48.6°

Chapter 16

Special Relativity

16.1 INTRODUCTION

In Chap. 12, we learned about the production of electromagnetic waves in space, and discovered that they were predicted to move with a speed of $c = 3 \times 10^8$ m/s—the speed of light. The question is: for which observer is this the velocity of light? For all other waves we know that the propagation velocity of the wave represents the velocity relative to the medium of the wave. For a transverse wave on a string the wave velocity is relative to the stationary string, for a water wave it is relative to the stationary water, for a sound wave it is relative to the air that carries the wave. For an observer on the shore of a river that is moving with a velocity V, a water wave traveling in the direction of the current will appear to move with a speed of $(v + V)$, where v is the speed of the wave relative to the stationary water, while a water wave traveling in the direction opposite to the current will move with a speed of $(v - V)$. In the case of an electromagnetic wave it was presumed that there was also such a medium, called the ether, which permeated all of space, and while otherwise undetected, served to carry the electromagnetic waves. Only a person at rest in the ether would measure c, and any other observer would measure a larger or smaller velocity depending on his motion through the ether. A determined effort was made to discover the supposed special reference frame in which the ether was at rest and the velocity was c.

Michelson had developed a very sensitive instrument based on the interferometer discussed in Chap. 15, capable of measuring very small changes in the velocity of light due to motion of an observer through the ether. The largest change in velocity for an observer on earth occurs when the earth moves from one side of its orbit around the sun to the other side, in which the velocity of the earth changes by twice its orbital velocity. Even this velocity change is only about 6×10^4 m/s, which is very small compared to $c = 3 \times 10^8$ m/s, but the effect on the velocity of light would be measurable with Michelson's instrument. No differences in velocity were observed, an unexplained result until Einstein developed his special theory of relativity. In this theory there is no ether that selects out one particular inertial reference frame as fundamental, and the velocity of light is the same for all observers in inertial reference frames moving at arbitrary uniform velocities relative to one another.

Before discussing the special theory of relativity we will analyze Michelson's experiment. A sketch of the interferometer is shown in Fig. 16-1 (see Chap. 15). A source of light shines (at a specific frequency), onto a "half-silvered" mirror, which allows half the light to be reflected and the other half to be transmitted. The reflected light travels to mirror 1, at a distance of L_1, and is reflected back. The transmitted light travels the same distance, $L_2 = L_1$ to mirror 2, and is also reflected back. Part of the two beams recombine and travel to the detector. If they are "in phase" ("out of phase") when they reach the detector they will constructively (destructively) interfere. Since the two paths have equal lengths, the light takes the same time to traverse each path if the velocity of light along each is the same. The two beams would then arrive in phase. If there is a difference in velocity of the light along the two paths, then the times of travel differ. If this difference is $\frac{1}{2}$ period ($T/2$), then the result will be destructive interference. When properly used a high resolution interferometer allows one to measure time differences of about $(\frac{1}{10})T$ for visible light. If one path of the interferometer is in the direction of the motion of the earth through the ether, and the other path perpendicular to that direction, then this sensitivity is just about sufficient to allow one to measure differences in travel time due to the earth's orbital velocity of 6×10^4 m/s.

Problem 16.1. A stationary observer sees light traveling to the right with the velocity c. It strikes a stationary mirror after traveling 2 m, and is reflected back, traveling at the same velocity, c.

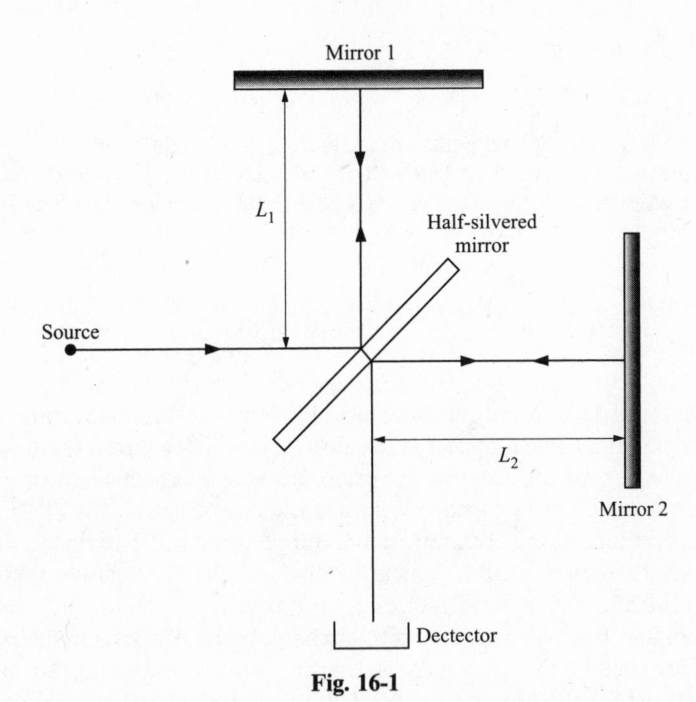

Fig. 16-1

(*a*) How long does it take for the light to travel back and forth according to the observer?

(*b*) If the velocity of light were equal to c in a medium moving to the right with a velocity of 3000 m/s, how long does it take for the light to travel back and forth, according to the observer?

(*c*) What is the difference between these two times?

Solution

(*a*) The time to travel each way is $L/c = (2 \text{ m})/c$. The total time is therefore $(4 \text{ m})/c = 1.33 \times 10^{-8}$ s, to three place accuracy.

(*b*) When traveling to the right, the light appears to the observer to have a speed of $c + 3000$, and when traveling to the left the velocity appears to be $c - 3000$. The time for a round trip is therefore $2/(c + 3000) + 2/(c - 3000) = 2[(c + 3000) + (c - 3000)]/[(c + 3000)(c - 3000)] = 4c/[c^2 - 3000^2] = 1.33 \times 10^{-8}$ s, to three place accuracy.

(*c*) As we saw above, the two results are the same to at least three place accuracy. To obtain the small difference we subtract algebraically and reduce the result. The difference between these this is $4c/[c^2 - 3000^2] - 4/c = (4/c)[c^2/(c^2 - 3000^2) - 1] = (4/c)\{c^2 - [c^2 - 3000^2]\}/(c^2 - 3000^2) = (4/c)(3000^2)/(c^2 - 3000^2) = 1.33 \times 10^{-18}$ s. This difference is 10^{-10} of the time in part (*a*). It would require a very sensitive detector to detect this difference.

Problem 16.2. In a Michelson interferometer, one path is in air throughout its length. In the other path we insert a length L, in which there is a vacuum. The light used has a wavelength of 500 nm. The index of refraction of air is 1.000,293. Assume the only difference in velocity is due to the different speeds of light in vacuum and in air.

(*a*) Calculate the period of the light used.

(*b*) Calculate the extra time spent on the all air path compared to the path with a length L of vacuum.

(*c*) What minimum length of vacuum is needed to get destructive interference?

Solution

(*a*) The period is given by $T = \lambda/c = 500 \times 10^{-9}/3 \times 10^{8} = 1.67 \times 10^{-15}$ s.

(*b*) Except for the length *L*, the two paths are identical, and the light spends the same time on each path. The only difference is that on the all air path the light travels back and forth along the length *L* of air, while on the other path the light travels back and forth in vacuum. The time spent in the air path is $2L/v = 2nL/c$, where *n* is the index of refraction in air. The time spent in vacuum is $2L/c$. The difference in time is $\Delta t = (2nL/c - 2L/c) = (2L/c)(n - 1) = L(1.95 \times 10^{-12})$.

(*c*) For the minimum length giving destructive interference the time difference, Δt, must equal $T/2 = 0.833 \times 10^{-15}$. Thus, $L(1.95 \times 10^{-12}) = 0.833 \times 10^{-15}$, or $L = 4.27 \times 10^{-4}$ m, or about 0.4 mm.

Michelson and Morley tried to find changes in the velocity of light traveling along perpendicular paths at different times of the year, and found no differences. They came to the conclusion that the velocity of light was always *c* for an observer on earth, no matter in which direction the observer was moving or, in other words, all observers measure the same speed of light, *c*.

This violates all our intuition and common sense. Suppose instead of light we applied this rule to a moving baseball. It is as if a man in a train, which is moving along the positive *x* axis at 20 m/s, throws a baseball in the same direction with a speed of 20 m/s relative to the train, and a man on the embankment watching the train and baseball zoom by measures the baseball's velocity as 20 m/s relative to the earth—the same as the speed of the train! However, not only is this concept—light traveling at the same velocity as measured by all observers—difficult to accept logically, it also leads to additional predictions which seemingly make even less sense, as we shall see shortly. It was Einstein who decided to investigate the logical consequences of assuming that light traveled at the same speed for all observers in inertial reference frames (i.e. observers traveling at uniform velocities relative to one another).

Einstein made two assumptions: first, that the laws of nature were identical for all inertial observers, as with Newton's laws of mechanics; second, that there was no special medium, or "ether", and that light waves traveled with the same speed in any direction in empty space as measured by any and all inertial observers. This included observers moving relative to each other while measuring the same light wave! We will study the consequences of these two assumptions, which form the basis of **Einstein's Special Theory of Relativity**.

16.2 SIMULTANEITY

We have always assumed that if one observer notes two events occurring simultaneously at different points in space, then any other observer will also note those events occurring simultaneously. If, however, all inertial observers measure the *same light beam* as traveling with velocity *c*, irrespective of their motion relative to each other, then this simultaneity assumption is not true.

Consider the apparatus shown in Fig. 16-2(*a*). An observer who is at rest with respect to this apparatus sees light emitted from a source halfway between the two mirrors, and traveling with the same velocity, *c*, in both directions. Since the distance to each mirror is identical, the light reaches the two mirrors simultaneously. We say that the two events, light reaching mirror 1 and light reaching mirror 2, occurred simultaneously for this observer. Now let us suppose that this apparatus, together with this observer, are on a space ship moving to the right with a velocity *V*, as measured by a stationary observer, *A* [see Fig. 16-2(*b*)]. Observer *A* sees the light emitted at the center of the apparatus as still traveling to both the right and left *with the same velocity*, *c*. Mirror 2 is moving away from the light with a velocity *V*, so the light has to travel further before reaching it and therefore takes longer to reach the mirror than if it were stationary. On the other hand mirror 1 is approaching the light, so the light travels a shorter distance and reaches this mirror sooner than for a stationary mirror. The light thus reaches mirror 1 before it reaches mirror 2 according to observer *A*, and the two events were *not* simultaneous, even though they were simultaneous according to *B*. In fact, according to *A*, event 1 (the light reaching mirror 1), occurs before event 2 (the light reaching mirror 2). If a third observer, *C*, were

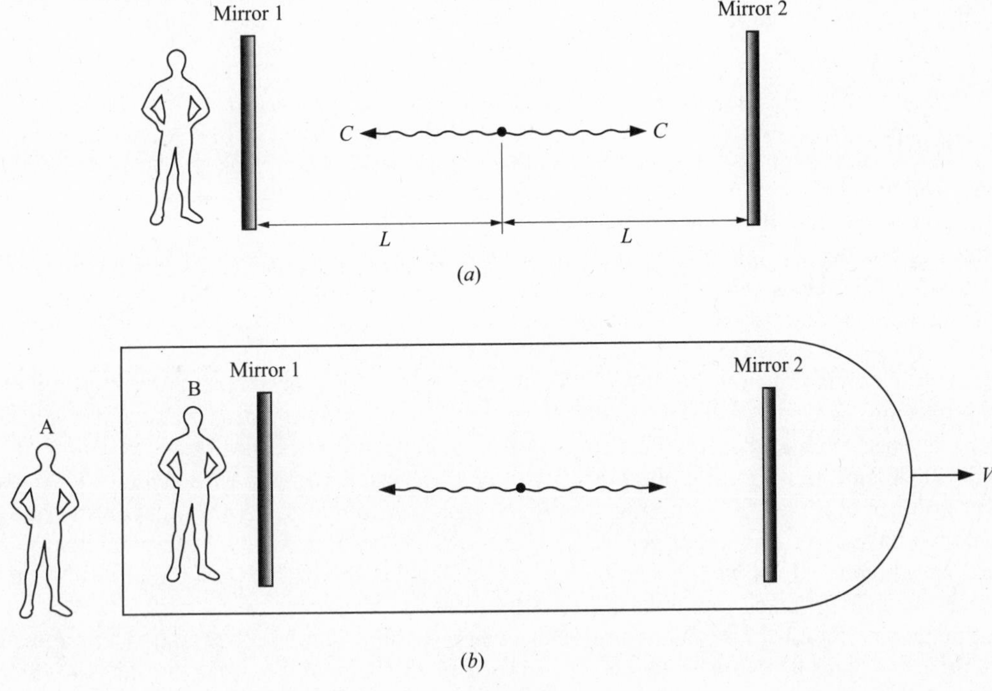

Fig. 16-2

in a frame of reference in which he/she saw the apparatus moving to the *left* with velocity V', the conclusion would be that event 1 occurred *after* event 2. These conflicting conclusions seem totally ridiculous, but follow from Einstein's hypothesis. Remember, however, that the difference in time between the events is hardly noticeable unless the velocity V is comparable to c. We have no common experience with such large velocities and we cannot therefore rule out the possibility that this disagreement about simultaneous events really occurs. In fact, the special theory of relativity has by now been amply verified by experiment, and we know that this is precisely what does happen.

Problem 16.3. Assume that, in Fig. 16-2(b), the distance from the light to each of the mirrors is $L_A = 2 \times 10^6$ m and the velocity V is 3×10^7 m/s, all according to A. Calculate the difference in time between event 1 (the light reaching mirror 1) and event 2 (the light reaching mirror 2).

Solution

To reach mirror 1, the light only has to travel a distance L_A minus the distance traveled by the mirror in the same time t_1. The light travels a distance ct_1 and the mirror travels a distance Vt_1. Thus $L_A - Vt_1 = ct_1$, or $L_A = (c + V)t_1$, and $t_1 = L_A/(c + V)$. Similarly, to reach mirror 2, in time t_2, the light has to travel L_A plus the distance traveled by mirror 2, Vt_2. Thus, $L_A + Vt_2 = ct_2$, and $t_2 = L_A/(c - V)$. The time difference between the events is $t_2 - t_1 = L_A[1/(c - V) - 1/(c + V)] = 2VL_A/(c^2 - V^2)$. Substituting the values of L_A and V into the equation, we get:

$$\Delta t = 2(3 \times 10^7)(2 \times 10^6)/[3 \times 10^8)^2 - (3 \times 10^7)^2] = 0.00135 \text{ s.}$$

The Δt is observably large because both L_A and V are large. As either of these quantities approaches zero, the time difference also approaches zero. Therefore, there never is a time difference at the same position, nor is there a time difference if the observers are not moving relative to each other.

This absence of simultaneity for events observed by an observer moving relative to the apparatus also leads to the prediction that clocks at different locations that are synchronized to the same time in one system will not be synchronized in another moving system. Whenever we think of the problem of simultaneity, it is appropriate to bring to mind this picture of the apparatus that we used in this discussion.

16.3 TIME DILATION

In this section we will again use a hypothetical apparatus to show that, in Einstein's theory, moving clocks run at a slower rate than stationary clocks, even if the clocks are identical. We will conceive of a clock whose operation depends on the motion of light and show that the time interval between successive "ticks" on the clock (its period) will be longer when observed by an observer, A, that sees the clock moving, than by an observer, B, at rest with respect to the clock. Thus observer A will say that the moving clock runs slow compared with an identical clock that is not moving.

Our special clock is shown in Fig. 16-3(*a*). Light starts at the bottom mirror and travels the distance L to the upper mirror where it is reflected back to the bottom mirror. The period of this clock is the time to travel once back and forth. This time will be $T = 2L/c$ for the stationary observer. We now assume this clock and its stationary observer, B, are in a spaceship, moving with velocity V to the right relative to another observer, A, as in Fig. 16-3(*b*). Using Newtonian logic, the light from the bottom mirror will still have a vertical component c, but will now, in addition, have a horizontal component V. The actual speed relative to A would then be $c' = (c^2 + V^2)$, and the path would appear as in the figure. The time for the light to reach the upper mirror and return to the lower mirror will not have changed, since that depends only on the vertical component which is still c. Thus both A and B will agree on the period of the clock. This is no longer true if we impose our Einsteinian idea that the speed of the light is the same to both observers. According to observer A, the light from the bottom mirror reaches the center of the upper mirror in a time $T'/2$, where T' is the period for the whole trip from bottom to top and back as measured by A. During the time $T'/2$ the mirror will have moved to the right a distance of $VT'/2$, and therefore the light will travel a diagonal distance $[L^2 + (VT'/2)^2]^{1/2}$ as shown in the figure. Since A sees the light moving with a speed of c [not $(c^2 + V^2)^{1/2}$], he concludes that $cT'/2 = [L^2 + (VT'/2)^2]^{1/2}$. Squaring both sides and multiplying by $(2/c)^2$ and rearranging terms, we finally find that $T'(1 - V^2/c^2)^{1/2} = 2L/c$. Recalling that $T = 2L/c$, we get $T' = T/(1 - V^2/c^2)^{1/2}$. Let us call the factor $[1/(1 - V^2/c^2)^{1/2}] = \gamma$

$$\gamma = [1/(1 - V^2/c^2)^{1/2}], \tag{16.1}$$

and we have

$$T' = \gamma T \tag{16.2}$$

The factor γ only makes sense if $V < c$, and we will have to show later on that this will always be the case. Then $(1 - V^2/c^2)^{1/2} < 1$, and $\gamma > 1$. Thus $T' > T$, and the period of the moving clock (B's clock observed by A) is longer than the period of the identical clock at rest (B's clock observed by B). If it takes longer between "ticks" on a moving clock, then that clock runs slow, and we have shown that moving clocks run slow. It can be shown that this consequence of Einstein's hypothesis is true of all

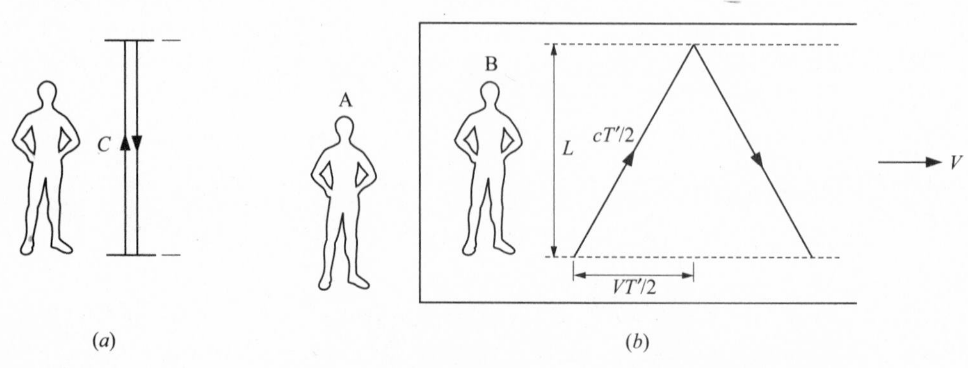

Fig. 16-3

moving clocks, including mechanical and electronic clocks. This includes biological clocks as well. Suppose a person who is stationary ages two years on his own stationary clock. When he observes an identical person moving with velocity V, he will say that that person's clocks run slow (slower heart rate, slower chemical reactions, etc.) and he will have aged less than two years. In a time interval of two years, while his own clock will have gone through $2/T$ "ticks", the moving clock will have gone through $2/T' = 2/\gamma T$ "ticks". Thus the moving person will have aged only $2/\gamma < 2$ years. It is important to note that increasing the period of moving clocks by the factor γ implies that the measured time interval between events decreases by the factor $1/\gamma$, so the server sees his moving counterpart as being aged less. It should be noted that the statement that the moving clock runs slow does not at all imply that there is any defect in the clock. On the contrary, we are talking about perfectly accurate clocks that keep perfect time. **Relativity** requires that such a perfect clock, if it is observed to be moving, will be observed to run slow compared to the identical clock at rest.

Problem 16.4. A person in a space ship, moving uniformly with a speed of $0.8c$, observes the earth. He notices that the earth rotates about its axis in a periodic fashion, and measures the period on his own clock.

(*a*) How long is this period, as measured by the spaceperson?

(*b*) After one week on the spaceperson's clock, how many rotations has the earth made on its axis, i.e. how many earth days have elapsed?

Solution

(*a*) The clock on the earth is a moving clock, from the point of view of the spaceperson. Therefore, it runs slow, and the period of any clock on the earth is increased by the factor γ. Thus the period of the earth's rotation is increased from one day to $\gamma = [1/(1 - v^2/c^2)]^{1/2} = [1/\{1 - (0.8c)^2/c^2\}]^{1/2} = 1/(1 - 0.64c^2/c^2)^{1/2} = 1/0.6 = 1.667$ days.

(*b*) In one week on the spaceperson's clock, seven of his days will have elapsed. Since the earth's clock runs slow, only $7/\gamma$ earth days will have elapsed, or $7(0.6) = 4.2$ days.

Problem 16.5. An observer on earth watches a spaceship traveling to a star that is 3 light years away according to the earthperson's ruler, as in Fig. 16-4. The spaceperson is traveling uniformly at a speed of $0.8c$. When the spaceship passes the earth, each observer sets his clock to zero.

(*a*) According to observer A, on earth, how long does it take the spaceship to reach the star?

(*b*) According to observer A, what time will have elapsed on the spaceperson's clock when he reaches the star? Will observer B, in the spaceship agree to this reading on his clock?

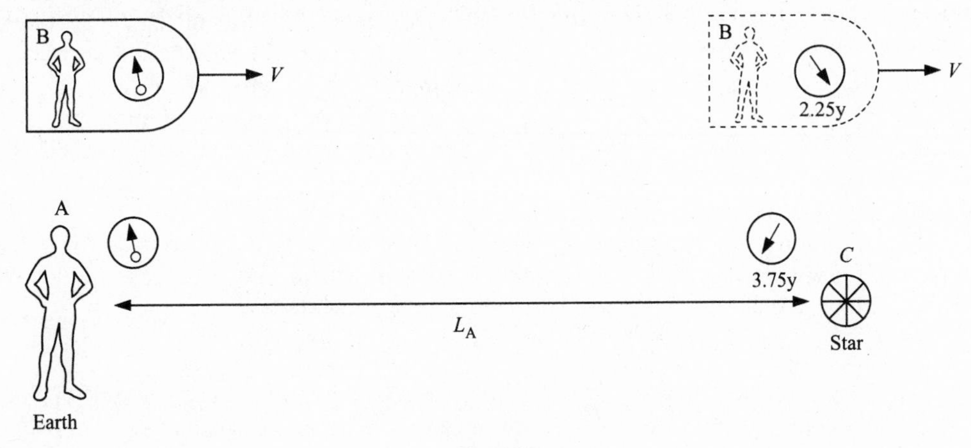

Fig. 16-4

(*c*) According to observer *B*, how much time will have elapsed on observer *A*'s clock until *B* meets the star?

Solution

(*a*) As long as observer *A* makes measurements in her own system, using her own rulers and clocks, she can use all the kinematics that we have previously learned. Thus the time to reach the star will equal L_A/V, where L_A is the distance to the star, and V is the speed of travel to the star. The distance is 3 light years, or three times the distance traveled by light in one year. The velocity is $0.8c$. Thus $t = 3$ light years/$0.8c$. Now one light year is c(1 year), so $t = 3c/0.8c$ years $= 3.75$ years.

(*b*) Observer *A* says that the clock of observer *B* runs slow, so less time will have elapsed on the spaceperson's clock. The factor $1/\gamma = 0.6$, so the time elapsed on *B*'s clock is $3.75(0.6) = 2.25$ years. When the spaceperson passes the star, one can take a picture of the clock, and everybody will have to agree to what is seen on this picture. Therefore *B* will also have to agree that his own clock reads only 2.25 years when he meets the star.

(*c*) From the point of view of *B*, 2.25 years have passed on his clock until he meets the star. He claims that his clock is stationary and the earth and star are moving with the velocity of $0.8c$ in the opposite direction. The star will have reached him after 2.25 years on *B*'s stationary clock. *B* considers the clocks on the earth and on the star as moving clocks, which run slow. Thus he will say that the time elapsed on *A*'s clock is just $2.25(0.6) = 1.35$ years.

Parts (*b*) and (*c*) of the previous problem present some interesting dilemmas. For starters observer *A* sees a space ship traveling at $0.8c$ through three light years and this takes 3.75 years, while observer *B* sees a star approaching at the same speed $0.8c$ and it takes only 2.25 years. This implies that observer *B* must believe the star was less than three light years away at $t = 0$. We will discuss this in the next section. In addition, observer *B* thinks that *A*'s clock has progressed through only 1.35 years while *A* knows his clock has progressed through 3.75 years. Since *A* and B are not at the same place, they cannot compare clocks at this time, and there is no definitive event (like the passing star) that allows a snapshot of *A*'s clock to be taken that both *A* and *B* will have to agree upon. We will see later that there is no contradiction in *A* and *B*'s conclusion regarding clock *A*'s reading. *A* says that *B* has aged less than *A* (2.25 rather than 3.75 years), while *B* says that *A* has aged less (1.35 rather than 2.25 years). This again is not a contradiction; but what if they somehow arrange to meet again and compare ages? For instance, the spaceship could reverse direction and return to earth. On both parts of the trip, *B* has the moving clock and ages more slowly, according to *A*, with the opposite statement from *B*. When they meet they can compare ages and they will have to agree on the result. This is known as the twin paradox (when *A* and *B* are identical twins), which we will discuss later. It will turn out that *B* is the one who has aged more slowly if they should meet.

Thus, for instance, a traveler to a distant star can reach the star after he ages only ten years, while the observer on earth, who sent him off on the trip, ages 100 years. The observer on earth is unlikely to be alive anymore when the traveler reaches the star, but the traveler will only be ten years older than when he left the earth. One might think that it is advantageous to travel at high speed since one lives longer. However, the traveler is not really gaining extra years. His heart beats the number of times that it should in ten years, he has only ten years worth of thoughts and other activities, he can't do anything more than one can do in ten years. He really is living for only ten years, even though the earth observer has aged, and lived, much more.

This time dilation effect is apparent only when the speed of travel is comparable to that of light. This is why the effect was not seen before it was predicted by Einstein. The next problem illustrates one of many ways in which this effect has by now been verified experimentally.

Problem 16.6. **Muons** are fundamental particles that are unstable and "decay" (break up into other particles) with a "half-life" of $\tau = 5 \times 10^{-6}$ s. This means that, on the average, half of them decay in this time. What is their half-life to an observer who sees them moving at a speed of $0.95c$?

Solution

The half-life, τ, is an internal clock of these particles (see Fig. 16-5). If the particles are moving, then a stationary observer sees this clock running slowly, and he will measure a half-life that is increased by the factor γ. Thus the half-life will be $\tau\gamma$. Now, $\gamma = [1/(1 - (0.95c)^2/c^2)^{1/2}] = 3.20$, and therefore the half-life that a stationary observer would measure is $5 \times 10^{-6}(3.20) = 1.6 \times 10^{-5}$ s. This effect has been amply verified by experiment and shows that this prediction is actually true.

Let us now turn our attention to the so-called "**twin paradox**". One twin remains on the earth, while the other one travels away at a speed V (see Fig. 16-6). After traveling a certain time the second twin turns around and returns to earth at the same speed. The twin on earth claims that the traveler's clock is running slow both on the trip away from the earth and on the trip back to the earth. Therefore, he claims that his brother will return younger than himself. However, the traveling twin will claim that *he* is stationary, and his earthbound brother is moving, first away from and then back to the spaceship. He therefore says that the earthperson's clock is running slow, and the earthperson will be the younger one when they meet again. Who is right? When they meet they can easily determine who is the younger one. The answer to the apparent paradox lies in the fact that, as we have already stated, our conclusions about time dilation are correct for *inertial* frame observers. The calculations have to be modified for

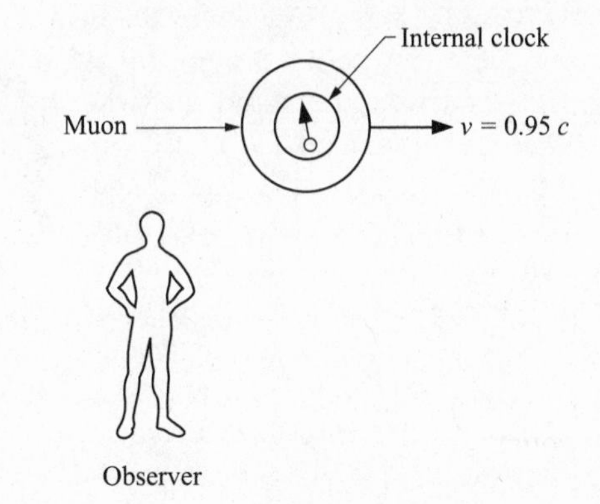

Fig. 16-5

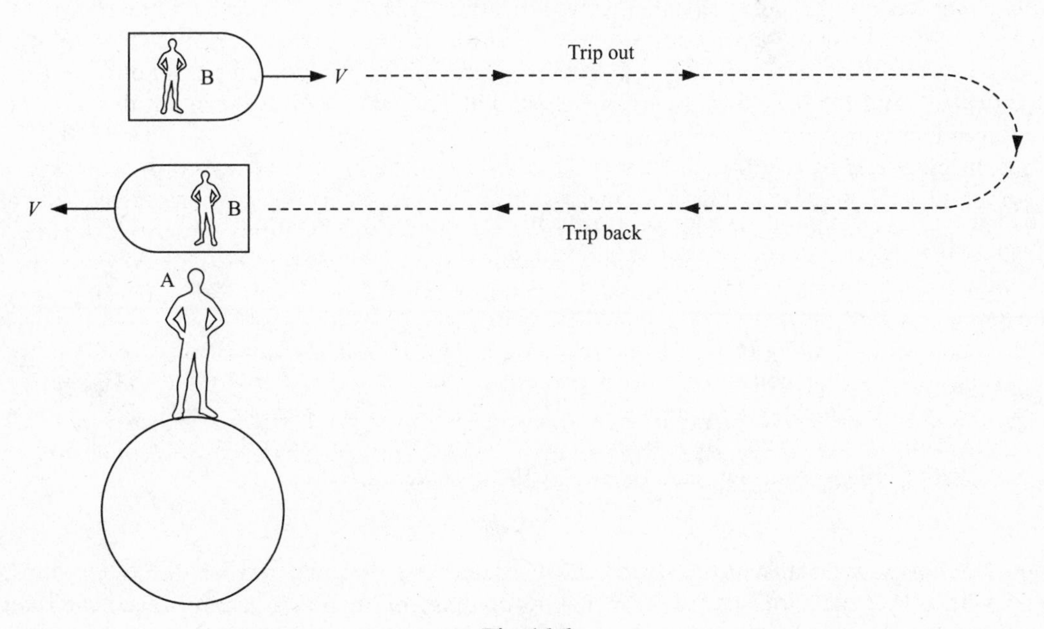

Fig. 16-6

observers that are in an accelerating reference frame. In order for the twins to meet again, one of them has to turn around and thereby change his vector velocity. In our case, the traveling twin is the one that turned around and therefore accelerated, so that his calculation has to be modified. The twin on earth was always inertial, so his calculation is correct, and the traveler will come back younger. The difference in age will be very small unless the speed is comparable to that of light, or unless the distance traveled is very large. In principle this effect occurs every time one gets on a train or plane, but the time difference is so small as to be virtually undetectable.

Let us return to the case of a traveler to a nearby star that we discussed in Problem 16.5 (see Fig. 16-4). There the traveler's clock read 2.25 y when he met the star. The earth observer says that the trip took 3.75 y, and the fact that the traveler's clock reads 2.25 y is due to the fact that the traveler has a moving clock that runs slow. What is the point of view of the traveler? He agrees that his own clock reads 2.25 y when he meets the star, but claims that his own clock is stationary. He will say that the earthperson's clock is moving, and runs slow, and that only 1.35 y have elapsed on the earthperson's clock. They are at different positions in space, so they cannot compare their clocks. However, we can look at the clock of an observer on the star, C. This observer is stationary relative to the earth, and his clock runs at the same rate as that of the earthperson. The traveler will claim that C's clock is a moving clock and runs slow. Now, A and C must agree on their observations, since they are not moving relative to each other, and, if they synchronized their clocks at the start, they will read the same time on both clocks. Thus, A and C will say that C's clock read 3.75 y when B arrives at the star. Since we can take a picture of the clock when they meet, there can be no disagreement on this point, and B will have to agree that C's clock reads 3.75 y. How can we understand this when B claims that the trip took only 1.35 y on A and C's clock? The answer is that although A and C say that their clocks are synchronized, B will, of necessity, say that they are not synchronized. Recall that when A and C synchronize their clocks, they simultaneously set their clocks to the same time. One way of doing this is to use light emitted from a point midway between A and C, and then set their clocks to zero when the light arrives simultaneously at the two locations. But B, watching this synchronization procedure, will say that both A and C were moving to the left (in the figure), so the light arrived at C before it arrived at A. Therefore C's clock was set to zero before A's clock, and it will always be ahead of A's clock by this amount. When one calculates this time difference, (as we will do in the next section), one finds it to be exactly 2.40 y. Therefore, B's explanation for the time he sees on C's clock is as follows. When B is at the position of A, A's clock read zero, and C's clock already read 2.40 y. The clocks of A and C run slow, and only 1.35 y elapsed until C reaches B. At that time, C's clock will therefore read $2.40 + 1.35 = 3.75$ y, as is indeed the case. We therefore see that the ideas of simultaneity (or lack thereof) are necessary to clarify some otherwise difficult predictions of time dilation. We will now use this same example to show another consequence of relativity, namely length contraction.

16.4 LENGTH CONTRACTION

Consider the same case of traveler B moving toward a star at a distance of L_A from the earth, as measured by the observer at A. We already showed that A will claim that the trip takes a time of L_A/V, while B claims that he meets the star after $L_A/\gamma V$. From the point of view of B, the star is moving to the left at a speed of V (see Fig. 16-7). If the time it takes for the star to reach B is $L_A/\gamma V$, then the distance that the star travels, according to B, is $V(L_A/\gamma V) = L_A/\gamma$. Thus B will say that the distance between the earth and the star has been contracted, and is reduced by the factor $1/\gamma$. For A the distance between A and C is fixed and not moving, and he measures L_A, while for B the distance, while fixed, is moving. This moving length is contracted, and becomes L_A/γ. This is another general principle of special relativity, that moving lengths contract, as measured by the observer at rest.

$$L' = L_A/\gamma \tag{16.3}$$

Actually, only the lengths in the direction of motion contract, not the lengths in the direction perpendicular to V. Thus A and B, who have identically built rulers and clocks, will each claim that the other's ruler is contracted and the other's clock runs slow.

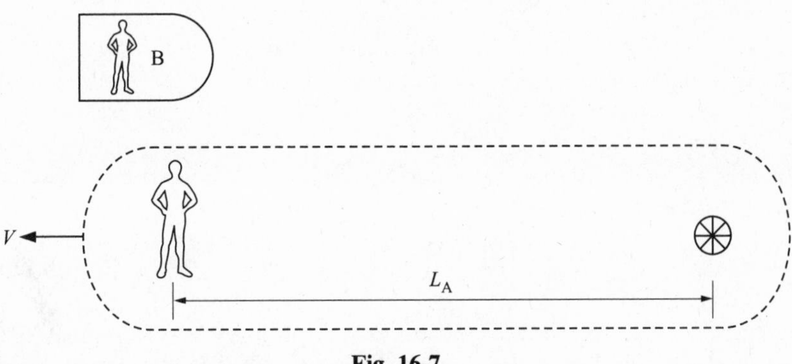

Fig. 16-7

Problem 16.7. A meter stick is moving at a speed of $0.8c$, according to observer B. What is the length of the meter stick as measured by B?

Solution

The problem refers to a "meter stick". This means that an observer at rest with respect to the meter stick measures a length of one meter. Observer B, who sees the "meter stick" moving, will say that the length of the stick is contracted, and will measure only $1/\gamma$ of its rest length. Now $\gamma = 1/[1 - (0.8c)^2/c^2]^{1/2} = 1/0.6$, and $1/\gamma = 0.6$. Thus, B will measure the length of the moving meter stick to be 0.6 m.

Problem 16.8. An observer measure the length of a moving space ship to be 150 m. The space ship is moving with a speed of $0.6c$, as measured by the observer. What is the length of the spaceship as measured by the spaceperson in the space ship?

Solution

The observer measures a contracted length, since, according to the observer, the space ship is moving. The length measured by the spaceperson is the length of the space ship at rest, which is greater than the moving length by the factor γ. The length of the ship at rest is called the "proper length", and is designated as L_0. Thus $L = L_0/\gamma$, or $L_0 = \gamma L = 150/0.8 = 187.5$ m, since $\gamma = 1/0.8$.

Problem 16.9. An observer at the bottom of a mountain measures the height of the mountain to be 1.8×10^2 m. Radioactive particles move downward at a speed of 1.5×10^8 m/s, and decay with a half-life of 2.2×10^{-8} s when they are at rest.

(*a*) According to the observer, how long does it take for the particles to reach the bottom of the hill?

(*b*) According to a second observer moving with the particles, how high is the hill?

(*c*) According to the second observer, how long does it take until they meet the bottom of the hill?

(*d*) According to the earth stationary observer, what is the half-life of the particles that he measures on his own clock, i.e. how long does it take for half the particles to decay?

Solution

(*a*) The observer sees particles moving with a speed of 1.5×10^8 m/s, and moving a distance of 1.8×10^2 m. The time this takes on his own clock is $1.8 \times 10^2/1.5 \times 10^8 = 1.2 \times 10^{-6}$ s.

(*b*) According to the second observer, the hill is a moving length, which is contracted by the factor $1/\gamma$. Thus $L = L_0/\gamma = 1.8 \times 10^2(0.866) = 156$ m.

(*c*) According to the second observer, the bottom of the hill is moving toward him with a speed of 1.5×10^8 m/s, and has to travel a distance of 156 m. The time is therefore $t = 156/1.5 \times 10^8 = 1.04 \times 10^{-6}$ s. This could also have been calculated using time dilation. According to the Earth-

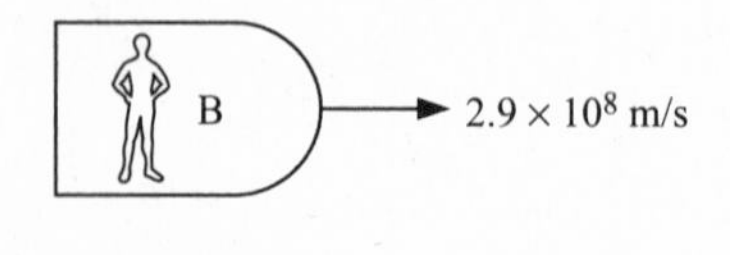

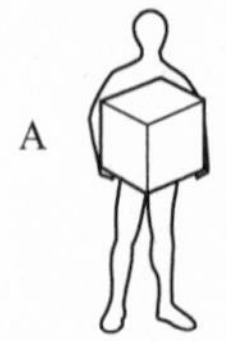

Fig. 16-8

bound observer, the clock on the particles is a moving clock, which runs slow. Therefore the time elapsed as measured on this moving clock in traveling down the hill is decreased by the factor $1/\gamma$. The time is therefore $1.2 \times 10^{-6}/\gamma = 1.2 \times 10^{-6}(0.866) = 1.04 \times 10^{-6}$ s.

(*d*) The half-life of the particles at rest is an intrinsic clock of the particles. According to the earthbound observer, this clock runs slow. The observer will therefore say that the time for particles to decay is increased by the factor γ. The half-life he measures is then $2.2 \times 10^{-8}\ \gamma = 2.2 \times 10^{-8}/0.866 = 2.54 \times 10^{-8}$ s.

Problem 16.10. Observer A holds a cube with each side having a length of 1.5 m. He is observed by a space traveler moving with a velocity of 2.9×10^8 m/s along the direction of one side of the cube, relative to observer A (see Fig. 16-8). What is the volume of the object as seen by observer A and as seen by observer B?

Solution

Observer A sees a cube with volume of $1.5^3\ \text{m}^3 = 3.375\ \text{m}^3$. Observer B sees the two perpendicular lengths of the object as having lengths of 1.5 m, since there is no length contraction except for parallel lengths. The parallel side is seen as contracted by the factor $1/\gamma = 0.256$. Thus, the volume of the object, which is no longer a cube as seen by B, is $1.5^2(0.256)(1.5) = 0.864\ \text{m}^3$.

Problem 16.11. A meter stick is moving at a speed of V parallel to a stationary wall (see Fig. 16-9). There is a hole in the wall of length 0.6 m. How fast must the stick be moving so that it could (in principle) fit through the hole in the wall?

Solution

The observer on the wall will see the meter stick as a moving length, which is contracted from one meter to $1/\gamma$ m. To fit through the hole, this length must equal 0.6 m. Thus, $0.6 = 1/\gamma = (1 - v^2/c^2)^{1/2}$. Then, $1 - v^2/c^2 = 0.36$, $(v/c)^2 = 1 - 0.36 = 0.64$, and $v = 0.8c$.

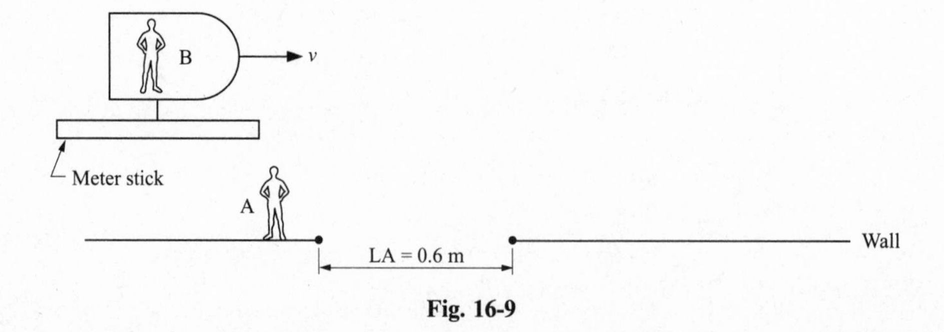

Fig. 16-9

16.5 LORENTZ TRANSFORMATION

We have learned in the previous sections, that what an observer measures in one inertial system is quite different than what a observer in another inertial system would measure. All of these differences should be obtainable if we knew how to relate an event at any point x, y, z, t in one system to the corresponding point x', y', z', t' in the other system. The equations that relate space–time points in one system to the other are called **transformation equations**. For Newtonian physics those transformation equations are called Galilean, while for Einstein's relativity theory they must be different and are called **Lorentz transformations**.

For simplicity let us consider two inertial frames of reference with a common x axis and parallel y axes, and with one frame (x', y', z', t') moving along the positive x axis of the other frame (x, y, z, t) with speed V. We assume that as the origins cross, clocks in both systems are set to zero.

Figure 16-10, shows two such inertial frames. Clearly,

$$x' = x - Vt, \qquad y' = y; \qquad z' = z, \qquad t' = t \tag{16.4}$$

corresponds to the **Galilean transformation equations**, which we intuitively understand. Thus, any event that takes place at x, y, z at time t in one frame takes place at x', y', z', t' in the other frame. For our case, $y' = y$, $z' = z$ and most obviously $t' = t$ since the clocks are assumed to always measure the same time. Along the x axis we clearly have x' less than x by the amount Vt that the primed system origin has moved to the right.

Given the peculiarities of the predictions of the theory of relativity it should be no surprise that the corresponding transformation equations will be quite different. Starting from Einstein's basic hypothesis and symmetry considerations, we can show that for the same two inertial frames given in Fig. 16.10 the Lorentz transformation equations are:

$$\begin{aligned} x' &= \gamma(x - Vt) \\ y' &= y; \qquad z' = z \\ t' &= \gamma(t - Vx/c^2), \end{aligned} \tag{16.5}$$

where $\gamma = 1/[1 - (V/c)^2]^{1/2}$, as defined earlier [Eq. (*16.1*)].

There are two main differences between these equations and the Galilean transformation equations, Eqs. (*16.4*). First, we have the factor γ in the x' equation. Second, the time t' no longer equals the time t. Instead, t' depends not only on the time t, but also on the position x. Just like x' depends on both x and t (even without relativity), t' now depends on both x and t as well. This means that the space coordinates and the time coordinates are not independent measurements if one transforms from one frame to another. We view x, y, z and t as the coordinates of an event in a space consisting of four dimensions, of which three (x, y, and z) are "spacelike", and one (t) is "timelike". In transforming from one frame to another space coordinates influence time coordinates and vice versa. This concept is developed more thoroughly in more advanced courses on relativity.

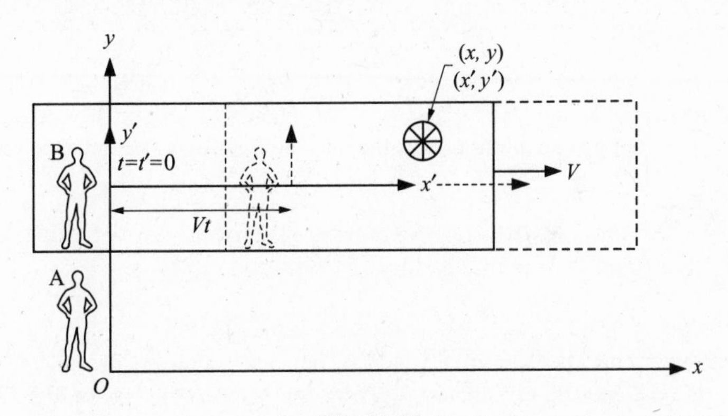

Fig. 16-10

We will now proceed to show that these equations provide for time dilation and for length contraction.

Problem 16.12. A bar is stationary relative to an observer in system B, and has a measured length of $L_0 = \Delta x' = x'_2 - x'_1$, as measured by B. System B is moving with velocity V relative to system A in a direction parallel to the bar. What length is measured by A for this moving bar?

Solution

When B measures the length of the bar what he actually does is to measure the position of the two ends of the bar, and he gets $L_0 = \Delta x' = x'_2 - x'_1$. We want to know the distance $x_2 - x_1 = L$ that is measured by A, who sees this bar as moving. When A measures the length of the bar what he actually does is to measure the position of the two ends of the bar at the same time from his point of view, i.e. a "snapshot" view. Thus, he measures x_2 and x_1 at the same time $t_2 = t_1$. Using the Lorentz equation for $x' = \gamma(x - Vt)$, and solving for x, we get $x = x'/\gamma + Vt$, and $x_2 - x_1 = (x'_2 - x'_1)/\gamma + V(t_2 - t_1)$. Since $t_2 = t_1$ for our case, we have $(x_2 - x_1) = (x'_2 - x'_1)/\gamma = L_0/\gamma = L$. This is just the length contraction that we developed previously, $L = L_0[1 - (V/c)^2]^{1/2}$.

It is important to note that the same length contraction can be obtained using the strange time equation in the Lorentz transformation. A knows that system B is traveling to the right with velocity V. He therefore decides to measure the length of the bar by determining the time t_A when the front end of the bar passes a given point x and the later time t_B when the back end passes the same point x. Then, $L = V(t_B - t_A)$. Observer B also sees these two events as a definitive measure of the length of the bar in his system, since point x, moving to the left from B's point of view, passes one end of the bar at t'_A and the second end at t'_B. Since x moves at speed V, B must have that $L_B = V(t'_B - t'_A)$. From the Lorentz transformation Eqs. (*16.5*) we have

$$t'_B = \gamma(t_B - Vx_B(c^2), \qquad t'_A = \gamma(t_A - Vx_A/c^2, \qquad \text{with} \qquad x_B = x_A = x.$$

So: $$t'_B - t'_A = \gamma(t_B - t_A), \qquad \text{and:} \qquad L_0 = V(t'_B - t'_A) = \gamma V(t_B - t_A) = \gamma L \rightarrow L = L_0/\gamma,$$

as before. We now turn to time dilation.

Problem 16.13. A clock that is stationary in system B has a period of T_0 when measured by B. Since system B is moving with velocity V relative to system A, what is the period that an observer in A will measure for this moving clock?

Solution

The clock is always at the same position in system B, since it is stationary in that system. To measure the period T_0 observer B will measure the difference in time between the first and second "tick" of the clock, at the same position x'. Thus $T_0 = t'_2 - t'_1$, where $x'_2 = x'_1$. The period observed by A for this moving clock is $t_2 - t_1$. If we solve the last Lorentz equation [Eq. (*16.5*)] for t in terms of t' and x, we get that $t = t'/\gamma + Vx/c^2$. Then $\Delta t = t_2 - t_1 = (t'_2 - t'_1)/\gamma + V(x_2 - x_1)/c^2$. But we known from the first Lorentz equation that $(x_2 - x_1) = (x'_2 - x'_1)/\gamma + V(t_2 - t_1) = V(t_2 - t_1)$ since $x'_2 = x'_1$. Therefore $\Delta t = T_0/\gamma + V^2\Delta t/c^2$, and rearranging, $\Delta t(1 - V^2/c^2) = T_0/\gamma = T_0(1 - V_2/c_2)^{1/2}$, or $\Delta t = \gamma T_0$. This is exactly what we know should be the case for time dilation, that the period of a moving clock is increased by the factor γ.

Problem 16.14. Show that the Lorentz transformation equations are the same as the Galilean transformation equations if the velocity V is much smaller than c.

Solution

The Lorentz equations are given by Eqs. (*16.5*). When $V \ll c$, $\gamma = 1/[1 - (V/c)^2]^{1/2}$ approaches 1. Then $x' = x - Vt$, to a very good approximation, giving the same result as for the Galilean transformation. Similarly, as long as $x < ct$ (the distance traveled by light during the time scale involved), then Vx/c^2

$< V(ct)/c^2 = Vt/c \ll t$, so Vx/c^2 can be ignored and $t' \approx \gamma t$. Since $\gamma \approx 1$ for our case, we have to a very good approximation that $t' = t$, as in the Galilean transformation.

16.6 ADDITION OF VELOCITIES

Without relativity we know how to evaluate the velocity of an object that is seen by one observer if we know the velocity of that object as seen by another observer (see e.g., Ibid., Sec. 3.5). In Fig. 16-11 we have an observer in system A and another observer in system B, moving to the right relative to A with velocity $V_{B/A}$. They both view an object C and measure its velocity relative to their own coordinate system. A measures the velocity as $v_{C/A}$ (read: v of C relative to A) and B measures the velocity as $v_{C/B}$. Without relativity we know that (for motion in one dimension)

$$v_{C/A} = v_{C/B} + V_{B/A} \tag{16.6a}$$

or, equivalently

$$v_{C/B} = v_{C/A} - V_{B/A} \tag{16.6b}$$

This is consistent with the Galilean transformation. In case any of these velocities is directed to the left, then that velocity will be negative in the equation. This is the equation for the addition of velocities without relativity. For relativity, these equations have to be modified. The new equations for relative motion in one dimension (where all velocities are algebraic, i.e. can be positive or negative) is:

$$v_{C/B} = (v_{C/A} - V_{B/A})/(1 - v_{C/A} V_{B/A}/c^2) \tag{16.7a}$$

recalling that $v_{B/A} = -v_{A/B}$, this can be re-expressed as:

$$v_{C/B} = (v_{C/A} + V_{A/B})/(1 + v_{C/A} V_{A/B}/c^2) \tag{16.7b}$$

Interchanging A and B leaves the equation valid as long as one puts in the correct signs, so we also have:

$$v_{C/A} = (v_{C/B} + V_{B/A})/(1 + v_{C/B} V_{B/A}/c^2) \tag{16.7c}$$

These results can be derived directly from the Lorentz transformation.

In these equations the velocities are positive to the right (along the positive sense of the x axis) and negative to the left. These equations effectively state that, for a relativistic calculation, one has to add the extra denominator factors. For Eq. (16.7c) this is $1/(1 + v_{C/B} V_{B/A}/c^2)$. This factor is essentially equal to one unless both $v_{C/B}$ and $V_{B/A}$ are not very small compared to c. In case one or both are small compared to c, the formula is essentially the same as the non-relativistic formula.

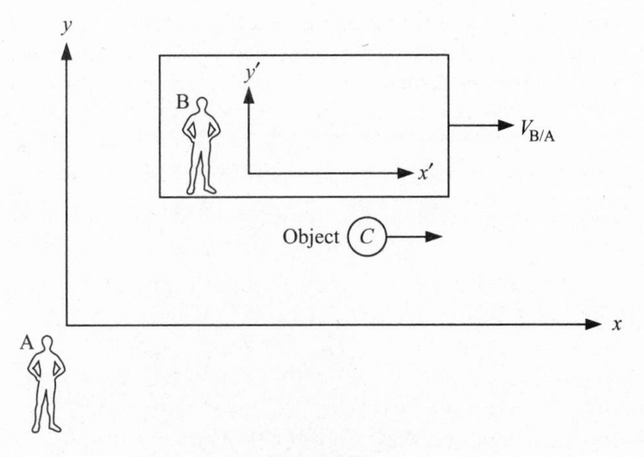

Fig. 16-11

Problem 16.15. System B is moving to the right relative to A with a velocity of $0.9c$. Observer B fires a rocket along the direction of this velocity with a speed of $0.8c$ relative to himself. What is the velocity of the rocket as measured by A, if the rocket moves (*a*) to the right; and (*b*) to the left?

Solution

(*a*) We use formula 34.7*c* to get $v_{C/A}$, with $v_{C/B} = 0.8c$ and $V_{B/A} = 0.9c$. Substituting in the equation we get $v_{C/A} = (0.8c + 0.9c)/(1 + 0.8c(0.9c)/c^2) = 1.7c/1.72 = 0.9883c$. We note that without relativity the result would have been $v_{C/A} = 1.7c$, which is greater than c. The additional factor introduced in the relativistic equation prevented $v_{C/A}$ from exceeding c. This is a general result of this equation for the addition of velocities. If the velocities in one system do not exceed c, then those in a transformed system will also not exceed c.

(*b*) Now we use the same formula, but $v_{C/B}$ is now $-0.8c$. Substituting in the equation we get $v_{C/A} = (-0.8c + 0.9c)/(1 + (-0.8c)(0.9c)/c^2) = 0.1c/(1 - 0.72) = 0.36c$.

Problem 16.16. System B is moving to the right relative to A with a velocity of V. Observer B sees a ray of light moving along the direction of this velocity with a speed of c relative to himself. What is the velocity of the light as measured by A, if the light moves to (*a*) the right; and (*b*) to the left?

Solution

We know that the answer must be that A also sees a velocity of c for the light. To verify this we use the formula 34.7*c* with $v_{C/B} = \pm c$. Thus $v_{C/A} = (\pm c + V_{B/A})/(1 \pm cV_{B/A}/c^2) = c(\pm c + V_{B/A})/(c \pm V_{B/A}) = \pm c$, as required.

It can be shown from the general form of Eqs. (*16.7a*), (*16.7b*) and/or (*16.7c*) that if the magnitudes of any two of the velocities are each less than c, then the magnitude of the third velocity must be less than c as well. This leads the way to the logical self-consistency of the claim that no object or signal can travel faster than c as seen by *any* observer.

Problem 16.17. An observer sees two objects moving away from each other. One object moves to the right with a velocity of 1.8×10^8 m/s, and the second object moves to the left with a velocity of 2.5×10^8 m/s.

(*a*) According to this observer, how fast are the two objects moving away from each other?

(*b*) According to an observer on one object, how fast is the other object moving away from him?

Solution

(*a*) Since we are not transforming from one system to another system we can calculate the relative velocity of the two objects without having to use relativity. According to the observer, the first object moves a distance of 1.8×10^3 m to the right every second, and the second object moves a distance of 2.5×10^8 m to the left every second. Therefore they move apart a distance of $(1.8 + 2.5) \times 10^8$ m every second, so their velocity of separation is 4.3×10^8 m/s. This is greater than c, but it is not a problem since no object itself is moving faster than c to this observer.

(*b*) If we take the observer on the second object (moving to the left) as observer A and the first observer as B, then $V_{B/A} = +2.5 \times 10^8$ m/s (B sees A moving to the left and A sees B moving to the right). Also, $v_{C/B} = +1.8 \times 10^8$ m/s. Substituting into Eq. (*16.7c*) we get $v_{C/A} = (1.8 \times 10^8 + 2.5 \times 10^8)/[1 + (1.8 \times 10^8)(2.5 \times 10^8)/c^2] = 2.867 \times 10^8$ m/s. This is less than c, as it must be.

16.7 RELATIVISTIC DYNAMICS

In the previous sections we have developed the kinematics of special relativity. We found that when one transforms from one frame to another, we have phenomena such as time dilation, length contraction, a change in the addition of velocities and inequivalence of simultaneity. We now investigate how this affects the dynamics of particles.

In mechanics we learned that the laws of physics that predict the behavior of particles were encompassed in **Newton's laws**. We derived the laws of conservation of energy and momentum from these basic concepts. Furthermore, Newton's laws and the conservation laws hold in all inertial reference frames. It is a basic tenet of special relativity that basic laws of physics should still be valid in any inertial reference frame and not be restricted to only one special reference frame. Given the peculiarities of the Lorentz transformation and the corresponding change in the law of addition of velocities this could only be true if Newton's laws were modified. Einstein showed that the appropriate modifications that would accomplish this task are the following:

(*a*) mass is not independent of velocity, but rather $m = \gamma m_0$. When the particle is at rest, $v = 0$, and $\gamma = 1$, and $m = m_0$. Therefore m_0 is called the rest mass of the particle.

$$m = \gamma m_0 \tag{16.8}$$

(*b*) The momentum still equals mv, but $mv = \gamma m_0 v$. This is a fairly easy new concept to accept, once one recognizes that the mass varies with velocity.

$$p = mv = \gamma m_0 v \tag{16.9}$$

(*c*) **Newton's second law** must be written as $F = \Delta p/\Delta t$, rather than $F = m\Delta v/\Delta t = ma$. Since m now changes with velocity, the two expressions are not equivalent, and the correct expression is

$$F = \Delta p/\Delta t \tag{16.10}$$

(*d*) The **total energy** of the particle is $mc^2 = \gamma m_0 c^2$. This energy is present even if the particle is not moving. When $v = 0$, the energy is $m_0 c^2$, which is called the rest mass energy of the particle. The kinetic energy of a particle is the *additional* energy that the particle has when it moves with velocity v, and equals $E_K = mc^2 - m_0 c^2 = (\gamma - 1)m_0 c^2$. This is a major modification, both in the introduction of a new rest mass energy and in the new formula for the kinetic energy, which is very different than the non-relativistic formula.

$$E = mc^2 = \gamma m_0 c^2 \tag{16.11}$$

$$E_K = mc^2 - m_0 c^2 = (\gamma - 1)m_0 c^2 \tag{16.12}$$

We will discuss these modifications in more detail in the following paragraphs. It should be noted that the only way we can determine whether these modifications are correct is to perform experiments to test their predictions. By now there is overwhelming evidence that nature behaves as predicted by these equations. Thus we have to modify our view of dynamics just as we had to modify our view of kinematics.

The relativistic mass of a particle varies as the velocity of the particle changes. This is in contrast to the non-relativistic case where the mass does not change. Nevertheless each particle has a unique mass that is inherent to that particle, and that is the rest mass. While the mass may change, the rest mass is an invariant quantity. Different observers will see a particle moving with different velocities and will therefore measure a different mass for the particle, but they will all measure the same rest mass. To avoid confusion we will always use the symbol m to mean the actual mass of the particle, and m_0 to mean the rest mass. This means that $m = \gamma m_0$. As the particle increases its speed, it becomes more massive and therefore harder to accelerate. The nearer one gets to the velocity of light, c, the larger γ will be and the larger m will be. In fact, as v approaches c, the mass approaches infinity, and we cannot increase its velocity further. It is this phenomenon that assures us that the velocity will never become equal to c, and that γ will not become infinite or imaginary.

As the velocity increases, the momentum increases for two reasons. The momentum is equal to mv, and both factors increase as v increases. For small v ($v \ll c$), γ is nearly equal to unity and hardly varies with v. This is the non-relativistic region where $m \approx m_0$, and the momentum $p \approx m_0 v$. As v gets very close to c, γ becomes very large, and v hardly increases anymore. The main change in p comes from the change in γ which is very sensitive to even slight changes in v. Then $p \approx mc = \gamma m_0 c$ since $v \approx c$. This is the extreme relativistic region.

Problem 16.18. A particle of rest mass 1.67×10^{-27} kg is moving with a velocity of $0.8c$. Calculate (*a*) the mass of the particle; (*b*) the momentum of the particle; (*c*) the rest mass energy of the particle; and (*d*) the kinetic energy of the particle.

Solution

(*a*) We first calculate the value of γ, which is needed in nearly all the calculations. $\gamma = 1/[1 - (v/c)^2]^{1/2} = 1/0.6 = 1.6667$. Then $m = \gamma m_0 = 1.67 \times 10^{-27}/0.6 = 2.78 \times 10^{-27}$ kg.

(*b*) The momentum is $mv = 2.78 \times 10^{-27}(0.8c) = 6.68 \times 10^{-19}$ kg · m/s.

(*c*) The rest mass energy of the particle is $m_0 c^2 = 1.67 \times 10^{-27}(3 \times 10^8)^2 = 1.50 \times 10^{-10}$ J.

(*d*) The kinetic energy is $E_K = (m - m_0)c^2 = (2.78 - 1.67) \times 10^{-27}(3 \times 10^8)^2 = 9.99 \times 10^{-11}$ J.

Problem 16.19. A particle of rest mass 1.67×10^{-27} kg is moving with a velocity of $0.0008c$. Calculate (*a*) the mass of the particle; (*b*) the momentum of the particle; (*c*) the rest mass energy of the particle; and (*d*) the kinetic energy of the particle.

Solution

(*a*) We first calculate the value of γ, which is needed in nearly all the calculations. $\gamma = 1/[1 - (v/c)^2]^{1/2} = 1/[1 - (0.0008)^2]^{1/2} = 1.000,000,32$. Then $m = \gamma m_0 = 1.67 \times 10^{-27}$.

(*b*) The momentum is $mv = 1.67 \times 10^{-27}(0.0008c) = 4.01 \times 10^{-22}$ kg · m/s.

(*c*) The rest mass energy of the particle is $m_0 c^2 = 1.67 \times 10^{-27}(3 \times 10^8)^2 = 1.50 \times 10^{-10}$ J.

(*d*) The kinetic energy is $E_K = (m - m_0)c^2 = (\gamma m_0 - m_0)c^2 = (1.000,000,32 - 1)1.67 \times 10^{-27}(3 \times 10^8)^2 = 4.81 \times 10^{-17}$ J.

Problem 16.20. Show that the answers to (*a*), (*b*) and (*d*) in the previous problem are the same as the non-relativistic results.

Solution

Since $\gamma \approx 1$, $m = m_0$, and $p = m_0 v$ as in a non- relativistic calculation. The kinetic energy without relativity would be $E_K = m_0 v^2/2 = 1.67 \times 10^{-27}(0.0008c)^2/2 = 4.81 \times 10^{-17}$ J, as in the relativistic calculation. It can be shown in general that, for $v \ll c$, the kinetic energy is the same whether calculated using relativity or without relativity. In this case essentially all the total energy is in the rest mass energy, since the kinetic energy is very small. Then we can justifiably do all our calculations without using the relativistic formulas.

Problem 16.21. A particle of rest mass 1.67×10^{-27} kg is moving with a velocity of $0.9998c$. Calculate (*a*) the mass of the particle; (*b*) the momentum of the particle; (*c*) the rest mass energy of the particle; and (*d*) the kinetic energy of the particle.

Solution

(*a*) We first calculate the value of γ, which is needed in nearly all the calculations. $\gamma = 1/[1 - (v/c)^2]^{1/2} = 1/[1 - (0.9998)^2]^{1/2} = 50.0$. Then $m = \gamma m_0 = 8.35 \times 10^{-26}$ kg.

(*b*) The momentum is $mv = 8.35 \times 10^{-26}(0.9998c) = 2.50 \times 10^{-17}$ kg · m/s.

(*c*) The rest mass energy of the particle is $m_0 c^2 = 1.67 \times 10^{-27}(3 \times 10^8)^2 = 1.50 \times 10^{-10}$ J.

(d) The kinetic energy is $E_K = (m - m_0)c^2 = (\gamma m_0 - m_0)c^2 = (50.0 - 1)1.67 \times 10^{-27}(3 \times 10^8)^2 = 7.36 \times 10^{-9}$ J. In this case the velocity is nearly c, and the total energy consists of nearly only kinetic energy, since the kinetic energy is much larger than the rest mass energy. This is the extreme relativistic case.

There are two essential points that are crucial to an understanding of relativistic mass and energy. The first point is that the **total energy** of a particle is essentially given by its **relativistic mass**. This is known as Einstein's famous equation for the equivalence of mass and energy, $E = mc^2$. All energy, including kinetic energy, is equivalent to mass. Therefore, whenever the energy of a particle is increased, its mass increases. If the particle is accelerated so that it moves with a larger velocity, its mass must increase to account for the added kinetic energy. A further analysis of relativistic dynamics shows that the same is true for any other kind of energy, such as potential energy, nuclear energy, etc. We therefore understand that the mass of a particle is no longer a constant under any and all circumstances as we thought it was before the theory of relativity.

The second point is that even the **rest mass** of a particle is equivalent to energy. While the rest mass of a particle is an intrinsic property of that particle, if the particle itself is a composite of smaller constituents (e.g. as molecule made up of atoms; an atom made up of protons, neutrons and electrons, etc.) then some of that rest mass energy can reappear in other forms when the particle breaks apart. This "**rest mass energy**" is in fact the sum total of all the energies of its constituents and can transform to other forms of energy. If we can cause a particle to disintegrate, with the consequent loss of rest mass energy, then that energy will have to appear in some other form to maintain conservation of energy (e.g., rest mass of smaller constituents, kinetic energy, potential energy). Furthermore, electromagnetic waves, which carry energy, must, according to $E = mc^2$, carry mass equivalent to that energy. This concept has enormous consequences which have been amply verified experimentally in many types of experiment, and which form the basis for nuclear reactors and nuclear weapons and explain the major source of energy of the stars. We will now explore some of these consequences.

Problem 16.22. Calculate the energy, in Joules and in eV, corresponding to the mass of (a) an electron; (b) a proton; (c) 1 g.

Solution

(a) The mass of an electron is 9.11×10^{-31} kg. The energy that this equals is $m_0 c^2 = 9.11 \times 10^{-31}(3 \times 10^8)^2 = 8.20 \times 10^{-14}$ J. Converting to eV by dividing by $[1.60 \times 10^{-19}]$ J/eV, gives 5.12×10^5 eV = 0.511 MeV.

(b) The mass of a proton is 1.67×10^{-27} kg. The energy that this equals is $m_0 c^2 = 1.67 \times 10^{-27}(3 \times 10^8)^2 = 1.50 \times 10^{-10}$ J. Converting to eV by dividing by 1.60×10^{-19} gives 9.39×10^8 eV = 939 MeV = 0.939 GeV.

(c) The rest mass energy of 1 g is $m_0 c^2 = (1.0 \times 10^{-3})(3 \times 10^8)^2 = 9 \times 10^{16}$ J $= 5.625 \times 10^{32}$ eV.

Problem 16.23. Calculate the velocity of (a) an 1.5 MeV electron (above rest mass energy); (b) a 1.5 MeV proton (above rest mass energy); and (c) a proton with a mass of three times its rest mass.

Solution

(a) The rest mass energy of an electron is 0.511 MeV, as calculated in the previous problem. We are told that the electron is a 1.5 MeV electron above rest mass energy. This means that the electron has been accelerated through the equivalent of a difference of potential of 1.5 MV, and therefore acquired an additional energy (kinetic energy) of 1.5 MeV. This kinetic energy is greater than the rest mass energy, and we must therefore use relativity to calculate the velocity.
Now, $E_K = (\gamma - 1)m_0 c^2 = 1.5$ MeV $= (\gamma - 1)(0.511$ MeV), which gives $\gamma = 1 + (1.5/0.511) = 3.935$. Using $\gamma = 1/(1 - v^2/c^2)^{1/2}$, we get $1 - v^2/c^2 = (1/\gamma)^2 = 0.0646$, and $v^2/c^2 = 0.9354$, $v = 0.967c = 2.90 \times 10^8$ m/s.

(*b*) The rest mass energy of a proton is 939 MeV. The kinetic energy of this proton is only 1.5 MeV, which is much less than its rest mass energy. Therefore we can calculate the velocity using non-relativistic formulas. The kinetic energy is 1.5 MeV $= 2.4 \times 10^{-13}$ J $= m_0 v^2/2 = 1.67 \times 10^{-27}\ v^2/2$. This gives $v^2 = 2.87 \times 10^{14}$, and $v = 1.70 \times 10^7$ m/s, which is small compared to c. (Doing the calculation relativistically gives $v = 1.69 \times 10^7$ m/s.)

Note. A numerical trick method for solving this problem is to remember 1.5 MeV $= m_0 v^2/2$ while 939 MeV $= m_0 c^2$. Dividing we get $2(1.5)/939 = v^2/c^2 \rightarrow v/c = 5.65 \times 10^{-2} \rightarrow v = 1.70 \times 10^7$ m/s.

(*c*) This case is one with a total energy of three times the rest mass energy, since the mass is three times the rest mass. Thus we must use relativistic formulas. $m = \gamma m_0 = 3m_0$, so $\gamma = 3$. Thus $1 - v^2/c^2 = 1/\gamma^2 = \frac{1}{9}$, and $v = 2.83 \times 10^8$ m/s.

The equivalence of mass and energy also applies to potential energy. For instance when an electron approaches a proton it loses potential energy. When this electron binds to a proton to form a hydrogen atom the two particles are bound together by an ionization energy, which is the amount of energy that must be supplied from the outside to separate the two particles from each other. Therefore, in the bound state, the two particles have less energy than in the separated state by the ionization energy of 13.9 eV. From the equivalence of mass and energy, the atom must have less mass than the two particles have when they are separated. The mass lost is the mass equivalent to 13.6 eV, or $\Delta m = 2.42 \times 10^{-35}$ kg. Since the mass of the hydrogen atom is 1.67×10^{-27} kg, this represents a percentage loss of mass of 1.45×10^{-6}%, which is just too small to detect directly. However, in nuclear reactions, the energy with which a proton is bound to a neutron is about 2.2 MeV, corresponding to a mass of 4×10^{-30} kg, or a loss in mass of 0.12%, which is easily measurable. In fact, whenever neutrons and protons are bound together in a nucleus, they lose a measurable amount of mass corresponding to the energy with which they are bound together, the "**binding energy**". This lost energy is known as the "**mass defect**". The energy lost in forming nuclei, or in transforming from one nucleus to another with a greater loss of mass, is the method by which energy is released in the sun and other stars, as well as in nuclear reactors and nuclear weapons. The amounts of energy released are enormous compared to the amounts released in chemical reactions. For chemical reactions the amounts released are comparable to molecular ionization energies of electron volts, whereas for nuclear reactions the amounts released are comparable to nuclear binding energies of MeVs. Thus, nuclear reactions are of order 1 million times more powerful than chemical reactions for the same number of atoms or molecules.

An example of the conversion of mass to energy is the annihilation of an electron. In general, electrons are stable and cannot decay, as we will see when we discuss particle physics. However, there is another particle, called a **positron**, which is essentially the same as an electron, except with positive charge. When an electron meets a positron they annihilate each other and both disappear. The resultant loss of mass is converted to energy in the form of two pulses of electromagnetic radiation (called **γ-rays**). These γ-rays can be detected, and their energy measured.

Problem 16.24. Suppose an electron at rest meets a positron also at rest and they annihilate each other producing two γ-rays of equal energy. How much energy does each γ ray have?

Solution

The rest mass energy of an electron is 0.511 MeV, as calculated in Problem 16.22. The positron has the same rest mass energy, and the total energy released is therefore 1.022 MeV. This is given, in equal amounts, to two γ-rays, and each will therefore have an energy of 0.511 MeV.

We stated previously that, with relativity, we must write Newton's second law in the form of $F = \Delta p/\Delta t = \Delta(mv)/\Delta t$. For a constant force, we can write $F\Delta t = \Delta(mv)$. Thus the momentum is changing at a constant rate. If the particle, of rest mass m_0 started at rest, then the change in mv will initially be confined to a change in v, as for a non-relativistic case. As v increases, the mass starts to increase as well, and further changes in p involve both m and v. After still more time, v approaches c and it will hardly change any more, so all the increase in p will take place as an increase in m. This can be seen in the following problem.

Problem 16.25. Suppose an electron is accelerated from rest by a constant force of 3×10^{-24} N. What is its momentum, its velocity and its mass after a time of (*a*) 1 s; (*b*) 10^2 s; (*c*) 10^4 s?

Solution

(*a*) For a steady force the momentum can be calculated using $\Delta p = p_f - p_i = p_f = F\Delta t$. Thus, after 1 s the momentum is $3 \times 10^{-24}(1) = 3 \times 10^{-24}$ kg · m/s. If we assume that we do not have to use relativity (we will have to check that v turns out to be much less than c, and that therefore this was a good assumption), then $3 \times 10^{-24} = 9.11 \times 10^{-31}\ v$, and $v = 3.3 \times 10^6$ m/s. Since this velocity is clearly much less than c, we are justified in not using relativistic formulas. Then $m \approx m_0 = 9.11 \times 10^{-31}$ kg.

(*b*) Again $p_f = F\Delta t = 3 \times 10^{-24}(10^2) = 3 \times 10^{-22}$. We cannot use non-relativistic formulas because then the velocity will be 100 times as big as in (*a*) and will be greater than c. Therefore we must use relativistic formulas. $p = mv = \gamma m_0 v = 3 \times 10^{-22}$. Then $v/(1 - v^2/c^2)^{1/2} = 3 \times 10^{-22}/9.11 \times 10^{-31} = 3.29 \times 10^8$. Squaring both sides gives $v^2/(1 - v^2/c^2) = 1.084 \times 10^{17}$, and $v^2 = 1.084 \times 10^{17}(1 - v^2/c^2)$. Rearranging gives $v^2(1 + 1.205) = 1.084 \times 10^{17}$, $v^2 = 4.916 \times 10^{16}$, $v = 2.22 \times 10^8$ m/s. Then $\gamma = 1.485$, and $m = 1.485 m_0 = 1.35 \times 10^{-30}$ kg.

(*c*) Again $p_f = F\Delta t = 3 \times 10^{-24}(10^4) = 3 \times 10^{-20}$. This is certainly a relativistic case, and we use the equations of relativity. Proceeding as in part (*b*) we get $v/(1 - v^2/c^2)^{1/2} = 3 \times 10^{-20}/9.11 \times 10^{-31} = 3.29 \times 10^{10}$. Squaring both sides gives $v^2/(1 - v^2/c^2) = 1.084 \times 10^{21}$, and $v^2 = 1.084 \times 10^{21}(1 - v^2/c^2)$. Rearranging gives $v^2(1 + 12{,}044) = 1.084 \times 10^{21}$, $v^2 = 8.9996 \times 10^{16}$, $v \approx 3 \times 10^8$ m/s $= c$. To calculate γ we must be careful, since $v \approx c$. Now $\gamma = 1/(1 - v^2/c^2)^{1/2}$, and we already saw that $v\gamma = 3.29 \times 10^{10}$. Since $v \approx c$ we can use $v = c$ in this last expression to give $c\gamma = 3.29 \times 10^{10}$, giving $\gamma = 1/(1 - v^2/c^2)^{1/2} = 109.7$. Then $m = \gamma m_0 = 109.7\ m_0 = 9.99 \times 10^{-29}$ kg.

If we had realized at the start that we were in the extreme relativistic region, we could have simplified our calculation by using the fact that $v \approx c$. Then $p = \gamma m_0 v = \gamma m_0 c = 3 \times 10^{-20}$, gives $\gamma = 109.8$, $m = \gamma m_0 = 9.99 \times 10^{-29}$ kg.

This problem illustrates that for small p ($p \ll m_0 c$) m remains constant at m_0 and the momentum changes because of changes in v, whereas for large $p (p \gg m_0 c)$ the velocity is approximately c and does not change much while the changes in p are due to changes in m.

In non-relativistic theory we have a relationship between the kinetic energy of a particle and its momentum: $E_K = (\frac{1}{2}) m_0 v^2$ and $p = m_0 v$, so $E_K = p^2/(2m_0)$. In a similar way, we can find a relationship between relativistic total (rest mass plus kinetic) energy and momentum. To obtain this relationship we note that $E = mc^2 = \gamma m_0 c^2$ and $p = mv = \gamma m_0 v$. Consider the expression $E^2 - p^2c^2 = \gamma^2 {m_0}^2 c^4 - \gamma^2 {m_0}^2 c^2 v^2 = \gamma^2 {m_0}^2 c^2 (c^2 - v^2) = \gamma^2 {m_0}^2 c^4 (1 - v^2/c^2)$. Recalling $\gamma^2 = 1/(1 - v^2/c^2)$ we have $E^2 - p^2c^2 = {m_0}^2 c^4$ or:

$$E^2 = p^2c^2 + {m_0}^2c^4 \tag{16.13}$$

This equation gives the general relation between energy and momentum for any particle of rest mass m_0.

Problem 16.26. For parts (*a*), (*b*) and (*c*) of Problem 16.25 use the solution for the momentum to find the energy E and mass m using the relativistic energy momentum equation.

Solution

(*a*) $E^2 = (3 \times 10^{-24}\ \text{kg} \cdot \text{m/s})^2 (3 \times 10^8\ \text{m/s})^2 + (9.11 \times 10^{-31}\ \text{kg})^2 (3 \times 10^8)^4 = 8.1 \times 10^{-31}\ \text{J}^2 + 6.72 \times 10^{-27}\ \text{J}^2 = 6.72 \times 10^{-27} \rightarrow E = 8.20 \times 10^{-14}$ J. $m = E/c^2 = 9.11 \times 10^{-31}$ kg.

(*b*) $E^2 = (3 \times 10^{-22})^2(3 \times 10^8)^2 + (9.11 \times 10^{-31})^2(3 \times 10^8)^4 = 8.1 \times 10^{-27} + 6.72 \times 10^{-27} = 1.48 \times 10^{-26} \rightarrow E = 1.22 \times 10^{-13}$ J. $m = E/c^2 = 1.35 \times 10^{-30}$ kg.

(*c*) $E^2 = (3 \times 10^{-20})^2(3 \times 10^8)^2 + (9.11 \times 10^{-31})^2(3 \times 10^8)^4 = 8.1 \times 10^{-23} + 6.72 \times 10^{-27} = 8.1 \times 10^{-23} \rightarrow E = 9 \times 10^{-12}$ J. $m = E/c^2 = 1.0 \times 10^{-28}$ kg.

Problems for Review and Mind Stretching

Problem 16.27. A man standing on the ground determines that a long piece of wood which is moving by him at a speed of $0.8c$ parallel to its length takes 10^{-7} s to pass by him.

(*a*) What length does he say the wood has?

(*b*) What length would be measured for the piece of wood by a person sitting on the wood?

Solution

(*a*) The wood travels a distance of $vt = 0.8c(10^{-7})$ from the time the front of the wood passes the observer until the time that the rear of the wood passes the observer. This is the length of the wood as measured by this observer. Thus $L = 24$ m.

(*b*) The person on the wood sees the wood as stationary, while the person on the ground sees the wood as moving. Therefore, the length in (*a*) is the (contracted) moving length $L = L_0/\gamma$. Thus $L_0 = \gamma L = [1/(1 - v^2/c^2)^{1/2}]L = [1/(1 - 0.8^2)^{1/2}]L = (24)/0.6 = 40$ m.

Problem 16.28. Observer A, is at rest relative to points (1) and (2). Another observer, B, is in a rocket moving with a velocity $V = 0.8c$ to the right relative to A, as in Fig. 16-12. According to B, the distance between the two points (1) and (2) is 4×10^9 m.

(*a*) According to A, what is the distance between points (1) and (2)?

(*b*) According to A, how long does it take for B to travel from (1) to (2)? How long does it take according to B?

(*c*) According to A, how long does it take for light to travel from point (1) to (2)? How long does it take according to B?

Solution

(*a*) According to A the distance is stationary and equals L_0. B measures a moving distance of $L = 4 \times 10^9$ m $= L_0/\gamma$, so $L_0 = (4 \times 10^9)/0.6 = 6.67 \times 10^9$ m.

(*b*) According to A, B is traveling a distance of 6.67×10^9 m with a speed of 2.4×10^8 m/s. The time that this takes is $6.67 \times 10^9/2.4 \times 10^8 = 27.8$ s. According to B, point (2) is approaching B with a speed of 2.4×10^8 m/s, but the distance that it has to travel is only 4×10^9 m. The time needed is

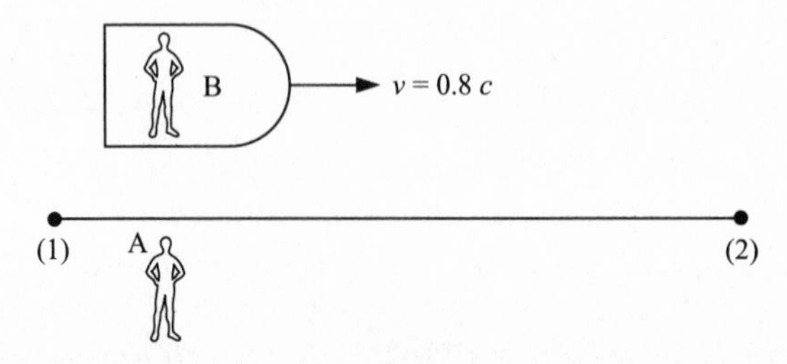

Fig. 16-12

$4 \times 10^9/2.4 \times 10^8 = 16.7$ s. We could have obtained this result by noting that, according to A, B's clock runs slow, and the time elapsed on his clock is smaller than on A's clock by the factor $1/\gamma$. The time on B's clock is therefore $27.8(0.6) = 16.7$ s.

(c) Light travels from (1) to (2) at the speed c, and the time needed to reach the stationary point (2), according to A, is $6.67 \times 10^9/3 \times 10^8 = 22.2$ s. According to B, the distance from (1) to (2) is 4×10^9 m, but point (2) is moving toward the light with speed v. The distance the light travels before reaching point (2) is (4×19^9 m $- vt$) where t is the time the light travels. Thus $ct = 4 \times 10^9 - vt$ or $vt + ct = 4 \times 10^9 = (2.4 + 3) \times 10^8 t$, and $t = 4 \times 10^9/5.4 \times 10^8 = 7.41$ s.

Problem 16.29. Observer A sees observer B moving to the right at a speed of $0.9c$, and observer C moving to the left with a speed of $0.95c$. Each observer has an identical meter stick in his possession.

(a) According to observer A, how long is the meter stick of (i) observer B; and (ii) observer C?

(b) According to observer B, how long is the meter stick of (i) observer A; and (ii) observer C?

(c) According to observer C, how long is the meter stick of (i) observer A; and (ii) observer B?

Solution

(a) (i) Observer A sees B moving at $V = 0.9c$, giving $\gamma = 2.294$. Observer A sees the meter stick of B as a moving length with $L = L_0/\gamma = 0.436$ m.

(ii) Observer A sees C moving at $V = -0.95c$, giving $\gamma = 3.203$. Observer B sees the meter stick of C as a moving length with $L = L_0/\gamma = 0.312$ m.

(b) (i) Observer B sees A moving at $V = -0.9c$, giving $\gamma = 2.294$. Observer B sees the meter stick of A as a moving length with $L = L_0/\gamma = 0.436$ m.

(ii) Using the relativistic formula for the addition of velocities, we calculate that B sees C moving at $V = (-0.9c - 0.95c)/[1 + (-0.9c)(-0.95c)/c^2] = -0.9973c$, giving $\gamma = 13.63$. Observer A sees the meter stick of C as a moving length with $L = L_0/\gamma = 0.073$ m.

(c) (i) Observer C sees A moving at $V = 0.95c$, giving $\gamma = 3.203$. Observer C sees the meter stick of A as a moving length with $L = L_0/\gamma = 0.312$ m.

(ii) Observer C sees B moving at $V = 0.9973c$, giving $\gamma = 13.63$. Observer C sees the meter stick of B as a moving length with $L = L_0/\gamma = 0.073$ m.

Problem 16.30. Two twins, one on earth and one on a spaceship moving at a speed of $0.995c$, meet after the second twin turns around and returns to earth at the same speed. How far away did he travel according to the twin on earth so that when he returns he is 15 years younger than the twin who remained on earth?

Solution

According to the twin on earth, the time he traveled back and forth is $2L/v$. During this time the spaceperson's clock ran slow, and he took only $2L/\gamma v$ on his clock. Since the earth twin did not accelerate, we can accept his calculation as correct. (There is a small correction for the acceleration period as seen by the earth twin. It is the accelerating twin who perceives an enormous speedup of time on earth during the acceleration.) Therefore their difference in age is $2L/v - 2L/\gamma v = (2L/v)(1 - 1/\gamma) = (2L/v)(\gamma - 1)/\gamma) =$ (15 years)(N), where N = number of seconds/year. Thus, $L = (15/2)Nv\gamma/(\gamma - 1) = (7.5N)(0.9995c)(31.63)/30.63 = 7.74\ Nc$. But $Nc = 1$ light year, so $L = 7.74$ light years.

Problem 16.31. Observer A sends a rocket away from himself at a speed of 2×10^8 m/s. The rocket is programmed to release a probe which will travel back to earth with a speed of 1.3×10^8 m/s. With what speed will the rocket have to launch the probe relative to the rocket (assume that the rocket's speed does not change during the release)?

Solution

If the rocket is system B, then $V_{B/A} = 2 \times 10^8$ m/s, and $v_{C/A} = -1.3 \times 10^8$ m/s. We seek $v_{C/B}$. Using the addition of velocity Eq. (*16.7b*),

$$v_{C/B} = (-1.3 \times 10^8 - 2 \times 10^8)/[1 + (-1.3 \times 10^8)(-2 \times 10^8)/c^2] = 2.56 \times 10^8 \text{ m/s}.$$

Problem 16.32. A rocket is launched from earth at time $t = 0$ from the origin with a speed of 2.75×10^8 m/s. A second rocket is launched after a time of 25 s with a speed of 2.95×10^8 m/s.

(*a*) According to a person on the first rocket, at what time and from what position was the second rocket launched? (Assume that the clocks in the first rocket were set to zero at the time of the launch.)

(*b*) According to a person on the earth, how long does it take for the second rocket to catch the first rocket?

Solution

(*a*) We call the earth system A and the first rocket system B. Then the second rocket was launched at $x = 0$ and $t = 25$. To get these coordinates as viewed by B, we use the Lorentz equations, to get $x' = \gamma(x - Vt)$, $t' = \gamma(t - Vx/c^2)$. Now $\gamma = 2.502$, giving

$$x' = 2.502[0 - 2.75 \times 10^8(25)] = -1.72 \times 10^{10} \text{ m}, \quad t' = 2.502[25 - 0] = 62.55 \text{ s}.$$

(*b*) According to person A, the first rocket is at a distance of $2.75 \times 10^8(25) = 6.875 \times 10^9$ m from the second rocket when the second rocket started. The second rocket is catching up to the first rocket at a speed of $(2.95 - 2.75) \times 10^8$ m/s, according to A. Thus, it takes $6.875 \times 10^9/0.2 \times 10^8 = 344$ s to catch the first rocket.

Problem 16.33. Referring to Problem 16.32, according to a person on the first rocket, at what time does the second rocket reach him?

Solution

We will do this calculation in two different ways. First, we calculate the time interval on A's clock from the launch of the first rocket (B) to the time when the second rocket reaches B. This time is 25 s + 344 s = 369 s. Since B's clock runs slow, the equivalent time interval on B's clock will be $369/\gamma = 147.5$ s.

The second method is to use the Lorentz equations. The launch of the first rocket is at $t = 0$ and the two rockets meet at $t = 369$ s, as in the first method. Since the second rocket traveled for 344 s from the origin of A at a speed of 2.95×10^8 m/s until reaching the first rocket, they meet at $x = 344(2.95 \times 10^8) = 1.015 \times 10^{11}$ m. Using the transformation equations, $t' = 0$ at the launch of the first rocket and the two rockets meet at

$$t' = (2.502)[369 - (1.015 \times 10^{11})(2.75 \times 10^8)/(9 \times 10^{16})] = 147.3 \text{ s},$$

which matches the first method to within rounding errors.

Problem 16.34. Two observers are moving from each other with a relative velocity V. They both observe a spaceship. According to the first observer, the spaceship is moving with a velocity of 2.3×10^8 m/s to the right. According to the second observer, the spaceship is moving to the left with a velocity of 1.9×10^8 m/s. What is the velocity of the second observer according to the first observer?

Solution

If we call the first observer A and the second observer B, we can substitute into Eq. (*16.7d*) and then solve for $v_{B/A}$. Doing this gives the equation

$$2.3 \times 10^8 = (-1.9 \times 10^8 + V_{B/A})/(1 + (-1.9 \times 10^8)V_{B/A}/c^2).$$

To solve, we multiply by the denominator to get

$$2.3 \times 10^8(1 + (-1.9 \times 10^8)V_{B/A}/c^2) = (-1.9 \times 10^8 + V_{B/A}).$$

Rearranging to bring $v_{B/A}$ to one side gives

$$(2.3 \times 10^8)(1.9 \times 10^8)V_{B/A}/9 \times 10^{16}) + V_{B/A} = 2.3 \times 10^8 + 1.9 \times 10^8 = 4.2 \times 10^8.$$

Then we get that

$$V_{B/A} = 4.2 \times 10^8/(1 + 0.485) = 2.83 \times 10^8 \text{ m/s}.$$

An alternative method is the following. Call the second observer C and the spaceship B. Then $v_{C/B} = -v_{B/C} = +1.9 \times 10^8$ m/s. Then from Eq. (*16.7d*)

$$v_{C/A} = (1.9 \times 10^8 + 2.3 \times 10^8)/[(1 + 1.9 \times 10^8)(2.3 \times 10^8)/c^2] = 2.83 \times 10^8 \text{ m/s}.$$

Problem 16.35. What speed does a particle have to have so that (*a*) its mass is three times its rest mass; (*b*) its kinetic energy is three times its rest mass energy; and (*c*) its momentum is $3m_0 c$?

Solution

(*a*) Since $m = \gamma m_0 = 3\, m_0$, we need $\gamma = 3 = 1/[1 - v^2/c^2]^{1/2}$, or $9[1 - v^2/c^2] = 1$, $v^2/c^2 = 1 - \frac{1}{9}$, $v = 0.943c$.

(*b*) Since $E_K = (m - m_0)c^2 = (\gamma - 1)m_0 c^2 = 3m_0 c^2$, therefore $\gamma = 4 = 1/[1 - v^2/c^2]^{1/2}$, or $16[1 - v^2/c^2] = 1$, $v^2/c^2 = 1 - \frac{1}{16}$, $v = 0.968c$.

(*c*) Since $p = mv = \gamma m_0 v = 3m_0 c$, therefore $\gamma v = 3c = v/[1 - v^2/c^2]^{1/2}$, or $[1 - v^2/c^2] = (\frac{1}{9})v^2/c^2$, $1 = v^2/c^2(1 + \frac{1}{9})$, $v^2/c^2 = 0.9$, $v = 0.949c$.

Problem 16.36. A person, A, on the ground sets up an x,y coordinate system, which is at rest relative to himself. Another person, B, in a bus moving with a velocity V relative to A sets up an x',y' coordinate system, which is at rest relative to B, as in Fig. 16-13. They both view the object, D, of rest mass 1.5 kg and measure its velocity. Person A measures a velocity of $0.8c$ to the right relative to himself, and person B measures a velocity of $(\frac{5}{13})c$ to the right relative to herself.

(*a*) Calculate the velocity of the bus, V, relative to A.

(*b*) Calculate the momentum of D, as measured by A.

(*c*) Calculate the total energy of D, as measured by A.

(*d*) Show that the relativistic energy momentum equation is satisfied in system A.

Solution

(*a*) We use the formula for the relativistic addition of velocities, $v_{D/A} = (v_{D/B} + V_{B/A})/(1 + v_{D/B}V_{B/A}/c^2)$. Here, $v_{D/A} = 0.8c$, $v_{D/B} = (\frac{5}{13})c$, and $V_{B/A} = V$. Substituting, we get $0.8c = (5c/13 + V)/(1 + (\frac{5}{13})cV/c^2)$. Thus, $0.8c((1 + (\frac{5}{13})V/c) = (\frac{5}{13})c + V$, or $V[1 - (\frac{5}{13})(0.8)] = [0.8 - (\frac{5}{13})]c$, or $V = 0.6c$.

(*b*) The momentum is $p = mv = \gamma m_0 v$, where $v = 0.8c$. This gives $p = (1/0.6)(1.5 \text{ kg})(0.8c) = 6.0 \times 10^8$ kg · m/s.

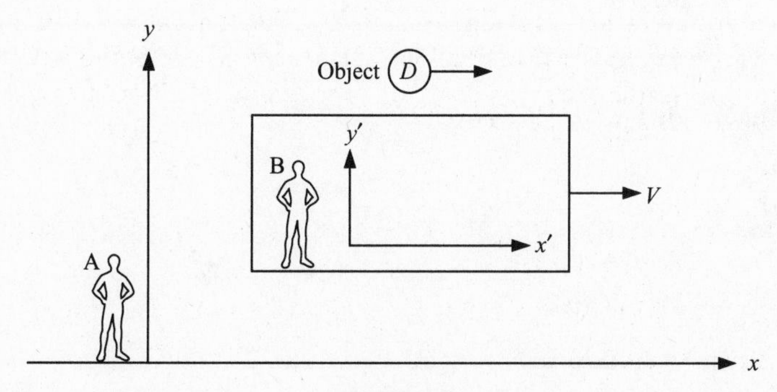

Fig. 16-13

(c) The total energy is $E = mc^2 = \gamma m_0 c^2 = (1/0.6)(1.5 \text{ kg})c^2 = (2.5 \text{ kg})c^2 = 2.25 \times 10^{17}$ J.

(d) $E^2 = 5.06 \times 10^{34}$ J^2; $p^2c^2 = (6 \times 10^8)^2(3 \times 10^8)^2 = 3.24 \times 10^{34}$ J^2; $m_0{}^2c^4 = (1.5 \text{ kg})^2(3 \times 10^8)^4 = 1.82 \times 10^{34}$ J^2; $p^2c^2 + m_0{}^2c^4 = 5.06 \times 10^{34} \rightarrow E^2 = p^2c^2 + m_0{}^2c^4$ as required.

Problem 16.37. Refer to Problem 16.36.

(a) At $t = 0$, the object D and the two origins just pass each other and A and B set their clocks at the origin to zero at that moment, i.e. $x = x' = 0$, $t = t' = 0$. After 2×10^{-7} s on A's clock, where is the object D according to A?

(b) Where is the object D according to B, and what is the time according to B?

(c) Use part (a) to deduce the velocity of D relative to B.

Solution

(a) According to A the distance covered in a time of 2×10^{-7} s is $vt = 0.8c(2 \times 10^{-7}) = 48$ m. Therefore the object is at $x = 48$, at $t = 2 \times 10^{-7}$ s.

(b) Using the Lorentz transformation equations, $x' = \gamma(x - Vt)$, $t' = \gamma(t - Vx/c^2)$, where $\gamma = 1/[1 - (V/c)^2]^{1/2}$, and $V = 0.6c$ from part (a) we get $\gamma = 1/0.8$, $x' = (1/0.8)[48 - 0.6c(2 \times 10^{-7})] = 15$ m, and $t' = (1/0.8)(2 \times 10^{-7} - 0.6c(48)/c^2) = 1.3 \times 10^{-7}$ s.

(c) $v_{D/B} = (x' - 0)/(t' - 0) = (15 \text{ m})/(1.3 \times 10^{-7} \text{ s}) = 1.154 \times 10^8 = 0.385c = (\frac{5}{13})c$, as required.

Problem 16.38. A space ship is moving away from an observer A, at a speed of $0.8c$. The spaceship has two guns, one a laser gun shooting rays of light, and one an ion gun shooting particles of rest mass 5×10^{-24} kg at a speed of $0.6c$ relative to the spaceship (see Fig. 16-14).

(a) When observer A watches the light ray that is shot from the laser gun, with what velocity does he see it move?

(b) When observer A watches the ion particles that are shot from the ion gun, with what velocity does he see them move?

(c) What is the momentum of an ion as measured by observer A?

(d) What is the kinetic energy of an ion as measured by observer B?

Solution

(a) The speed of light is c for all observers. A will see the light moving with that velocity.

(b) Here we must use the relativistic addition of velocities to get $v_{I/A} = (v_{I/B} + v_{B/A})/(1 + v_{I/B}v_{B/A}/c^2) = (0.6c + 0.8c)/(1 + 0.48) = 0.946c$.

(c) We know that $p = mv = \gamma m_0 v$. Here $v = 0.946c$, and $m_0 = 5 \times 10^{-24}$ kg. Calculating $\gamma v = v/[1 - v^2/c^2]^{1/2} = 2.92c$, and $p = 4.38 \times 10^{-15}$ kg · m/s.

(d) We know that the kinetic energy is $(m - m_0)c^2 = (\gamma - 1)m_0 c^2$. Here we must use the velocity of the ion as measured by B, $v_{I/B} = 0.6c$, since we seek the kinetic energy that is measured by B. Thus $\gamma = 1/0.8$, and $E_K = (1.667 - 1)(5 \times 10^{-24})(c^2) = 1.125 \times 10^{-7}$ J.

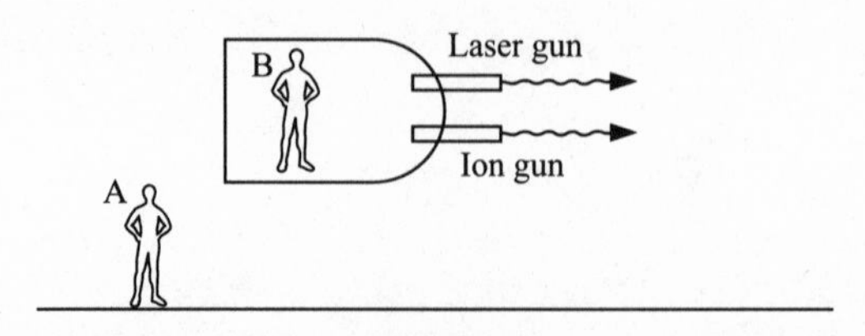

Fig. 16-14

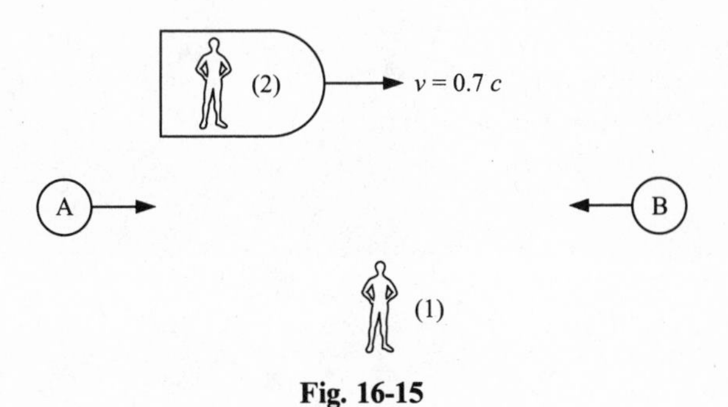

Fig. 16-15

Problem 16.39. An observer sees two objects, A and B, each of rest mass 2×10^{-23} kg, approach each other, each having a velocity of $0.6c$ (see Fig. 16-15). They collide and, during the collision, each loses some rest mass (which converts to kinetic energy), so that each has a rest mass of 1.5×10^{-23} kg after the collision. The objects each move apart with a velocity v after the collision.

(*a*) What is the momentum of each object before the collision?

(*b*) What is the kinetic energy of each object before the collision?

(*c*) What is the total energy of each object before the collision?

(*d*) Use conservation of energy to calculate the velocity of each object after the collision.

(*e*) Redo part (*c*) using the relativistic energy momentum equation.

Solution

(*a*) The momentum is given by $p = \gamma m_0 v$. For each object, $\gamma = 1/0.8$, so $p = (2 \times 10^{-23}\ \text{kg})(0.6c)/0.8 = 4.5 \times 10^{-15}\ \text{kg} \cdot \text{m/s}$.

(*b*) The kinetic energy for each object is $(\gamma - 1)m_0 c^2 = 4.5 \times 10^{-7}$ J.

(*c*) The total energy of each object is $\gamma m_0 c^2 = (2 \times 10^{-23}\ \text{kg})c^2/0.8 = 2.25 \times 10^{-6}$ J.

(*d*) The final total energy of each object is $m_f c^2$. By symmetry each object is moving away from the observer with the same velocity, v, after the collision, each object has the same energy, and conservation of energy means that this energy is the same as before the collision. Recalling that after the collision the rest mass has changed to $m_0 = 1.5 \times 10^{23}$ kg, we have for each object $E_f = \gamma_f m_{0f} c^2 = 2.25 \times 10^{-6}$ J. Solving for γ_f we get $\gamma_f = 2.25 \times 10^{-6}/(1.5 \times 10^{-23})c^2 = 1.667$, and, when solving for v we get $v = 0.8c = 2.4 \times 10^8$ m/s.

(*e*) $E^2 = p^2 c^2 + m_0 c^4 = (4.5 \times 10^{-15}\ \text{kg} \cdot \text{m/s})^2(3 \times 10^8\ \text{m/s})^2 + (2 \times 10^{-23}\ \text{kg})^2(3 \times 10^8\ \text{m/s})^4 = 5.06 \times 10^{-12}\ \text{J}^2 \rightarrow E = 2.25 \times 10^{-6}$ J.

Problem 16.40. Refer to Problem 16.39.

(*a*) If another observer is moving to the right with a velocity of $0.7c$ (according to the original observer), and observes the collision, what velocity does the new observer measure for A and for B before the collision?

(*b*) What total energy does this second observer see for each object before the collision?

(*c*) Use the addition of velocities to calculate the velocity of each object after the collision according to the second observer.

Solution

(*a*) To get the initial velocities as seen by the second observer, we use the addition of velocity formula, $v_{A/2} = (v_{A/1} + v_{1/2})/(1 + v_{A/1} v_{1/2}/c^2) = (0.6c - 0.7c)/(1 - 0.42) = -0.172c = -5.17 \times 10^7$ m/s. Similarly, $v_{B/2} = (v_{B/1} + v_{1/2})/(1 + v_{B/1} v_{1/2}/c^2) = (-0.6c - 0.7c)/(1 + 0.42) = -0.915c = -2.75 \times 10^8$ m/s.

(*b*) The total energy of each particle is $\gamma_i m_0 c^2$. Now $\gamma_{iA} = 1/(1 - v_{iA}{}^2/c^2)^{1/2} = 1/(1 - 0.172^2)^{1/2} = 1.0152$, and $\gamma_{iB} = 1/(1 - v_{iB}{}^2/c^2)^{1/2} = 1/(1 - 0.915^2)^{1/2} = 2.479$. Then, using $m_{0i} = 2 \times 10^{-23}$, $E_{KA} = 1.83 \times 10^{-6}$ J, and $E_{KB} = 4.462 \times 10^{-6}$ J.

(*c*) Using the addition of velocity equation and the results of part (*d*) of Problem 16.43, we get for the final velocities, $v_{A/2} = (v_{A/1} + v_{1/2})/(1 + v_{A/1} v_{1/2}/c^2) = (-0.8c - 0.7c)/(1 + 0.56) = -0.9615c = -2.885 \times 10^8$ m/s. Similarly, $v_{B/2} = (v_{B/1} + v_{1/2})/(1 + v_{B/1} v_{1/2}/c^2) = (0.8c - 0.7c)/(1 - 0.56) = 0.227c = 6.82 \times 10^7$ m/s.

Supplementary Problems

Problem 16.41. The two arms of a Michelson interferometer are each length 2.5 m. In one arm the medium is air of index of refraction $n_a = 1.000,293$. In the other arm, a length of 1.1 m is vacuum, and the rest is air.

(*a*) How much extra time is spent by light in going back and forth in the first path than in the second path?

(*b*) How much extra length would one have to add to the second arm so that the light spends the same time on both paths?

Ans. (*a*) 2.15×10^{-12} s; (*b*) 3.22×10^{-4} m = 0.322 mm

Problem 16.42. In Fig. 16-2(*b*), the light reaches mirror 1 a time of 0.02 s before it reaches mirror 2, according to observer *A*. The velocity *V* is 2×10^8 m/s.

(*a*) How far apart are the points, as measured by *A*?

(*b*) How far apart are the points, as measured by *B*?

(*c*) How long does it take for the light to reach each mirror, according to *B*?

Ans. (*a*) 2.5×10^6 m; (*b*) 3.35×10^6 m; (*c*) 5.59×10^{-3} s

Problem 16.43. Two points in the universe are 10^{10} m apart, as measured by an observer at rest relative to the points. If someone wants to travel from one point to the other in 25 s (on the traveler's clock), how fast will he have to travel?

Ans. 2.4×10^8 m/s

Problem 16.44. A person travels between two points in the universe, and it take him 4000 s on his clock to make the trip. His velocity, as measured by a person resting on one of the points is 0.6*c*.

(*a*) How far apart are the points, as measured by the traveler?

(*b*) How far apart are the points, as measured by the person resting on one of the points?

(*c*) How long does the trip take, as measured by the observer resting on one of the points?

Ans. (*a*) 7.2×10^{11} m; (*b*) 9×10^{11} m; (*c*) 5000 s

Problem 16.45. A man, *B*, in a rocket ship is traveling from the earth to a star at the speed of 0.8*c*, as measured by a man, *A*, on earth, who is at rest relative to the earth and the star. On *B*'s clock, it takes 3 years to meet the star.

(*a*) How long does *A* say that it took for *B* to reach the star?

(*b*) What is the distance from *A* to the star according to *B*, and what is that distance according to *A*?

(*c*) If man *A* claims that his clock, and the clock on the star are synchronized (i.e. they read the same time), which clock is ahead according to *B*?

Ans. (*a*) 5 y; (*b*) 2.4 light years = 2.27×10^{16} m, 4 light years = 3.78×10^{16} m; (*c*) on the star

Problem 16.46. A weight on a spring on earth is oscillating with a frequency of 3×10^3 Hz, as measured by an observer on earth. An observer in a rocket ship is moving at a speed of 2×10^8 m/s relative to the earth. What does he measure for the frequency of this oscillation?

Ans. (*a*) 2.24×10^3 Hz

Problem 14.47. Mesons decay with a half-life of 2×10^{-8} s when they are stationary. An observer on earth sees them traveling with a speed of $0.95c$.

(*a*) What is their half-life, according to this observer?

(*b*) How far do the mesons travel before half of them decay, according to this observer?

(*c*) How far does the earth travel before half of them decay, according to an observer at rest relative to the mesons?

Ans. (*a*) 6.41×10^{-8} s; (*b*) 18.3 m; (c) 5.7 m

Problem 16.48. The edge of the observable universe is believed to be about 10^{10} light years away. A spaceman wants to travel to that edge during his lifetime of 50 y.

(*a*) How fast will he have to travel?

(*b*) How far away is the edge of the universe to the observer traveling with that speed?

Ans. (*a*) $c(1 - 1.25 \times 10^{-17})$; (*b*) 50 light years

Problem 16.49. A UFO is traveling across the sky at a speed of $0.4c$. A person on earth measures the length of the UFO to be 0.3 m. What length is measured for the UFO after it lands?

Ans. 0.33 m

Problem 16.50. A space traveler wants to travel to a star that is 50 light years away, as measured by an observer on earth. The space traveler says that the star is only 22 light years away.

(*a*) How fast is the traveler moving relative to the earth?

(*b*) How long does the traveler say that it took to reach the star?

Ans. (*a*) $0.898c$; (*b*) 24.5 light years

Problem 16.51. Two persons on earth are separated by a distance of 3×10^8 m. The first person fires a rocket toward the second person at a speed of $0.8c$. When the second person receives the rocket he retaliates by firing a rocket back to the first person at a speed of $0.9c$. A spaceperson, traveling in the direction from the first to the second person at a speed of $0.3c$ observes the exchange. When the spaceperson (B) passes the first person (A) they both set their times to zero, and they are at their origins. The first rocket is fired at that time, $t = 0$.

(*a*) According to A, at what time did the rocket reach the second person?

(*b*) According to A, at what time did the second rocket reach him?

(*c*) According to B, at what time and at what position did the rocket reach the second person?

(*d*) According to B, at what time and at what position did the second rocket return to A?

Ans. (*a*) 1.25 s; (*b*) 2.36 s; (*c*) $x' = 1.97 \times 10^8$ m, $t' = 1.00$ s; (*d*) $x' = -2.23 \times 10^8$ m, $t' = 2.47$ s

Problem 16.52. Observer A measures the distance between two stationary points to be 5×10^8 m. He notices that each of the points sends out a beam of light at time $t = 0$. Observer B is moving in the direction from the first to the second point at a speed of $0.7c$, and passes the first point at time $t' = t = 0$.

(*a*) According to A, at what time did the light reach a point halfway between the points (i.e. when the beams meet)?

(*b*) According to B, at what time was the light emitted from the second point?

(*c*) According to B, at what time did the beams meet?

Ans. (*a*) 0.833 s; (*b*) $t' = -1.633$ s; (*c*) $t' = 0.350$ s

Problem 16.53. An observer sees three objects moving to the right. He measures their velocities to be $v_1 = 0.5c$, $v_2 = 0.63c$ and $v_3 = 0.99c$.

(*a*) What is the velocity of 2 relative to 1?

(*b*) What is the velocity of 1 relative to 3?

(*c*) What is the velocity of 3 relative to 2?

Ans. (*a*) $0.19c$; (*b*) $-0.97c$; (*c*) $0.957c$

Problem 16.54. An electron moves with a speed of $0.8c$.

(*a*) What is the total energy of the electron?

(*b*) What is the kinetic energy of the electron?

(*c*) What is the momentum of the electron?

Ans. (*a*) 1.365×10^{-13} J; (*b*) 5.46×10^{-14} J; (*c*) 3.64×10^{-22} kg · m/s

Problem 16.55. An observer, A, sees an object, of rest mass 0.06 kg, moving to the right with a velocity of $0.8c$. A different observer sees the same object moving to the left with a velocity of $0.6c$.

(*a*) According to A, what is the momentum of the object?

(*b*) According to B, what is the kinetic energy of the object?

(*c*) What is the velocity of B relative to A?

Ans. (*a*) 2.4×10^7 kg · m/s; (*b*) 1.35×10^{15} J; (*c*) $0.946c$

Problem 16.56. Electron A moves to the right relative to the observer, and has a total energy of 1.35×10^{-13} J. Another electron, B, moves to the left with a speed of 2.8×10^8 m/s relative to the observer.

(*a*) What is the kinetic energy of electron A?

(*b*) What is the velocity of electron A?

(*c*) What is the momentum of electron A?

(*d*) What is the velocity of electron B as measured by electron A?

Ans. (*a*) 5.31×10^{-14} J; (*b*) 2.38×10^8 m/s; (*c*) 3.57×10^{-22}; (*d*) -2.98×10^8 m/s

Problem 16.57. A particle has a rest mass of 5×10^{-23} kg, and is initially at rest. The particle explodes into two particles of equal mass, and in the process, a fraction of the initial rest mass is destroyed. The lost rest mass is converted into kinetic energy of the remaining particles. The total kinetic energy of the two particles after the explosion is 1.1×10^{-7} J.

(*a*) What was the total initial energy of the particle?

(*b*) What is the rest mass energy of each of the remaining particles? What is the rest mass of each of the remaining particles?

(*c*) What is the velocity of each of the remaining particles?

Ans. (*a*) 4.5×10^{-6} J; (*b*) 2.20×10^{-6} J, 2.44×10^{-23} kg; (*c*) 6.59×10^7 m/s

Problem 16.58. A particle, of rest mass 2×10^{-20} kg, is accelerated from rest by a constant force of 6×10^{-11} N. After how much time will the particle have (*a*) a momentum of 6×10^{-14}; (*b*) a velocity of $0.6c$; (*c*) a kinetic energy of 3×10^{-3} J; and (*d*) a mass of 5×10^{-20} kg?

Ans. (*a*) 1.0×10^{-3} s; (*b*) 0.075 s; (*c*) 0.297 s; (*d*) 0.229 s

Problem 16.59. Two particles approach each other as in Fig. 16-16. One particle has a rest mass of m_0, and is moving to the right with a speed of $0.6c$. The other particle has a rest mass of $0.5m_0$ and is moving to the left with a speed of $0.8c$. They collide and form a new particle with a rest mass of m_0' moving with a velocity of v'.

(*a*) In terms of m_0, what is the total momentum of the system before the collision?

(*b*) In terms of m_0, what is the total energy of the system before the collision?

(*c*) Using conservation of momentum, what is the momentum of the particle, $p = mv'$, after the collision?

(*d*) Using conservation of energy, what is the total energy of the particle, $E_T = mc^2$, after the collision?

(*e*) Using the fact that $m'cv'/m'c^2 = v'/c$, where m' is the relativistic mass of m_0', calculate the velocity of the particle after the collision.

(*f*) Calculate the rest mass, m_0', after the collision.

Ans. (*a*) $m_0 c/12$; (*b*) $(25/12)m_0 c^2$; (*c*) $m_0 c/12$; (*d*) $(25/12)m_0 c^2$; (*e*) $c/25$; (*f*) $2.08m_0$

Problem 16.60. Refer to Problem 16.36.

(*a*) Calculate the momentum of D as measured by B.

(*b*) Calculate the energy of D as measured by B.

(*c*) Is the relativistic energy–momentum equation: $E^2 = p^2c^2 + m_0{}^2c^4$ true in B's frame of reference?

Ans. (*a*) 1.875×10^8 kg · m/s; (*b*) 1.46×10^{17} J; (*c*) yes

Problem 16.61. Refer to Problem 16.40 and the perspective of the second observer.

(*a*) Find the momentum of objects A and B before the collision, and the total momentum before the collision.

(*b*) Find the momentum of objects A and B after the collision, and the total momentum after the collision.

(*c*) Find the total energy of each object after the collision.

(*d*) Comparing (*a*) and (*b*), is momentum conserved in this collision?

(*e*) Comparing (*c*) to Problem 16.44(*b*), is energy conserved in this collision?

Ans. (*a*) -1.05×10^{-15} kg · m/s, -1.363×10^{-14} kg · m/s, -14.70×10^{-15} kg · m/s; (*b*) -1.575×10^{-14} kg · m/s, 1.050×10^{-15} kg · m/s, -14.70×10^{-15} kg · m/s; (*c*) 4.912×10^{-6} J, 1.386×10^{-6} J; (*d*) yes; (*e*) yes

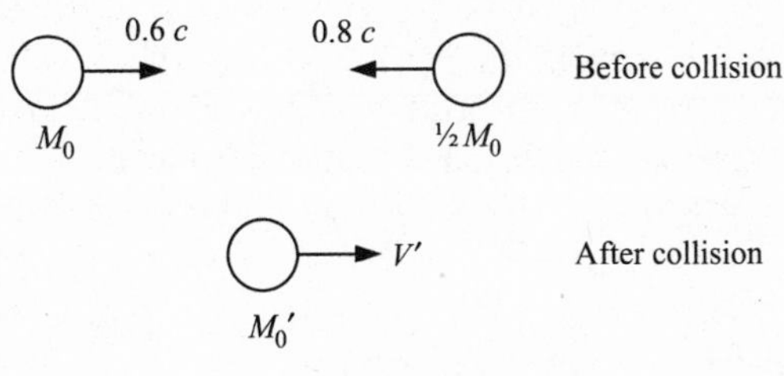

Fig. 16-16

Chapter 17

Particles of Light and Waves of Matter: Introduction to Quantum Physics

17.1 INTRODUCTION

In the last chapter, we learned that, in trying to understand the meaning of the predicted velocity of electromagnetic waves in space, we were led to the development of the theory of relativity. In this chapter we will learn that, in trying to understand the nature of light waves (electromagnetic waves), in certain situations, the question of whether light consists of particles or waves has to be reopened, helping to lead the way to the development of the theory of quantum mechanics.

Before we can reasonably discuss whether light behaves like a particle or like a wave, we have to understand what we mean by particle behavior and what we mean by wave behavior. We will discuss four areas in which particle behavior differs from wave behavior.

Localization

When we think of a particle, we think of a quantity of mass located at a given position in space that is localized. An extended object such as a rigid body can be composed of many point particles each localized at a given point in space, each with a known velocity and acceleration. There is certainly no possibility of considering any one particle as being simultaneously at many different positions in space.

In contrast, a wave is definitely not localized in space, but rather is a disturbance that is spread over a fairly sizable area. For a pure sinusoidal wave the wave is, strictly speaking, spread out from plus to minus infinity. Even for a pulse wave, there is a definite extent in space at any instance. One generally thinks of a wave as being a disturbance located simultaneously at many points of space.

It is therefore clear that the particle and wave picture differ on the question of localization.

Interference and Diffraction

When two waves meet in a region of space in which a positive displacement of one wave is canceled by a negative displacement of the second wave we have destructive interference. In regions in which the positive displacements reinforce we can get double the amplitude of each individual wave. Since the intensity of the wave is proportional to the square of the displacement, the energy density is increased by a factor of 2^2, or four, giving more energy than that produced by the two waves individually. The total energy over all space will, of course, still be equal to the sum of the energy from the two individual waves, but some regions will have a reduced energy and others will have an increased energy. A particle could never give us such interference since it is localized. Therefore, if one detects interference or diffraction phenomena, it is a clear indication that we are dealing with waves.

Energy Flow

Both waves and particles carry energy and momentum. But there is a big difference in the mechanism of that energy flow. In the case of a particle beam, the point particle carries the energy and momentum, and each particle in the beam carries its own energy. Once the particle passes an area, all of its energy has been transported past that area. The intensity of the energy flow is given by the average energy carried through a perpendicular unit area per second which is just the average number of particles passing the unit area per second times the energy of each particle. However, the flow of energy is not uniform or continuous. As one reduces the number of particles that are flowing, the energy becomes less and less continuous, with bigger gaps between the time that a particle carries energy through the area and the time that the next particle brings in additional energy. Thus the energy comes in discontinuous pulses, with each pulse bringing a discrete amount of energy. If all the particles in our beam have equal energy, then the energy passing a unit area in any time interval will be a whole number multiple of that particle energy.

A wave, when it carries energy and momentum, carries this energy in a continuous fashion. As the wave passes through the area the energy passes through the area smoothly with the rate changing as the displacement changes, but without discontinuous jumps in energy flow, even if the energy is greatly decreased. For a traveling wave of given amplitude the total energy passing a unit area in a given time will vary continuously with the time, unlike the particle beam. This difference leads to the idea of quantization.

Quantization

If light behaved as a particle, it would be easy to understand that light energy could be transported in integral multiples of some fundamental energy associated with that particle; each particle carries a "**quantum**" of energy. It would be much harder to understand how a continuously flowing wave could transport such quantized units of energy.

Before the 20th century, the preponderance of evidence led physicists to believe that light (and other frequency electromagnetic waves) were indeed waves. During the beginning of the 20th century, more and more experiments pointed to a particle picture for light, and some experiments even showed both particle and wavelike properties. We will now examine some of these experiments.

17.2 LIGHT AS A WAVE

In Chap. 12 we discussed the electromagnetic theory which predicted that electromagnetic waves exist in nature. Light was found to have all the properties predicted by this theory, including the correct speed, c, and the correct transverse polarization. It was quickly universally accepted that light was indeed an example of electromagnetic waves, as described by Maxwell's equations.

The wave nature of light is experimentally demonstrated by all the interference and diffraction phenomena discussed in Chap. 15. There we discussed the case of a double slit, in which a single light wave is incident on two slits each of width a separated by a small distance, d, as in Fig. 17-1. We assume $a \ll d$. Light reaches the screen which is located at a large distance, L, from the slits. The light going through the two slits travels further from the lower slit to the point, P, than does light from the upper

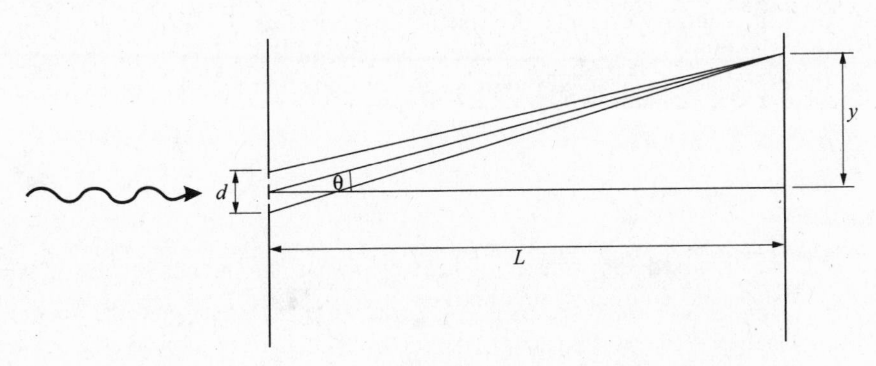

Fig. 17-1

slit. This gives rise to interference between the light going through the two slits, which is seen as a pattern of fringes on the screen. Since $a \ll d$ there will be many fringes within the broad central maximum of the single slit diffraction pattern. At some points on the screen, there is complete destructive interference between the two waves, resulting in no light hitting the screen at those points. We understand this because we assume that light is a wave which simultaneously goes through both slits. If light went through only one slit, we would still get the single slit diffraction pattern on the screen with a wide central maximum, but not the interference fringes of the double slit. If one were to first close one slit, letting light through the second slit, and a short time later close the second slit letting light through the first slit, the cumulative light reaching the screen would not correspond to the fringe distribution of light seen when light goes through both slits. This is easy to understand on the basis of light being a wave, as was described in Chap. 15.

If light were made up of a beam of particles impinging on the two slits, each particle would go through either one slit or the other slit. One cannot say that a particle goes simultaneously through both slits. Therefore, the particles that go through one slit will distribute themselves on the screen in the same manner whether or not the second slit is open. They are not even aware of the existence of a second slit. Since the distribution doesn't change as a consequence of both slits being open, the distribution of particles from one slit will overlap the distribution from the other slit on the screen. There is no way to explain the presence of interference fringes and their existence seemed to altogether eliminate the possibility that light consists of particles.

If light is an electromagnetic wave it will transport energy at a rate given by the intensity, I, of the light, which is the amount of energy transported through a unit area per second. I is proportional to the square of the electric field in the wave, and is given by I_{AV} = average power/area = $c^2\varepsilon_0 E_0{}^2/2$. This energy is transported continuously (albeit time varying with the periodicity of the wave) as the wave moves past the area, and not in sporadic bursts. As the wave impinges on a material, its energy can be partially or totally absorbed by the atoms or molecules of the material with a corresponding reduction in the wave amplitude. For matter to absorb a given amount of this energy requires that the wave impinge for at least the length of time needed for the required amount of energy to be deposited in the material. The same is, of course, true for the transfer of momentum to the material.

To summarize, if light is a wave, then we expect to see the interference and diffraction effects that actually take place in nature. Furthermore, we expect these waves to transport energy and momentum in a continuous fashion, and as energy is absorbed we expect the amplitude to continuously decrease until it reaches zero when all the energy is absorbed. The fact that we do indeed see interference effects for light seems to be irrefutable proof that light acts as a wave.

17.3 LIGHT AS PARTICLES

During the first part of the 20th century, several experiments with light gave strong evidence that light behaves as particles. Although the experiments on interference show that light behaves as a wave, we cannot disregard the contrary evidence of these new experiments. To resolve this discrepancy physicists developed the quantum theory that we will discuss later in this chapter. At this stage, we want to discuss some of the experiments which led physicists to conclude that light has particle—like properties.

Black Body Radiation

In Chap. 17.3, we discussed the transfer of heat by radiation. In that section it was noted that electromagnetic radiation was emitted by any body at a temperature T, in accordance with the **Stefan–Boltzmann law**,

$$H = A\varepsilon\sigma T^4 \tag{17.1}$$

The energy emitted per second is proportional to the **emissivity**, ε $(0 < \varepsilon < 1)$, to the surface area of the emitter, and to the absolute temperature, T, to the fourth power, T^4. A perfect emitter, where ε is equal to one, is called a **black body**, since it is also a perfect absorber and completely absorbs all incident radiation, giving it the appearance of black which represents the absence of light. Scientists checked whether the properties of this radiation agree with predictions of Maxwell's equations. It was easy to show that the emitted radiation had the usual wave properties and that the dependence of the total emitted radiation on T^4 is consistent with these equations. However, this radiation is emitted over a continuous spectrum of wavelengths (or frequencies), and the study of the dependence of the intensity of the black-body radiation on frequency led to contradictions. If one plots the rate of energy emitted as a function of frequency at a particular temperature, as determined by experiment, one gets a curve such as the one shown in Fig. 17-2(*a*). The vertical is actually a plot of energy rate per unit frequency interval as determined for each frequency so that the total area under the curve represents the total energy emitted. The highest rate of energy radiated is at a frequency f_m, as shown in the figure. If we plotted the corresponding curve at higher temperatures the frequency f_m would move to higher values [Fig. 17-2(*b*)]. At room temperature ($T = 300$ K) f_m is in the infrared, and at $T = 10{,}000$ K the f_m is in the visible region. The hotter the temperature the "bluer" is the light, while a cooler object is "redder". The overall intensity of the light is affected by the temperature even more strongly, with the height of the energy distribution rising rapidly with temperature. This leads to the overall T^4 dependence of the light intensity, as we stated previously. To explain this radiation energy distribution from Maxwell's equations, one argues as follows. All material consists of charged particles, which are moving around or vibrating in the material. As they accelerate in their motions, they radiate due to their acceleration. As the material gets hotter there is more thermal motion, and consequently more radiation. When one applies this idea and tries to calculate the dependence of the radiation frequency at a given temperature, one finds good agreement for low frequencies (long wavelengths), but there is no maximum frequency, f_m, predicted, and the theory actually implies that the energy continues to rise as f increases leading to absurd results. All attempts to explain the entire curve of Fig. 17-2 using Maxwell's equations proved unsuccessful, and some new idea was clearly necessary. This idea was supplied by Max Planck in 1900. He noted that one could reproduce the experimental curve if one assumes that the energy of the electromagnetic radiation at a particular frequency f was restricted to integral multiples of hf, where h is a constant now known as **Planck's constant**. Thus the energy at a frequency f is limited to certain values, given as integral multiples of a basic quantum of radiation at that frequency, hf. It is as if the light came as discrete particles, each having an energy hf. (These particles or "quanta" of light were later verified to exist and are called photons, and all photons of a given frequency have the same energy, hf.) With this assumption, Planck was able to theoretically produce the same curve as was determined by experiment, and to derive a value for h (now known as Planck's constant) as $h = 6.625 \times 10^{-34}$ J $\cdot$ s. This was the first serious experimental evidence that light behaves as particles. Further evidence was not long in coming.

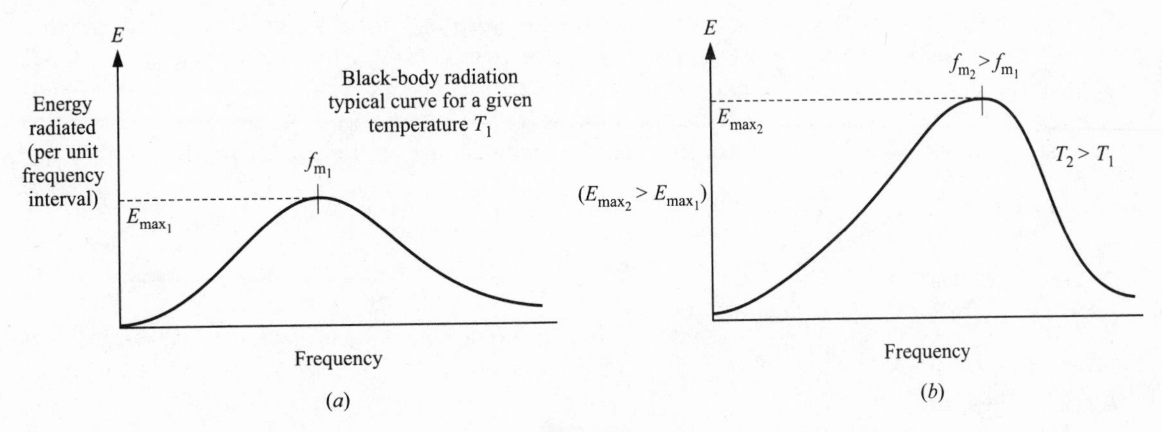

Fig. 17-2

Problem 17.1. Assuming that photons obey Planck's energy rule, $E_{ph} = hf$. At what frequency will photons have the following energies, and for each case where in electromagnetic spectrum are these photons: (*a*) 0.010 eV; (*b*) 3.0 eV; (*c*) 10 eV; (*d*) 1.0×10^4 eV; (*e*) 1.0×10^9 eV?

Solution

(*a*) The energy of a photon is $E = hf$, or $f = E/h$. To convert from eV to Joules we divide by 1.6×10^{-19}. Thus, $f = 0.010(1.6 \times 10^{-19})/6.625 \times 10^{-34} = 2.42 \times 10^{12}$ Hz. This corresponds to a wavelength of $\lambda = c/f = 3.0 \times 10^8/2.42 \times 10^{12} = 1.24 \times 10^{-4}$ m = 124 μm, which is in the infrared.

(*b*) Again to convert from eV to Joules we divide by 1.6×10^{-19}. Thus, $f = 3.0(1.6 \times 10^{-19})/6.625 \times 10^{-34} = 7.25 \times 10^{14}$ Hz. $\lambda = c/f = 3.0 \times 10^8/7.25 \times 10^{14} = 4.14 \times 10^{-7}$ m = 414 nm, which is in the visible.

(*c*) The energy of a photon is $E = hf$, or $f = E/h$. To convert from eV to Joules we divide by 1.6×10^{-19}. Thus, $f = 10(1.6 \times 10^{-19})/6.625 \times 10^{-34} = 2.42 \times 10^{15}$ Hz. This corresponds to a wavelength of $\lambda = c/f = 3.0 \times 10^8/2.42 \times 10^{15} = 1.24 \times 10^{-7}$ m = 124 nm, which is in the ultraviolet.

(*d*) Here, $f = 1.0 \times 10^4(1.6 \times 10^{-19})/6.625 \times 10^{-34} = 2.42 \times 10^{18}$ Hz. $\lambda = c/f = 3.0 \times 10^8/2.42 \times 10^{18} = 1.24 \times 10^{-10}$ m = 0.124 nm, which is in the "soft" X-ray region (i.e. low frequency end of the X-ray spectrum).

(*e*) Here, $f = 1.0 \times 10^9(1.6 \times 10^{-19})/6.625 \times 10^{-34} = 2.42 \times 10^{23}$ Hz. $\lambda = c/f = 3.0 \times 10^8/2.42 \times 10^{23} = 1.24 \times 10^{-15}$ m = 1.24×10^{-6} nm, which is in the "very hard" X-ray and gamma ray region.

Photo-Electric Effect

The **photo-electric effect** provided the most convincing evidence that light must be considered to consist of particles. The detailed experimental data from this effect is unable to be understood on the basis of a wave picture for light and requires one to accept the particle nature of light.

In the photo-electric effect, a beam of light is shone on a metal, with the result that electrons are emitted from the metal. This occurs because energy of the light is transmitted to the electrons in the metal in sufficient quantity to break the electrons free from the bonds that hold them in the material. This general fact can be understood whether light is a particle or a wave, since in both cases the light brings energy to the metal.

The electrons that are emitted from the material have a certain amount of kinetic energy after the emission. If one sends in light of at a definite frequency, f, the number of electrons emitted with a given velocity can be experimentally measured over the range of observed velocities, and the result can be plotted. This is done in Fig. 17-3(*a*) where the kinetic energy is plotted on the horizontal and the

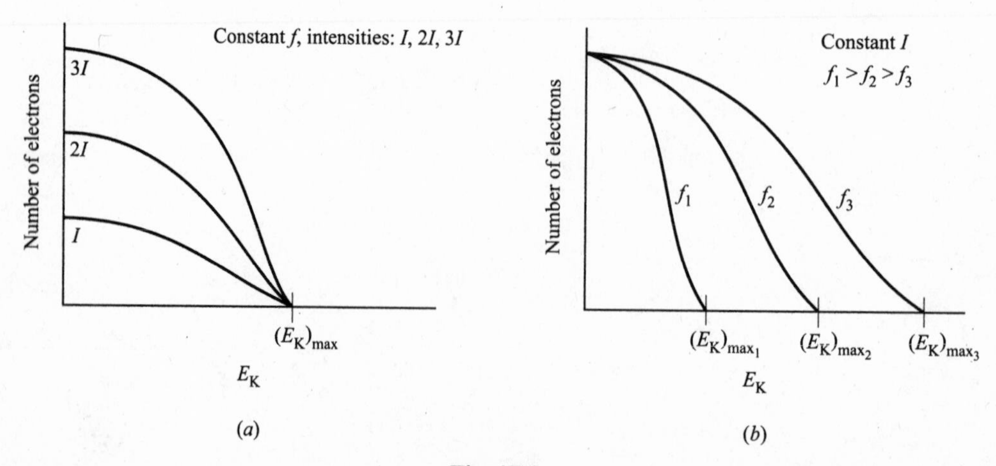

Fig. 17-3

number of electrons (per unit velocity) is plotted on the vertical. The area under the curve then represents the total number of electrons emitted. Three plots are shown for one frequency at three different intensities of the light. This plot has two very important features. First, there is a maximum kinetic energy that the electrons can have. Second, this maximum kinetic energy does not depend on the intensity of the incident light. Both of these features are difficult to understand on the basis of a wave picture of light. As we already mentioned previously, a wave transmits energy on a continuous basis, and the amount of energy accumulated in the material depends on how long one has light striking the metal. Presumably, after sufficient time, enough energy has accumulated on an individual electron to allow it to break out of the material. If the electron stays in the material longer, it should accumulate more energy, and then emerge with more kinetic energy. Why should there be a maximum kinetic energy that can be accumulated by the electron? And even more difficult is the question of why this maximum energy does not increase with the intensity of the incident light. If the intensity increases, the energy is being supplied more rapidly to the metal and one would expect the electrons to be able to accumulate more energy. Even more astonishing is the fact that the maximum kinetic energy depends on the frequency irrespective of the intensity. Furthermore, the relationship between the maximum kinetic energy and the frequency is a very simple linear relationship, shown in Fig. 17-4 for different target molecules. We notice from this figure that for each metal there is a critical frequency, f_c, below which there are no photo-electrons. How can we understand this phenomenon? Why should light with a low frequency be unable to bring in sufficient energy to emit electrons, no matter what intensity is used. As shown this minimum frequency depends on the metal being used (as well as on the state of its surface). Another remarkable feature of these plots is that the slope of the straight lines for all the metal surfaces are the same and given by $(\Delta E/\Delta f)$ = slope = h, where h turns out to be Planck's constant!

Einstein realized that, if one accepts Planck's hypothesis that light comes in quantized amounts of energy given by hf, then all these experimental facts are easily understood. On a particle picture of light, one interprets this experiment as follows. At the frequency, f, particles of light (photons), each having an energy of hf are incident on the material. When energy is absorbed by an electron all the energy of the photon has to be absorbed at once, since the photon cannot have any energy other than hf. This energy is then available to the electron in its attempt to escape from the metal. It can lose some of the energy it garnered in breaking loose from its bonds to atoms and in collisions with atoms on its way to the surface, and it has to lose a certain amount of additional energy at the surface in order to break out. This latter amount of energy is called the **work function** of the material, and is similar to the ionization energy needed for an electron to escape from an atom except that it now represents the minimum amount of energy needed to escape from the material surface. The kinetic energy remaining to the electron is just the energy it received (hf), minus the energy it lost on the way to the surface (E_s), and

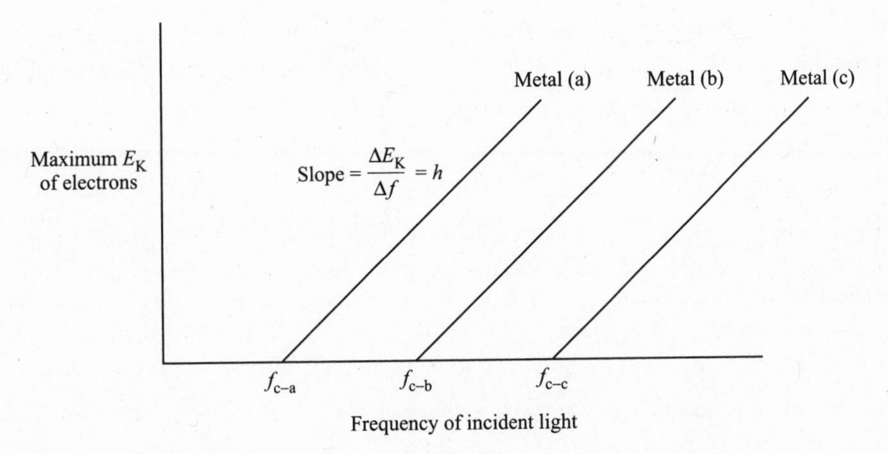

Fig. 17-4

minus the energy needed to break out of the surface, the work function, W. Thus,

$$E_K = hf - E_s - W \tag{17.2}$$

The maximum kinetic energy will occur if the energy lost to reach the surface is minimum, i.e. for $E_s = 0$ (the case of no collisions and an electron initially unbound to any atoms). Then

$$(E_K)_{max} = hf - W \tag{17.3}$$

This is just the straight line relationship seen experimentally, and shown in the graph in Fig. 17-4. We can see the linear relationship between $(E_K)_{max}$ and f, and identify h as the slope of the curve. A crucial test of this theory is whether the h one determines experimentally from this experiment is identical to the h determined from black body radiation. The two values of h are indeed the same strengthening the argument that the energy of light at any frequency f is quantized in units of hf.

Interpreting Eq. (*17.3*), we can understand why maximum kinetic energy does not depend on the intensity of the light. Increasing intensity, only increases the number of photons incident on the metal, but does not increase the energy of each photon. To increase the energy of the emitted electrons, one must increase the energy of the incident photons, which can be done only by increasing the frequency. To understand why there is a minimum frequency below which no photo-electrons are emitted, we note that if $hf < W$, electrons cannot break through the surface. $hf_{min} = W \rightarrow (E_K)_{max} = 0$ is thus the threshold condition. Unless hf is bigger than W, no electron will be emitted with any kinetic energy (except for the extremely unlikely case that two photons simultaneously strike the same electron). The cutoff frequency is just $f_{min} = W/h$ varies with material since W varies with material. Indeed the cutoff frequency provides a convenient method for measuring the work function of different materials.

Despite the beautiful correlation between the photon hypothesis and the strange results represented in Figs. 17-3 and 17-4, there were still scientists who were skeptical because of the contradiction to the wave theory of light. They argued that there might be some unknown mechanism to explain the observed phenomena without the radical photon hypothesis. Perhaps the particular nature of solids lent themselves to these strange results. Nonetheless, there was one final observation that convinced almost all that only Einstein's hypothesis would work. This last experiment was as follows. Suppose one suddenly opens a shutter to allow very low intensity light to be incident on a metal; how long does one have to wait for the first photo-electron to appear? If the light comes in as a wave, one should have to wait a minimum amount of time until sufficient energy has accumulated to dislodge the electron. This time could be calculated and was sufficiently large to be measurable for low intensity light. Clearly, this time should increase as the intensity decreases, since then the energy is being supplied at a slower rate. When those experiments were performed only the number of photo-electrons decreased with decreasing intensity, but some were always emitted almost instantaneously. The conclusion was inescapable: it is essentially impossible to understand how electrons can be emitted immediately, if light acts as a wave.

Problem 17.2. Light is incident on a metal with a work function of 3.0 eV.

(*a*) What is the minimum frequency of light that will result in the emission of photo-electrons?

(*b*) What is the minimum wavelength of that light?

(*c*) What is the minimum frequency of light that will result in the emission of photo-electrons with a kinetic energy of 2.0 eV?

Solution

(*a*) The minimum frequency is one where the energy of a photon just equals the work function. This is equivalent to saying that $(E_K)_{max} = 0$. Using the relationship that $E_{photon} = hf_c$, and equating this to W, we get that $f_c = E_{photon}/h = 3.0(1.6 \times 10^{-19})/6.625 \times 10^{-34} = 7.25 \times 10^{14}$ Hz.

(*b*) The wavelength is given by: $\lambda = c/f = (3.0 \times 10^8)/7.25 \times 10^{14} = 4.14 \times 10^{-7}$ m $= 414$ nm.

(*c*) The minimum frequency is one where the energy of the photon is equal to the work function plus the 2.0 eV kinetic energy. This is equivalent to using Eq. (*17.3*) with $(E_K)_{max} = 2.0$ eV. We get that $hf = (3.0 + 2.0)$ eV $= 5.0(1.6 \times 10^{-19})$, or $f = 1.21 \times 10^{15}$ Hz.

Problem 17.3. When light of wavelength 450 nm is incident on a certain metal, electrons are emitted with kinetic energies up to 2.0 eV.

(*a*) What is the work function of the metal?

(*b*) What is the minimum frequency needed to emit any photo-electrons?

Solution

(*a*) Using Eq. (*17.3*), we get $(E_K)_{max} = hf - W = hc/\lambda - W$, or $W = hc/\lambda - (E_K)_{max} = [6.625 \times 10^{-34}(3.0 \times 10^8)/450 \times 10^{-9}] - 2.0(1.6 \times 10^{-19}) = 1.22 \times 10^{-19}$ J $= 0.76$ eV.

(*b*) This frequency is one for which photons have an energy equal to the work function, or $hf = W = 1.22 \times 10^{-19}$ J. Then, $f = 1.22 \times 10^{-19}/6.625 \times 19^{-34} = 1.84 \times 10^{14}$ Hz.

Problem 17.4. A beam of light of frequency $f = 1.8 \times 10^{18}$ Hz has an intensity of $I = 1.3 \times 10^{-12}$ W/m^2.

(*a*) What is the average time between successive photons?

(*b*) Assuming a wave picture of light and assuming that a given electron can absorb energy from a cross-sectional area of incoming light comparable to the area of a typical atom, $A \approx 10^{-20}$ m^2, how long will it take for an electron to absorb enough energy to escape a work function of $W = 3.0$ eV?

Solution

(*a*) Each photon has an energy of $hf = 6.625 \times 10^{-34}(1.8 \times 10^{18}) = 1.19 \times 10^{-15}$ J. If $\Delta n/\Delta t$ are the number of photons incident on a unit area per second, then the intensity of the light is $I = (\Delta n/\Delta t)hf$. The average time between successive photons hitting a unit area is $\Delta t/\Delta n = hf/I = 1.19 \times 10^{-15}/1.3 \times 10^{-12} = 9.17 \times 10^{-4}$ s.

(*b*) The energy accumulated must be $W = (3.0 \text{ eV})(1.6 \times 10^{-19} \text{ J/eV}) = 4.8 \times 10^{-19}$ J. This energy equals IAt where t is the needed time. Thus, $(4.8 \times 10^{-19} \text{ J}) = (1.3 \times 10^{-12} \text{ W/m}^2)(1.0 \times 10^{-20} \text{ m}^2)t \rightarrow t = 3.6 \times 10^{13}$ s.

Note. Even if we assume a single electron can absorb light from a cross-sectional area of 10,000 atoms, $A \approx 10^{-16}$ m^2, the result would still be $t = 3.6 \times 10^9$ s, corresponding to hundreds of years! In fact, electrons are emitted within thousandths of a second as can be seen from part (a).

We have seen that the details of the photo-electric effect can be explained very simply by using the assumption of Planck that the energy of electromagnetic waves is quantized in multiples of hf. The reverse process, which we shall see is used for the production of X-rays, also gives evidence for the particle nature of light.

Production of X-rays

X-rays, which were discovered by **Roentgen** in 1895, can be produced by taking energetic electrons (1000 or more electron volts), and letting them strike a metal plate. This is essentially the reverse of the photo-electric effect, where electromagnetic rays incident on a metal produced electrons. Here, electrons striking the metal produce electromagnetic radiation. Maxwell's equations predict that any charged particle that is accelerated (or decelerated) should radiate electromagnetic waves. When electrons hit a metal plate, they decelerate as they are stopped in the metal, and therefore the wave theory of light predicts the emission of electromagnetic radiation. This radiation is called **Bremsstrahlung radiation,** which, in German, means "braking" radiation. However, when we consider the details of the radiation emitted we are, once again, led to the conclusion that this radiation acts like particles.

The electrons are generally produced in an evacuated tube, and are emitted from a filament with little initial kinetic energy. They are then accelerated through a potential difference, V, to a high kinetic energy given by eV where e is the magnitude of the negative charge on the electron, before they strike the metal plate (Fig. 17-5). In electron volts, this kinetic energy is just numerically the potential difference V through which they have fallen. When these electrons stop in the metal, they emit electromagnetic radiation over a large range of wavelengths. The surprising experimental fact is that there is a minimum wavelength, λ_{min}, that can be emitted for each V. This minimum wavelength varies inversely with V, but is not affected by the electron current density (electrons/unit area/unit time) j hitting the plate at a given V. A plot of the intensity emitted as a function of wavelength λ, is given in Fig. 17-6 for one V, but for three different electron current densities. If one plots $1/\lambda_{min}$ *vs.* V, one gets the graph in Fig. 17-7. Since this gives a straight line, it shows that $1/\lambda_{min}$ is proportional to V.

To understand these results, we again think of the electromagnetic waves as photons of energy $hf = hc/\lambda$. When an electron strikes the metal it releases kinetic energy due to decelerating collisions within the metal. Depending on the collision the electron can release any fraction of its kinetic energy in the form of a photon. However, the maximum amount of energy that can be released is the entire kinetic energy of the electron, eV which corresponds to a photon of frequency $f_{max} = eV/h$, or a wavelength, $\lambda = hc/eV$. Then $1/\lambda_{min} = (e/hc)V$. Then, not only is there a minimum wavelength, but $1/\lambda_{min}$ is proportional to V, and the slope of the graph of $1/\lambda_{min}$ *vs.* V should be e/hc. This is exactly what is found experimentally. These results cannot be explained by a wave theory of light. Actually, when one produces X-rays in this manner, one often finds that the curve of X-ray intensity *vs.* λ has sharp spikes

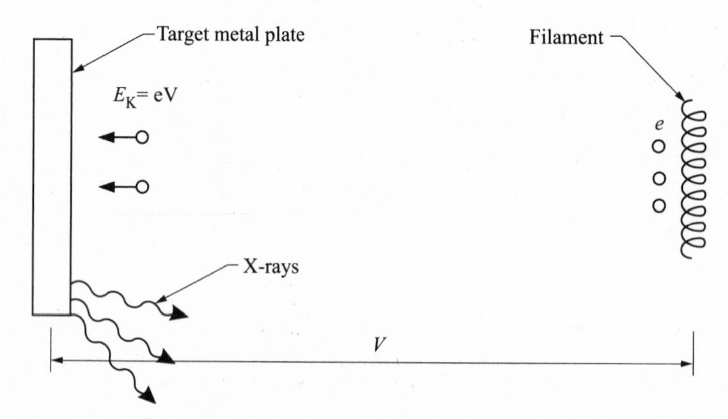

Fig. 17-5

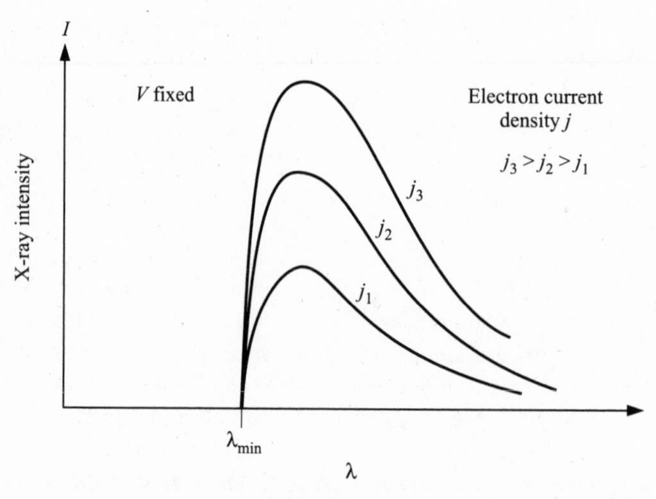

Fig. 17-6

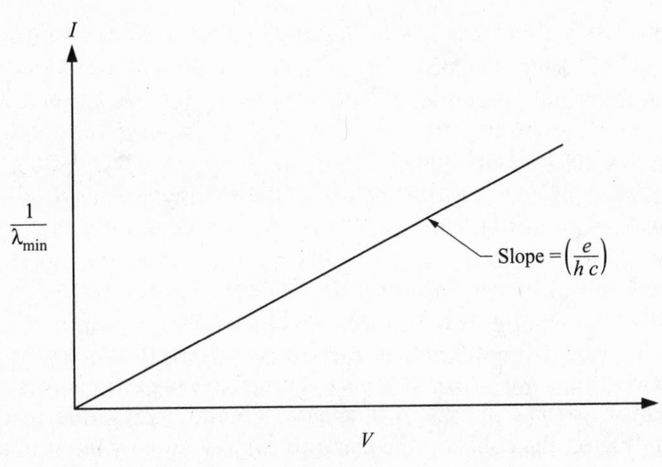

Fig. 17-7

at specific wavelengths that are characteristic of the material bombarded by the electrons. These spikes are due to an effect that we will discuss later, but the spikes all appear at wavelengths longer than λ_{min} for the given voltage. If one reduces the voltage so that λ_{min} gets longer than the wavelength of these characteristic spikes, then they disappear.

> ***Note.*** The plot in Fig. 17-7 appears to go through the origin. This is not quite true because the electrons actually gain a small energy in addition to eV as they enter the surface, namely the work function energy, W, discussed in the photoelectric effect. Thus, we should have $\lambda_{min} = hc/(eV + W)$. Since eV for X-rays is $>10^3$ eV while $W \approx 1$ eV, W is usually negligible compared to eV, and is ignored.

Problem 17.5. Electrons are accelerated through a difference of potential of $V = 10$ kV.

(*a*) What kinetic energy, E_K, do the electrons have as they strike the metal plate?

(*b*) What is the minimum wavelength, λ_{min}, X-ray that is emitted?

Solution

(*a*) $E_K = eV = 10$ keV (in electron volts). Then, 10 keV $= 10 \times 10^3(1.6 \times 10^{-19})$ J $= 1.6 \times 10^{-15}$ J.

(*b*) λ_{min} corresponds to the maximum frequency, which occurs when all the electron's kinetic energy goes to the photon. $\lambda_{min} = (hc/e)/V$, and substituting in the values of h, c and e, we get: $\lambda_{min} = 1.24 \times 10^{-6}$ J · m/s · C)/(10^4 kV) $= 1.24 \times 19^{-10}$ m.

Problem 17.6. If one wishes to produce X-rays with a wavelength of 1.0×10^{-12} m, what minimum difference of potential is needed in the X-ray tube? What kinetic energy does this correspond to in electron volts?

Solution

The X-rays will have a maximum energy of hc/λ. To get this energy, the electrons must have at least this energy, so $eV_{min} = hc/\lambda$. Then $V_{min} = hc/e\lambda = 1.24 \times 10^6$ V $= 1.24$ MV. The electrons must have a minimum kinetic energy of 1.24 MeV.

Compton Scattering

When energetic photons such as X-rays are incident on a material, they must interact with the electrons in the material in a way in which both energy and momentum are conserved. For electrons that are loosely bound to atoms, the collision can be thought of as a free collision between two isolated objects: the photon and the electron, and both momentum and energy are conserved as in an elastic collision between two objects on a frictionless surface. Such an interaction between light and matter is called **Compton scattering**. Here, as we shall see, the photon can scatter at an angle, θ, from the incident direction. The electron flies off as well, receiving some momentum and energy from the photon. To be able to apply the principles of conservation of energy and momentum to the collision between the photon and electron, we need a formula for the momentum of the photon. To develop this formula, we will use two approaches. First, we note that for electromagnetic waves, the average energy transported per unit area/unit time is given by the magnitude of the **Poynting vector**, $P = E_0 B_0/2\mu_0$ discussed in Chap. 12. The momentum transported by the electromagnetic wave per unit area/unit time was shown to equal P/c. If the energy and momentum is carried by identical photons then we would have to conclude that the same relation holds for each photon's energy and momentum, i.e. its momentum is equal to its energy divided by c, or $p = hf/c = h/\lambda$. The direction of p would be the direction in which the photon is traveling. The second approach uses the concept of relativistic mass that we used to get the momentum for electromagnetic waves originally. Since photons travel at speed c they would have infinite energy if they had a non-zero rest mass: $E = \gamma m_0 c^2$, but $\gamma = \infty$ for $v = c$, so $E = \infty$ which is non-physical. We therefore assume $m_0 = 0$. From our energy momentum equations we then have $E^2 = p^2c^2 + m_0{}^2c^4 \to E^2 = p^2c^2 \to E/c = p$, as required. Then $E = hf \to p = E/c = hf/c = h/\lambda$, the same as our earlier result. Note that a photon, while not having rest mass m_0 may still be thought of as having a relativistic mass given by $hf = mc^2$ and momentum given by $p = mc$. Thus for our photon we have:

$$E = hf \qquad \text{and} \qquad p = h/\lambda \tag{17.4}$$

We now analyze the collision between a photon, of wavelength λ, striking an initially stationary electron, as in Fig. 17-8. After the collision, the electron moves away with a velocity v at an angle β from the original direction, and a photon moves away with wavelength λ', at an angle ϕ below that direction. Before the collision the total energy was $hf = hc/\lambda$ for the photon, and $m_0 c^2$ for the electron. Thus $E_i = hc/\lambda + m_0 c^2$. The total energy after the collision is hc/λ' for the photon and mc^2 for the electron, where $m = \gamma m_0$, with m_0 the electron rest mass, $\gamma = (1 - v^2/c^2)^{-1/2}$, and v the velocity of the electron. Equating the energy before and after the collision, we get

$$hc/\lambda + m_0 c^2 = hc/\lambda' + \gamma m_0 c^2 \tag{17.5}$$

Before the collision the momentum was h/λ, all in the x direction. After the collision, the photon has momentum of h/λ' at an angle ϕ, and the electron has momentum $\gamma m_0 v$ at an angle β with the x axis.

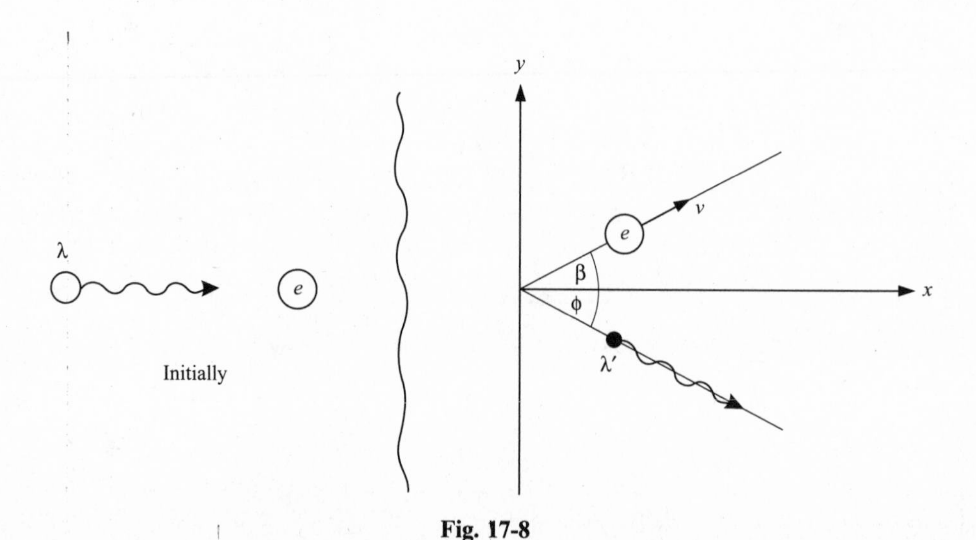

Fig. 17-8

For the x and y directions we then have, using conservation of momentum,

$$h/\lambda = (h/\lambda') \cos \phi + \gamma m_0 v \cos \beta, \text{ in the } x \text{ direction,} \tag{17.6}$$

and

$$0 = -(h/\lambda') \sin \phi + \gamma m_0 v \sin \beta, \text{ in the } y \text{ direction.} \tag{17.7}$$

Eqs. (*17.5–7*) are three equations in four unknowns: ϕ, λ', β, v. For a given photon scattering angle we can solve for the other three unknowns at that angle. The result for λ' is (after some messy algebra)

$$\lambda' = \lambda + (h/m_0 c)(1 - \cos \phi), \tag{17.8}$$

or

$$\lambda' - \lambda = \lambda_c(1 - \cos \phi), \tag{17.9}$$

where $\lambda_c = h/m_0 c = 2.43 \times 10^{-12}$ m, is called the **Compton wavelength**.

This equation tells us that the photon after the collision has a longer wavelength, or less energy than the incident photon. When ϕ is zero, there is no scattering, and $\lambda' = \lambda$. When $\phi = \pi$, $\lambda' - \lambda = 2\lambda_c$, and there is the maximum change in energy. Since $\lambda_c = 2.43 \times 10^{-12}$ m, the change in wavelength will only be noticeable for incident wavelengths shorter than about 10^{-10} m, which is in the X-ray region.

Problem 17.7. An X-ray, of wavelength 1.0×10^{-11} m is Compton scattered at an angle of 85°.

(*a*) What is the wavelength of the scattered X-ray and how much energy has the photon lost?

(*b*) What is the mass of the electron after the collision?

Solution

(*a*) Using Eq. (*17.9*), we get that $\lambda' - \lambda = \lambda_c(1 - \cos \phi)$, or $\lambda' = \lambda + \lambda_c(1 - \cos \phi) = 1.0 \times 10^{-11} + 2.43 \times 10^{-12}(1 - \cos 85°) = 1.22 \times 10^{-11}$ m. The energy of a photon is $hf = hc/\lambda$, so $\Delta E = hc(1/\lambda - 1/\lambda') = 3.61 \times 10^{-15}$ J = 22.6 keV.

(*b*) Since the photon lost 3.61×10^{-15} J of energy, the electron must have gained this amount of kinetic energy. Then its mass must have increased by $\Delta E/c^2 = 4.01 \times 10^{-32}$ kg, and its mass would be $m_0 + \Delta m = 9.11 \times 10^{-31}$ kg $+ 4.01 \times 10^{-32}$ kg $= 9.51 \times 10^{-31}$ kg.

Problem 17.8. An X-ray with a wavelength of 3.0×10^{-13} m is Compton scattered in a metal. The electron has a kinetic energy of 0.90 MeV after the collision. What is the wavelength and direction of the scattered photon?

Solution

The electron gained 0.90 MeV of energy, so the photon lost that amount of energy. Therefore $hc/\lambda' = hc/\lambda - 0.90$ MeV. Solving for λ', we get $\lambda' = 3.83 \times 10^{-13}$ m. Using Eq. (*17.9*), we get that $\lambda' - \lambda = \lambda_c(1 - \cos \phi)$, or $(\lambda' - \lambda)/\lambda_c = 1 - \cos \phi$, which gives $\phi = 15°$.

When one experimentally observes Compton scattered photons at any angle, one finds, in addition to the longer wavelength photons, some photons at the original wavelength. This corresponds to the case where the photon scatters off an electron that stays bound in the atom in the collision process. This is equivalent to the photon elastically scattering off the whole atom. Eqs. (*17.8*) and (*17.9*) still hold but the mass m_0 is now the rest mass of the atom as a whole, which is thousands of times larger than that of the electron. Thus $\lambda - \lambda'$ is negligible compared to λ for almost all X-rays.

17.4 MATTER WAVES

In the previous two sections we have shown that experiments indicate that electromagnetic waves exhibit both wavelike and particle like behavior. It is therefore not unreasonable to wonder whether particles like electrons, which clearly exhibit particle properties of localization (not being spread over a large region of space) and of quantization (charge, rest mass), also exhibit wavelike properties of interference and diffraction. DeBroglie speculated as to the wavelike properties one might assign to particles, and in particular, what wavelength and frequency should a particle have, if it has a rest mass m_0 and is moving with a velocity v? To conform with the case of electromagnetic waves, DeBroglie hypothesized that the frequency and wavelength should be determined by the same basic relations used for photons. Thus he defined the wavelength λ_{DB} (now known as the **DeBroglie wavelength**) and the frequency f_{DB} by:

$$\lambda_{DB} = h/p, \tag{17.10}$$

where $p = mv$ with $m = \gamma m_0$, and

$$f_{DB} = E/h, \text{ where } E = mc^2 \text{ is the total energy of the electron} \tag{17.11}$$

These are the same relationships between wavelength and momentum, and between frequency and energy that exist for the photon. However, it is definitely *not true* that $\lambda_{\Delta B} f_{DB} = c$ nor the particle velocity v. In this regard, these "**matter waves**" do not behave like electromagnetic waves. They are a new type of "matter" wave that we will discuss further in the next sections.

If DeBroglie's speculation is true, then the electrons represented by these waves should exhibit interference and diffraction appropriate to the wavelengths associated with the electron. Let us determine the magnitude of these wavelengths for typical electrons.

Problem 17.9. Determine the DeBroglie wavelength of electrons with a velocity of (*a*) 1.0×10^3 m/s; (*b*) 1.0×10^6 m/s; (*c*) 1.0×10^8 m/s; (*d*) $0.99c$.

Solution

(*a*) The wavelength is given by $\lambda = h/p = h/mv = h/\gamma m_0 v$. We must therefore calculate the momentum of this electron before we can calculate its wavelength, using relativity when necessary. We do not have to use relativity as long as the velocity is much smaller than c, which we can take as $v \ll 0.1c$. This is clearly the case for $v = 1.0 \times 10^3$ m/s. Here, $p = 9.1 \times 10^{-31}(1.0 \times 10^3) = 9.1 \times 10^{-28}$, and $\lambda = 6.625 \times 10^{-34}/9.1 \times 10^{-28} = 7.28 \times 10^{-7}$ m $= 0.728$ μm.

(*b*) Again, $v = 1.0 \times 10^6 \ll c$, so $p = 9.1 \times 10^{-31}(1.0 \times 10^6) = 9.1 \times 10^{-25}$, and $\lambda = 6.625 \times 10^{-34}/9.1 \times 10^{-25} = 7.28 \times 10^{-10}$ m $= 0.728$ nm.

(*c*) With $v = 1.0 \times 10^8$ m/s, we must use relativity. Remembering that $\gamma = 1/(1 - v^2/c^2)^{1/2}$, we find that $\gamma = 1/(1 - 0.111)^{1/2} = 1.06$. Then $p = 1.06(9.1 \times 10^{-31})(1.0 \times 10^8) = 9.65 \times 10^{-23}$, and $\lambda = 6.625 \times 10^{-34}/9.65 \times 10^{-23} = 6.86 \times 10^{-12}$ m.

(*d*) Here, $\gamma = 1/[1 - (0.99)^2]^{1/2} = 7.09$, $p = 1.92 \times 10^{-21}$, and $\lambda = 3.46 \times 10^{-13}$ m.

Problem 17.10. Determine the DeBroglie wavelength of electrons with kinetic energies of (*a*) 10 eV; (*b*) 1.0 keV; (*c*) 1.0 MeV.

Solution

The wavelength is given by $\lambda = h/p = h/mv = h/\gamma m_0 v$. Obtaining v is easy if we do not have to use relativity. We initially assume that relativity is not needed, calculate v, and if $v \ll c$, we accept the answer. If v is close to c, we will recalculate using relativity. In general we will find that if the kinetic energy is much less than the rest mass energy, which is 0.511 MeV, then relativity is not needed.

(*a*) For $E_K = 10$ eV $= (1/2)mv^2 = 1.6 \times 10^{-18}$ J, we get $v^2 = 3.52 \times 10^{12}$, and $v = 1.88 \times 10^6$ m/s. Thus $v \ll c$, $p = m_0 v = 1.71 \times 10^{-24}$, $\lambda = h/p = 3.88 \times 10^{-10}$ m.

(*b*) For $E_K = 1.0 \times 10^3$ eV, we get $v^2 = 3.52 \times 10^{14}$, and $v = 1.88 \times 10^7$ m/s, which is still small, compared to c. $p = m_0 v = (9.1 \times 10^{-31})(1.88 \times 10^7) = 1.71 \times 10^{-23}$. Finally, $\lambda = 3.88 \times 10^{-11}$ m.

(*c*) For $E_K = 1.0 \times 10^6$ eV, we certainly will have to use relativity, since $E_K > m_0 c^2$. Then, $E_K =$

$(m - m_0)c^2 = (\gamma - 1)m_0 c^2 = (\gamma - 1)(0.511 \times 10^6) = 1.0 \times 10^6$, giving $\gamma = 1 + 1.0 \times 10^6/0.511 \times 10^6 = 2.957$, and $v = 0.94c = 2.82 \times 10^8$. Then, $p = \gamma m_0 v = 7.60 \times 10^{-22}$, and $\lambda = 8.72 \times 10^{-13}$ m.

We see that the wavelengths are of the order of 10^{-10} m for electrons with kinetic energies of 10–1000 eV. This is just the size of the spacing of atoms in solids, and the scattering of waves of these wavelengths from solids should give diffraction effects, if the electrons really have these wavelike properties. When experiments were performed using crystals as diffraction gratings, this diffraction was indeed seen. Similar predictions for the wavelengths of more massive particles, such as protons and neutrons, have also been experimentally verified.

These "electron waves" can be used to develop microscopes and other instruments that mimic their optical counterparts, if one can develop the appropriate "lenses" and detectors. The electron microscope is, by now, a well-developed instrument for making measurements of very small dimensions. As we learned in Chap. 15, diffraction effects limit the smallest sizes that can be resolved in microscopes. Generally, one cannot measure sizes smaller than the wavelength of the waves being used. With electron waves we can get very small wavelengths, and can measure down to the atomic level.

Problem 17.11. An electron microscope is to be used to measure atoms whose dimensions are 3.5×10^{-10} m. For accuracy, one must use waves with wavelengths ten times smaller than the sizes being measured. What is the minimum energy that the electrons need in order to be usable in this application?

Solution

The maximum wavelength we can use is $\lambda = 3.5 \times 10^{-11}$ m. Since the wavelength is given by $\lambda = h/p$, we need a momentum of $p = h/\lambda = 6.625 \times 10^{-34}/3.5 \times 10^{-11} = 1.89 \times 10^{-23}$. Without using relativity, this gives $v = p/m_0 = 2.08 \times 10^7$ m/s which is less than $0.1c$ and justifies not using relativity. Then the kinetic energy is $(1/2)m_0 v^2 = 1.97 \times 10^{-16}$ J $= 1.23$ keV. This is the minimum energy required for the electrons in the microscope.

Problem 17.12. A photon and an electron each have an energy of 1300 eV. What is the wavelength of each of them?

Solution

For both a photon and an electron, the wavelength is equal to h/p. However, p is not the same for each of them. For a photon, $p = h/\lambda = hf/c = E/c$, or $\lambda = hc/E$. Thus, $\lambda_p = 6.625 \times 10^{-34}(3.0 \times 10^8)/(1300)(1.6 \times 10^{-19}) = 9.56 \times 10^{-10}$ m.

For an electron, we must first calculate v. Assuming that relativity is not required, we can use $(1/2)mv^2 = 1300(1.6 \times 10^{-19})$, giving $v = 2.14 \times 10^7$ m/s, which is much less than c and justifies not using relativity. Then $p = m_0 v = 9.1 \times 10^{-31}(2.14 \times 10^7) = 1.95 \times 10^{-23}$, and $\lambda_e = 3.41 \times 10^{-11}$ m.

17.5 PROBABILITY AND UNCERTAINTY

We have learned in the previous sections, that both electromagnetic waves and particles have properties that resemble waves and properties that resemble particles. This dilemma led to the development of quantum mechanics, and the development of concepts far different from those previously held.

To understand the fundamental concept introduced by quantum mechanics, let us once again turn to the case of a double slit, with normally incident plane wave light passing through the slits and forming an interference pattern on a distant screen. The screen intensity pattern is shown in Fig. 17-9(*a*). Let us try to understand what is seen on the screen if we assume that light consists of particles of photons. As one moves along the screen away from the center ($y = 0$), one sees a series of fringes, as shown in Fig. 17-9(*b*). On a photon picture what actually happens is that many photons pass through the slits, and distribute themselves on the screen, forming the pattern seen. Where there is a bright

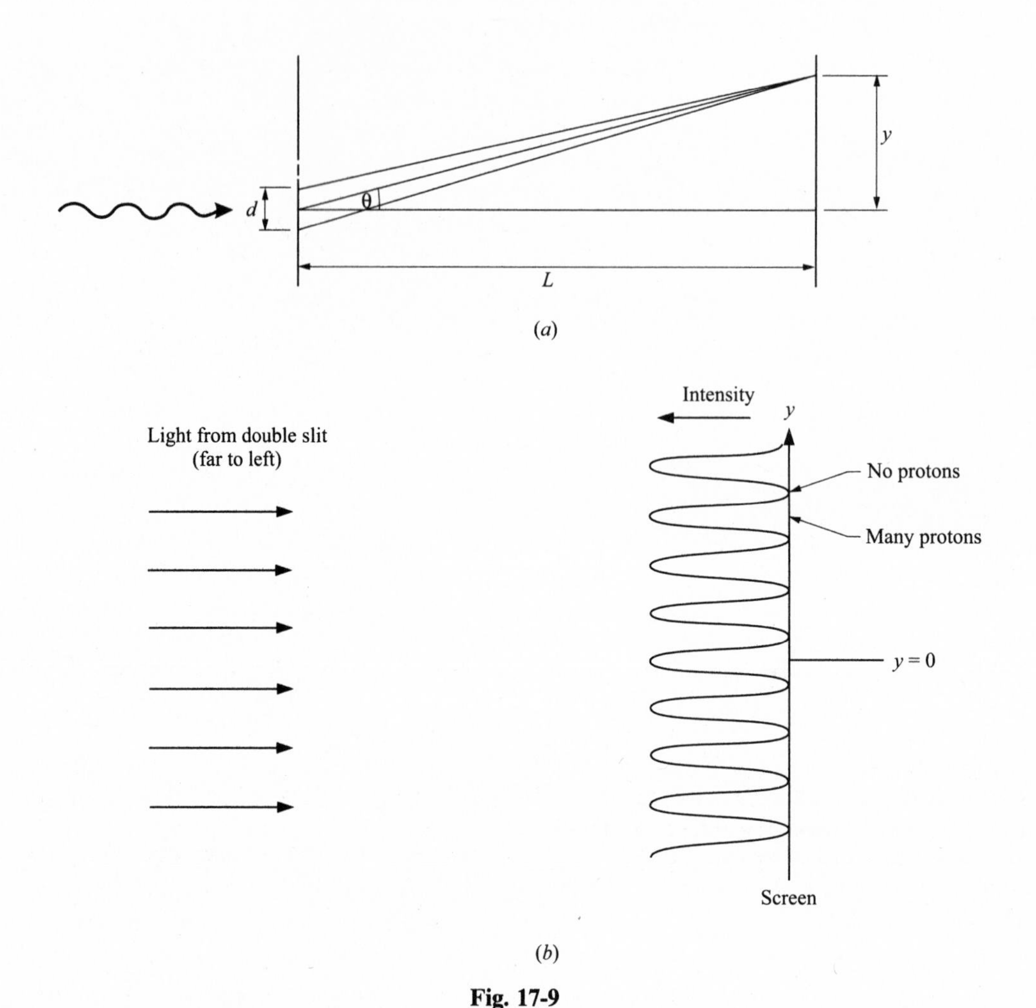

Fig. 17-9

fringe, many photons have arrived, and where there is no light , no photons have arrived. The pattern actually gives one the distribution of the numbers of photons on the screen. Any one photon arrives at only one spot on the screen, and the pattern that one sees is the average of very large numbers of photons. If one were to reduce the intensity of the light so that only a few photons are passing through the slits at any time, then the individual arrival of the photons on the screen can actually be detected, thus clearly demonstrating the particle nature of light.

Problem 17.13. Sunlight at the earth's surface has an intensity of approximately $I = 1.4\ \text{kW/m}^2$. If the mean wavelength of the light is $\lambda = 500$ nm, how many photons pass through a square meter per second?

Solution

The energy of a photon is $hf = hc/\lambda$. If the energy passing one square meter per second (the intensity) is I, then the number of photons passing that square meter per second is $I/hf = (1.4 \times 10^3\ \text{W/m}^2)(500 \times 10^{-9}\ \text{m})/[(6.625 \times 10^{-34}\ \text{J} \cdot \text{s})(3.0 \times 10^8\ \text{m/s}) = 3.52 \times 10^{21}\ \text{photons/m}^2 \cdot \text{s}$.

Since the intensity pattern is actually a pattern of the distribution of photons, the pattern must reflect the probability that a photon reaches a certain point on the screen when both slits are open. One cannot predict where any individual photon will go, but one can predict with great certainty what the pattern will be after large numbers of photons have contributed to the pattern. This concept, that the only thing that we can predict is the probability of a photon's location, is a basic concept of quantum mechanics. It is manifested in the "uncertainty principle" that we will discuss later on.

Pursuing our analysis of the two slit fringe distribution, our photon model requires that the intensity of light distribution reflect an underlying probability distribution for the photons. As we learned in Chap. 12, the intensity is proportional to the square of the "displacement" of the electromagnetic wave, its electric field. For a double slit, Maxwell's equations predict an electric field distribution corresponding to two interfering electromagnetic waves. The square of this electric field distribution is proportional to the intensity of the wave at any point, and thus is proportional to the probability that a photon arrives at any point in space. The electric field itself is thus proportional to a "**probability amplitude**" the square of which gives the probability. Furthermore this probability amplitude exhibits the properties of interference that are normally associated with waves.

Thus in the case of a photon, **Maxwell's wave equation** gives the probability amplitude for finding a photon. To apply this idea to electrons, one needs a different wave equation that will predict the probability amplitude distribution of electrons. Such an equation was developed by Schroedinger, and is known as **Schroedinger's equation**, for a non-relativistic electron. For a relativistic electron, the appropriate equation is called **Dirac's equation**. The exact form of the equation does not concern us at this stage, since we need more advanced mathematics to make use of the details of the equations. The central idea however is that there is *some* wave equation which has to be solved to predict the behavior of a particle. When this equation is solved for a particular case, the resulting wave (called a wave function) gives an amplitude which, when squared, is proportional to the probability distribution of the particles. This is the identical concept that we used for photons. Particles have to be dealt with in the same manner as electromagnetic waves, except that the wave equation is different.

We are now faced with the following question. In the introduction we stated that one of the characteristics of a particle is that it can be localized. This means that at a given instant we can consider it to be at a certain point in space. A wave, on the other hand is necessarily spread out over a large area. How can we reconcile this? We have actually come across "localized" waves before. When we give a long rope a single snap the pulse is fairly well localized over some spread Δx at any given instant. The same can be true of a "pulse" of electromagnetic radiation. It can be shown that such pulses can be obtained by superposing a large number of regular traveling waves of different wavelengths. Such a pulse is called a "wave packet". In Chap. 2 we discussed the "beats" that are produced when just two waves of different but very nearby wavelengths (λ_1, λ_2 with $\Delta\lambda = |\lambda_1 - \lambda_2| \ll \lambda_1$ or λ_2) are combined. The result was a wave of almost the same wavelength as the original waves but with a slowly varying amplitude of frequency $\Delta f = f_1 - f_2$. Hence, the rapidly varying wave λ has superposed a longer wavelength wave corresponding to the frequency Δf. The result was a traveling wave consisting of a series of wave packets like the one in Fig. 17-10, one following the other. If one adds many more waves with wavelengths between λ_1 and λ_2, one can eliminate the wave everywhere except in the region of one packet. Such a packet traveling through space indeed resembles a localized particle. The mathematical

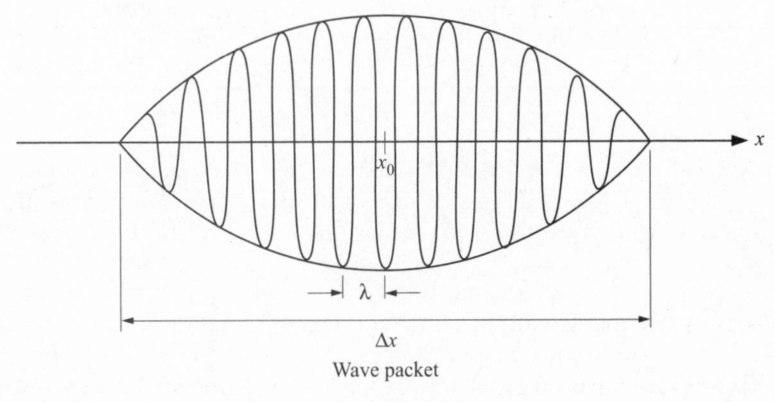

Wave packet

Fig. 17-10

subject of Fourier analysis addresses the relationship between the spatial size Δx of such a wave packet and the range of wavelengths $\Delta\lambda$ necessary to form it. The length of a packet, Δx, is large when the many wavelengths that are added together are all in a close range $\Delta\lambda$ of each other, and Δx gets smaller as the wavelengths involved are spread over a broader range $\Delta\lambda$. Thus, to build a packet with a small Δx requires a large spread in $\Delta\lambda$, whereas a packet with large Δx can be constructed from nearby wavelengths, and $\Delta\lambda$ is small. Since the momentum of a particle is given by h/λ, a large range of λ implies a large range in the momentum, Δp. Thus to get a small Δx, we need a large Δp. Using Fourier analysis, we can show that for motion along the x axis, Δx and Δp are related by:

$$\Delta x\ \Delta p \geq h/2\pi = \hbar, \qquad (\hbar = \text{“}\hbar\text{ bar”} = h/2\pi) \tag{17.12}$$

This is known as the **uncertainty principle**. It states that if one wants to describe a particle (at a given time) as being localized in a region Δx (i.e. have a spatial uncertainty Δx), then that particle must have an uncertainty in its x direction momentum, Δp, which is at least as large as $\hbar/\Delta x$. Similarly, if one wants the x direction momentum to be known to within Δp, then there must be an uncertainty in the position of the particle which is at least as large as $\Delta x = \hbar/\Delta p$. Of course, if p is uncertain, so is v and the kinetic energy. This has important consequences which we shall now explore.

Problem 17.14. A proton has a mass of 1.67×10^{-27} kg and is as close to motionless as possible. What minimum uncertainty in its momentum and in its kinetic energy must one have, if one confines the proton to a region of (*a*) 1.0 mm; (*b*) an atom of length 5.0×10^{-10} m; and (*c*) a nucleus of length 5.0×10^{-15} m?

Solution

(*a*) The uncertainty principle requires that $\Delta p \geq \hbar/\Delta x$. Thus $\Delta p \geq [(6.625 \times 10^{-34}\ \text{J}\cdot\text{s})/2\pi)]/(1.0 \times 10^{-3}\ \text{m}) = 1.05 \times 10^{-31}\ \text{kg}\cdot\text{m/s}$. Now $\Delta p = m_0\,\Delta v$, as long as relativity can be ignored. Then $\Delta v \geq \hbar/m_0\,\Delta x = (6.625 \times 10^{-34}/2\pi)/[1.67 \times 10^{-27}(1.0 \times 10^{-3})] = 6.31 \times 10^{-5}$ m/s. Since the only reason the proton is not motionless is the uncertainty principle, $\Delta v \approx (v - 0) = v$ is the minimum bound on the uncertainty in the proton's velocity. The minimum uncertainty in its kinetic energy is then $(1/2)m_0 v^2 = 0.5(1.67 \times 10^{-27})(6.31 \times 10^{-5})^2 = 3.32 \times 10^{-36}\ \text{J} = 2.08 \times 10^{-17}$ eV.

(*b*) Using $\Delta x = 5.0 \times 10^{-10}$ in the same equations as in (*a*) we get $\Delta p = 2.11 \times 10^{-25}$ and $\Delta v = 126$ m/s, $E_K = 1.33 \times 10^{-23}\ \text{J} = 8.32 \times 10^{-5}$ eV.

(*c*) Using $\Delta x = 5 \times 10^{-15}$ in the same equations as in (*a*), we get, $\Delta p = 2.11 \times 10^{-20}$, $\Delta v = 1.26 \times 10^{7}$ m/s, $E_K = 1.33 \times 10^{-13}\ \text{J} = 8.32 \times 10^{5}$ eV. This proton is still not relativistic (the rest mass energy is 938 MeV), but its kinetic energy is significant, nearly 1 MeV.

We see that if we are considering a particle, such as a proton, which we know is confined to a small region, then quantum mechanics requires that such a particle cannot be described as having a precise momentum, even momentum zero, but rather as having a range of possible momenta. This means that particles of matter, even if the temperature is at absolute zero where there is no thermal energy, must still have an average kinetic energy related to this range of momenta. This kinetic energy at a temperature of absolute zero, is called the "**zero point energy**", and there is no way to avoid having this minimum amount of energy.

A manifestation of this idea is the fact that electrons cannot be confined in a nucleus (10^{-14} m). Since the mass of an electron is so small, the zero point kinetic energy $= (\Delta p)^2/2m_e$ is much larger than the Coulomb binding energy of an electron with the particles in a nucleus. Even confining an electron to an atom, whose size is $\approx 10^{-10}$ m, requires a zero point kinetic energy of ≈ 4 eV. This is comparable to the potential energy of an electron at this distance from a proton, and is consistent with the kinetic energy of an electron moving in a circle of this radius about the proton.

Problem 17.15. Consider an electron that is at a distance of 1.0×10^{-10} m from a proton.

(*a*) What is the zero point energy of an electron confined within this distance?

(*b*) What is the electric potential energy between the electron and proton?

(*c*) If the electron is moving in a circle about the proton at this radius, what is its kinetic energy?

(*d*) Is it possible to have an uncertainty in momentum but an exact energy?

Solution

(*a*) To get the zero point energy we use the uncertainty principle, $\Delta x\, \Delta p \geq \hbar$. Then $\Delta p = (6.625 \times 10^{-34}/2\pi)/1.0 \times 10^{-10} = 1.05 \times 10^{-24}$ kg · m/s, $\Delta v = \Delta p/m_0 = 1.05 \times 10^{-24}/9.1 \times 10^{-31} = 1.15 \times 10^6$ m/s, $E_K = (\frac{1}{2})m_0 v^2 = 6.02 \times 10^{-19}$ J $= 3.76$ eV (using $v \approx \Delta v$).

(*b*) The electrical potential energy is $[1/(4\pi\varepsilon_0)qq'/r = -9 \times 10^9\ (1.6 \times 10^{-19})^2/1.0 \times 10^{-10} = (-2.30) \times 10^{-18}$ J $= -14.4$ eV.

(*c*) If the electron is moving in a circle around the proton, then the centripetal force must equal the electrical force between the electron and proton, or, $mv^2/r = (\frac{1}{4}\pi\varepsilon_0)qq'/r^2$, or $mv^2 = [1/4\pi\varepsilon_0]qq'/r = 2.3 \times 10^{-18}$ J from part (*b*). Then $E_K = (\frac{1}{2})mv^2 = 1.15 \times 10^{-18}$ J $= 7.20$ eV.

(*d*) Yes. For the case of part (c) the momentum is still uncertain because the velocity in the x direction still varies from $+v \rightarrow -v$, having a Δv and $m\Delta v$ uncertainty even if E_K is definite.

So far we have assumed that the particle or photon is confined in the x direction. Actually, the particle or photon is in three dimensional space and it is easy to generalize the uncertainty principle and to include these other dimensions as well. The result is the following set of equations.

$$\Delta x\, \Delta p_x \geq h/2\pi = \hbar,$$
$$\Delta y\, \Delta p_y \geq h/2\pi = \hbar, \qquad (17.13)$$

and

$$\Delta z\, \Delta p_z \geq h/2\pi = \hbar$$

It can be shown that there is an uncertainty principle linking time and energy as well:

$$\Delta t\, \Delta E \geq h/2\pi = \hbar \qquad (17.14)$$

This latter equation means that if one knows for sure that an event occurred within a limited time interval Δt, then there must be an uncertainty in the energy involved of at least $\hbar\Delta t$. For instance, if a photon is created by some interaction within a short time Δt, then the energy of this created photon is uncertain by the amount $\Delta E \geq \hbar/\Delta t$, and consequently, the frequency is uncertain by the amount $\Delta f = \Delta E/h \geq 1/(2\pi\Delta t)$ or $\Delta f\, \Delta t \geq 1/2\pi$.

Problem 17.16. A photon is emitted in a process characterized by a time interval T. This means that the photon is most likely to be emitted some unpredictable instant within that time interval. What is the uncertainty in the energy and in the frequency of the emitted photon if T equals (*a*) 1.0 millisecond (ms $= 10^{-3}$ s); (*b*) 1.0 microsecond (μs $= 10^{-6}$ s); (*c*) 1.0 nanosecond (ns $= 10^{-9}$ s); (*d*) 1.0 picosecond (ps $= 10^{-12}$ s); (*e*) 1.0 femtosecond (fs $= 10^{-15}$ s); (*f*) 1.0×10^{-20} s?

Solution

(*a*) We use the uncertainty principle for time, $\Delta E = \hbar\Delta t$, and $\Delta f = 1/(2\pi\Delta t)$. For Δt we substitute T, and get $\Delta E = 6.625 \times 10^{-31}$ J, $\Delta f = 159$ Hz.

(*b*) Using $\Delta t = 1.0 \times 10^{-6}$, we get $\Delta E = 1.055 \times 10^{-28}$ J, $\Delta f = 1.59 \times 10^5$ Hz.

(*c*) Using $\Delta t = 1.0 \times 10^{-9}$, we get $\Delta E = 1.055 \times 10^{-25}$ J, $\Delta f = 1.59 \times 10^8$ Hz $= 159$ Mhz.

(*d*) Using $\Delta t = 1.0 \times 10^{-12}$, we get $\Delta E = 1.055 \times 10^{-22}$ J, $\Delta f = 1.59 \times 10^{11}$ Hz $= 159$ Ghz.

(*e*) Using $\Delta t = 1.0 \times 10^{-15}$, we get $\Delta E = 1.055 \times 10^{-19}$ J $= 0.659$ eV, $\Delta f = 1.59 \times 10^{14}$ Hz.

(*f*) Using $\Delta t = 1.0 \times 10^{-20}$, we get $\Delta E = 1.055 \times 10^{-14}$ J = 65.9 keV, $\Delta f = 1.59 \times 10^{19}$ Hz.

This uncertainty in energy also means that for short intervals of time there may be an apparent violation of conservation of energy. For instance, a particle may be trapped inside an enclosure requiring a certain energy to get out, as in Fig. 17-11. In this schematic diagram, the particle is at an energy above zero, inside the equivalent of a well, requiring an activation energy E_a to climb out of the well. Once it gets out it will be able to descend to lower energy, ending up with less potential, and more kinetic energy at the bottom. Without the uncertainty principle of quantum mechanics, the thermal energy is unlikely to enable the particle to surmount the barrier and leave the well. But quantum mechanics allows for the possibility that, for the short time needed to travel from the inside to the outside of the well, the energy uncertainty will be big enough to surmount the barrier and it will escape from the well. At the end there will, of course, be conservation of energy. This process is called tunneling, since the particle appears to have dug a hole through the wall of the well and emerged on the outside. This process actually occurs in the radioactive decay of nuclei, and in certain semiconductor devices called tunnel diodes, as well as in many other types of experimental situations.

The uncertainty principle can also be used to give an intuitive explanation for the diffraction pattern from a narrow slit. Let us assume that particles with mass m and a velocity v impinge along the x axis on a slit of width w as in Fig. 17–12. As the particles pass through the slit their position along the y axis is constrained to be within the width of the slit, or $\Delta y = w$. This means that we cannot assume that the particles are only moving in the x direction, since this constraint on y implies an uncertainty in p_y. Thus the particles will be moving in the y direction as well as in the x direction after passing through the slit, with a velocity v_y that is of order $\Delta v_y = (1/\mathrm{m})\Delta p_y \approx (1/m)(\Delta y) = \hbar/mw$. Particles with small v_y will reach a screen at a distance L from the slit at points near the center of the pattern, while particles with larger v_y will be further away from this central line. Thus the diffraction is spread on the screen. This also explains why a small slit causes more spreading, since then Δy is smaller and there must be particles with larger v_y due to the uncertainty principle.

Our discussion of the ideas of quantum mechanics is far from complete. We have discussed the fact that particles of matter and waves of light actually do not act as either pure particles or pure waves. They are "quantized" as one would expect of particles, but their motion and location is determined by a probability wave, a "**wave function**", that can only predict the probability distribution for finding the particle or quantum. This probability wave is the solution of some specific wave equation, and exhibits all the interference and diffraction effects normally associated with waves. The actual probability is proportional to the square of the waves. This formulation gives rise to the uncertainty principle, limiting our ability to simultaneously know precisely the position and the momentum of a quantum, and having other consequences in our understanding of physical law. The formulation of the theory has succeeded

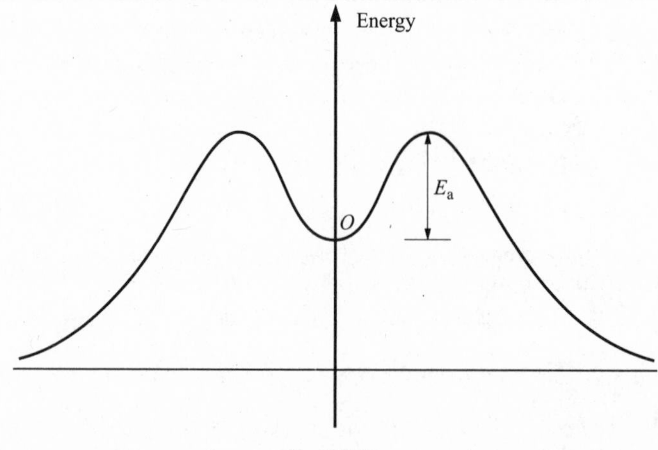

Fig. 17-11

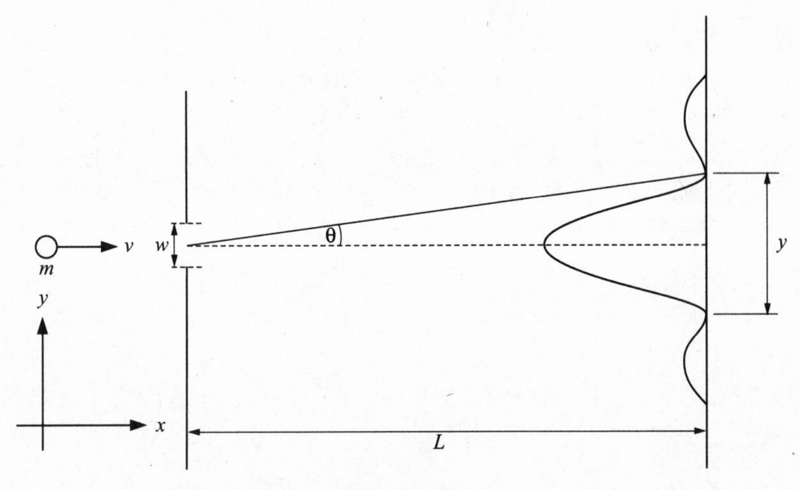

Fig. 17-12

in giving us an understanding of atomic, molecular and solid-state phenomena to a very high degree of accuracy. In the next chapter we will develop some of these applications, and explain our present understanding of these areas.

Problems for Review and Mind Stretching

Problem 17.17. Light impinges on a metal which has a work function of 2.0 eV, and electrons are emitted with velocities up to a maximum of 6.0×10^6 m/s.

(*a*) What is the frequency of the light?

(*b*) What is the minimum frequency of light needed to produce *any* photoelectrons?

Solution

(*a*) We use Eq. (*17.3*), $(E_K)_{max} = hf - W$. Here, $(E_K)_{max} = (1/2)m_0 v_{max}^2 = 0.5(9.1 \times 10^{-31})(6.0 \times 10^6)^2 = 1.64 \times 10^{-17}$ J = 102 eV, and W = 2.0 eV. Then $hf = 104$ eV $= 1.67 \times 10^{-17}$ J, and $f = 2.52 \times 10^{16}$ Hz.

(*b*) The minimum frequency needed to produce any photoelectron is one that is just large enough to get the electron out of the metal with no kinetic energy, so that $hf = W$. Thus $f = 2(1.6 \times 10^{-19})/6.625 \times 10^{-34} = 4.83 \times 10^{14}$ Hz.

Problem 17.18. Electrons are moving with a velocity of 4.0×10^6 m/s.

(*a*) What is the DeBroglie wavelength of these electrons?

(*b*) If these electrons had started from rest, through what potential difference would they have had to be accelerated to attain their velocity?

Solution

(*a*) The wavelength is given by $\lambda = h/mv = 6.625 \times 10^{-34}/(9.1 \times 10^{-31})(4.0 \times 10^6) = 1.82 \times 10^{-10}$ m.

(*b*) The kinetic energy of these electrons is $(\frac{1}{2})(9.1 \times 10^{-31})(4.0 \times 10^6)^2 = 7.28 \times 10^{-18}$ J = 45.5 eV. Then these electrons must have been accelerated through a potential difference of 45.5 V.

Problem 17.19. Referring to Problem 17.18:

(*a*) If the electrons are used to produce X-rays, what is the shortest wavelength they could produce? (Ignore work function effects.)

(*b*) If the electrons are known to be located within a region of 6.6×10^{-5} m, what is the minimum uncertainty in their velocity?

Solution

(*a*) The minimum wavelength X-ray has the maximum energy, which equals all the kinetic energy of the electron. Thus $hf_{max} = hc/\lambda_{min} = 7.28 \times 10^{-18}$, or $\lambda_{min} = 6.625 \times 10^{-34}(3.0 \times 10^8)/7.28 \times 10^{-18} = 2.73 \times 10^{-8}$ m.

(*b*) Using the uncertainty principle, $\Delta x\, \Delta p = \Delta x\, \Delta(mv) \geq \hbar$, we get $\Delta v = \hbar/m\Delta x = 1.76$ m/s.

Problem 17.20. We want to use X-rays which have a wavelength of 2.1×10^{-11} m. Will we be able to obtain such X-rays from a machine in which the electrons are accelerated through a difference of potential of 50 kV?

Solution

If the machine accelerates electrons through a difference of potential of 50 keV, then the minimum wavelength that can be produced is one that has an energy of 50 keV. Thus, $\lambda_{min} = hc/E = (6.625 \times 10^{-34})(3.0 \times 10^8)/(50 \times 10^3)(1.6 \times 10^{-19}) = 2.48 \times 10^{-11}$ m. Since this is bigger than 2.1×10^{-11} m, the machine will be unable to produce such X-rays.

Problem 17.21. An X-ray, of wavelength 3.0×10^{-10} m, is Compton scattered from a free electron initially at rest. The difference in wavelength between the incident and the scattered photon is 2.1×10^{-12} m.

(*a*) Which photon has the longer wavelength?

(*b*) Through what angle is the photon scattered?

(*c*) What is the kinetic energy of the scattered electron? What is its velocity?

(*d*) What is the direction of the velocity of the electron?

Solution

(*a*) Since the photon loses energy, the scattered photon has less energy than the incident photon. Therefore the scattered photon has the longer wavelength.

(*b*) Using Eq. (*17.9*), $\lambda' - \lambda = \lambda_c\,(1 - \cos\phi)$, we get $2.1 \times 10^{-12} = 2.43 \times 10^{-12}\,(1 - \cos\phi)$, or $1 - \cos\phi = 0.864$, $\cos\phi = 0.136$, $\phi = 82.2°$.

(*c*) The electron gains kinetic energy equal to the loss of energy of the photon. The energy of the photon is hc/λ for the incident photon, and hc/λ' for the scattered photon. The energy lost is therefore $\Delta E = hc(1/\lambda - 1/\lambda') = (hc/\lambda\lambda')(\lambda' - \lambda) = 6.625 \times 10^{-34}(3.0 \times 10^8)(2.1 \times 10^{-12})/(3.0 \times 10^{-10})^2 = 4.64 \times 10^{-18}$ J $= 29.0$ eV. This is the kinetic energy of the electron. Since this energy is much less than the rest mass energy of the electron, the electron is non-relativistic. Then it has a velocity given by $E_K = (1/2)m_0 v^2$. Then $v = 3.19 \times 10^6$ m/s.

(*d*) We use conservation of momentum in the direction perpendicular to the initial path of the photon. In that direction there was no momentum before the scattering, and therefore the momentum of the electron and photon after the scattering must cancel in this direction. Therefore the electron must have a component of momentum in this direction equal to that of the photon. The photon has momentum of $(h/\lambda')\sin\phi$ in this direction, and the electron has $mv \sin\beta$ in this direction. Thus, $(6.625 \times 10^{-34})(\sin 82.2)/(2.979 \times 10^{-10}) = (9.1 \times 10^{-31})(3.19 \times 10^6)\sin\beta$. Solving, we get $\sin\beta = 0.076$, $\beta = 4.9°$.

Problem 17.22. A crystal serves as a diffraction grating with a spacing of 2.3×10^{-10} m. Particles are diffracted from the grating and the first diffraction maximum occurs at an angle of 36°.

(*a*) What is the DeBroglie wavelength of the particles?

(*b*) If the particles are electrons, what energy do they have? What velocity?

(*c*) If the particles are neutrons, what energy do they have? ($M_n \approx 1840m_e$)

Solution

(*a*) In a diffraction grating the maxima occur at angles for which $d \sin \phi = n\lambda$. For the first maximum we get $d \sin \phi = \lambda = (2.3 \times 10^{-10})(\sin 36°) = 1.35 \times 10^{-10}$ m.

(*b*) The DeBroglie wavelength of a particle is related to its momentum by $p = h/\lambda$. Thus the momentum of the particles is $p = (6.625 \times 10^{-34})/(1.35 \times 10^{-10}) = 4.90 \times 10^{-24}$ kg · m/s. For an electron this means that the velocity is $v = p/m = 4.90 \times 10^{-24}/9.1 \times 10^{-31} = 5.39 \times 10^6$ m/s, which is non-relativistic. Then the energy is $E_K = (1/2)m_0 v^2 = 1.32 \times 10^{-17}$ J = 82.5 eV.

(*c*) For neutrons, the momentum would also be 4.90×10^{-24}. Noting $E_K = p^2/2m$, we get $E_K = 7.18 \times 10^{-21}$ J = 0.045 eV.

Problem 17.23. A simple harmonic oscillator consists of a mass, m, attached to a spring with a force constant k, and can oscillate at a frequency of f, given by $f = [1/(2\pi)]\sqrt{k/m}$. Use the uncertainty principle to show that it must have some energy even at a temperature of 0 K, and estimate the minimum amount of energy that it must have.

Solution

The energy of a simple harmonic oscillator is given by $E = (\frac{1}{2})kx^2 + (\frac{1}{2})mv^2$. For this energy to be zero, we would have to have both x and v equal to zero. But we know from the uncertainty principle that if x is very small then v cannot be restricted to a fixed value, certainly not to zero. Alternatively if v is known to be very small then we cannot restrict x to be zero. Therefore we must have energy in this oscillator, either from the potential energy due to a non-zero x or from the kinetic energy due to a non-zero v.

To estimate the minimum energy that we must allow for the oscillator, we use the uncertainty principle, $\Delta x\, \Delta p = \Delta x(m\Delta v) \geq \hbar$. Assume that we know Δx. Then $\Delta v \geq \hbar/m\Delta x$, and the energy can be written as $E \geq (1/2)[k(\Delta x)^2 + m(\hbar/m\Delta x)^2]$. We want to find the minimum value for this energy. If Δx is small, then the first term is small but the second term is large. If Δx is large, then the first term is large and the second term is small. Clearly we must choose Δx somewhere in between, perhaps at the value which will result in equal contributions from each term. (In fact, we can show mathematically that the minimum occurs at just that value.) If we choose to equalize the two terms, then $k(\Delta x)^2 = m(\hbar/m\Delta x)^2$, or $(\Delta x)^4 = (\hbar/m\omega)^2$, where $\omega^2 = \sqrt{k/m}$. Then $(\Delta x)^2 = \hbar/m\omega$, and $E_{min} = (1/2)[k\hbar/m\omega](2)$, since the two terms are equal. This gives $E_{min} = \hbar\omega = hf$. We therefore predict that the "zero point energy" of an oscillator is approximately hf by using the uncertainty principle.

Note. An accurate calculation, using the tools of quantum mechanics, predicts a "zero point energy" of $(\frac{1}{2})hf$, in good agreement with our approximation.

Supplementary Problems

Problem 17.24. The intensity of the radiation from a radio station at '1010 on your dial" is 5.3×10^3 W/m² · s. [Hint: AM stations are in kHz.]

(*a*) What is the energy of a single photon of that radio station?

(*b*) How many photons pass through an area of 1 m² per second from this radiation?

Ans. (*a*) 6.69×10^{-28} J; (*b*) 7.92×10^{30} photons/s

Problem 17.25. A beam of photons consists of pulses of duration 15 ms. During each pulse, the average power was 0.60 W, and the wavelength of the photons was 1.06 μm.

(*a*) What is the energy of one photon in the pulse?

(*b*) What is the total energy carried by each pulse?

(*c*) How many photons are contained in each pulse?

(*d*) What is the minimum uncertainty in the energy of each photon?

Ans. (*a*) 1.875×10^{-19} J; (*b*) 9×10^{-3} J; (*c*) 4.8×10^{16} photons; (*d*) 7.0×10^{-33} J

Problem 17.26. A microwave point source emits spherically symmetric electromagnetic radiation at a wavelength of 63.5 μm. The source emits this radiation at a steady rate of 6.0×10^4 W.

(*a*) How many photons per second pass through a unit area at a distance of 400 m?

(*b*) How many photons per second pass through a unit area at a distance of 4.0×10^4 m?

(*c*) How many photons per second pass through a unit area at the distance of the moon, i.e. at a distance of 3.8×10^8 m?

Ans. (*a*) 9.54×10^{18} photons/m$^2 \cdot$ s; (*b*) 9.54×10^{14} photons/m$^2 \cdot$ s; (*c*) 1.06×10^7 photons/m$^2 \cdot$ s

Problem 17.27. A material has a work function of 3.2 eV.

(*a*) What is the maximum wavelength that can be incident on this material and produce photo-electrons with a velocity of 5×10^5 m/s?

(*b*) If light at a wavelength of 3.8×10^{-7} m is incident on the material, what is the maximum velocity of emitted photo-electrons?

Ans. (*a*) 3.18×10^{-7} m; (*b*) 1.56×10^5 m/s

Problem 17.28. A certain material emits photo-electrons only if the incident radiation has a frequency greater than 1.5×10^{16} Hz.

(*a*) What is the work function for this material?

(*b*) What is the minimum frequency needed to produce photo-electrons with a kinetic energy of 2.5 eV?

Ans. (*a*) 62.1 eV; (*b*) 1.56×10^{16} Hz

Problem 17.29. Photons with a frequency of 3.2×10^{15} Hz produce photo-electrons with a maximum kinetic energy of 6.0 eV when incident on a certain material.

(*a*) What is the work function for this material?

(*b*) What is the minimum frequency of electromagnetic radiation needed to produce photo-electrons with a kinetic energy of 1.5 eV?

Ans. (*a*) 7.25 eV; (*b*) 2.11×10^{15} Hz

Problem 17.30. Photons, of wavelength 1.3×10^{-7} m are incident on a material that has a work function of 1.9 eV. An electron in the material absorbs a photon, and then loses 25% of its acquired energy to collisions until it reaches the surface of the material and is emitted. What is the kinetic energy of this photo-electron when it emerges?

Ans. 5.27 eV

Problem 17.31. Neutrons have a kinetic energy of 0.013 eV ($M_n \approx 1840\ m_e$).

(*a*) What is the DeBroglie wavelength of the neutrons?

(*b*) If these neutrons are diffracted by a crystal with an interatomic spacing of 3.1×10^{10}, at what angle will the first maximum occur?

(*c*) If one of these neutrons lost all of its kinetic energy and emitted a photon, what wavelength would the photon have?

Ans. (*a*) 2.51×10^{-10} m; (*b*) 54.1°; (*c*) 9.56×10^{-5} m

Problem 17.32. An X-ray beam is Compton scattered through an angle of 11°. The emerging electrons have a kinetic energy of 0.20 MeV.

(*a*) What is the change in the wavelength of the photons?

(*b*) What is the change in the frequency of the photons?

(*c*) What was the wavelength of the original photon?

Ans. (*a*) 4.46×10^{-14} m; (*b*) 4.83×10^{19} Hz; (*c*) 5.04×10^{-13} m

Problem 17.33. In a Compton scattering experiment, the emerging electron (initially at rest) has an energy of 0.32 MeV, and the scattered photon has an energy of 1.55 MeV.

(*a*) What is the energy of the incident photon?

(*b*) What is the change in the wavelength of the photon?

(*c*) At what angle does the scattered photon emerge?

Ans. (*a*) 1.87 MeV; (*b*) 1.37×10^{-13} m; (*c*) 19.3°

Problem 17.34. A photon has a wavelength of 2.3×10^{-13} m, and is Compton scattered.

(*a*) At what scattering angle is the wavelength of the scattered photon a maximum?

(*b*) What is the maximum wavelength of the scattered photon?

(*c*) What is the maximum kinetic energy of the scattered electron?

Ans. (*a*) 180°; (*b*) 5.09×10^{-12} m; (*c*) 5.16 MeV

Problem 17.35. A positron (electron anti particle) and an electron approach each other along a path, each having a kinetic energy of 0.35 MeV. They collide and annihilate each other producing two photons moving in opposite directions. [positrons and electrons have the same rest mass].

(*a*) What is the energy of each photon?

(*b*) What is the wavelength of each photon?

(*c*) What is the momentum of each photon?

Ans. (*a*) 0.86 MeV; (*b*) 1.44×10^{-12} m; (*c*) 4.59×10^{-22}

Problem 17.36. A neutron, with a kinetic energy of 0.038 eV, is diffracted off a crystal, and the first diffraction maximum occurs at $\phi = 44°$.

(*a*) What is the DeBroglie wavelength of the neutron?

(*b*) What is the interatomic spacing of the crystal?

Ans. (*a*) 1.47×10^{-10} m; (*b*) 2.12×10^{-10} m

Problem 17.37. Particles are incident on a double slit with a spacing of 3.0×10^{-7} m. At what angle will the first maximum occurs in the interference pattern if the particles are (*a*) electrons with a kinetic energy of 9.2 eV; (*b*) neutrons with a kinetic energy of 0.008 eV; and (*c*) electrons with a velocity of 2.7×10^4 m/s?

Ans. (*a*) 0.077°; (*b*) 0.061°; (*c*) 5.16°

Problem 17.38. An electron is confined to a box with one relatively short side of width 1.0×10^7 m. Use the uncertainty principle for x to calculate (*a*) the minimum velocity to which such an electron can be constricted; (*b*) the minimum kinetic energy to which the electron can be constricted, i.e. its zero point energy; and (*c*) the temperature at which the thermal kinetic energy would equal this zero point energy. [Hint: Recall $E_K = (1/2)kT$ for one-dimensional motion.]

Ans. (*a*) 1.16×10^3 m/s; (*b*) 6.11×10^{-25} J; (*c*) 0.089 K

Problem 17.39. The ionization energy for a neutron in a certain nucleus is 0.23 MeV. How small must the nucleus be so that the zero point energy equals the ionization energy of the neutron?

Ans. 9.5×10^{-15} m

Problem 17.40. A wave packet exists in the region between $x \pm \Delta x/2$, with a momentum between $p_{av} \pm \Delta p/2$. If $x = 3.5 \times 10^{-6}$ m, $p_{av} = 8.3 \times 10^{-26}$ kg · m/s, and $\Delta x = 2.2 \times 10^{-8}$ m, use the uncertainty principle to calculate (*a*) Δp; (*b*) p_{max} and p_{min}; (*c*) λ_{max} and λ_{min}; and (*d*) $\Delta\lambda$.

Ans. (*a*) 4.79×10^{-27} kg · m/s; (*b*) 8.54×10^{-26} kg · m/s, 8.06×10^{-26} kg · m/s; (*c*) 8.22×10^{-9} m, 7.76×10^{-9} m; (*d*) 4.6×10^{-10} m

Problem 17.41. An unstable particle called η^0 has a lifetime of approximately 10^{-18} s. Use the uncertainty principle to calculate (*a*) the uncertainty in the mass of the particle, Δm; and (*b*) if the rest mass is 9.8×10^{-28} kg, what fraction of the rest mass is Δm?

Ans. (*a*) 1.17×10^{-33} kg; (*b*) 1.12×10^{-6} ($\approx$ 1 part per million)

Chapter 18

Modern Physics: Atomic, Nuclear and Solid-State Physics

18.1 INTRODUCTION

In the last chapter, we learned the basic concepts of quantum mechanics. These concepts were required in order to be able to explain the experimental phenomena associated with photons such as the photoelectric effect and with the wavelike properties of particles such as electron diffraction. One of the early successes of this new theory was in explaining the properties of atoms. An atom is composed of a positively charged nucleus consisting of protons and neutrons and a surrounding shell of negatively charged electrons which have about 2000 times less mass than the protons and neutrons of the nuclei. It was conjectured that the attractive electrical force between the nucleus and the electrons would result in a planetary system similar to the planetary system created by the gravitational force in our solar system. However it is easy to show that such a system based on electromagnetism is not stable since the orbiting electrons accelerate emitting radiant energy and standard electromagnetic theory was unable to find a solution to this problem. Niels Bohr showed that if one makes certain ad hoc assumptions about quantization, similar to those suggested by Planck for the black-body radiation, he could indeed explain some detailed phenomena associated with atoms containing a single electron. This theory is known as the theory of the **Bohr atom**, and it provides a framework for thinking about atoms. Although we now know that it disagrees with quantum theory which was developed more fully after Bohr proposed his theory of the atom, it helps us understand some of the basic ideas of atomic physics.

To understand why there is a problem of stability we will consider what would happen if an electron were to move in a circle about a nucleus. In Fig. 18-1 we show such a hypothetical atom. A negatively charged electron is circling the positively charged nucleus at radius, r, and with velocity, v. Since a particle moving in a circle has acceleration (centripetal acceleration), there must be a force supplying this acceleration (the centripetal force). The electrical force between the nucleus and the electron will be able to do this, just as the gravitational force between the sun and the earth can provide the centripetal force to keep the earth in its orbit about the sun. There is, however, one major difference. The earth is uncharged, while the electron is charged. Electromagnetic theory predicts that any charged particle that accelerates, such as a particle does when it moves in a circular orbit, must radiate electromagnetic waves, thus losing energy. While we know that atoms do radiate at certain special wavelengths, and it is tempting to speculate that this comes from the electron's acceleration in its circular motion, the fact that it loses energy in the process means that it will spiral into the nucleus as its energy

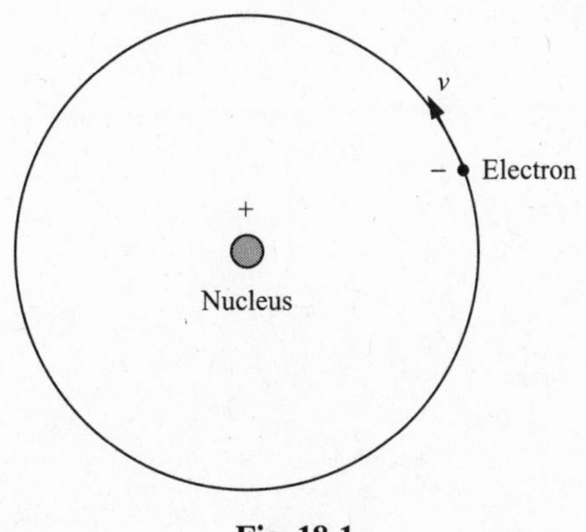

Fig. 18-1

decreases, and the orbit will be unstable. In fact, one can calculate that the atom should decay in a very short time, and there should therefore actually be no atoms in existence. Bohr developed a framework to help understand why the atoms do not spontaneously radiate away energy, and this framework carries over into quantum mechanics.

18.2 ATOMIC PHYSICS

The Bohr Atom

Let us consider in some detail the picture of an atom consisting of a nucleus of charge Ze, where Z is the number of protons in the nucleus, and an electron of charge $-e$ and mass m moving in a circle of radius r and with velocity v around this nucleus. This is shown in Fig. 18-1. The momentum of the electron is mv, and therefore the DeBroglie wavelength is h/mv. One can imagine a wave drawn around the circle representing the path of the electron, with this wavelength, as in Fig. 18-2. In order to obtain a stable wave (a standing wave) around this path, the circumference must be an integral multiple of the wavelength, or

$$2\pi r = n\lambda_{DB} = nh/mv \tag{18.1}$$

Then, mvr = angular momentum of electron $= n(h/2\pi) = n\hbar$, or

$$L = \text{angular momentum} = n\hbar \tag{18.2}$$

Bohr made the assumption that the angular momentum is quantized in multiples of $\hbar$, and that the only possible allowed orbits for the electron were ones with such an angular momentum. In that case, the atom would usually be unable to radiate energy, since radiating energy might require the electron to move to an orbit which did not satisfy this quantization condition. Only if the electron radiates away the exact amount of energy needed to place the electron in another allowed orbit can the atom radiate. We will soon calculate the energy of the electron in each of these allowed orbits, but it should be clear that other than the possibility of radiative transitions between the discrete orbits, the electrons are in stable repetitive motion. Even when the electrons jump between orbits, emitting definite amounts of radiative energy, they will ultimately reach the lowest energy orbit that is consistent with the angular momentum quantization. Electrons in this lowest "ground state" are truly stable and if undisturbed will continue circulating forever. Obviously, the Bohr hypothesis violates the predictions of classical electromagnetic theory. It should be noted that the quantum theory also predicts the existence of a ground state, and other allowed states of specified energy, but the existence of these states and energies are grounded in the solution of the Schroedinger equation, rather than in Bohr's ad hoc assumption regard-

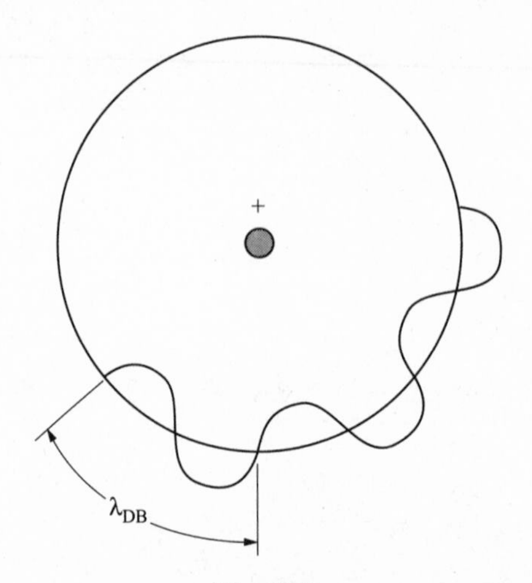

Fig. 18-2

ing standing waves. Bohr's concepts were revolutionary at the time and represented a major breakthrough that helped lead the way to the full quantum theory. Here we explore the Bohr atom more closely. We usually picture the allowed states for the atom in an energy level diagram, as in Fig. 18-3. Here, the energy is zero if the nucleus and the electrons are not bound to each other. This is the **ionized atom**. The **ground state** is the lowest allowed energy (E_1) bound state, and is negative relative to an electron at rest at infinity, as is required for a bound system. Other levels are labeled by a numerically ascending index, E_2, E_3, etc. The atom can make a transition by moving from one allowed level to another, but not by losing or gaining any amount of energy that is not consistent with the energies of these allowed levels, thus assuring conservation of energy. This basic concept of transitions was introduced by Bohr as a second hypothesis and carries over into the quantum theory as well.

If we consider a Bohr atom, with a nucleus of Z protons (plus some neutrons) of charge Ze, and *one* electron, with charge $-e$ and mass m, moving in a circle of radius r and with velocity v around the nucleus, we know that the electrical force kZe^2/r^2 ($k = 1/(4\pi\varepsilon_0)$ must equal the centripetal force mv^2/r. Assuming that the angular momentum, L, is quantized, so $L_n = mvr = n\hbar$, we have two equations in the two unknowns v and r for each value of n. We can then solve for v and r, as well as for the total energy of the orbit with that value of n. The result, in terms n and the various fundamental constants (e, c, m, h, ε_0) are:

$$v = kZe^2/n\hbar \tag{18.3}$$

$$r = (h^2/4\pi^2mke^2)(n^2/Z) \tag{18.4}$$

$$E_T = -RZ^2/n^2, \qquad \text{where} \qquad R = 2\pi^2mk^2e^4/h^2 \equiv \textbf{Rydberg constant} \tag{18.5}$$

Problem 18.1. Using the formulas developed, substitute numbers for the fundamental constants, and calculate (*a*) the energy of the *n*th level; (*b*) the radius of the *n*th level; (*c*) the energy of the lowest energy level for hydrogen ($Z = 1$); (*d*) the radius of the lowest level for hydrogen ($Z = 1$); (*e*) the energy of the level for hydrogen with $n = 3$, and its radius; and (*f*) the energy and radius of the ground state for singly ionized (one electron) helium ($Z = 2$).

Solution

(*a*) Substituting in Eq. (*18.5*), we get for the energy of the state with a given n,

$$E_n = -(2.18 \times 10^{-18})Z^2/n^2 \text{ Joules,} \tag{18.6a}$$

or

$$E_n = -(13.6)Z^2/n^2 \text{ electron volts} = -RZ^2/n^2 \tag{18.6b}$$

(with R = "Rydberg energy" = 13.6 eV)

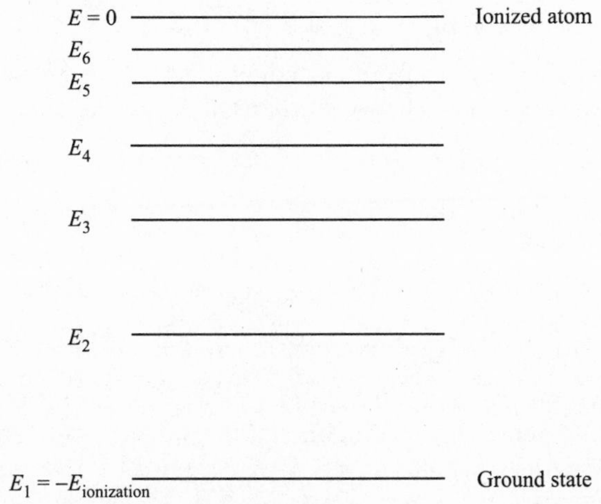

$E = 0$ ——— Ionized atom
E_6 ———
E_5 ———
E_4 ———
E_3 ———
E_2 ———
$E_1 = -E_{\text{ionization}}$ ——— Ground state

Fig. 18-3

(*b*) Substituting in Eq. (*18.4*), we get

$$r_n = (5.29 \times 10^{-11})n^2/Z \text{ m} = a_0 n^2/Z \tag{18.7}$$

(with a_0 = "Bohr radius" = 5.29×10^{-11} m)

(*c*) The lowest energy occurs if the energy is most negative. This occurs for $n = 1$. Thus, the ground state energy is $E_1 = -2.18 \times 10^{-18}$ J $= -13.6$ eV.

(*d*) The radius of the ground state ($n = 1$) is $r_1 = a_0 = 5.29 \times 10^{-11}$ m.

(*e*) Using $n = 3$ and $Z = 1$, we get $E_3 = -13.6/3^2 = -1.51$ eV, and $r_3 = (5.29 \times 10^{-11}\text{ m})(3^2) = 4.76 \times 10^{-10}$ m.

(*f*) Using $n = 1$ and $Z = 2$, we get $E_1 = -13.6(2^2) = -54.4$ eV, and $r_1 = (5.29 \times 10^{-11}\text{ m})/2 = 2.65 \times 10^{-11}$ m.

The Bohr theory of the atom thus predicts the existence of quantized energy levels, each associated with a distinct label, *n*, which is called the quantum number of that level. In those levels the atom cannot radiate away any energy unless, after losing an amount of radiant energy, the atom finds itself in another allowed energy level. Thus the atom in an upper energy level E_u when it radiates energy must lose just enough energy to bring it to the energy of a lower energy level E_l. In general, this means that $E_{rad} = E_u - E_l$. In such a transition between allowed orbits, the radiation is typically emitted as a single photon of electromagnetic energy. Since $E_{rad} = hf$ for a photon, only specific frequencies of radiation (and correspondingly, only specific wavelengths) can be radiated by an atom. These frequencies and wavelengths are given by:

$$f_{ul} = (E_u - E_l)/h, \tag{18.8a}$$

where f_{ul} is the frequency for the transition from the upper to the lower level, and

$$1/\lambda_{ul} = (E_u - E_l)/hc \tag{18.8b}$$

Formulas (*18.8*) depend only on the concept of energy levels and do not depend on the method of arriving at the energies of each level. Thus, these equations are valid for all atoms, even with many electrons, if one can find a method to calculate the energy levels of those atoms. They carry over into the correct quantum mechanical treatment of the atom, where the values of the energy levels are now calculated using the correct quantum mechanical equations.

For the Bohr atom with one electron, the allowed transitions are from states of high *n* to a lower *n*. If one starts at an upper level, *m*, and ends at a lower level, *n*, then

$$1/\lambda_{mn} = (E_m - E_n)/hc = [(2.18 \times 10^{-18})Z^2/hc][1/n^2 - 1/m^2] \text{ in inverse meters}$$
$$= [1.097 \times 10^7 Z^2(1/n^2 - 1/m^2)] \text{ m}^{-1} \tag{18.9}$$

All the wavelengths resulting from transitions ending in the $n = 1$ ground state form a series which, for hydrogen, is called the **Lyman series**. The wavelengths in this series are

$$1/\lambda_{m1} = 1.097 \times 10^7(1/1^2 - 1/m^2) \text{ m}^{-1} \text{ (Lyman series)} \quad (m = 2, 3, 4, \ldots) \tag{18.10a}$$

Similarly those ending at $n = 2$ form the **Balmer series**, and those ending at $n = 3$ form the **Paschen–Bach** series. Their wavelengths are given by

$$1/\lambda_{m2} = 1.097 \times 10^7(1/2^2 - 1/m^2) \text{ m}^{-1} \text{ (Balmer series)} \quad (m = 3, 4, 5, \ldots) \tag{18.10b}$$

$$1/\lambda_{m3} = 1.097 \times 10^7(1/3^2 - 1/m^2) \text{ m}^{-1} \text{ (Paschen–Bach series)} \quad (m = 4, 5, \ldots =) \tag{18.10c}$$

These transitions are shown on the energy level diagram in Fig. 18-4. Because the $n = 1, 2, 3$ energy levels are so well separated from each other in energy the wavelengths for the lines in each series all correspond to a range of energies small compared to the lowest energy in that series, but are quite distinct from the wavelengths of the other series. That is, the wavelengths for these series do not overlap—a fact that is not true for such series in general. These formulas for the wavelengths are

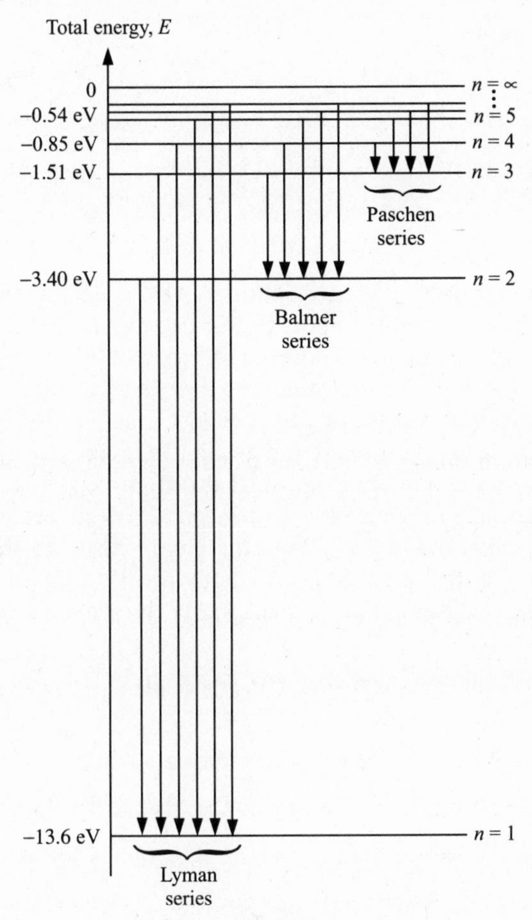

Fig. 18-4

dependent on the details of the Bohr theory of the atom and presumably would have to be modified when the full quantum mechanical theory is applied. One of the great successes of the Bohr theory was that these formulas indeed gave answers which were in very good agreement with experiment. In fact, the quantum mechanical treatment gives identical formulas in a first approximation for these one electron atoms. The agreement that Bohr was able to show between his theory and these experimental observations showed that his approach was in the right direction.

Problem 18.2. Consider a hydrogen atom.

(*a*) Calculate the first three wavelengths in the Lyman series. In what part of the spectrum are these wavelengths?

(*b*) Calculate the first three wavelengths in the Balmer series. In what part of the spectrum are these wavelengths?

Solution

(*a*) Using Eq. (*18.10a*), we get

$$1/\lambda_{21} = 1.097 \times 10^7(1/1^2 - 1/2^2) = 8.23 \times 10^6, \qquad \lambda_{21} = 1.215 \times 10^{-7}\ \text{m} = 121.5\ \text{nm}$$

$$1/\lambda_{31} = 1.097 \times 10^7(1/1^2 - 1/3^2) = 9.75 \times 10^6, \qquad \lambda_{31} = 1.026 \times 10^{-7}\ \text{m} = 102.6\ \text{nm}$$

$$1/\lambda_{41} = 11.097 \times 10^7(1/1^2 - 1/4^2) = 1.028 \times 10^7, \qquad \lambda_{41} = 9.72 \times 10^{-8}\ \text{m} = 97.2\ \text{nm}$$

These are all in the ultraviolet region of the spectrum.

(*b*) Using Eq. (*18.10b*), we get

$$1/\lambda_{32} = 1.097 \times 10^7(1/2^2 - 1/3^2) = 1.52 \times 10^6, \qquad \lambda_{32} = 6.56 \times 10^{-7}\ \text{m} = 656\ \text{nm}$$

$$1/\lambda_{42} = 1.097 \times 10^7(1/2^2 - 1/4^2) = 2.06 \times 10^6, \qquad \lambda_{42} = 4.86 \times 10^{-7}\ \text{m} = 486\ \text{nm}$$

$$1/\lambda_{52} = 1.097 \times 10^7(1/2^2 - 1/5^2) = 2.30 \times 10^6, \qquad \lambda_{52} = 4.34 \times 10^{-7}\ \text{m} = 434\ \text{nm}$$

These are all in the visible part of the spectrum.

Just as emission of radiation is restricted to photons whose energies are equal to the differences between energy levels of the atom, so too the atom can absorb radiation only for those wavelengths that have energies that will result in the atom moving to another allowed energy level. If the atom starts in the ground state, E_1, then the only photons that can be absorbed are those whose energies equal $E_n - E_1$, where E_n are the energies of states at higher energies than E_1. For instance, consider the energy level for an atom that is given in Fig. 18-5. Here the ground state has energy -25 eV, and levels 2, 3, 4, and 5 are at higher energies by the amounts shown. If the atom begins in the ground state, then only photons of energy 6 eV (to level 2), or 10 eV (to level 3), or 12 eV (to level 4), etc. can be absorbed. To ionize the atom, that is to remove the electron from the atom—to break the bond—requires a photon with an energy of at least 25 eV. If the photon has energy greater than 25 eV, then the excess energy over 25 eV will be given to the electron in kinetic energy after the atom has been ionized.

Problem 18.3. Consider the atom whose energy level diagram is given in Fig. 18-5. Suppose that the atom starts in level 3.

(*a*) What is the shortest wavelength photon that the atom can emit?

(*b*) What is the longest wavelength photon that it can absorb (starting in level 3)?

(*c*) What is the lowest frequency photon that can ionize the atom (starting in level 3)?

(*d*) If the atom starts in the ground state what energy photon is needed to ionize the atom *and* give the emitted electron a kinetic energy of 5 eV?

Solution

(*a*) The shortest wavelength corresponds to the greatest energy. Thus we want the atom to make a transition to the lowest possible energy level, which is the ground state. The energy difference between level 3 and the ground state is 10 eV, and therefore this will be the energy of the emitted photon with the smallest wavelength. The wavelength is given by $\lambda = c/f = hc/\Delta E = (6.625 \times 10^{-34}\ \text{J} \cdot \text{m})(3.0 \times 10^8\ \text{m/s})/(10\ \text{eV})(1.6 \times 10^{-19}\ \text{J/eV}) = 1.24 \times 10^{-7}\ \text{m}$.

(*b*) The longest wavelength corresponds to the smallest energy. Therefore the atom will make a transition to a higher level than level 3, but to one that is closest in energy, i.e. to level 4. The energy difference between levels 3 and 4 is 2 eV, and the wavelength corresponding to this energy is 6.21×10^{-7} m.

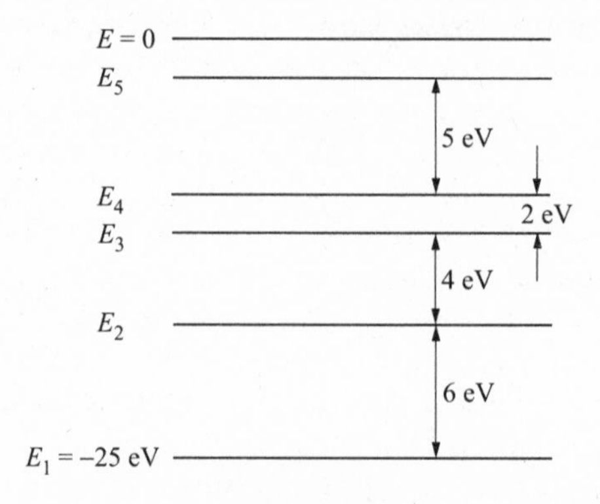

Fig. 18-5

(*c*) To ionize the atom starting in level 3 we have to supply the energy needed to go from level 3 to an energy of zero. The energy of level 3 is 10 eV above the ground state, whose energy is given as -25 eV. Thus the energy of level 3 is -15 eV, and 15 eV is needed to ionize the atom from level 3. The photon needed for this has a frequency of $f = (15 \text{ eV})(1.6 \times 10^{-19} \text{ J/eV})/(6.625 \times 10^{-34} \text{ J} \cdot \text{s}) = 3.62 \times 10^{15}$ Hz.

(*d*) From the ground state, 25 eV is needed to ionize the atom. To give the emitted electron a kinetic energy of 5 eV, we need an additional 5 eV of energy. Thus the photon must have an energy of $25 + 5 = 30$ eV.

The Bohr theory was successful in giving a rationale for the stability of atoms, and it correctly predicted the wavelengths of radiation from single electron atoms. The theory further showed that the radiation from atoms in general was determined by differences in energy levels of the atoms. However, his theory was unable to be generalized to atoms with more than one electron, and was unable to explain why the energy levels of atoms should be quantized in the first place. Additionally, while using the wave nature of electrons to generate the angular momentum quantization condition, it still basically treats the electron as a classical particle traveling in definite circular orbits (like a planet) about the nucleus. Nevertheless the picture is a useful one in enabling us to think about the process that leads to an atom that has quantized energy and angular momentum levels. We will now proceed to a qualitative examination of a true quantum mechanical treatment of an atom.

The Quantum Theory of the Atom

We learned that in the quantum theory the behavior of particles was determined by the wave-function solutions of Schroedinger's equation, which represent the probability amplitude for the particle, and when squared, yield the probability that a particle is at a given point in space at a given time. To understand the atom, one has to solve this equation for the case of an electron around a nucleus. We are not going to solve the equation here, but rather we will show what major consequences arise from any such solution.

The Schroedinger equation is a wave equation, similar to the equations that give rise to other types of waves, such as waves on a cord, sound waves, water waves and electromagnetic waves. In the case of a wave on a cord, we showed that there are both traveling solutions and standing wave solutions which we called "**stationary solutions**". The same is true of Schroedinger's equation, where the stationary solutions (as hinted at by the Bohr theory) correspond to the bound states of the atom. For the case of our cord of length L, the standing wave solutions could be distinguished from one another by one index, for instance the number of nodes or antinodes in the standing wave. For each number, the properties of the wave, particularly the frequency, are determined. There are solutions only for specific frequencies, designated as f_n. Suppose instead of a cord one now examines the property of waves on a taut sheet of material such as the surface of a drum. Again there are stationary solutions (with definite nodes and antinodes) at certain allowed frequencies. In this case however, these waves will have nodes and antinodes in two independent directions of the plane. To specify the solution that we are considering we must label it with two index numbers, n and m, one for each direction. Thus, the allowed frequencies are now designated as $f_{n,m}$. If we now move to the solution of a wave equation in three dimensions we find, once again, that stationary solutions exist for only certain frequencies. These solutions consist of waves with nodes and antinodes in three directions. The solutions must now be labeled with three indices, such as n, l and m, and we will get only certain allowed frequencies, designated as $f_{n,l,m}$. We know from DeBroglie's hypothesis that matter waves of a particular frequency f have energy equal to hf. This remains true in quantum mechanics and we therefore see that the only possible energies that can be found for bound states, or stationary solutions, of Schroedinger's equation for an atom in three dimensions are special values designated as $E_{n,l,m}$. This is the origin of the fact that only certain energy levels are permitted for the atom and this explains the basic idea that originated with Bohr that electrons can radiate only frequencies that lead to transitions between allowed energy levels.

The exact formulas for the energies allowed and the meaning of the indices n, l, and m, require an actual solution of the wave equation. These solutions have been obtained exactly for an atom with one electron, and in approximation for atoms with more than one electron. We will discuss the results of these solutions in detail further on but we will mention right now that for an atom with one electron the energies obtained in the first approximation are just those of the Bohr atom.

Although bound state solutions of the Schroedinger equation require only three labels in three dimensions, it was experimentally determined from the details of the radiation emitted by these atoms, that a fourth label was required as well. These labels are called "**quantum numbers**" and the accepted notation for these quantum numbers in a one electron atom are n, l, m_l and m_s. This fourth quantum number arises from a purely quantum mechanical concept, the intrinsic spin of the electron.

The Quantum Numbers for a One Electron Atom

The most important quantum number in determining the energy of the atom is the number n, called the principal quantum number, which is associated with the distance, r, between the nucleus and the electron. This is reasonable since we know that the energy is determined in classical mechanics only by this distance r. This number takes on integral values beginning with 1, so $n = 1, 2, 3, 4, \ldots \infty$. Again this is reasonable since r has a minimum value of zero and no maximum value, so n has a lower bound of 1 but no upper bound. Indeed n can be identified with the energy quantum number in the Bohr theory, [Eq. (*18.6b*)], and when we neglect the small effect of the other quantum numbers, the Schroedinger energy levels are given by the same equations as for the Bohr atom:

$$E_{n,\,l,\,m_l,\,m_s} = -(2.18 \times 10^{-18})Z^2/n^2 \text{ Joules, or} \tag{18.11a}$$

$$E_{n,\,l,\,m_l,\,m_s} = -(13.6)Z^2/n^2 \text{ electron volts} \tag{18.11b}$$

Note that, unlike the Bohr atom, in the full quantum mechanics, the quantum number n does not tell you anything about the angular momentum of the electron. This number is related to the distance (actually mean distance) of the electron from the nucleus and has nothing to do with a circulation around the nucleus which can give rise to an angular momentum. Indeed, the concept of a circulating electron in an orbit as prescribed by the Bohr theory, has *no validity* in quantum mechanics and angular momentum in quantum mechanics cannot be given a simple visual picture as in the Bohr theory.

The quantum number, l, is related to one of the angular coordinates. It is called the angular momentum quantum number, since it determines the magnitude of the orbital angular momentum of the electron about the nucleus. For a given n, the allowed values of l are restricted to integers in the range from zero to $(n - 1)$. Also the orbital angular momentum, L, is given by

$$L = (h/2\pi)\sqrt{(l)(l+1)}, \qquad l = 0, 1, 2 \ldots (n-1) \tag{18.12}$$

The above tells us that the allowed values of l are restricted by n. Thus for $n = 1$, l can have values from 0 to $(n - 1) = 0$, or the only possible value for l is zero. For other n we can carry out similar calculations.

Problem 18.4. Calculate the allowed values for the angular momentum quantum number l if the principal quantum number n is (*a*) $n = 2$; (*b*) $n = 3$; (*c*) $n = 4$.

Solution

(*a*) The value of l ranges from 0 to $(n - 1) = (2 - 1) = 1$. Thus $l = 0, 1$.

(*b*) The value of l ranges from 0 to $(n - 1) = (3 - 1) = 2$. Thus $l = 0, 1, 2$.

(*c*) The value of l ranges from 0 to $(n - 1) = (4 - 1) = 3$. Thus $l = 0, 1, 2, 3$.

In the case of multi electron atoms ($Z > 1$), in addition to the Coulomb energy between each electron and the nucleus, there is Coulomb energy due to the repulsive force between electrons. The allowed energies for the electrons can then become very complex. To a first approximation, however, we can often assume that the dominant effect is due to the nucleus (with a modification of Z perhaps) so that the electrons can be treated as if they were in a single electron atom. In this approximation all electrons with a common principle quantum number, n, are said to be in the same shell. All electrons with $n = 1$ are said to be in the K shell, and all have essentially the same energy. All electrons with the $n = 2$ are said to be in the L shell, with $n = 3$ they are in the M shell, with $n = 4$ in the N shell, etc. As we stated previously, the angular momentum is determined by the quantum number l (not n). All electrons with the same quantum number l, have the same angular momentum. Although the correct formula for the angular momentum is $L = \sqrt{l(l+1)}[h/2\pi]$, we often say that the "electron has angular momentum l", meaning that we can calculate the correct value when l is given. It should be noted that in multi electron atoms the energy levels of an electron are dependent somewhat on l as well as on n, but the formulas are quite complex. In general, n is still the dominant factor in the energy of an electron.

Problem 18.5. Calculate the true allowed values of the angular momentum if the principal quantum number n is (*a*) $n = 1$; (*b*) $n = 2$; (*c*) $n = 3$.

Solution

(*a*) The value of l ranges from 0 to $(n - 1) = (1 - 1) = 0$. Thus $l = 0$. With $l = 0$, $L = \sqrt{0(1)} = 0$. Thus the only possible value for the angular momentum is zero. We see from this that the picture of an electron circling a nucleus is incorrect, since any such electron would have to have angular momentum.

(*b*) The value of l ranges from 0 to $(n - 1) = (2 - 1) = 1$. Thus $l = 0, 1$. Therefore L can be either zero (when $l = 0$), or $(\sqrt{2})(h/2\pi)$ (when $l = 1$).

(*c*) The value of l ranges from 0 to $(n - 1) = (3 - 1) = 2$. Thus $l = 0, 1, 2$. Now there are three possible values for the angular momentum. $L = 0$ (when $l = 0$), $L = (\sqrt{2})(h/2\pi)$ (when $l = 1$) and $L = (\sqrt{6})(h/2\pi)$ (when $l = 2$).

Any electron with $l = 0$ is called an s electron, and has no angular momentum. When $l = 1$, the angular momentum is $(\sqrt{2})(h/2\pi)$, and such an electron is called a p electron. When $l = 2$, the angular momentum is $(\sqrt{6})(h/2\pi)$, and such an electron is called a d electron. An electron with $l = 3$ is called an f electron, with $l = 4$ is called a g electron, etc.

Problem 18.6. For a multi-electron atom (*a*) what shell contains s and p electrons, but none with higher l; and (*b*) what is the minimum value of n so that d electrons are also found?

Solution

(*a*) To get only s and p electrons we must have $l = 0$ and 1, but no larger value of l. The maximum value of l for a given n is $(n - 1) = 1$, giving $n = 2$, the L shell. Note that the K shell is eliminated since it cannot have p electrons ($n = 1 \rightarrow l_{max} = 0$).

(*b*) To get d electrons we must have n big enough that l can equal 2. This requires n to be at least $n = 3$. For greater n there will also be d electrons, but for $n < 3$ there are no d electrons. Thus $n = 3$ is the minimum value of n for which d electrons exist.

The third quantum number, m_l is related to the second angular coordinate. For a given l the values of m_l are restricted to integer values ranging from $-l$ to $+l$. This quantum number is often called the magnetic quantum number because the energy of an electron depends on m_l in the presence of a magnetic field. For a given l there are $(2l + 1)$ possible values for m_l.

Problem 18.7. What are the possible values for m_l if the electron (*a*) is an s electron; (*b*) has an angular momentum quantum number of 1; (*c*) has an actual angular momentum of $(\sqrt{6})(h/2\pi)$; and (*d*) is in the L shell?

Solution

(*a*) If the electron is an s electron, then $l = 0$. The m_l ranges from -0 to $+0$, which means that the only allowed value for m_l is zero.

(*b*) If the electron has $l = 1$, then m_l ranges in integral steps from -1 to $+1$, which means that the only allowed values for m_l are -1, 0, or $+1$. There are thus three possible values for m_l as expected, since $(2l + 1) = 2(1) + 1 = 3$.

(*c*) If $L = (\sqrt{6})(h/2\pi) = [\sqrt{(l+1)}][h/2\pi]$, then $l = 2$. For $l = 2$, the allowed values of m_l are $-2, -1, 0, 1, 2$. These five values $[2(2) + 1) = 5]$ are the only possible values for m_l when $l = 2$.

(*d*) If the electron is in the L shell then $n = 2$. For $n = 2$, there are two possible values for l, namely $l = 0$ or 1. For $l = 0$, only $m_l = 0$ is permitted, while for $l = 1$ we can have $m_l = -1$, 0 or $+1$. Thus the only allowed values for m_l are 0 and ± 1.

The magnetic quantum number has only a very small effect on the energy of the quantum state, except when a magnetic field is existent. However there is significance to this quantum number even in the absence of a magnetic field. Suppose we choose a certain direction as the z axis. (Usually this is an arbitrary choice, but in the presence of a magnetic field we typically choose the z axis along the direction of the field.) As defined in Beginning Physics I, Sect. 10.4, angular momentum is a vector quantity whose direction (for simple symmetric bodies) defines the axis about which the body is rotating. As we have seen, the quantum number l specifies the magnitude of the angular momentum. The quantum number m_l helps to pin down the allowed directions for the angular momentum by specifying the z components, thus restricting **L** to certain directions relative to this z axis. This is called space quantization, since the direction of **L** is restricted to certain directions in space. Each value of m_l is associated with a possible direction in space by the following rule. The component of the angular momentum along the z axis must equal $m_l(h/2\pi)$. For $l = 1$, this is illustrated in Fig. 18-6. If $l = 1$, there are three possible values for m_l, -1, 0, and $+1$. Thus L_z can equal $-(h/2\pi)$, 0, or $+(h/2\pi)$, even though $|\mathbf{L}| = (\sqrt{2})(h/2\pi)$ for all three cases. This is shown in the figure where the length of **L** is drawn on the circle with radius $(\sqrt{2})(h/2\pi)$. For $L_z = +h/2\pi$ $(m_l = +1)$, the component along the z axis is positive and of length $h/2\pi$. For $l = 0$, we must have $L_z = 0$, so the vector **L** is perpendicular to the z axis and the component along z is zero. For $m_l = -1$, the component along the z axis is negative, and of length $h/2\pi$, as shown. The angle, θ, between **L** and the z axis is given by $\cos\theta = L_z/L = m_l/\sqrt{(l)(l+1)}$.

If we have an s electron, for which $l = 0$, there is no angular momentum and therefore no component along z. For a d electron there are five possible values for m_l, and therefore five possible spatial orientations for **L** relative to the z axis.

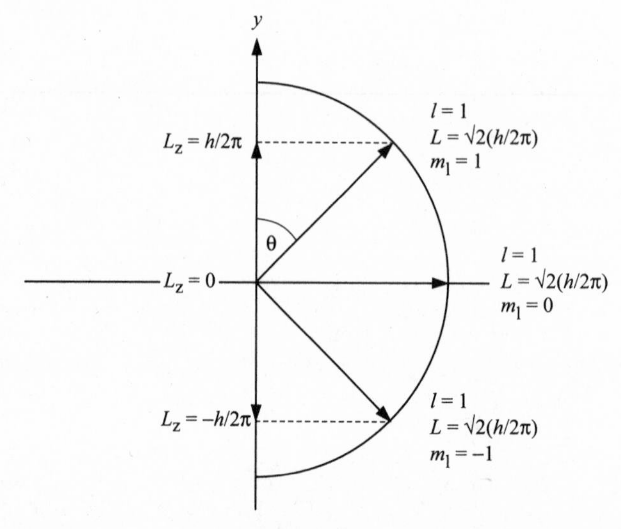

Fig. 18-6

Problem 18.8. For a d electron, draw all the possible directions for L_z relative to the z axis.

Solution

For a d electron $l = 2$, and the possible values of m_l are -2, -1, 0, 1, and 2. The length of the angular momentum vector is $(\sqrt{6})(h/2\pi)$, and the possible values for L_z are $-2(h/2\pi)$, $-1(h/2\pi)$, 0, $1(h/2\pi)$ and $2(h/2\pi)$. This is drawn in Fig. 18-7.

To simplify these drawings we often use the approximation that $L = l(h/2\pi)$, rather than the more exact value of $L = [\sqrt{(l)(l+1)}](h/2\pi)$. Then for $m_l = l$, $L_z = L$, and the vector **L** would point along the positive z axis. In this approximation, the maximum positive or negative values of m_l always correspond to the angular momentum pointing either parallel or antiparallel to the z axis. This is shown in Fig. 18-8 for $l = 1$, $l = 2$ and $l = 3$. It is again important to reiterate that although angular momentum remains an important physical quantity in quantum mechanics, one cannot identify this quantity with the concept of a circulating electron! Our intuition no longer holds sway in the realm of the atom!

From these considerations we see that the shell with a particular value of n consists of several different states of nearly the same energy. It is important to know how many levels there are in each shell even if they really were at the same energy, as we shall see when we discuss atoms with many

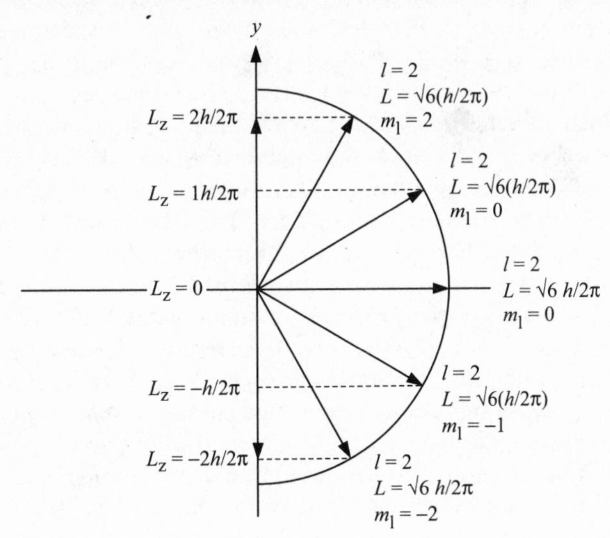

Fig. 18-7

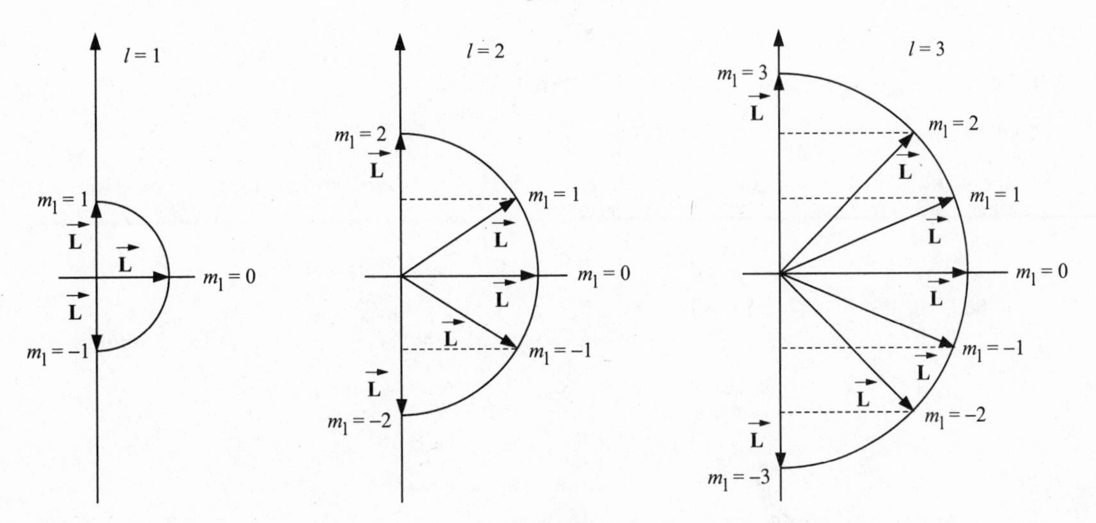

Fig. 18-8

electrons. In many cases, however, the energy of the levels depends somewhat on l and on m_l, as well as on n. Since electrons can make transitions only between allowed energy levels, this allows transitions with slightly different energies to occur. The additional wavelengths allowed give rise to what is called the fine structure in the lines, and one should be able to determine how many different states are available for the initial and for the final states by examining the fine structure. When one does this one discovers that there are more states available to the electron than is given by the three quantum numbers n, l, m_l and m_s. We need an additional quantum number in order to get agreement between the theory and experiment. This additional quantum number does not arise from the motion of an electron in three dimensions, but rather arises from the intrinsic properties of the electron itself. If one assumes that the electron has an intrinsic spin (an internal spin about its own axis) which is quantized, then the atom has additional angular momentum due to the intrinsic spin. This assumption was an ad hoc assumption until quantum mechanics was made to conform to the theory of relativity and the Schroedinger equation was replaced by the Dirac equation. The relativistic equation actually predicted the correct value for the electron spin. We now know that all electrons (as well as protons and neutrons and many other particles) have an internal angular momentum of magnitude $[\sqrt{(\frac{1}{2})(\frac{1}{2}+1)}](h/2\pi)$ (called the spin angular momentum), which is intrinsic to the electron just as its charge and its mass are. It cannot be increased or decreased any more than the charge or rest mass can. As in the case of the orbital angular momentum, the spin angular momentum is also quantized in space, with its component along the z direction being restricted to $\pm\frac{1}{2}(h/2\pi)$. We assign a quantum number m_s, which is $\pm\frac{1}{2}$, to these two possible states. We call the state with $m_s = +\frac{1}{2}$ one in which the spin is along the axis and the state with $m_s = -\frac{1}{2}$ one in which the spin is opposite to the axis. This gives us two additional quantum values for each orbital state, and we now have to designate each energy level by four quantum numbers, n, l, m_l and m_s. Because the electron is charged, the spinning electron produces a magnetic field which in turn impacts slightly on the energy levels of the electrons. With the addition of this spin, the theory agrees with experiment to a remarkable degree. Of course there is also spin in the nucleus, but this results in much smaller magnetic energy changes for the electrons than is caused by the spin of the electron. However, even these small changes can be seen and give rise to the "hyperfine" structure of the radiated lines, as well as to the NMR (**nuclear magnetic resonance**) effects which are currently used extensively in medicine. We will not discuss these nuclear effects further.

We can now draw the energy level diagram for a single electron in an atom, specifying all the states that can exist. For simplicity we will first address the case of only one electron in the atom and then turn to the multi electron atom. Even in the one electron atom there is fine structure of the energy levels because the electrons with different l values have different magnetic properties and interact with the spin magnetic field of the electron differently. Furthermore if we assume the atom is placed in an external magnetic field the energy levels will depend slightly on m_l and m_s as well.

Problem 18.9. List all the possible states in the K shell and in the L shell.

Solution

For the K shell, $n = 1$. The only possible value for l is then zero, and m_l must also be zero. For m_s we can have $\pm\frac{1}{2}$. Thus the K shell has two states with

$$n = 1, \qquad 1 = 0, \qquad m_l = 0, \qquad m_s = \pm\tfrac{1}{2}$$

For the L shell, $n = 2$. Then l can be 0 or 1. If $l = 0$, we must have $m_l = 0$, but $m_s = \pm\frac{1}{2}$. If $l = 1$, then m_l can be -1, 0 or $+1$, and $m_s = \pm\frac{1}{2}$. This gives us a total of eight states which can be listed as follows:

$$n = 2, \qquad 1 = 0, \qquad m_l = 0, \qquad m_s = \pm\tfrac{1}{2} \text{ (two states)}$$

$$\left.\begin{array}{lll} l = 1, & m_l = -1, & m_s = \pm\frac{1}{2} \\ & m_l = 0, & m_s = \pm\frac{1}{2} \\ & m_l = +1, & m_s = \pm\frac{1}{2} \end{array}\right\} \text{(six states)}$$

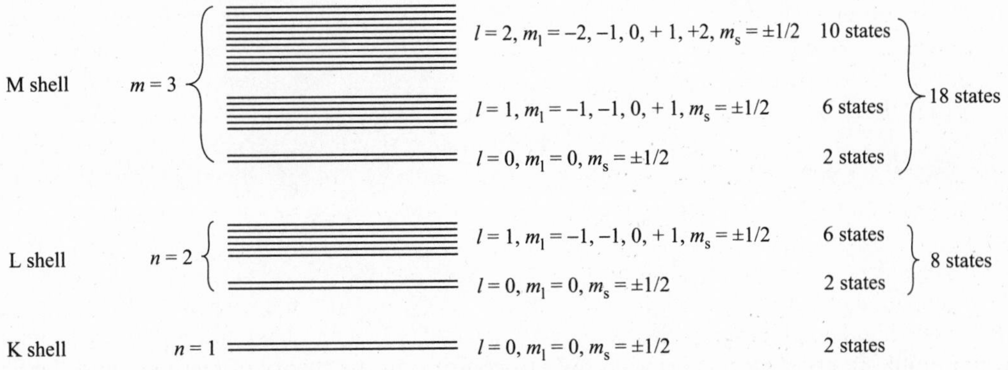

M shell $m = 3$
$l = 2, m_l = -2, -1, 0, +1, +2, m_s = \pm 1/2$ 10 states
$l = 1, m_l = -1, -1, 0, +1, m_s = \pm 1/2$ 6 states
$l = 0, m_l = 0, m_s = \pm 1/2$ 2 states
18 states

L shell $n = 2$
$l = 1, m_l = -1, -1, 0, +1, m_s = \pm 1/2$ 6 states
$l = 0, m_l = 0, m_s = \pm 1/2$ 2 states
8 states

K shell $n = 1$
$l = 0, m_l = 0, m_s = \pm 1/2$ 2 states

Fig. 18-9

We say that the L shell has two sub shells, one with $l = 0$, and the other with $l = 1$. Using this same procedure for the other shells, one arrives at the energy level diagram shown in Fig. 18-9 (assuming a small external magnetic field).

The procedure we have described has led us to the conclusion that the energies of electrons in an atom are quantized, that each state is characterized by four quantum numbers and that radiation can be emitted only if the energy of the photon equals the difference between the energies of two different levels. When one actually solves the Schroedinger equation to obtain the predicted energies (putting in the electron spin in an ad hoc way) the agreement between theory and experiment is excellent.

The Wave Function of a One Electron Atom

When we solve Schroedinger's equation, we get stationary solutions for only certain allowed energies. The solution to the equation at one of these allowed energies is the "wave function" for the electron in this state. We learned previously that the meaning of this wave function was that it gave us the probability for finding the electron at various positions in space and time. Let us examine what kind of distribution these solutions give for the electron's position at some fixed time.

For s electrons we find that the distribution is spherically symmetric, having the same value at any angle, and depending only on distance from the nucleus. The probability is zero at $r = 0$ and at $r = \infty$, and peaks somewhere in between. One can calculate an average value for the radial distance of the electron from the nucleus from this probability. In Fig. 18-10, we plot the probability distribution for s electrons for a few different values of n, and indicate the average value of the electron's distance. This distance is given in terms of the Bohr radius $a_0 = 5.29 \times 10^{-11}$ m, which is the radius of the lowest level orbit in the Bohr atom. We see that the electron cannot be considered to be located at any specific position in space, but rather there is an "**electron probability cloud**" in which the electron is concentrated. For $n = 1$, the maximum probability location for the electron is $r = a_0$ while the average position of the electron is $1.5a_0$. The total probability of finding the electron at any radial distance is the area under the curve between $r = 0$ and $r = \infty$, and must equal 1 since the electron is definitely *somewhere* in this range. For larger n, there are nodes in the radial electron probability distribution reminiscent of standing waves on a string. The values of the average radial position of the electron are shown in the figure for $n = 2, l = 0$ and $n = 3, l = 0$.

For p electrons, the distribution depends on angle as well as on the distance from the nucleus. The average value for r is not too different than for $l = 0$, but the electron cloud now has an angular dependence. One way of viewing this angular distribution for the $l = 1$, $m_l = 0$ state is shown in Fig. 18-11. The probability of finding the electron at an angle is proportional to the length from the origin to the surface shown, at that angle. The peak probability is at 0° (and 180°) while the probability is zero at 90°. While this seems to give unique status to the z axis it can be shown that by taking suitable combinations of the three $l = 1$ states one can generate the same distribution about the x axis or y axis. Similar distributions can be plotted for other values of n and l. The important point is that the location of the electron must be considered as a probability distribution, and cannot be thought of as located at

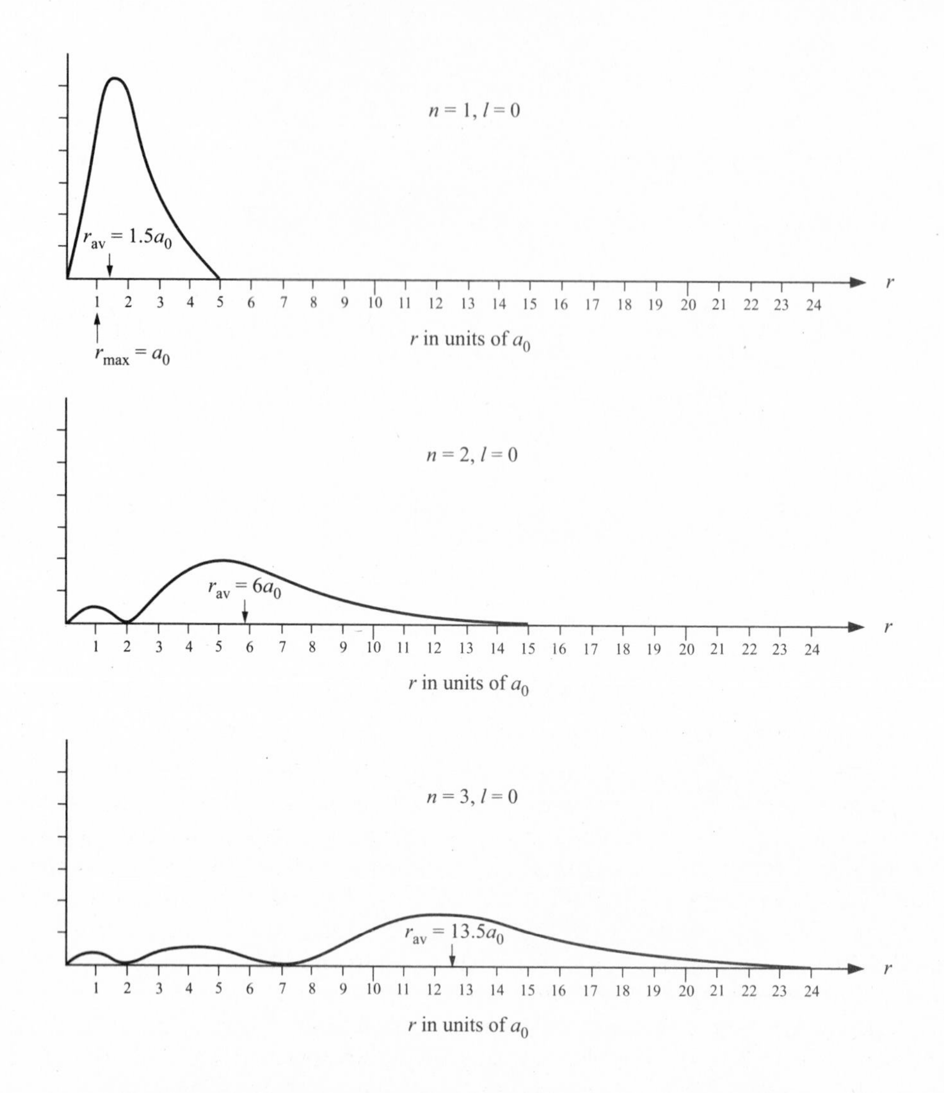

Schroedinger predicted radial probability distribution for a one electron atom.

Fig. 18-10

a particular orbit at a given time. Similarly the velocity or momentum of the electron also must be considered a probability distribution, so the electron does not have a particular momentum in a particular orbit at a given instant of time.

Many Electron Atoms and the Pauli Exclusion Principal

We now turn our attention to the case of atoms with more than one electron. There still is an electrical interaction between each electron and the nucleus which is the same as for only one electron. However, as noted earlier, there is also an electrical interaction between the electrons themselves. This complicates the problem tremendously, especially if there are many electrons. Fortunately, it is possible to obtain some important features of the solution by using a very crude approximation. If we assume that the effect of all electrons can be considered as modifying the force due to the nucleus by providing some shielding of the large positive charge on the nucleus, then we can calculate the energy levels for each electron using this new, more complicated force. The result is an energy level diagram similar to that for only one electron, except that the energy now depends more strongly on the value of l and not only on the value of n. Each electron, however, has the same energy level diagram. In this model we

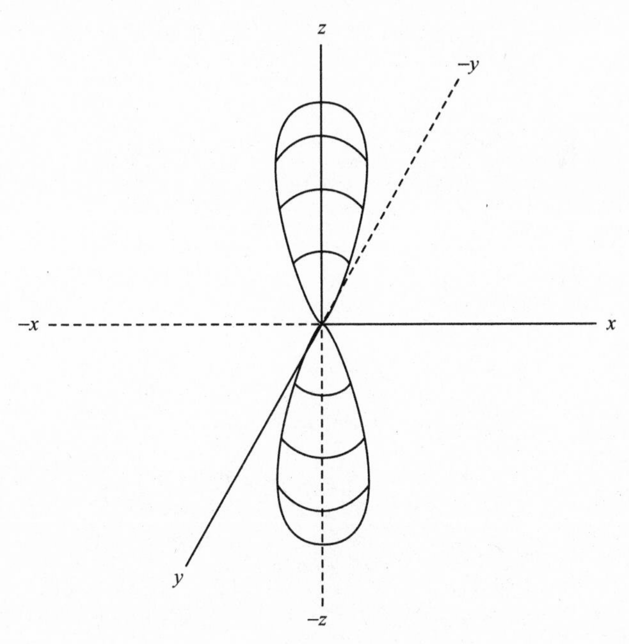

Fig. 18-11

would expect that in the most stable configuration all the electrons would be in the lowest possible energy level, namely the K shell. Any electron in a higher level could radiate its excess energy and transfer to this lowest state. However, this is not what is found experimentally. The most stable, lowest energy state includes electrons in a number of different energy levels. To explain this discrepancy, **Pauli** proposed his "**exclusion principle**" which states that no more than one electron in an atom can have the same quantum numbers n, l, m_l and m_s. We now know that this principle applies to any particle that has a spin of $\frac{1}{2}$ (or $\frac{3}{2}$, $\frac{5}{2}$, etc.). In that case, only one electron can have the quantum numbers for the lowest energy level. The next electron will have to go into the next level, and subsequent electrons into ever higher levels. To see how this occurs, we make use of Fig. 18-9. We see that there are two levels in the K shell, and the first two electrons will go there. The third electron will have to go to the L shell which is at a much higher energy. Here too there are two states in the s subshell, and then there are six states in the p subshell. After a total of ten electrons have been added (two in the K shell and eight in the L shell), one must add the next electron to the M shell which is at a still higher energy. This process continues until one exhausts all the electrons of the atom. The actual energy separations of the levels are not quantitatively shown in Fig. 18-9. The separation between shells and between subshells within shells depends sensitively on the number of electrons in the atom. When there are just enough electrons to fill a shell the energies of these electrons are particularly low, giving the atom great stability. Whenever there is only one electron in a shell, that electron is not held too tightly, and can easily be ionized from the atom. Whenever there is just one electron missing to complete a shell, it is very stabilizing for the atom to capture one more electron (from a nearby atom) to complete the shell. These general ideas can explain many of the features of the periodic table. Atoms with closed shells neither want to lose or accept electrons, and these atoms are therefore chemically inert. Atoms with one more electron than is needed to fill a shell lose electrons readily in chemical reactions and these atoms correspond to the alkali metals. Atoms with one electron less than a full shell (or full sub p shell) gain electrons readily in chemical reactions and these atoms are called the halogens. The following table shows this in some detail (only n, l values are given).

Beyond this point the situation becomes more complex. One would expect that the next energy levels to be occupied are those with $n = 3$ and $l = 2$. However the energy of the ten d electrons turns out to be greater than the energy of the two electrons with $n = 4$ and $l = 0$. Therefore these two s states are filled next, followed by the ten d electrons with $n = 3$ and the six p electrons with $n = 4$. These

Table 18.1. Properties of the Light Elements

Shell	n	l	Element	no. of total electrons	Property
K	1	0	Hydrogen	1	
	1	0	Helium	2	Inert gas (full K shell)
L	2	0	Lithium	3	Alkali metal (one L shell s electron)
	2	0	Beryllium	4	
	2	1	Boron	5	
	2	1	Carbon	6	
	2	1	Nitrogen	7	
	2	1	Oxygen	8	
	2	1	Fluorine	9	Halogen (one L shell p electron needed for completion)
	2	1	Neon	10	Inert gas (full L shell)
M	3	0	Sodium	11	Alkali metal (one M shell s electron)
	3	0	Magnesium	12	
	3	1	Aluminium	13	
	3	1	Silicon	14	
	3	1	Phosphorus	15	
	3	1	Sulfur	16	
	3	1	Chlorine	17	Halogen (one M shell p electron needed for completion)
	3	1	Argon	18	Inert gas (full s and p subshells)

eighteen states close out the next shell with krypton, an inert gas with 36 electrons. Bromine with one less electron ($n = 4$, $l = 1$) is a halogen, and rubidium with one more electron ($n = 5$, $l = 0$) is an alkali metal. In this manner one can explain most of the chemical properties of the elements. Detailed calculations have made it possible to obtain very good agreement between theory and experiment.

Problem 18.10. List the electrons that fill the ground state of the following atoms: (*a*) magnesium with 12 electrons; (*b*) arsenic with 33 electrons; (*c*) iron with 26 electrons; and (*d*) rubidium with 37 electrons.

Solution

(*a*) The first two electrons have $n = 1$, $l = 0$, and are therefore 1s electrons. The next two electrons have $n = 2$, $l = 0$ and are 2s electrons. The next six electrons have $n = 2$, $l = 1$ and are 2p electrons. This adds up to ten electrons. The remaining two electrons are 3s electrons ($n = 3$, $l = 0$). This can be summarized as $(1s)^2(2s)^2(2p)^6(3s)^2$, where the superscript represents the number of electrons in the (nl) subshell.

(*b*) Proceeding as in part (*a*), the first 12 electrons are $(1s)^2(2s)^2(2p)^6(3s)^2$. The next six electrons are 3p electrons. The next electrons are 4s, rather than 3d, since they turn out to have lower energy. This adds two more electrons (4s). We then add the ten 3d electrons, for a total of 30 electrons. The remaining three electrons are 4p electrons, giving the structure as: $(1s)^2(2s)^2(2p)^6(3s)^2(3p)^6(4s)^2(3d)^{10}(4p)^3$.

(*c*) For this case we begin as in parts (*a*) and (*b*) for the first 20 electrons, giving $(1s)^2(2s)^2(2p)^6(3s)^2(3p)^6(4s)^2$. The remaining six electrons are 3d electrons, and the final structure is: $(1s)^2(2s)^2(2p)^6(3s)^2(3p)^6(4s)^2(3d)^6$.

(*d*) The first thirty electrons are the same as in case (*b*), giving $(1s)^2(2s)^2(2p)^6(3s)^2(3p)^6(4s)^2(3d)^{10}$. This is followed by the six 4p electrons. The remaining electron will be the 5s electron, which has less energy than the 4d electrons (and certainly less than the 4f electrons). The final structure therefore is: $(1s)^2(2s)^2(2p)^6(3s)^2(3p)^6(4s)^2(3d)^{10}(4p)^6(5s)^1$.

X-Ray Lines

With this information, we can now also explain a special feature of the X-ray spectrum when electrons are stopped in a material. We discussed the origin of the continuous X-ray spectrum in the last chapter. There the energy from a decelerating electron is converted directly to a photon. The incident electron can lose energy in a non-radiative manner also. It could transfer energy by a "collision" to one of the bound electrons in the material, resulting in the removal of that electron from the atom. If that removed electron is one of the inner electrons (for instance, in the K or L shell), then there will temporarily be a "hole" in the electron structure. To fill this hole, one of the outer electrons (for instance, from the M or N shell), could transfer to this state, leaving a hole in the outer state instead. By dropping into the inner state the electron loses energy, and that energy is radiated away by a photon with a wavelength corresponding to the difference in energy between the two states. If the lower state was in the K shell, then the radiation is called K radiation (K_α or K_β etc. depending on the initial state): if the lower state was in the L shell, then the radiation is L radiation, etc. For each material, only particular wavelengths are produced, those corresponding to the difference in energy between the levels for that particular material. These wavelengths are often in the X-ray region, especially for materials with many electrons. These characteristic wavelengths then appear in the X-ray spectrum in addition to the continuous spectrum due to deceleration of the incoming electrons which is essentially the same for all materials.

There is a special feature of this radiation. The inner (K shell) electrons are hardly affected by the presence of the outer electrons in terms of the energy of their level. Their energy can be calculated as being that of a single electron atom with the nucleus having a charge of $(Z - 1)e$, where we use $(Z - 1)$ because there is another electron in the K shell which shields out one positive charge. Then the energy of the level would be proportional to $(Z - 1)^2$, as we saw for a single electron atom previously. As the atomic number of the atom (Z) increases, the energy of the emitted X-ray should increase and be proportional to $(Z - 1)^2$. This was indeed found experimentally to be the case even before the origin of these X-rays was understood. Indeed, Mosely was able to use this information to correct some errors in the ordering of the periodic table, and to ensure that the table was correctly ordered in terms of increasing Z.

Problem 18.11. Assuming that the energy level of a K shell electron in a lead atom ($Z = 82$) is given by that of a one-electron atom with atomic number $(Z - 1)$, calculate: (*a*) the energy of the K electron; (*b*) the wavelength emitted if a passing free electron with almost no initial kinetic energy falls into this level; and (*c*) the ratio of this wavelength to the corresponding one for tin ($Z = 50$).

Solution

(*a*) The energy is given by Eq. (*18.6a*), $E_n = -(2.18 \times 10^{-18})Z^2/n^2$ Joules, with $Z = (82 - 1)$, and $n = 1$. Thus $E = -1.43 \times 10^{-14}$ J.

(*b*) The wavelength is given by $E = hc/\lambda = E_i - E_f = 0 - (-1.43 \times 10^{-14})$. Thus, $\lambda = 6.625 \times 10^{-34}(3.0 \times 10^8)/1.43 \times 10^{-14} = 1.39 \times 10^{-11}$ m.

(*c*) Since the energy of the K level is proportional to $(Z - 1)^2$, the wavelength is proportional to $(Z - 1)^{-2}$. Thus $\lambda_{Pb}/\lambda_{Sn} = (50 - 1)^2/(82 - 1)^2 = 0.366$.

Lasers

We have seen how the ideas of quantum mechanics have enabled us to understand many of the phenomena associated with atomic physics. Indeed, all of atomic physics can be well understood in great detail using the equations of quantum mechanics. This is also true of solid-state physics, which we will discuss later on in this chapter. We can use these concepts to understand the operation of lasers, which can produce very intense and focused light, whose frequency is restricted to a very narrow range. Most lasers are made of solid-state materials, but the principles can be understood from our previous discussion.

Suppose that we have an atom that has an energy level structure similar to the one shown in Fig. 18-12. The electrons in all the atoms would tend to be found in the "**ground state**", in which the electrons fill the various levels from the bottom up, and thus has the lowest overall energy, E_0. If the electrons were somehow removed to a higher state we say that the atom is in an "**excited state**" with energy E_1. When such an electron transitions back to its original ground state, a photon is emitted of energy $(E_1 - E_0)$. Often such an excited state electron will rapidly transition downward emitting the photon of frequency $f_1 = (E_1 - E_0)/h$. This is called "**spontaneous emission**". In some cases, however, the transition downward is not particularly favored to occur and, if undisturbed, the atom can stay in the excited state for a relatively (on atomic time scales) long time. This is called a metastable state. It was shown already by Einstein, that if such an atom is in state (1) and a photon at frequency f_1 passes by, this radiation could stimulate the electron to make the transition back to the ground state, by emitting a second photon at frequency f_1 that is in phase with the first photon. Such a process is called "**stimulated emission**". The new photon also travels in the same direction as the initial photon. If there were many atoms in level (1), then each of the photons could stimulate other emissions which would stimulate even more emission causing a cascade of photons at the same frequency. This radiation would be especially intense since the radiation from all the independent atoms are in phase with each other. This is the basic concept needed for understanding the operation of the laser. Somehow we produce a group of atoms that are in a metastable state (1). Then, when light of frequency f_1 is produced spontaneously by just one atom, it will stimulate an avalanche of other atoms to simultaneously produce more radiation at the same frequency, in phase with the original radiation and traveling in the same direction. The resulting radiation can not only be very intense but also very sharply focused along the direction of photon travel.

To provide a group of atoms in metastable level (1), we often make use of level (2). We shine light of frequency $f_2 = (E_2 - E_1)/h$ onto the atoms. When this is absorbed by the atoms in the ground state, they are excited into level (2). We say that the atoms are "pumped" into level (2). The atoms in level (2) can lose a small amount of energy and transfer into level (1), either radiatively or otherwise. Thus we are continuously providing atoms in level (1), without using any radiation of frequency f_1 to produce those states. When sufficient numbers of these atoms appear in level (1), one of them will spontaneously transition to the ground state setting off the cascade for laser operation. Naturally, it is not trivial to find atoms with the right level structure to work properly, but sufficient numbers of systems have been found in the visible, infrared and ultraviolet regions that lasers are quite common nowadays. There are many different types of level structure that are used, and many different ways to fill the metastable state, but the basic idea is similar to the case of the level structure and pumping mechanism used in our illustration. In the microwave and radio regions, similar sources were discovered and are called masers. Actually the very first demonstrations of the laser principle was with a maser.

Problem 18.12. A laser is used in eye surgery to attach a retina, and has a wavelength of 514 nm.

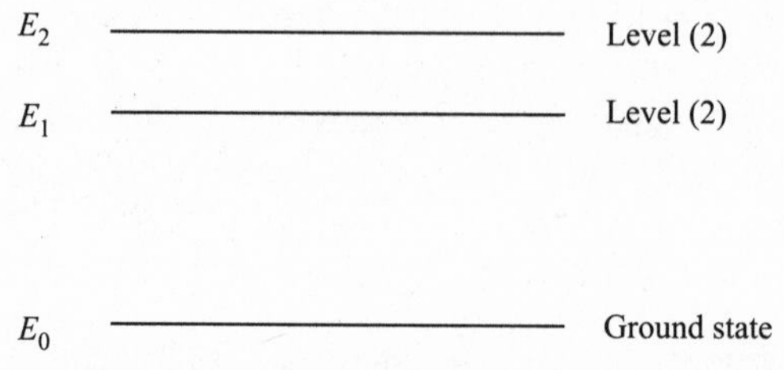

E_2 ———————— Level (2)

E_1 ———————— Level (2)

E_0 ———————— Ground state

Fig. 18-12

(*a*) What is the energy difference, in eV, between the levels that produce the stimulated radiation?

(*b*) If the laser beam has a power of 2 W and is turned on for 0.10 s, how many photons are emitted by the laser?

Solution

(*a*) The energy difference is given by $\Delta E = hf = hc/\lambda = (6.625 \times 10^{-34}\ \text{J} \cdot \text{s})(3.0 \times 10^8\ \text{m/s})/(514 \times 10^{-9}\ \text{m}) = 3.87 \times 10^{-19}\ \text{J} = 2.42\ \text{eV}$.

(*b*) Each photon has an energy of 3.87×10^{-19} J as we calculated in (*a*). Thus there are $(2\ \text{J/s})/(3.87 \times 10^{-19}\ \text{J/photon}) = 5.17 \times 10^{18}$ photons/s emitted. During a time of 0.10 s, the number of photons emitted is 5.17×10^{17} photons.

18.3 NUCLEI AND RADIOACTIVITY

The Nucleus

The properties that we discussed in atomic physics were due to the electromagnetic interactions of the electrons and nucleus in an atom. The properties of the wave functions and energy levels of these negatively charged electrons in the atom are responsible for the chemical properties of materials. The central role of the nucleus is to provide the positive charge necessary to bind the electrons in the atom. In addition the nucleus provides most of the mass of the atoms and the same atom (i.e. the atomic nucleus Z) can have different mass nuclei called isotopes, which only minimally affect chemical properties but significantly affect the weights of the chemicals. Quite apart from chemical properties, however, the nucleus itself is the source of remarkable properties, including being the storehouse of vast quantities of energy, and we now turn our attention to its study. The nucleus while the source of most of the mass of the atom occupies only a very small part of the volume of the atom, as was shown experimentally by Rutherford. The nucleus consists of particles called nucleons, and we now know that nucleons are either protons or neutrons. Protons have a positive charge, of the same magnitude as an electron, and have a mass of 1.673×10^{-27} kg, roughly 2000 times as massive as an electron. The neutron is uncharged, and has a mass which is only slightly larger than that of a proton. Since the nucleus consists of only protons and neutrons, it is useful to measure the mass of the nucleus in a unit in which these individual nucleons have a mass of approximately one. The atomic mass unit (u) is defined to make the mass of a carbon atom that has six protons and six neutrons in its nucleus, surrounded by electrons, have a mass that is equal to exactly 12 u. $1\ \text{u} = 1.661 \times 10^{-27}$ kg. In this unit, the mass of a proton is 1.0072765 u and the mass of a neutron is 1.0086649 u. An electron has a mass of 0.0005485799 u, so that a normal hydrogen atom (proton plus electron) has a mass of 1.0078250 u. Nuclei, which must consist of integral numbers of protons and neutrons, thus have masses that are close to integral multiples of u. The "**mass number**" of the nucleus equals the number of nucleons (protons and/or neutrons) in the nucleus and is, of course, the integral approximation to the mass in units of u. The "**atomic number**" of the nucleus, Z, equals the number of protons in the nucleus, and the number of neutrons will equal $N = A - Z$. Atoms with nuclei of the same Z have the same number of electrons surrounding the nucleus, and therefore have the same chemical properties. The nuclei may, as noted, have different numbers of neutrons. Nuclei with the same Z and different N are called **isotopes**. Thus, hydrogen normally has a nucleus of one proton. However, an isotope, "heavy" hydrogen, also exists, and has a nucleus containing one neutron in addition to the proton. This nucleus is called a deuteron, and the atom (deuteron plus electron) is called deuterium. Water that is made from deuterium is called "**heavy**" **water**. A third isotope of hydrogen also exists, tritium, containing two neutrons, but it is unstable and decays. We generally designate a nucleus (or the specific isotopic atom) by giving Z and A, and write it as $_ZX^A$, where X is the chemical symbol for the atom. For purposes of the periodic table the atom is characterized by giving the atomic number Z.

Problem 18.13. Complete the table by calculating the number of neutrons, and by using the periodic table to get the atomic symbol.

Atomic number	Mass number	Neutron number	Symbol
1	1	0	$_1H^1$
1	2	1	$_1H^2$
2	4		
2	3		
3	7		
4	9		
6	12		
11	23		
18	40		
26	57		
39	89		
73	181		
79	197		
82	207		
86	222		
92	238		
92	235		

Solution

To calculate the neutron number we merely calculate $A - Z$.

Atomic number	Mass number	Neutron number	Symbol
1	1	0	$_1H^1$
1	2	1	$_1H^2$
2	4	2	$_2He^4$
2	3	1	$_2He^3$
3	7	4	$_3Li^7$
4	9	5	$_4Be^9$
6	12	6	$_6C^{12}$
11	23	12	$_{11}Na^{23}$
18	40	22	$_{18}Ar^{40}$
26	57	31	$_{26}Fe^{57}$
39	89	50	$_{39}Y^{89}$
73	181	108	$_{73}Ta^{181}$
79	197	118	$_{79}Au^{197}$
82	207	125	$_{82}Pb^{207}$
86	222	136	$_{86}Rn^{222}$
92	238	146	$_{92}U^{238}$
92	235	143	$_{92}U^{235}$

From this table we see that for small atomic number, Z, the number of neutrons and the number of protons are nearly equal. However, as Z increases, the number of neutrons increases even faster, so that there are many more neutrons than protons if Z is large. How do we understand this phenomenon? To answer this question we first have to understand why neutrons and protons are bound together in the nucleus in the first place. Obviously, there must be some force that attracts these particles to each other. This force is not one of the forces that we have already discussed. The gravitational force, which is attractive, is much too small to hold the particles together. The electrical force is repulsive for the protons and is zero for the neutral neutrons. So neither of these forces will be able to form a nucleus. We are forced to conclude that there is another force, which we call the nuclear force (or the *strong interaction*), that holds the nucleons together. When we investigate the properties of this force, we find that it is very large, but only at very small distances. The force is called a "short range" force, which is zero unless the particles are within a distance of about 10^{-15} m from each other. Within that small distance, the force is essentially the same for protons and for neutrons. For such a strong force it is very difficult to do exact calculations of the energy levels that quantum mechanics predicts, especially since the force is more complex than the electromagnetic forces. However, some general ideas can be easily seen. If we use the same approach as for atoms, we expect that the "orbital" motion of the particles about each other will produce a series of levels. Because of the **Pauli exclusion principle**, each of these levels will be able to contain only two protons and two neutrons corresponding to the two spin directions. If all the forces between protons and between neutrons were identical, then the energy levels for protons would be the same as for neutrons, and might look like the simplified schematic of Fig. 18-13. From an energy stability point of view, as we fill these levels we first fill the lowest levels, E_0 and E'_0 with two neutrons and two protons. Then we add equal numbers of neutrons and protons to the next levels, E_1 and E'_1. This explains why nuclei tend to have equal numbers of neutrons and protons. However, we know that the positive charge on a proton causes a repulsion between protons, which gets bigger as the number of protons increases. While the attractive power of the nuclear force is far greater than the coulomb repulsion for two protons, as Z increases the coulomb force is no longer negligible. Therefore, the protons will not be bound as tightly as the neutrons, which means that their energy levels are nearer to the ionization (or dissociation) level. The energy levels would then look more like Fig. 18-14. Now, as we fill the states the neutrons will be more tightly bound. When we reach level 7, the energy of the proton's seventh level is higher than the eighth level of neutrons. Thus, we will have a more stable nucleus by adding extra neutrons before adding more protons. This is why there will be more neutrons than protons as Z increases. In Fig. 18-15, we show a plot of Z vs. N of the nuclei, and the deviation from $Z = N$ at high values of Z is clear.

Binding Energies

When nucleons are held together in a nucleus, they are bound to each other by the nuclear force. It requires a certain amount of energy (called its **binding energy**) to remove a nucleon from the nucleus (similar to the ionization energy for an electron in an atom). The energy required to separate all the nucleons from each other is the total binding energy (B.E.) of the nucleus. For a nucleus with A nucleons the average energy needed to remove one nucleon is (B.E.)/A, or the binding energy per nucleon. When one separates all the nucleons, one must supply this amount of energy per nucleon.

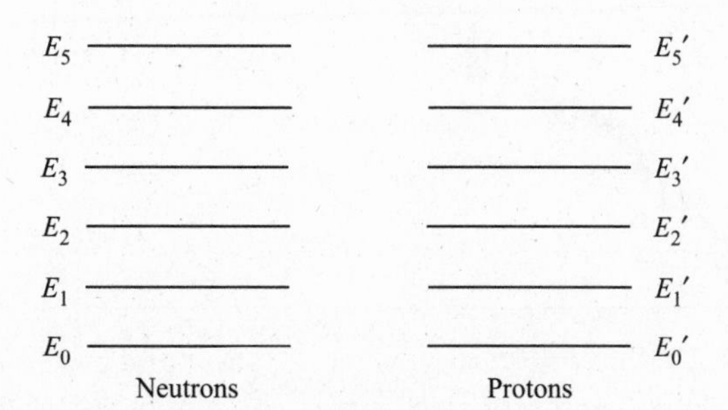

Fig. 18-13

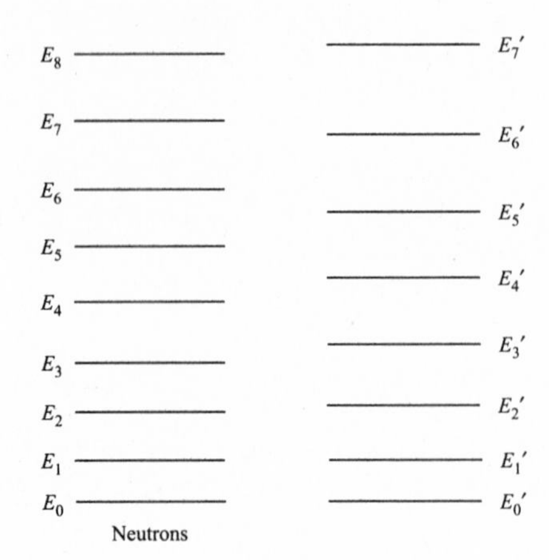

Fig. 18-14

When one takes separated nucleons and forms a nucleus there must be a release of this same amount of energy per nucleon. Thus the total binding energy, B.E., is also the amount of energy released when one forms the nucleus from individual nucleons. The greater the binding energy, the more energy has been released. If we consider the separated nucleons to have zero energy, then the bound nucleus would have, as expected, a negative energy of magnitude equal to the binding energy.

We learned in the chapter on relativity that energy is equivalent to mass. If one loses energy, then one has equivalently lost mass. The amount of mass lost, Δm, must equal the amount of energy lost, ΔE, divided by c^2, since $E = mc^2$. We therefore expect that when we take a proton and a neutron and form a deuteron, we will have less mass in the deuteron than the sum of the mass of the proton and neutron. The same will be true for any other nucleus: it will have less mass than the sum of the masses of its constituents. This "lost" mass is called the "mass defect", and must be equal to the (B.E.)$/c^2$. If we measure the mass of the nucleus, we can calculate the mass defect by comparing this mass to the sum of the masses of its constituents. This will then give us the binding energy of the nucleus. If m_X is the

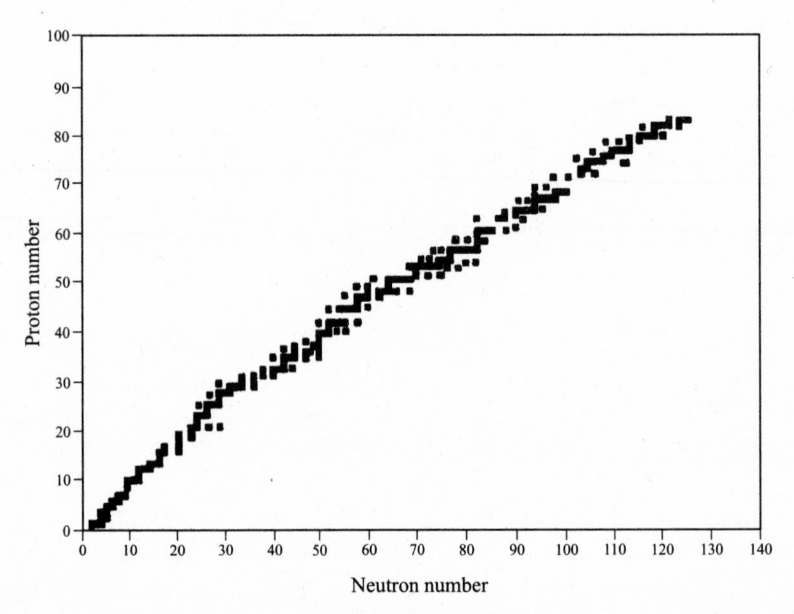

Fig. 18-15

atomic mass of an atom $_ZX^A$ and m_H the mass of a hydrogen atom (proton plus electron) then:

$$\Delta m = Z(m_H) + (A - Z)(m_n) - m_X, \text{ for an atom } _ZX^A \qquad (18.13a)$$

Note that by including the electron mass in m_H we have Z electron masses and Z proton masses in the first term on the right of the equation. Since m_X is an atomic mass it also has Z electrons in it, and when we subtract m_X we are left with the difference in mass between the nucleons (Z protons and $(A - Z)$ neutrons and the mass of the bare nucleus). [We neglect here the slight mass difference due to the small binding energies of the electrons.]

$$\text{B.E.} = \Delta m(c^2) \qquad (18.13b)$$

The binding energy can be separately measured by experimentally determining how much energy we must supply to separate the nucleons in a nucleus from each other. When this is done, we get excellent agreement.

Problem 18.14. The atomic mass of a neutron is 1.0086649 u, and that of $_1H^1$ (a proton plus electron) is 1.0078250 u. Calculate the mass defect for the listed atoms, whose masses are given below.

(*a*) $_1H^2$, with a mass of 2.014102 u.

(*b*) $_2He^3$, with a mass of 3.016029 u.

(*c*) $_2He^4$, with a mass of 4.002603 u.

(*d*) $_6C^{12}$, with a mass of 12.00000 u.

(*e*) $_{26}Fe^{57}$, with a mass of 56.935396 u.

(*f*) $_{92}U^{238}$, with a mass of 238.050786 u.

Solution

(*a*) The mass defect, Δm, is obtained by adding together the masses of all the constituents, and then subtracting the mass of the nucleus. For $_1H^2$, the sum is $1(m_n) + 1(m_H) = 1.0086649 + 1.0078250 = 2.0164899$. Then, the mass defect is $2.0164899 - 2.014102 = 0.002388$ u.

(*b*) Proceeding in the same manner, we get $\Delta m = 1(1.0086649) + 2(1.0078250) - 3.016029 = 0.008286$ u.

(*c*) Again, $\Delta m = 2(1.0086649) + 2(1.0078250) - 4.002603 = 0.030377$ u.

(*d*) $\Delta m = 6(1.0086649) + 6(1.0078250) - 12.000000 = 0.098939$ u.

(*e*) $\Delta m = 31(1.0086649) + 26(1.0078250) - 56.935396 = 0.536666$ u.

(*f*) $\Delta m = 146(1.0086649) + 92(1.0078250) - 238.050786 = 1.934189$ u.

Problem 18.15. Determine the binding energy per nucleon, in MeV, for the cases in Problem 18.14.

Solution

(*a*) The binding energy in Joules is $(\Delta m)c^2$, where Δm is in kg. If $\Delta m = 1$ u, the binding energy is $(1.661 \times 10^{-27}\ \text{kg})(9.0 \times 10^{16}\ \text{m}^2/\text{s}^2) = (1.493 \times 10^{-10}\ \text{J})/1.602 \times 10^{-19}\ \text{eV} = 932$ MeV (where we have used more accurate values for c^2 and J/eV than previously). Therefore, the conversion from a mass in *u* to MeV is 932. For $_1H^2$, the mass defect was 0.002388 u, and this corresponds to a binding energy of (0.002388 u)(932 MeV/u) = 2.226 MeV. Since there are two nucleons in this nucleus, the binding energy per nucleon is 1.11 MeV.

(*b*) Here the mass defect is 0.008286 u, giving a binding energy of 932(0.008286) = 7.72 MeV, and a binding energy per nucleon of 7.72/3 = 2.57 MeV.

(*c*) Here the mass defect is 0.030377 u, giving a binding energy of 932(0.030377) = 28.31 MeV, and a binding energy per nucleon of 28.31/4 = 7.07 MeV.

(*d*) Here the mass defect is 0.098939 u, giving a binding energy of 932(0.098939) = 92.21 MeV, and a binding energy per nucleon of 92.21/12 = 7.68 MeV.

(*e*) Here the mass defect is 0.536666 u, giving a binding energy of 932(0.536666) = 500.02 MeV, and a binding energy per nucleon of 500.02/57 = 8.77 MeV.

(*f*) Here the mass defect is 1.934195 u, giving a binding energy of 932(1.934189) = 1802.66 MeV, and a binding energy per nucleon of 1802.66/238 = 7.57 MeV.

Problem 18.16. The binding energy per nucleon for $_{82}Pb^{208}$ is 7.93 MeV/nucleon. What is the atomic mass of this atom?

Solution

The total binding energy of the atom is 208(7.93) = 1649.44 MeV. This means that the mass defect is 1649.44/932 = 1.76979 u. Using Eq. (*18.13a*), we get that the mass of the lead is $m =$ 126(1.0086640) + 82(1.0078250) − 1.76979 = 207.964 u.

Problem 18.17. Use the data given in Problem 18.14 for the isotope $_1H^2$. From energy considerations, calculate the minimum photon frequency necessary to separate the neutron and proton. This process is known as photo-disintegration.

Solution

The energy needed to dissociate this nucleus is equal to the binding energy of the nucleus. We calculated the binding energy [Problem 18.16(*a*)] to be 2.23 MeV. Therefore, the photon must have an energy of at least 2.23 MeV. (In order to conserve momentum also, the energy will have to be slightly greater than this.) A photon with an energy of 2.23 MeV has a frequency of $f = E/h = (2.23 \times 10^6 \text{ eV})(1.6 \times 10^{-19} \text{ J/eV})/(6.625 \times 10^{-34} \text{ J} \cdot \text{s}) = 5.39 \times 10^{20}$ Hz.

The above problems illustrated how we can get information on how strongly nucleons are bound within a nucleus. If the binding energy is large, then a lot of energy has been released in forming the nucleus, and the nucleons are held very tightly. If the binding energy is small, then the nucleons are only weakly bound together. It may be possible to change that nucleus into a different nucleus with more binding energy, thus releasing more energy. To see if this is possible, it is useful to plot the binding energy per nucleon vs. A, and to analyze the result. In Fig. 18-16, we plot (B.E.)/A vs. A. We note that there is a maximum (B.E.)/A of 8.77 MeV/nucleon, which occurs for Fe^{57}. This particular nucleus has nucleons which are more tightly bound than in any other nucleus. If the nucleons in other nuclei would be able to rearrange themselves into Fe^{57}, they would each release more energy until their new binding energy increases to 8.77 MeV/nucleon. This can occur if the light elements ($A < 57$) fuse together to form nuclei with larger A, in a process called fusion. This is the source of energy of the sun and the other stars, as well as of the hydrogen bomb. It is hoped to someday be able to use this source of energy in a controlled manner on earth by combining hydrogen atoms into helium atoms which have larger binding energies. One can also release energy by converting very heavy nuclei (such as uranium) into lighter nuclei which have a larger (B.E.)/A, thus releasing energy. This process is known as fission, and is the source of energy of nuclear reactors, as well as of the atomic bomb.

Problem 18.18. $_2He^4$ has a binding energy per nucleon of 7.0 MeV/A, while $_1H^2$ (deuterium) has a binding energy per nucleon of 1.11 MeV/A. If one fuses two atoms of deuterium into one atom of He^4, how much energy is released?

Solution

Before the fusion the total binding energy was 2(1.11) = 2.22 MeV. After the fusion, the binding energy is 4(7.0) = 28 MeV. An additional energy of (28 − 2.22) ≈ 25.8 MeV was therefore released in this fusion.

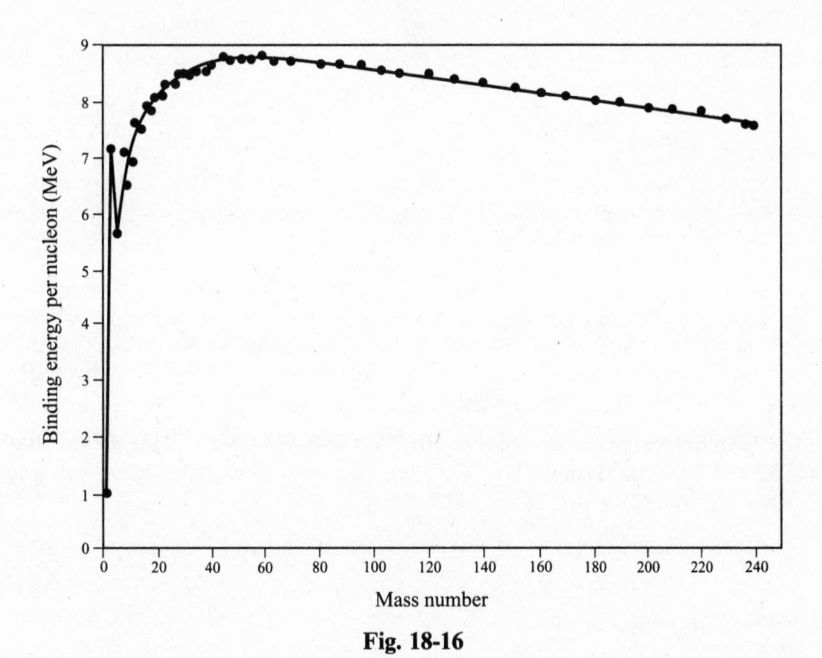

Fig. 18-16

Radioactivity

So far we have discussed the properties of stable nuclei. Many nuclei, including those occurring naturally, are unstable, and they decay to another nucleus in a characteristic manner. In this section, we will discuss the general properties of radioactive decay, and in the next section we will discuss the properties of certain specific types of decay.

Each particular nuclear decay is characterized by a "**half-life**", τ, which is the average time that it takes for half of the nuclei to decay. Although one cannot predict which individual atom will have its nucleus decay at any time, only half the original nuclei, N_0, still remain after a time τ, or $N_0/2$ nuclei. If one waits an additional interval of time τ, then half of these decay, and we will have only $(N_0/2)/2 = N_0/2^2$ remaining. After another τ, the remaining number is $N_0/2^3$. In general, after a time of $m\tau$, the remaining number is $N_0/2^m$. This is plotted in Fig. 18-17. Clearly, N/N_0 drops exponentially as $2^{-(t/\tau)}$, and it is not hard to show that we can re-express this exponential in terms of the natural

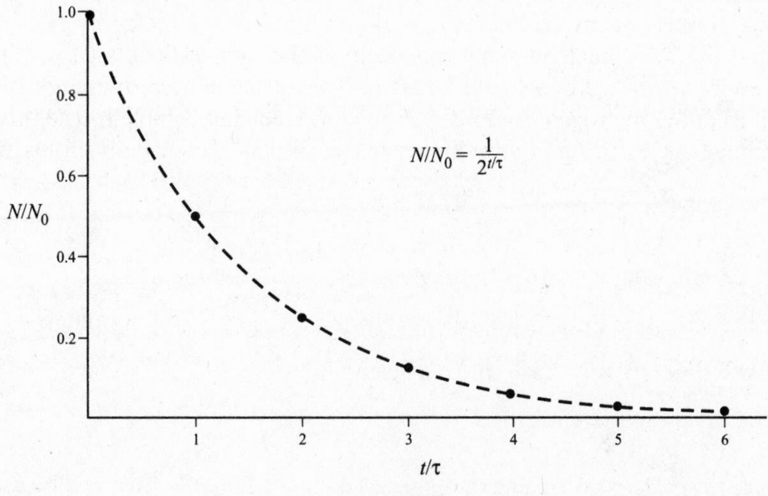

Fig. 18-17

exponent e. The result is:

$$N = N_0 e^{-\lambda t}, \qquad \text{with } \lambda = (\ln 2)/\tau. \tag{18.4}$$

This is an exponential decay, with decay constant λ. There is another fundamental way to arrive at this same equation. If one assumes that the number of particles that decay during a short time Δt (short compared with the half-life) is proportional to the number of particles present at that time, then we can write

$$\Delta N/\Delta t = -\lambda N \tag{18.15}$$

with λ some positive constant and where the minus sign means that N is decreasing. This equation can be solved using calculus, and the result is Eq. (*18.14*). The quantity, $|\Delta N/\Delta t|$, represents the number of decays per second at any given time. This quantity is called the activity of the radioactive material, and is measured in terms of **Becquerel** (Bq), which is one decay per second. A more common unit for activity is the **Curie** (Ci), which is 3.70×10^{10} Bq.

Problem 18.19. A certain material has a decay constant of $4.2 \times 10^3\ \text{s}^{-1}$, and starts with 3.0×10^7 atoms.

(*a*) What is the half-life of the material?

(*b*) What is the initial activity of the material?

(*c*) How many atoms remain after a time of 1.2×10^{-3} s?

(*d*) What is the activity of the sample after 1.2×10^{-3} s?

Solution

(*a*) Using Eq. (*18.14*), $\tau = (\ln 2)/\lambda = 0.693/4.2 \times 10^3 = 1.65 \times 10^{-4}$ s.

(*b*) Using Eq. (*18.15*), $\Delta N/\Delta t = -\lambda N$, we get that the activity $|\Delta N/\Delta t| = 4.2 \times 10^3(3.0 \times 10^7) = 1.26 \times 10^{11}\ \text{Bq} = 3.4$ Ci.

(*c*) Using Eq. (*18.14*), $N = N_0\ e^{-\lambda t}$, we get $N = 3.0 \times 10^7 \exp(-4.2 \times 10^3 \times 1.2 \times 10^{-3}) = 3.0 \times 10^7 \exp(-5.04) = 1.94 \times 10^5$ atoms.

(*d*) Using Eq. (*18.15*), $\Delta N/\Delta t = -\lambda N$, we get that the activity, $|\Delta N/\Delta t| = 4.2 \times 10^3(1.94 \times 10^5) = 8.16 \times 10^8\ \text{Bq} = 0.022$ Ci.

Problem 18.20. A certain sample has 5.6×10^8 particles. After a time of 25 s, only 0.70×10^8 particles remain. What is the half-life for this decay?

Solution

We first calculate $N/N_0 = 0.70 \times 10^8/5.6 \times 10^8 = 1/8$. There are now two ways to proceed to solve this problem. First, we note that $1/8 = 1/2^3$, which means that three half-lives have elapsed. Then $25 = 3\tau$, and $\tau = 25/3 = 8.33$ s. The second method uses Eq. (*18.14*), $N/N_0 = e^{-\lambda t}$, $0.125 = e^{-\lambda t}$, $\ln(0.125) = -\lambda t = -2.079$, $= 2.079/25 = 0.832 = (\ln 2)/\tau$, $\tau = 8.33$ s.

Problem 18.21. A certain sample has 6.4×10^7 particles, and a half-life of 3.5 s.

(*a*) What is the decay constant for this decay?

(*b*) At what time will there be only 1.3×10^7 particles left?

(*c*) What is the activity after a time of 10 s?

Solution

(*a*) We know that $\lambda = (\ln 2)/\tau$. Thus $\lambda = 0.693/3.5 = 0.198\ \text{s}^{-1}$.

(*b*) Using Eq. (*18.14*), $N = N_0\,e^{-\lambda t}$, we get $N = 1.3 \times 10^7 = 6.4 \times 10^7 \exp(-0.198t)$. Then, $\exp(-0.198t) = 0.203$, $\ln(0.203) = -1.59 = -0.198t$, $t = 8.05$ s.

(*c*) The activity, $|\Delta N/\Delta t| = +\lambda N$. We must first calculate $N = N_0 \exp(-\lambda t) = 6.4 \times 10^7 \exp(-0.198 \times 10) = 8.84 \times 10^6$. Then $|\Delta N/\Delta t| = 0.198(8.84 \times 10^7) = 1.75 \times 10^6$ Bq $= 4.73 \times 10^{-5}$ Ci $= 47.3\ \mu$Ci.

One of the applications of naturally occurring radioactivity is in dating of geological or archeological samples. If one assumes that one knows the initial composition of a material in terms of the nuclei present, then, if some of the nuclei are radioactive we can determine the age of the sample by measuring the composition at the present time. The age is given by the equation $N/N_0 = \exp(-\lambda t)$, or $t = \ln(N_0/N)/\lambda$. Unfortunately, one does not generally know N_0. What can be done is illustrated by the following problem.

Problem 18.22. The isotope of carbon, C^{14}, is radioactive, with a half-life of 5730 years, while C^{12} is stable. Initially the ratio of C^{12} to C^{14} is 8.3×10^{11}. After a time, t, the ratio has increased to 9.1×10^{12}. Determine the time t.

Solution

Let N_{12} equal the number of C^{12} nuclei originally present. Since this isotope is stable, we will still have the same number after the time t. Using the ratios of C^{12} to C^{14} given for the initial and final time, we see that the initial number of C^{14} nuclei, N_0, was $N_{12}/8.3 \times 10^{11}$, and the final number of C^{14} nuclei present, N, is $N_{12}/9.1 \times 10^{12}$. Thus, $N/N_0 = (8.3 \times 10^{11})/(9.1 \times 10^{12}) = 0.0912$. But we also have $N/N_0 = \exp(-\lambda t)$. Now $\lambda = 0.693/\tau = 0.693/5730$ y $= 1.210 \times 10^{-4}$ y^{-1}. Thus, $\ln(0.0912) = \lambda t \rightarrow -2.39 = -1.21 \times 10^{-4}$ t, or $t = 19{,}790$ y.

This technique would allow us to date organic materials on the basis of the following assumptions: (1) The ratio of C^{12}/C^{14} in the atmosphere has remained essentially stable over geologic time; (2) As an organic substance grows and absorbs carbon, the ratio absorbed is the same as the ratio in the sea of air above; (3) Once the organic material (e.g. a tree) dies there is no more carbon absorbed or emitted chemically, so that the only change in composition comes from the radioactive decay of the carbon 14. We can usually compensate for uncertainties in these assumptions, and in many cases we can check the accuracy of age determinations by using additional radioactive decays of other isotopes. This has become a standard technique, although it requires careful measurements as well as thorough analysis to be sure that it is legitimately applied in each case.

Natural Decay Processes

In a natural decay process, the nucleus spontaneously emits some particle and becomes transformed into some other nucleus. In this process the conservation laws of nature must all be satisfied, and the possible modes of decay are restricted by this condition. Indeed, by examining what possible processes do not occur, we are led to new conservation laws that are needed to explain why they do not occur. The same is true of other nuclear reactions, such as fission, fusion, inelastic collisions, etc. In this section we will discuss the three main decay modes which are found to occur naturally, and study their characteristics with these conservation laws in our mind.

There are three conservation laws which we know from our study of mechanics: conservation of energy, of linear momentum and of angular momentum. In addition, we learned in the chapter on electricity about the law of conservation of charge. All of these conservation laws must be satisfied in any decay that occurs in nature. In addition, we find in the naturally occurring decays that the total number of protons and neutrons (nucleons) must be the same before and after a decay. A proton or a neutron can change into the other, but not into electrons or photons, for instance. This conservation law is just a special case of the more general law of conservation of "**baryons**", which include a whole host of exotic particles in addition to protons and neutrons. As long as the total number of baryons is conserved, any particle that is a baryon is allowed to change into a different baryon (provided that all the other conservation laws are also satisfied). For the decays we will discuss in this section, the only baryons that are involved are the nucleons, namely the proton and the neutron.

There are three main naturally occurring decays, which are called α, β and γ. In α-decay the particle emitted from a larger nucleus is the nucleus of a helium atom, in β-decay the particle emitted is an electron and in γ-decay the particle emitted is a photon.

For **α-decay** the decay process can be written as follows:

$$_ZX^A \rightarrow {}_{Z-2}D^{A-4} + {}_2\text{He}^4 \tag{18.16}$$

Here, the nucleus X has Z protons and $(A - Z)$ neutrons, for a total of A nucleons. After the decay the "daughter" (D) nucleus has $(Z - 2)$ protons and a total of $(A - 4)$ nucleons. The alpha particle has two protons and a total of four nucleons. Thus, the number of protons after the decay is $(Z - 2) + 2 = Z$, as before the decay. This guarantees conservation of charge. Similarly, the total number of nucleons after the decay is $(A - 4) + 4 = A$, as it was before the decay. This guarantees conservation of baryon number. It is easy to see that in Eq. (*18.16*) this means that if the total of the subscripts before the decay equals the total afterwards, then charge is conserved. Also, if the total of superscripts are equal before and after the decay, then the number of nucleons (baryons) is conserved.

We now turn our attention to conservation of energy. We know from relativity that energy and mass are equivalent, so we must be sure to also include the energy due to rest mass. If the rest mass on the left side of the equation is greater than the rest mass on the right-hand side, this means that some rest mass has been "lost" and converted into kinetic energy of the decay products. If the rest mass is smaller on the left side than on the right side, then the energy after the decay would be greater than before the decay, which cannot happen unless energy is added by some other source beforehand. We therefore can conclude that a particle will decay by itself only if the decay products have less rest mass than the original particle. The excess mass of the original particle is given as kinetic energy to the decay products. We can write this in the form of an equation:

$$m_X c^2 = (m_\text{D} + m_\text{He})c^2 + E_\text{K} \tag{18.17a}$$

or

$$E_\text{K} = (m_X - m_\text{D} - m_\text{He})c^2 \tag{18.17b}$$

If we recall that the equivalent energy of 1 u of rest mass equals 932 MeV, the kinetic energy in MeV will equal 932 times the mass differences in u. Thus,

$$E_\text{K} = (m_X - m_\text{D} - m_\text{He})(932 \text{ Mev/u}), \text{ with all } m \text{ in u.} \tag{18.17c}$$

Problem 18.23. The isotope of uranium, ${}_{92}\text{U}^{238}$, is radioactive, and decays via emission of an α particle into thorium. The masses of the particles are $m_\text{U} = 238.0508$ u, $m_\text{Th} = 234.0436$ u and $m_\text{He} = 4.0026$ u.

(*a*) Write the symbolic equation for the decay.

(*b*) How much kinetic energy is given to the decay products?

Solution

(*a*) From conservation of charge we know that thorium must have 90 protons, and from conservation of baryons we know that thorium must have 234 nucleons. Thus the equation becomes:

$${}_{92}\text{U}^{238} \rightarrow {}_{90}\text{Th}^{234} + {}_2\text{He}^4$$

(*b*) The kinetic energy is equal to the energy equivalent of the loss in rest mass, as given in Eq. (*18.17c*). Thus $E_\text{K} = (238.0508 - 234.0436 - 4.0026)(932) = 4.3$ MeV.

Problem 18.24. Referring to Problem 18.23, and using the equation for the kinetic energy given to the decay products, express this kinetic energy in terms of the binding energies of the particles involved rather than in terms of the masses of the particles.

Solution

We start with Eq. (*18.17c*), $E_K = (m_X - m_D - m_{He})(932)$ MeV, with all m in u. We now write the masses in terms of the binding energies using Eq. (*18.13*), $\Delta m_X = Z(m_H) + (A - Z)(m_n) - m_X$, and (B.E.) $= \Delta m_X(c^2)$. Thus $m_X = Z(m_H) + (A - Z)(m_n) - (\text{B.E.})_X/c^2$. Alternatively, with masses in u and energy expressed in equivalent mass units in u, this can be written as $m_X = Z(m_H) + (A - Z)(m_n) - (\text{B.E.})_X$.

Then
$$m_U = 92(m_H) + (238 - 92)(m_n) - \text{B.E.}_U$$
$$m_{Th} = 90(m_H) + (234 - 90)(m_n) - \text{B.E.}_{Th}$$
$$m_{He} = 2(m_H) + (4 - 2)(m_n) - \text{B.E.}_{He}$$

We use this to find that $(m_U - m_{Th} - m_{He}) = \text{B.E.}_{Th} + \text{B.E.}_{He} - \text{B.E.}_U$, since the masses of the protons and the neutrons cancels. This is because Z constant and A constant implies that there is no change in either the number of protons or in the number of neutrons. Reexpressing the binding energies in energy units (e.g., MeV) we have for the kinetic energy released in the decay: $E_K = \text{B.E.}_{Th} + \text{B.E.}_{He} - \text{B.E.}_U = \text{B.E.}_f - \text{B.E.}_i = \Delta(\text{B.E.})_X$. In any decay or reaction where the number of protons and the number of neutrons individually does not change, the energy released is just equal to the net *increase* in binding energy after the decay.

We see from the above that the kinetic energy of the decay particles can be determined from a knowledge of the masses or binding energies of the particles involved. This kinetic energy is distributed among the decay products. In the case of alpha decay the energy is shared by the daughter nucleus and the α-particle. From energy considerations alone, we cannot determine how much energy is given to the daughter nucleus and how much to the α-particle. However, using the other conservation law, conservation of momentum, will enable us to assign a unique energy to each of these particles.

We consider the case of a nucleus that spontaneously decays while at rest. In that case, there is no momentum before the decay. Conservation of momentum requires that there be no momentum after the decay as well. After the decay, two particles are moving, and each has momentum. For the total momentum to be zero afterwards requires that these two particles are moving in opposite directions with equal momentum. We can write this as an equation where v is the magnitude of the velocity of each particle,

$$m_D v_D = m_\alpha v_\alpha \tag{18.18}$$

If we combine this with the equation for conservation of energy, $E_K = E_{KD} + E_{K\alpha}$, we have two equations in the two unknowns, v_D and v_α. The kinetic energy is expressed in terms of v either with the relativistic or the non-relativistic relation, as required. Solving those two equations gives us the velocity and the energy of each of the decay products for this decay.

Problem 18.25. For the decay of uranium in Problem 18.23, calculate the velocity and the energy of the α-particle.

Solution

We showed in Problem 18.23 that the kinetic energy of the decay products was 4.3 MeV. This is much less than the rest mass energy of the decay products, so we will be able to use non-relativistic equations for the kinetic energy. Thus $E_K = 4.3\ \text{MeV} = (\frac{1}{2})m_{Th}v_{Th}^2 + (\frac{1}{2})m_{He}v_{He}^2$. Conservation of momentum means $m_{Th}v_{Th} = m_{He}v_{He}$. From this equation we get that $v_{Th} = m_{He}v_{He}/m_{Th}$. Substituting this into the energy equation gives: $4.3\ \text{MeV} = (1/2)m_{Th}(m_{He}v_{He}/m_{Th})^2 + (\frac{1}{2})m_{He}v_{He}^2 = (\frac{1}{2})m_{He}v_{He}^2\{1 + m_{He}/m_{Th}\} = 4.3 \times 10^6(1.6 \times 10^{-19})\ \text{J} = (\frac{1}{2})m_{He}v_{He}^2(1 + 4.00/234) = (\frac{1}{2})(4.00 \times 1.66 \times 10^{-27})v_{He}^2(1.017)$, giving $v_{He}^2 = 2.037 \times 10^{14}$, and $v_{He} = 1.43 \times 10^7$ m/s. The kinetic energy of the alpha particle is $(\frac{1}{2})m_{He}v_{He}^2 = 6.76 \times 10^{-13}$

J = 4.22 MeV. It is not surprising that almost all the kinetic energy is carried off by the α-particles given that thorium is so massive composed to the α.

For each type nucleus that undergoes α-decay, all the α-particles are released at the same energy. In the case of U^{238} that energy is 4.22 MeV. In the case of Bi^{212} that energy is 6.09 MeV. This uniqueness is due to the fact that there are only two decay products, so that the equations of conservation of energy and momentum are sufficient to uniquely solve for the two unknown velocities. If there were three decay products, then the two conservation law equations could not determine unique values for the three unknown velocities, and the kinetic energy will be distributed among the three products over a range of values.

In **β-decay**, an electron is spontaneously emitted from a nucleus. The decay products that we see are thus a daughter nucleus plus an electron. If this were all that took place, then the reaction equation would be:

$$ {}_{z}X^{A} \rightarrow {}_{z+1}\mathrm{D}^{A} + {}_{-1}\mathrm{e}^{0} \tag{18.19a}$$

where X is the radioactive nucleus, D is the nucleus to which X decays (the daughter nucleus), and e is the electron. We have conservation of nucleons (baryons) since the electron is not a baryon and there are equal numbers of nucleons, A, in both the decaying nucleus and in the daughter nucleus. Conservation of charge occurs because the daughter nucleus has one extra positive charge (one extra proton) and the negative charge of the electron compensates for this. Since the daughter nucleus has an extra proton, but the same number of nucleons, it has one less neutron. The β-decay is therefore effectively a process in which one neutron in the nucleus converts into a proton plus an electron. In fact, free neutrons are themselves radioactive, decaying via β-decay into a proton and electron with a half-life of 12 minutes. We will examine the case of the β-decay of the neutron as the prototype for all β-decays. The equation for this decay can be written as:

$$ {}_{0}\mathrm{n}^{1} \rightarrow {}_{1}\mathrm{p}^{1} + {}_{-1}\mathrm{e}^{0} \tag{18.20a}$$

Problem 18.26. From conservation of energy, determine the kinetic energy available to the decay products when a neutron at rest undergoes β-decay.

Solution

For conservation of energy we require that the rest mass energy of the neutron equal the rest mass energy of the decay products (proton plus electron) plus the kinetic energy given to the products. This can occur only if the rest mass of the neutron is larger than the combined rest mass of a proton and electron. We note that a proton plus free electron has essentially the same rest mass as a hydrogen atom, ${}_{1}H^{1}$. (The binding energy of the electron in the hydrogen atom corresponds to an equivalent mass of less than 3/100,000 of the rest mass of an electron!) For the neutron decay, we can then say that the rest mass of the neutron must equal the rest mass of a hydrogen atom plus the sum of the kinetic energies given to the proton and electron that fly off. The kinetic energy given to the proton and electron would then equal

$$E_K = (m_N - m_{H)}(932)\mathrm{MeV}, \text{ with the masses in u.} \tag{18.21a}$$

Using the values for the masses we get $E_K = (1.008{,}665 - 1.007{,}825)(932) = 0.783$ MeV. This energy is distributed among the decay products.

We showed previously, for the case α of decay, that if there are only two decay products, then conservation of momentum, combined with the conservation of energy, gives unique values for the velocity and kinetic energy of the two products. Thus, the outgoing electron in β-decay of the neutron should always have the same kinetic energy. When the experiment is performed, we find that the energy of the electron ranges from zero to nearly 0.783 MeV. Furthermore, if we apply conservation of momentum, and note that the initial momentum was zero (the neutron was at rest), the final momentum will also have to be zero. If there are only two decay products their momenta will have to be not only equal in magnitude but opposite in direction. When we observe the decay products (proton and electron) we find that they do not usually travel in opposite directions. Thus both conservation of energy and conservation of momentum are violated if there are only two particles after the decay. Lastly, it turns out that such a decay also violates conservation of angular momentum (we will not discuss the details). Unless scientists were prepared to abandon the three greatest conservation laws known to man their only possible conclusion was that there must be at least one more particle among the decay products, which for some reason had eluded detection in studies of β-decay, and this particle was called the **neutrino**. Using conservation laws scientists were able to deduce some of the properties that this elusive particle must have. To conserve charge in the decay, the neutrino must be neutral. To conserve baryons, the neutrino cannot be a "baryon". To conserve energy, the neutrino has no rest mass, or at best very little mass, as can be shown by detailed careful measurements of the energy and momentum of the decay products. To conserve angular momentum, the neutrino must have a spin of $\frac{1}{2}$. Arguments such as these were used to determine many other properties of the neutrino as well. Direct observation of the neutrino was accomplished only about 20 years after its existence was predicted from experiments on β-decay and it had all the properties predicted. Its discovery gave the scientific community a collective sigh of relief and was a great triumph for the validity of the great conservation laws of physics. The elusiveness of the neutrino is a consequence of the fact that it interacts weakly with other matter, thus not providing easy clues to its whereabouts. The symbol we use for the neutrino is ν, and the correct equation for β-decay becomes

$$_{Z}X^{A} \to {}_{Z+1}\mathrm{D}^{A} + {}_{-}\mathrm{e}^{0} + \bar{\nu} \qquad (18.19b)$$

which replaces incorrect Eq. (*18.19a*) and for the neutron, this becomes

$$_{0}\mathrm{n}^{1} \to {}_{1}\mathrm{p}^{1} + {}_{-1}\mathrm{e}^{0} + \bar{\nu} \qquad (18.20b)$$

which replaces incorrect Eq. (*18.20a*).

The energy released as kinetic energy is distributed among the three decay products, and given that the neutrino has essentially zero rest mass, [Eq. (*18.21a*)] for the neutron decay remains correct. For the more general decay of atomic nucleus X into daughter nucleus, D, Eq. (*18.19b*) we have

$$E_{\mathrm{K}} = (m_X - m_{\mathrm{D}})(932)\ \mathrm{MeV, with\ the\ masses\ in\ u} \qquad (18.21b)$$

where m_X is the atomic mass of atom X and m_{D} is the atomic mass of daughter atom D. Using the atomic masses keeps proper track of the total number of electrons before and after the decay since the D atom has one more electron than the X atom.

Problem 18.27. $_{5}\mathrm{B}^{12}$ decays via β-decay. The atomic mass of this isotope of boron is 12.014,354 u.

(*a*) Write down the reaction equation for this decay.

(*b*) What kinetic energy is given to the decay products?

Solution

(*a*) Using Eq. (*18.19b*) we get:

$$_{5}\mathrm{B}^{12} \to {}_{6}\mathrm{C}^{12} + {}_{-1}\mathrm{e}^{0} + \bar{\nu}$$

(*b*) The kinetic energy of the decay products is the difference between the atomic masses of the parent and daughter nuclei. The mass of the boron was given as 12.014,354 u, and the mass of C^{12} is exactly 12.000,000 u (this isotope was used to define the unit). Thus $E_{\mathrm{K}} = (12.014,354 - 12.000,000)(932)\mathrm{MeV} = 13.4\ \mathrm{MeV}$.

We can ask about the possibility that a proton decay into a neutron. Since the proton has less rest mass than a neutron, this cannot happen for a free proton because of conservation of energy. In a nucleus, however, it is possible that a nucleus with an extra neutron and one less proton has less mass than the original nucleus, so energy conservation would allow this to occur. However, conservation of charge requires that if a proton becomes an neutron, another positively charged particle must also be created. Indeed, a particle which is essentially identical to an electron, except for its positive charge was predicted and discovered, and given the name positron. This positron is the "antiparticle" of the electron, and is also called a positive β-particle (β^+), or a positive electron (e^+). We can therefore also have positron decay, for which the reaction is:

$$_ZX^A \rightarrow {}_{Z-1}D^A + {}_1e^0 + \nu \tag{18.22}$$

It should be noted that the neutrino also has an anti-neutrino counterpart. By convention, the particle emitted in Eq. (*18.22*) is called the neutrino and the particles emitted in regular β decay, Eqs. (*18.19b*) and (*18.20b*) are labeled anti-neutrinos. We will not discuss this process any further and instead turn to the final form of spontaneous nuclear decay.

This third general category of decay (**γ-decay**) is one in which the emerging particle is a "γ-ray", which is a very energetic electromagnetic wave (a photon). In this decay there is no change in either Z or A, since the photon (γ-ray) is uncharged and not a "baryon". This means that the effect of the emission of a γ-ray is merely to carry away energy from the nucleus. We view this in the same manner as for atomic physics where a photon carries away energy as the electrons make transitions between different levels. Similarly, the nucleus has energy levels, and a photon is emitted when the nucleus makes a transition to a lower level.

We can summarize the results of our analysis of the three types of decay in terms of the new nuclei they produce. For α-decay there is a decrease of 4 in A and a decrease of 2 in Z. For β-decay there is a decrease of 0 in A and an increase of ± 1 in Z (depending on whether it is an electron or a positron that is emitted). For γ-decay there is no change in either A or Z. Such decaying nuclei are called "radioactive", the degree of radioactivity depending on the half-life for the decay. Highly radioactive means short half-life, weakly radioactive means long half-life. We see, however, that the only possible change in A that can occur is a decrease of 4. There is no restriction on Z since changes of ± 1 can occur. The restriction on the change in A means that there are four distinct groups of radioactive nuclei characterized by their mass number. In each group the value of A differs by multiples of 4 (238, 234, 230, 226, ... 186, ... etc.). Within each group it is possible for any nucleus to be connected to another one via a series of decays, but there is no possibility, using α-, β-, or γ-decay to connect nuclei from different groups. Thus there occur in nature four "decay series", and three of these are still present. The fourth does not contain any nucleus with a half-life long enough so that any atoms remain in nature, having all decayed since the formation of the earth (although they can be created in nuclear collisions). The other three are headed by a nucleus with a long half-life.

Problem 18.28. The last, and lightest, nucleus in each of the four decay series is a different isotope of lead, all four of which are stable. They are Pb^{206}, Pb^{207}, Pb^{208}, and Pb^{209}. Which will be the isotope reached if one starts with $_{92}U^{238}$?

Solution

The final nucleus must have a mass number, A, that differs from 238 by a multiple of 4. The four isotopes have A of 206, 207, 208, and 209, which differ from 238 by 32, 31, 30, and 29, respectively. Only 32 is divisible by 4, which means that the stable nucleus to which U^{238} decays is Pb^{206}.

Induced Reactions

In addition to the naturally occurring nuclear transformations via radioactivity, it is possible to induce nuclear transformations by bombarding the nucleus with another particle or nucleus. For instance, one can strike the nucleus with a proton, or a neutron, or an α-particle, or a γ-ray, etc. The result of such a collision could be a new nucleus plus the emission of a different particle or particles. In all such collisions, called nuclear reactions, the outcomes are restricted by all of the conservation laws that we discussed previously.

Problem 18.29. A nuclear reaction is induced when an α-particle strikes a nucleus of N^{14}, joining to form a new nucleus, and a proton is emitted in the reaction.

(*a*) How many protons are contained in the new nucleus that is formed?

(*b*) What is the mass number for the new nucleus?

(*c*) Write down the reaction equation for this case.

Solution

(*a*) We use conservation of charge to determine the number of protons in the new nucleus. Before the collision, we had seven protons in the nitrogen nucleus ($Z = 7$) and two in the α-particle for a total of nine positive charges. After the collision we have one free proton and therefore the new nucleus must have eight protons and therefore is an oxygen nucleus ($Z = 8$).

(*b*) We use conservation of nucleons (baryons) to determine the mass number of the new nucleus. Before the collision we had 14 nucleons in the nitrogen and four in the α-particle for a total of 18 nucleons. After the collision there is one free nucleon (the proton) and therefore the oxygen nucleus must have 17 nucleons.

(*c*) The reaction equation is therefore:

$$_2He^4 + {}_7N^{14} \rightarrow {}_1H^1 + {}_8O^{17}$$

This reaction is often written as $_7N^{14}(\alpha, p)_8O^{17}$, which is interpreted to mean that a nitrogen nucleus is struck by an α-particle, emitting a proton (α in, proton out) and producing an oxygen nucleus.

Problem 18.30. Complete the reaction, write down the reaction equation and state what is taking place in the following reactions.

(*a*) $_5B^{10}(n, \alpha)$ ___

(*b*) $_{12}Mg^{25}(\gamma, p)$ ___

(*c*) $_6C^{13}(p, \gamma)$ ___

Solution

(*a*) We use conservation of charge and nucleons to determine the number of protons and nucleons in the new nucleus. Before the collision, we had five protons in the boron nucleus and none in the neutron for a total of five positive charges. After the collision we have two protons in the α-particle and therefore the new nucleus must have three protons and therefore is a lithium nucleus ($Z = 3$). Similarly we get that $A = 7$. The reaction is one in which an neutron strikes a boron nucleus producing a lithium nucleus and an α particle. The reaction equation is:

$$_0n^1 + {}_5B^{10} \rightarrow {}_2He^4 + {}_3Li^7, \qquad \text{or} \qquad {}_5B^{10}(n, \alpha)_3Li^7$$

(*b*) Using the same reasoning we deduce that the new nucleus is $_{11}Na^{24}$. The reaction is one in which a γ-ray strikes a magnesium nucleus producing a nucleus of sodium and a free proton. The reaction equation is:

$$\gamma + {}_{12}Mg^{25} \rightarrow {}_1H^1 + {}_{11}Na^{24}, \qquad \text{or} \qquad {}_{12}Mg^{25}(\gamma, p)_{11}Na^{24}$$

(*c*) Again from charge and baryon conservation we determine that the new nucleus is $_7N^{14}$, which means that the reaction was one in which a proton strikes a carbon nucleus producing nitrogen and a γ-ray.

The reaction equation is:

$${}_1H^1 + {}_6C^{13} \rightarrow \gamma + {}_7N^{14}, \qquad \text{or} \qquad {}_6C^{13}(p, \gamma){}_7N^{14}$$

We now turn to special kinds of reactions called fission and fusion reactions.

Fission Reactions

We noted earlier that the binding energy per nucleon is higher for A in the range of 60 than it is for a very large value of A, as in uranium. Thus, if a uranium nucleus would "**fission**", i.e. break apart into two smaller nuclei, then each nucleon on average would gain binding energy and this additional binding energy would be released in the form of kinetic energy given to the fission products. Although such fission does occur spontaneously, i.e. without any outside assistance, it occurs very rarely. The reason for this rarity is that in order to fission into two smaller nuclei, the atom must first stretch apart, and the energy of this stretched state is higher than the original energy. This is not energetically possible (at least in Newtonian physics) unless it somehow received enough energy from an outside source to overcome this barrier. Since quantum mechanics relies on waves for describing the probabilities of particle motions, and waves can penetrate thin barriers, we can show that there is a slight probability that the uranium can fission even if it does not get the energy needed to surmount the barrier. This process is called "tunneling", since we think of the particle as if it had dug a tunnel through the wall of the barrier. For nuclear fission, however, this is very rare. In order to help the uranium nucleus to fission, we can send in additional energy via a photon (photofission) or a neutron. Then the nucleus has enough energy to surmount the barrier and it will immediately break apart. Often, the products of fission include, in addition to fairly massive nuclei, other more elemental particles such as neutrons.

Problem 18.31.

(*a*) Complete the reaction equation for the fission process:

$${}_0n^1 + {}_{92}U^{235} \rightarrow {}_{\_\_\_}?^{141} + {}_{36}Kr^{-} + 3{}_0n^1$$

(*b*) Explain why extra neutrons are likely to be emitted in a fission process as they are in the example shown.

(*c*) If the average binding energy per nucleon is 7.6 MeV/A for uranium, and is 8.5 MeV/A for the fission fragments, how much energy is given to the fission products in this process? Assume the incoming neutron that induces the fission is very slow moving.

Solution

(*a*) We use conservation of charge to determine the number of protons in the nucleus. Before the fission, we had 92 protons and after the fission we have 36 protons in the krypton. Therefore the other nucleus must have 56 protons and therefore is a barium nucleus ($Z = 56$). Similarly, using conservation of nucleons, we originally have 236 nucleons (235 in the uranium plus one free neutron). After the fission we have 141 nucleons in the barium and three free neutrons for a total of 144. The krypton nucleus must therefore have 92 nucleons. The reaction equation is therefore:

$${}_0n^1 + {}_{92}U^{235} \rightarrow {}_{56}Ba^{141} + {}_{56}Kr^{92} + 3{}_0n^1$$

(*b*) We learned previously that as Z increases, the number of neutrons in a stable nucleus increases even more quickly. Thus when one doubles the number of protons one more than doubles the number of neutrons. Similarly if one halves the number of protons one needs less than half the number of neutrons with which one started. A fission process is like dividing the nucleus into half, and there will then be more neutrons available than are needed for the new nuclei. These excess neutrons are generally released in a fission process. Actually, the fission fragments themselves usually still have too many neutrons, and therefore are radioactive. They generally decay by changing a neutron into a proton (β-decay), and those decay products sometimes are also β-emitters or emit a neutron (delayed neutrons).

(c) In this fission process we have the same number of protons before and after the fission as well as the same number of neutrons, i.e. no nucleon changes into a different nucleon. Since we are told that the incoming neutron's kinetic energy is negligible, the total energy we have before the fission is the rest mass of the uranium and the rest mass of the neutron. This will equal the rest mass of all the nucleons in the uranium nucleus and the extra neutron, minus the binding energy of the nucleons in the uranium. Similarly, after the fission the total rest mass energy will equal the rest mass of all the protons and neutrons in the products, minus the binding energy of the nucleons in the Ba and the Kr. To get the energy released, we have to subtract the rest mass energy after the fission from the rest mass energy before the fission. The rest masses of the protons and neutrons before and after will cancel, and we are left with $E_{released} = (-\text{B.E.}_{\text{U}}) - (-\text{B.E.}_{\text{Ba}} - \text{B.E.}_{\text{Kr}}) = \text{B.E.}_{\text{Ba}} + \text{B.E.}_{\text{Kr}} - \text{B.E.}_{\text{U}} = 8.5\ \text{MeV}(141 + 92) - 7.6\ \text{MeV}(235) = 194.5\ \text{MeV}$. Thus, each fission process releases approximately 200 MeV of energy.

The energy released is huge compared to the energy released in a chemical reaction of two atoms or two molecules, such as the explosion of TNT. In chemical reactions the energy released per molecular interaction is typically a few eV, which is 10^8 times smaller than the above fission reaction. This is why nuclear energy is much more potent than chemical energy. The energy released can be used destructively in an atom bomb, or constructively in a nuclear reactor. Whether for a bomb or to generate thermal energy we must arrange to have many nuclei fission in order to produce a large total amount of energy. This is made possible by using the extra neutrons that are released in each fission to initiate a new fission. The process is known as a "**chain reaction**". If the chain is not controlled it could produce a large, nearly instantaneous release of energy as in a bomb. If it is controlled, the energy release can be gradual, as in a reactor.

Problem 18.32. A fission reaction produces 200 MeV for each fission.

(a) How many fissions per second are required to produce energy at the rate of 100 MW?

(b) How much mass of uranium is being split every second?

Solution

(a) To get 100 MW, we must produce 100 MJ of energy every second. This is 6.25×10^{26} eV/s $= 6.25 \times 10^{20}$ MeV/s. Since each fission produces 200 MeV we must have $6.25 \times 10^{20}/200 = 3.13 \times 10^{18}$ fissions/s.

(b) Each fission involves one uranium nucleus of approximate mass 235 u of uranium. This is a mass of $235(1.66 \times 10^{-27}\ \text{kg}) = 3.90 \times 10^{-25}$ kg. Thus, every second there are fissions involving $(3.92 \times 10^{-25}\ \text{kg})(3.13 \times 10^{18}\ \text{fissions}) = 1.23 \times 10^{-6}$ kg of uranium.

Note. A more accurate figure would be obtained using the actual atomic mass of U^{235}, which accounts for the difference of proton and neutron masses from 1 u, accounts for the binding energy, and includes the mass of the electrons in the U^{235} atom.

In practice, if one wants to build a reactor (or a bomb) one must have the U^{235} isotope for the fission process. Natural uranium contains mainly U^{238} (99.3%), with only 0.7% of U^{235}. The likelihood of fission occurring in U^{238} (we call this likelihood the "**cross-section**") is too small to use as a practical material. Only U^{235} (and a new nucleus, produced in reactions, plutonium, Pu^{239}) has a sufficiently large cross-section for use in a reactor. It is very difficult to separate U^{235} from U^{238} since both isotopes have the same chemical properties and they differ only slightly in mass. This difficulty is a key factor that limits the ability for making nuclear weapons to a few highly industrialized countries. When one has "enriched" the uranium so that it contains about 3% of U^{235}, one can use this mixture as the fuel for a reactor. The Pu^{239} that can also be used for a fission reactor is produced in the reaction of U^{238} with a neutron. When one adds a neutron to U^{238}, the new nucleus, U^{239} decays via β-decay to Np^{239}, which decays further via β-decay to Pu^{239}. The reactions are

$$_0n^1 + {_{92}U^{238}} \rightarrow {_{92}U^{239}} \rightarrow {_{93}Np^{239}} + \beta^- + \bar{\nu} \rightarrow {_{94}Pu^{239}} + \beta^- + \bar{\nu}$$

The plutonium produced in this reaction can be chemically separated from the uranium and used as fuel for a fission reactor (or bomb). Since all reactors contain U^{238}, all reactors produce some Pu during their operation. The possibility of the proliferation of Pu is a source of great concern. Some reactors are built to specifically take advantage of the production of new fuel in the form of Pu. One can build a reactor that produces more fuel (in the form of Pu) than it consumes (in the form of U^{235}). These reactors are called "**breeder**" **reactors**.

The fission reaction that is induced by the incoming neutron produces extra neutrons, with a large amount of kinetic energy, which we want to use to induce further fissions in a chain reaction. However, only slow neutrons are efficient in inducing fission, so we first have to slow down the neutrons. This is done by a "**moderator**", which is a material with which the neutrons collide, and to which they transfer their energy. We know that energy transfer is most efficient if an energetic object collides with another body of roughly the same mass (e.g. one pool ball gives all its energy to another pool ball that is initially at rest in a head on collision, but loses very little energy when colliding with the wall of the table). Therefore, the moderator should be made of a material which contains atoms nearly the same mass as a neutron. The closest material in mass would be hydrogen (a proton), which is plentiful in water. However, a proton has a tendency to absorb the neutron and form a deuteron ($_1H^2$), which removes the neutron from play instead of slowing it down, thus inhibiting the chain reaction tremendously. Therefore, "**heavy**" **water**, already made from deuterium is a much better alternative moderator. Many reactors use carbon or sodium as the moderator. No matter what the process, if the excess neutrons that are produced in a fission induce, on average, less than one new fission, then the reactor is "sub-critical" and will not produce a chain reaction. If the average number of fissions induced by the extra neutrons is just one, the reactor is critical, and a chain reaction will be sustained. If more than one fission is induced, on average, then the reaction will increase rapidly in number, possibly leading to an explosion (bomb). The number of induced fissions produced is controlled by the use of "**control rods**" made of material that absorbs neutrons. These are automatically adjusted to keep the energy production at the intended level. Another control mechanism that prevents a properly constructed reactor from getting out of control is the fact that as the reactor core overheats, the moderator will boil away and this will automatically reduce the number of slow neutrons and hence the fission reaction. There is therefore no danger that reactors will explode as a nuclear bomb. What are the problems of reactors?

We learned that the fission fragments produced in the fission reaction are radioactive. Many of the fission fragments have half-lives that are very long, e.g. thousands of years. Thus, a nuclear reactor continually produces radioactive nuclei whose radioactivity persists for a long time. This "exhaust" material must be collected and stored in a safe environment so that the radioactive material cannot escape into the soil, water or air. It is very difficult to find a storage method that can maintain safety over decades or even centuries and that is politically acceptable to everybody. This is one of the most serious problems associated with nuclear reactors. Another serious problem is the possibility of an accident in a reactor that would release radioactivity into the environment. Although many safeguards are built into reactors to avoid this release, there is always a possibility that the safeguards are not sufficient. In the USA the most serious accident that occurred was the near meltdown at Three Mile Island. This effectively destroyed the reactor, but, fortunately, very little radioactivity was released into the atmosphere. However, the accident in Chernobyl, in the USSR, was much more serious, and a huge amount of radioactivity was released, which traveled in the atmosphere to many parts of Europe as well as in the USSR. These dangers have to be assessed in the context of alternative sources of power. It is well known that electric generators that use coal and oil as their source of energy damage the environment with acid rain, and with an increase in the temperature of the planet due to the "**greenhouse effect**", as well as depleting a non-renewable resource. The alternative of reducing the use of energy is a difficult one to implement, although efforts are being made to limit the use of energy to the greatest possible acceptable extent. This is a problem that is not likely to be solved easily. One great hope for a solution would provide almost inexhaustible supplies of "clean" energy is nuclear fusion.

Fusion Reactions

Just as it is possible to release energy by splitting heavy nuclei, it is also possible to generate energy by combining together light nuclei. The binding energy per nucleon of light nuclei is smaller than that of nuclei with mass numbers near 70, and fusing those light nuclei together increases the average binding energy per nucleon, thus releasing energy. This process is known as fusion, and is the source of energy of the sun (and other stars) as well as of the H-bomb.

Problem 18.33. A possible fusion reaction combines two neutrons and two protons into an α-particle. The mass of an α-particle (actually of the ${}_2\mathrm{He}^4$ atom) is 4.002,603 u, while that of a neutron is 1.008,665 u and of ${}_1\mathrm{H}^1$ is 1.007,825 u.

(*a*) What is the binding energy of the α-particle?

(*b*) How much energy is released in this fusion process?

(*c*) How many fusions are required per second to generate a power of 100 MW?

(*d*) How much mass of "fuel" is involved per second in generating this 100 MW?

Solution

(*a*) The binding energy is the energy equivalent of the loss in mass as one fuses together the particles into the nucleus. The mass deficit Δm is $(2m_N + 2m_H - m_{He}) = 2(1.008{,}665 + 1.007{,}825)$ u $- 4.002{,}603$ u $= 0.030{,}377$ u $= 28.3$ MeV.

Note. Again, using atomic masses, the masses of the electrons cancel out (two hydrogen − one helium) so we get the same results as using the masses of the nuclei alone.

(*b*) Since the initial constituents are two free neutrons and two free protons, the energy released is just the binding energy of the α-particle. Thus, each fusion releases 28.3 MeV.

(*c*) To get 100 MW, we must produce 100 MJ of energy every second. This is 6.25×10^{26} eV/s $= 6.25 \times 10^{20}$ MeV/s. Then we must have $(6.25 \times 10^{20}\ \mathrm{MeV/s})/(28.3\ \mathrm{MeV/fusion}) = 2.19 \times 10^{19}$ fusions/s.

(*d*) Each fusion involves two neutrons and two hydrogen atoms, or 4.03 u of fuel. This corresponds to $4.03(1.66 \times 10^{-27}\ \mathrm{kg}) = 6.69 \times 10^{-27}$ kg. Thus, every second there are fusions involving $(6.69 \times 10^{-27}\ \mathrm{kg})(2.19 \times 10^{19}\ \mathrm{fusions/s}) = 1.47 \times 10^{-7}$ kg of hydrogen and neutron fuel.

It is also possible to fuse deuterons together to form helium, or to use a tritium atom and a proton. Actually this is what is done in a H-bomb, but in the sun the process used to generate energy combines four protons. The conservation of charge requires that electrons be emitted in the same process.

Problem 18.34. A possible fusion reaction combines four protons into an α-particle. The mass of an α-particle (actually of ${}_2\mathrm{He}^4$) is 4.002,603 u, while that of ${}_1\mathrm{H}^1$ is 1.007,825.

(*a*) Write down the reaction equation, and show how it conserves charge and baryons.

(*b*) How much energy is given to the products in this fusion process?

Solution

(*a*) We start with the basic facts that are given, that four protons combine into one He nucleus. This would be written as $4\,{}_1\mathrm{H}^1 \rightarrow {}_2\mathrm{He}^4$. However, this does not satisfy conservation of charge, since there are four positive charges before the fusion and only two after the fusion. To conserve charge we must add two positrons to the products of the reaction. This can be viewed as if two protons had converted to neutrons and positrons within the nucleus, as happens in β^+-decay. In that decay we learned that a neutrino is also emitted. Therefore we will also get two neutrinos (in addition to the positrons) in this fusion process. The final reaction equation is therefore:

$$4\,{}_1H^1 \rightarrow {}_2He^4 + 2\beta^+ + 2\nu.$$

This reaction also conserves baryons, since there are four baryons in the four protons, and four baryons in the helium. The positrons and neutrinos are not baryons.

(*b*) The binding energy is the energy equivalent of the loss in mass as one fuses together the particles into the nucleus. The mass deficit Δm of the nucleus is $(4m_H - m_{He}) = 4(1.007{,}825) - 4.002{,}603 = 0.028{,}697$ u = 26.7 MeV. However, additional rest mass is created in the form of the two positrons, each with rest mass energy 0.511 MeV. The neutrinos have no rest mass. The energy available as kinetic energy of the products is thus $26.7 - 2 \times 0.511 = 25.7$ MeV.

We see from this problem that neutrinos are released in the fusion process. These neutrinos hardly interact with any material since they are not charged, and since they do not bind to nucleons. They interact so weakly with matter that they can penetrate the entire interior of the earth without interacting with any matter. However, since there are so many neutrinos released by the sun in its fusion process, we can build large, massive detectors and expect a few of this huge number to interact. This is how neutrinos were finally detected. There are several neutrino detectors around the world which are used to detect neutrinos from the sun, from other astronomical events, and from nuclear reactors. This effort has been so refined that we now know that there are three different types of neutrinos (and associated antineutrinos), only one type of which is involved in β-decay.

If one could build a controlled fusion reactor, we would be able to generate energy by combining the very abundant hydrogen and/or deuterium found in water, and create helium, a very useful, yet environmentally harmless inert gas. No radioactive byproducts would be produced. This would be an excellent way to help alleviate the problems associated with energy generation that were mentioned in the previous section. However, it is very difficult to fuse these particles together, since the electrical force of the positive charges on the protons causes them to repel each other. Unless the protons are moving with a very large kinetic energy, they will not approach each other closely enough to be able to interact and fuse. Thus, for the last fifty years, we have been unable to build a practical fusion reactor (called a thermonuclear reactor), despite intensive effort. Much progress has been made but there is as yet no assurance that such a reactor will become a practical machine.

Elementary Particles

The particles that we have been considering at this stage are the proton and neutron (which are baryons), the electron, positron and neutrino (which we will find are all in a new class called leptons), and the photon. Positrons were predicted by Dirac after he showed that the correct set of quantum wave equations that are consistent with relativity predicted the existence of an "antiparticle" for each particle. These antiparticles are nearly identical with their particles and have either the same or the opposite properties of their particle. Thus a positron has the same mass as the electron, but the opposite charge. Similarly, the antiproton has the same mass as the proton, but the opposite charge (the proton has positive and the antiproton has negative charge). Additionally, the antiproton is a member of the baryon family, but counts as a missing baryon, i.e. if you add together a proton and an antiproton you have a net zero number of baryons. We say that the antiproton has a baryon number of -1. The same is true of an antineutron, which has a baryon number of -1, and the same mass as the neutron. The properties of particles and antiparticles are always such that they can annihilate each other if they meet, with their entire rest masses being converted into energy (usually γ-rays), without violating any conservation law. Similarly, pairs of particle–antiparticle can be produced directly from the conversion of γ-ray energy into the pair (usually in the presence of some heavy object such as a nucleus) without violating any conservation laws. The γ-ray must, of course, have at least an energy equal to the combined rest mass energy of the created particles.

Problem 18.35.

(*a*) What minimum amount of energy is needed to create an electron–positron pair?

(*b*) What minimum amount of energy is needed to create a proton–antiproton pair?

Solution

(*a*) The energy needed is the rest mass energy of an electron plus a positron. Since they both have the same rest mass, this will equal twice the rest mass energy of an electron. We calculated in Problem 16.22 that the rest mass energy of an electron is 0.511 MeV. Therefore, we need 1.022 MeV to create an electron–positron pair.

(*b*) Using the same reasoning as in (*a*), the energy we need is twice the rest mass energy of a proton. This is [see Problem 16.22(*b*)]: $E = 2(0.938\text{ GeV}) = 1.876\text{ GeV}$. In practice, if the pair is formed from a photon or the kinetic energy of moving particles, we need more energy in order to conserve both momentum and energy.

The photon is a particle that turns out to be its own antiparticle. This means that there is no distinction between particle and antiparticle photons, since both have the same charge (zero), rest mass (zero), baryon number (zero), spin (one), etc. The photon has another important property, arising from its connection with Maxwell's equation. When a full quantum theory for electromagnetism was developed it turned out that there was a strong connection between the interactions of charged particles and photons. Using the concepts of quantum mechanics it was possible to show that the electromagnetic interactions (i.e. Coulomb's law) can be described in terms of the "virtual" exchange of photons. By this we mean that the electromagnetic interaction between charged particles can be considered a consequence of a possible continuous creation of photons by one charge and absorption by the other charge. Since these photons are not free to travel away, they are called **virtual photons** to distinguish them from photons that carry energy through space (**electromagnetic wave photons**). Thus the photon can be considered the carrier of the electric and magnetic force between charged particles. The full theory of these interactions is called quantum electrodynamics. As we will see, the idea that one type of particle is the carrier of the force between other particles can be carried over to other forces as well.

As a result of observations of cosmic rays (particles which enter our atmosphere from extraterrestrial space), a new particle was discovered, and is now called a muon, and given the symbol μ^-. This particle has spin $\frac{1}{2}$ and a negative charge e. It has a mass of $207m_e$, but in other respects it behaves essentially the same as an electron. It is sometimes called a heavy electron. Neither the electron (or its antiparticle, the positron), nor the μ^- (or its antiparticle, the μ^+) can be bound in the nucleus. This is because these particles (and the neutrinos as well) are not affected by the force that holds nucleons together in the nucleus. That nuclear force produces what we call the "strong interaction" and the electrons, muons and neutrinos are not affected by the strong interaction. They are, however affected by a much weaker force, giving rise to the "weak interaction", which also affects baryons. Since it is weak, it is not strong enough to bind any of these particles into the nucleus, and any processes using this weak force have much longer half lives than those involving the much stronger "strong interaction". These weakly interacting particles form a "family" of their own, and are called leptons. Just as there is law of conservation of baryons, there is also a law of conservation of leptons. As in the case of baryons, we assign a "**lepton**" number of $+1$ to a lepton, and of -1 to an antilepton. Thus the difference between a neutrino and an antineutrino is that they have opposite lepton numbers. This also explains why β-decay requires an antineutrino rather than a neutrino. In the example of β-decay that we discussed, $\text{n} \rightarrow \text{p} + \text{e} + \bar{\nu}$, we have no leptons before the decay. After the decay there is an electron, which has a lepton number of one. To conserve leptons, there must also be an antilepton present, with lepton number -1. This is an antineutrino. To complete the lepton family we must add another particle, similar to the electron and muon, called the τ particle with a rest mass of $3490m_e$. Also, there are actually three kinds of neutrinos (and their associated antineutrinos), called ν_e, ν_μ and ν_τ, each associated respectively with an electron, a muon and a τ-particle. These six particles (e, μ, τ, ν_e, ν_μ and ν_τ), together with their antiparticles comprise the leptons. The muons and the τ-particle are unstable, and decay to other particles of smaller rest mass. The muon, since it is charged, decays to an electron of the same charge (μ^- to an electron, and μ^+ to a positron), plus the appropriate neutrino and antineutrino.

Problem 18.36. For the decay of a μ^- into an electron, neutrino and antineutrino, show that this conserves charge, baryons, leptons, and mass. The decay is:

$$\mu^- \rightarrow e^- + \bar{\nu}_e + \nu_\mu$$

Solution

Before the decay there is one negative charge, as there is after the decay, giving conservation of charge. There are no baryons before or after the decay, so we have conservation of baryons. There is one lepton before the decay (the muon) and a net of one lepton after the decay ($+1$ for the e, $+1$ for the ν_μ neutrino, and -1 for the $\bar{\nu}_e$ antineutrino). Thus we have conservation of leptons. The rest mass before the decay is $207m_e$ (the mass of the μ). After the decay there is rest mass only in the electron. The decrease in rest mass is converted into kinetic energy of the decay products. Thus we have conservation of energy.

In the above analysis we assumed that the lepton numbers of the electron and electron neutrino and the muon and muon neutrino were separately conserved. This requirement explains why the muon does not decay into an electron plus a photon. This process would satisfy all the conservation laws, including lepton number. The decay into electron plus photon is ruled out by the more stringent lepton conservation law. The new law states that among the leptons, there are three subclasses, which must each be separately conserved. The three subclasses consist of (1) the electron and its neutrino, (2) the muon and its neutrino, and (3) the τ-particle and its neutrino. The τ-particle also decays into a lighter lepton (muon or electron) of the same charge plus the appropriate neutrino and an antineutrino.

The six leptons that we have discussed are considered to be among the fundamental particles that cannot be subdivided further. The photon is also a fundamental particle. The next group of particles that we discuss, the **mesons**, are strongly interacting like the baryons, but have full integer instead of half integer spin. Like the baryons, mesons were once considered to be fundamental particles. Now mesons (and baryons) are thought of as composed of more fundamental particles called quarks. The difference between baryons and mesons lies in the different combinations of quarks that are involved. We will discuss the properties of quarks shortly.

Mesons are particles that consist of one quark and one antiquark. They have electric charges of either zero or ± 1. They are neither baryons nor leptons and there is no law that their number must be conserved. They often have 0 spin angular momentum, but may have other integral (not half-integral) amounts of spin. All mesons are affected by the strong interaction, as are the baryons. We call any particle that participates in the strong interaction a **hadron**. Thus mesons and baryons are both hadrons.

The least massive meson is the **pion** (π^0, π^- or π^+), which has a mass of $\approx 270m_e$. The $\pi^\pm$ are particle antiparticle pairs, and the π^0 is its own antiparticle. The π^0 decays very rapidly into two photons. Since there are no leptons involved in the decay, the weak interaction is not required and the decay is caused by the much stronger (and therefore faster acting) strong and/or electromagnetic interaction. However, in the decay of the $\pi^\pm$, we must have a charged particle as one of the decay products to conserve charge. The only charged particles lighter than the pion are leptons, and therefore the weak interaction is involved. Thus the decay of the $\pi^\pm$ is much slower, having a half-life of $\approx 10^{-8}$ s (compared with 10^{-16} for π^0). These pions decay into a muon and a neutrino, or, more rarely, an electron and a neutrino.

Problem 18.37. A π^- decays into a muon and an antineutrino. Show that this conserves charge, baryons, leptons, and mass.

Solution

The decay is: $\pi^- \rightarrow \mu^- + \bar{\nu}_\mu$. Before the decay there is one negative charge, as there is after the decay, giving conservation of charge. There are no baryons before or after the decay, so we have conservation of baryons. There are no leptons before the decay and a net of zero leptons after the decay ($+1$ for the μ and -1 for the $\bar{\nu}_\mu$ antineutrino). Thus we have conservation of leptons. The rest mass before the decay is $270m_e$ (the mass of the π). After the decay there is rest mass only in the μ. The decrease in rest mass is converted into kinetic energy of the decay products. Thus we have conservation of energy.

The next massive meson is the **kaon**, K^0, and $K^\pm$. The $K^\pm$ are particle–antiparticle pairs. There are actually two K^0 particles which combine to form particle—antiparticle pairs. The mass of the K particle is $\approx 970m_e$. The kaons decay into less massive pions and/or leptons. The half-life is $\approx 10^{-8}$ s. It is hard to understand why this particle should have such a long half-life. After all, the kaon participates in the strong interaction, and can decay into pions that also participate in the strong interaction. There is a similar problem with the decay of a massive baryon called the Λ (lambda)-particle, which is often created together with the kaon. The solution that was conceived is that these particles (as well as some other new particles) have a new quantum number, called "strangeness", which must be conserved in an interaction involving nuclear forces (strong interaction). Strangeness is a new property that, like electric charge, baryon number and lepton number, can be both positive and negative. A kaon is given a strangeness of $+1$, and a Λ-particle is given a strangeness of -1. Pions and nucleons have strangeness 0. Since the kaon is the least massive "strange" particle, it cannot conserve strangeness in its decay to a less massive particle. Therefore the decay cannot be a process governed by the strong interaction, which requires conservation of strangeness. It can, however take place via the weak interaction, which does not require conservation of strangeness. Since this interaction is weak the decay has a relatively long half life. Conservation of strangeness introduces a new concept insofar as it is not an absolute conservation law. It is required for the strong interaction (and the electromagnetic interaction), but not for the weak interaction.

The next group of particles is the baryon family. The baryons are each made up of three quarks. We have already met the proton and the neutron. The proton is stable because it is the lowest mass baryon, and conservation of baryons prevents it from decaying to a meson or lepton. The neutron does decay to a proton plus leptons (electron and antineutrino), which requires the weak interaction and is therefore slow. The next baryon that was discovered is the Λ-particle, which has a mass of $2182m_e$, and a charge of 0. It has enough mass so that the Λ^0 should be able to decay into a proton plus a π^- (or neutron plus π^0) and should therefore decay very rapidly. In fact, it has a comparatively longer half-life than expected, one of $\approx 10^{-10}$ s. The answer developed is the same as for the kaon, namely that the Λ-particle has a "strangeness" number of -1 (its antiparticle has a strangeness of $+1$), and any decay via the strong interaction requires conserving strangeness. If it could decay into a kaon (which also has strangeness) then the decay would be much faster. However the mass of a kaon plus proton is larger than the mass of the Λ-particle. Conservation of energy therefore precludes a decay into a kaon. The Λ-particle must therefore decay via the weak interaction, which does not require conservation of strangeness, and it has a comparatively long half-life. The next group of baryons are the three Σ particles, $\Sigma^\pm$ and Σ^0 (and their three antiparticles). These particles have a mass only slightly larger than a Λ, and the same strangeness as the Λ^0. The Σ^0 can decay into a Λ^0 and a photon, satisfying all the conservation laws. This decay does not involve leptons and proceeds via the strong and electromagnetic interactions. The half-life is correspondingly small, around 10^{-20} s. The charged Σ-particles however must decay into a charged particle, either a proton or a charged pion. Neither of these has strangeness, so strangeness cannot be conserved. This requires the slower weak interaction, and their half-lives are $\approx 10^{-10}$ s. The next baryons are the two Ξ-particles, the Ξ^0 and Ξ^-, with antiparticles Ξ^0 and Ξ^+. Their mass of $\approx 2580m_e$ would allow decay into a Λ and a π. Since they have a half-life of $\approx 10^{-10}$ s, the decay must be mediated by the weak interaction. If the Ξ had a strangeness of 1, the above decay could proceed via the strong interaction and would be fast. The only reason for the slow decay must be that the strangeness of the Ξ-particle is -2! Indeed this is the case. The last baryon we will discuss is the Ω^- particle, which was predicted to exist after the existence of quarks was postulated. This particle has a long half-life because it is the least massive baryon with a strangeness of -3. As we see, there is a seemingly inexhaustible array of oddball particles, and the conservation laws begin to seem arbitrary and capricious. The quark model, however, brings order and reason back into the scheme, as we shall soon see.

Problem 18.38. A Λ-particle is often created together with a kaon when a high energy pion collides with a proton in the following reaction:

$$\pi^- + p^+ \rightarrow \Lambda^0 + K^0$$

(*a*) Show that this reaction can satisfy conservation of charge, baryons, leptons, and strangeness, as required for the strong interaction.

(*b*) The rest masses of the particles are:

$$m_\pi = 273m_e$$
$$m_p = 1836m_e$$
$$m_\Lambda = 2184m_e$$
$$m_K = 974m_e$$

What is the least amount of energy that the pion must have in order to conserve energy? (Actually more energy will be required since the reaction products will have to have momentum, and therefore kinetic energy, in order to conserve momentum also.)

Solution

(*a*) Before the decay there is one negative charge and one positive charge for a net charge of zero. After the reaction, neither particle has a charge giving conservation of charge. There is one baryon before the reaction (the proton), and one baryon after the reaction (the Λ-particle), so we have conservation of baryons. There are no leptons before the reaction and no leptons after the reaction. Thus we have conservation of leptons. Before the reaction we had a strangeness of zero, since a proton and a pion are not particles with strangeness. After the reaction, the Λ-particle has a strangeness of -1 and the kaon has a strangeness of $+1$, giving a net strangeness of zero, as before the reaction.

(*b*) The rest mass before the reaction is $273m_e + 1836m_e = 2109m_e$ (the mass of the π and p). After the reaction there is rest mass of $2184m_e + 974m_e = 3158m_e$ (the masses of the Λ and K). The extra rest mass energy after the collision must be supplied by the kinetic energy of the pion. The minimum energy required is $(3158 - 2109)m_e = 1049m_e = 536$ MeV.

With all these particles that have been discovered (there are many more than we have discussed) we wonder which are **fundamental particles**, and which can be considered as **composites** of more fundamental particles. At present we believe that the photon and the six leptons (and their antiparticles) are all fundamental particles. However, the mesons and the baryons are viewed as composites of more fundamental particles, called **quarks**. We know of six different quarks (and their corresponding antiquarks). The six quarks are called the "up", "down", "strange", "charm", "top" and "bottom" quarks. All six have been detected indirectly by identifying the properties of mesons containing these quarks. Each quark has a charge of $-\frac{1}{3}$ or $+\frac{2}{3}$, a baryon number of $\frac{1}{3}$ (antiquarks have charge of $+\frac{1}{3}$ or $-\frac{2}{3}$ and a baryon number of $-\frac{1}{3}$) and a spin of $\frac{1}{2}$. The charges are: up (u) = $\frac{2}{3}$; down (d) = $-\frac{1}{3}$; strange (s) = $-\frac{1}{3}$; charm (c) = $\frac{2}{3}$; bottom (b) = $-\frac{1}{3}$; and top (t) = $\frac{2}{3}$. Only the s ("strange") quark has a non-zero strangeness (of -1). Similarly, the charm, top and bottom quarks each have a separate unique quantum number, which, like strangeness is conserved in strong interactions. All mesons are composed of one quark and one antiquark, and all baryons are composed of three quarks. Thus, for example: proton = uud; neutron = udd; Λ = uds; $\pi^+ = u\bar{d}$; $\pi^- = \bar{u}d$; $K^+ = u\bar{s}$, where a bar on top indicates an antiparticle.

Problem 18.39. By combining the appropriate number of quarks, show that:

(*a*) All mesons have a baryon number of zero.

(*b*) All mesons have a charge of ± 1 or 0.

(*c*) All baryons have a baryon number of 1.

(*d*) All baryons have a charge of -1, 0, $+1$ or $+2$.

Solution

(*a*) Mesons consist of one quark and one antiquark. All quarks have a baryon number of $\frac{1}{3}$ and all antiquarks have a baryon number of $-\frac{1}{3}$. Combining one quark and one antiquark gives a total baryon number of zero.

(*b*) Mesons consist of one quark and one antiquark. All quarks have a charge of $(-\frac{1}{3})$ or $+\frac{2}{3}$, and all antiquarks have a charge of $\frac{1}{3}$ or $-\frac{2}{3}$. If one combines a quark of charge $-\frac{1}{3}$ with an antiquark of charge $\frac{1}{3}$, or if one combines a quark of charge $\frac{2}{3}$ with an antiquark of charge $-\frac{2}{3}$, then the meson will have a charge of 0. If one combines a quark of charge $-\frac{1}{3}$ with an antiquark of charge $-\frac{2}{3}$, the meson will have a charge of -1. If one combines a quark of charge $\frac{2}{3}$ with an antiquark of charge $\frac{1}{3}$, the meson will have a charge of $+1$. These are the only possibilities.

(*c*) All baryons consist of three quarks. Each quark has a baryon number of $\frac{1}{3}$, so the baryon will have a baryon number of 1.

(*d*) All baryons consist of three quarks. All quarks have a charge of $-\frac{1}{3}$ or $+\frac{2}{3}$. The possible combinations are:

$$(-\tfrac{1}{3}) + (-\tfrac{1}{3}) + (-\tfrac{1}{3}) = -1$$

$$(-\tfrac{1}{3}) + (-\tfrac{1}{3}) + (+\tfrac{2}{3}) = 0$$

$$(-\tfrac{1}{3}) + (+\tfrac{2}{3}) + (+\tfrac{2}{3}) = +1$$

$$(+\tfrac{2}{3}) + (+\tfrac{2}{3}) + (+\tfrac{2}{3}) = +2$$

These are the only possibilities giving the charges expected.

Note. All *antibaryons* will have charges of -2, -1, 0 or $+1$.

All known mesons and baryons (there are many more than the ones that we described earlier) can be classified using their quark composition. The quarks are considered to be fundamental particles in the same way that leptons are fundamental. However, the leptons interact only by means of the weak and the electromagnetic interactions, whereas the quarks also interact by means of the strong interaction. It should be noted that conservation of charge, baryon number, strangeness, etc. in strong interactions can be understood in the context of the quarks in the initial system, which are grouped into mesons and baryons, rearranging themselves into different groups in the final system, but requiring that the quark quantum numbers before and after must still add up to the same values.

We mentioned previously that the electromagnetic interaction can be considered as arising from the interchange of "virtual" photons. In a similar manner, physicists now believe that the weak interaction can be considered as arising from the virtual interchange of other particles, called "vector bosons", and named $W^{\pm}$ and Z^0. These particles have all been detected. [The weak interaction and the electromagnetic interaction have now been shown to be two aspects of the common "electroweak" interaction, which involves virtual interchanges of the four fundamental particles, γ (photon), $W^{\pm}$ and Z^0.] The same type of theory is also being applied to the strong interaction. Here we have a force between any of six quarks. The particles that are being interchanged are called "**gluons**" since they are responsible for "gluing together" the quarks. In a full theory each quark has a three valued index called "color" charge. The color charge is a quantity comparable to electric charge in electrodynamics, except that there are two types of electric charges and three types of color charges. This theory of the strong interaction is called "**chromodynamics**" because the label "color" has been commonly used for this index. Neither quarks nor gluons have ever been seen alone. It is now believed that quarks can only exist in bound states so that one will never see a free quark. This is a consequence of the nature of the strong force provided by the gluons to the quarks. In contrast, the electromagnetic force allows charged particles to exist in both bound states (electrons in an atom) or free states (electrons or protons moving through space). Much of our present understanding of elementary particles and the world built out of them is explained by a comprehensive theory called the "Standard Model". Even so, much more needs to be understood, such as the reason the fundamental particles have the rest masses that they do, and whether all the forces of nature can be understood to be different aspects of the same interaction.

18.4 SOLID-STATE PHYSICS

Another area in which the concepts of quantum mechanics have made a major impact is in the area of solid-state physics. This refers to the study of materials in which the atoms are firmly fixed in place and not free to move as they are in gases and liquids. The simplest case is the situation when the atoms are arranged in a fixed symmetric pattern which is repeated throughout the solid. Such a system is called a crystal. The results of analyzing this special case can be generalized to more complicated cases of solids with varying degrees of disorder (amorphous solids) and even to liquids. Together all the classes of material are called "condensed matter".

We have seen that atoms are described by quantum mechanics in terms of energy levels. Each energy level can hold two electrons, one for each spin direction. If one has two, or more, atoms that are far apart, then each has its own identical energy level structure, and electrons can be specified as being in a particular level of a particular atom. As one brings two atoms closer together, the wave functions begin to overlap, and the electron can no longer be considered as confined to only one atom. The result is that the individual levels of each of the two overlapping atoms become two closely spaced shared levels in each of which the electron is shared by the two atoms. As one adds more and more atoms, the energy levels split into more and more of these closely spaced levels (3 atoms, 3 levels; 4 atoms, 4 levels; etc.), and are best described as an energy "band" with allowed energies extending from the bottom to the top of the band. Again, each level in a band allows for two electrons, one for each spin direction. The number of possible electrons in each band is therefore equal to twice the number of atoms that created this band. Each energy level of the individual atoms become converted to a band able to accommodate two electrons per atom. This status is depicted in Fig. 18-18, where the original energy levels are shown as they convert into bands. Electrons are contributed to these bands from all the atoms in the solid. The first electron from each atom is contributed to the lowest band 1, and will use up half of the states in that band. The second electron (with opposite spin) on each atom also goes into band 1, filling that band. The third electrons fill half of band 2, and the fourth electrons fill that band. It is therefore clear that if there are an odd number of electrons per atom, then the last band to be filled will not be full. If there are an even number of electrons per atom, then one expects the highest band to be filled. However, if the bands overlap, as do bands 4 and 5 in Fig. 18-18, then the highest band containing electrons may not be filled even for an even number of electrons per atom. There is an "energy gap" between some of the bands which, in Fig. 18-18 is large between bands 1 and 2, and is small between bands 3 and 4. At the energies within these gaps, there cannot be any electrons in the solid. This basic idea permits us to understand why certain solids act as insulators, while others act as conductors, and still others act as "**semiconductors**". When one imposes a voltage between the ends of a wire, an electric field tries to accelerate the electrons in the material. This acceleration adds small amounts of energy to the electrons. If there are energy states available to the electrons at this small added energy, then the electrons can be accelerated by the electric field, and the material will be a conductor. This is the case if the highest band containing electrons is not filled. The electrons that can move as a result of the imposed voltage are called "**free**" **electrons**, and the band in which they move is called the conduction band. The lower bands are called **valence bands**. If the band is filled, then there are no energy levels available for the highest energy electrons, and they will be unable to accelerate. We will then have an

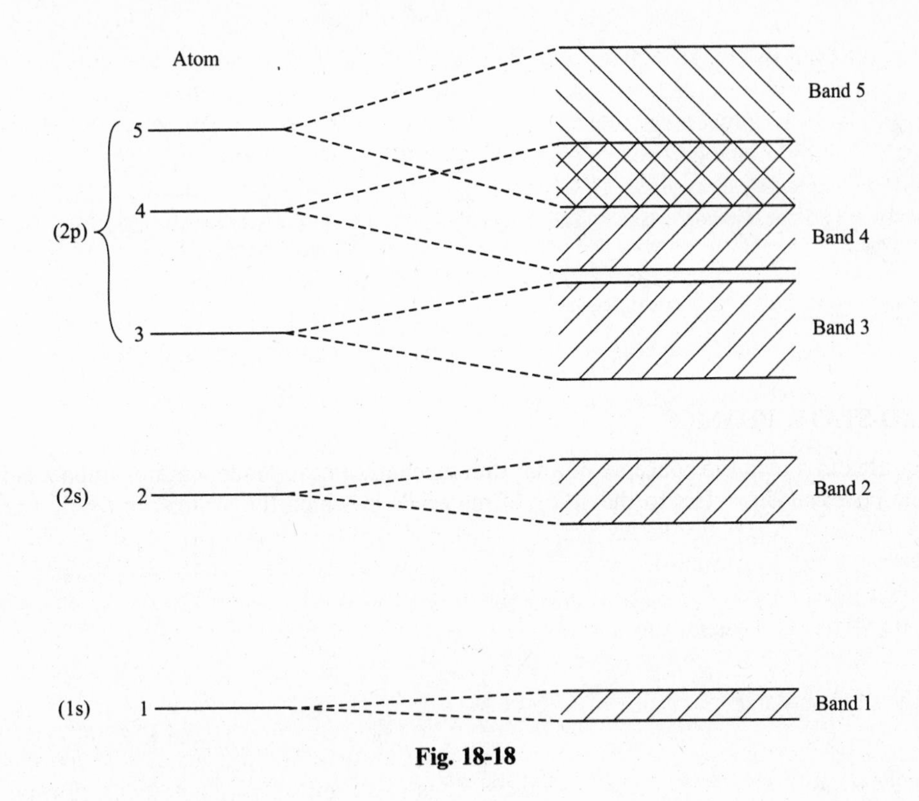

Fig. 18-18

insulator, which has a filled upper valence band. If one can excite electrons from this filled valence band to the empty next higher band, that band would constitute a conduction band, and these excited electrons can move in an electric field. Additionally, there is now room in the valence band for electrons to accelerate into the vacated level left by the excited electron. A vacated level leaves a net positive charge in the region where the electron was excited and conducted away. Under the external electric field another electron in that same level can hop to the new location, neutralizing that location, but leaving a net positive charge where it came from. The net effect is as if a positive charge moved in the direction opposite to that of the electrons. We describe the apparent motion of a positive charge as follows: when an electron excites into the conduction band we say that there is a positive "hole" created in the valence band and that the positive holes act just like positively charged particles and provide conduction in this band, just as the negative electrons provide conduction in the conduction band. The material will be conducting as long as these excited electrons and the holes they leave behind persist. This type of excitation can occur if the temperature is sufficiently high that the thermal energy of the electrons allows them to jump to the conduction band. For a large energy gap this can occur only at very high temperatures, but if the gap is small it occurs at normal temperatures. Materials such as these, with small energy gaps, are called semiconductors. They can change from insulators to conductors as the temperature increases. In practice, we can get electrons in the conduction band of these semiconductors, or holes in their valence band in a different manner. If one inserts an "impurity" into the material, that is, one inserts atoms which have one extra electron than the atoms of the base material, then there will be one more electron per impurity atom than is needed to fill the valence band. These added electrons can partially fill the conduction band, and provides conduction appropriate to negative electrons. We call this material an **n-type semiconductor**. On the other hand, if one inserts an impurity with one less electron than the host material, then there will be one state in the valence band that may be unfilled. This "hole" acts to provide conduction appropriate to a positive charge, and this semiconductor is called **p-type**. We see that we have great flexibility in producing the type of material we want by "doping" the base semiconductor with the appropriate type and amount of impurity. The most common semiconductor in current use is silicon.

It is also possible to excite electrons in an insulator or semiconductor to the conduction band by absorbing a photon. The incident photon that is absorbed will make it possible for the material to conduct electricity. This process is the basis for constructing photodetectors, since the absorption of a photon is signaled by a change in the conductivity of the material. It is also the basis for converting solar energy into electrical energy. Of course, only photons with sufficient energy (large enough frequency) can excite these electrons, so different materials will be sensitive to different wavelengths.

Problem 18.40. Silicon has a band gap of 1.1 eV.

(*a*) At what temperature is the average thermal energy of the electrons equal to the band gap energy?

(*b*) At what maximum wavelength can photons excite electrons from the valence band into the conduction band?

(*c*) What is the maximum band gap that one can have in a material if one wants to build a detector for infrared light at 9.5 μ?

Solution

(*a*) The thermal energy of a free electron equals $(\frac{3}{2})kT$. The band gap is $1.1\text{ eV} = 1.1(1.6 \times 10^{-19})\text{ J} = (\frac{3}{2})(1.38 \times 10^{-23}\text{ J/K})T$, giving $T = 8.5 \times 10^3\text{ K}$.

(*b*) The energy of the photon is hc/λ. This will equal or exceed the band gap energy if λ is less than or equal to: $\lambda = hc/E_{\text{gap}} = (6.625 \times 10^{-34}\text{ J}\cdot\text{s})(3.0 \times 10^8\text{ m/s})/(1.1\text{ eV})(1.6 \times 10^{-19}\text{ J/eV}) = 1.13 \times 10^{-6}\text{ m} = 1.13\ \mu$.

(*c*) Again we use that $\lambda = hc/E_{\text{gap}}$ and solve for $E_{\text{gap}} = hc/\lambda = 6.625 \times 10^{-34}(3.0 \times 10^8)/(9.5 \times 10^{-6}\text{ m}) = 2.09 \times 10^{-20}\text{ J} = 0.13\text{ eV}$.

Semiconductor devices are presently the components of all electronics systems, including radios, TV, radar and computers. There are many different ways in which semiconductors can be used to provide the properties needed to be useful in electronic applications. All these devices make use of the fact that one can make materials in which one or more parts are n-type and other parts are p-type. By applying voltages to different parts one can vary the current flow through the material. Some simple junctions function as diodes which conduct current primarily in only one direction. Some function as transistors that have the same properties as the old vacuum tubes and can amplify a signal greatly. Some act as data storage elements that can take on one of two different states, that can be viewed as "zero" and "one", and are used in computers extensively. Some act as lasers, others as light emitting diodes. Devices have been built as photovoltaic cells, and as photodetectors. With modern technology it is possible to build millions of these elements in a small volume. This is known as building integrated circuits, and for even more densely packed devices, as VLSI (very large scale integration). There is no aspect of modern day technology that does not make use of some of these devices.

Problems for Review and Mind Stretching

Problem 18.41. Consider the case of a Bohr atom of hydrogen. Assume that the radiation that is emitted is the result of transitions between the bound levels or from $E = 0$ (the minimum energy of an unbound, or ionized, electron) to one of the levels.

(*a*) What is the minimum and the maximum wavelength of the light emitted in the first series ending on the $n = 1$ level?

(*b*) What is the minimum and the maximum wavelength of the ight emitted in the second series ending on the $n = 2$ level?

(*c*) What is the minimum and the maximum wavelength of the light emitted in the third series ending on the $n = 3$ level?

(*d*) What is the minimum and the maximum wavelength of the light emitted in the fourth series ending on the $n = 4$ level?

(*e*) Between which series is there overlap in wavelength?

Solution

(*a*) We use Eq. (*18.9*), $1/\lambda_{mn} = (E_m - E_n)/hc = 1.097 \times 10^7 Z^2(1/n^2 - 1/m^2)\ \text{m}^{-1}$, with $n = 1$ and $Z = 1$. The maximum wavelength occurs for $m = 2$ (the nearest level corresponding to the least energy photon), and the minimum wavelength is for $m = \infty$ (the $E = 0$ state). Thus $1/\lambda_{max} = 1.097 \times 10^7(1 - \frac{1}{2}^2) = 8.23 \times 10^6\ \text{m}^{-1}$, or $\lambda_{max} = 1.22 \times 10^{-7}$ m. Furthermore, $1/\lambda_{min} = 1.097 \times 10^7(1 - 0) = 1.097 \times 10^7\ \text{m}^{-1}$, or $\lambda_{min} = 9.11 \times 10^{-8}$ m. Thus, the wavelength ranges from 91.1 to 122 nm.

(*b*) We use Eq. (*18.9*), $1/\lambda_{mn} = (E_m - E_n)/hc = 1.097 \times 10^7 Z^2(1/n^2 - 1/m^2)\ \text{m}^{-1}$, with $n = 2$ and $Z = 1$. The maximum wavelength occurs for $m = 3$ (the nearest level corresponding to the least energy photon), and the minimum wavelength is for $m = \infty$ (the furthest away in energy). Thus $1/\lambda_{max} = 1.097 \times 10^7(\frac{1}{2}^2 - \frac{1}{3}^2) = 1.523 \times 10^6\ \text{m}^{-1}$, or $\lambda_{max} = 6.56 \times 10^{-7}$ m. Furthermore, $1/\lambda_{min} = 1.097 \times 10^7(\frac{1}{4}) = 2.74 \times 10^6\ \text{m}^{-1}$, or $\lambda_{min} = 3.64 \times 10^{-7}$ m. Thus, the wavelength ranges from 364 to 656 nm.

(*c*) We again use Eq. (*18.9*) with $n = 3$ and $Z = 1$. The maximum wavelength occurs for $m = 4$ and the minimum wavelength is for $m = \infty$. Thus $1/\lambda_{max} = 1.097 \times 10^7(\frac{1}{3}^2 - \frac{1}{4}^2) = 5.33 \times 10^5\ \text{m}^{-1}$, or $\lambda_{max} = 1.875 \times 10^{-6}$ m. Furthermore, $1/\lambda_{min} = 1.097 \times 10^7(\frac{1}{9}) = 1.22 \times 10^6\ \text{m}^{-1}$, or $\lambda_{min} = 8.20 \times 10^{-7}$ m. Thus, the wavelength ranges from 820 to 1875 × nm.

(*d*) We again use Eq. (*18.9*) with $n = 4$ and $Z = 1$. The maximum wavelength occurs for $m = 5$ and the minimum wavelength is for $m = \infty$. Thus $1/\lambda_{max} = 1.097 \times 10^7(\frac{1}{4}^2 - \frac{1}{5}^2) = 2.47 \times 10^5\ \text{m}^{-1}$, or $\lambda_{max} = 4.05 \times 10^{-6}$ m. Furthermore, $1/\lambda_{min} = 1.097 \times 10^7(\frac{1}{16}) = 6.86 \times 10^5\ \text{m}^{-1}$, or $\lambda_{min} = 1.46 \times 10^{-6}$ m. Thus, the wavelength ranges from 1460 to 4050 nm.

(*e*) We tabulate the minimum and maximum wavelengths for each series.

Series	λ_{min} (nm)	λ_{max} (nm)
1	91.1	122
2	364	656
3	820	1875
4	1460	4050

The minimum of each higher series is above the maximum of the previous series except for the last series. Thus series 3 and 4 overlap.

Problem 18.42. Certain atoms in the ground state can absorb radiation at wavelengths of 150 nm, 450 nm, 600 nm and 700 nm.

(*a*) From this data deduce the energies of some of the levels of this atom above the ground state.

(*b*) After absorbing radiation at the given wavelength the atom can radiate at these same wavelengths as well as at additional wavelengths. What are the additional wavelengths?

Solution

(*a*) The atom will have energy levels above the ground state with energies equal to the energy of the photon that is absorbed. The longest wavelength corresponds to the nearest energy level, which we call E_1. The energies that can be absorbed are $E = hc/\lambda$. Using the wavelengths given, these energies are:

$$E_4 = 1.325 \times 10^{-18}\ \text{J} = 8.28\ \text{eV}$$
$$E_3 = 4.42 \times 10^{-19}\ \text{J} = 2.76\ \text{eV}$$
$$E_2 = 3.31 \times 10^{-19}\ \text{J} = 2.07\ \text{eV}$$
$$E_1 = 2.84 \times 10^{-19}\ \text{J} = 1.77\ \text{eV}.$$

These are the energies of the levels above the ground state.

(*b*) If the atom is in one of these states and makes a transition to the ground state it will radiate at the wavelengths given. However, the atom in level 4 can also make a transition to level 3, 2, or 1, the one in level 3 can make a transition to levels 2 or 1, and the one in level 2 can make a transition to level 1. The wavelengths emitted in these transitions correspond to the energy differences between the levels. These are:

$$4 \to 1,\ \lambda = hc/(E_4 - E_1) = hc/(13.25 - 2.84) \times 10^{-19} = 1.91 \times 10^{h7}\ \text{m}$$
$$4 \to 2,\ \lambda = hc/(E_4 - E_2) = hc/(13.25 - 3.31) \times 10^{-19} = 2.00 \times 10^{-7}\ \text{m}$$
$$4 \to 3,\ \lambda = hc/(E_4 - E_3) = hc/(13.25 - 4.42) \times 10^{-19} = 2.25 \times 10^{-7}\ \text{m}$$
$$3 \to 1,\ \lambda = hc/(E_3 - E_1) = hc/(4.42 - 2.84) \times 10^{-19} = 1.26 \times 10^{-6}\ \text{m}$$
$$3 \to 2,\ \lambda = hc/(E_3 - E_2) = hc/(4.42 - 3.31) \times 10^{-19} = 1.79 \times 10^{-6}\ \text{m}$$
$$2 \to 1,\ \lambda = hc/(E_2 - E_1) = hc/(3.31 - 2.84) \times 10^{-19} = 4.23 \times 10^{-6}\ \text{m}$$

Problem 18.43. The characteristic K_α X-rays from W (tungsten, $Z = 74$) have a wavelength of 2.18×10^{-11} m. What is the characteristic wavelength of the same line from the next element Re (Rhenium, $Z = 75$)?

Solution

The characteristic X-rays for high Z atoms have energies proportional to $(Z - 1)^2$, or wavelengths proportional to $1/(Z - 1)^2$. Thus $\lambda_{Re}/\lambda_W = (73/74)^2 = 0.973$, and $\lambda_{Re} = 0.973(2.18 \times 10^{-11}\ \text{m}) = 2.12 \times 10^{-11}$ m.

Problem 18.44. The binding energy of an α-particle is 28.24 MeV. The mass of a neutron is 1.008,665 u and of ${}_1H^1$ is 1.007,825 u.

(*a*) What is the mass of ${}_2He^4$ in u?

(*b*) What minimum frequency photon is needed to completely dissociate an α-particle into its nucleons?

(*c*) If the α-particle is broken up into two deuterons (${}_1H^2$), 23.86 MeV must be supplied. What is the binding energy of a deuteron?

(*d*) If one wanted to get fusion power by creating α particles from deuterons, how many α particles would one have to form per second to get 2 W of power?

Solution

(*a*) We know that the binding energy is the difference between the mass of the constituents and the mass of the nucleus, converted to energy. Thus, converting to atomic mass units, B.E. = (28.24 MeV/(932 MeV/u) = $(2m_n + 2m_H - m_{He})$. Thus $28.24/932 = 2(1.008{,}625) + 2(1.007{,}825) - m_{He}$. This results in $m_{He} = 4.002{,}60$ u.

(*b*) The energy we need is 28.24 MeV. This requires a photon of minimum frequency $hf = 28.24\ \text{MeV} = 28.24 \times 10^6(1.6 \times 10^{-19})$ J, or $f = 6.82 \times 10^{21}$ Hz.

(*c*) The reaction is $E_K + {}_2He^4 \to 2\,{}_1H^2$. Since there are two protons and two neutrons on each side, their masses will cancel, leaving only the binding energies. Then $E_K - \text{B.E.}_{He} = -2\ \text{B.E.}_H = 23.86\ \text{MeV} - 28.24\ \text{MeV} = -4.38$ MeV. Then $\text{B.E.}_H = 4.38/2 = 2.19$ MeV.

(*d*) If one fuses two deuterons into He, one releases 23.9 MeV of energy. For a power of 2 W one must generate 2 J/s = $(2/1.6 \times 10^{-19})$ eV/s = 1.25×10^{13} MeV/s. Since each fusion releases 23.9 MeV, we need $1.25 \times 10^{13}/23.9 = 5.23 \times 10^{11}$ fusions/s.

Problem 18.45. A certain sample has 5.1×10^{11} particles which decay with a half life of 1.5×10^{-2} s.

(*a*) What is the initial activity of the sample?

(*b*) How many particles have decayed after a time of 2.0×10^{-6} s?

(*c*) How many particles have decayed after a time of 5.2×10^{-2} s?

Solution

(*a*) The activity, $|\Delta N/\Delta t|$, is λN, where $\lambda = (\ln 2)/\tau$. Thus, initially, the activity is $|\Delta N/\Delta t| = [0.693/(1.5 \times 10^{-2}\text{ s})][5.1 \times 10^{11}] = 2.36 \times 10^{13}$ decays/s $= 6.38 \times 10^{2}$ Ci.

(*b*) As long as the time during which the decays occur is small compared to the half life, the number of particles in the sample remains essentially constant. Then we can use $\Delta N = \lambda N \Delta t = (2.36 \times 10^{13}\text{ s}^{-1})(2 \times 10^{-6}\text{ s}) = 4.7 \times 10^{7}$ particles.

(*c*) In this case the time during which decays occur is not much less than the half life, and we cannot use $\Delta N = \lambda N \Delta t$ since N changes during the time we are considering. We therefore calculate the number of particles remaining after the time t, using the equation $N = N_0 2^{(-t/\tau)} = N_0 e - {}^{\lambda t} = 5.1 \times 10^{11}$ exp $[-(46.21)(-5.2 \times 10^{-2}\text{ s})] = 4.61 \times 10^{10}$ particles. The number that have decayed is therefore $5.1 \times 10^{11} - 4.6 \times 10^{10} = 4.64 \times 10^{11}$ particles.

Problem 18.46. The β-decay of ${}_{15}P^{32}$ has a half life of 14.3 days and 1.72 MeV is released in the decay. ${}_{15}P^{32}$ has a mass of 31.973,907 u.

(*a*) Write down the reaction equation for the process.

(*b*) What is the mass of the daughter atom?

(*c*) If one has 2.3×10^{-11} kg of phosphorus, what is the activity of the sample?

Solution

(*a*) ${}_{15}P^{32} \rightarrow {}_{16}S^{32} + \beta^- + \bar{\nu}$.

(*b*) Using conservation of energy, $m_P = m_S + E_K$. Therefore, $m_S = 31.973{,}907 - 1.72/932 = 31.973{,}907 - 0.001{,}83 = 31.97{,}206$ u.

(*c*) The activity equals $\Delta N/\Delta t = \lambda N = 0.693\ N/\tau$. We are given the mass, m, of our sample and we need the number, N, of atoms this corresponds to. If one has N atoms, then there are $n = N/N_A$ moles, where N_A is Avogadro's number. Each mole has a mass of $M = 31.97$ g $= 0.031{,}97$ kg. Thus, $n = m/M = (2.3 \times 10^{-11}\text{ kg})/(0.031{,}97\text{ kg}) = N/(6.02 \times 10^{23}\text{ atoms/mol}) \rightarrow N = 4.33 \times 10^{14}$ atoms. The activity is $0.693(4.33 \times 10^{14})/[(14.3\text{ d})(24\text{ h/d})(60\text{ min/h})(60\text{ s/min})] = 2.43 \times 10^{8}$ Bq $= (2.43 \times 10^{8}/3.7 \times 10^{10})$ Ci $= 6.6 \times 10^{-3}$ Ci.

Problem 18.47. The Ω^--particle can decay into a Λ^0 and a K^-: $\Omega^- \rightarrow \Lambda^0 + K^-$. Show that this conserves (*a*) charge, (*b*) energy, (*c*) baryons, and (*d*) leptons.

(*e*) Show that the strangeness changes by one. What does this imply about the decay?

Solution

(*a*) There is one negative charge before the decay and one after the decay, conserving charge.

(*b*) The mass of the Ω^- is $3272 m_e$ while the masses of the Λ^0 and K^- after the decay are $(2184 + 967)m_e = 3151 m_e$. Thus the initial mass is greater than the final mass and the extra mass is converted into kinetic energy. This will satisfy conservation of energy.

(*c*) There was one baryon before the decay (the Ω^-) and there is one baryon after the decay (the Λ^0). The kaon is not a baryon. Therefore, we have conservation of baryons.

(*d*) There are no leptons before or after the decay, so leptons are conserved.

(*e*) The Ω^- has a strangeness of -3, while the Λ^0 and the K^- each have a strangeness of -1. Thus the total strangeness after the decay is -2, which is one less than the strangeness before the decay. The decay cannot occur via the strong interaction since strangeness is not conserved, and we therefore expect the half life to be relatively long.

Supplementary Problems

Problem 18.48. A hydrogen atom is in the level $n = 2$.

(*a*) What wavelength photon is needed to cause a transition to $n = 3$?

(*b*) What maximum wavelength is needed to ionize the atom from $n = 2$?

Ans. (*a*) 6.56×10^{-7} m; (*b*) 3.65×10^{-7} m

Problem 18.49. A hydrogen atom is in a state with $n = 3$, $l = 2$, $m_l = -1$, $m_s = \frac{1}{2}$. Assume that the energy of a hydrogen atom state is given by $E = -13.6\ \text{eV}/n^2$.

(*a*) What is the orbital angular momentum of the atom in this state?

(*b*) If the atom changes to the state $n = 1$, $l = 1$, $m_l = 0$, $m_s = \frac{1}{2}$, what is the frequency of the photon emitted?

(*c*) List all the possible states of the atom with $n = 3$.

(*d*) If light of wavelength 200 nm is absorbed by the atom (with $n = 3$), what is the kinetic energy of the emitted electron?

Ans. (*a*) $1.49 \times 10^{-34}\ \text{kg} \cdot \text{m}^2/\text{s}$; (*b*) 2.93×10^{15} Hz;
(*c*) $l = 0$; $m_l = 0$; $m_s = \pm\frac{1}{2}$
$l = 1$; $m_l = 0, \pm 1$; $m_s = \pm\frac{1}{2}$
$l = 2$; $m_l = 0, \pm 1, \pm 2$; $m_s = \pm\frac{1}{2}$
(a total of 18 states);
(*d*) 4.70 eV

Problem 18.50. Write down all the possible states of an f electron (for a given *n*).

Ans. $l = 3$; $m_l = 0, \pm 1, \pm 2, \pm 3$; $m_s = \pm\frac{1}{2}$ (a total of 14 states)

Problem 18.51. An atom has a ground state at an energy of 25 eV below ionization. Other levels are at -1.5, -6, and -10 eV. The atom is in the state at -6 eV.

(*a*) What wavelengths can be emitted by the atom?

(*b*) What wavelengths can be absorbed by the atom?

Ans. (*a*) 3.11×10^{-7} m, $6.53 \times 10^{-7=8}$ m; (b) 2.76×10^{-7} m, 2.07×10^{-7} m and below.

Problem 18.52. A certain atom is in the state with $n = 4$ and $m_l = -1$.

(*a*) What are the possible values of *l* for this state?

(*b*) What are the possible values of angular momentum for this state?

(*c*) What are the possible angles that the angular momentum can make with the *z* axis for this state?

Ans. (*a*) $l = 1$, 2 or 3; (*b*) $L = 1.49 \times 10^{-34}$, 2.58×10^{-34}, or $3.65 \times 10^{-34}\ \text{kg} \cdot \text{m}^2/\text{s}$; (c) 135°, 114° or 106°

Problem 18.53. List the electrons that fill the ground state of the following elements: (*a*) silicon with 14 electrons; (*b*) titanium with 22 electrons; (*c*) copper with 29 electrons; and (*d*) selenium with 34 electrons.

Ans. (*a*) $(1s)^2(2s)^2(2p)^6(3s)^2(3p)^2$; (*b*) $(1s)^2(2s)^2(2p)^6(3s)^2(3p)^6(4s)^2(3d)^2$; (*c*) $(1s)^2(2s)^2(2p)^6(3s)^2(3p)^6(4s)^2(3d)^9$; (*d*) $(1s)^2(2s)^2(2p)^6(3s)^2(3p)^6(4s)^2(3d)^{10}(4p)^4$

Problem 18.54. What must be the difference in energy of the levels responsible for producing the laser light in the case of: (*a*) an ultraviolet laser at 100 nm; and (*b*) an X-ray laser at 1.0 nm?

Ans. (*a*) 12.4 eV; (*b*) 1.24 keV

Problem 18.55. An X-ray laser operates at a wavelength of 1.0 nm. How many photons are emitted per second if the power of the laser is: (*a*) 2.0 W; (*b*) 10 W; and (*c*) 10 kW?

Ans. (*a*) 1.01×10^{16}/s; (*b*) 5.03×10^{16}/s; (*c*) 5.03×10^{19}/s

Problem 18.56. Lead, with atomic number 82, has isotopes with mass number (*a*) 204, (*b*) 206, (*c*) 207, (*d*) 208, (*e*) 210 and (*f*) 214. Determine the number of neutrons, of protons and the symbol for each of these isotopes.

Ans. (*a*) $N = 122$, $Z = 82$, ${}_{82}Pb^{204}$; (*b*) $N = 124$, $Z = 82$, ${}_{82}Pb^{206}$; (*c*) $N = 125$, $Z = 82$, ${}_{82}Pb^{207}$; (*d*) $N = 126$, $Z = 82$, ${}_{82}Pb^{208}$; (*e*) $N = 128$, $Z = 82$, ${}_{82}Pb^{210}$; (*f*) $N = 132$, $Z = 82$, ${}_{82}Pb^{214}$.

Problem 18.57. What is the binding energy per nucleon for each of the isotopes in the previous problem, if the atomic masses of the isotopes are: $m_{204} = 203.973{,}037$ u; $m_{206} = 205.974{,}455$ u; $m_{207} = 206.975{,}885$ u; $m_{208} = 207.976{,}641$ u; $m_{210} = 209.984{,}178$ u; $m_{214} = 213.999{,}764$ u? (The mass of a neutron is 1.008,665 u and of ${}_1H^1$ is 1.007,825 u.)

Ans. (*a*) 7.88 MeV/A; (*b*) 7.88 MeV/A; (*c*) 7.87 MeV/A; (*d*) 7.87 MeV/A; (*e*) 7.84 MeV/A; (*f*) 7.78 MeV/A

Problem 18.58. The binding energy for an electron in a hydrogen atom is 13.6 eV. The binding energy for a nucleon in a nucleus of heavy water is 2.25 MeV. What fraction of the original mass of the constituents is lost in forming the atom and in forming the nucleus?

Ans. (*a*) 1.4×10^{-6}%; (*b*) 0.12%

Problem 18.59. At a certain time, a sample is decaying at the rate of 9.4×10^4 Ci. The decay constant is measured to be $1.3 \times 10^4\ s^{-1}$.

(*a*) How many particles are there in the sample at this time?

(*b*) At what time will the activity be reduced to $\frac{1}{2}$ of its initial value?

Ans. (*a*) 2.7×10^{11} particles; (*b*) 5.3×10^{-5} s

Problem 18.60. Particles decay with a half life of 155 years. How long does it take till only 0.10% remain?

Ans. 1.54×10^3 y

Problem 18.61. The half life for the α-decay of Bi^{212} is 60.6 s. How much mass of bismuth is needed for an activity of 3.3×10^{-3} Ci?

Ans. 3.76×10^{-15} kg

Problem 18.62. Element X with 180 nucleons, decays via α-emission, and releases 4.1 MeV in the process. The half life for the decay is 3.0×10^4 s.

(*a*) If one started with 10^{25} nuclei of X, how many will remain after 1.2×10^5 s?

(*b*) The binding energy of X is 5.9 MeV/A, and of the α-particle is 2.5 MeV/A. What is the binding energy per nucleon of the daughter nucleus?

Ans. (*a*) 6.25×10^{23}; (*b*) 5.95 MeV/A

Problem 18.63. The energy released in the α-decay of $_{84}Po^{210}$ is 5.3 MeV. The mass of this isotope of polonium is 209.982,876 u and the mass of a helium atom ($_2He^4$) is 4.002,603 u.

(*a*) Identify the daughter nucleus.

(*b*) What is the mass of the daughter isotope?

Ans. (*a*) $_{82}Pb^{206}$; (*b*) 205.9746 u

Problem 18.64. The nucleus of $_3Li^7$ has a mass defect of 0.0422 u. (The mass of a neutron is 1.008,665 u and of $_1H^1$ is 1.007,825 u.)

(*a*) What is the binding energy per nucleon for this nucleus?

(*b*) What is the mass of $_3Li^7$?

(*c*) A photon of large enough frequency can cause the emission of a neutron from $_3Li^7$. What is the resulting nucleus? If the isotope of this nucleus has an atomic mass of 6.015 u, what is the minimum frequency of the photon needed to cause the neutron emission?

Ans. (*a*) 5.62 MeV/A; (*b*) 7.0159 u; (*c*) $_3Li^6$, 1.75×10^{21} Hz

Problem 18.65. $_{83}Bi^{214}$ can decay to $_{82}Pb^{210}$ either by α-emission followed by β-emission or in the reverse order. Write the reaction equation for both paths.

Ans. (*a*) $_{83}Bi^{214} \rightarrow {}_{81}Tl^{210} + \alpha$, $_{81}Tl^{210} \rightarrow {}_{82}Pb^{210} + \beta + \bar{\nu}$;
(*b*) $_{83}Bi^{214} \rightarrow {}_{84}Po^{214} + \beta + \bar{\nu}$, $_{84}Po^{214} \rightarrow {}_{82}Pb^{210} + \alpha$

Problem 18.66. A slow neutron causes a nucleus of $_{94}Pu^{239}$ to fission, producing two neutrons and two identical fragments. The plutonium has a binding energy per nucleon of 7.62 MeV/A, and 207 MeV of kinetic energy is given to the fission fragments and outgoing neutrons.

(*a*) Fill in the missing information for X in the reaction equation, $_{94}Pu^{239} + {}_0n^1 \rightarrow 2_ZX^A + 2_0n^1$.

(*b*) What is the B.E./A for the isotope X?

Ans. (*a*) $_{47}Ag^{119}$; (*b*) 8.52 MeV/A

Problem 18.67. A fission reactor, using U^{235}, produces energy at the rate of 160 MW. Each fission release energy of 208 MeV. How much uranium is used during a period of three years?

Ans. 177 kg

Problem 18.68. A nuclear reaction converts nuclide X into nuclide Y in the reaction X(n, α)Y. If the incident neutron has an energy of 1.2 MeV, the products have a kinetic energy of 5.3 MeV. The binding energy of the α-particle is 28.3 MeV. What is the difference in binding energy of the two nuclides, B.E.$_X$ − B.E.$_Y$?

Ans. 24.2 MeV

Problem 18.69. In the fusion reaction, $_2He^3 + {}_2He^3 \rightarrow {}_2He^4 + 2_1H^1$, calculate how much energy is released and given to the products as kinetic energy. (Atomic masses are: $_2He^3$ = 3.016,030 u, $_2He^4$ = 4.002,603 u, $_1H^1$ = 1.007,825 u.)

Ans. 12.86 MeV

Problem 18.70. Which conservation laws are violated in each of the following hypothetical reactions?

(*a*) $p \rightarrow n + e^{+} + \nu$

(*b*) $\pi^{-} \rightarrow 2\gamma$

(*c*) $n \rightarrow K^{0} + \pi^{0}$

(*d*) $\Lambda^{0} \rightarrow n + \pi^{0}$

(*e*) $\pi^{+} \rightarrow \mu^{+} + e^{+} + e^{-}$

Ans. (*a*) energy; (*b*) charge; (*c*) baryons; (*d*) strangeness; (*e*) leptons

Problem 18.71. In the strong interaction: $\pi^{-} + p^{+} \rightarrow \Lambda^{0} + K^{0}$

(*a*) What are the quark combinations before the reaction?

(*b*) What are the quark combinations after the reaction?

(*c*) Noting that the quantum numbers of a quark–antiquark pair cancel each other out, what are the "net" quarks before and after the reaction?

Ans. (*a*) $(\bar{u}d)$, (uud); (*b*) (uds), $(d\bar{s})$; (*c*) u,d,d before and after

Problem 18.72. A crystalline material turns from insulating to conducting when bathed in light of wavelength less than 260 nm. What is the band gap between the valence band and the conduction band?

Ans. 4.78 eV

Index